# International Conference on Technological Advancements in Materials, Design, Manufacturing and Energy Sectors (ICTAMDMES'20)

IOP Conference Series: Materials Science and Engineering
Volume 923

Chennai, India
20 - 21 February 2020

ISBN: 978-1-7138-4763-2
ISSN: 1757-8981

**Printed from e-media with permission by:**

Curran Associates, Inc.
57 Morehouse Lane
Red Hook, NY 12571

**Some format issues inherent in the e-media version may also appear in this print version.**



Institute of Physics
Dirac House, Temple Back
Bristol BS1 6BE UK

Phone: 44 1 17 929 7481
Fax: 44 1 17 920 0979

techtracking@iop.org

**Additional copies of this publication are available from:**

Curran Associates, Inc.
57 Morehouse Lane
Red Hook, NY 12571 USA
Phone: 845-758-0400
Fax: 845-758-2633
Email: curran@proceedings.com
Web: www.proceedings.com

# TABLE OF CONTENTS

IOP Conf. Series: Materials Science and Engineering **923** (2020) 011001 doi:10.1088/1757-899X/923/1/011001

**Abstract**

International Conference on " Technological Advancements in Materials, Design, Manufacturing and Energy Sectors" (ICTAMDMES 2020) has been held successfully on February 20$^{th}$ and 21$^{st}$, 2020 at St. Joseph's College of Engineering, Chennai, Tamilnadu, India. International Conference on Technological Advancements in Materials, Design, Manufacturing and Energy Sectors (ICTAMDMES 2020) aims to provide a forum for researchers, practitioners, and professionals from the industry, academia and government to discourse on research and development, professional practice in mechanical engineering and material research. It is one of the leading international conferences for presenting novel and fundamental advances in the fields of mechanical engineering, manufacturing and energy sectors. It also serves to foster communication among researchers and practitioners working in a wide variety of scientific areas with a common interest in improving mechanical engineering.

# Peer review declaration

All papers published in this volume of *IOP Conference Series: Materials Science and Engineering* have been peer reviewed through processes administered by the Editors. Reviews were conducted by expert referees to the professional and scientific standards expected of a proceedings journal published by IOP Publishing.

- **Type of peer review:** Single-blind

  All the submitted manuscripts in this volume of IOP Conference Series: Materials Science and Engineering underwent a single blind peer review processes administered by the proceedings editors. Reviews were conducted by expert referees related to the professional and scientific standards expected of a proceedings journal published by IOP publishing.

- **Conference submission management system:**

  Conference submission management system was done through EasyChair.

- **Number of submissions received:**

  A total of 252 manuscripts received from various themes related to the conference.

- **Number of submissions sent for review:**

  After the manuscripts were received, initial screening were made by the conference editors and a total of 160 submissions were found to be suitable for the review process, which then underwent a single blind peer review process.

- **Number of submissions accepted:**

  After single blind peer review process by the referees assigned by the conference administrators, 80 manuscripts were accepted for presentation and publication for conference proceedings.

- **Acceptance Rate (Number of Submissions Accepted / Number of Submissions Received X 100):**

  Acceptance Rate = (80/252) * 100 = 31.74%

- **Average number of reviews per paper:**

  Each manuscript accepted for review were assigned two referees for completely reviewing.

IOP Conf. Series: Materials Science and Engineering **923** (2020) 011002 doi:10.1088/1757-899X/923/1/011002

- **Total number of reviewers involved:**

  A total of 40 reviewers were involved from various institutions relating to the conference theme for reviewing the manuscript.

- **Any additional info on review process:** -

- **Contact person for queries:**

  For any queries, the conference organizing committee members listed below can be contacted.

1. **Dr.Vaddi Seshagiri Rao,**
   Principal
   St.Joseph's College of Engineering
   Old Mamallapuram Road, Chennai-119
   TamilNadu, India
   E-mail: stjosephsconference@gmail.com

2. **Dr.K.M.Kumar,**
   Associate Professor
   Department of Mechanical Engineering
   St.Joseph's College of Engineering
   Old Mamallapuram Road, Chennai-119
   TamilNadu, India
   E-mail: stjosephsconference@gmail.com

IOP Conf. Series: Materials Science and Engineering **923** (2020) 012001 doi:10.1088/1757-899X/923/1/012001

# Design and analysis of connecting rod using composite materials (AL7075, AL6061, AL7075+SiC, AL6061+SiC)

**Mohanakrishnan R[1], Iniyan K[2], Sabarish Kumar P[3], Arun Vasantha Geethan K[4]**
[1,2,3,4] Department of Mechanical Engineering, St. Joseph's Institute of Technology, Chennai ,Tamilnadu, India.

E-mail:rmkrishnan98@gmail.com,iniyan19@gmail.com,pskgps@gmail.com

**Abstract**. Connecting rod is a vital part of an internal combustion engine which forms a link between piston and the crankshaft. Due to huge amount of forces acting on the piston owing to high engine load and rpm, cyclic compressive and tensile forces act on the connecting rod and results in fatigue and failure. So, to in order avoid this, the existing connecting rod is to be replaced by composite materials. Because these materials have good mechanical properties such as wear resistance, hardness and high tensile strength. A 3D model of connecting rod is modeled using Solidworks and FEA analysis is carried out by ANSYS 18.2. In this paper analysis is performed on the connecting rod using AL7075, AL6061, AL7075+10%SiC and AL6061+10%SiC to find out the best material for construction of connecting rod.
*Keywords*: Connecting rod; ANSYS; AL7075; AL6061; AL7075+SiC; AL6061+SiC

## 1. Introduction

India is one of the world's largest automobile manufacturers. Two wheelers account for around 80% of the vehicle sales. The important part of an automobile is the internal combustion engine. It consists of piston, connecting rod, crankshaft, and crank. The connecting rod forms a link between piston and crankshaft, the function of connecting rod is to transmit force acting on the piston to the crankshaft. The connecting rod is a slider and crank mechanism which converts the reciprocating motion of the piston into the rotational motion of the crankshaft. Connecting rod has two ends, one is pin end and the other is crank end. Pin end is attached with piston by gudgeon pin and crank end is attached to the crank pin by a crankshaft.

So, it is essential to rectify the problems for the better utilization of the engine and better life and performance of the engine. Usually the connecting rods are made up of forged steel, cast iron. The main disadvantage of using these materials is that they have higher density, which leads to produce to heavier engine components and motorcycle. So, a less dense material should be used for the manufacture of the connecting rod in order to reduce weight and inertia forces. In this case, Aluminium is chosen for the manufacture of the connecting rod due to low density.

Aluminium and its alloy possess low density, high thermal conductivity and high resistance to corrosion. These materials have low melting point and high ductility. The mechanical properties can be improved by alloying and cold working. The major alloying elements are zinc, magnesium, silicon, copper etc. Recent attention has been given to the alloys of aluminium and other low-density materials as engineering materials for applications in aerospace, automobile and defense etc.

Aluminum is added with composite materials to make metal matrix composites. These materials improve the mechanical properties of the connecting rod. The most commonly used composite material is Silicon Carbide

Silicon carbide (10%SiC), which is also known as carborundum, is semiconductor containing silicon and carbon. It exists in nature as the extremely rare mineral moissanite. Synthetic 10%SiC powder has been mass-produced since 1893 for use as an abrasive material

Grains of silicon carbide can be bonded together by the process called sintering to form very hard ceramics that are widely used in applications requiring high endurance such as car brakes, car clutches and ceramic plates in bulletproof vests, cutting tools, grinding materials.

## 2. Problem specification

The aim of the paper is to design, analyze and determine the mechanical properties, feasibility of the connecting rod made of Al6061, Al7075, Al6061+10%SiC and Al7075+10%SiC. 3D model of the connecting rod was made using Solidworks 2016. Model was imported into ANSYS 18.2 for analysis. Model was analyzed to find the stress, strain, and the deformation.

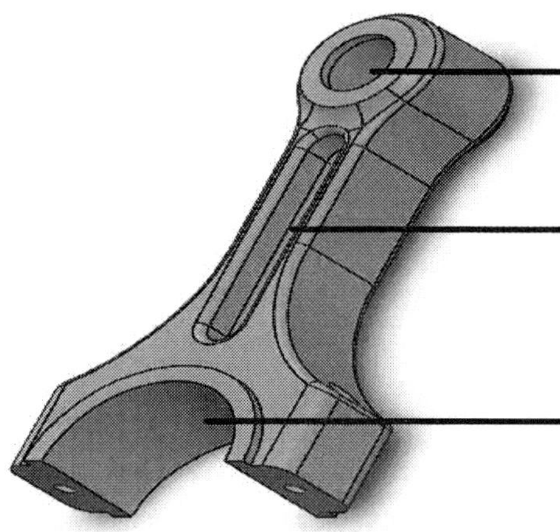


**Figure 1**: 3D model of connecting rod

## 3. Design of connecting rod

A connecting rod is an engine component which is subjected to alternating direct compressive and tensile forces. Due to fact that the compressive forces are predominant over the tensile forces, the cross section of the connecting rod is designed as a strut and the Rankine Gordon formula is employed to find the dimensions of the cross section. A connecting rod subjected to an axial load may buckle in x-axis as neutral axis within the plane of motion of the rod, or y-axis may be a neutral axis. For buckling analysis, the connecting rod is assumed to be fixed at both ends. The connecting rod is to be designed to be strong in both the axes.
According to Rankin formulae,
$W_{cr}$ About x-axis,

$$= \frac{\sigma_c * A}{1+a\left[\frac{L}{k_{xx}}\right]^2} = \frac{\sigma_c * A}{1+a\left[\frac{l}{k_{xx}}\right]^2} \quad [\therefore \textit{for both ends hinged } L = l]$$

$W_{cr}$ About y-axis,

$$= \frac{\sigma_c * A}{1+a\left[\frac{L}{k_{yy}}\right]^2} = \frac{\sigma_c * A}{1+a\left[\frac{l}{2k_{yy}}\right]^2} \quad [\because for\ both\ ends\ fixed\ L = l/2]$$

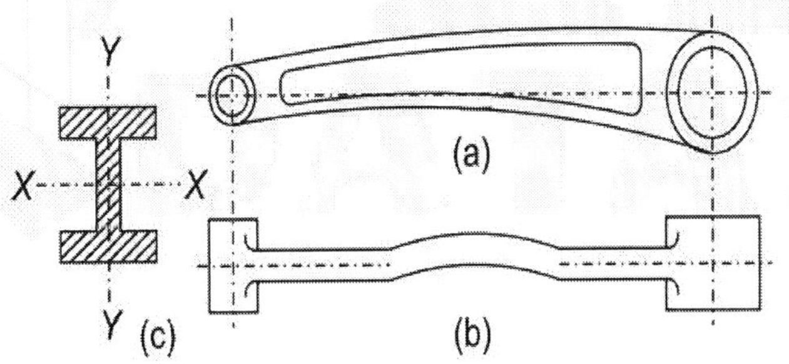


**Figure 2**: Buckling on applying load

The loads must be equal in order to resist buckling in the both axes.

$$= \frac{\sigma_c * A}{1+a\left[\frac{L}{k_{xx}}\right]^2} = \frac{\sigma_c * A}{1+a\left[\frac{l}{2k_{yy}}\right]^2} \quad [\text{or}]$$

$$\left[\frac{l}{k_{xx}}\right]^2 = \left[\frac{l}{2k_{yy}}\right]^2$$

$$K^2{}_{xx} = 4K^2{}_{yy} \ [\text{or}] \ I_{xx} = 4I_{yy}$$

If I xx > 4*Iyy, then buckling will occur about the Y-axis and if I xx < 4*Iyy, then buckling occurs about the X-axis. But practically Ixx is kept slightly lesser than 4 Iyy. A value between 3 and 3.5 is taken for it. In practice the connecting rod is only designed to withstand buckling about x axis, since the buckling about y axis is lesser than x axis by 4 times.

Cross sectional area= $2[4t^2] + 3t^2 = 11t^2$

Moment of inertia about the X-axis $= 2[4t^2] + 3t^2 = 11t^2$

Moment of inertia about x-axis,

$$I_{xx} = \frac{1}{12}[4t\,(5t^3) - 3t\,(3t^3)] = \frac{419}{12}t^4$$

Moment of inertia about y-axis,

$$I_{yy} = \frac{2}{12} * 4t^4 + \frac{1}{12} * 3t^4 = \frac{131}{12}t^4$$

$$\frac{I_{xx}}{I_{yy}} = 3.2$$

The value of Ixx/Iyy is found out to be between 3 and 3.5. Hence the chosen proportions are appropriate

IOP Conf. Series: Materials Science and Engineering **923** (2020) 012001 doi:10.1088/1757-899X/923/1/012001

## 4. Pressure calculation of 100cc engine

**Table 1:** Specifications of 100cc Engine

| | |
|---|---|
| **Displacement** | 100cc |
| **Bore** | 50mm |
| **Stroke** | 49.5mm |
| **Length of rod** | 99mm |
| **Max power** | 5500w @ 8000rpm |
| **Max torque** | 8Nm |
| **Compression ratio** | 9:1 |
| **Density of petrol** | 737.23 kg/m3 |

*4.1 Calculation of gas pressure:*

PV=MRT (Ideal Gas equation)

Where,

P- Pressure

V- Volume in $m^3$

R- Gas constant in $Jmol^{-1}K^{-1}$

T- Temperature in K

M- Mass in kg

M = ρ*V = 737.23 * 9.7*$10^{-5}$ = 0.071kg

Where,

ρ = Density of petrol

V = volume of the cylinder

R = $R^*/M_{wt}$ = (8.314/114.28) = 72.786

$R^*$= Gas constant

$M_{wt}$= molecular weight of petrol

P = MRT/V = (0.071*72.786*300)/9.72*$10^{-5}$

= 15.95MPa.

*4.2 Calculation of Force due to pressure:*

$F_P$= P*A

Where,

P- Gas pressure in MPa

A-Area in $mm^2$

$F_P$ = 15.95 * (π/4) * $50^2$

= 3.13KN

*4.3 Calculation of Inertia force:*

$F_I$ = mr $w^2(\cos\theta + \cos 2\theta/\text{n})$

Where,

$F_I$- Inertia force

$w$- Angular velocity

w = 2πN/60 = 837.76

$F_I$ = 0.2488*1.25*24.75$10^{-3}$ $*$ $837.76^2$

= 5232.6N

IOP Conf. Series: Materials Science and Engineering **923** (2020) 012001 doi:10.1088/1757-899X/923/1/012001

## 5. Design calculations of existing connecting rod

$$W_{cr} = \frac{\sigma_c * A}{1+a\left[\frac{L}{k_{xx}}\right]^2}$$

Where,

$W_{cr} = W_{cr}$ * FOS = 156509.375N
L = length of connecting rod =125mm
d=bore diameter = 50mm
A=area of cross section = $11t^2$
$\sigma_c$ = 572MPa
$k_{xx}$ = 1.78t

On substituting these values, we get
t = 5mm, B=20mm, H=25mm

Thickness at crank end =1.5*H = 31.25 mm
Thickness at piston end = 0.75*H = 18.75 mm

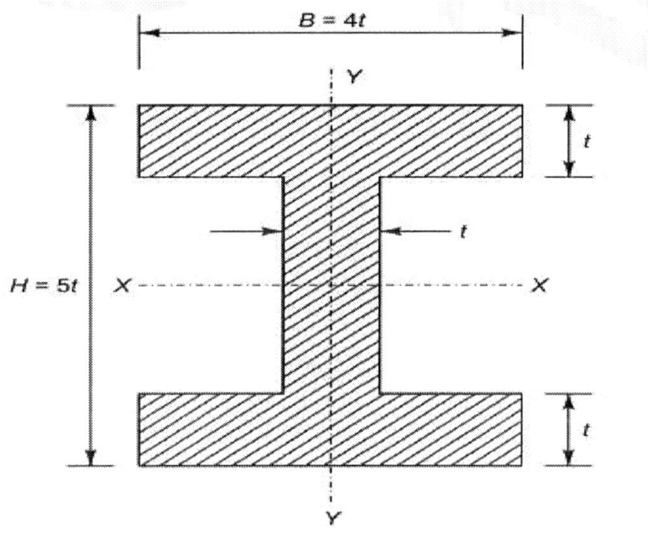


**Figure 3**: Cross section of the shaft section of the connecting rod

## 6. Results and discussion

### *6.1 Stress*

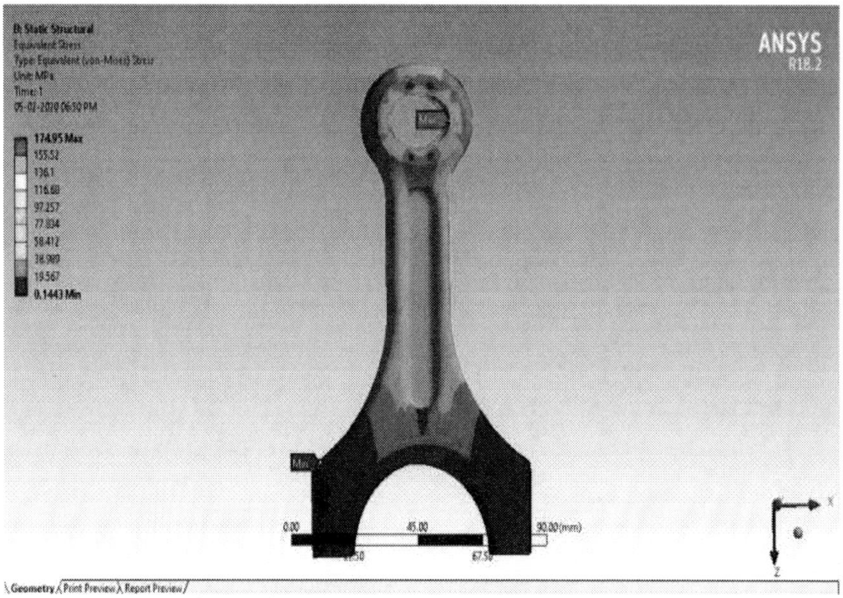


**Figure 4**: AL6061 under load showing stress

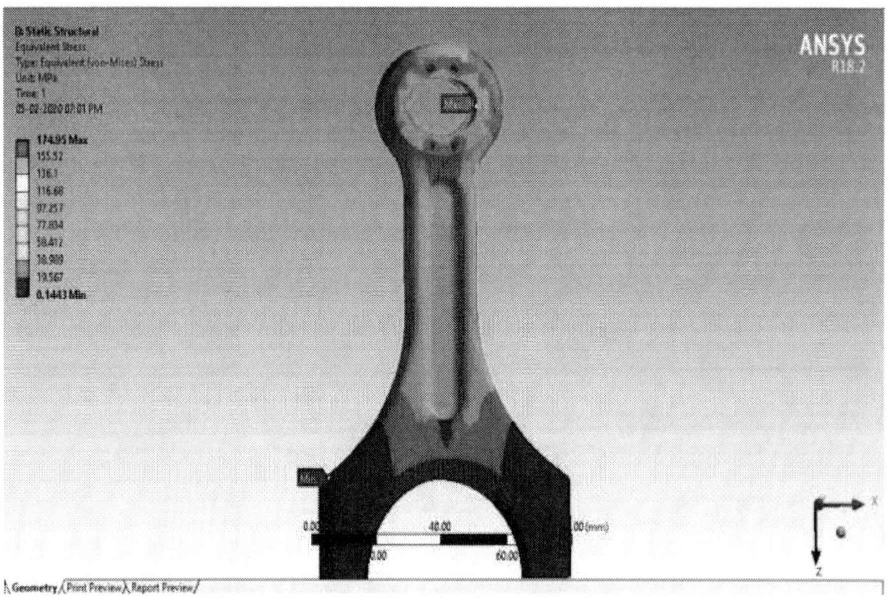


**Figure 5**: AL7075 under load showing stress

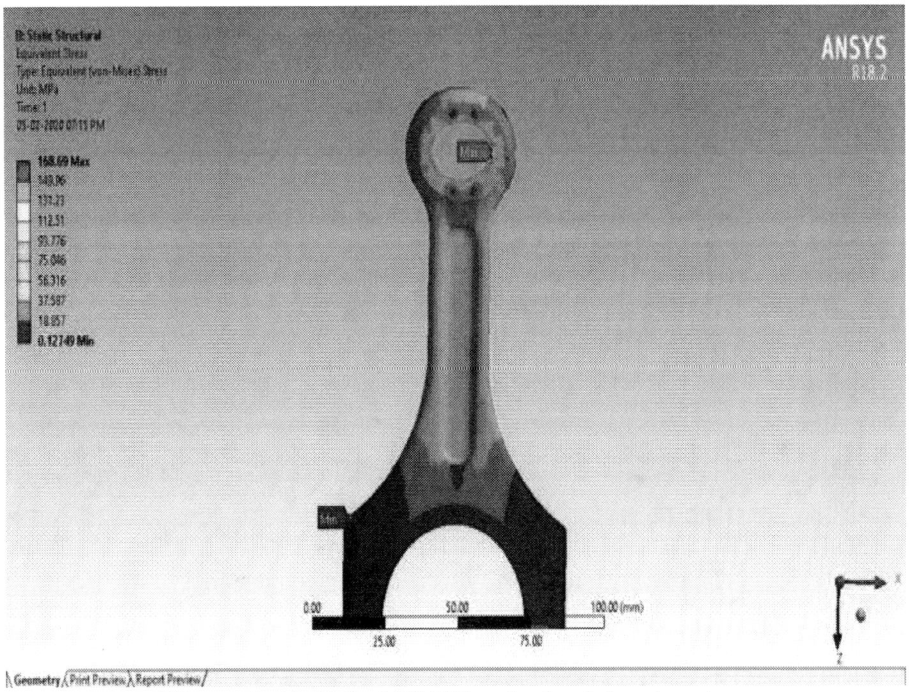


**Figure 6**: AL6061+10%SiC under load showing stress

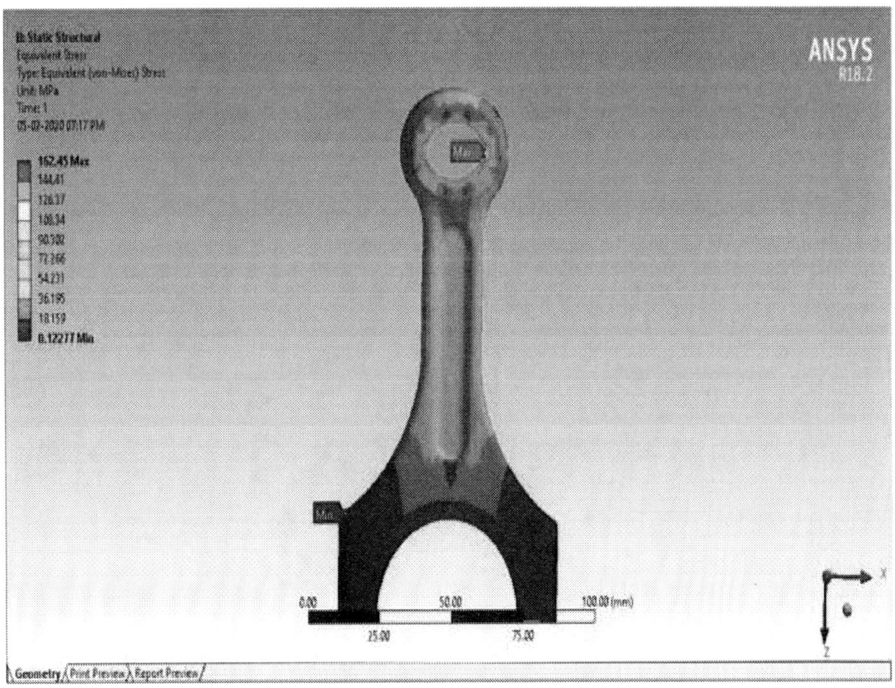


**Figure 7**: AL7075+10%SiC under load showing stress

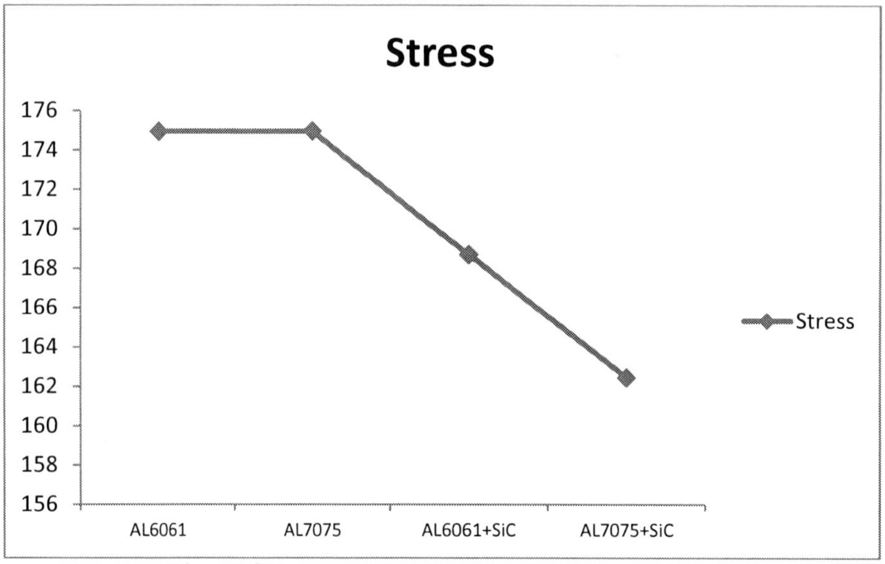


**Figure 8**: Comparison of stress acting on materials

### *6.2 Elastic Strain*

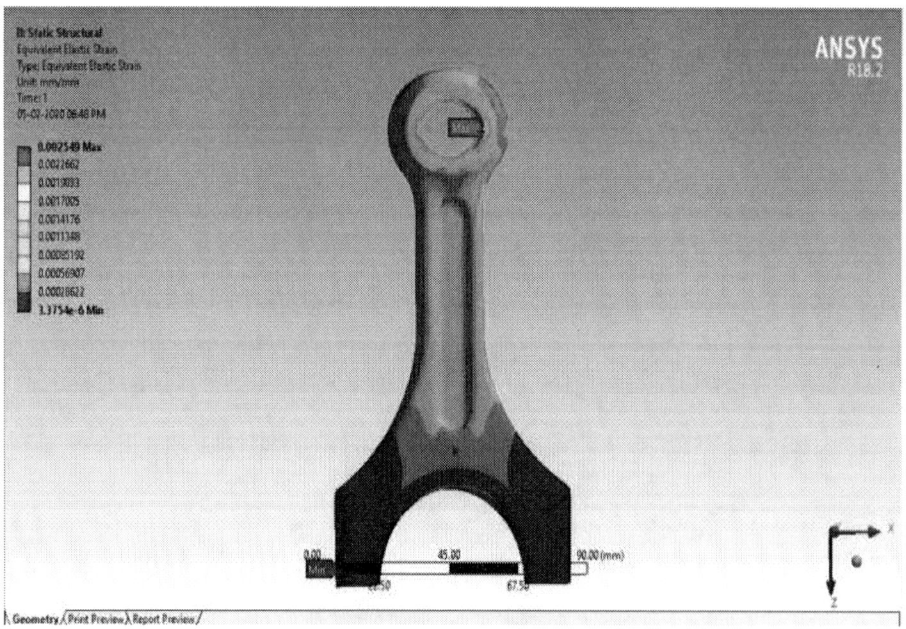


**Figure 9**: AL6061 under load showing elastic strain

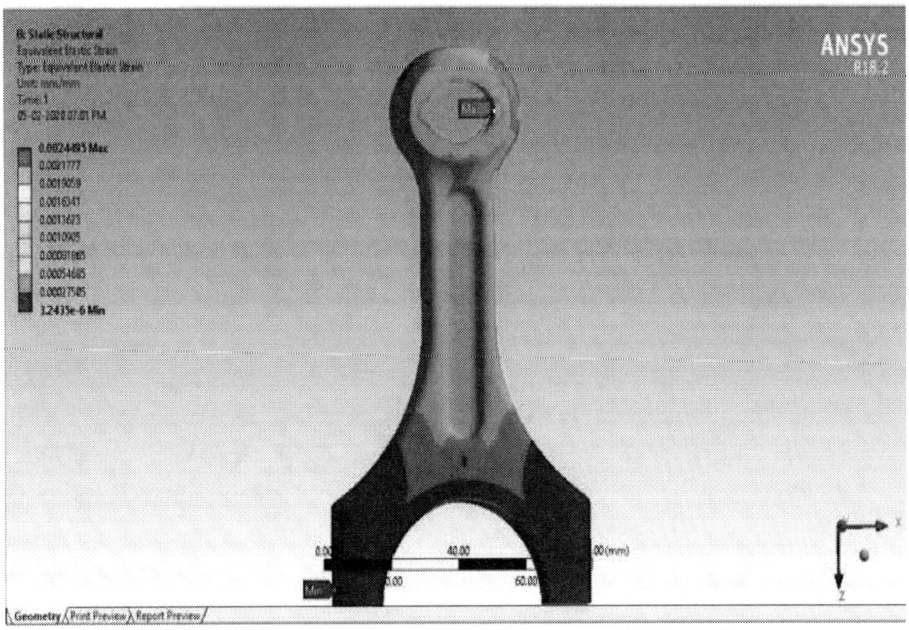


**Figure 10**: AL7075 under load showing elastic strain

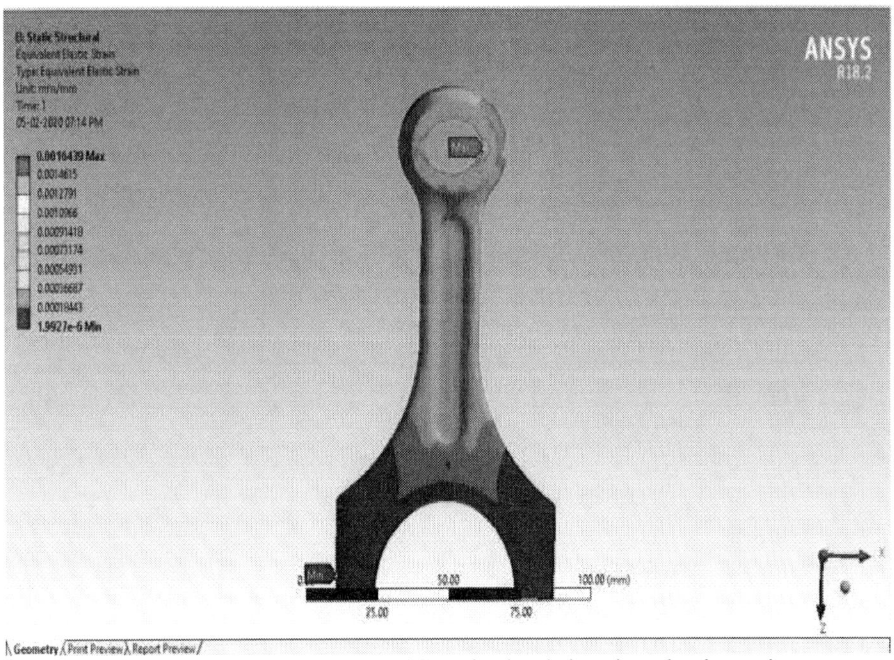


**Figure 11**: AL6061+10%SiC under load showing elastic strain

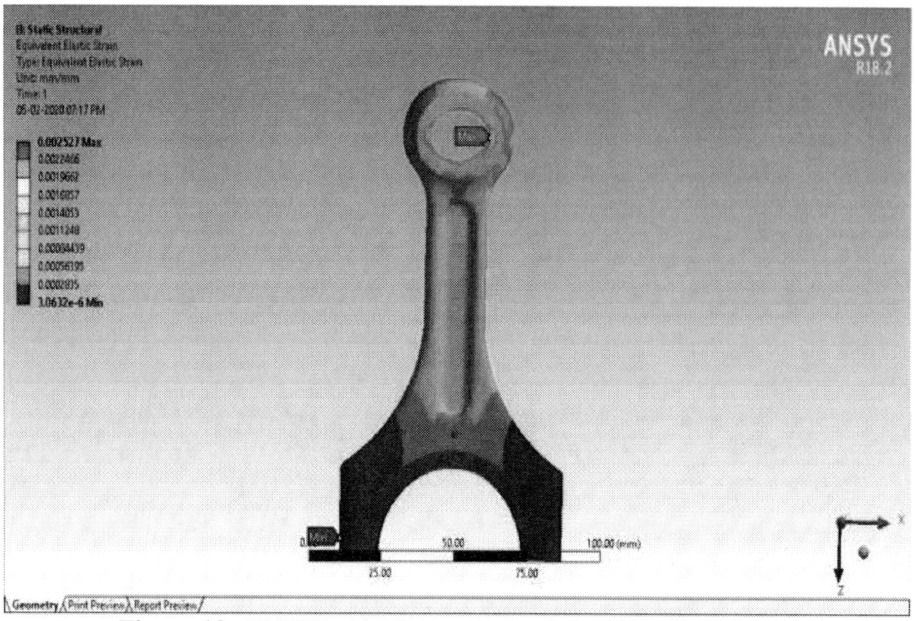


**Figure 12**: AL7075+10%SiC under load showing elastic strain

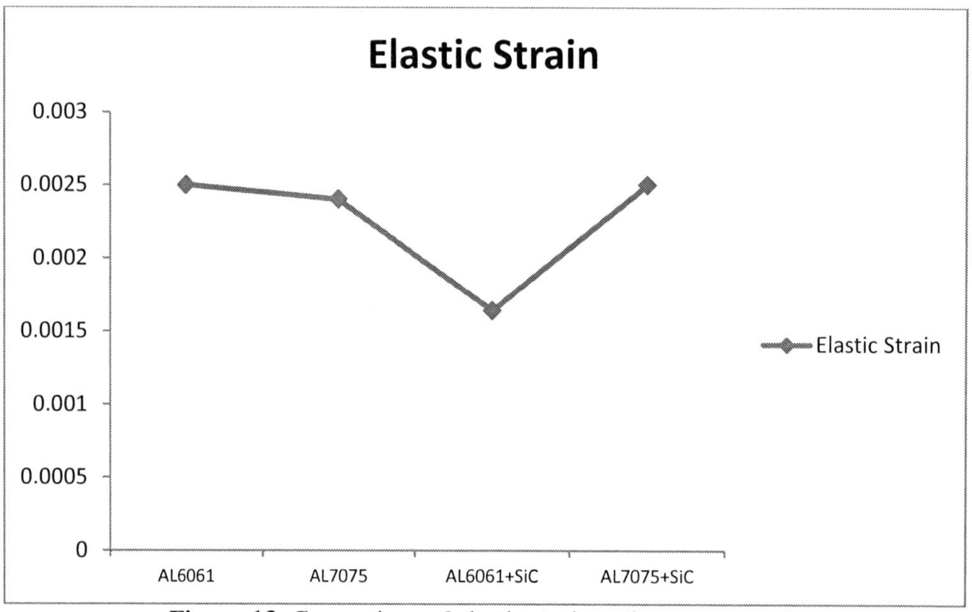


**Figure 13**: Comparison of elastic strain acting on materials

### 6.3 Deformation

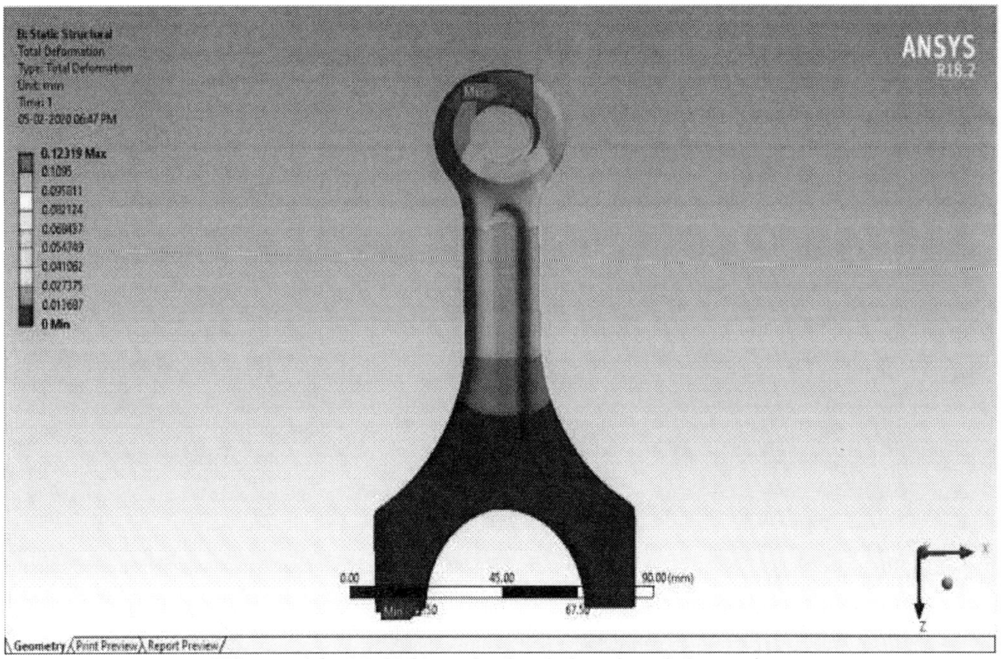


**Figure 14**: AL6061 under load showing deformation

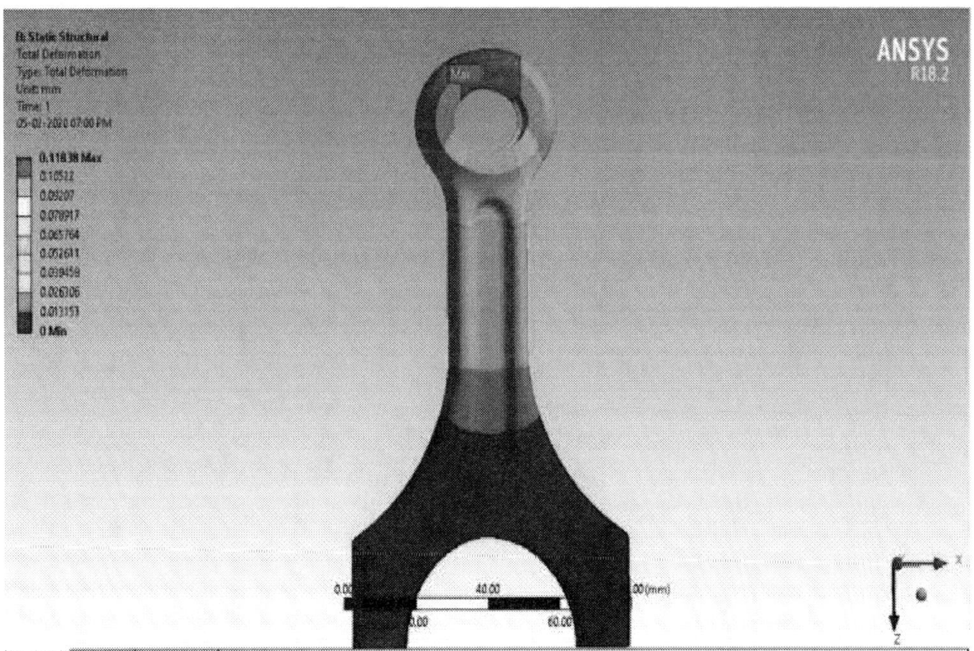


**Figure 15**: AL7075 under load showing deformation

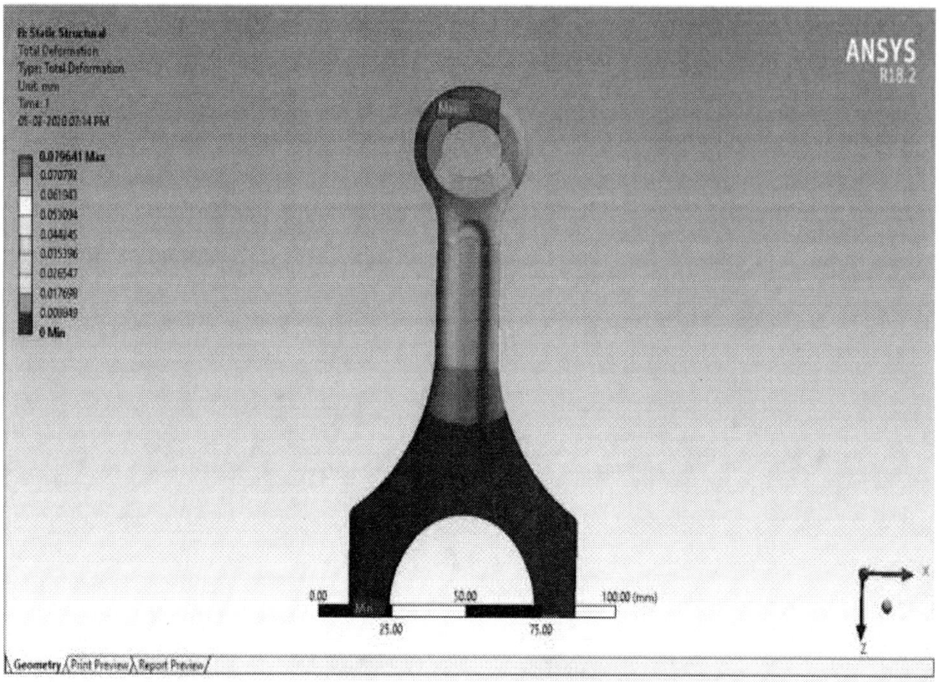


**Figure 16**: AL6061+10%SiC under load showing deformation

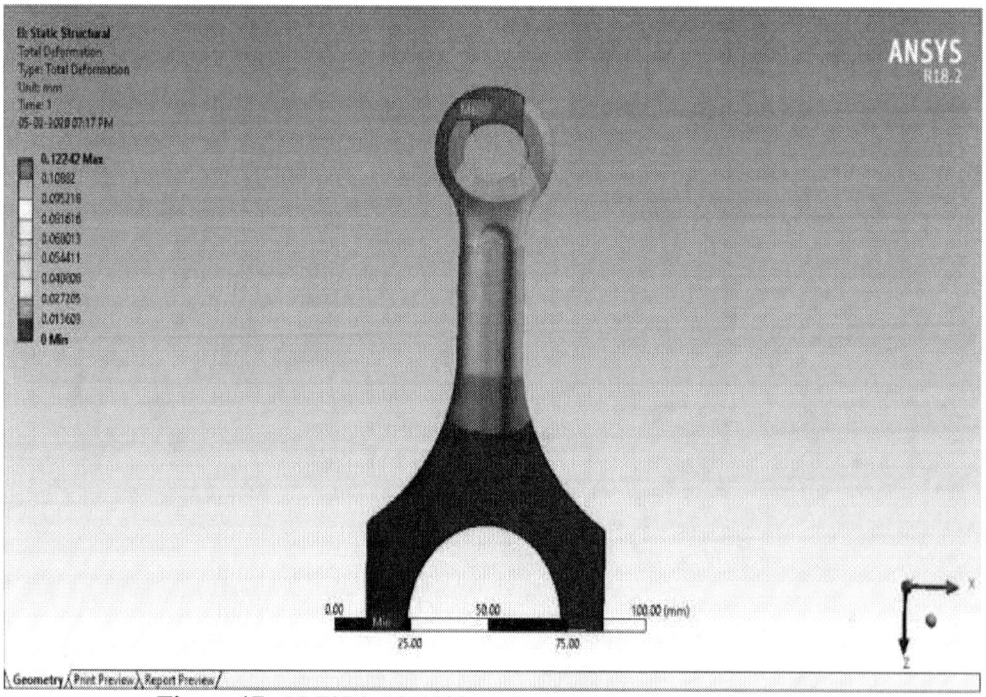


**Figure 17**: AL7075+10%SiC under load showing deformation

IOP Conf. Series: Materials Science and Engineering **923** (2020) 012001 doi:10.1088/1757-899X/923/1/012001

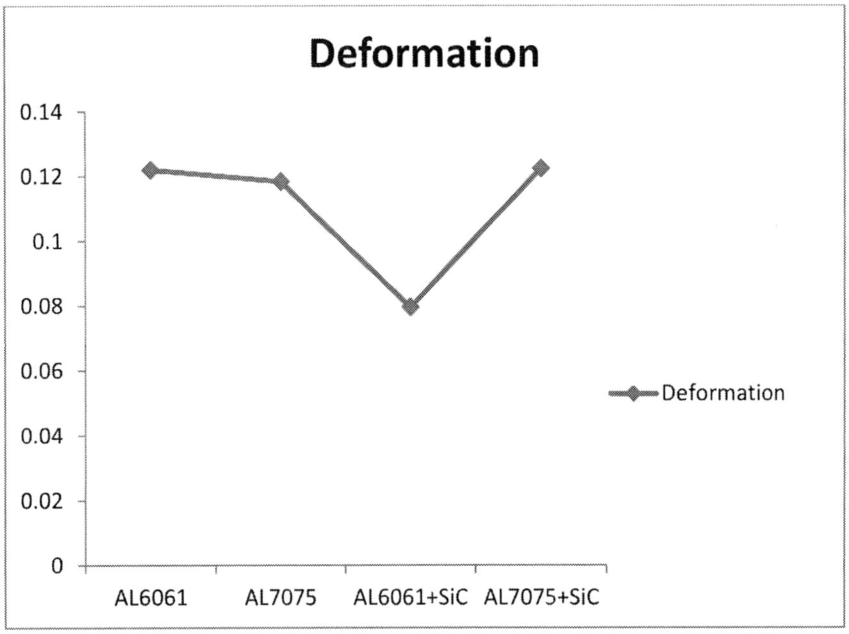


**Figure 18**: Comparison of deformation acting on materials

**Table 2**: Overall values of mechanical properties

| Materials | Stress (MPa) | Strain | Deformation (mm) |
|---|---|---|---|
| **AL6061** | 174.95 | 0.0025 | 0.1219 |
| **AL7075** | 174.95 | 0.0024 | 0.1183 |
| **AL6061+10%SiC** | 168.69 | 0.00164 | 0.0796 |
| **AL7075+10%SiC** | 162.45 | 0.0025 | 0.1224 |

## 7. Material cost calculation

Volume of connecting rod = $8.844294*10^{-5}$

**Table 3**: Mass of materials

| MATERIAL | MASS (kg) |
|---|---|
| **AL6061** | 0.2387 |
| **AL7075** | 0.2485 |
| **AL6061+10%SiC** | 0.2441 |
| **AL7075+10%SiC** | 0.2512 |

Mass of AL6061 rod = $\rho*V = 2700*8.844294*10^{-5}$
$= 0.2387$kg

Mass of AL7075 rod = $\rho$*V = 2810*8.844294*$10^{-5}$
= 0.2485kg
Mass of AL6061+SiC rod = $\rho$*V = 2760* 8.844294*
= 0.2441kg
Mass of AL7075+SiC rod = $\rho$*V = 2840* 8.844294*$10^{-5}$
= 0.2512kg
AL6061 = ₹ 215/ Kilogram
AL7075 = ₹600/ Kilogram
Silicon Carbide Powder = ₹85/ Kilogram
10%SiC alloy so 90% aluminium and 10%SiC is
Cost of 1 kg of 7075-SiC will be = 600*0.9 + 85*0.1
= ₹548.5
Cost of 1kg of 6061-SiC will be =215*0.9+85*0.1
= ₹202

*(The above values are taken from www.indiamart.com)*

Cost of connecting rod is
AL6061 = 0.2387*215 = ₹51.32
AL7075 = 0.2485*600 = ₹149.1
AL6061+10%SiC = 0.2441*202 = ₹49.3
AL7075+10%SiC = 0.2512*548.5 = ₹137.78

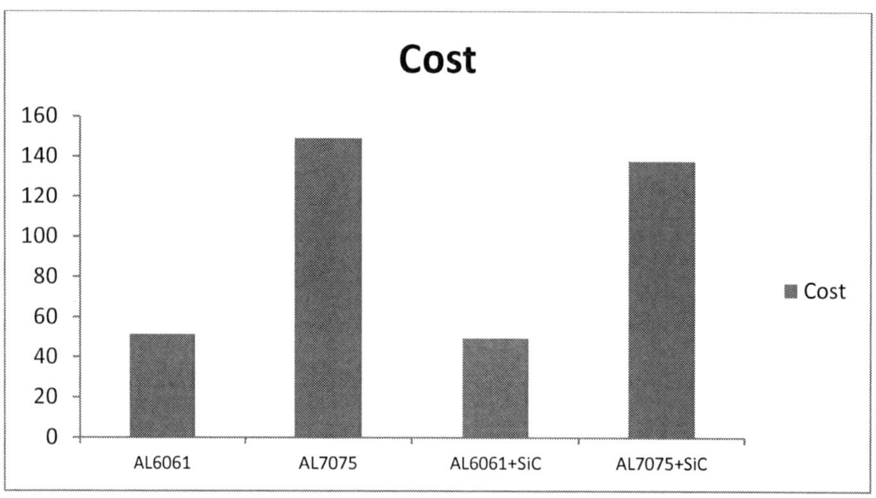


**Figure 19**: Material cost comparison

## 8. Conclusion

On comparing the various result data from the FEA analysis method, we find

- The elastic strain on AL7075 alloy is lesser than AL6061 Alloy by 4% and the total deformation of Al 7075 is lesser than AL6061 Alloy by 3%.
- When comparing the stress of AL7075+10%SiC alloy with AL7075 alloy we find that the stress reduction about 7% and increase in elastic strain by 4% and deformation by 3.5%.
- When comparing AL7075+10%SiC alloy with AL 6061+10%SiC alloy we see that AL7075+10%SiC has a lower stress by 3.7% but higher elastic strain and deformation of 3.5%.

We choose **AL6061+10%SiC** as the alternate material because it is able to withstand the applied load without much deformation or strain and also the fact that AL6061+10%SiC is also lighter in weight compared with AL7075+10%SiC and is also relatively cheap market price.

**References**

[1] M. Srinivas Rao, A. Girish Kumar, Dr.D. Vijayaganapathy.2018 *International Journal of Pure and Applied Mathematics* **118 (24)** 1-7
[2] Dr. B S N Murthy, K. Adarsh Kumar, Mohammed Abdul Shafeeq, S Sai Sundara Praveen.2017 *International Journal of Engineering Research & Technology* 7 **(3)** 1-5
[3] Sneha G.Pawar, Dhanashri A. Jagatap, Sujit S. Malgave.2017 *International Journal of Engineering Development and Research* **5 (4)** 54-56
[4] Leela Krishna Vegi, Venu Gopal Vegi.2013 *International Journal of Scientific & Engineering Research* **4 (6)** 2081-2090
[5] Vikas Singh, Sumit Verma, Harish Chandra Ray, Vishal Bharti, Abinesh Bhaskar.2017 *International Journal Research in Applied Science and Engineering Technology* **5 (5)** 1152-1178
[6] K. Arun Vasantha Geethan .2012 *International Journal of Applied Environmental Sciences*, **7** 77-92
[7] P. Sabarish Kumar, K. Karuppasamy, A.M.B. Rakheeb Basha, A.P.Vetrivel.2015 *International Journal of Applied Engineering Research* **10 (66)** 284-288
[8] Paul A. Rometsch, Yyong Zhang, Steven knight.2014 *Transactions of non-ferrous metals society of china* 1-4
[9] Sebastian Antony, Arjun A, Shinos T K, Anoop P.2016 *International Journal of Engineering Research & Technology* **5 (10)** 188-192
[10] Professor Gurunath V Shinde, Professor Vinayak D Yadav, Professor V M Jamadar, Professor G S Jadhav, Dr S M Sawant.2014 *International Journal of Engineering Research & Technology* **3 (2)**
[11] Amit Kumar B Solanki, Jimmy Maiya, Neetin Mishra Anand Rathod.2018 *International Journal of Mechanical and Production Engineering* **6 (1)** 53-56

IOP Conf. Series: Materials Science and Engineering **923** (2020) 012002 doi:10.1088/1757-899X/923/1/012002

# Enhancement of ergonomic wearable chair with voice controlled using Raspberry pi

**V P F Christina*, D Delicia Elisheba*, Y Britto*, Kevin Franklin* and Dr D Raveena Judie Dolly***
*Department of ECE, Karunya Institute of Technology and Sciences, Coimbatore.

*E-mail:cvpf@karunya.edu.in

**Abstract.** Ergonomics is the designing of products and systems that suit a range of people who use them. The Ergonomically designed wearable chair ensures the ability of the user to sit and stand anywhere. It provides the convenience and required posture for their workplace thereby overcoming disabilities caused due to long hours of standing and sitting. This device increases the efficiency of the user by reducing tiredness, pain in the lower back, leg and ankle region. Enhancing the wearable device with voice control will facile the operation of the system and provide automatic control over the device. The chair is wearable due to its exoskeleton-like design and portable due to its lightweight arrangement. This paper elaborately explains the design, mechanism, and application of the wearable ergonomic chair with voice automation. Future developments and challenges are also discussed in this paper.
*Keyword:* ergonomic, wearable, chair, voice, control, raspberry pi 3

## 1. Introduction

It is a known fact that necessity is the starting point for all inventions. Energy is the fuel for productivity. In today's fast-growing world, a person spends most of their energy at work. Sometimes they tend to work overtime. Putting for long hours in their work tends to cause physical as well as mental problems. Sitting and standing postures for long periods come under sedentary behaviors. This behavior leads to health issues like back pain, musculoskeletal disorder, cardiovascular diseases, etc.; also It is a proven fact that sitting and standing for too long can reduce life expectancy. Such kind of problems affects the productivity of the organization where they are working. A recent survey among the workers shows that 61 percent of them are working below their energy level; this is mainly due to prolonged standing. Many industries have been trying to solve this, by replacing the human force with machines. But with this replacement, the industries faced even greater problems such as maintenance, power shortage, etc.

It nearly took 20 years for researchers to seek a solution to solve the productivity problem in industries. Many solutions were brought forward by them but the problem still exists. So the engineers paved a way to an invention i.e. an intermediate device that helps the workers in both sitting and standing hence resulting in increased productivity. With the introduction of this invention, the purpose of conventional chairs has been rendered null and void. In this paper, we had renowned our invention with more sophistication by incorporating the voice module which enables the person to sit/stand easily., This device is useful for surgeons who have to perform surgery for a long duration of time where both their hands will be busy. It is also useful for factory/industry workers who perform most of their tasks in a standing position. Since it's an exoskeleton device it doesn't occupy much space like our conventional chairs. This device is power efficient, lightweight portable and walk able. In this paper, we have enhanced our base model [1], to further increase the features of the ergonomic device to adapt to the current trends of the society. Some of the health disorders and the proposed idea have been explained in detail below.

## 2. Musculoskeletal disorder

Operation Theater is a terrifying experience not only for patients but for surgeons too. They also experience a dreadful situation during surgery time. Out of 10, 9 surgeons experience musculoskeletal disorder shown in Fig 1; this is due to a continuous standing position during the surgery time [2]. There are cases where the surgeons have to perform surgery for more than 12 hrs. This leads to many health and mental illnesses. Since our energy is directly related to our production rate, to work efficiently one needs energy, decreases in time, lower the energy rate that in response lower the productivity. When it comes to surgeons it is very important to maintain the energy level, because they are lifesavers. They play a Nobel position in this society.

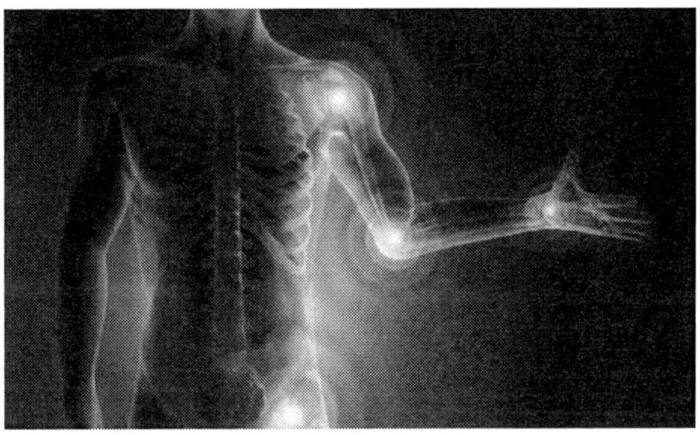

**Figure 1 :** Musculoskeletal disorder [10].

They are given a status next to god, so as a citizen and as an engineer it's everyone's responsibility to find a solution for their problem. So this device that we have designed will be used for the people who stand for a very long time especially for the people who want to switch their positions between standing and sitting, It's pretty much helpful for the surgeons because most of the time they have to use both their hands to perform surgeries in such circumstances since our device is voice automated it will be very useful for them. The main advantage of our device is that it doesn't occupy much space and it is a wearable, portable device so the user can walk easily. By using our device an individual's energy expectancy can be maintained so that the productivity and efficiency of a person will be high for example if person works on an average of 8 hrs per day without our device will work for more than 2 hrs by using our device because their energy have been saved by using this device.

## 3. Varicose Disease

It is a disorder which happens in the vein near the superficial surface of the skin shown in fig2. This type of disorder is mainly due to high blood pressure Blood flows to the heart through a valve, when this valve becomes weak the blood will start accumulated in the vein which in turn swells the vein. The valve becomes weak when a person sits or stands for a very long time. This pools the blood which is flowing through the leg. This type of disease affects your day today activity [3].

IOP Conf. Series: Materials Science and Engineering **923** (2020) 012002 doi:10.1088/1757-899X/923/1/012002

### *3.1 Symptoms of varicose disease*

- Rashes
- Change of skin color
- Inflammation on the leg
- Burning sensation on the leg
- Aching of leg etc…
- Itching on the areas around the leg

### *3.2 Prevention*

One can prevent varicose vein by improving the circulatory system and muscular tone. It's very difficult for the blood to flow against the gravity. Which consequently pool the blood around the affected areas? There are some other ways in which we can prevent this disease.

- Regular exercise
- Changing the sitting and standing position often
- Avoid wearing heels and high shoe

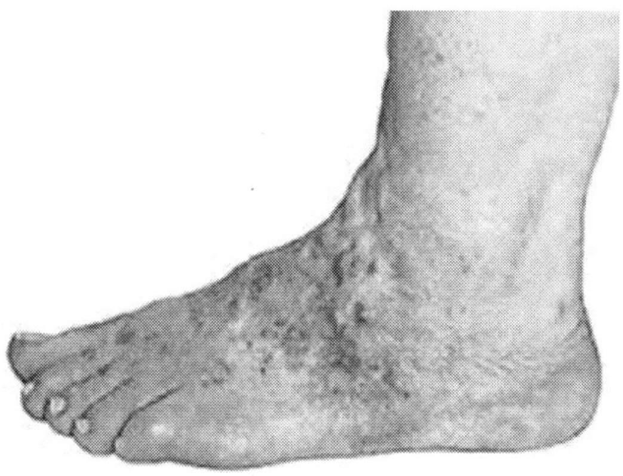

**Figure 2:** Varicose veins [11].

## 4. Base Model Working Principle

Our base model is a flexible, portable, wearable device which covers the hip, back of the leg to our foot. User has to wear the device and shoes then they have to adjust the belts that are tied tightly, once all these process are done the person have to give a physical pressure to supporting rod that is placed in the butt region shown in Fig3.

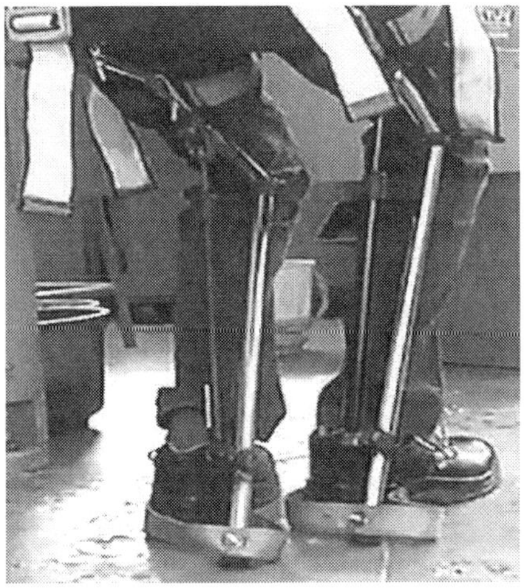

**Figure 3:** chair-less chair.

There will be a piston attached to the leg like structure which is made up of stainless steel to sit, so while applying the pressure the piston will be compressed and the person will be able to sit, and for standing the person has to release the pressure from the supporting rod which is placed in the butt region so that the piston will expand and the person will be able to stand. With a support of our earlier designed model, we taught of enhancing the model with voice automations and other stuffs, the idea and working principle of our enhanced model is discussed below.

### *4.1 Working Principle of Enhanced Model*

The enhanced version of our base model is also an exoskeleton, wearable, flexible device with voice automation to make it more easier and useful for the people who are standing for a very long time especially this voice automated chair will be useful for the surgeons [4], because there are situations where both their hands will be busy during the operation, so voice automated ergonomic chair will be helpful in such circumstances.

Once the device is booted by the battery, analog input i.e. (a continuous signal with a time varying feature) is given to the voice recognition module, this analog input is then converted into digital signals (in the form of 1's and 0's) These signals are then transmitted to the raspberry pi via a USB TTL module, it is a protocol converter which is used to convert the data to other communication standards. The signals are processed inside the raspberry pi (a versatile single board computer) and then the motors are controlled accordingly, using the relay controller, with respect to these the ergonomic device will be able to sit and stand according to the user's desire.

IOP Conf. Series: Materials Science and Engineering **923** (2020) 012002 doi:10.1088/1757-899X/923/1/012002

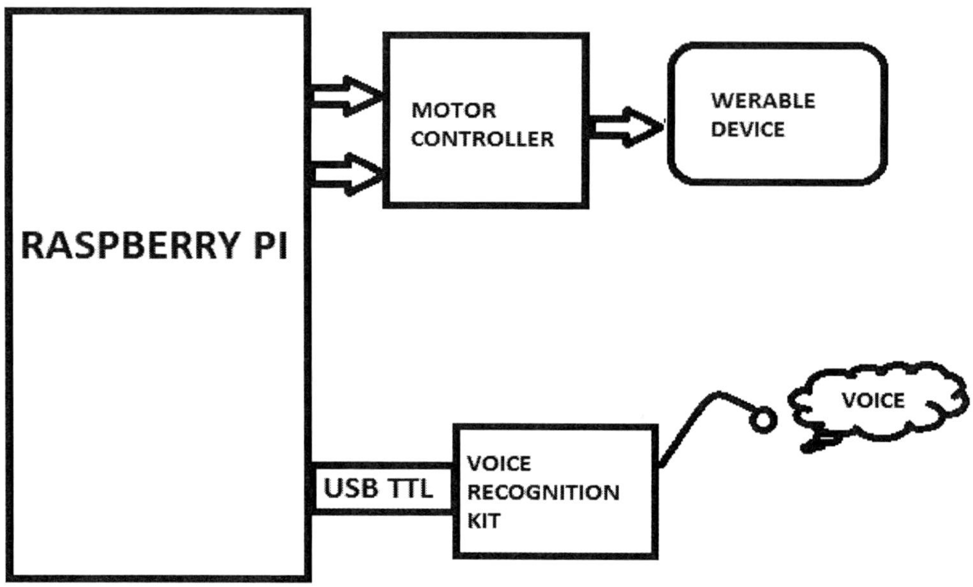


**Figure 4:** Block Diagram of Ergonomic wearable chair.

## 5. Construction of the Ergonomic Wearable Chair

The construction of the ergonomic wearable chair consists of three parts:

I. The skeletal design
II. Voice recognition System

### *5.1 The Skeletal Design*

The strength and balance of the wearable chair rely on the skeletal design of the device. The base of the skeleton is constructed with graded aluminum alloys used in aircraft. These alloys are light in weight and provide rigidity and durability. The positioning of the aluminum rods is very significant because of their major role in balancing the whole model. The skeletal design comprises two legs each containing two aluminum rods connected with the help of nuts and bolts. The connection between the two rods will bend exactly in the human knee position, making the device operate like an exoskeleton. So, when the user wants to sit or stand the rods are arranged in such a way it bends and straightens like the user's leg [5].

The linear actuator is the main component of this ergonomic wearable chair. The design consists of a linear actuator which has only two movements forward and backward, this is controlled by giving the DC power supply to the polarities of the motor.

The linear actuator is placed right behind the skeletal structure. When the actuator is energized forward the rod arrangement is expanded to make a person stand and when it is energized backward the arrangement is contracted to make a person sit. The linear actuator can carry up to 120kgs of load and requires a 12v DC power supply [9]. The belts are the next component in the design that is attached to the skeletal structure of the wearable chair to keep the device intact to the user's body. The belts are placed at the hip, thigh and ankle region to create a steady balance. The hip region's belt holds the whole device to the body and contains spaces to place the batteries to run the actuators and electronic modules to automate the device with voice control. The ankle region's belt support will keep the device intact to the user by not letting the lower rod arrangements to leave away from the user's circumference. If required extra straps or belts can be attached from the lower body to the shoulders providing extra support.

### *5.2 Voice recognition System*

It is a compact user-friendly device, it depends totally on a speaker to recognize the voice, and all types of sounds can be incorporated into the kit. Before recognizing the voice the user has to instruct or train the kit, the commands that we are instructing will be stored in a large set of the library [6]. It consists of the USB TTL module, which is just a USB (Universal Serial Bus) to serial TTL (Transistor Transistor Logic) converter which is built based on IC PL2303. This module can be used in all types of the operating system [7]. It can be used in all type of development boards to serially transmit through USB; this device does not require any external power source because it pulls the power from the pi itself.

The next component is the Raspberry pi which is manufactured in two ways they are A and B, A type of pi doesn't have an Ethernet port and it consumes less power whereas B type pi has an Ethernet. It is a minicomputer in series, it does all the functions that a computer does, it requires a SD card in which we have to install the NOOBS/RASPBEIAN and the pi is powered up by a USB it consumes a power of 3-5 V.CPU which is the brain pi is responsible for carrying the instructions through operations, it uses ARM11 processor [8]. GPU is a chip present in pi which speeds up the image calculation. A connection with raspberry pi to detect voice command is shown in Fig 5.

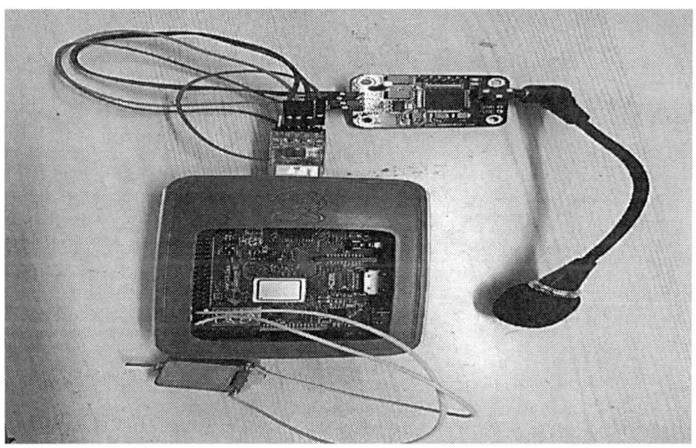

**Figure 5**: Connections with raspberry pi to detect voice command.

## 6. Comparison of Base and Enhanced Model

There are some major changes that have been implemented on our base model, which improves the efficiency and comfort of the user. Hence, the difference between the base and enhanced model are explained in the following table 1.

**Table 1 :** Comparison of base and enhanced model.

| CONTENTS | BASE MODEL | ENHANCED MODEL |
|---|---|---|
| Type of device | Mechanical | Electrical |
| Weight | 6 Kgs | 4 Kgs(excluding batteries) |
| `Material Used | Stainless steel | Aluminum graded alloy |
| Compressing component | Hydraulic piston | Linear Actuator |
| Accessibility | Can walk up to few meters | Can walk easily |
| Compressing type | Physical pressure | Automatic |
| Balancing material | Painters belt | Lase like straps |
| Shoes | Customized shoes are required | Can use their own shoes |
| Power consumption | No power required | Up to 12 v |

## 7. Applications

- Doctors/Surgeons
- Factory Workers
- Applicable for patients suffering from Spastic Diplegia
- Can be used in Physiotherapeutic Exercises
- Industrial Employees
- Useful for people working in assembly lines and distributed systems
- Applicable in the field of hair-care services
- Designers and Artists
- It is also helpful for people working in the food industry such as chefs, etc.

## 8. Challenges

There are not many challenges in this proposed model. The main challenge is the cost of the device since it will be more expensive than the conventional chairs. In this model, the fame will be light weight but with the inclusion of batteries it will get little heavier. Finally, the most basic challenge is the stability of the model, due to the placement of the linear actuator the balancing capacity of the device will be less.

## 9. Conclusion

"Too much of anything is good for nothing" is a quote that suites the condition where people work long hours standing or sitting. This proposed model will ensure that the above criterion could be managed. Invention is the key to change; this proposal can bring a change in every possible workplace by increasing the efficiency and convenience of the worker.

## 10. Acknowlegement
We would like to acknowledge program in global surgery and social health of harvard medical School, USA for providing financial support to implement this project.

## REFERENCES

[1] Delicia Elisheba D, Britto Y and Christina V.P.F, Chair-Less Chair for Lumbar Pain Reduction, International Journal of Mechanical Engineering and Technology9 (11), 2018, pp. 500–507.
[2] Guo HR, Chang YC, Yeh WY, Chen CW, Guo YL. Prevalence of musculoskeletal disorder among workers in Taiwan: a nationwide study. Journal of occupational health. 2004;46(1):26-36.
[3] Bahk JW, Kim H, Jung-Choi K, Jung MC, Lee I. Relationship between prolonged standing and symptoms of varicose veins and nocturnal leg cramps among women and men. Ergonomics. 2012 Feb 1;55(2):133-9.
[4] Halim I, Omar AR, Saman AM, Othman I. A review on health effects associated with prolonged standing in the industrial workplaces. IJRRAS. 2011 Jul;8(1):14-21.
[5] https://www.researchgate.net/publication/312368279_A_review_on_health_effects_associated_with_prolonged_standing_in_the_industrial_workplaces
[6] https://pdfs.semanticscholar.org/e644/fee06fbed7ff76997ab36ad14d051a573400.pdf
[7] Porter EW, inventor; L&H Holdings USA Inc, assignee. Voice recognition system. United States patent US 4,829,576. 1989 May 9.
[8] Yue CZ, Ping S. Voice activated smart home design and implementation. In2017 2nd International Conference on Frontiers of Sensors Technologies (ICFST) 2017 Apr 14 (pp. 489-492). IEEE.
[9] https://www.firgelliauto.com/blogs/news/how-does-a-linear-actuator-work
[10] https://www.google.com/search?q=musculoskeletal+disorders&source=lnms&tbm=isch&sa=X&ved=2ahUKEwjdo7v6-ZHoAhWyzDgGHXixBPgQ_AUoAXoECBEQAw
[11] https://www.google.com/search?q=varicose+veins&tbm=isch&ved=2ahUKEwjpxvv7-ZHoAhXFAXIKHfX9CXQQ2-cCegQIABAA&oq=varicose+&gs_l=img.1.0.0l10.49605.52925..54517...0.0..0.348.2946.2-1j8......0....1..gws-wiz-img.......0i67j0i131.sM-p55qeLO4&ei=gJpoXumGDcWDyAP1-6egBw

IOP Conf. Series: Materials Science and Engineering **923** (2020) 012003 doi:10.1088/1757-899X/923/1/012003

# A Numerical Investigation of Automotive Lambda Sensor to Improve the Life Span of the Sensor using CFD

**S Aravind[1], P Ragupathi[1], D Sathish Kumar[1], G Vignesh[1]**

[1]Department of Mechanical Engineering, Karpagam Academy of Higher Education, Coimbatore-641021, Tamil Nadu, India

aravind.s@kahedu.edu.in

**Abstract**. The modern technology has tremendously improved the controlling technologies for fuel injection system and the emission of various harmful greenhouse gases. For detecting oxygen level in the combustion chamber and exhaust gases, a sensor named lambda sensor is used in most of the automobiles. In this paper, a numerical simulation of this lambda sensor was carried out to improve the life span of the sensors protection wall by changing its design with the help of computational fluid dynamics techniques. The simulation results showed that exhaust gas entering the sensor with a pressure range reduced by about 1.41% in comparison to its prevailing design. At the central zone of the sensor, the pressure value reduced by 9.81% with the design modification performed in the protection layer of the lambda sensor. Similarly, the exhaust gas leaving the sensor showed a 6.45% pressure drop in par with the prevailing model. The velocity and temperature of the exhaust gas were the other parameters studied. The simulation results of improved protection layer design when compared to the prevailing model indicated that, the exhaust gas velocity at the inlet and central zone of the sensor was reduced by 19.37% and 17.1% respectively. The exhaust gas temperature showed a drop of 15% in the central zone of the sensor for the new improved model. Thus, from the simulation results it can be stated that, the design modification done in the protection layer of the lambda sensor will positively improvise the life span of the sensor from present life time of 2.5 Lakh kilometres or 15 years.

## 1. Introduction

In the modern world, automotive revolution is going on in a fast pace. The automotive sectors are trying their best to put forward innovations in automobiles. In order to automate the processes in the automobiles, it needs the help of engineering technology. It uses many types of sensors for controlling the processes in the automobiles. It controls processes like engine performance, parking automatically and cruise control are some of the applications where it needs the help of sensors [1,2,4-6]. One such sensor is the oxygen sensor, which measures the amount of oxygen in the fluid that is under investigation. The lambda sensor with zirconia ceramic layered sensing element with a tinny layer of platinum were mutually found in exhaust and reference region was used in this investigation [6]. It is available in both heated and unheated forms. The sensor measures the change between the amount of oxygen in the exhaust gas and air. Excess oxygen request was seen for rich mixtures. This request develops a voltage in the sensor, because of transference of oxygen particles over the sensor. Mixtures of lean nature resulted in voltage reduction due to overabundance of oxygen [4-5]. Data on oxygen fixation is directed to the engine administration system or engine control unit (ECU), this alters the measure of fuel infused into the engine to make up for abundance air or fuel. The engine control unit endeavors to keep up, all things considered, by a proper specific air–fuel proportion by translating the data it picked up from the oxygen sensor. Such a valuable sensor helps in expanding the utilization of fuel proficiently [1-6]. Many research works have been carried out in studying and analyzing such lambda sensors [4-15]. Tiffee et al. [4] described various types pf oxygen sensors for emission control and reduce the fuel consumption by controlling the air fuel ratio. They elucidated about the lambda sensors with for DI engines with $\lambda = 1$ and other sensors with $\lambda > 1$. Shuk et al. [7] tested oxygen sensors with different electrode materials and found that oxide electrodes showed fast response and long-term stability for varying oxygen (O2) concentration. Antonio et al. [10] studied the application of automotive lambda sensor for measuring the concentration of O2 in the exhaust flue gases in industrial combustion system. They suggested two different methods for use of lambda sensors in the industry. Lima et al. [8] reported the application of titanium dioxide thin film gas sensor for measurement of lambda and analyzed to check its response for different O2 concentration in the exhaust gases. Izu et al. [9] investigated the resistive O2 sensor using ceria-zirconia as the sensing element and it acted as the temperature recompensing material. They noticed that sensor elements made of such material was independent of the temperature for wide range of 773 to 1073 K. Klett et al. [13] in their work numerically and experimentally tested the lambda sensor with double protection tube for its heat transfer, flow pattern and temperature field. They found that the projected outcomes of numerical model were in phase with the experimental outcomes. This helped to find the aging and heating power of the lambda sensor. The performance of such lambda sensors can also be studied effectively by creating numerical models using some of the commercially available numerical tools. Since it reduces the experimentation cost and time. Some of the researchers put in their interest in modeling the lambda sensor using CFD tools. Bruck et al. [15] conducted numerical and experimental investigation on lambda sensors to find the mass transfer of the exhaust gases flowing through them. Experimental setup based on similarity theory was modelled and the same was numerically modelled in CFD software. This model projected the flow outline and species transference inside the shield of the lambda sensor. Zhang et al. [11] performed a numerical investigation using CFD tool in an exhaust system. This work provided the guidelines for the application of CFD tools in various zones of the exhaust system. This helps in building up of new concepts and modifying the existing design for optimum performance. Chen et al. [12] performed CFD analysis to model the flow field in an oxygen sensor to envisage the convection coefficient, poisoning and swapping time. The CFD investigation results were found to be in par with the tested outcomes.

Many works have been done in lambda sensors but the thought of improvising the life time of the sensor was not done. So, in this present numerical investigation, design of the protection barrier of the lambda sensor is modified and analyzed under the real-time observations to extend the life time. The existing and new design of the protection wall was modeled and analyzed using CFD package.

## 2. Numerical analysis

### *2.1. Working of Lambda sensors*

The ALS is introduced in the engines exhaust gas framework at a location where the engines exhaust temperature range is quite good, which is important for proficient sensor working. The sensor projects into the exhaust gas, outline being with the end goal that one cathode surface is encompassed by the exhaust gas and the other is associated with the environment. Indeed, even under abundance fuel conditions, there is constantly residual oxygen in the spark ignition engine's exhaust gas (at lambda = 0.94, approx. 0.101 to 0.301% by volume). Amid progress from a lean mixture (high lingering oxygen level) to a rich mixture (low leftover oxygen level), the remaining oxygen level diminishes suddenly by a power of ten in the stoichiometric location (lambda = 1) of air-fuel mixture. This outcomes in the abrupt rise in voltage of the sensor which yields at lambda = 1. The voltage range of the sensor and its inward resistance rely on temperature. Dependable closed loop control happens at temperatures over 360°C (unheated sensor) or more 140°C (warmed sensor) [6]. The characteristics of the Lambda sensor working at 600°C for Rich mixture is explained in figure 1 (air insufficiency) and Lean mixture (air excess). The engine (controlled framework), the Lambda sensor (measuring component), the controller in the engine control unit, and the injectors (actuators) are in closed loop control. The variable under control in the exhaust gas is the leftover oxygen [5]. The objective is to produce an ideal air-fuel mixture by modifying the fuel quantity (controlled variable) infused by the injectors is done by the control loop system. The closed loop control naturally considers uncommon modes, for example, commence, increasing speed, and full load. In the Lambda Sensors, the outer wall is protecting core part or heart of lambda sensors from the high temperature and high velocity exhaust gas. Two wall protections and three wall protection systems are available in the lambda sensors. These walls are reducing the temperature, pressure, velocity and other parameters in the exhaust gas flow. The figure 2 shows the SI engine with exhaust system with catalytic converter and position of the lambda sensor [5].

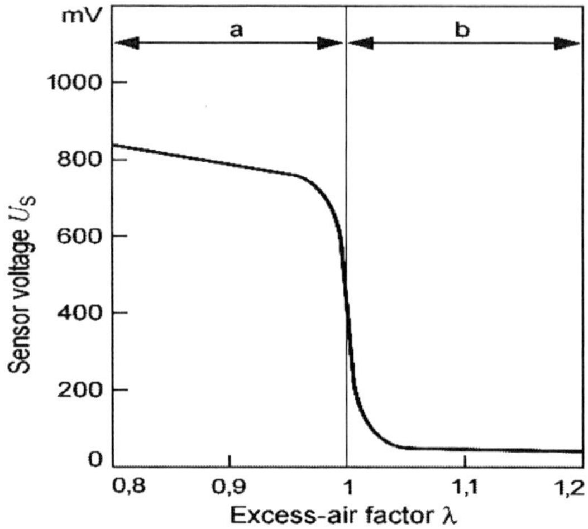


**Figure. 1** Voltage characteristics of the Lambda sensor at working temperature of 600°C for a. Rich mixture (insufficient air) b. Lean mixture (excess air)

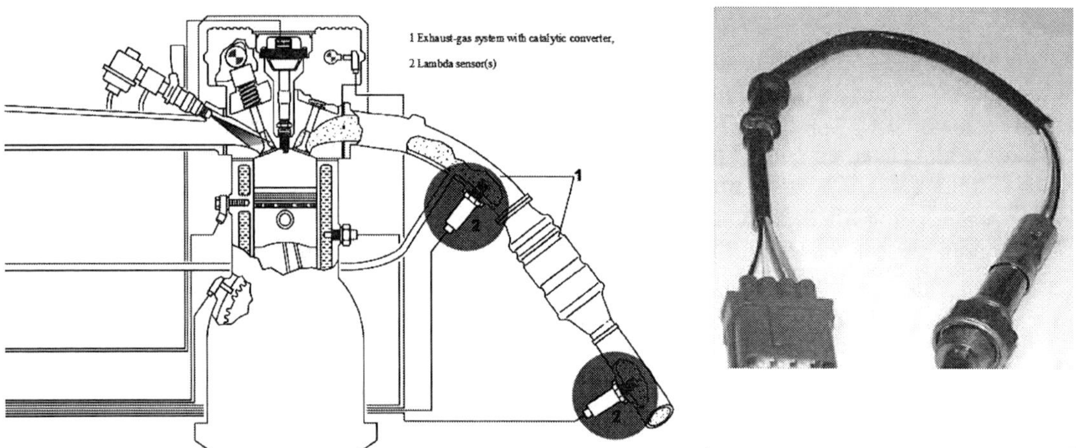


a) b)

**Figure. 2** a) Engine with exhaust-gas system b) Lambda sensor

*2.2. Computational fluid dynamics (CFD)*

CFD confines over area of interest, with stated (known) circumstances on the bounded area. It is a numerical tool for simulating the behaviour of systems involving fluid stream, heat transfer and other associated physical processes with the help of computers. It works by resolving the fluid flow equations starting with equations of conservation of mass, momentum and energy [16-18]. These are given below from eq. 1, 2a-2c and 3. The CFD became a vital tool in practically every branch of fluid dynamics, from aerospace applications to weather forecast. Different stages in Fluent Analysis are,

1. Pre-processing: problem formulation, governing equations and constraints in the boundary; Discretization.
2. Solving: Numerical solution of the given equation.
3. Post- processing: Plot and analyze the results.

IOP Conf. Series: Materials Science and Engineering **923** (2020) 012003 doi:10.1088/1757-899X/923/1/012003

*2.2.1 Governing equations.*

The equations governing the flow are obtained by the application of physical principle [16-18]. The most important equations governing the fluid dynamics are namely:

a. Mass is Conserved (Continuity equation)

The integral form of the continuity equation:

$$\frac{\partial}{\partial t}\iiint \rho dv + \iint \rho V.ds = 0 \tag{1}$$

b. Newton's second law, F=ma (Momentum equation)

X-component

$$\rho\frac{Du}{Dt} = -\frac{\partial p}{\partial x} + \frac{\partial \tau xx}{\partial x} + \frac{\partial \tau yx}{\partial y} + \frac{\partial \tau zx}{\partial z} + \rho f_X \tag{2}$$

Y-component

$$\rho\frac{Dv}{Dt} = -\frac{\partial p}{\partial y} + \frac{\partial \tau xy}{\partial x} + \frac{\partial \tau yy}{\partial y} + \frac{\partial \tau zy}{\partial z} + \rho f_Y \tag{3}$$

Z-component

$$\rho\frac{Dw}{Dt} = -\frac{\partial p}{\partial z} + \frac{\partial \tau xz}{\partial x} + \frac{\partial \tau yz}{\partial y} + \frac{\partial \tau zz}{\partial z} + \rho f_Z \tag{4}$$

c. Energy is conserved (First law of thermodynamics equation)

$$\rho\frac{D}{Dt}\left(e+\frac{V^2}{2}\right) = \rho\dot{q} + \frac{\partial}{\partial x}\left(k\frac{\partial T}{\partial x}\right) + \frac{\partial}{\partial y}\left(k\frac{\partial T}{\partial y}\right) + \frac{\partial}{\partial z}\left(k\frac{\partial T}{\partial z}\right) - \frac{\partial(up)}{\partial x} - \frac{\partial(vp)}{\partial y}$$
$$-\frac{\partial(wp)}{\partial z} + \frac{\partial(u\tau xx)}{\partial x} + \frac{\partial(u\tau yx)}{\partial y} + \frac{\partial(u\tau zx)}{\partial z} + \frac{\partial(v\tau xy)}{\partial x} + \frac{\partial(v\tau yy)}{\partial y}$$
$$+\frac{\partial(v\tau zy)}{\partial z} + \frac{\partial(w\tau xz)}{\partial x} + \frac{\partial(w\tau yz)}{\partial y} + \frac{\partial(w\tau zz)}{\partial z} + \rho f_V \tag{5}$$

*2.3 Methodology*

Life time of the Lambda (Oxygen) sensors depends on the Exhaust gas flow parameters (Temperature, Pressure, Velocity) Starting conditions, Usage of gears, Exhaust gas containment and others. It is known that by reducing the temperature at the sensing element area we can reduce the pressure by Gas eqn. PV=RT which leads to the extended life time of Lambda sensors. By changing the protection layer design, exhaust gas temperature can be reduced by making an obstacle in the flow passage. To perform that break, there is a need to change the design in existing protection wall [11]. The exhaust gas flow inside the sensor need to be determined for the impact of new design on temperature loss. The figure 3 shows the prevailing design of the double wall and triple wall protection lambda sensors.

Figure 3. prevailing design of double wall and triple wall protection lambda sensor

## 3. Results and discussions

The CFD analysis of the existing double wall protection lambda sensor was performed and the figures 4-6 shows the pressure, velocity and temperature distribution respectively.

*3.1. Double wall protection lambda sensor*

The exhaust gas at high temperature enters the sensor through the protection wall and passes over the sensing element. The pressure inside the sensing chamber and the velocity distribution is shown in the figure 4 and 5. The Figure 6 shows the distribution of temperature inside the sensor when the hot exhaust gas enters.

The exhaust gas maximum pressure is 2.05 kPa at the entry of exhaust gas. The pressure at the core sensing part is around 2.0 kPa and the minimum pressure is 1.98 kPa at exit of exhaust gas from the lambda sensors. The maximum velocity in the exhaust gas is 89 m/s at the entry of exhaust gas. The velocity at the core sensing part is around 49 m/s.

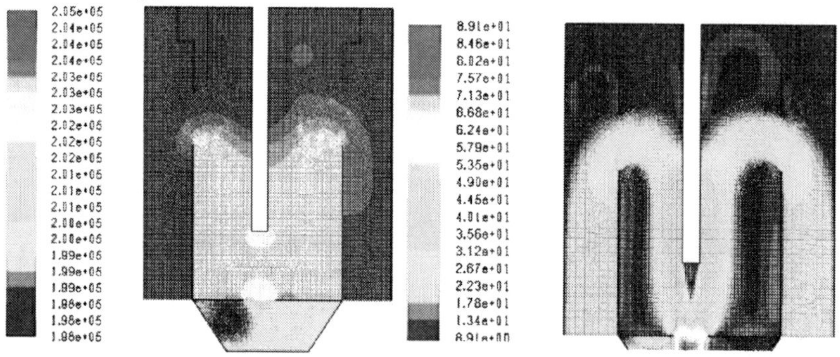


Figure 4, 5. Pressure and velocity distribution of double wall protection lambda sensor

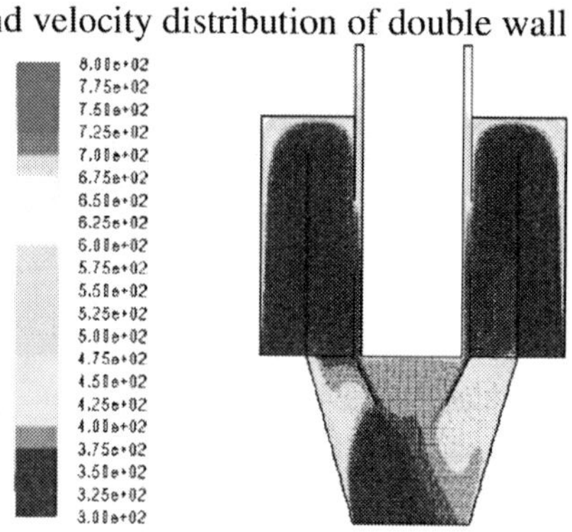


Figure 6. Temperature distribution of double wall protection lambda sensor

The exhaust gas maximum temperature is 800°C at the entry of exhaust gas. The temperature at the core sensing part is around 675°C.

### 3.2. *Triple wall protection lambda sensor*

The exhaust gas at high temperature enters the sensor through the protection wall and passes over the sensing element. The pressure inside the sensing chamber and the velocity distribution is shown in the figure 7 and 8. The figure 9 shows the distribution of temperature inside the sensor when the hot exhaust gas enters. The exhaust gas maximum pressure is 2.23 kPa at the entry of exhaust gas. The pressure at the core sensing part is around 2.2 kPa and the minimum pressure is 1.97 kPa at exit of exhaust gas from the lambda sensors. The maximum velocity in the exhaust gas is 90 m/s at the entry of exhaust gas. The velocity at the core sensing part is around 35m/s.

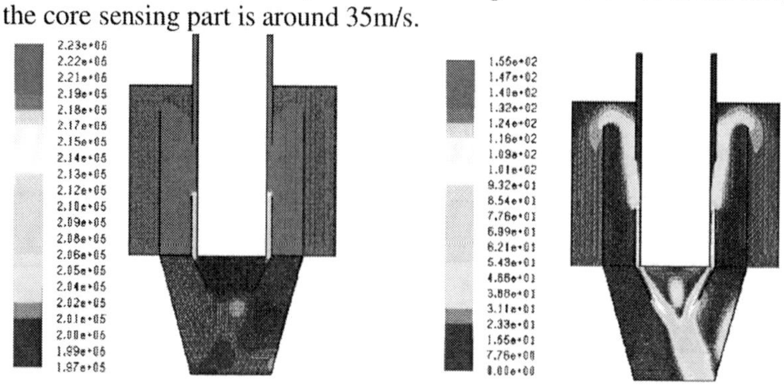


**Figure 7. & Figure 8.** Pressure and Velocity distribution of double wall protection lambda sensor

The exhaust gas maximum temperature is 800°C at the entry of exhaust gas. The temperature at the core sensing part is around 700°C.

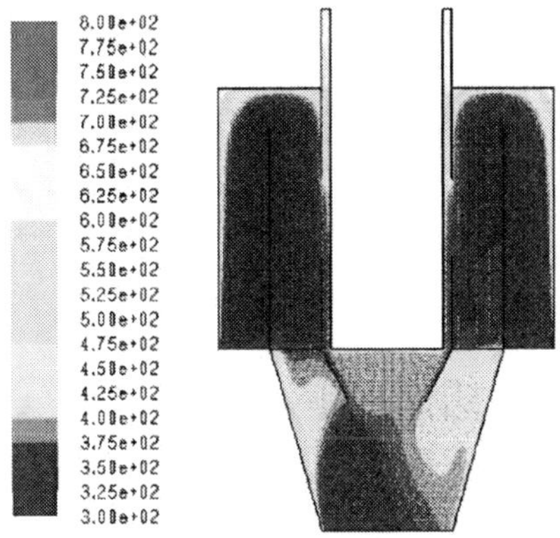


**Figure 9.** Temperature distribution of double wall protection lambda sensor

### *3.3. Modified double wall protection lambda sensor*

The exhaust gas at high temperature enters the sensor through the protection wall and passes over the sensing element. The pressure inside the sensing chamber and the velocity distribution is shown in the figure 10 and 11. The figure 12 shows the distribution of temperature inside the sensor when the hot exhaust gas enters. The exhaust gas maximum pressure is 2.20 kPa at the entry of exhaust gas. The pressure at the core sensing part is around 2 kPa and the minimum pressure is 1.85 kPa at exit of exhaust gas from the lambda sensors. The maximum velocity of the exhaust gas is 75 m/s at the entry of exhaust gas. The velocity at the core sensing part is around 30 m/s.

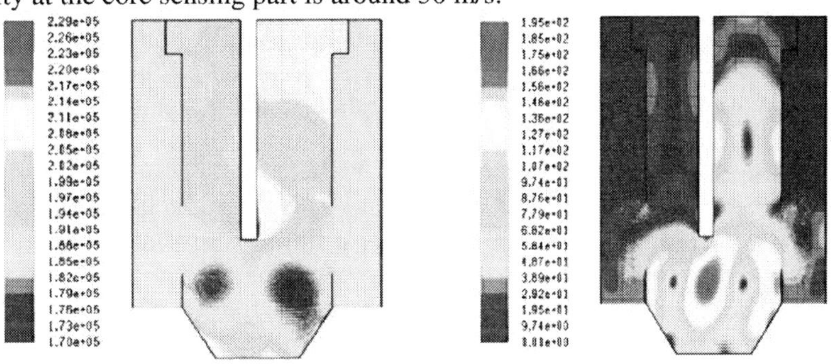


**Figure 10. and Figure 11.** Pressure and Velocity distribution of modified double wall protection lambda sensor

The maximum temperature exhaust gas is 800° Cat the entry of exhaust gas. The temperature at the core sensing part is around 610° .C

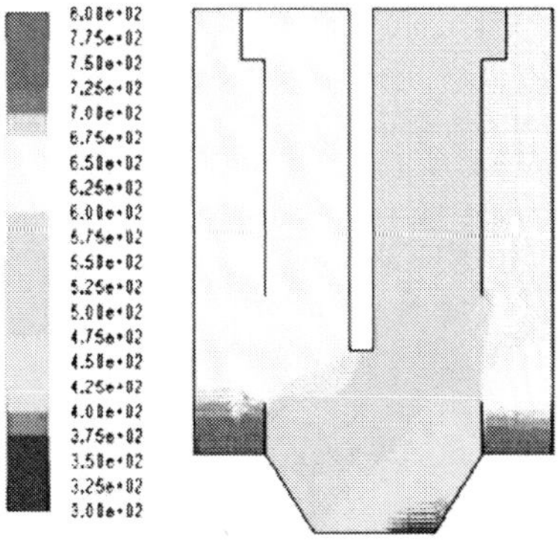


**Figure 12.** Temperature distribution of modified double wall protection lambda sensor

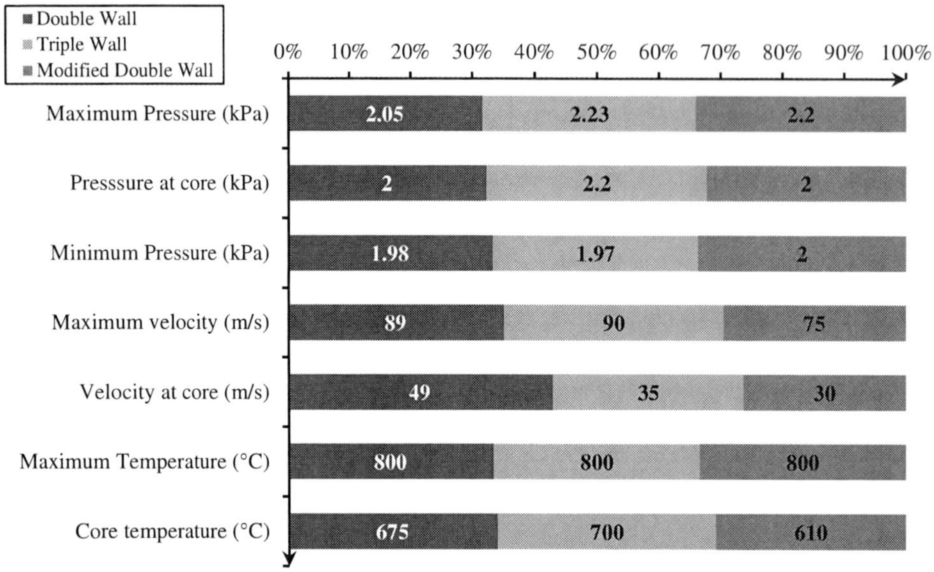


**Figure. 13** comparison charts of pressure, velocity and temperature of the existing and modified protection wall.

The comparison of the pressure, velocity and temperature distribution of existing protection wall and modified protection wall is shown in the chart form in figure. 13.The analysis result shows that, the modified double wall protection lambda sensor is better than the existing ones. The pressure distribution of modified double wall protection at the core of the sensor reduced by about 10% when compared to triple wall protection. This is due to the positioning of the exhaust gas entry to the core somewhere at the middle of the sensor wall module [11, 12, 15]. The exhaust gas velocity entering the center of the modified double wall protection sensor reduced by 63.3% and 16.6% when compared to the triple and double wall protection sensor. This reduction in the velocity distribution in the modified design is due to the presence of the inlet for the entry of exhaust gas at somewhere middle. This acts as a block for the streamlined flow of the exhaust gas [5, 8,11]. The temperature distribution at the core of the modified double wall protection sensor decreased by about 14.75% and 10.65% when compared to the double and triple wall protection sensor. This reduction in temperature distribution of the modified design is due to the delayed contact between the wall and the core area of the sensor [10, 13, 14].

## 4. Conclusion

The outcomes of the investigation show that the design modification made in the protection layer of the lambda sensor performs more effectively when compared to the prevailing design. The simulation results show the exhaust gas pressure at the entry level has reduced by about 1.35% in par with the prevailing design. There is a reduction of about 10.1% in the pressure at the central zone of the lambda sensor by the design modification made in the protection layer. The exhaust gas pressure at the exit of the sensor also show a reduction by 6.4% due to the design alteration made. The design of protection layer of the lambda sensor was modified in this investigation because of this, the exhaust gas velocity at the inlet region and central region reduced by 20.5% and 16.5% respectively. The third parameter under study was the exhaust gas temperature which got reduced by 15% at the central zone of the sensor. Thus, the simulation results of prevailing protection layer design and the modified design were compared and found that the modified design showed appreciable reduction in the all the three parameters like exhaust gas pressure, velocity and temperature. This would certainly increase the life of the oxygen sensor which is around 15 plus years for existing model to around 17 plus years or even more.

**References**

[1] Gibson, R.W., Kumar, R.V. and Fray, D.J., 1999. Novel sensors for monitoring high oxygen concentrations. Solid State Ionics, 121(1-4), pp.43-50.

[2] Auckenthaler, T.S., Onder, C.H. and Geering, H.P., 2002. Control-oriented investigation of switch-type air/fuel ratio sensors. IFAC Proceedings Volumes, 35(1), pp.331-336.

[3] Ivers-Tiffée, E., Härdtl, K.H., Menesklou, W. and Riegel, J., 2001. Principles of solid state oxygen sensors for lean combustion gas control. Electrochimica Acta, 47(5), pp.807-814.

[4] Riegel, J., Neumann, H. and Wiedenmann, H.M., 2002. Exhaust gas sensors for automotive emission control. Solid State Ionics, 152, pp.783-800.

[5] Moos, R., Rettig, F., Hürland, A. and Plog, C., 2003. Temperature-independent resistive oxygen exhaust gas sensor for lean-burn engines in thick-film technology. Sensors and Actuators B: Chemical, 93(1-3), pp.43-50.

[6] Xu, Y., Zhou, X. and Sorensen, O.T., 2000. Oxygen sensors based on semiconducting metal oxides: an overview. Sensors and Actuators B: Chemical, 65(1-3), pp.2-4.

[7] Shuk, P., Jantz, R. and Guth, H.U., 2012. OXYGEN SENSOR WITH ADVANCED OXIDE ELECTRODE MATERIALS. International Journal on Smart Sensing & Intelligent Systems, 5(1).

[8] de Lima, L.C., Carmona, H.A., da Silva, C.V. and Junior, F.C., 2011. Oxygen Excess Control of Industrial Combustion Through the Use of Automotive Lambda Sensor. Int. Trans. J. Eng. Manag. Appl. Sci. Technol, 3(2), pp.365-373.

[9] Izu, N., Nishizaki, S., Shin, W., Itoh, T., Nishibori, M. and Matsubara, I., 2009. Resistive oxygen sensor using ceria-zirconia sensor material and ceria-yttria temperature compensating material for lean-burn engine. Sensors, 9(11), pp.8884-8895.

[10] de Souza Sobrinho, A.S., Junior, F.S.C. and de Lima, L.C., 2012. Monitoring industrial combustion through automotive oxygen sensor. International Transaction Journal of Engineering, Management, & Applied Sciences & Technologies, 3(2), pp.203-205.

[11] Zhang, X. and Romzek, M., 2008. Computational fluid dynamics (CFD) applications in vehicle exhaust system (No. 2008-01-0612). SAE Technical Paper.

[12] Chen, D.K., Jeswani, P. and Li, J.Z., 2000. Optimization of oxygen sensor. SAE transactions, pp.848-856.

[13] Klett, S., Piesche, M., Weyl, H., Wiedenmann, H.M., Schneider, U. and Neumann, H., 2004. Numerical and experimental analysis of the 3D flow-pattern in exhaust gas sensors. SAE transactions, pp.108-123.

[14] Li, C., Xia, J.Y., Chu, J., Wang, Y.H., Zhuang, Y.P. and Zhang, S.L., 2013. CFD analysis of the turbulent flow in baffled shake flasks. Biochemical engineering journal, 70, pp.140-150.

[15] Brück, M., Mader, G., Piesche, M. and Klett, S., 2007. Numerical and Experimental Analysis of the Mass Transfer in Exhaust Gas Sensors (No. 2007-01-1144). SAE Technical Paper.

[16] Tu, J., Yeoh, G.H. and Liu, C., 2018. Computational fluid dynamics: a practical approach. Butterworth-Heinemann.

[17] Jamshed, S., 2015. Using HPC for Computational Fluid Dynamics: A Guide to High Performance Computing for CFD Engineers. Academic Press.

# IOT IN CAD 4.0 WITH AUGMENTED REALITY GLASS

**Nigal Ashik P.A[1], Suseendhar.P[2] Manoj.N[3], Subashchandrabose.P[4]**

1, 3, 4 Department of Mechanical Engineering, Jai Shriram Engineering College, Tirupur, Tamil Nadu, India

2 Department of Electrical and Electronics Engineering, Jai Shriram Engineering College, Tirupur, Tamil Nadu, India

G-mail: nihalashik@gmail.com

**Abstract**. This product absorbs the power of our smart phone to endow with the augmented reality (AR) experience. The sensor data fusion and camera allow the software to accurately recognize and trail the surroundings. Radial target image can be easily identified and trailed with the help of our smartphone, thereby creating a link between the holographic image and the target image is possible. This software allows us for the viewing of our own 3D CAD models in augmented reality. The software makes it achievable to sight any CAD models in accurate three dimensions (3D), to size, and in the real context of our own situation. Through mirror, lightweight stereoscopic lenses, and combiner glass, stereo image can be focussed on display before our eyes. As a result, virtual 3D holographic image is generated, which can be viewed from any angle.

## 1. Introduction

Computer Aided Design (CAD) is the tool, which plays main vital role in every mechanical and automobile organization. Every product should be designed by design engineer as per organization requirement. But visualizing the CAD structure in a computer screen is not sufficient for uneducated peoples and also it is traditional method. Augmented reality gives better solution of these problems. Augmented reality (AR) is a type of technology which makes blend between a real world and computer graphics coexisting in the same place. It has proven in many different areas, such as engineering, medical and education. There are some smart glasses gained attention by most of the peoples who loves AR experience. In any mechanical organization customer satisfaction is not only depends upon quality, time and budget of the components. Now a day's customers expecting from organization how they different from others. Those cultures will be come when, through only these types of technical interactions with our customers and their feedbacks. When compared to smart glasses, augmented reality glasses much cheaper than that. Augmented reality glasses give better experience of our own customers. Here both software (UNITY software) and hardware (AR glass) gives better AR experience in many industrial and non-industrial fields. Here it is manufactured only for mechanical and automobile organization's development. But it is also used for some other fields such as civil construction, architecture planning, education and medical. In civil construction it is used for constructional design and showing that to clients in a virtual 3D view for customer satisfaction. The organization will be developed through these types of activities. In architectural it is similar to constructional design but here we can place our Computer Aided Design (CAD) structure in real and we can transport or change the colour of the particular product as per the requirement. In medical field it is used as a tool for patient education, both for treatment and disease prevention. Through this patient can learn about taking medicines or contacting doctors when specific symptoms occur on their body. Most of the students having smart phone now a day. So, there is no need to any new equipment for AR experience. They just installed this software in their smart phone is enough to achieve AR experience.

## 2. Literature review

### *2.1. Mehdi Mekni and E Lemieux, 2014.*
The author explained about augmented and virtual reality in a mobile technology. These projects used for various applications in all over the domain such as medical, military, robotics, games, entertainment, manufacturing etc., here the author conclude as it is a future based project in architecture, engineering and construction.

### *2.2. Petar Pejić and Taško Rizov, 2014.*
Augmented reality gives rapid development in manufacturing industry. For example, author gives real time pictures of the actual application in both mechanical and architectural field. Here the author says it is an emerging technology in education. Students can easily identify the actual 3D structure of 2D diagram which is being the textbook also they can get real time experience.

### *2.3. Jorge Bacca and Silvia Baldiris, 2016*
Author referred 32 studies published in indexed journals and he gives an overall literature of the augmented reality and virtual reality in education. Both AR and VR gives better experience to the students who want to know about actual structure of what they are studying in their textbook. Author describes the level of augmented reality in education in the period of 2014.

### *2.4. Philipp A. Rauschnabel and Alexander Brem, 2016*
Here author compared augmented reality in mobile technology and wearable gadgets such as Microsoft hololens, google glass and smart glass. AR gadgets gives better experience than the mobile technology when application like a mechanical, architecture and medical. By using this method customer intention can be reduced.

### *2.5. Abrar Omar Alkhamisi and Muhammad Mostafa Monowar, 2013*
Author describes augmented reality growth and future applications in a various field. Study explains about how to get customer attention, fulfilling their expectation and economic growth compared to competitive companies in the market. Also, it deals about various applications such as medical, education, news and sports.

## 3. Augmented reality and virtual reality

### *3.1. Augmented reality (AR)*
AR is a field of view (FOV) where it mixes of both real world and virtual world through software. It gives better augmented experience of software user's when compared to virtual reality. It is the current research of all over the software developing organizations. Through this we can scale, rotate any designed CAD part as per our requirement.

### *3.2. Virtual reality*
Virtual reality is the field of view (FOV) where it consists only graphical world through software. It is the traditional method of viewing any CAD parts with the help of computers. It is only possible when user know about that particular software. Best example of virtual reality is games. Every game developed under the virtual reality platform.

## 4. Smart glass

It is a type of glass which is used to provide better augmented experience for users. There are few glasses developed for AR experience such as, google AR glass, AR Microsoft hololens and some other AR glasses like a game glass.

### *4.1. Google glass*

It is a device which can be wearable by peoples. It is accessed by human's voice or touch control. It can be identified by GPS location also it gives current status of flight, train, time, temperature and location.

### *4.2. Microsoft hololens*

Compared to google glass, it gives real time AR experience of the actual object in a particular place. Industrial peoples referring hololens compared to google glass for purchasing.

### *4.3. AR Glass*

It is not like a smart glass, but it is cheaper than that cause of there is no need of importing programs into this. It has basic construction and simple principle of working to achieve AR experience.But it needs smart phone to achieve AR vision as shown in figure. Here it gives very much support to augmented reality view for users.

## 5. Procedure of software development

### *5.1. CAD Design*

Computer aided design is the use of computers to create, modify or analysis and design. There are number of designing software's available for both constructional and mechanical fields. But augmented software can read any 3D CAD models as well as poly modal objects. Through this we can achieve augmented experience when we designed in any CAD software.

### *5.2. FBX Conversion*

This software can accept any 3D object which is designed in any designing software. But the main condition is importing file format should be in a film box (FBX) format. In figure. We can easily understand the basic difference between field vision of both CAD and FBX format files. Because of unity is game developing software and it can read only FBX, 3Dmax and some other formats.

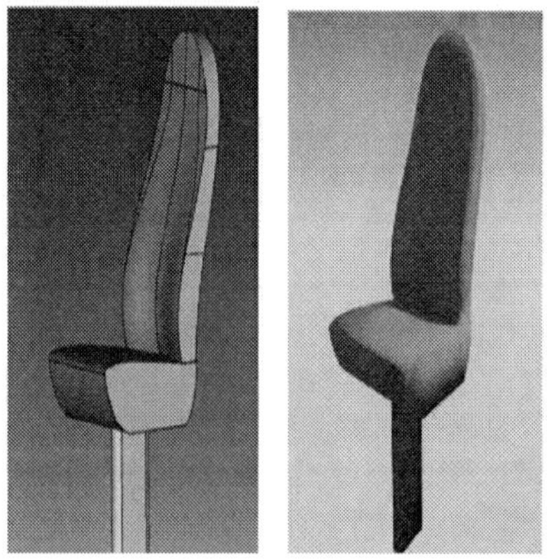

Figure 1. converted file from CATPart to FBX

IOP Conf. Series: Materials Science and Engineering **923** (2020) 012004 doi:10.1088/1757-899X/923/1/012004

### *5.3. Development of license and target image through Vuforia developer*
Here the proccss of buying license and target image which is user will scan in future for occurring holographic image. Vuforia developer website providing unity license and database for user's target image. Actual screen shot of Vuforia website as shown in figure. Here the format of image file (JPG) converted to unity file format. Then we can import both our license and target data base easily in a unity software.

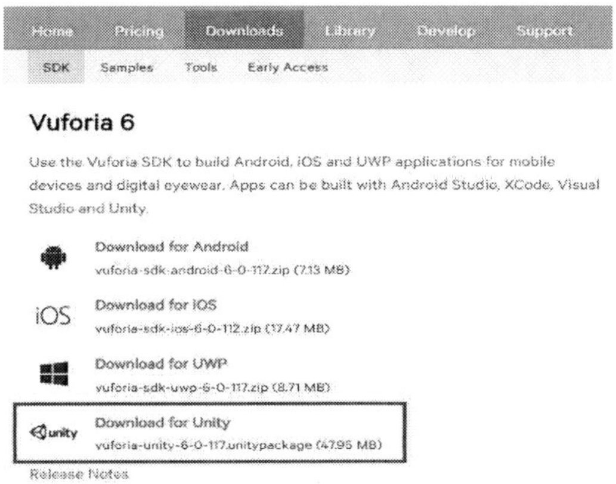


Figure 2. screen shot of Vuforia developer website

### *5.4. Unity software*
Here only the maximum process taken for achieving augmented reality experience in user's smart phone. First, we have to switch the platform either android, iOS, mac or windows. Next step is importing both license and target image into the software. After completion of importing file we can import our CAD file. Check whether target image can read by unity software or not. Then develop unity software for corresponding CAD modals. Actual screen shot of unity software as shown in figure. At finally install software in user's smart phone and the augmented view CAD modal will be show in real world.

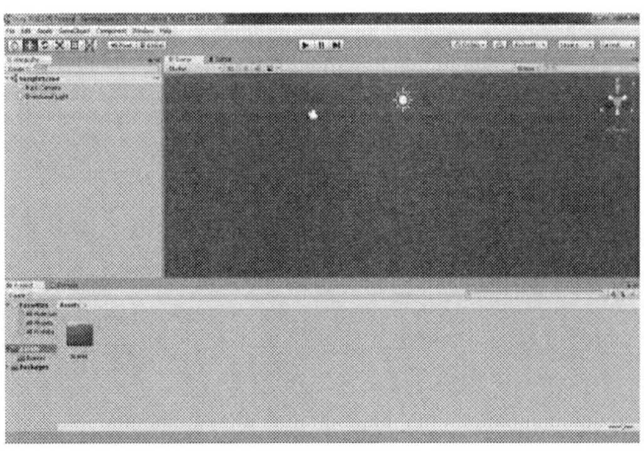

Figure 3. Screen shot of unity software

## 6. Principle of AR glass working
The angle between Smartphone's display and mirror is 40o inclination. The Fresnel lens placed right angle (horizontal) to the mirror. The combiner glass placed same as like that of mirror but in opposite side. First, the Smartphone's vision passed on the mirror. The wave of Smartphone is wide in initial condition. Now the wide wave reflects on the Fresnel lens and the wave becomes straight with each

other. Thus, the straight wave passed on the combiner glass now the clear augmented experience we can be attained.

## 7. Conclusion

Through these types of activity, we can improve our customers in our organization, knowledge in education, information about true 3Dview of organs in medical field and make our own construction and present that in front of our customer in augmented view. In future everywhere artificial intelligence will be come. That time we should know about these types of concepts and softwares.

## References

[1] Abrar Omar Alkhamisi and Muhammad Mostafa Monowar'Rise of Augmented Reality: Current and Future Application Areas',International Journal of Internet and Distributed Systems, 2013, 1, 25-34.

[2] Bacca, J., Baldiris, S., Fabregat, R., Graf, S., and Kinshuk'Augmented Reality Trends in Education', Educational Technology & Society, 17 (4), 133–149.

[3] Petar PEJIĆ, Taško RIZOV, Sonja KRASIĆ and Bojana STAJIĆ (2014) 'Augmented reality application in engineering',SMAT2014-AEM37.

[4] Philipp A. Rauschnabel, Alexander Brem and Young K. Ro (2016) 'Augmented Reality Smart Glasses: Definition, Conceptual Insights, and Managerial Importance',research gate,Article · July 2015

[5] Mehdi Mekni and Andr´e Lemieux (2014) 'Augmented Reality: Applications, Challenges and Future Trends',ISBN: 978-960-474-368-1

IOP Conf. Series: Materials Science and Engineering **923** (2020) 012005 doi:10.1088/1757-899X/923/1/012005

# Secondary tire inflation system for blown out tire

**S Premnath[1], S Sudharsan[2] and A V J Nithishvardan[2]**
[1] Associate Professor, Department of Automobile Engineering, Sri Venkateswara College of Engineering, Sriperumbudur, Tamil Nadu, India
[2] Student, Department of Automobile Engineering, Sri Venkateswara College of Engineering, Sriperumbudur, Tamil Nadu, India

**Abstract.** Automobile is one of the greatest inventions of mankind which helps people to migrate from one place to another in a short span of time with comfort. In the last few decades, many innovative technologies have been emerging. One such technology is Self Tire Inflation System. This system will monitor the tire pressure continuously and inflates or deflates the tire depending upon the road conditions. But there are many chances for the tire to get blown out, hence the plan of implementing an additional tire-tube setup in the tire has been brought up, which will act as a secondary tube until the primary tube gets blown out. This tube setup makes use of the Self Tire Inflation System and immediately inflates when the primary tire-tube setup gets damaged. This will act as an immediate remedy for reducing the accidents that are caused due to sudden blown out of tire, especially for the Off-Road vehicles and military vehicles that roll their wheels on tough terrain and extreme weather conditions.
*Keywords:* tire pressure, secondary tire inflation system, butterfly valve and nozzle

## 1. Introduction

Automobile has become an integral part of our life today. Humans are completely reliable on automobiles for the transportation purposes. The demand for automobiles has been increasing day by day. Many automobile industries are competing with each other to launch new and advanced vehicles focusing more on safety and comfort of the customers. There are many innovative technologies to improve the safety system in a vehicle. One such upcoming technology is the Automatic Tire Inflation system. This system will continuously monitor the tire pressure and also inflate or deflate the tire depending upon the road condition. In the situations where the tire suddenly gets blown out, making the driver lose the control over the steering wheel and ultimately leading to accident. So, to avoid this risk an additional feature has been implemented along with the tire inflation system. This additional feature helps in reducing the time to repair the tire and simultaneously reduces the manual effort. This system helps in improving fuel economy of the vehicle as well.

The best application of such automatic tire inflation system is used in military vehicles, where the land conditions keep continuously changing and the probability of tire damage is very high. Such vehicles are designed to operate in extreme weather condition such as heavy rainfall, snowfall, deserts storms, etc.

- Rathi et al (2017) states that the only part of a vehicle which is not getting automated is tires. From this research work it was known that a slight drop in pressure, will affect the tire wear, safety, gas mileage and whole vehicle's performance. In order to neglect the unwanted stress and time spent for inflating the tires, an Automatic Air Inflation system was designed. The system states that the tires are optimally inflated at all running conditions.[1]

- Ibrahim savas et al. (2016). This research is aimed to develop an automatic tire pressure control system for control over the slippage in order to achieve better the tractive efficiency of tractors. The developed design of the system was a flexible structure which could be adapted to all brand and their model of tractors. With the help automatic tire pressure control system, it is aimed, to reduce fuel consumption , to increase the efficiency of drawbars and to rule out negative traffic effects which will occur on the field.[2]

- Kaczmarek et al. (1984) stated that "One of the most effective and well proven systems that have been adapted to wheeled tactical vehicles to improve the overall vehicle mobility is CTI". However, after World War II no serious consideration of the benefits of CTI occurred until the early 1980's,the period when most of the military vehicles produced in the United States were equipped with CTI. [3]

- Burase et al. surveyed on Automobile Air Inflating System. This research paper ,comprises of a new approach for automobile, a self-inflating tire that guarantees the tires to be properly inflated at all times. This system uses portable compressor that will supply air to all the four tires through hoses and a rotary joint fixed between the wheel spindle and wheel hub at each wheel. The rotary joints effectively allow air to be delivered to the required under-inflated tire. This system acknowledges an essential improvement in gas mileage, tire wear reduction, and an increase in handling and tire performance in diverse conditions.[4]
- Hemantsoni and S. Adakmol et al focused on automatic tire inflation system and he introduced the centralized compressor-based system with the aim of improving mileage, tire life and car safety. Using this system air is delivered to all four tires of the vehicle using hoses and a rotary joint constrained between the wheel spindle and wheel hub at every wheel. This is one of the most beneficial project for society as the system automatically re-inflates all the four tires whenever required under all environmental and road conditions. [5]

- P. Omprakash et al. worked on the mechanism for air refilling system. The aim of this project was to introduce a system that can be used in any type of tire either tube or tubeless. With the intention of providing tight connection during rotation of tire, the system uses the permanent connection between valve and hoses so that it can be lose only when substituting the tire by the driver.[6]

The above Literatures served as an eye opener for new safety systems for automotive vehicles.

## 2. Materials and methods

### *2.1 Design of secondary tire-tube setup*

- The design in this system is quite easy to understand and this setup has been made with very few and intricate modifications which has been theoretically out shown a good result. The former method of Automatic Tire Inflation System, had proven to be a good solution for few decades.

- But for the wheels to roll normally in any extreme situations, the changes can be adapted to the system by providing a Secondary Tire-Tube setup inside the original tire with a lesser dimension. The pressure is maintained at a pre-set amount by the usage of the air compressors. A pair of nozzles, a pair of butterfly valves, pressure sensors and the pressure regulators are used to support this new setup. This system can even help in car to move in extreme terrains, even after the damage of the outer tire. This system is of high advantage for the Off- Road vehicles, providing the traveller no worry about the tire and the terrain.

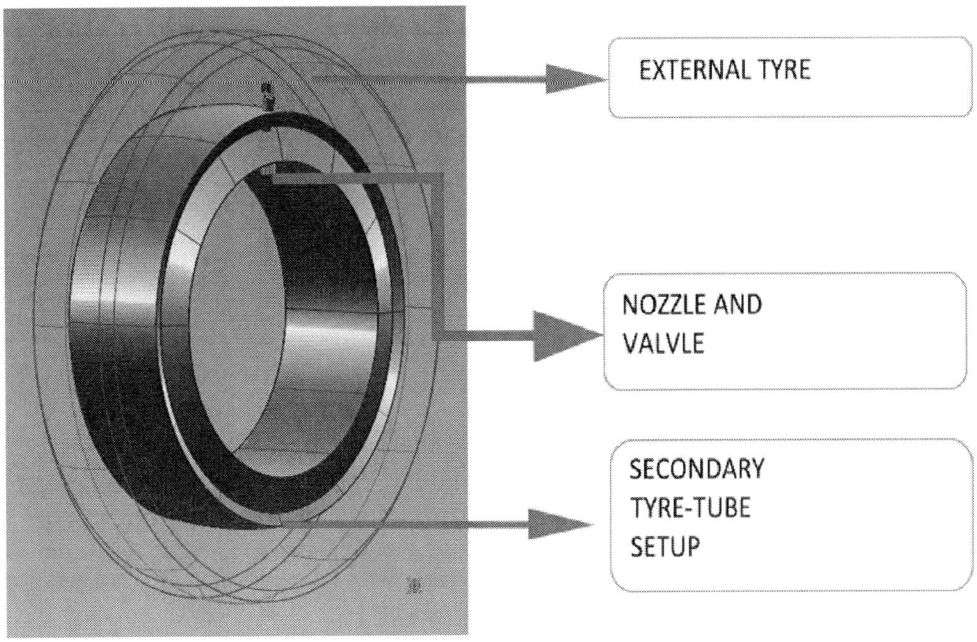


Figure 2.1 Design of complete secondary tire-tube setup

*2.1.1 Components required*

1. Air Compressor- Pressurises the air
2. Pressure Sensor - Monitors the pressure changes and signals ECU
3. Two Nozzles - To change the velocity of feed air
4. Two Butterfly Valves - Closing and opening of air passage
5. Secondary Tire-Tube setup
6. Rubber material - Silica or Carbon Black

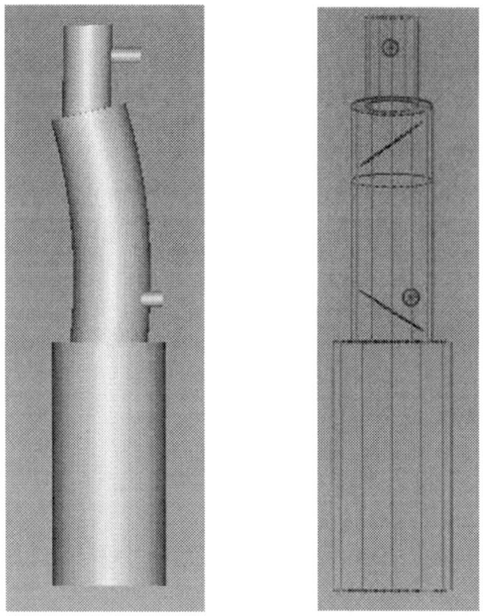

Figure 2.2 Air compressor with nozzle and butterfly valves, pressure sensor at the end points

Figure 2.3 External and internal tire with nozzle position

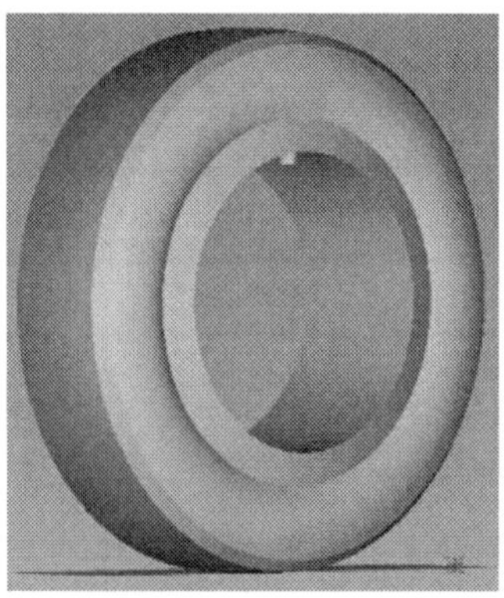

Figure 2.4 External tire and tube setup

IOP Conf. Series: Materials Science and Engineering **923** (2020) 012005 doi:10.1088/1757-899X/923/1/012005

*2.2 Working*

The working or the workflow can be categorized depending upon their condition

i. Pressure Reduction
ii. Puncture in the external tire-tube setup

*2.2.1 Pressure reduction:*

This phenomenon is quite familiar, where the pressure in the tubes go down the pre-set or the recommended level. The pressure decrease in the tube is sensed by the pressure sensor which send the data to the ECU, which in turn signals and activates the air compressor to feed the compressed air into the tube. The air compressor takes in atmospheric air and compresses it to the required pressure that has to be fed into the tube, thus ensuring proper tire pressure and this is achieved by the movement of the butterfly valves and the nozzles which are in the flow path of air.

*2.2.2 Puncture in the external tire-tube setup:*

This situation has higher chances to occur in the Off-Road and military vehicles whose wheels rolls on tough terrains and extreme weather conditions. When travelling in such tough terrains, the spokes or the rocks or any obstacle may tend to damage the external tube thus leading to the sudden fall in pressure which is sensed by the pressure sensor and signalled to the ECU.

This action is followed by the closing of the butterfly valve at the external tube, thus causing the air to flow to the external tube. Due to the position of butterfly valve in the inner side in open condition, the compressed air is let into the secondary tube setup which is raised to the pressure little lesser than the recommended value of pressure.
The punctured tube shrinks to stick on to the inner side of the external tire, simultaneously the inner tire expands along with the tube and makes contact with the external tire's inner surface. This system of using a Secondary Tire-Tube inflation system can serve as a temporary solution but for a quite longer period, until the original tube and tire is replaced.

*2.3 Conditions to be maintained*

1.The outer tube must be maintained at a pressure (-5 psi) than the recommended value

2.The inner tube must be maintained at a pressure three-fourth of the pre-set value of pressure

3.The proper working of the valves must be ensured

4.The Aurdino coding must be precise to achieve accurate results

*2.4 Components and Design Setup:*

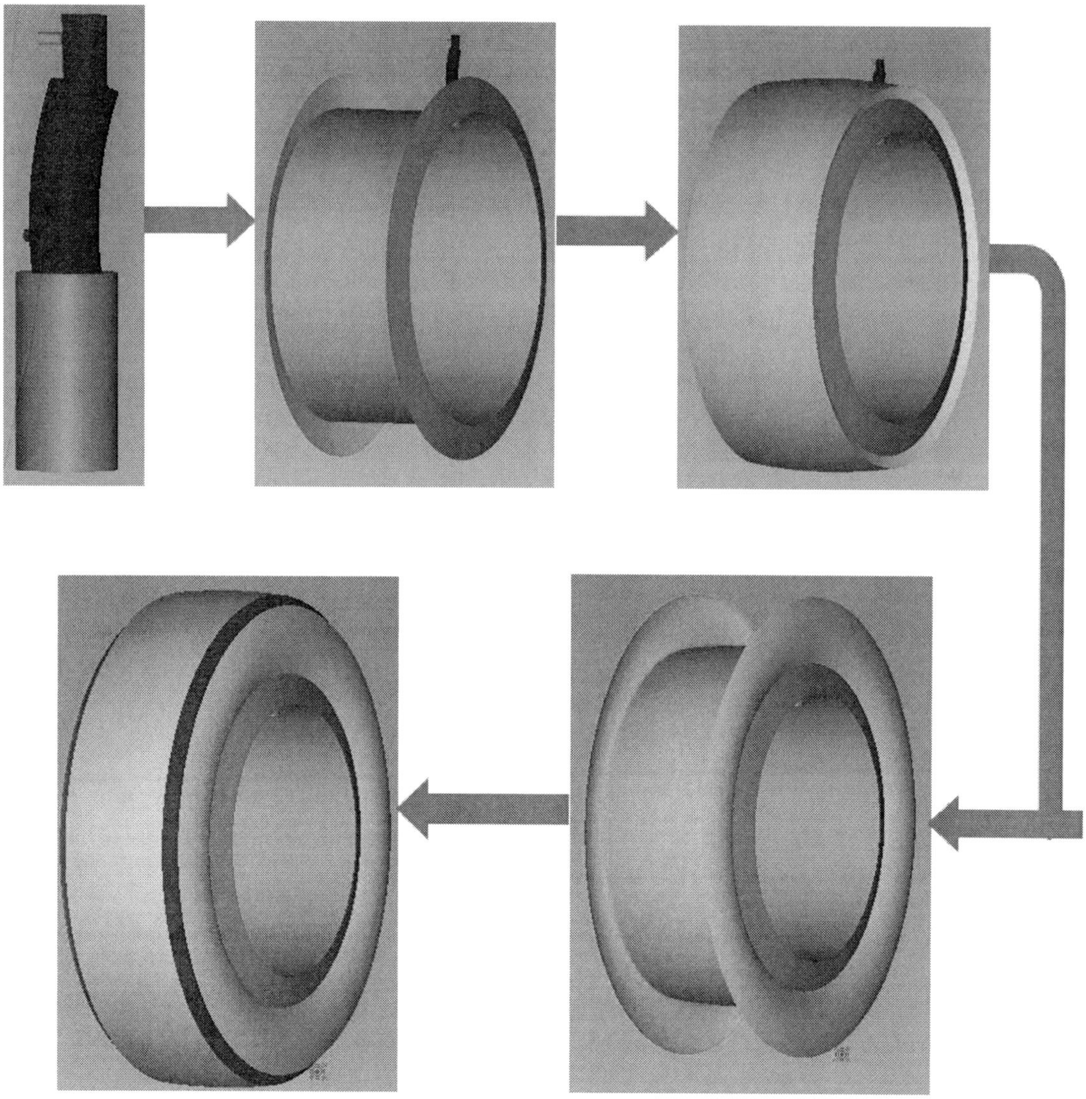

Figure 2.5 Design flow of secondary tire inflation system

IOP Conf. Series: Materials Science and Engineering **923** (2020) 012005 doi:10.1088/1757-899X/923/1/012005

## 3. Results and conclusion

The tire pressure is an important parameter in any automobile. This pressure also decides the mileage of the vehicle. Any variation in the tire pressure may lead to lack of control over stability of the vehicle and ultimately leading to accident. However, to avoid such accidents, the idea of Self Tire Inflation System has been brought up. In the situations where tire gets blown out, there is a need for immediate solution. The Secondary Tire Inflation system, which is the solution, will inflate the secondary tube when the primary tire-tube blown out and feeds air pressure to the wheel, thus ensuring instant stability to the vehicle. Thus, this system acts as an immediate remedy to avoid the accidents or loss of life due to tire blown outs, also reduces the manual work and avoids the necessity to repair the blown-out tire. Hence, this system can be applied especially to the Off-Road vehicles and also to military vehicles.

**4. Future scope:** The present research work can be extended by employing the design of experiment (DOE) testing along with Internet of Things (IoT).

## 5. References

[1] Rathi K, Shitole A, Shelke A, Kalhapure S, and Pujari B R 2017 Automatic Central Air Inflation System, *IOSR Journal of Mechanical and Civil Engineering*

[2] Ibrahim Savas, Birol, Yenal, Bahattin, Yilmaz and Ali,
Automatic Tire Pressure Control System to improve the tractive efficiency of tractors, *Journal of Agricultural Machinery Sciences*, 2014, **Vol 10(3)**, 253-259

[3] Kaczmarek R W (1984): Central Tire Inflation Systems (CTIS) – a means to enhance vehicle mobility In: Proceedings of the *8th International Conference of the International Society for Terrain-vehicle Systems*, Cambridge, June 1, 1984: 1255–1271

[4] I Burase S Kamble, A Patil and A Kharat A Survey on Automatic Air Inflating System for Automobile, *International Journal of Innovative Research in Science, Engineering and Technology*, **Vol 5(10)**, 2016, ISSN: 2319- 8753

[5] Hemant Soni, Pratik Golar, Ashwin Kherde, Design Of Automatic Tire Inflation System, **Vol-01**, Issue 4/April 2014 ISSN: 2347-5420

[6] P Omprakash, T Senthil Kumar, M A R S- Mechanised Air Refilling System

[7] http://www ijser org>Automatic Tire Inflation System

# Simulation of Transient Liquid Phase Bonding Process and the Influence of Interlayer Thickness

**B.Benita**

Assistant Professor, Department of Computer Science & Engineering, Francis Xavier Engineering College, Tirunelveli-627002

**D.S.Samuvel Prem Kumar**

Asst.Professor, Department of Mechanical Engineering, JP College of Engineering, Tenkasi

**R.Pravin**

Under Graduate Student, Department of Mechanical Engineering,
Mepco Schlenk Engineering College (Autonomous), Sivakasi.

**N.Samuel Dinesh Hynes**

Independent Researcher, & Former Part Time Lecturer, Department of Computer Science & Engineering, Government College of Engineering, Tirunelveli-627007.

**J.Angela Jennifa Sujana**

Associate Professor,Department of Information Technology,Mepco Schlenk Engineering College (Autonomous),Sivakasi.

**Abstract**. Ceramic/metal joints are possible with silid state joining techniques. Transient liquid phase bonding is a solid state joining process which is choice process. In this paper, numerical simulation of TLP bonding of ceramic/metal is done by Finite Element simulation. The finite element analysis of the TLP process provides the temperature and residual stress distribution during the bonding. Aluminium sheet as interlayer having high wettability is used. Transient thermal analysis is performed and discussed.

**Keywords**. TLP; Temperature Distribution; Transient Liquid Phase Bonding; Interlayer Thickness; Dissimilar Joints.

## 1. Introduction

In the present manufacturing scenario, the need for joining of dissimilar materials is drastically increasing and it can't be ignored. Various conventional and unconventional joining processes are explored to find the joining possibility. Among them, transient liquid phase bonding is an effective process to join dissimilar materials with high integrity. The major drawback of the transient liquid phase bonding process is the thermal mismatch between the base materials. Transient Liquid Phase

Bonding (TLPB) is applied to join materials of various combinations such as metal-metal, metal-ceramics and ceramics-ceramics, required in automotive and electronic industries [4,5].

The interlayer material is one of the important process parameter and it is placed between the base metals to be jointed [6-9]. The major consideration for the interlayer material is that it should possess a lower melting point than the substrate/base material. The whole specimen is placed in a furnace and heated to a temperature below the melting point of the substrate material and the assembly is held at the bonding temperature for a specific time period for isothermal solidification [10-13]. The resultant joint possesses a higher re-melting temperature than the initial melting temperature.

Friction stud welding [14-33], Friction Plug Welding [34-38], Diffusion bonding [39, 40], Friction Welding [41-55] Friction Drilling [56-67], and Friction Riveting [68-70] are few processes employed for joining of dissimilar metals. TLB technique overcomes most of the difficulties encountered by all these techniques. Arafin & Medraj (2006) said that the holding time decreases by increasing the bonding temperature & decreasing the interlayer thickness, using interlayers. Experimental work had good agreement with numerical simulation results generated through finite element based software [9].

## 2. Numerical Analysis

It is widely known that thermal mismatch is a major factor in determining joint mechanical behaviour because this may lead to the formation of residual stress at the bonded interfaces. These stresses are pre-induced on any stresses generated during subsequent thermal or mechanical loading, and thus strongly influence the strength of the joint. A three dimensional transient, isotropic solid with heat source finite element model was developed to simulate the transient liquid phase bonding process in Graphite/Copper using the commercial code ANSYS 15.0. As a first step in the analysis the transient temperature field T which is a function of time t and the spatial coordinates (x, y, z), is solved. Modelling heat evolution between the coil and bonded sheet is an important step in understanding how it affects material flow within and surrounding the joint. To compensate for the lack of a predicted temperature field, measured temperature values from the literature [6] are used to construct an approximate temperature field for the TLPB process. This temperature field is then used as input for the structural model for the same TLPB process. As such, a problem geometry that accommodates the simulation of the preceding TLPB test is used.

Table 1: Material Properties

| S.No | Material | Density Kg/m3 | Thermal Conductivity W/mK | Specific Heat J/Kg/K |
|---|---|---|---|---|
| 1 | Graphite | 2490 | 8.7 | 771 |
| 2 | Aluminium | 2700 | 173 | 896 |
| 3 | Copper | 8960 | 398 | 384.56 |

### *2.1. Assumptions*

- The ambient temperature was assumed to be 20°C.
- Only thermal loading is applied.
- Bonding Temperature is kept constant as 700°C.
- Thickness of the interlayer is varied as 0.1mm,0.2mm and 0.3mm
- Uniform cooling and dissimilar materials are perfectly bonded at the interface. The properties of Graphite, Aluminium and Copper are presented in table 1.

### *2.2.Material Modelling*

The sheets of Graphite and Copper having dimensions of 0.015m*0.015m*0.06m is modelled. The Aluminium foil having dimensions of 0.015m*0.015m*0.001m issued as interlayer for the Graphite/Copper joint. Aluminium metal is taken as interlayer due to its high wettability and also it has a melting point of 660°C, which is lower than the melting point of the base materials. Figure 1 shows the modelled geometry and Table 1 gives the properties of materials.

SOLID70 which is a three-dimensional thermal solid is used as the element type for thermal analysis. SOLID70 has a three dimensional thermal conduction capability. The element possess eight nodes with a single degree of freedom and temperature, at each node. The element is applicable to a three dimensional, steady-state, or for transient thermal analysis. The element also can compensate for mass transport and heat flow from a constant velocity field. If the model which has a conducting solid element is also to be analyzed structurally, that element must be replaced by an equivalent structural element such as SOLID185. A free surface of the element (that is, not adjacent to another element and not subjected to a boundary constraint) is assumed to be adiabatic. Thermal transients having a fine integration time step and a severe thermal gradient at the surface will also be requiring a fine mesh at the surface. This element is not having the mass transport or fluid flow options. SOLID185 Structural Solid is suitable for modelling general 3-D solid structures. It allows for prism, tetrahedral, and pyramid degenerations when used in irregular regions. Various element technologies such as B-bar, uniformly reduced integration, and enhanced strains are supported. Figure (2) shows the meshes generated for the present simulation. The shape of the mesh element is hexahedral shape. The meshes are composed of a total number of 2230 nodes and 10640 solid elements.The boundary conditions applied are convection coefficient of 1.33W/m2k around the surface of the plate and surface load of 901W/m2around the interface.

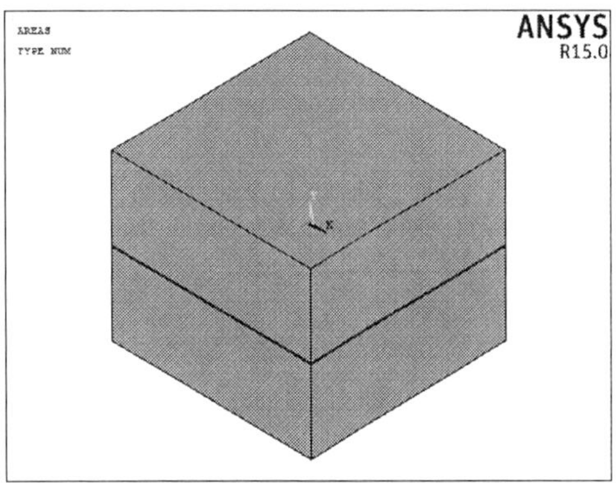


Figure 1: Modelled Geometry

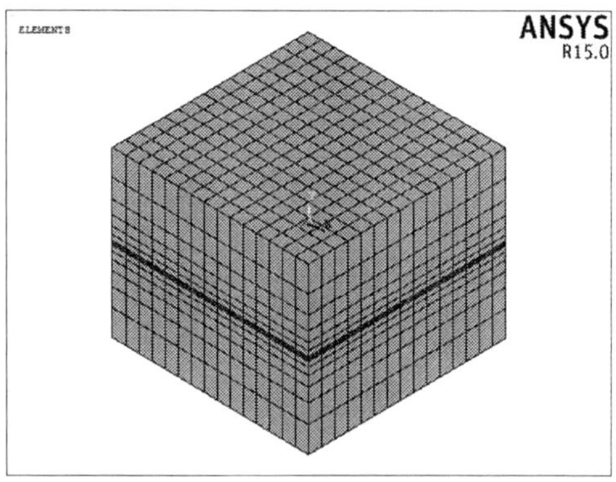


Figure 2: Meshed Model

As the model is a symmetric geometry, it possess isotropic property. The mapped mesh is used to mesh the model. The fine mesh is obtained. The meshed model is shown in Figure 2.

## 3. Results and Discussion

The above numerical simulation is carried out to determine the temperature distribution of TLP joint for 700°C bonding temperature under various interlayer thickness as 0.1mm, 0.2mm and 0.3mm. The

IOP Conf. Series: Materials Science and Engineering **923** (2020) 012006 doi:10.1088/1757-899X/923/1/012006

result obtained shows that a high level of temperature distribution is obtained at 0.3mm interlayer thickness for the bonding temperature of700°C. Since the temperature distribution is high at 0.3mm interlayer thickness, the stress created at that point is more which affects the strength of the TLP joint. Therefore it is preferable to go for a bonding temperature of 700°C for graphite/copper joint with aluminium interlayer of 0.1mm interlayer thickness to get higher joint strength. Figure 3 shows the temperature distribution for 0.1 mm thick aluminium interlayer at 700°C. The minimum temperature value 88.95°C is obtained at Graphite side. This is because of the low thermal conductivity of the graphite. As the temperature at the Graphite is very much less than the melting point of Graphite, the properties of Graphite is not affected in this process.

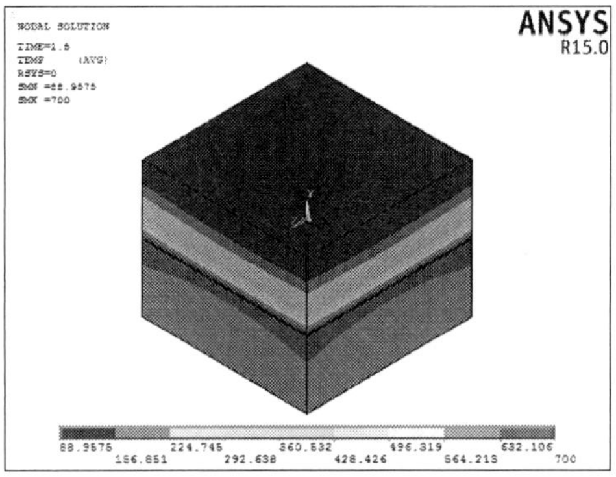


Figure 3: Temp distribution at IT=0.1mm Interlayer Thickness

Similarly, Figure 4 and Figure 5 shows the temperature distribution for 0.2 mm and 0.3mm thick aluminium interlayer at 700oC. The minimum temperature value for those models are also obtained at Graphite side. At this point, the temperature at the copper side is nearly around 600oC. This is because of the high thermal conductivity of the copper. As the temperature at the copper is nearly the Glass Transition Temperature of copper, the possibility of affecting the properties of copper is more. To avoid that a cooling system is provided at the copper side in this process.

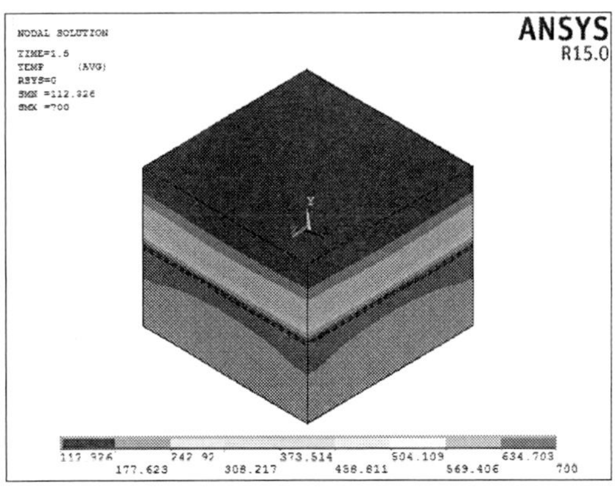


Figure 4: Temp distribution at IT=0.2mm Interlayer Thickness

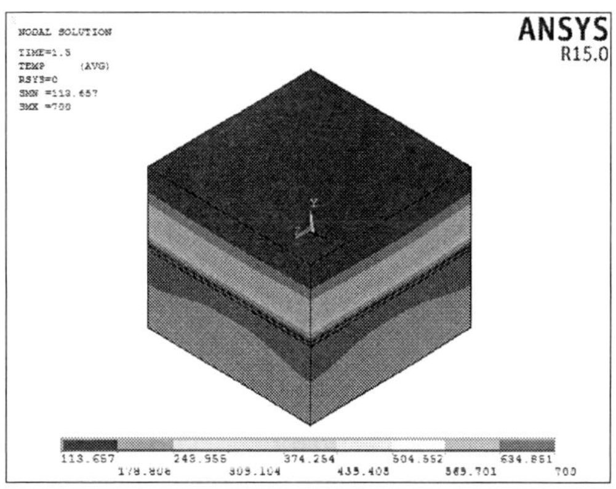


Figure 5: Temp distribution at IT=0.3mm Interlayer Thickness

## 4. Conclusion

The major problem encountered during transient liquid phase bonding process is the thermal mismatch that occurs during bonding. The thermal mismatch is the melting point difference between graphite and copper. This defect can be reduced by introducing aluminium interlayer. The main consideration about the interlayer is that it should possesshigh wettability and a melting point lower than the base materials. When an interlayer had been introduced in between the specimen, joining had been successively achieved. Numerical investigation shows that the interlayer acts as aninterface of joint for the two materials. Because of this tendency, thermal mismatch becomes minimum and hence thermal stress becomes minimum. It is observed that when 0.1 mm thick aluminium sheet is used as an interlayer, thermal mismatch is minimized. This could eventually reduce the residual stress and thereby increase the joint efficiency between graphite and copper joints.

## References

[1] M.A.Arafin,M.Medraj,D.P.Turner,P.Bocher "Transient liquid phase bonding of Inconel 718 and Inconel 625 with BNi-2: Modeling and experimental investigations" Journal of Materials Science and Engineering A 447(2007) 125-133.
[2] Rajesh N et al (2014) "Ultrasonic evaluation of friction stud welded AA6063/AISI1030 steel joints" Mater Des 62:118–123.
[3] Rajesh Jesudoss Hynes N, ShenbagaVelu P, "Simulation of friction welding of alumina and steel with an aluminium interlayer", Int J AdvManufTechnol, Springer, DOI 10.1007/s00170-015-7874-8.
[4] Grant O.Cook.III.CarlD.Sorensen "Overview of transient liquid phase and partial transient liquid phase bonding" Journal of Material Science (2011) 46:5305-5323 DOI 10.1007/s10853-011-5561-1.
[5] HongshengChen,ChongshengLong,Tianguo Wei Gao,HongxingXiao,Le Chen "Effect of Ni interlayer on partial transient liquid phase bonding of Zr-Sn-Nb alloy and stainless steel" Materials and Design 60 (2014) 358-362.
[6] ZhihongZhong,ZhangjianZhou,ChangchunGe "Brazing of doped graphite to Cu using stress relief interlayers" Journal of Materials Processing Technology 209 (2009) 2662-2670.
[7] Jin-Woo Park and Thomas W.Eagar "Application of Transient Liquid Phase Bonding of Microelectronics and MEMS packaging" Department of Materials Science and Engineering,Massachusetts Institute of Technology,Cambridge,MA 02139.
[8] Rajesh Jesudoss Hynes N et al," Numerical Simulation of Heat Flow in Friction Stud Welding of Dissimilar Metals", Arabian Journal for Science and Engineering, 2014, 39, 3217-3224.
[9] V.Jalilvand,H.Omidvar,M.R.Rahimipour,H.R.Shakeri "Influence of bonding variables on transient liquid phase bonding behavior of nickel based superalloy IN-738LC" Materials and Design 52 (2013) 36-46.
[10] Rajesh N et al (2014) "Ultrasonic evaluation of friction stud welded AA6063/AISI1030 steel joints" Mater Des 62:118–123.
[11] Rajesh Jesudoss Hynes N, ShenbagaVelu P, "Simulation of friction welding of alumina and steel with an aluminium interlayer", Int J AdvManufTechnol, Springer, DOI 10.1007/s00170-015-7874-8.
[12] Rajesh Jesudoss Hynes N et al," Numerical Simulation of Heat Flow in Friction Stud Welding of Dissimilar Metals", Arabian Journal for Science and Engineering, 2014, 39, 3217-3224.
[13] V.Jalilvand,H.Omidvar,M.R.Rahimipour,H.R.Shakeri "Influence of bonding variables on transient liquid phase bonding behavior of nickel based superalloy IN-738LC" Materials and Design 52 (2013) 36-46.
[14] Rajesh N et al (2014) "Investigation on joining of aluminum& mild steel by friction stud welding" MaterManuf Process 27:1409–1413.
[15] Rajesh et al (2014) "Numerical simulation on joining of ceramics with metal by friction welding technique" Int J Mod PhysConfSer 22:190–195.
[16] N. Rajesh Jesudoss Hynes, P. Nagaraj, J.A.J.Sujana, "Ultrasonic Evaluation of Friction stud welded AA6063/AISI1030 steel joints," Materials & Design, Elsevier publications, Volume 62, pp 118–123, October 2014
[17] Rajesh Jesudoss Hynes N, Nagaraj P, Palanichamy R, Arumugham CAK & Angela JennifaSujana J, "Numerical Simulation of heat flow of Friction Stud Welding of Dissimilar metals," The Arabian Journal for Science and Engineering, Springer Publications, DOI 10.1007/s13369-013-0932-3, 2014, Volume 39, Issue 4, pp 3217–3224
[18] Rajesh Jesudoss Hynes N, Nagaraj P & Angela JennifaSujana J , "Mechanical Evaluation and Microstructure of Friction welded Aluminium-Mildsteel joints," The Arabian Journal for Science and Engineering Springer Publications, DOI 10.1007/s13369-014-1082-y, June 2014, Volume 39, Issue 6, pp 5017–5023.
[19] Rajesh Jesudoss Hynes N, Nagaraj P &MebySelvaraj R (2013) , "Finite Element based Thermal Modelling of Friction Welding of Dissimilar Materials," International Journal of Modern Physics Conference Series,ISSN: 2010-1945 DOI: 10.1142/S201019451301012X, vol. 22, pp.196–202.
[20] Rajesh Jesudoss Hynes N, Nagaraj P &VivekPrabhu M (2013), "Evaluation of Bending Strength in Friction Welded Alumina/Mild Steel Joints by Applying Factorial Technique," International Journal of Modern Physics Conference Series,ISSN: 2010-1945, DOI: 10.1142/S201019451301012X, vol. 22, pp.184-189.
[21] Rajesh Jesudoss Hynes N, Nagaraj P & Joshua Basil S (2013), "Numerical Simulation on Joining of Ceramics with Metal by Friction Welding Technique," International Journal of Modern Physics Conference Series,ISSN: 2010-1945, DOI: 10.1142/S2010194513010118, vol. 22, pp.190-195.
[22] Rajesh Jesudoss Hynes N, Nagaraj P and Thanga Kumar P, "Thermal Modeling of Friction Plug Welding," International Journal of Applied Engineering Research, ISSN 0973-4562 Vol. 9 No.26 (2014) pp. 9031-9033.
[23] Rajesh Jesudoss Hynes N, Nagaraj P and Tharmaraj R, "Prediction of Thermal Profile During Friction Stud Welding of Aluminium - Mild Steel Joints," International Journal of Applied Engineering Research ISSN 0973-4562 Volume 10, Number 8 (2015), pp. 6107-6110.
[24] Rajesh Jesudoss Hynes N, Nagaraj P & Prakash P, "Numerical Modeling of Temperature Distribution in Friction Stud Welding of Dissimilar Metals," International Journal of Applied Engineering Research, ISSN 0973-4562 Vol. 9 No.26 (2014) pp. 9023-9027.

[25] Rajesh Jesudoss Hynes N, Nagaraj P & Prakash P, "Mathematical Model to Predict Heat Flow in Underwater Friction Stud Welding," Advanced Materials Research, doi:10.4028/www.scientific.net/AMR.984-985.596, Vols.984-985 (2014), pp.596-599.
[26] Rajesh Jesudoss Hynes N, Nagaraj P &Thanga Kumar P, "Mathematical Modeling of Friction Plug Welding with Preheating effect," Advanced Materials Research, doi:10.4028/www.scientific.net/AMR. 984-985.600, Vols. 984-985 (2014), pp.600-603.
[27] Rajesh Jesudoss Hynes N, Nagaraj P & Prakash P, "Thermal Analysis on Joining of Dissimilar Metals by Friction Stud Welding," Advanced Materials Research, doi:10.4028/www.scientific.net/AMR.984-985.592, Vols. 984-985 (2014), pp.592-595.
[28] Rajesh Jesudoss Hynes N, Tharmaraj R, "One Dimensional Thermal Model for Heat Flow in Friction Stud Welding," Journal of Thermal Engineering and Applications. 2015; 2(2), pp. 22–27.
[29] N Rajesh Jesudoss Hynes and P ShenbagaVelu,"Thermo Mechanical Modelling of Friction Welding using Finite Element Method,"European Journal of Advances in Engineering and Technology, 2015, 2(7), pp. 29-32.
[30] N. Rajesh Jesudoss Hynes, P. ShenbagaVelu, "Simulation on Friction Welding Of MgAZ31 / AA 6061 T6 Joints," Recent Advances in Computer Science, Recent Advances in Computer Engineering Series, ISSN: 1790-5109, Volume 32, 2015, pp. 524-528.
[31] N. Rajesh Jesudoss Hynes, R. Tharmaraj, P. ShenbagaVelu, and R. Kumar, "Finite element based simulation on friction stud welding of metal matrix composites to steel,"American Institute of Physics (AIP) Conference Proceedings, doi: 10.1063/1.4946607, ISSN: 0094243X, Volume 1728 (2016); pp.020556-1 to 020556-4.
[32] R. Meby Selvaraj, and N. Rajesh Jesudoss Hynes, Finite element approach in thermal modelling of friction stud welding, AIP Conference Proceedings 2142, 110006 (2019); https://doi.org/10.1063/1.5122466
[33] R. Sankaranarayanan, and N. Rajesh Jesudoss Hynes, Friction riveting for joining of wide range of dissimilar materials, AIP Conference Proceedings 2142, 150004 (2019); https://doi.org/10.1063/1.5122553
[34] N. Rajesh Jesudoss Hynes, P. ShenbagaVelu, R. Tharmaraj, R. Kumar, "Numerical investigation on friction welding of alumina / AA 6063 T6 joints," American Institute of Physics (AIP), doi: 10.1063/1.4946601, ISSN: 0094243X, Conference Proceedings, Volume 1728, (2016); pp 020550-1 to 020550-5.
[35] N. Rajesh Jesudoss Hynes, M. AbeyramNithin, "Simulation on friction taper plug welding of AA6063-20Gr metal matrix composite", American Institute of Physics (AIP) Conference Proceedings, doi: 10.1063/1.4946607, ISSN: 0094243X, Volume 1728, (2016); pp 020556-1 to 020556-4.
[36] N. Rajesh Jesudoss Hynes, R.Kumar, J. Angela JennifaSujana, "Modelling of Process Parameters of Friction Stud Welding Using Fuzzy Logic System", International Journal of Advanced Engineering Technology-IJAET, E-ISSN 0976-3945, Volume 8, Issue 1, 2016; pp. 413-417.
[37] N. Rajesh Jesudoss Hynes, P. ShenbagaVelu, "Simulation of friction welding of alumina and steel with aluminum interlayer," The International Journal of Advanced Manufacturing Technology, DOI:10.1007/s00170-015-7874-8, ISSN 0268-3768, October 2017, Volume 93, Issue 1–4, pp 121–127.
[38] Rajesh Jesudoss Hynes N, Angela JennifaSujana J &Karuppasamy P, "Simulation of Friction Stud Welding Process with an inter-metallic layer," International Journal of Applied Engineering Research, ISSN 0973-4562 Vol. 9 No.26 (2014) pp. 9028-9030.
[39] N. Rajesh Jesudoss Hynes, P. ShenbagaVelu, R. Kumar, M. Karthick Raja, "Investigate the influence of bonding temperature in transient liquid phase bonding of SiC and copper", Ceramics International, Volume 43, Issue 10, July 2017, Pages 7762–7767.
[40] N. Rajesh Jesudoss Hynes, M. Karthick Raja, "Numerical simulation on influence of bonding temperature in transient liquid phase bonding,"American Institute of Physics (AIP) Conference Proceedings, doi: 10.1063/1.4946594, ISSN: 0094243X, Volume 1728, (2016); pp 020543-1 to 020543-4.
[41] P. ShenbagaVelu, N. Rajesh Jesudoss Hynes,"Numerical Analysis of Friction Welded Titanium Joints," Journal of Achievements in Materials and Manufacturing Engineering, Vol. 76, Issue 1 (2016), pp. 26-29.
[42] Dr.N.Rajesh Jesudoss Hynes, Mr.ShenbagaVelu, "Effect of rotational speed on Ti-6Al-4V-AA 6061 friction welded joints," Elsevier, Journal of Manufacturing Processes, DOI: 10.1016/j.jmapro.2018.02.014; Volume No: 30, Issue No: -1; pp. 288-297.
[43] Rajendran, T.P., Hynes, N.R.J., Nikolova, M.P. et al. Influence of heat treatment on friction-welded joints made of high-carbon high-chromium tool steel/low-carbon steel for tooling applications. J Braz. Soc. Mech. Sci. Eng. 42, 87 (2020).
[44] PackiarajRajendran, T., Rajesh Jesudoss Hynes, N. & Christopher, T. Characterization of high-carbon high-chromium tool steel/low-carbon steel friction-welded joints for industrial tooling applications. J Braz. Soc. Mech. Sci. Eng. 40, 316 (2018).
[45] Dr.N.Rajesh Jesudoss Hynes, Mr.ShenbagaVelu, "Friction push plug welding in airframe structures using Ti-6Al-4V plug," Springer, Journal of the Brazilian Society of Mechanical Sciences and Engineering, DOI: 10.1007/s40430-018-1088-6; Volume No: 40, Issue No: 3; pp. 158-165.
[46] Mr.ShenbagaVelu, Dr.N.Rajesh Jesudoss Hynes, "Microstructural and Mechanical properties on Friction Welding of dissimilar metals used in motor vehicles," Materials Research Express, DOI: 10.1088/2053-1591/aaabe6; Volume No: 5, Issue No: 2; pp. 26521-26532.
[47] N. Rajesh Jesudoss Hynes, R. Kumar, R. Tharmaraj, P. ShenbagaVelu, "Production of aluminium metal matrix composites by liquid processing methods," American Institute of Physics (AIP) Conference Proceedings, doi: 10.1063/1.4946609, ISSN: 0094243X, Volume 1728, (2016); pp 020558-1 to 020558-5.
[48] Dr.N.Rajesh Jesudoss Hynes, Mr.VivekPrabhu&Dr.P.Nagaraj, "Joining of hybrid AA6063-6SiCp-3Grp composite and AISI 1030 steel by friction welding," Elsevier, Defence Technology, DOI: dx.doi.org/10.1016/j; Volume No: 13, Issue No: 1; pp. 338-345.
[49] MebySelvaraj R, Rajesh Jesudoss Hynes N, "Assessment of Influencing Factors on Mechanical and Electrical Properties of Al/Cu Joints," DOI: https://doi.org/10.1063/1.5033163; AIP Conference Proceedings, Volume 1953 D, Issue No: 1; pp. 1300191-1300195.
[50] Rajesh Jesudoss Hynes N, Raja S, "Experimental Study on Joining of AA6063 and AISI 1040 steel," DOI: https://doi.org/10.1063/1.5033164; AIP Conference Proceedings, Volume 1953 D, Issue No: 1; pp. 1300201-1300204.

[51] ShenbagaVeluP, Rajesh Jesudoss Hynes N, "Numerical Modeling of Friction welding of Bi-metal joints for Electrical applications," DOI: https://doi.org/10.1063/1.5033272; AIP Conference Proceedings, Volume 1953 D, Issue No: 1; pp. 1400971-1400975.

[52] Rajendran, T.P., Hynes, N.R.J., Nikolova, M.P. et al. Influence of heat treatment on friction-welded joints made of high-carbon high-chromium tool steel/low-carbon steel for tooling applications. J Braz. Soc. Mech. Sci. Eng. 42, 87 (2020).

[53] Packiaraj Rajendran, T., Rajesh Jesudoss Hynes, N. & Christopher, T. Characterization of high-carbon high-chromium tool steel/low-carbon steel friction-welded joints for industrial tooling applications. J Braz. Soc. Mech. Sci. Eng. 40, 316 (2018).

[54] Rajesh Jesudoss Hynes N, Nagaraj P & Angela Jennifa Sujana J (2013),"Controller for friction stud welding machine," IEEE Conference proceedings on Energy Efficient Technologies for sustainability, pp.879-882.

[55] Rajesh Jesudoss Hynes N, Nagaraj P & Angela Jennifa Sujana J (2013), "Analytical Modeling of Temperature profile in friction stud welding of dissimilar metals," IEEE Conference proceedings on Energy Efficient Technologies for sustainability, pp.827-829.

[56] N. Rajesh Jesudoss Hynes, M. V. Maheshwaran, "Numerical analysis on thermal drilling of aluminum metal matrix composite,"American Institute of Physics (AIP) Conference Proceedings, doi: 10.1063/1.4946597, ISSN: 0094243X, Volume 1728 (2016); pp.020546-1 to 020556-4.

[57] N.Rajesh Jesudoss Hynes, M. Muthukumaran, N. Rakesh and C.K.Gurubaran, "Numerical Analysis in Friction Drilling Of AISI 1020 Steel and AA 6061 T6 Alloy," Recent Advances in Environmental and Earth Sciences and Economics, Energy, Environmental and Structural Engineering Series, ISSN: 2227-4359, Volume 39, 2015, pp. 145-149.

[58] Mr.R.Kumar, Dr.N.Rajesh Jesudoss Hynes, "Finite-element simulation and validation of material flow in thermal drilling process,"Springer, Journal of the Brazilian Society of Mechanical Sciences and Engineering, DOI: 10.1007/s40430-018-1091-y; Volume No: 40, Issue No: 3; PageNo: 162-172.

[59] N. Rajesh Jesudoss Hynes et al, "Optimum Bushing Length in Thermal Drilling of Galvanized Steel Using Artificial Neural Network Coupled with Genetic Algorithm,"Materiali in tehnologije / Materials and technology, ISSN 1580-2949, doi:10.17222/mit.2016.290, 51 (2017) 5, pp. 813–822.

[60] Dr.N.Rajesh Jesudoss Hynes, R.Kumar, "Simulation and Experimental Validation of Al7075-T651 Flow Drilling Process," Journal of the Chinese Society of Mechanical Engineers, 2017, Vol.38, No.4, pp. 413-420.

[61] Mr.R.Kumar, Dr.N.Rajesh Jesudoss Hynes, "Process optimization for maximizing bushing length in thermal drilling using integrated ANN-SA approach," Journal of the Brazilian Society of Mechanical Sciences and Engineering, DOI: 10.1007/s40430-017-0820-y; Volume No: 39, Issue No: 12; pp. 5097-5108.

[62] Kumar R, Rajesh Jesudoss Hynes N, "Numerical Analysis of Thermal Drilling Technique on Titanium sheet metal," DOI: https://doi.org/10.1063/1.5033158; AIP Conference Proceedings, Volume 1953 D, Issue No: 1; pp. 1300141-1300144.

[63] Vijayabaskar P, Rajesh Jesudoss Hynes N, "Simulation of Friction Stir Drilling Process," DOI: https://doi.org/10.1063/1.5033284; AIP Conference Proceedings, Volume 1953 D, Issue No: 1; pp. 1401091-1401094.

[64] Vignesh N J, Rajesh Jesudoss Hynes N, "Thermal Analysis of Friction Riveting of Dissimilar Materials," DOI: https://doi.org/10.1063/1.5033285; AIP Conference Proceedings, Volume 1953 D, Issue No: 1; pp. 1401101-1401104.

[65] R Kumar, N Rajesh Jesudoss Hynes, J Angela Jennifa Sujana, "Multi-objective optimization of green technology thermal drilling process using grey-fuzzy logic method", Journal of Cleaner Production (SCI Journal, IF 7.100), Volume 236, Available online 21 July 2019.

[66] R. Kumar, N. Rajesh Jesudoss Hynes, Anish Khan, Simulation of thermo-mechanical behaviour of friction drilling process, Int. J. Computational Materials Science and Surface Engineering, Vol. 9, No. 1, 2020, pp. 70-84.

[67] R.Kumar, N. RajeshJesudoss Hynes, "Thermal drilling processing on sheet metals: A Review", Elsevier, International Journal of Lightweight Materials and Manufacture, Available online 19 August 2019.

[68] Vignesh N J, Rajesh Jesudoss Hynes N, "Thermal Analysis of Friction Riveting of Dissimilar Materials," DOI: https://doi.org/10.1063/1.5033285; AIP Conference Proceedings, Volume 1953 D, Issue No: 1; pp. 1401101-1401104.

[69] Sankaranarayanan R, Rajesh Jesudoss Hynes N, "Prospects of Joining Multi-Material Structures," DOI: https://doi.org/10.1063/1.5033165; AIP Conference Proceedings, Volume 1953 D, Issue No: 1; pp. 130021-130025.

[70] R. Sankaranarayanan, and N. Rajesh Jesudoss Hynes, Friction riveting for joining of wide range of dissimilar materials, AIP Conference Proceedings 2142, 150004 (2019); https://doi.org/10.1063/1.5122553

# Static Analysis of Aluminum 6063 Alloy for Steering Knuckle Application in Student Formula Car

**GAshwinPrabhu[1],K Muninathan[2],O Lokesh Kanna [3],G Monish[4]and SRArun [5]**

[1,2]Faculty of Department of Mechanical Engineering, St. Joseph's College of Engineering, Chennai -119

[3,4,5]Department of Mechanical Engineering, St. Joseph's College of Engineering, Chennai -119

[1]E-mail: ashwin.prabhu1990@gmail.com

**Abstract:**Steering knuckle plays an important role in connecting the chassis of the vehicle with suspension, brakes, steering and the whole wheel assembly. It plays a major role in transferring the load from the road to the suspension and vice versa. The knuckle comes under the division called unsprung mass in a vehicle system. The knuckle should be light in weight and should also possess superior strength. We have chosen an aluminum alloy commercially known as AL6063. This alloy is good for structural applications and also cheap than compared to other aluminum alloys. The knuckle is designed using a CAD software. The stress analysis is done applying the required constraints. The knuckle is designed by optimizing the shape and also changing the material from its traditional counterparts hence this would produce a better overall enhanced performance. A reduction in weight will improve the mileage of the vehicle in consideration.

***Keywords*:** Steering knuckle, chassis, unsprung mass.

## 1. Introduction

Steering knuckle is an important and non-standard product which connects the wheel assembly to the chassis. It has the suspension lower and upper arms, brakes rotors and calipers, wheels hub and bearings, steering tie rods, axles (if it is in the driving wheels) respective ball joints, all of these are connected to it. Hence it is an important part in a vehicle system. The steering knuckle also transfers the load from the tires when it comes in contact with and bumps or dips in the road surface. The knuckle should be stiff and light in weight to have an optimal performance. The knuckle weight will affect the suspension response, and also the handling of the vehicle. All racing cars are demanded to be lighter so to be more efficient and to be handled better. There are various forces that act on the steering knuckle, predominantly the forces due to acceleration and braking, force due to steering or pivoting force, suspension loads, and torsional force when the axle spins and the frictional force. The knuckle helps to intermediate these forces and transfer them accordingly. The car knuckle is designed by first obtaining all the parameters, the design of the knuckle depends on the suspension geometry, steering geometry, weight of the vehicle.

IOP Conf. Series: Materials Science and Engineering **923** (2020) 012007 doi:10.1088/1757-899X/923/1/012007

## 2. Objectives

- To minimize the weight of the individual part.
- The improve machinability.
- To improve the design.
- To make the design simpler.
- To increase the load bearing characteristics.

## 3. Methodology

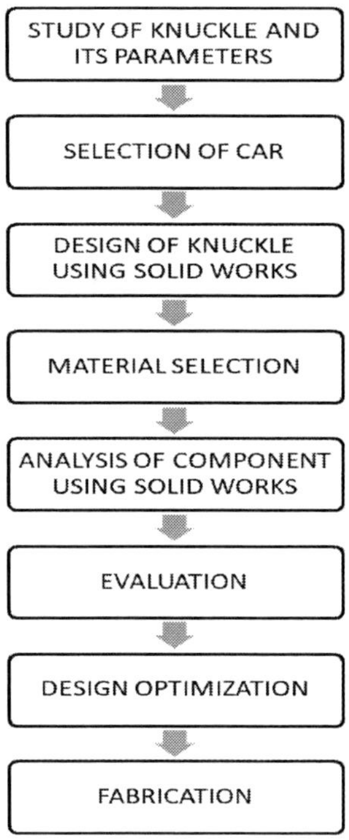


**Figure 1** Methodology

## 4. Material Selection

The material chosen is Aluminum 6063. This alloy of aluminum is cheaper and accessible. The alloy is predominantly used for structural applications. The alloy is made from a combination of many metals the base being aluminum.

**Table 1.** Properties of AL6063

| Properties of AL6063 | Value |
|---|---|
| Density | 2.69 g/cm3 |
| Youngs modulus | 68.3 Gpa |
| Tensile strength | 190 Mpa |
| Poisson ratio | 0.3 |

| | |
|---|---|
| Yield strength | 2.1e8 |

**Table 2.** Composition of alloy

| **Component** | **Weight %** |
|---|---|
| Al | 97.5 |
| Mg | 0.45 – 0.90 |
| Fe | 0.35 |
| Si | 0.2 – 0.6 |
| Cr,Cu,Ti,Zn, Mn | 0.1 |

## 5. Design of Steering Knuckle.

The knuckle is designed with many parameters in mind. The car is the main body hence based on the car the knuckle is shaped. The weight plays a major role while designing the knuckle. The two main parameters that shape the knuckle are the Steering axis inclination and the Castor angle. The steering axis inclinations is also called as the king pin inclination.

The various loads that act on the knuckle are given below

**Table 3.** Dimensions of the car

| **Parameter** | **Values** |
|---|---|
| Weight | 360 kg |
| Track width | 52 inches |
| Wheel base | 81.2 inches |
| Ground clearance | 2.5 inches |
| Centre of gravity height | 12.5 inches |
| Roll centre height | 3.76 inches |

## 6. Loading conditions

We have taken three conditions, they are

1. Bump Load -- 671 N
2. Braking load -- 178.32 Nm
3. Steering load – 50N

But for analysis purpose we take a much more value

1. Bump load – 5g force
2. Braking load -- 3g force
3. Steering load – 1.5g

## 7. Result

### *7.1 Knuckle design*

**Figure 2**. Knuckle Design

### *7.2 Bump Load*

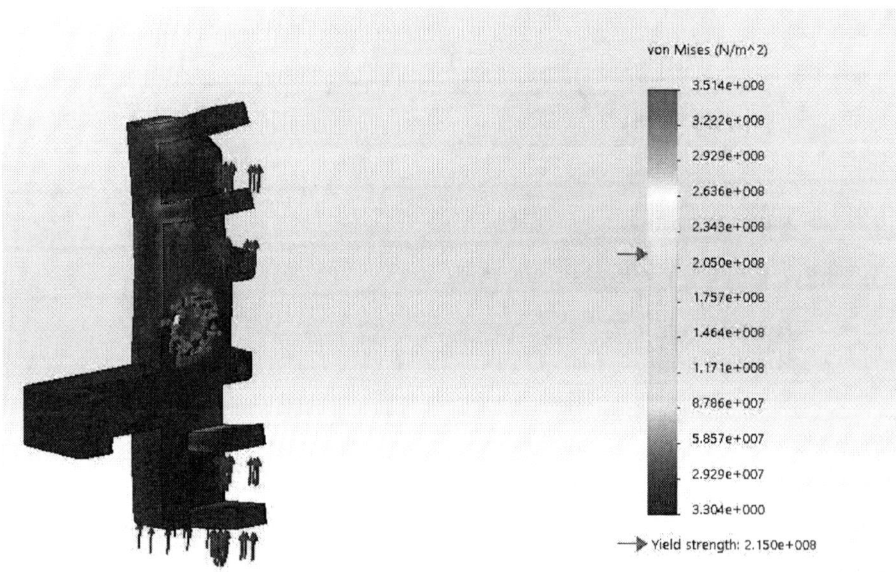

**Figure 3**. Bump Load

*7.3 Braking load*

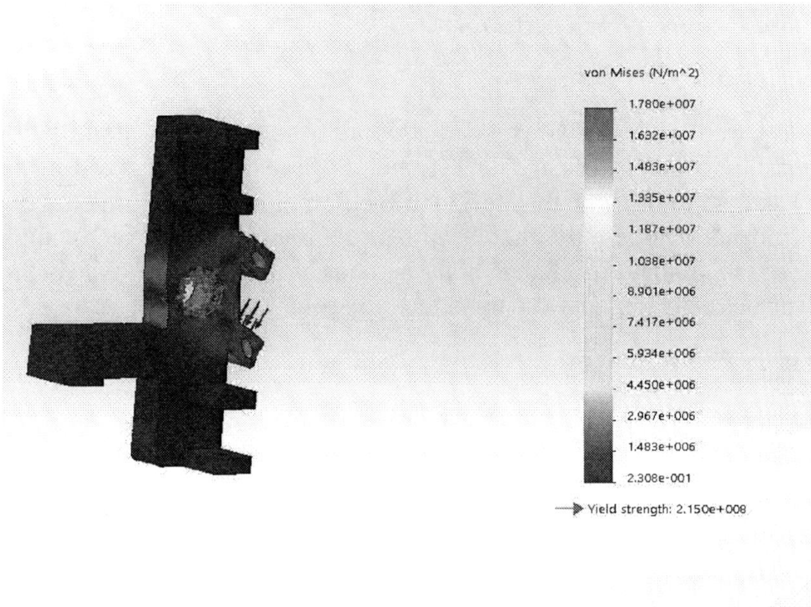


**Figure 4**. Breaking Load

*7.4 Steering Load*

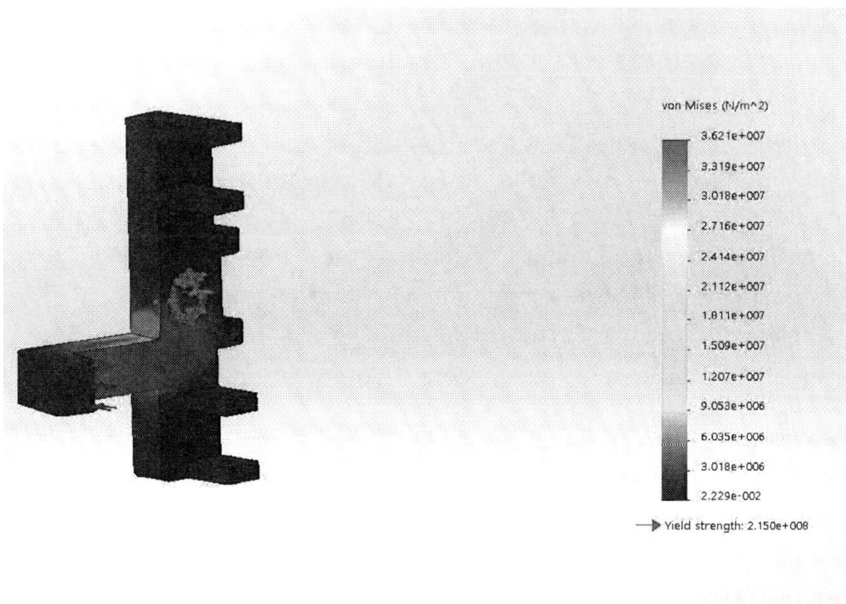

**Figure 5**. Steering Load

The material and design are capable of withstanding the impact loads. Hence the material can be used for this application. The factor of safety for the component is also in the preferable scale. Hence this has also reduced the weight of the individual component whichimproves the overall performance.

## 8. Conclusion

The reduction in weight will ultimately improve the handling and suspension response, most of all increasing the efficiency of the car in terms of fuel. Going forward we are trying to implement this design into commercial cars, aluminum does not corrode like its ferrous counterparts, hence it is more practical in everyday applications. Shape optimization and material change both methods have been employed in this project hence we can achieve greater cost saving in the long run.

## References

[1] Sharma M P, Mevewala D S, Joshi H, & Patel D A 2014 Static analysis of steering knuckle and its shape optimization Journal of Mechanical and Civil Engineering 34-38
[2] PurushottamDumbre, Prof Mishra A K, Aher V S, Swapnil S Kulkarni structural Analysis Of Steering Knuckle For Weight Reduction International Journal of Advanced Engineering Research and Studies EISSN2249–8974
[3] Babu B, Prabhu M, Dharmaraj P & Sampath R 2014 Stress analysis on steering knuckle of the automobile steering system International Journal of Research in Engineering Technology 3(3) 363-366
[4] Razak I H A, Yusop M Y M, Yusop M S M, Hashim M F Modelling, Simulation and Optimization Analysis of Steering Knuckle Component for Race Car IJRET: International Journal of Research in Engineering and Technology eISSN:23191163 pISSN: 2321-7308 Volume: 03 Issue: 11 Nov 2014.
[5] Wan Mansor Wan Muhamad, EndraSujatmika, Hisham Hamid, & Faris Tarlochan Design Improvement of Steering Knuckle Component Using Shape Optimization International Journal of Advanced Computer Science Vol. 2 No. 2 Pp. 65-69 Feb 2012
[6] William F Milliken and Douglas L Milliken A text book of Race Car Vehicle Dynamics.
[7] K. Muninathan, R RThirumaniraj, and M A Suryanarayanan, "Heat Transfer Modeling and Analysis Of Annular Diffuser with and without Double Tapered Struts", 2018 IEEE International Conference on System, Computation, Automation and Networking (ICSCA), Pondicherry, 2018, pp.1-5. doi: 10.1109/ICSCAN.2018. 8541227.

IOP Conf. Series: Materials Science and Engineering **923** (2020) 012008 doi:10.1088/1757-899X/923/1/012008

# STUDY OF CUTTING FORCES AND PREDICTION OF SURFACE QUALITY ANALYSIS USING NEURAL NETWORK MODEL, SUPPORT VECTOR REGRESSION MODEL BY VARIOUS TEXTURED TOOL CONDITION FOR Ti-6Al-4V ALLOY

D. Arulkirubakaran[1]*, R. MalkiyaRasaline Prince[2], R. Monish kumar[3], S. Aravinthkumar[4], C. Andrew Joshva[5]

[1,2,3,4,5] Department of Mechanical Engineering, Karunya Institute of Technology and Sciences, Coimbatore - 641114

arulkirubakaran@karunya.edu,russelmecher@gmail.com,monishr@karunya.edu,aravi1999kumar@gmail.com, andrewc@karunya.edu.in

**ABSTRACT:** To evaluate the performance of the textured tool on surface quality with three different types of the textured pattern using Wire-Cut Electrical Discharge Machining (W-EDM) on tungsten carbide cutting tools with two different groove depth dimensions 100 µm& 200µm respectively and the tools are coated with both TiN and TiAlN using Physical Vapour Deposition (PVD) technique. Surface roughness is predicted using the Support Vector Regression, multilayer Artificial Neural Network model (ANN) model. ANN training is carried out with a pure line transfer function and backpropagation algorithm. Easy off machining and good surface finish are achieved through TiAlN coated tool with linear texture along the perpendicular to chip flow direction than the tools considered for experimental and predicted conditions.

**KEYWORDS:** Surface roughness, coated tool, friction, ANN, SVR.

## 1. INTRODUCTION:

Modified tools surface on the rake face and flank face of cutting tool has a higher impact on the minimizing the friction between the contact surfaces. Also, it the lubrication between the interfaces. Turning of titanium alloy material undergoes various difficulties are stress induction during the formation of chip, larger friction coefficient between the tool-chip interface, larger temperature rise [1, 2]. The cutting tool wear behaviors were analyzed with various coated wear-resistant films of TiN, TiAlN, and ZrC etc. and their combinations improve the tool wear characteristics during machining of Hard-to-cut materials [3-9]. Modified tool (textured) surface has been introduced to improve the load capacity, coefficient of friction and wear resistance and there are attributed physical mechanisms such as wear debris entrapment through the hydrodynamic effect [10-18]. Machining impact with textured tools on surface roughness was examined with a coated tool for AISI 1030 steel with various experimental results. The experimental studies were analyzed on Artificial Neural Networks (ANN) for the different set of parameters and the results are compared by statistical error analyzing method [19,20]. Similar kinds of investigations were made by various authors [21, 22 and 23]. In the present study, three different types of micro surface texture at two depths 100µm and 200 µm on top surface (rake face) toolareproduced by Wire-EDM (W-EDM) and coated with two TiN&TiAlN. Machining tests are conducted on Titanium Grade 5 (Ti-6Al-4V) alloy. Two different models are used for the comparison of surface roughness such as Artificial Neural Network (ANN), Support Vector

Regression (SVR) or Support Vector Machine (SVM) for analyzing any of the combinations cutting parameters.

## 2. EXPERIMENTAL METHOD

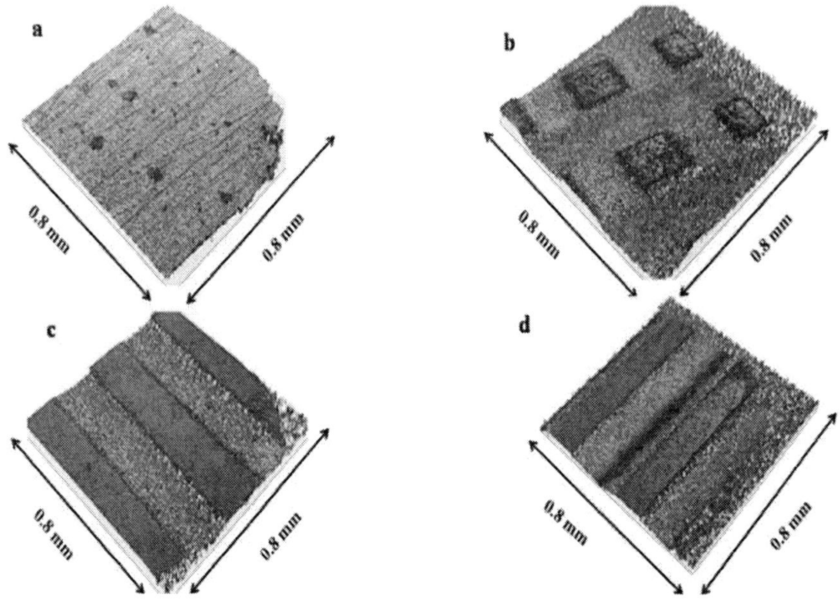


**Figure 1**:3D confocal topography images for a) NT, b) CR.T,c) PR.T, and d) PE.T [23].

Titanium alloy (Ti6Al4V) work material with 25 mm diameter and the 150 mm length rod is selected and the chemical element proportion is provided in Table 1. The Electrical Discharge Machining (EDM) is used for producing the linear textures and cross textures on the cutting tool as shown in figure.1. The texture groove dimensions are 250 μm width and 100 μm, 200 μm by W-EDM process. Titanium experiments are performed by CNC turning (LEADWELL-Fig.2) turning center with CNMA120408 (uncoated, coated carbide) inserts for executing the machining operation. The PCLNL 1616 H12 type tool holder with a rake angle of -6° was used for the turning purpose. The physical and mechanical properties of tungsten carbide tool as shown in Table.2.The coatings of TiN and TiAlN are processed using Physical Vapour Deposition (PVD). The machining parameters are chosen as 60, 90,120 m/min with a constant depth of cut of 0.5mm and feed rate of 0.1 mm/rev and using molybdenum disulfide ($MoS_2$) mixedwith gear oil. The Kistler 9257B piezo-electric type 3-component dynamometer was used for the measurement of forces for the different types of tools.The surface roughness measurement was measured using the Mitutoyo SJ411 surface roughness tester.

**Figure 2:** CNC Machine setup

| | |
|---|---|
| Aluminium | 5–6.75% |
| Vanadium | 3.5–4.5% |
| Carbon | <0.1% |
| Iron | <0.3% |
| Oxygen | <0.2% |
| Nitrogen | <0.05% |
| Hydrogen | <0.015% |
| Titanium | Balance |

**Table 1:**Chemical Composition of Ti-6Al-4V

| **Composition (wt. %)** | **WC + 6%Co** |
|---|---|
| Density(g/cm$^3$) | 14.5 |
| Hardness(GPa) | 16.0 |
| Toughness(MPa.m$^{1/2}$) | 14.8 |
| Thermal Conductivity(W/(m.K)) | 75.4 |
| Flexural Strength (MPa) | 2000 |

**Table 2:**Chemical composition and properties of WC/Co material

IOP Conf. Series: Materials Science and Engineering **923** (2020) 012008 doi:10.1088/1757-899X/923/1/012008

## 3.RESULT AND DISCUSSIONS

### 3.1 EFFECT OF TEXTURES ON CUTTING FORCES

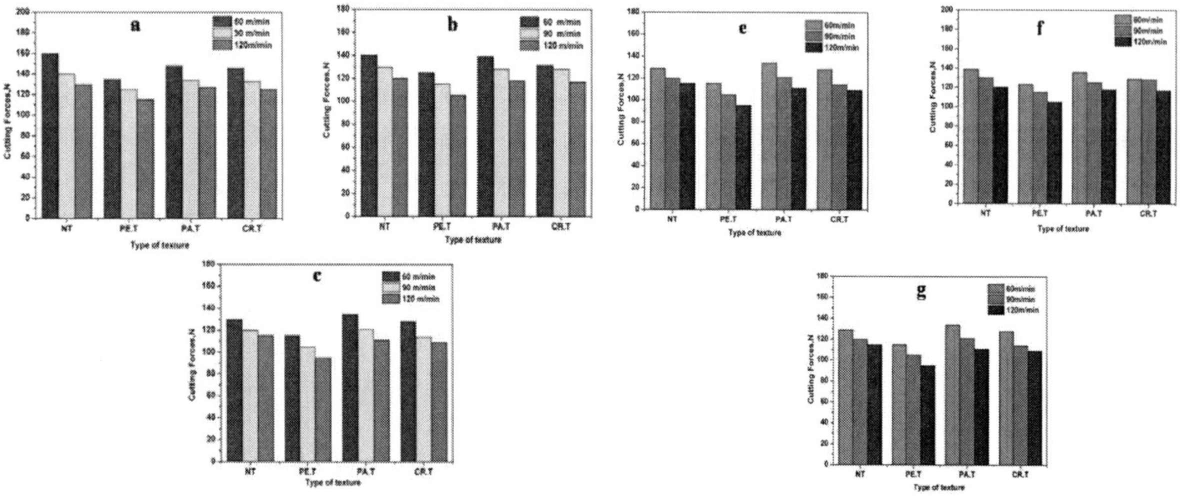


**Figure 3:**Cutting forces for various textured tools (100µm) (a) Uncoated, (b) TiN (c) TiAlN

**Figure 4:**Cutting forces for various textured tools (200µm) (e) Uncoated, (f) TiN (g) TiAlN

Figure 3 and Figure 4 shows the impact of fabricated cutting tools with conventional tool (without groove) on cutting forces for the considered groove depths and tools at 60,90 and 120 m/min with turning of Ti-6Al-4V alloy. From figure 3, the larger force of 160 N was observed for the conventional(non-textured) tool and the lesser force of 130 N was noticed with TiAlN coated conventional (non-textured) tool at minimum considered cutting speed(60m/min). However, at 120m/min similar behavior was noticed and the least force was observed as 109N than the other types of conventional(non-textured) tools considered for a groove depth of 100 µm. Similarly, the reduction in cutting forces was noticed with considered textured tools. The 6-12 % reduction of cutting forces were found in textured tools with conventional( non-textured) tools for all the considered conditions. Moreover, the least force was observed with TiAlN coated perpendicular textured tool for all the considered cutting conditions. The evidence for the diminishing effect in cutting forces due to the formation of a thin film which reduces the shear strength of the material work-piece top surface and presence $MoS_2$. Also, it induces the capillary action and acts as a transporting medium between the contact surfaces. Similar profound was observed for the grooved depth of 200 µm.

### 3.2. EFFECT OF TEXTURE ON SURFACE ROUGHNESS

The functional performance and quality of the machining are assessed by the surface finish produced by the machined work-piece. The surface finish was considered based on the degree of damage on the peak and valley height on the machined surface. The measured surface finish was represented in figure 7. The $R_a$ value was considered for the mean value of 30 measured points, at different positions of the workpiece by using the surface roughness apparatus. The surface roughness was found to be less for the TiAlN perpendicular textured tool with an average equal to 0.208µm. The predictions were analyzed using ANN and SVR. The various errors are Mean Absolute Deviation (MAD), Mean Square Error (MSE) and Mean

Absolute Percentage Error (MAPE). The factors for the machining experimentation were shown in section 2.

### 3.3. PREDICTION OF SURFACE ROUGHNESS THROUGH ANN AND SVR

#### 3.3.1. NEURAL NETWORK

Neural network models is a simple mathematical model that is used to predict surface roughness through MATLAB software with the help of the Back Propagation Neural Network (BPNN) algorithm. Figure 5 shows the developed neural network model.

#### 3.3.2. DESIGNING THE MODEL

The experiments were conducted according to the four different input parameters coating, groove depth, texture, and speed. The different levels of each parameter are shown in below table 3.

| Factors | 1 | 2 | 3 | 4 |
|---|---|---|---|---|
| Groove Size | 100 | 200 | - | - |
| Speed | 60 | 90 | 120 | - |
| Coating | NC | TiN | TiAlN | - |
| Texture | NT | PET | PAT | CRT |

**Table 3:** Experimental level and Parameters

#### 3.3.3.CHOICE OF THE NUMBER OF HIDDEN NEURONS

The selection of the optimal number of hidden neurons (N) is the necessary challenging tasks in ANN modeling. Simon Haykin [21] has stated that N should lie somewhere between 2 and infinity. Similarly, Hecht Nelisen [22] utilize the Kolmogorov's theorem to calculate the number of hidden neurons N=2(n+1) where n is the number of input neurons. In general, this value must be defined based on the requirement of the problem. The choice of the high value of neurons (N) can be reduced the error which causes the complexity of the problem. Thus an optimal N value must be considered which gives a tolerable error associated with ANN.

#### 3.3.4. CHOICE OF PARAMETERS

The main important parameters to be considered are the learning rate and the momentum factor. The learning rate is very small, the computations become slow, thereby takes much time to complete. The learning rate is largerthe model becomes unstable (oscillatory). In order to control the learning rate without disturbing the stability of the model the momentum factor.The values of the learning rate and momentum factor should lie between 0 & 1.

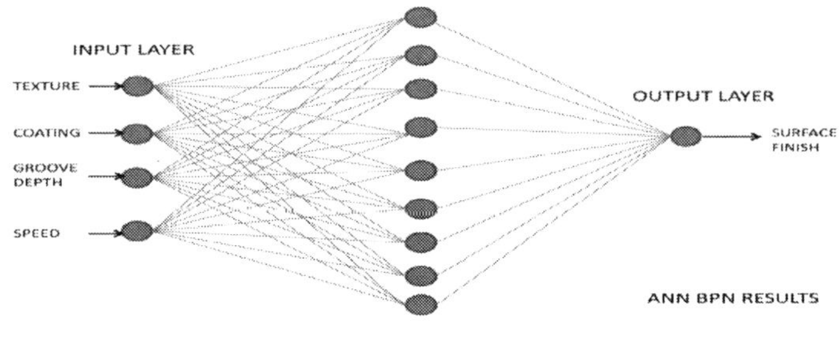


**Figure 5:.** Neural Network model

### 3.3.5. EVALUATION OF THE MODEL

The evaluation of the model is done using measured values of the least square error with the training data which is fed into the model. The training is stopped, when the least square value of error has been obtained without vary much deviation of least square error value for any number of iterations. The final model was developed based on regression value and accuracy of the regression plot for any given set of input parameters. The Mean Square Error (MSE) is used to validate the performance of ANN data.

### 3.3.6.ANN MODEL RESULT

The ANN model is developed and trained, a set of test data was given as input to the developed model. The output from the model is obtained and tabulated shown in Table 4. The average sum of mean square error by this method is found to be equal to 0.276.

| Experimental Output | ANN Output | Error | Mean Absolute Deviation | Mean Square Error | MAPE |
|---|---|---|---|---|---|
| 0.309 | 0.633 | -0.324 | 0.324 | 0.105 | 1.049 |
| 0.394 | 1.535 | -1.141 | 1.141 | 1.302 | 2.896 |
| 1.582 | 1.200 | 0.382 | 0.382 | 0.146 | 0.241 |
| 0.361 | 0.654 | -0.293 | 0.293 | 0.086 | 0.812 |
| 0.627 | 0.953 | -0.326 | 0.326 | 0.106 | 0.520 |
| 0.663 | 0.720 | -0.057 | 0.057 | 0.003 | 0.086 |
| 1.535 | 0.744 | 0.791 | 0.791 | 0.626 | 0.515 |
| 1.200 | 1.440 | -0.240 | 0.240 | 0.058 | 0.200 |
| 0.554 | 1.100 | -0.546 | 0.546 | 0.298 | 0.986 |
| 0.753 | 0.933 | -0.180 | 0.180 | 0.032 | 0.239 |
| MEAN | | | 0.428 | 0.276 | 0.754 |

**Table 4:** ANN Outputs

### 3.3.7. SUPPORT VECTOR REGRESSION MODEL (SVR)

The SVR model used to find out the real number valuesfor the large combination of possibilities. In this operation, the tolerance ε is considered based on the trial and error method to the SVR for the machining operation. However, the main motive is to minimizing the error and increase the margin by compensated error.

### 3.3.8 SVR MODEL PERFORMANCE AND RESULTS

Based on the SVR model formation, the set of developed and trained data's was given as input to the developed model. The output from this model is obtained through the SVR and it is tabulated in Table 5. The average sum of mean square error for this method is found to be 0.044.

| Experimental Output | SVR Output | Error | Mean Absolute Deviation | Mean Square Error | MAPE |
|---|---|---|---|---|---|
| 0.309 | 0.043 | 0.266 | 0.266 | 0.071 | 0.861 |
| 0.394 | 0.193 | 0.201 | 0.201 | 0.040 | 0.510 |
| 1.582 | 1.579 | 0.003 | 0.003 | 0.000 | 0.002 |
| 0.361 | 0.374 | -0.013 | 0.013 | 0.000 | 0.036 |
| 0.627 | 0.648 | -0.021 | 0.021 | 0.000 | 0.033 |
| 0.663 | 0.875 | -0.212 | 0.212 | 0.045 | 0.320 |
| 1.535 | 1.489 | 0.046 | 0.046 | 0.002 | 0.030 |
| 1.200 | 1.145 | 0.055 | 0.055 | 0.003 | 0.046 |
| 0.554 | 0.232 | 0.322 | 0.322 | 0.104 | 0.581 |
| 0.753 | 1.174 | -0.421 | 0.421 | 0.177 | 0.559 |
| MEAN | | | 0.156 | 0.044 | 0.298 |

**Table 5:** SVR Outputs

The performance of the SVR model can be predicted by the performance plots against the test data fed into the model. This plot is shown in figure 6, where the test output is plotted against the number of inputs

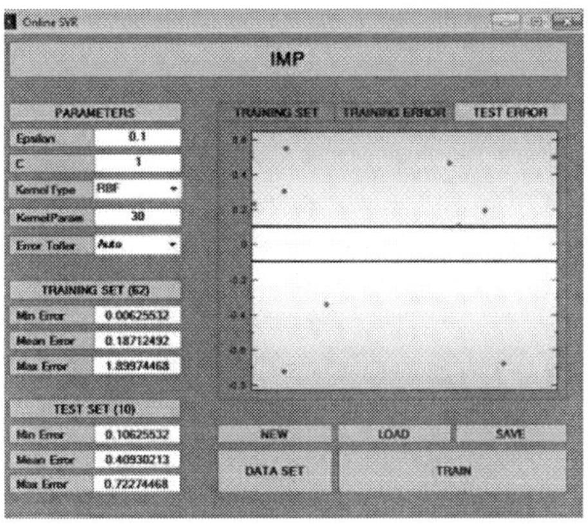


**Figure 6:** Performance plot of SVR

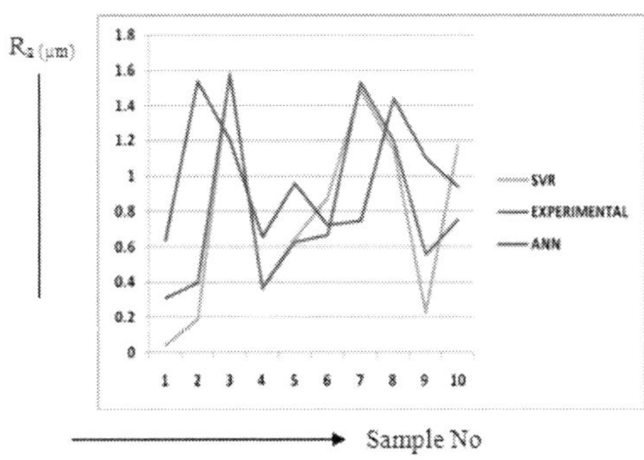


**Figure 7:** Comparison of ANN, SVR and Experimental Results

## 4. COMPARISON OF RESULTS

The test data is given as input into the two models considered for the evaluation of predicted outputs from the two models. From the comparison plot, the output of the two models compared with the actual experimental output. It was found that the SVR plot is giving a much accurate prediction compared to the ANN plot as shown in figure.7. and the comparison between the error values is shown in Table 6. Based on the comparison plot, the SVR model is better in predicting the output as compared to ANN for MSE, MAD,and MAPE errors.

| COMPARISON | MAD | MSE | MAPE |
|---|---|---|---|
| ANN | 0.428 | 0.276 | 0.754 |
| SVR | 0.156 | 0.044 | 0.298 |

**Table 6:** Comparison of Error

## 5. CONCLUSIONS:

The effect of the different combinations of texture, coating and groove depth for the evaluation of surface roughness and cutting force for the turning of titanium alloy was investigated. Conclusions are shown below:

- The perpendicular textured tool with TiAlN coating enhances machinability aspects for the groove depth of 100 µm.
- It reduces the cutting force by 6-12% with non-textured tool due to combined effect coating and reduction in tool-formation of chip interface friction.
- The SVR and ANN models were developed to predicted surface roughness values for the considered input conditions. Based on the predicted surface roughness value, the SVR model performed better than the ANN model for considered input conditions
- The comparative study shows the different types of errors for the SVR model and ANN models. The SVR give better result than ANN models which ensures the degree of accuracy is high in the SVR model than the ANN model.

## REFERENCES:

[1]. Wu H,Guo L 2014 Machinability of titanium alloy TC21 under orthogonal turning process, *Materials and Manufacturing Processes***29 (11–12)** 1441–1445

[2] Ai X 2004 High Speed Machining Technology first ed*National Defense Industry Press Beijing*

[3] Liu J Y, Han R D, Sun Y F 2005 Research on experiments and action mechanism with water Vapor as coolant and lubricant in Green cutting *Machine Tools and Manufacture***45**687–694

[4] Deng J X, Liu J H, Zhao J L, Song W L, Niu S2008 Friction and wear behaviors of the PVD ZrN coated carbide in sliding wear tests and in machining processes*Wear***264** 298–307

[5] Klocke F, Krieg T 1999 Coated tools for metal cutting – features and applications*CIRP Annals Manufacturing Technology* **48**515–525

[6] Kalss W, Reiter A, Derflinger V, Gey C, Endrino J L 2006 Modern coatings in high-performance cutting applications Refractory Metals and Hard Materials **24**399–404

[7] Ezugwu E O, Okeke C I 2001 Tool life and wear mechanisms of TiN coated tools inan intermittent cutting operation Material Processing Technology 11610–15

[8] Su G S, Liu Z Q 2012 Wear characteristics of nano TiAlN-coated carbide tools inultra-high speed machining of AerMet100Wear289 124–131

[9] Long Y, Zeng J J, Yu D H, Wu S H 2014 Microstructure of TiAlN and CrAlN coatingsand cutting performance of coated silicon nitride inserts in cast iron turning Ceramic International40889–894

[10] Long Y, Zeng J J, Wu S H 2014 Cutting performance and wear mechanism ofTi–Al–N/Al–Cr–O coated silicon nitride ceramic cutting insertsCeramic International40 9615–9620.

[11] Brizmer V, Kligerman Y, EtsionI 2003A laser textured parallel thrust bearing Tribology Transactions 46397–403

[12] Vrbka M, Samanek O, Sperka P, Navrat T, Krupka I, Hartl M 2010 Effect of surface texturing on rolling contact fatigue within mixed lubricated non-conformalrolling/sliding contacts, Tribology International43 1457–1465

[13] Pettersson U, Jacobson S 2003 Influence of surface texture on boundary lubricatedsliding contactsTribology International36 857–864

[14] Borghi A, Gualtieri E, Marchetto D, Moretti L, Valeri S 2008 Tribological effects ofsurface texturing on nitriding steel for high-performance engine applicationsWear 265 1046–1051

[15] Rapoport L, Moshkovich A, Perfilyev V, Lapsker I, Halperin G, Itovich Y, Etsion I 2008 Friction and wear of MoS2films on laser textured steel surfaces Surface Coatings and Technology202 3332–3340.

[16] Sugihara T, Enomoto T 2009 Development of a cutting tool with nano/micro- textured surface:improvement of anti-adhesive effect by considering the texture patterns Precision Engineering33 425–429.

[17] Toshiyuki Obikawa, AkihiroKamio,HidemitsuTakaok ,AkiraOsada 2011 Micro-texture at the coated tool face for high performance cutting Machine Tools and Manufacture51 966–972

[18] Koshy P, Tovey J 2011Performance of electrical discharge textured cutting toolsCIRP Annals-Manufacturing Technology60 153–156

[19] MuammerNalbant, HasanGokkaya, IhsanToktas, Gokhan Sur 2009 The experimental investigation of the effects of uncoated, PVD- and CVD-coated cemented carbide inserts and cutting parameters on surface roughness in CNC turning and its prediction using artificial neural networks Robotics and Computer-Integrated Manufacturing25211–223

[20]IlhanAsilturk , HarunAkkus 2011 Determining the effect of cutting parameters on surface roughness in hard turning using the Taguchi methodMeasurement441697–1704

[21] Simon Haykin 1999Neural networks a comprehensive foundationPrentice Hall International IncNew Jersey USA 161–175.

[22]Hecht-Nielsen R. 1990Neuro Computing Addison-Wesley Publishing Company Inc

[23] Arulkirubakaran D,Senthilkumar V, Vijay Kumawat 2016Effect of micro-textured tools on machining of Ti–6Al–4V alloy: Anexperimental and numerical approach Refractory Metals and Hard Materials54165-177.

IOP Conf. Series: Materials Science and Engineering **923** (2020) 012009 doi:10.1088/1757-899X/923/1/012009

# Split-Point and Attribute-Reduced Classifier Approach for Fault Diagnosis of Wind Turbine Blade through Vibration Signals

**A Joshuva *, M Arjun, B Sujith Adhithya, B Akash and S Abdul Wahaab**

Centre for Automation & Robotics (ANRO), Hindustan Institute of Technology and Science, Chennai 603103, Tamilnadu, India.

*Email: joshuva1991@gmail.com

Abstract. This study proposes a data processing and analysis of wind turbine blade faults using split-point and attribute-reduced classifier (SPAARC) through statistical-machine learning approach. In this study, the fault like erosion, hub-blade loose connection, pitch angle twist, bend and crack faults have been simulated and the vibration data has been taken using a piezoelectric accelerometer. With the recorded data, statistical features where extracted and with the extracted features were used to classify the fault condition on the wind turbine blade through SPAARC. The classification accuracy was found to be 85.67% and validated through 10-fold-cross-validation.

*Keywords:* Turbine blade fault; SPAARC; J48; statistical; 10-fold cross classification

## 1. Introduction

The International Renewable Energy Agency (IREA) has estimated that the operation and maintenance costs of wind turbines compensate for a substantial 20-25% of the total electricity costs leveled. For on- and off-shore activities, servicing is directly responsible for up to 80% of these costs [1-3]. Corrective or unplanned servicing, which results in replacements if the pieces become broken, can be very costly, in particular when major components failure like wind turbine blade. The wind turbine is especially deficient in its blades due to various factors. Such corners are the central segment for the extraction of wind energy. The turbine must be shut down and the physical examination must be done to identify defects in wind turbine cuts. This will cause a massive reduction in the production of productivity and will require a high labor charge [4, 5].

Innovative iterative non-linear filter and morphological analysis, Liu and Zhang conducted research on naturally weakened wind turbine blade with fault detection [6]. The goal is to eliminate heavy noise from iterative nonlinear filters and to isolate poor vibrating vibrations. In order to diagnose a bearing defect in the frequency domain, a morphological transformations dependent envelope approach is then used. The findings of the diagnostics indicate that the method introduced can be an important diagnostic tool for very weak layers of blades and is preferable to conventional failure diagnostic procedures. Tu et al., [7] have been working to assess blade damages from portable devices by using wind turbine noise.

This article is intended to use the wind turbine noise for the blade inspection process and to conduct on-site wind farm inspections by utilizing built portable devices. The downside of this device is that during service it may check wind turbine blades.

Liu et al., [8] performed a study in an observational wavelet thresholding system on the vibration analysis of large wind turbine fault detection blades. Failure of blade coats will substantially reduce energy output, so the detection of defective blade costly maintenance and accidental loss is important to avoid. However, the key problems in diagnosing low speed blade are that several noise disruptions obscure the poor fault vibration signals and the effective vibrate data are very limited. Freeman et al., [9] performed the study of the fault signature of the marine hydrokinetic turbine blade with the help of continuous wavelet transform. This paper discusses the effects of hydrodynamic asymmetry problems on marine kinetic turbines and the difficulties faced by the non-intrusive identification and condition control approaches when trying to identify such defects.

Chen et al., [10] presented a study on the algorithm to find errors in deep learning on variable-speed wind turbines. To evaluate the signal characteristics of the fault a long-term memory model (LSTM) is developed. The focus system for improving efficiency is introduced in the LSTM. The results of the simulation show that the proposed approach can detect an imbalance error with more than 98 percent accuracy which demonstrates the effectiveness of the proposed approach on the imbalance detection of wind turbine sheets. Various works were done utilizing recreation investigation of issue and structure examination of wind turbine cutting edge and different parts of the wind turbine like gearbox; be that as it may, just a not many in the exploratory examination for wind turbine sharp edge was done [11-14]. Figure 1 shows the methodology of the work done.

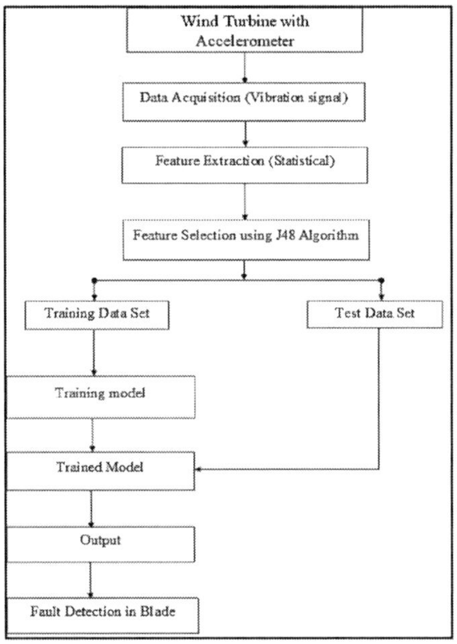


**Figure 1.** Methodology

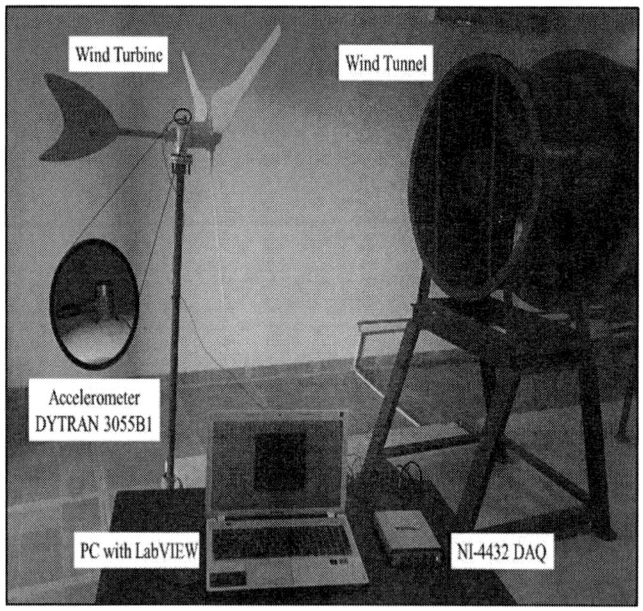


**Figure 2.** Experimental Setup

## 2. Experimental Studies

The trial arrangement, faults, and working technique are explained in detail in Joshuva and Sugumaran (2020) [15]. Figure 2 shows the experimental setup. The frequency of parameters used in the calculation was 12,000 Hz, with 10,000 information for each signal. The DAQ (NI-USB 4432) has been coupled with the Piezoelectric Accelerometer (DYTRAN 3055B1), which was used to store data. Figure 3 and Figure 4 shows the simulated fault on the blade and its vibration pattern.

IOP Conf. Series: Materials Science and Engineering **923** (2020) 012009 doi:10.1088/1757-899X/923/1/012009

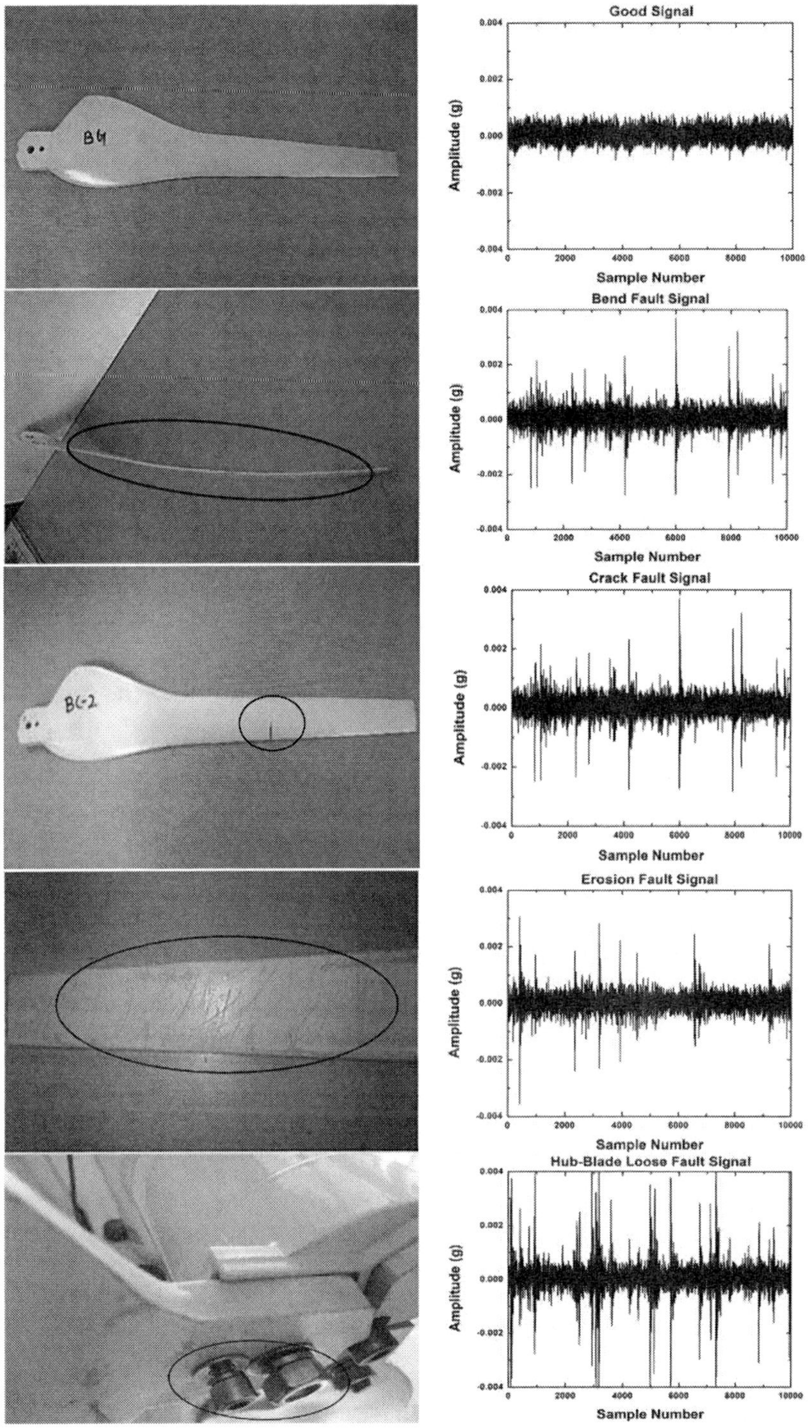

IOP Conf. Series: Materials Science and Engineering **923** (2020) 012009 doi:10.1088/1757-899X/923/1/012009

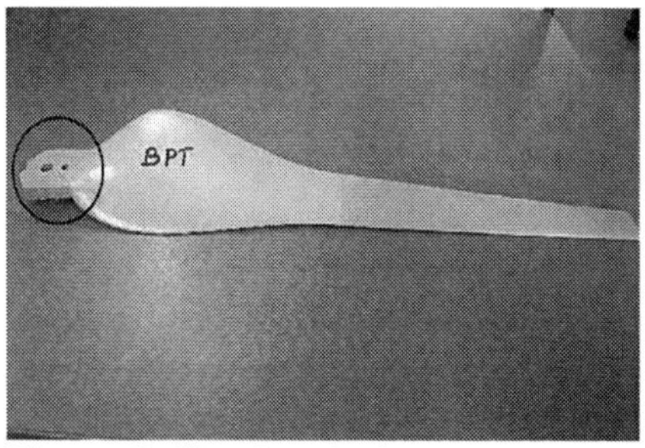


**Figure 3.** Simulate faults

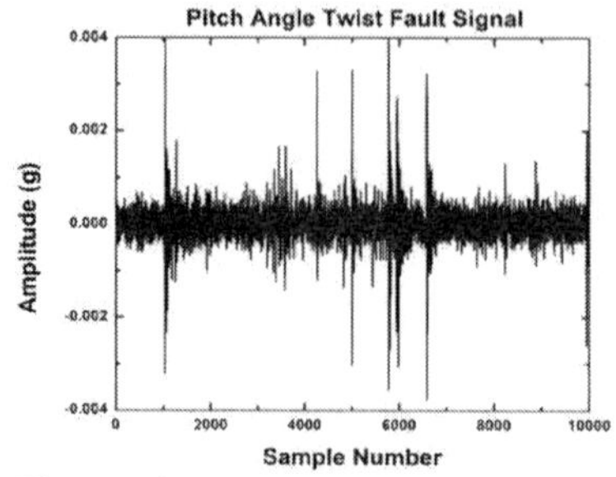


**Figure 4.** Vibration signal

## 3. Feature Extraction

In this study, descriptive statistical features have been extracted from the reordered signal and the procedure for feature extraction has been provided in [16]. The statistical features like mode, minimum, standard error, maximum sum, mean, median, range, skewness, kurtosis, standard deviation and sample variance.

## 4. Feature Selection

Feature Selection is the process where the user automatically or manually select those features which contribute most to the prediction variable or output [17]. Having irrelevant features in the data can decrease the accuracy of the models and make your model learn based on irrelevant features. In this study, J48 decision tree algorithm has been used as feature selection tool. The detailed procedure and the selection of technique has been provided in [18].

## 5. Feature Classification

SPAARC (Split-Point and Attribute-Reduction Classifier) [19] is a single tree classification algorithm based on the popular CART (Classification and Regression Tree) algorithm. It incorporates techniques for reducing the computational load, improving processing time whilst minimising effects on classification accuracy. These include 'split-point sampling' that reduces the number of split-points used when testing the suitability of an attribute at each node in the tree, plus 'node-attribute sampling', whereby a subset of attributes is tested at each alternate horizontal tree node level. Tests revealed the combined effect over a number of training datasets is the reduction in model-build time of as much as 69%. This reduced run-time is an ideal feature well suited to constrained devices such as smartphones. The detailed information about SPAARC was provided in [20].

## 6. Result and Discussion

For good condition cutting edge and other defective wind turbine blade conditions [21], the vibration signals were reported. Maximum set of 600 items; 100 of these were good-fit blade tests. In the case of other failures, such as bending of the blade, corrosion, blade crack, hub-blade loose blade, pitch angle twist, 100 tests have been obtained for each case. Statistical features have been identified as attributes and are used for the classification The required performance of the algorithm will be the corresponding state of the classified data [22]. The sum of instances per leaf and the amount of data for the reduced error cutting was kept at 50 in order to choose 4 predominant features of the J48 decision tree algorithm. The batch size and debug 'real' default parameters for SPAARC have now been set as standard. Figure 5 shows the tree flow of the SPAARC classifier with the features. The values in the bracket represents the number of correctly classified instance to the incorrectly classified instances. Figure 6 shows the 10-fold-cross-validation result for SPAARC [23]. Out of 600 instances, 514 (85.67%) instances have been correctly classified and remaining 86 (14.33%) instances are misclassified. The other statistical measures like kappa statistic, mean absolute error, root mean squared error, relative absolute error, root relative squared error values are present (Figure 3).

Table 1 represents the confusion matrix for SPAARC. The first row of the confusion matrix, the primary component (area (1,1)) speaks to the quantity of accurately classified having a place with the equivalent which represents good condition (Class A). The subsequent component (area (1,2)) talks

IOP Conf. Series: Materials Science and Engineering **923** (2020) 012009 doi:10.1088/1757-899X/923/1/012009

about the quantity of good cases of bent fault erroneously assigned (bend) (Class B). The third component (area (1,3)) discusses the amount of good cases of crack loss (crack) that have been erroneously delegated (Class C). The fourth component (area (1,4)) speaks of the volume of positive events wrongly called loose fault situations of hub-blade (loose) (Class D). The fifth component (area (1,5)) speaks to the quantity of good occurrences that were inaccurately named pitch angle twist fault condition (PAT) (Class E). The sixth component (area (1,6)) speaks to the quantity of good examples that were mistakenly delegated erosion fault condition (erosion) (Class F) [24].

```
Median < 6.56E-5
|   Standard Error < 3.98E-6
|   |   Standard Deviation < 2.81759E-4: Bend(3.0/0.0)
|   |   Standard Deviation >= 2.81759E-4: Erosion(90.0/8.0)
|   Standard Error >= 3.98E-6: PAT(98.0/7.0)
Median >= 6.56E-5
|   Sample Variance < 2.595E-7
|   |   Standard Error < 3.29E-6
|   |   |   Mean < 9.48E-5: Bend(76.0/2.0)
|   |   |   Mean >= 9.48E-5: crack60(3.0/0.0)
|   |   Standard Error >= 3.29E-6
|   |   |   Mean < 7.485E-5
|   |   |   |   Mean < 6.705E-5: Erosion(3.0/0.0)
|   |   |   |   Mean >= 6.705E-5: Bend(9.0/1.0)
|   |   |   Mean >= 7.485E-5: crack60(90.0/13.0)
|   Sample Variance >= 2.595E-7
|   |   Sample Variance < 4.6700000000000004E-7
|   |   |   Minimum < -0.0082975
|   |   |   |   Kurtosis < 30.11124363: Good(10.0/0.0)
|   |   |   |   Kurtosis >= 30.11124363
|   |   |   |   |   Standard Error < 5.375E-6: crack60(3.0/0.0)
|   |   |   |   |   Standard Error >= 5.375E-6: Loose(11.0/3.0)
|   |   |   Minimum >= -0.0082975
|   |   |   |   Kurtosis < 20.699946755
|   |   |   |   |   Range < 0.01264: Loose(9.0/1.0)
|   |   |   |   |   Range >= 0.01264: Good(7.0/0.0)
|   |   |   |   Kurtosis >= 20.699946755: Loose(64.0/8.0)
|   |   Sample Variance >= 4.6700000000000004E-7: Good(71.0/10.0)
```

**Figure 5.** Tree flow of SPAARC

```
Number of Leaf Nodes: 15

Size of the Tree: 29

Time taken to build model: 0.61 seconds

=== Stratified cross-validation ===
=== Summary ===

Correctly Classified Instances         514               85.6667 %
Incorrectly Classified Instances        86               14.3333 %
Kappa statistic                          0.828
Mean absolute error                      0.0676
Root mean squared error                  0.2049
Relative absolute error                 24.3208 %
Root relative squared error             54.968  %
Total Number of Instances              600
```

**Figure 6.** 10-fold-cross-validation

**Table 1.** Confusion matrix for SPAARC

| Blade Class | Class A | Class B | Class C | Class D | Class E | Class F |
|---|---|---|---|---|---|---|
| Class A | **79** | 0 | 1 | 20 | 0 | 0 |
| Class B | 0 | **84** | 7 | 0 | 0 | 9 |
| Class C | 0 | 8 | **89** | 3 | 0 | 0 |
| Class D | 18 | 0 | 7 | **75** | 0 | 0 |
| Class E | 0 | 0 | 0 | 0 | **98** | 2 |
| Class F | 0 | 4 | 0 | 0 | 7 | **89** |

**Table 2.** Class-wise accuracy of SPAARC

| Blade Class | TP | FP | PRE | REC | F-M | MCC | ROC | PRC |
|---|---|---|---|---|---|---|---|---|
| Class A | 0.79 | 0.03 | 0.81 | 0.79 | 0.80 | 0.76 | 0.94 | 0.79 |
| Class B | 0.84 | 0.02 | 0.87 | 0.84 | 0.85 | 0.83 | 0.95 | 0.80 |
| Class C | 0.89 | 0.03 | 0.85 | 0.89 | 0.87 | 0.84 | 0.96 | 0.80 |
| Class D | 0.75 | 0.04 | 0.75 | 0.75 | 0.75 | 0.71 | 0.92 | 0.73 |
| Class E | 0.98 | 0.01 | 0.93 | 0.98 | 0.95 | 0.94 | 0.98 | 0.89 |
| Class F | 0.89 | 0.22 | 0.89 | 0.89 | 0.89 | 0.86 | 0.96 | 0.83 |

Thus, the subsequent column speaks to the second condition i.e bend condition. The third line speaks to the information focuses for the third condition, for example crack condition. The fourth line speaks to the information focuses for fourth condition, for example hub-blade loose fault condition. The fifth column speaks to the information focuses for the fifth condition, for example pitch angle twist fault condition. The sixth line speaks to the information focuses for the sixth condition, for example erosion fault condition. In the confusion matrix, the diagonal components speak to the accurately grouped examples and the others are misclassified instances [25]. Table 2 provides comprehensive reliability in class via the true positive rate (TP), false positive rate (FP), F-measures (F-M), Matthews coefficient of correlation (MCC), precision (PRE), area of the receiver's operating characteristics (ROC), recall (REC), region of precision recall curve (PRC). The true positive (TP) value should be almost 1 in a correct classification and the false positive (FP) should be less than 0. Figure 5 reveals that the TP average in most classes is near 1 and that the FP rate is close to 0. This supports that in Table 1, the outcome shown by the confusion matrix.

## 7. Conclusion

The wind turbine is extremely important in the day-to-day production of wind power. This study aims to provide the fault classification on wind turbine blade using split-point and attribute reduced classifier (SPAARC). The SPAARC has been validated through 10-fold-cross-validation and the correctly classified instance was found to be 85.67%. This classifier has the minimal mean absolute error (0.0676). SPAARC may therefore practically be used for wind turbine blade condition monitoring to identify conditions of the blade failure.

## REFERENCES

[1] Fogaing MB, Gordon H, Lange CF, Wood DH and Fleck BA 2019 A review of wind energy resource assessment in the urban environment *InAdvances in Sustainable Energy* **2019** 7-36.

[2] Liu Z and Zhang L 2020 A review of failure modes, condition monitoring and fault diagnosis methods for large-scale wind turbine bearings *Measurement* **149** 107002.

[3] Ti Z, Deng XW and Yang H 2020 Wake modeling of wind turbines using machine learning *Applied Energy* **257** 114025.

[4] Du Y, Zhou S, Jing X, Peng Y, Wu H and Kwok N 2020 Damage detection techniques for wind turbine blades: a review *Mechanical Systems and Signal Processing* **141** 106445.

[5] Zuo H, Bi K and Hao H 2020 A state-of-the-art review on the vibration mitigation of wind turbines *Renewable and Sustainable Energy Reviews* **121** 109710.

[6] Liu Z and Zhang L 2019 Naturally damaged wind turbine blade bearing fault detection using novel iterative nonlinear filter and morphological analysis *IEEE Transactions on Industrial Electronics* **30**.

[7] Tu TH, Lo FC, Liao CC, Chung CF and Chen RC 2019 Using wind turbine noise to inspect blade damage through portable device *InINTER-NOISE and NOISE-CON Congress and Conference Proceedings 2019* **259** 1787-91.

[8] Liu Z, Zhang L and Carrasco J 2020 Vibration analysis for large-scale wind turbine blade bearing fault detection with an empirical wavelet thresholding method *Renewable Energy* **146** 99-110.

[9] Freeman B, Tang Y and VanZwieten J 2019 Marine hydrokinetic turbine blade fault signature analysis using continuous wavelet transform. *In2019 IEEE Power & Energy Society General Meeting (PESGM) 2019* **4** 1-5.

[10] Chen J, Hu W, Cao D, Zhang B, Huang Q, Chen Z and Blaabjerg F 2019 An imbalance fault detection algorithm for variable-speed wind turbines: a deep learning approach *Energies* **12** 2764.

[11] Joshuva A, Aslesh A and Sugumaran V 2019 State of the art of structural health monitoring of wind turbines *International Journal of Mechanical and Production Engineering Research and Development* **9** 95-112.

[12] Jiménez AA, Zhang L, Muñoz CQ and Márquez FP 2020 Maintenance management based on machine learning and nonlinear features in wind turbines *Renewable Energy* **146** 316-28.

[13] Vilar AÁ, Xydis G and Nanaki EA 2020 Small wind: a review of challenges and opportunities *InSustaining Resources for Tomorrow* 185-204.

[14] Movsessian A, Garcia D and Tcherniak D 2020 Adaptive feature selection for enhancing blade damage diagnosis on an operational wind turbine *InProceedings of the 13th International Conference on Damage Assessment of Structures* 594-605.
[15] Joshuva A and Sugumaran V 2020 A lazy learning approach for condition monitoring of wind turbine blade using vibration signals and histogram features *Measurement* **152** 107295.
[16] Agrawal B, Dubey S and Dixit M 2019 Standard statistical feature analysis of image features for facial images using principal component analysis and its comparative study with independent component analysis *InInternational Conference on Intelligent Computing and Smart Communication* 1027-48.
[17] Nagra AA, Han F, Ling QH, Abubaker M, Ahmad F, Mehta S and Apasiba AT 2020 Hybrid self-inertia weight adaptive particle swarm optimisation with local search using C4.5 decision tree classifier for feature selection problems *Connection Science* **32** 16-36.
[18] Sahu SK and Shrivas AK 2020 Comparative study of classification models with genetic search based feature selection technique *InCognitive Analytics: Concepts, Methodologies, Tools, and Applications* 773-83.
[19] Denison DG, Mallick BK and Smith AF 1998 A bayesian CART algorithm *Biometrika* **85** 363-77.
[20] Yates D, Islam MZ and Gao J 2018 SPAARC: a fast decision tree algorithm. *InAustralasian Conference on Data Mining* **28** 43-55.
[21] Solimine J, Niezrecki C and Inalpolat M 2020 An experimental investigation into passive acoustic damage detection for structural health monitoring of wind turbine blades *Structural Health Monitoring* **3** 1475921719895588.
[22] Wu R, Zhang D, Yu Q, Jiang Y and Arola D 2019 Health monitoring of wind turbine blades in operation using three-dimensional digital image correlation *Mechanical Systems and Signal Processing* **130** 470-83.
[23] Jung K, Bae DH, Um MJ, Kim S, Jeon S and Park D 2020 Evaluation of nitrate load estimations using neural networks and canonical correlation analysis with k-fold cross-validation *Sustainability* **12** 400.
[24] Braun T, Spiliopoulos S, Veltman C, Hergesell V, Passow A, Tenderich G, Borggrefe M and Koerner MM 2020 Detection of myocardial ischemia due to clinically asymptomatic coronary artery stenosis at rest using supervised artificial intelligence-enabled vectorcardiography-a five-fold cross validation of accuracy *Journal of Electrocardiology* **8**.
[25] Hasnain M, Pasha MF, Ghani I, Imran M, Alzahrani MY and Budiarto R 2020 Evaluating trust prediction and confusion matrix measures for web services ranking *IEEE Access* **8** 90847-61.

IOP Conf. Series: Materials Science and Engineering **923** (2020) 012010 doi:10.1088/1757-899X/923/1/012010

# Rough Set Theory Based Blade Condition Classification on Wind Turbine through Statistical Features

**A Joshuva*, K Rakesh Kumar, G S Sriram Gangadhar, S S Dhanush, M Arjun**

Centre for Automation & Robotics (ANRO), Hindustan Institute of Technology and Science, Chennai 603103, Tamilnadu, India.

*Email: joshuva1991@gmail.com

Abstract. Wind energy has become a one of the alternative energy source due to fossil fuel crisis. These wind energies are being harvested from the wind through wind turbines. These wind turbines are subjected to various environmental factors and prone to severe vibration on blade. This vibration lead to the catastrophic calamities and cause severe capital loss and wind production loss. This study proposes a data processing and analysis of wind turbine blade faults using rough set theory based feature classification. The feature extraction (statistical features) and the feature selection (J48 decision tree algorithm) methods were used to identify the best features for fault classification. Using rough set theory, with five statistical features, 75.5% of classification accuracy have been obtained for the fault identification on wind turbine blade.

*Keywords:* Turbine blade fault; rough set theory; J48; statistical; 10-fold cross classification

## 1. Introduction

Due to total environmental pollution, wind energy has become the most important resource in the present period [1]. Wind power is an excellent renewable energy source and a normal generation alternative. The wind turbine is used for electricity transformation increasing wind energy. The capacity, efficiency, protection, durability, viability, and life-cycle of turbines should be improved to make wind power even more important [2]. The wind turns the rotor post and spins the shaft within the wind turbine. The sensitive, lengthy and versatile thin boundary blades are part of the most incredibly complex power and the friction on the straight edge blades will easily be carried on [3]. The strained and dangerous vibration induces rotor-edge pulverization and harm [4]. The blade is one of the most important components that is often damaged. The vibration of the blade when operating is hard to figure due to its massive development and working condition [5]. The study uses vibration source and rough set theory to diagnose damage on wind turbine blades.

Hsu et al., [6] carried out a research on diagnostics and automated management of wind turbines by way of computational process control and machine learning. This research has studied sensor data by combining elements in the maintenance test list in data mining systems with practitioners' insight. The findings show a high degree of precision: 92.68% for the software Decision Tree and 91.98% for the Random Forest Software. The study shows that wind turbine faults can be observed and components replacement requirements can be forecast by data mining and simulation. Rizk et al., [7] also performed hyperspectronic imagery used to identify damage to wind turbine blades and icing. In this analysis it is important to use the hyperspectral imaging device for detecting harm to the wind turbine blade and for icing. Within this study, the harm types, triggers and methods used for identification are mentioned specifically. In addition, current issues and successful attempts to examine real harm identification to turbine blades are addressed. The results showed that surface and subsoil defects and icing incidents could be observed in their early occurring phases through hyperspectral imaging.

Ghanbarpour et al., [8] conducted dependable energy extraction in wind turbines using a concept of predictive failure resistance. The point of this article is to achieve control objectives in the event of partial load area disturbances, instability, sensor and actuator faults for wind turbines. The goal is to map the best generator speed by harvesting maximal energy in wind turbine control systems in the partial load region. Jiang et al., [9] proposed to operate on the main fault position in wind turbine systems for graphically semi-supervised temporal deep learning. A different approach to identifying the central failure chain in wind turbine systems utilizing unlabeled data is presented in this paper. The first approach is to train and derive high-level functions from the data is a deep learning algorithm, a stacked sparse auto-encoder. A semi-monitored time study algorithm is then used to generate a pseudo-labeled data set with an unlabeled data set.

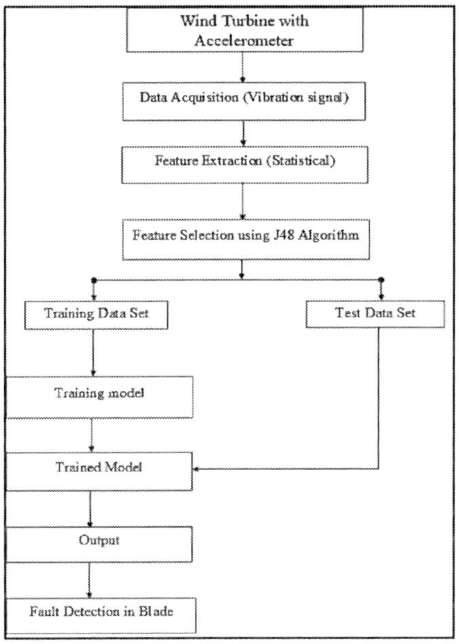


**Figure 1.** Methodology

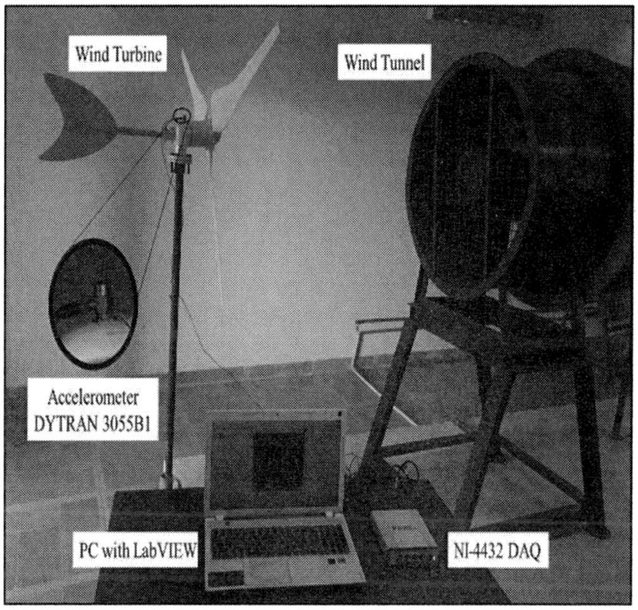


**Figure 2.** Experimental Setup

Various works were done utilizing recreation investigation of issue and structure examination of wind turbine cutting edge and different parts of the wind turbine like gearbox; be that as it may, just a not many in the exploratory examination for wind turbine sharp edge was done [10]. For the discovery of faults in cutting edges of wind turbine, the machine learning method was considered; however, use was limited on drawing. The analysis was performed under a very limited system of imperfections [11]. This is particularly valid on account of fault finding of the wind turbine cutting edge. Consequently, there is a solid need to plan an issue determination framework that can deal with numerous flaws in wind turbine edges utilizing rough set theory draws near. This study aims to detect distinctive cutting-edge problems with a rough set theory and statistical analysis. The methodology of the work carried out is shown in Figure 1.

## 2. Experimental Studies

IOP Conf. Series: Materials Science and Engineering **923** (2020) 012010 doi:10.1088/1757-899X/923/1/012010

The trial arrangement, faults, and working technique are explained in detail in Joshuva and Sugumaran (2020) [12]. Figure 2 shows the experimental setup. The sampling frequency used in the assessment was 12,000 Hz and each signal has a length of 10,000 data. Piezoelectric accelerometer (DYTRAN 3055B1) was used along with DAQ (NI-USB 4432) for recording the data. Figure 3 and Figure 4 shows the simulated fault on the blade and its vibration pattern.

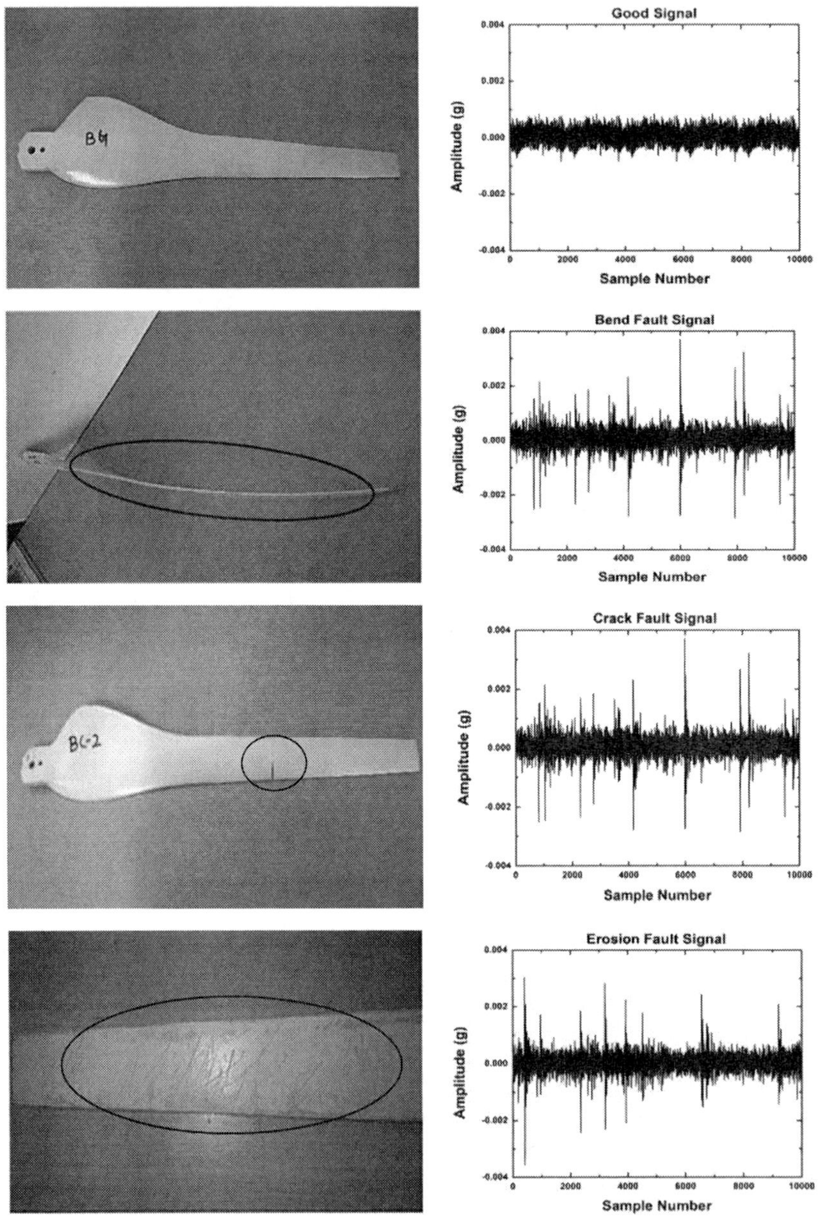

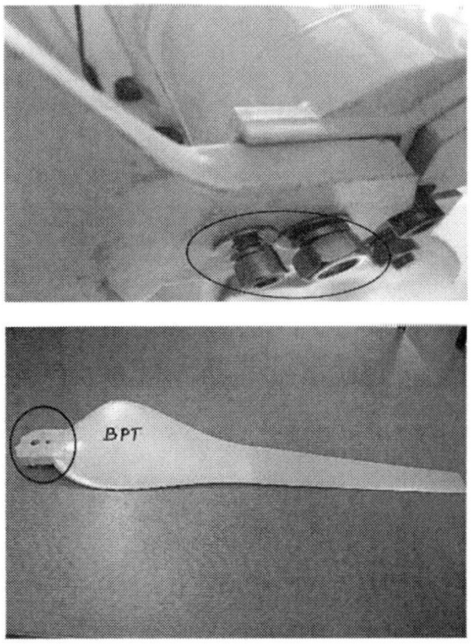


Figure 3. Simulate faults

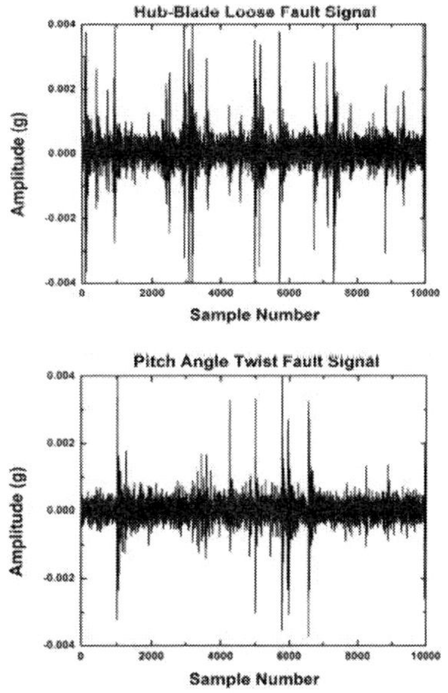


**Figure 4.** Vibration signal

## 3. Statistical based Feature Extraction

In this study, descriptive statistical features have been extracted from the reordered signal and the procedure for feature extraction has been provided in [13]. The statistical features like mode, minimum, standard error, maximum sum, mean, median, range, skewness, kurtosis, standard deviation and sample variance.

## 4. J48 Decision Tree based Feature Selection

Feature Selection is the process where the user automatically or manually select those features which contribute most to the prediction variable or output. Having irrelevant features in the data can decrease the accuracy of the models and make your model learn based on irrelevant features. In this study, J48 decision tree algorithm has been used as feature selection tool. The detailed procedure and the selection of technique has been provided in [14].

## 5. Rough Set Theory based Feature Classification

The notion of rough set theory (RST) was introduced by Pawlak [15]. It is a principle that emerges from basic research into the conceptual features of information systems. Rough-set principle in comparison repositories was the technique for the retrieval or exploration of information. It is closely related to fuzzy theory. One can use rough set approach to discover structural relationship within imprecise and noisy data [16]. Rough sets and fuzzy sets are complementary generalizations of classical sets. The approximation spaces of rough set theory are sets with multiple memberships, while fuzzy sets are concerned with partial memberships. Soft Computing includes along with rough sets, at least fuzzy logic, neural networks, probabilistic reasoning, belief networks, machine learning, evolutionary computing, and chaos theory. The working procedure of rough set theory is given in [17].

## 6. Result and Discussion

For good condition blade and other wind turbine blade faults the vibrating signals were reported. A minimum of 600 samples, including 100 from a good condition weapon, have been obtained. In fact,

100 samples of every state were tested for other faults including blade bending, corrosion, blade crack, hub blade loose links, twisting of the pitch. The statistical attributes were derived as features and are used for classification purposes [18]. The relevant performance of the algorithm is the corresponding state of the classified data. For choosing four main features of the J48 decision tree algorithm, the total number of instances and the amount of data used on reduced error pruning were 50. The parameters for RST such as batch size (100) and "real" debug choice were configured as usual. The 10-fold cross-validation result for RST [19] is shown in figure 5.

For 600 instances, 453 (75.5%) were correctly categorized and 123 (20.5%) remaining instances became incorrectly classified with 24 (4%) unclassified instances. The other statistical measures like kappa statistic, mean absolute error, root mean squared error, relative absolute error, root relative squared error values are present (Figure 5) [20]. Table 1 represents the confusion matrix for rough set theory. The primary component (area (1,1)) speaks to the quantity of accurately classified having a place with the equivalent (Class A). The following component (area (1.2)) refers to the number of good instances of bend failure (bend) (Class B). The third component (area (1,3)) refers to the number of good instances in which crack (crack) was erroneously delegated (Class C). The fourth component (area (1,4)) refers to the quantity of good instances which were wrongly named loose (loose) hub blade defects (Class D). The fifth component (area (1,5)) speaks to the quantity of good occurrences that were inaccurately named pitch angle twist fault condition (PAT) (Class E). The sixth component (area (1,6)) speaks to the quantity of good examples that were mistakenly delegated erosion fault condition (erosion) (Class F) [21].

```
Time taken to build model: 0.64 seconds

=== Stratified cross-validation ===
=== Summary ===

Correctly Classified Instances        453              75.5    %
Incorrectly Classified Instances      123              20.5    %
Kappa statistic                         0.7436
Mean absolute error                     0.0712
Root mean squared error                 0.2668
Relative absolute error                26.6927 %
Root relative squared error            73.0653 %
UnClassified Instances                 24                4     %
Total Number of Instances             600
```

**Figure 5.** 10-fold-cross-validation result for rough set theory

Thus, the subsequent column speaks to the second condition i.e bend condition. The third line speaks to the information focuses for the third condition, for example crack condition. The fourth line speaks to the information focuses for fourth condition, for example hub-blade loose fault condition. The fifth column speaks to the information focuses for the fifth condition, for example pitch angle twist fault condition [22]. The sixth line speaks to the information focuses for the sixth condition, for example erosion fault condition. In the confusion matrix, the diagonal components speak to the accurately grouped examples and the others are misclassified instances. Table 2 [23] demonstrates the thorough precision of the section. The class-specific consistency refers to real positive (TP) performance, false positives (FP), precision (PRE), recall (REC), F-Measure (F-M), Matthews coefficient of correlation (MCC), receiver area (ROC) area, PRC area [24]. For a better categorizer, the true positive (TP) value

IOP Conf. Series: Materials Science and Engineering **923** (2020) 012010 doi:10.1088/1757-899X/923/1/012010

should be almost 1 in a correct classification and the false positive (FP) should be less than 0. Table 2 reveals that the rate of TP in most groups is similar to 1 and the rate of FP almost 0. It means that the finding shown in Table 1 is a confusion matrix [25].

**Table 1.** Confusion matrix for rough set theory

| Blade Class | Class A | Class B | Class C | Class D | Class E | Class F |
|---|---|---|---|---|---|---|
| Class A | **76** | 1 | 0 | 20 | 0 | 0 |
| Class B | 2 | **78** | 8 | 0 | 0 | 9 |
| Class C | 4 | 7 | **61** | 11 | 7 | 2 |
| Class D | 22 | 1 | 6 | **62** | 2 | 0 |
| Class E | 1 | 0 | 2 | 1 | **91** | 5 |
| Class F | 0 | 2 | 2 | 0 | 8 | **85** |

**Table 2.** Class-wise accuracy of rough set theory

| Blade Class | TP | FP | PRE | REC | F-M | MCC | ROC | PRC |
|---|---|---|---|---|---|---|---|---|
| Class A | 0.78 | 0.06 | 0.72 | 0.78 | 0.75 | 0.70 | 0.85 | 0.59 |
| Class B | 0.80 | 0.02 | 0.87 | 0.80 | 0.83 | 0.80 | 0.87 | 0.72 |
| Class C | 0.66 | 0.03 | 0.77 | 0.66 | 0.71 | 0.66 | 0.78 | 0.53 |
| Class D | 0.66 | 0.66 | 0.66 | 0.66 | 0.66 | 0.59 | 0.77 | 0.47 |
| Class E | 0.91 | 0.03 | 0.84 | 0.91 | 0.87 | 0.84 | 0.93 | 0.78 |
| Class F | 0.87 | 0.03 | 0.84 | 0.87 | 0.85 | 0.85 | 0.82 | 0.90 |

## 7. Conclusion

This study aims to provide the fault classification on wind turbine blade using rough set theory (RST). The RST has been validated through 10-fold-cross-validation and the correctly classified instance was found to be 75.5%. This classifier has the minimal mean absolute error (0.0712). Hence rough set theory for condition monitoring of the wind turbine blade can be employed practically to identify the conditions in blade.

## REFERENCES

[1] Wu X, Hu Y, Li Y, Yang J, Duan L, Wang T, Adcock T, Jiang Z, Gao Z, Lin Z and Borthwick A 2019 Foundations of offshore wind turbines: a review *Renewable and Sustainable Energy Reviews* **104** 379-93.

[2] Ti Z, Deng XW and Yang H 2020 Wake modeling of wind turbines using machine learning *Applied Energy* **257** 114025.

[3] Fogaing MB, Gordon H, Lange CF, Wood DH and Fleck BA 2019 A review of wind energy resource assessment in the urban environment. *InAdvances in Sustainable Energy* 7-36.

[4] Madi E, Pope K, Huang W and Iqbal T 2019 A review of integrating ice detection and mitigation for wind turbine blades *Renewable and Sustainable Energy Reviews* **103** 269-81.

[5] Igwemezie V, Mehmanparast A and Kolios A 2019 Current trend in offshore wind energy sector and material requirements for fatigue resistance improvement in large wind turbine support structures–a review *Renewable and Sustainable Energy Reviews* **101** 181-96.

[6] Hsu JY, Wang YF, Lin KC, Chen MY and Hsu JH 2020 Wind turbine fault diagnosis and predictive maintenance through statistical process control and machine learning *IEEE Access* **8** 23427-39.

[7] Rizk P, Al Saleh N, Younes R, Ilinca A and Khoder J 2020 Hyperspectral imaging applied for the detection of wind turbine blade damage and icing *Remote Sensing Applications: Society and Environment* **5** 100291.
[8] Ghanbarpour K, Bayat F and Jalilvand A 2020 Dependable power extraction in wind turbines using model predictive fault tolerant control *International Journal of Electrical Power & Energy Systems* **118** 105802.
[9] Jiang N, Hu X and Li N 2020 Graphical temporal semi-supervised deep learning–based principal fault localization in wind turbine systems *Proceedings of the Institution of Mechanical Engineers, Part I: Journal of Systems and Control Engineering* **12** 0959651819901034.
[10] Joshuva A, Aslesh A and Sugumaran V 2019 State of the art of structural health monitoring of wind turbines *International Journal of Mechanical and Production Engineering Research and Development* **9** 95-112.
[11] Xu D, Liu PF, Chen ZP, Leng JX and Jiao L 2020 Achieving robust damage mode identification of adhesive composite joints for wind turbine blade using acoustic emission and machine learning *Composite Structures* **236** 111840.
[12] Joshuva A and Sugumaran V 2020 A lazy learning approach for condition monitoring of wind turbine blade using vibration signals and histogram features *Measurement* **152** 107295.
[13] Anh NX, Nataraja RM and Chauhan S 2020 Towards near real-time assessment of surgical skills: A comparison of feature extraction techniques *Computer Methods and Programs in Biomedicine* **187** 105234.
[14] Singh J, Bagga S and Kaur R 2020 Software-based prediction of liver disease with feature selection and classification techniques *Procedia Computer Science* **167** 1970-80.
[15] Pawlak Z 1998 Rough set theory and its applications to data analysis *Cybernetics & Systems* **29** 661-88.
[16] Tay FE and Shen L 2003 Fault diagnosis based on rough set theory *Engineering applications of artificial intelligence* **16** 39-43.
[17] Suraj Z 2004 An introduction to rough set theory and its applications *ICENCO* **3** 80.
[18] Lin Z and Liu X 2020 Wind power forecasting of an offshore wind turbine based on high-frequency SCADA data and deep learning neural network *Energy* **24** 117693.
[19] Huo Y, Xin L, Kang C, Wang M, Ma Q and Yu B 2020 SGL-SVM: a novel method for tumor classification via support vector machine with sparse group Lasso *Journal of Theoretical Biology* **486** 110098.
[20] Xu J, Zhang Y and Miao D 2020 Three-way confusion matrix for classification: a measure driven view *Information Sciences* **507** 772-94.
[21] Meehan AJ, Latham RM, Arseneault L, Stahl D, Fisher HL and Danese A 2020 Developing an individualized risk calculator for psychopathology among young people victimized during childhood: a population-representative cohort study *Journal of affective disorders* **262** 90-8.
[22] Chicco D and Jurman G 2020 The advantages of the matthews correlation coefficient (MCC) over F1 score and accuracy in binary classification evaluation *BMC genomics* **21** 6.
[23] Düntsch I and Gediga G 2020 Indices for rough set approximation and the application to confusion matrices *International Journal of Approximate Reasoning* **118** 155-72.
[24] Leong S, Zhao Y, Ribeiro-Rodrigues R, Jones-López EC, Acuña-Villaorduña C, Rodrigues PM, Palaci M, Alland D, Dietze R, Ellner JJ and Johnson WE 2020 Cross-validation of existing signatures and derivation of a novel 29-gene transcriptomic signature predictive of progression to TB in a Brazilian cohort of household contacts of pulmonary TB *Tuberculosis* **7** 101898.
[25] Zeng G 2020 On the confusion matrix in credit scoring and its analytical properties *Communications in Statistics-Theory and Methods* **49** 2080-93.

IOP Conf. Series: Materials Science and Engineering **923** (2020) 012011 doi:10.1088/1757-899X/923/1/012011

# Investigation of material impact on Truck frame design through Structural Analysis

**Padmanabhan S*, Sri Charan Undamatla and Harsha Vardhan Reddy**
School of Mechanical and Construction, Vel Tech Rangarajan Dr. Sagunthala R&D Institute of Science and Technology, Avadi, Chennai, India.

Email: padmanabhan.ks@gmail.com

**Abstract**. The truck chassis core purpose is to support the payload due to the various components located on it. The chassis was encountered to dynamic vibration due to the road irregularity and excitation of vibrating mechanisms, when the vehicle is moving on the road. Chassis helps to keep the automobile rigid and stiff. This paper investigates the material impact on the truck frame design. Two different materials were analysed on the heavy truck frame through finite element analysis. The Al alloy and Structural steel materials were applied on for channel cross-section of the truck frame. The results were compared and the best truck chassis material was identified with lower stress and deflection.

## 1. Introduction

Many investigators in the automotive engineering took the challenge to involve in the design, development and production of chassis. The chassis was encountered to dynamic vibration and forces due to the road irregularity and self-excitation of vibrating mechanisms, when the vehicle is moving on the road. Chassis will experience large amplitude of oscillations which may produce extreme bending and failure on it. Studies are on to obtain an optimized chassis design. To investigate the vibrations, the torsional stress and modal factors were analysed experimentally and then the results were validated with finite element model. Thus chassis was optimized with the enhanced structural stiffness. The operating competences of autos, an excited vibration frequency and dynamic model of the frame was investigated [1]. The base frame helps in supporting the different parts of the Truck. It has to resist the sudden impacts, jerks, twists and other stresses that are obtained during braking, acceleration, shocking road conditions. This carries the maximum load for all the operating conditions. The base frame acts as the spine of a heavy vehicle [2]. Finite element analysis techniques are used to analyse the structural exploration of truck chassis on base frames much easily [3]. Effects of frame geometry on collisions and structure of the ladder frame when subjected to various loads. It states that Frame geometry plays a major role during the collisions and various impacts [4].

An ideal automotive chassis are manufactured using Low weight and High Strength Carbon fibres. They made a structural analysis of the truck chassis frame for carbon fibres and steel. They analysed stresses, strain and various deformations and compared between Steel and carbon fibres [5]. The major issue in the automotive sector is weight reduction. They worked on TATA 2515EX which is a heavy vehicle chassis. They analysed it with different composite materials which are subjected to similar pressure as that of chassis made of steel [6]. In the traditional design process, the chassis design focused on the strength and to enhance the stiffness of the structure. A very slight attention to the weight of the chassis. This design technique includes the adding the cross members to the present chassis structure to intensify its torsional stiffness. As this step, the heaviness of the chassis increases. This will leads to reduction in fuel efficiency and increases the material cost due to the addition of cross members. Hence, in the modern design, chassis with adequate strength and nominal stiffness is essential. This paper presents the study of material influence in the frame design on truck chassis. Two different materials were analysed on the heavy truck frame through finite element analysis. The al alloy and structural steel materials were applied on for channel cross-section of the truck frame. The results were compared and the best truck chassis material was identified with lower stress and deflection.

## 2. Frame Design

The chassis consists of longitudinal and cross members on its structure. Longitudinal members are with c- section. The typical model of chassis is shown in Figure 1. Dimensions from the literature is the basis for our models [7]. The frame is designed to handle a weight of 3 tons from the body and 0.6 ton of Engine.

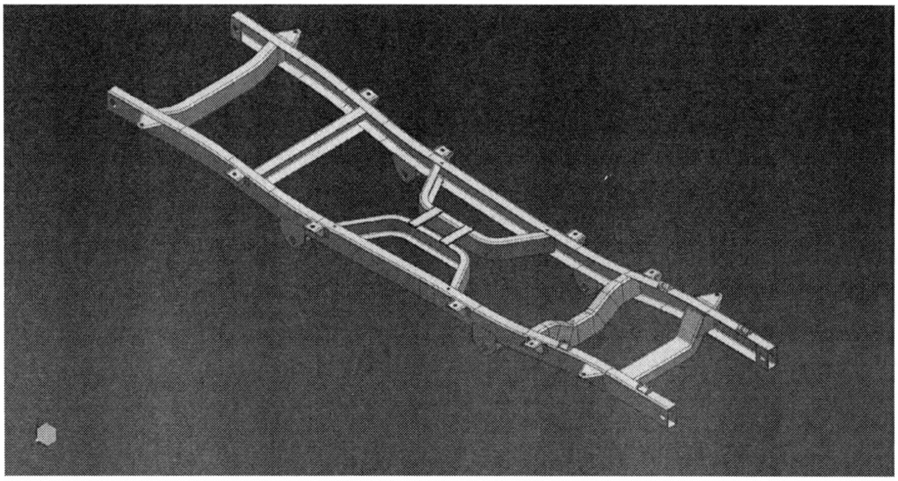

**Figure 1.** Chassis with C- Section

Top view of chassis is shown in Figure 2. Width at the front is less compared to the width at the rear. When the wheels are steered, they should not interfere with the chassis and this is the reason why less width is maintained in the front. Width at the front – 710 mm; Width at the rear – 880 mm.

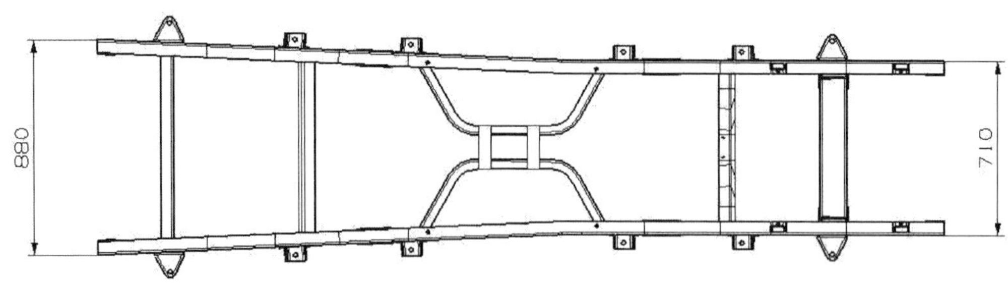


**Figure 2.** Plan of Chassis

Elevation of the chassis is shown in Figure 3. The wheelbase is nothing but the dimension between wheel centres and it is 2350 mm. The overall length of chassis is based on various factors which are out of scope.

IOP Conf. Series: Materials Science and Engineering **923** (2020) 012011 doi:10.1088/1757-899X/923/1/012011

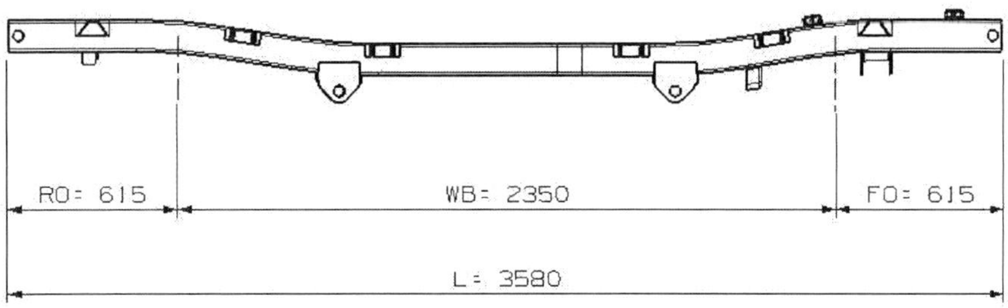


**Figure 3.** Front View of Chassis

In the steel manufacturing industry, AISI 4130 alloy is called Structural Steel. This is generally used in the fabrication of structures. The property of two different frame material is tabulated in Table 1.

**Table 1.** Properties of Frame materials

| Design Property | Structural Steel | Al Alloy |
|---|---|---|
| Young's Modulus, N/mm$^2$ | 207000 | 70000 |
| Poison's Ratio | 0.3 | 0.3 |
| Yield Strength, N/mm$^2$ | 910 | 500 |
| Ultimate Tensile Strength, N/mm$^2$ | 1030 | 550 |

The Finite Element Method used in an engineering as computational tool to achieve an estimated results for a bounded problems. It uses the different boundary constraints for every analysis. The free-free boundary limit will be applied with no constraint is applied to chassis, when it tested for a normal mode. At the same time, real operating condition was applied to the chassis at the static analysis. At the suspension mountings, the pinned boundaries are applied to the chassis because at these mountings won't allow any translation but permit only rotation about its axis.

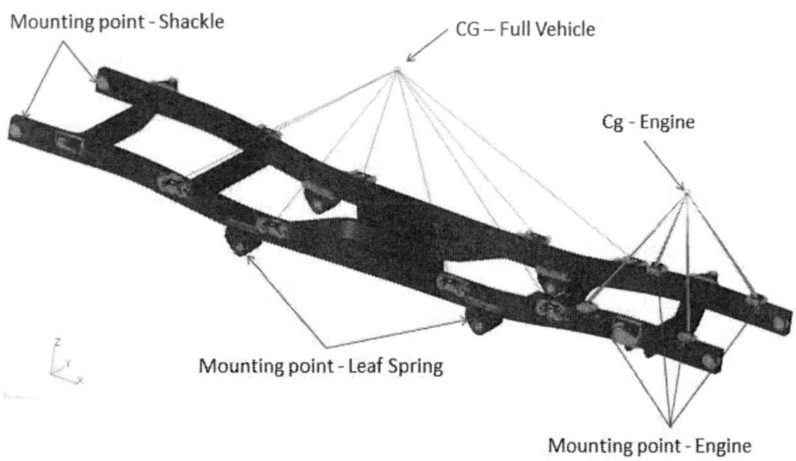


**Figure 4.** Finite Element Model

The model has meshed and the meshed model is shown in Figure 4. Shackle mounting point at 4 locations is arrested with 6 DOF. Leaf spring mounting points at 4 locations are arrested with 6 DOF. This is known as solution sequence 101. There will be reaction force at shock absorber mounting points. We can consider 3 as FOS while designing the Chassis. A load of 0.6 ton is assumed for Engine. This load is equally distributed to the 4 mounting locations.

## 3. Result and Discussion

Plot results for Max Principal, Max Shear & Von Mises Stresses can be easily seen for the al alloy in Figure 5 to 7. Von Mises stress is one of the parameters considered while designing any structure. Stress values obtained are compared with the allowable stress limits.

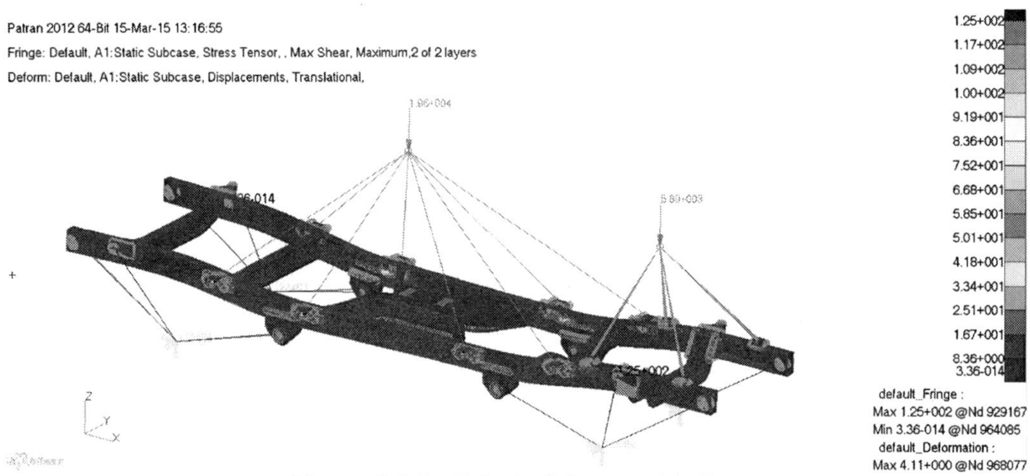


**Figure 5.** Max Principal Stress – Al alloy

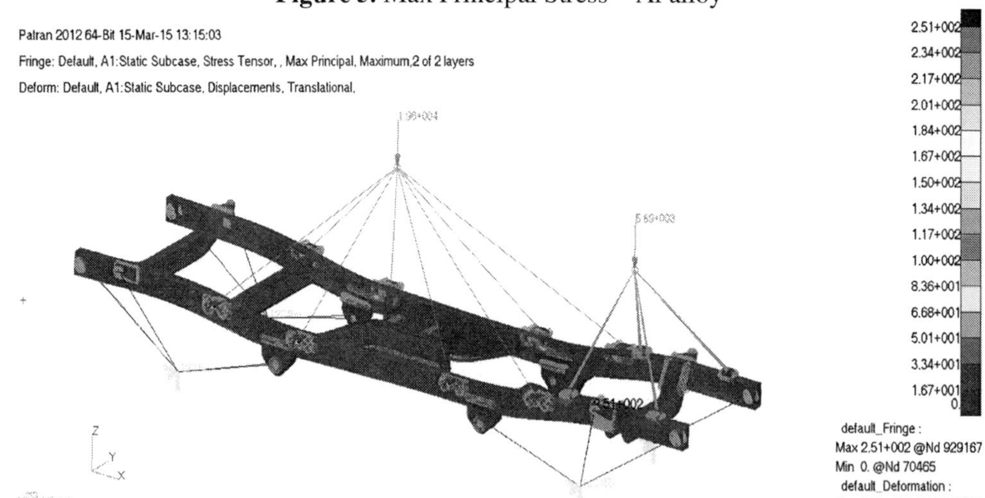


**Figure 6.** Max Shear Stress – Al alloy

This key point is one of the reasons that may cause the loss of exhaustion. The existing truck chassis is fitted with stiffeners for modification and study. Initially the model's thickness, where the highest deflection occurs in bending analysis, has been increased to a certain value with appropriate limit. And at the middle of the wheel foundation, one more cross beam was added to add stiffness to the platform. In order to reinforce and improve the chassis stiffness as well as the overall chassis performance, a series of modifications and tests were carried out by adding the stiffener [7].

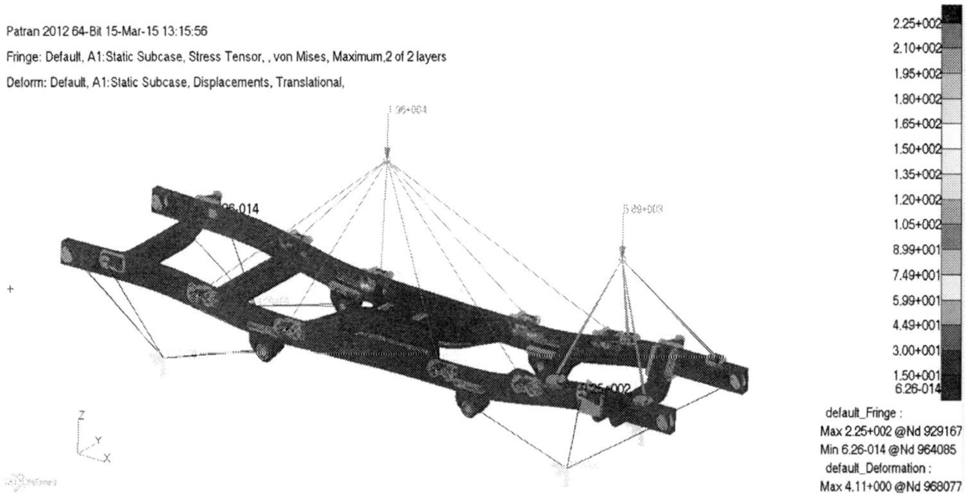


**Figure 7.** Von Mises Stress – Al alloy

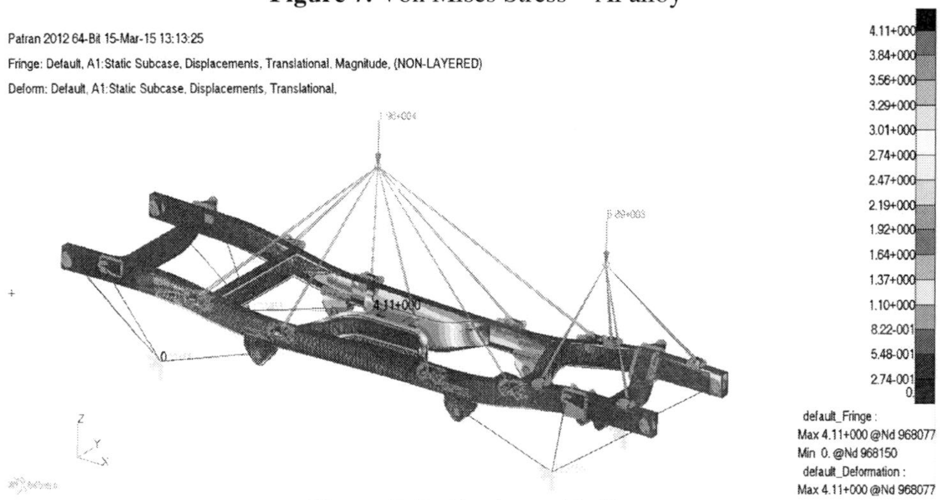


**Figure 8.** Deflection – Al alloy

Maximum deflection is noticed near one of the Body mounting points as shown in the Figure.8. Deflection value obtained is verified with the allowable limits obtained from experiments. The results of aluminium alloy on the truck frame is tabulated in Table 2.

**Table 2.** Stress values obtained for Al alloy

| Description | UTS (N/mm$^2$) | Values Obtained (N/mm$^2$) | RF | Deflection (mm) |
|---|---|---|---|---|
| Max Principal Stress | 550 | 251 | 2.2 | |
| Max Shear stress | 330 | 125 | 2.6 | 4.1 |
| Von Mises stress | 550 | 225 | 2.4 | |

The material changed to Structural Steel, and it was the most used widely in the design of an automobile frame. The plots at the made for Max Principal, Max Shear & Von Mises Stresses and deflection in Figure 9 to 12.

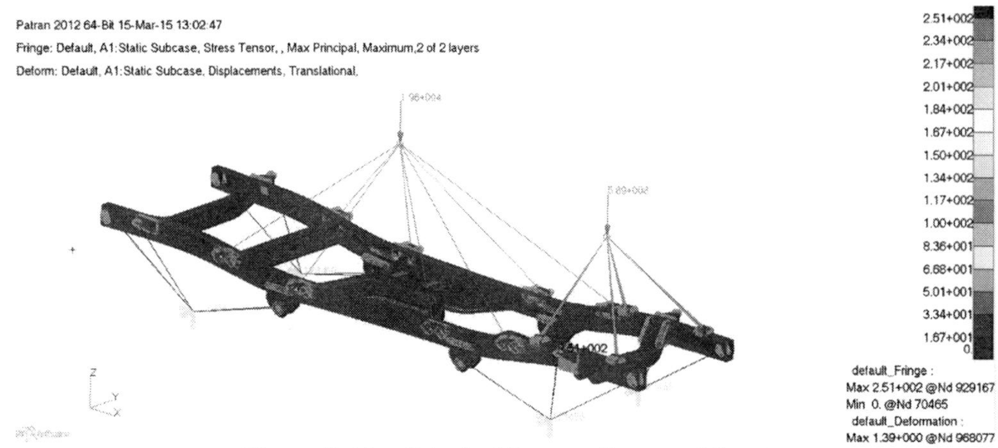


**Figure 9.** Max Principal Stress – Structural Steel

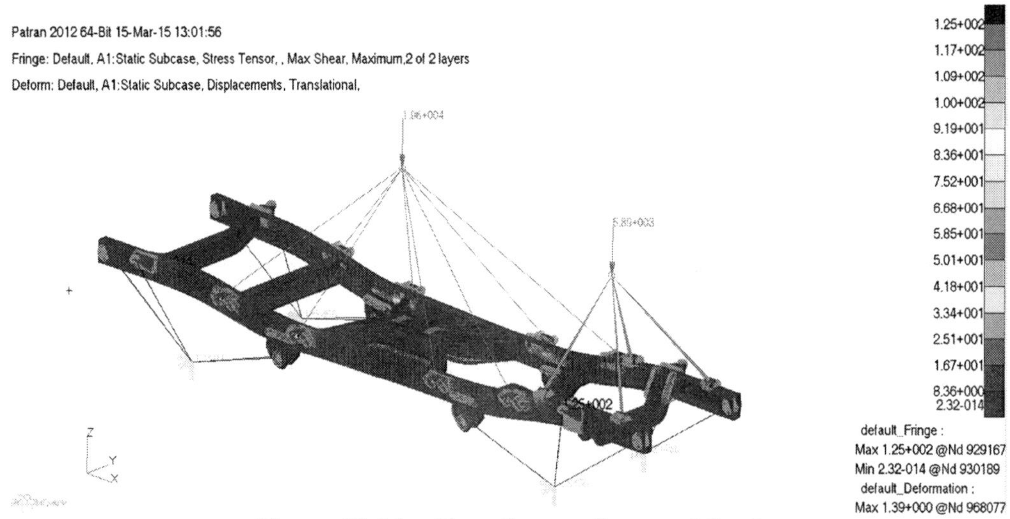


**Figure 10.** Max Shear Stress – Structural Steel

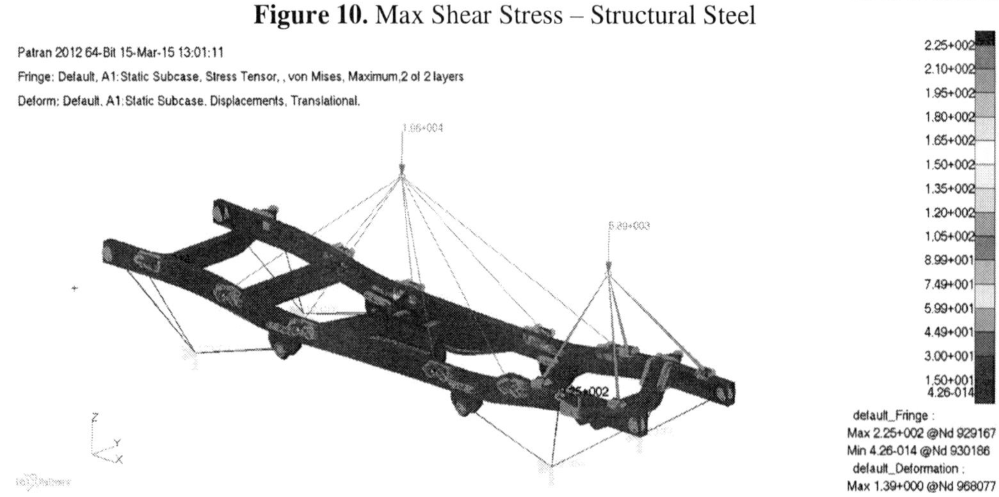


**Figure 11.** Von Mises Stress – Structural Steel

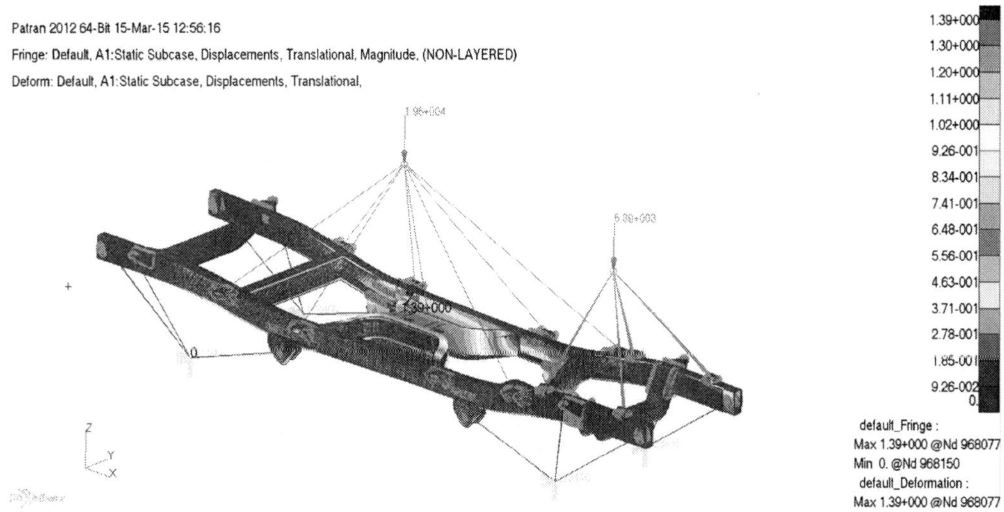


**Figure 12.** Deflection – Structural Steel

Analyses the stress state and performs the static strength test and dynamic strength check of the whole fracturing truck chassis, measures the tension of 15 chassis measuring points under the totally charged static condition and the correct front wheel lifting condition respectively and figures out the distribution of the frame stress and strain. The results show that the frame's maximum stress position under migration condition lies at the root of the tank's rear support [8].

**Table 3:** Stress values obtained for Structural Steel

| Description | UTS ($N/mm^2$) | Values Obtained ($N/mm^2$) | RF | Deflection (mm) |
|---|---|---|---|---|
| Max Principal Stress | 1030 | 251 | 4.1 | |
| Max Shear stress | 618 | 125 | 4.9 | 1.4 |
| Von Mises stress | 1030 | 225 | 4.6 | |

The comparison of aluminium steel and structural steel on the C-section truck frame, structural steel having higher results than the al alloy. The defection was also proven with minimum deflection. The comparison table was presented in table 4.

**Table 4.** Comparison of RF for Al Alloy and Structural Steel

| Description | Al Alloy | Structural Steel |
|---|---|---|
| RF 1 | 2.2 | 4.3 |
| RF 2 | 2.6 | 5.1 |
| RF 3 | 2.4 | 4.8 |
| Deflection | 4.1 | 0.6 |

## 4. Conclusion

Automotive chassis structure ensures less noise, vibration and harshness on the vehicle. This paper presents the study of material influence in the frame design on truck chassis. Two different materials were analysed on the heavy truck frame through finite element analysis. The al alloy and structural steel materials were applied on for channel cross-section of the truck frame. An evaluation of different two materials under this scenario shows that structural steel material- has potential for the frame structure than a simple change of materials. The FEA analysis was contacted with the same type of load and results were compared for best truck chassis material identification with lower stress and deflection. From the results, it is found that structural steel has better RF values when compared to Al alloy. It is also found that structural steel has less deflection than that of Al alloy.

## References

[1] Yuan Ren, Yongchang Yu, Binbin Zhao, Chuanhui Fan and He Li 2017 Finite Element Analysis and Optimal Design for the Frame of SX360 Dump Trucks *Procedia Engineering* **174** 2017 638-647

[2] Vijayan S N 2015 Design and Analysis of Automotive Chassis Considering Cross *International Journal of Current Research* **7** 15697-15701

[3] Prabhakaran S, Padmanabhan S and Joel C 2018 Design optimization and structural analysis of automotive chassis *International Journal of Mechanical and Production Engineering Research and Development* **8** 934-942

[4] Balaji G, Annamalai K and Duc Pham 2017 An experimental and numerical scrutiny of crashworthiness variables for square column with V-notch and groove initiators under quasi-static loading *Cogent Engineering* **4**

[5] Salvi Gauri Sanjay, Kulkarni Abhijeet, Gandhi Pratik Pradeep and Baskar P 2014 Finite Element Analysis of Fire Truck Chassis for Steel and Carbon Fiber Materials: *Journal of Engineering Research and Applications* **4 7** 69-74

[6] Chandra M R, Sreenivasulu S and Hussain S A 2012 Modeling and Structural Analysis of Heavy Vehicle Chassis Made of Polymeric Composite Material by Three Different Cross Sections, *International Journal of Modern Engineering. Research* **2 4**

[7] Goolla Murali, Subramanyam B and Dulam Naveen 2013 Design Improvement of a Truck Chassis based on Thickness *Altair Technology Conference Sigma Consultancy India* 1-6

[8] Lei L , Bin Zhang and Chunming Fan 2020 *IOP Conf. Ser.: Mater. Sci. Eng.* **751** 012057

IOP Conf. Series: Materials Science and Engineering **923** (2020) 012012 doi:10.1088/1757-899X/923/1/012012

# Design and Structural Analysis of Truck Frame

**Padmanabhan S, Harsha Vardhan Reddy and Sri Charan Undamatla**
School of Mechanical and Construction, Vel Tech Rangarajan Dr. Sagunthala R&D Institute of Science and Technology, Avadi, Chennai, India.

Email: padmanabhan.ks@gmail.com

**Abstract**. The chassis construction is the larger element in every automobile, it integrates the heavy vehicle main systems such as the power train, drive axles, suspension and trailer. The vehicle structural shape dependent on this skeleton, it absorbs the impacts from front, side and overturn effects. There are many industrialized sectors using this truck as means of transportation for logistics, agricultures, factories and other industries. In this paper, two different material was adopted on the C-section truck frame design. Mild steel and Titanium alloy were structurally analysed on the heavy truck frame through finite element analysis. The results were compared and the best truck chassis material was identified with lower stress and deflection.

## 1. Introduction

An automotive chassis will go through high amplitude of vibrations on running due to the road irregularity, which will lead to the notable changes in the structural elements. The chassis vibration will roots the stress formation at specific parts with maximum magnitude. It also creates weakness of the vehicle structure and formation of noise due to looseness of mechanical parts and leads to vehicle uncomforted. To rectify the above difficulties, the study of static and dynamic characteristics of vehicle chassis is essential. The road irregularity causes the structural vibration, which is the vital factors that will sources the fatigue failure. The amount of the stress excited on vibration can forecast the lifecycle of the automotive truck chassis [1]. The resonance may occur, when the high amplitude of vibrations of the vehicle structure matches with the amplitude of the external excitation, which will produce the unnecessary deflections on structure and failure on chassis [2]. The modal investigation of the vehicle frame will help to calculate the natural frequency and modal shapes of the parts. The rigidity of the system can be evaluated and resonance might be bypassed [3]. Structural optimization analysis with the constraints on the vehicle chassis can be used to investigate maximum shear stress and deflection at the maximum load condition [4]. The design analysis of truck frame using finite element method was useful to identified the performance of the chassis system, as its design strength cannot be found using empirical formulas [5]. In automotive sector, modal analysis is used in effective way to study the modal behaviour of car chassis. It is used to collect the important data to the inspect vehicle dynamic behaviour [6].

The amount stress developed on the chassis used to examine the lifecycles of the truck chassis. The precision of estimated lifespan of the vehicle chassis is subjected to the outcome of stress analysis [7]. On the investigation load sharing on the vehicle chassis is not constant across its total span area, so that to the concentration of load it can be vary the area of ladder chassis. On this analysis, the effect of decrease in cross-section area with limitations of various stresses and deflection, decrease in the

IOP Conf. Series: Materials Science and Engineering **923** (2020) 012012 doi:10.1088/1757-899X/923/1/012012

area will make difference in the quantity of material used for ladder chassis [8]. In this paper, two different material was adopted on the C-section truck frame design. Mild steel and Titanium alloy were structurally analysed on the heavy truck frame through finite element analysis. The results were compared and the best truck chassis material was identified with lower stress and deflection.

## 2. Truck Frame Design

The ladder frame chassis (Figure 1) is the oldest and simplest frame in all sectors. It contains of two proportioned beams and rails, running on the entire length of the vehicle. Many transverse cross-members are connected with that and mostly used on the trucks.

**Figure 1.** Ladder Frame chassis

Chassis consists of longitudinal members and cross members. Longitudinal members are with c-section. The typical model of chassis is shown in Figure 2. Dimensions on basis of literature for our models [1]. The frame is designed to handle a weight of 3 tons from the body and 0.6 ton of Engine.

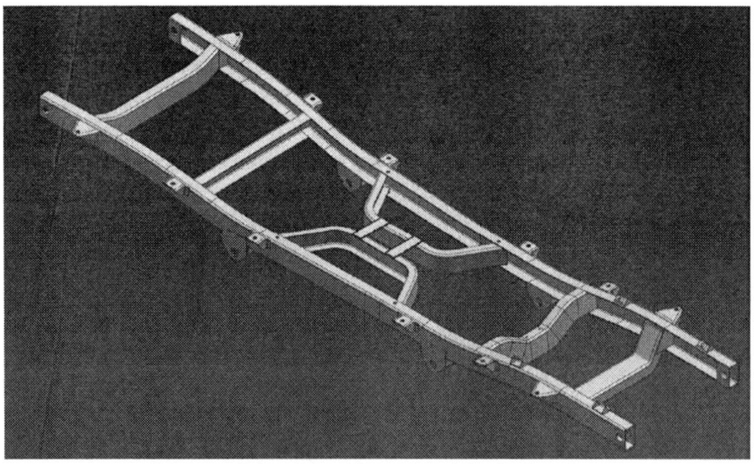

**Figure 2.** Ladder Chassis with C- Section

Top view of chassis is shown in Fig 3. Width at the front is less compared to the width at the rear. When the wheels are steered, they should not interfere with the chassis and this is the reason why less width is maintained in the front. Width at the front – 710 mm; Width at the rear – 880 mm.

IOP Conf. Series: Materials Science and Engineering **923** (2020) 012012 doi:10.1088/1757-899X/923/1/012012

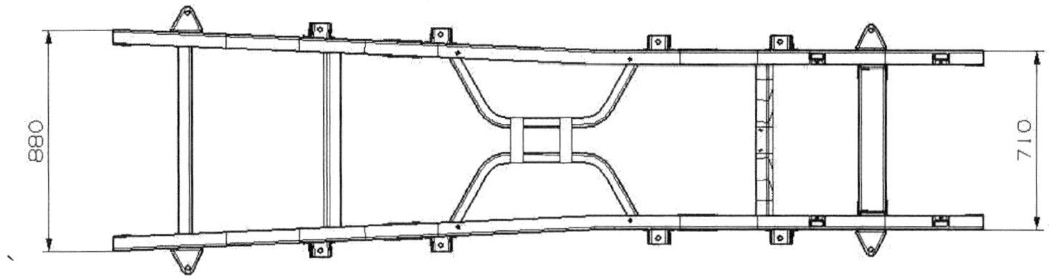


**Figure 3.** Top View of Chassis

Front view of the chassis is shown in Fig 4. Wheel base is the dimension between wheel centres. Wheel Base – 2350 mm ; length – 3580 mm .

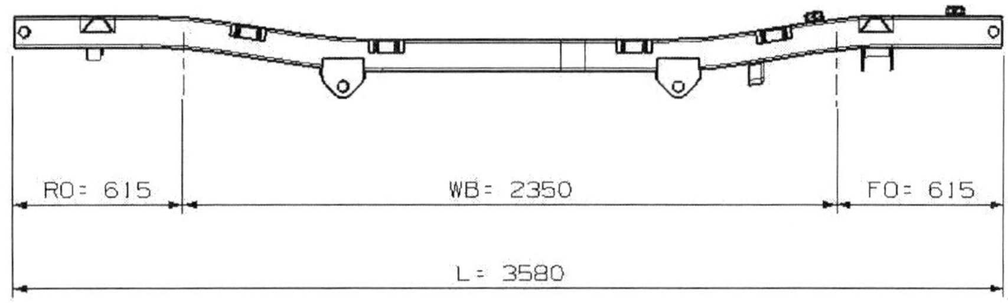


**Figure 4.** Front View of Chassis

The main influences of selecting material especially for frame is wide variation of characteristics such as mechanical and thermal resistance, easiness of production and durability. The first choice for the frame material with these features, Steel is leading choice. There was numerous progresses in irons and steels over the previous decades that through the steel stronger, harder and light-weight and refining other performance features. Titanium has an suggestion with space-technology and is observed by several people as an "eventual" material. It has a density roughly half that of steel, and also a slight over half of the stiffness value. It's a like condition with regards to ultimate and yield strengths. Mild steel and Titanium alloy were selected for the analyse. The property of two different frame material is tabulated in Table 1.

**Table 1.** Properties of Frame materials

| Design Property | Mild Steel | Al Alloy |
|---|---|---|
| Young's Modulus, N/mm$^2$ | 200000 | 120000 |
| Poison's Ratio | 0.3 | 0.3 |
| Yield Strength, N/mm$^2$ | 225 | 450 |
| Ultimate Tensile Strength, N/mm$^2$ | 500 | 970 |

## 3. Results and Discussion

The key factor in today's automobile truck manufacturing is to overwhelmed the growing needs for advanced performance, minimum weight, and extended life of automobile parts, all are should be at a realistic price and in a affordable time. Structural chassis analysis using Finite Element Method is to trace the important area which has the maximum stress concentration. The road irregularity causes the structural vibration, which is the vital factors that will sources the fatigue failure. The amount of the stress excited on vibration can forecast the lifecycle of the automotive truck chassis.

IOP Conf. Series: Materials Science and Engineering **923** (2020) 012012 doi:10.1088/1757-899X/923/1/012012

The chassis was modelled and meshed by using 2D – 4 noded – Quad – Iso mesh solid elements. Shackle mounting point at 4 locations is arrested with 6 DOF. Leaf spring mounting points at 4 locations are arrested with 6 DOF. This is known as solution sequence 101. There will be reaction force at shock absorber mounting points. We can consider 3 as FOS while designing the Chassis. A load of 0.6 ton is assumed for Engine. This load is equally distributed to the 4 mounting locations. Plot results for Max Principal, Max Shear & Von Mises Stresses can be easily seen for the Ti alloy in Figure 5 to 7. Von Mises stress is one of the parameters considered while designing any structure. Stress values obtained are compared with the allowable stress limits.

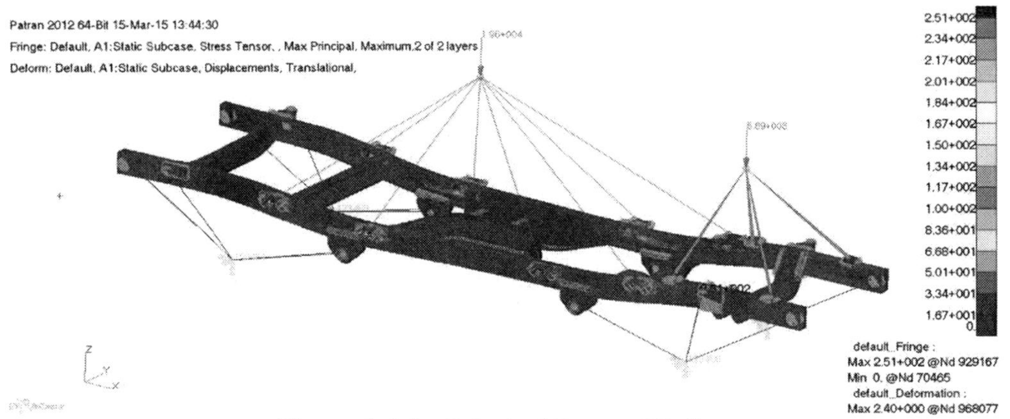


**Figure 5.** Max Principal Stress – Ti alloy

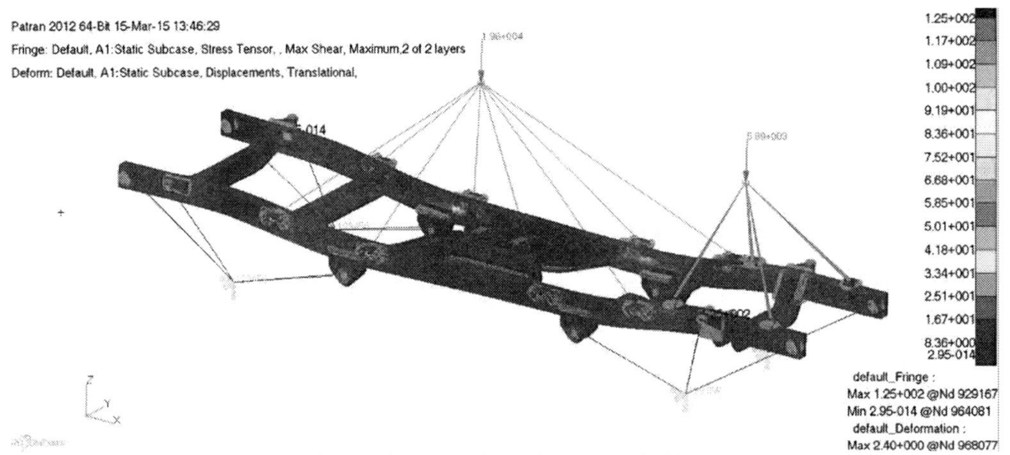


**Figure 6.** Max Shear Stress – Ti alloy

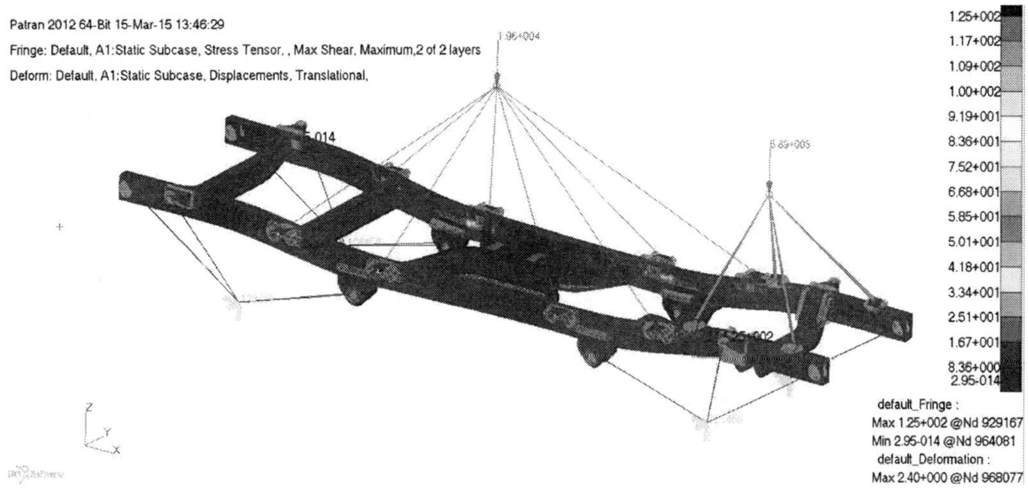


**Figure 7.** Von Mises Stress – Ti alloy

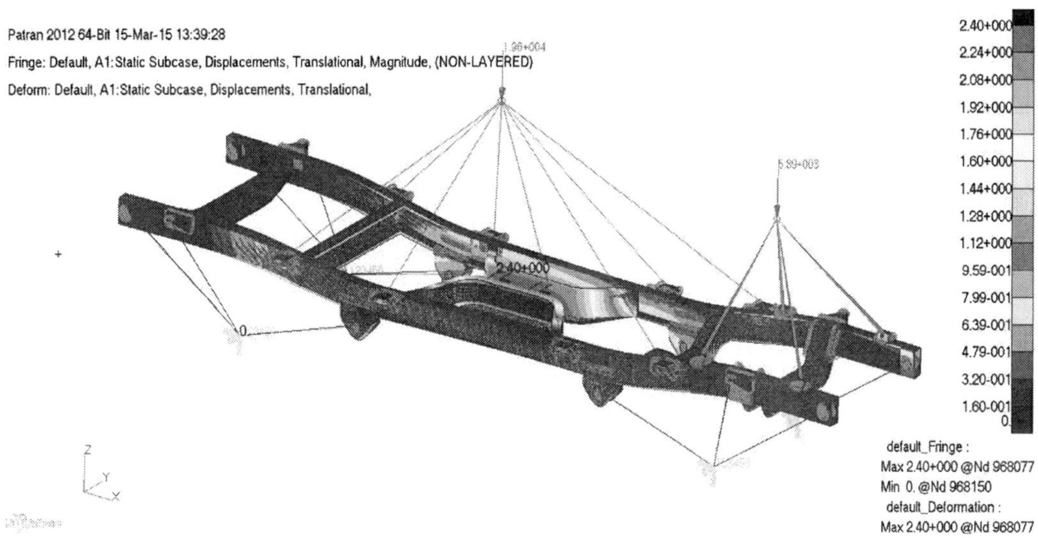


**Figure 8.** Displacement – Ti alloy

**Table 2:** Stress values obtained for Ti alloy

| Description | UTS ($N/mm^2$) | Values Obtained ($N/mm^2$) | RF | Deflection (mm) |
|---|---|---|---|---|
| Max Principal Stress | 970 | 251 | 3.9 | |
| Max Shear Stress | 582 | 125 | 4.6 | 2.4 |
| Von Mises Stress | 970 | 225 | 4.3 | |

This main argument is one of the factors that may cause an fatigue failure. The existing truck chassis is for modification and study equipped with stiffeners. Initially, the thickness of the model, where the highest deflection occurs in bending analysis, was increased to a given value with appropriate limit. Then one more cross beam was added at the centre of the wheel frame to give steadiness to the base. To strengthen and improve the rigidity of the chassis as well as the overall performance of the chassis, a series of modifications and tests were carried out by adding the stiffener.

The material changed to Mild Steel, and it was the most used widely in the design of an automobile frame. The plots at the made for Max Principal, Max Shear & Von Mises Stresses and deflection in Figure 9 to 12.

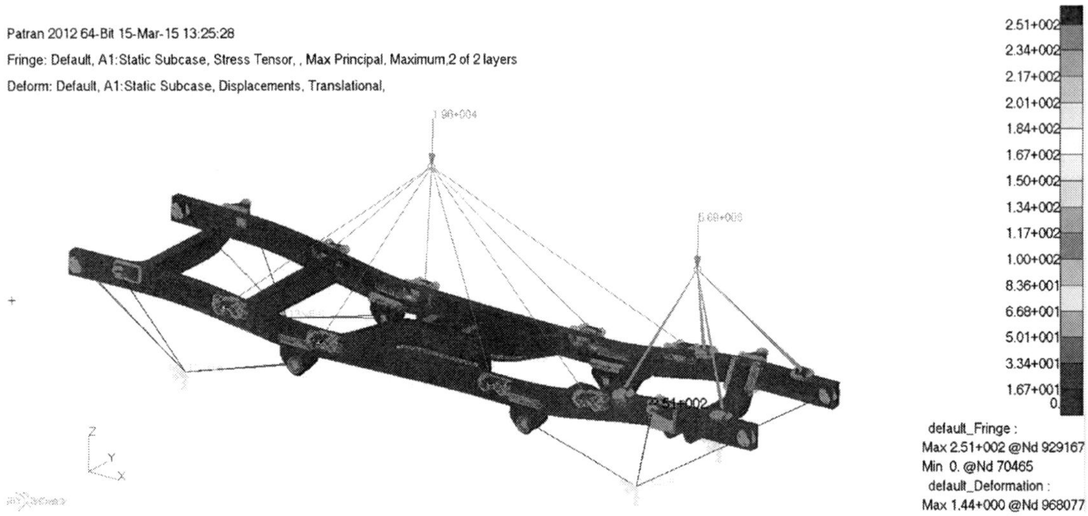


**Figure 9.** Max Principle Stress – Mild Steel

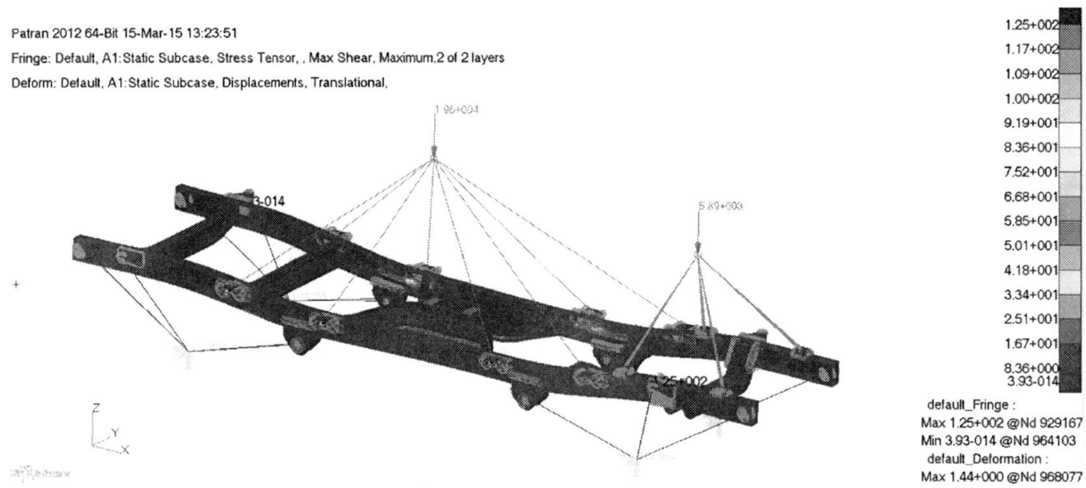


**Figure 10.** Max Shear Stress – Mild Steel

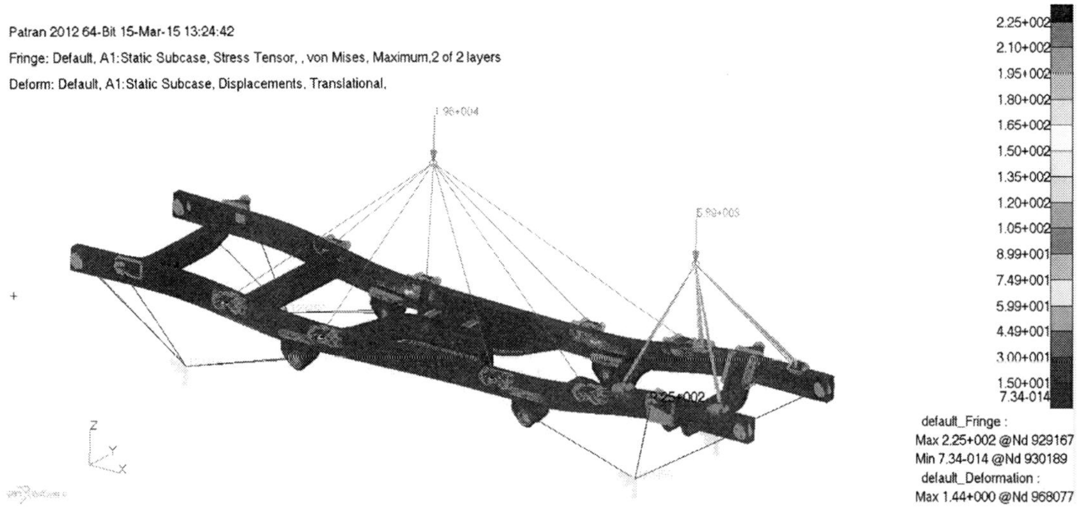


**Figure 11.** Von Mises Stress – Mild Steel

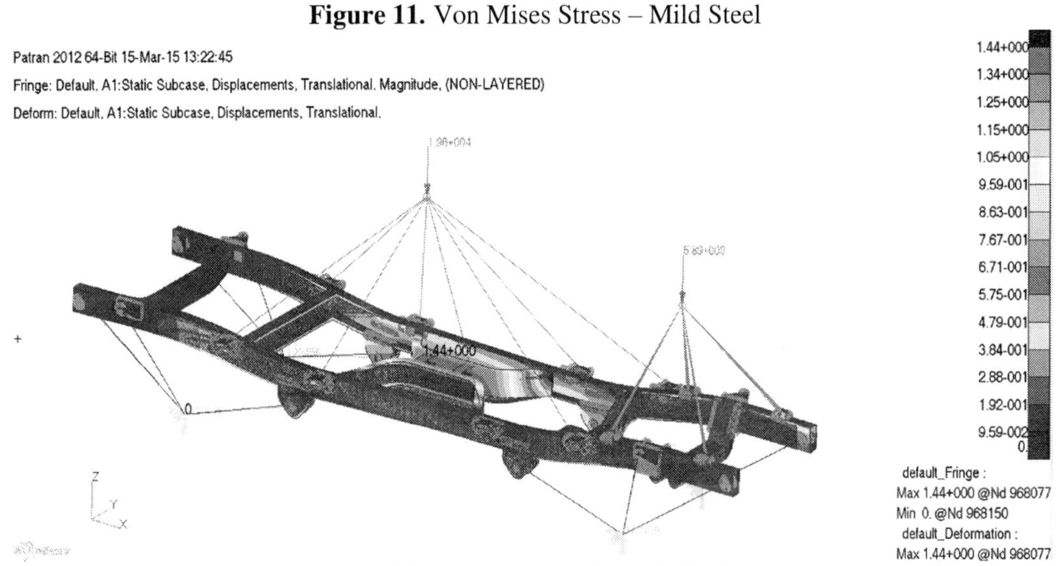


**Figure 12.** Deflection – Mild Steel

The stress state is analyzed and the entire fracturing truck chassis is tested and its static strength tested, the tension of 15 chassis points is measured under the statical condition fully charged and the right front wheel lifting conditions are calculated and the stresses and stress distribution are determined.

**Table 3.** Stress values obtained for Mild Steel

| Description | UTS ($N/mm^2$) | Values Obtained ($N/mm^2$) | RF | Deflection (mm) |
|---|---|---|---|---|
| Max Principal Stress | 500 | 251 | 2.1 | |
| Max Shear stress | 300 | 125 | 2.4 | 1.4 |
| Von Mises stress | 500 | 225 | 2.2 | |

The comparison of Mild steel and Titanium alloy on the C-section truck frame, Titanium alloy having higher results than the Mild Steel. Where as in the defection mild steel proven with minimum deflection. The comparison table was presented in table 4.

**Table 4.** Comparison of RF for Mils Steel and Ti alloy

| **Description** | **Mild Steel** | **Ti alloy** |
|---|---|---|
| RF 1 | 2.1 | 3.9 |
| RF 2 | 2.4 | 4.6 |
| RF 3 | 2.2 | 4.3 |
| Deflection | 1.4 | 2.4 |

## 4. Conclusion

The selection of vital materials for an automobile is the major and utmost significant aspect for automotive design. The huge availability of materials can be selected on the heavy vehicle frame and chassis, but the main challenge for its design constraints. The most significant conditions that a material choice should meet the light-weight capacity, design safety, commercial effectiveness, recyclability and lifespan considerations. In this paper, two different material was adopted on the C-section truck frame design. Mild steel and Titanium alloy were structurally analysed on the heavy truck frame through finite element analysis. The structural analysis of Mild steel and Titanium alloy on the C-section truck frame, Titanium alloy having higher results than the Mild Steel. Where as in the defection mild steel proven with minimum deflection.

## References

[1] Goolla Murali, Subramanyam B and Dulam Naveen 2013 Design Improvement of a Truck Chassis based on Thickness *Altair Technology Conference Sigma Consultancy India* 1-6

[2] Rao S S 2004 *Mechanical Vibration* Pearson Education Press

[3] Madhu and Venugopal 2014 Static Analysis, Design Modification and Modal Analysis of Structural Chassis Frame *International Journal of Engineering Research and applications* **4** 6-10

[4] Rajasekar K and Saravanan R 2014 Literature Review on Chassis Design of On-Road Heavy Vehicles *International Journal of Innovative Science, Engineering & Technology* **1** **7**

[5] Prabhakaran S, Padmanabhan S and Joel C 2018 Design optimization and structural analysis of automotive chassis *International Journal of Mechanical and Production Engineering Research and Development* **8** 934-942

[6] Bobrowskii A V, Zotov A V , Rastorguev D A, Gorokhova D A and Ugarova L A 2019 *IOP Conf. Ser.: Mater. Sci. Eng.* **537** 032078

[7] Patel Vijay Kumar V and Patel 2012 Structural Analysis of Automotive Chassis Frame and Design Modification for Weight Reduction *International Journal of Engineering Research and Technology* **1** **3**

[8] Abhishek Singh , Vishal Soni and Aditya Singh 2014 Structural Analysis of Ladder chassis for higher strength. *International Journal of Emerging technology and Advanced Engineering* **4** 253-259

IOP Conf. Series: Materials Science and Engineering **923** (2020) 012013 doi:10.1088/1757-899X/923/1/012013

# Identification and Analysis of Blood Group with Digital Microscope Using Image Processing

**Jayakumar P[1], Padmanabhan S[2], Suthendran K[3], Yeddu Nitish Kumar[2], Mada Sujith[2]**
[1]Research Scholar, Sathyabama Institute of Science and Technology, Chennai, India.
[2]School of Mechanical and Construction, Vel Tech Rangarajan Dr. Sagunthala R&D Institute of Science and Technology, Avadi, Chennai, India.
[3]School of Computing, Kalasalingam Academy of Research and Education, India.

Emaill: padmanabhan.ks@gmail.com

**Abstract.** Blood group identification is an essential examination for transfusion and also for lot of health check procedures. Conventional micro plate system is painstaking. In this study, an automatic moving conveyor belt setup was developed to assist the diagnosis procedure. Blood group identification is done based on the images acquired through light microscopes. A blood distribution on polystyrene coated bio paper and the captured image data has been analysed using Image Acquisition software. The image morphological patterns were analysed to obtain the identification blood group by capturing its microstructure. At end of each test, based on the clumping and its image morphology of the corresponding colour coding was analysed to identify the blood groups. The focus on the present study is to investigate the chance of quick and precise, sensitivity and concurrent identification of blood group will weigh up its usefulness in the routine use.

*Keywords: Blood group identification, Image processing, Automatic Mechine*

## 1. Introduction

Blood, fluid is an essential one to all life and it ships the nutrients and oxygen to human cells [1]. In addition, it also ropes the body heat & pH value, immunological operation [1, 2]. Blood component is alienated into two set they are cells and plasma. It contains red cells (RBC), leucocytes and platelets. The first one, RBC is the chief oxygen transporter, secondly the colourless leucocytes are concerned to immunity system and lastly, the platelets are tiny cells which are capable of forming clots to stop continuous bleeding. Blood plasma is the medium which ships hematopoietic cells, alimentary and various hormones to the needy cells. When exsanguinations happen then the transfusion is much crucial. Blood excerption is an alternative solution for the mislaid blood by saline. However, the RBC has to be bringing back for transfusions with suitable practice. Ahead of performing a secured blood transfusion, certain pre-transfusion tests must be undertaken, namely [2, 4]; carried out in test tube, plate, micro-plates, or gel separation [2-4].

The test tube and plate methods are results in a reduced amount of accuracy than the remaining two tests. But they are cheaper and superficial. Micro-plate or gel separation test is the powerful test for blood typing and it is mandatory for secure blood transfusions. Further, it is important to ensure that the selected blood type is suitable for the receiver [5-12]. During emergency, the adequate time is unavailable to do blood typing so they are choosing blood type O negative (universal donor) [1-3]. Sometimes, this may result into critical issues for the victim. It is imperative to computerize the blood typing [5-12]. The precision of blood typing is crucial and it can elude severe troubles in transfusions. The direct and reverse tests are in practice. In direct test, antigens presence is examined. In the other test the measurement of antibodies is done. Nowadays, Blood typing is done in either manual or automated [13,14].

In manual testing, the results are available in time. However, the human assistance is much needed to conduct the test. Based on ocular inspection, the agglutinated non-agglutinated samples are identified and typing is done. Therefore, the test result is depending on the examiner who involved in the process. So, there may be chance for mistyping [13]. In automation, such human dependency is ignored. Therefore, there are fewer chances for wrong typing [21-22]. Though it is having merit they are slow, high cost and the devices are large in dimensions. Here, the blood is assorted with reagents in order to find the agglutinations. In market, numerous apparatus are available which identifies typing automatically [12-14], Groupamatic, Auto-Grouper, Olympus PK 7200, Ortho AutoVue Innova System, Techno Twin Station and Immucor Galileo. But till now, no system conveys the result accessible in time for emergency condition [15-19].

While the microscopy method has fine sensitivity and authorize type of detection by image differentiation and detection and counting process is highly precise. Dissimilar cells in digital microscope images can be separated by a spatial investigation and intensity details of the image. There is a speedy and simple way of retrieving analysing a set of images for potential indication as with a automated method. Thus the novel system proposed in this paper, automatically execute the blood type determination process. Hence it can eradicate human mistake in an undersized period. Further, it is suitable during abnormal state. In slide test, the developed software does image processing techniques to identify blood type. The test results are captured by a digital microscope which has a colour image poised of quad tasters of blood & reagent. Such image is considered as an input to image acquisition software. This method intends to create an unique model which is suitable to carry out necessary pre-transfusion analysis for a safe and sound blood transfusion automatically.

## 2. Experimental Details

The blood of the patient is taken in a slide to respond with the well-known antibodies A, B & D followed agglutination response was recorded. Blood was poured on polystyrene coated bio paper.

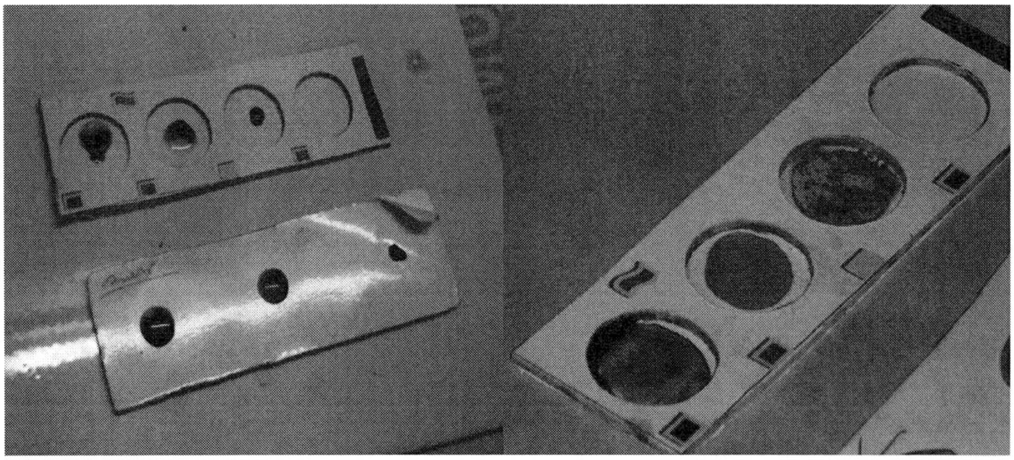

**Figure 1.** (a) Blood sample on polystyrene coated bio paper (b) Blood group testing slide with different colour marked on its bottom for colour coding.

The polystyrene coated bio paper can easily paste and remove from the bottom glass plate as shown in figure 1(a). The bottom of the each blood specimen the colour was marked with different colouring agent figure 1(b) for avoiding manual error.

The design of automatic machine for blood group identification as shown in figure 2. The conveyor belt was connected with motor and it operated horizontally. The test slide move on the conveyor belt when entered in to the digital microscope place it on automatically and the image will be scanned. Digital microscopy is highly sensible with range of 50 X (model name BW908/BW908C) is fixed vertically, where the image of the blood sample captured. The process time for analysis of multi plate (maximum 4 plates at the same time) was 30 sec. The captured image data was transferred in to computer where the image has been analysed by the combined system with IMAQ Vision software and image processing techniques. At end of the each test the based on the clumping and its image morphology of the corresponding colour coding as a result of the blood groups.

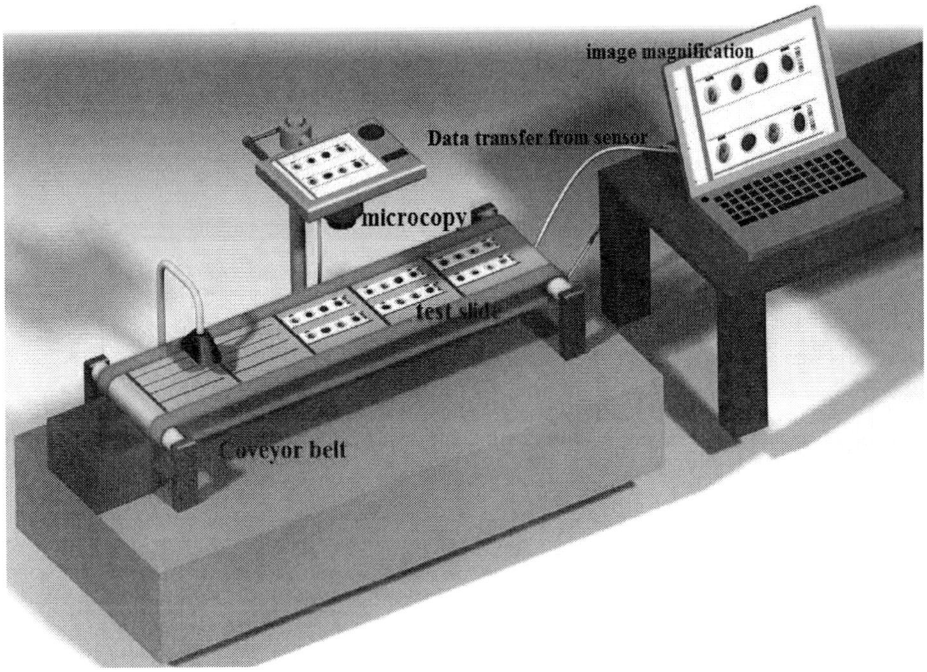


**Figure 2.** Image of automatic mechine for blood identification

## 3. Result and Discussions

If there is a presence of antigen in blood cell related to the counteragent, it results in amalgamation or clustering. When the blood amalgamates by way of anti-A antibody, then it is A type. Similarly when amalgamation by means of anti-B, then it is B type. Conversely if amalgamation is observed among both anti-A and -B then it is AB and no amalgamation reported as O type. In addition, if amalgamation happens through the f anti-RhD antibody, it is reported as positive and otherwise negative. Then, the features such as textural and colour features are extracted and stored in the database. Figure 3. Shows that an image of the digital microscope on blood clumping image capturing and its magnification.

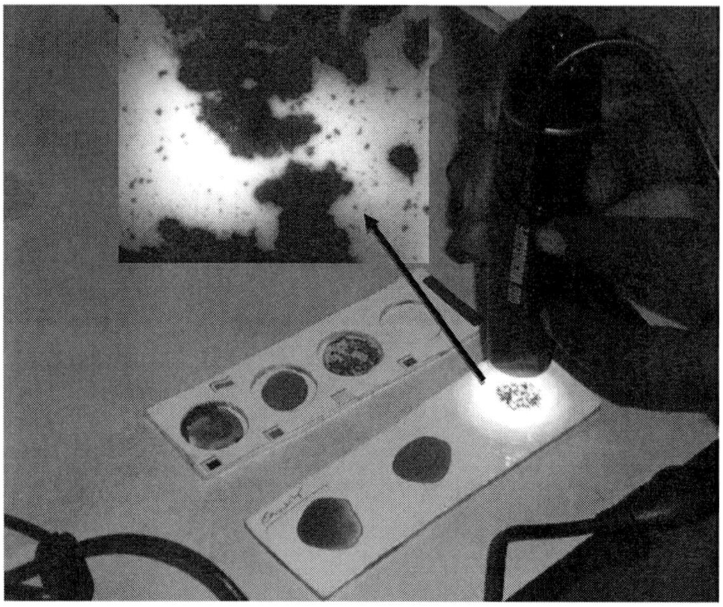

**Figure 3.** Image of the digital microscope on blood clumping image capturing and its magnification.

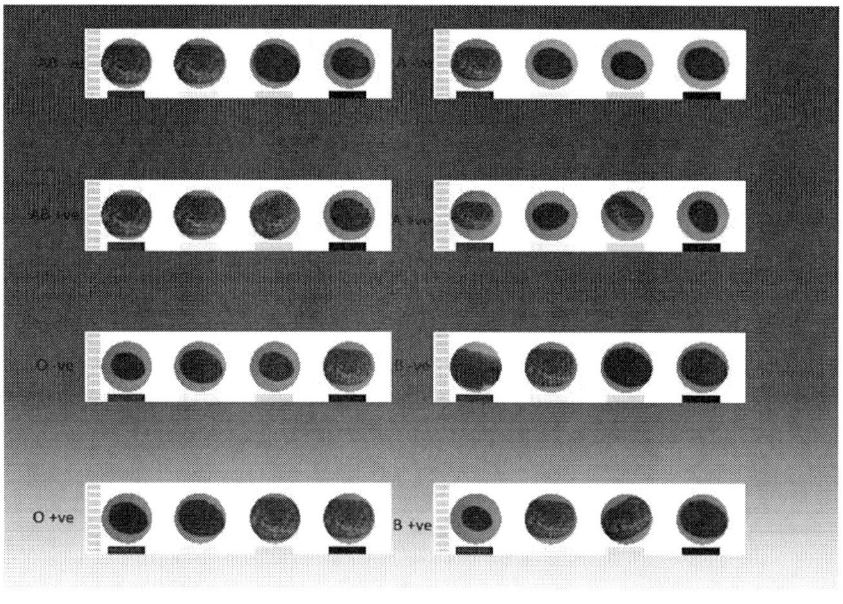


**Figure 4.** Images of blood groups on adding antigens.

The results of blood groups as shown in figure. 4 and scattering and non-scattering component was cleared from the above figure. From above image of the blood scattering indicates the reaction between antigens on corresponding to the blood groups. Microstructure and the size precision from the image indicate it's one of the excellent diagnostic tool for measurement of the blood types. Similar type of agglutination results was explained by Yaw-Jen Chang et al. [20].

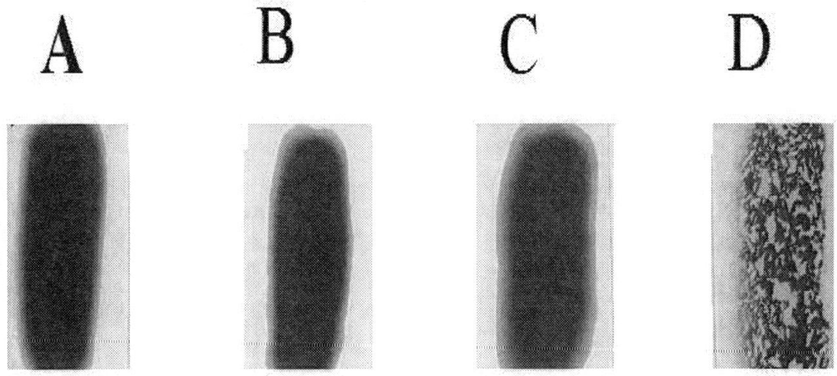


**Figure 5.** Scattered image of O negative Reagent of (a) anti-A (b) anti-B (c) Anti-AB and (d) anti-D.

Result of O negative blood group is presented in above figure 5. It clearly shows the amalgamation takes place only in sample d and not in samples a, b and c. It confirms the nonappearance of antigens AB in the analysed blood sample. Further, the amalgamation is occurred by means of anti-D reagent. Hence, it is proved the existence of antigens Rh in the analysed blood sample. So the results of the above work can be reported as O negative blood type.

## 4. Conclusion

The above experiment and its results give an effective way of blood type detection and the blood scattering image automatically enable the types of blood groups. During emergency, such way of blood type detection is highly usable. It also helps the admin to choose well-suited blood for transfusion, avoiding feasible blood incompatibilities. In future, it can be scaled down to make it handy.

## References

[1] Seeley R R, Stephens T and Tate P 2003 *Anatomy and Physiology* McGraw- Hill
[2] Davies A, Blakeley A and Kidd C 2001 *Human Physiology* Churchill Livingstone
[3] Rogers K 2011 *Blood Physiology and Circulation* Britannica Educational Publishing
[4] Klein HG and Anstee DJ 2005 *Mollisons Blood Transfusion in Clinical Medicine* Blackwell Science Ltd
[5] Brown M R and Crim P 2007 Organizing the Antibody Identification Process *Clin Lab Sci* **20** 122-126
[6] Myhre B A and McRuer D 2000 Human error a significant cause of transfusion mortality *Transfusion* **40** 879-885
[7] Petaja J, Andersson S and Syrjala M 2004 A simple automatized audit system for following and managing practices of platelet and plasma transfusions in a neonatal intensive care unit *Transfus Med* **14** 281-288.
[8] Delamaire M 2005 Automation of the immunohematology laboratory *Transfus Clin Biol* **12** 163- 168
[9] Henneman E A, Avrunin G S, Clarke, Osterweil L J, Andrzejewski C, Merrigan K, Cobleigh R, Frederick K, Katz-Bassett E and Henneman P L 2007 Increasing patient safety and efficiency in transfusion therapy using formal process defmitions *Transfus Med Rev* **21** 49-57
[10] Wagar E A, Tamashiro L, Yasin B, Hilborne L and Bruckner D A 2006 Patient safety in the clinical laboratory: a longitudinal analysis of specimen identification errors *Arch Pathol Lab Med* **130** 1662-1668

[11] Turner C L, Casbard A C, Murphy M F 2003 Barcode Technology:its role in increasing the safety of blood transfusion *Transfusion* **43** 1200-1209

[12] Callum J L, Kaplan H S, Merkley L L, Pinkerton P H, Fastman B R, Romans R A, Coovadia A and Reis M D 2001 Reporting of near-miss events for transfusion medicine: improving transfusion safety *Transfusion*, **41** 1204-1211

[13] Blood Policy and technology 1985 *Congress,Office of Technology Assessment* Washington

[14] Sturgeon P 2001 Automation: its introduction to the field of blood group serology *Immunohematology Journal of Blood Group Serology and Education* **17** 4

[15] Coakly W A 1981 *Handbook of Automated Analysis* Mercel Dekker pp 6l

[16] Ewing G W 2004 *Analytical Instrumentation Handbook* New York: Marcel Dekker pp 152

[17] Dada A, Beck D and Schmitz G 2007 Automation and Data Processing in Blood Banking using the Ortho AutoVue Innova System *Transfusion Medicine Hemotherapy* **34** 341-346

[18] Shin S Y, Kwon K C, Koo S H, Park J W, Ko C S, Song J H and Sung J Y 2008 Evaluation of two automated instruments for pre-transfusion testing: AutoVue Innova and Techno TwinStation *Korean j Lab Med* **3** 214-220

[19] Wittmann G, Frank J, Schram W and Spannagl M 2007 Automation and Data Processing with the Immucor Galileo system in a University Blood Bank *Transfusion Medicine Hemotherapy* **34** 347-352

[20] Yaw-Jen Chang, Ching-Wei Huang, Zh-Wei Chen and Jiun-Yi Chou 2011 Design of Microfludic Chip for Blood Typing System, *Proceedings of 2011 International Conference on System Science and Engineering* China

[21] Pooja Prakash M, Fepslin and Sreeraj R 2018 Combined Cryptography And Digital watermarking For Secure Transmission of Medical Images in EHR Systems *International Journal of Pure and Applied Mathematics* **118** 265-269

[22] Fepslin, Suthendran K and Arumugam S 2017 A Novel Reversible Data Hiding Method in Teleradiology to Maximize Data Capacity in Medical Images *Springer LNCS*, **10398** 55-62

IOP Conf. Series: Materials Science and Engineering **923** (2020) 012014 doi:10.1088/1757-899X/923/1/012014

# Experimental Analysis of Al-359 reinforced with $B_4C$ and $Al_2O_3$ composite

**G AshwinPrabhu[1], K Muninathan[2], R Raja Srinivasan [3], S Jayaganesan[4] and M Adithiya Kumar[5]**

[1,2]Faculty of Department of Mechanical Engineering, St. Joseph's College of Engineering, Chennai -119
[3,4,5]Department of Mechanical Engineering, St. Joseph's College of Engineering, Chennai -119

[1]E-mail: ashwin.prabhu1990@gmail.com

**Abstract:** Now a days composite materials plays a vital role in various aspects of engineering such as in the development of light weight and high strength components. The composite material's strength depends upon the type and the reinforcement. In composite analysis, the influence of $B_4C$ particles on the mechanical behaviour and tribological behaviour of Al-359 with Al2O3 composites has been studied. The manufacturers of automobiles have been showing an interest and focusing on the wear resistance and other properties for improvement. There are various characteristics of aluminium such as high thermal conductivity, low wear rate, high specific heat capacity and low density when compared to conventional cast iron. Here composites with aluminium oxide (Al2O3) and boron carbide (B4C) are used as reinforcement.

***Keywords:*** Aluminium, light weight, strength, reinforcement, cost reduction, resistance

## 1. Introduction

Today most of the research has been done in the improvement of the material and its standard. Everyone in the society who has been doing their research has been analyzing using different compositions. A composite material plays a vital role in the field of engineering. A composite material is the combination of two or more materials which provides a different property such as its physical and chemical properties are different when combined. Depending upon the composition the characteristics differ accordingly. The classification of composite are done on the basis of the material matrix such as Metal Matrix Composites (MMC), Ceramic Matrix Composites (CMC) and Polymer Matrix Composites (PMC). The application of composite materials has seen a significant increase in the recent years. Improvements have been in the sector of manufacturing so as to achieve the best outcome and reduction in the cost of the product. Composite material is needed widely for the production of light materials and shock absorbent materials. The raw materials used in the work are: -

1. Al-359
2. Boron Carbide ($B_4C$)
3. Aluminium Oxide ($Al_2O_3$)

IOP Conf. Series: Materials Science and Engineering **923** (2020) 012014 doi:10.1088/1757-899X/923/1/012014

### 1.1 Al-359

Aluminium is the type of chemical element which is produced from bauxite. It has a wide range of applications because of low density and the ability of resistance to corrosion. The composition of Al-359 is aluminium (90.42%), silicon (9.04%) and magnesium (0.64%). The number "359" represents the grade of the aluminium. Among various cast aluminum alloys, Al-359 composition has comparatively high amounts of silicon in it. The characteristics of silicon are used for the improvement of casting fluidity and lower melting temperature. Silicon also has the effect of strengthening.

### 1.2 Boron Carbide ($B_4C$)

Boron carbide is one of the significant components utilized in the composites. Boron carbide has been used in various things such as in tank armor, bulletproof vests and in the various industrial applications. The Vickers hardness of boron carbide is greater than 30Gpa. After cubic boron nitride and diamond, boron carbide is the hardest material known among other materials. The person named Henri Moissan synthesized boron carbide and the process involved in the synthesis is by the reduction of boron trioxide either with the carbon or magnesium in the presence of carbon in electric arc furnace. When carbon is present the reaction occurs above the melting point of $B_4C$ where a large amount of carbon monoxide is liberated. The reaction of the process is $2\ B_2O_3 + 7\ C \rightarrow B_4C + 6\ CO$.

### 1.3 Aluminium Oxide ($Al_2O_3$)

Aluminium oxide is the chemical formulation of aluminium and oxygen. $Al_2O_3$ is one of the components used in the aluminium metal in the form of abrasive to improve the hardness of the material and is also used as a refractory material for high melting point. It is also amphoteric in natur. There are various applications of aluminium oxide such as in composite fibres, abrasives, glass etc.

## 2. Literature Survey

In composite materials, Metal matrix composites (MMC) is becoming a large leading material. Because of their highly excelling engineering properties particle reinforced aluminium MMCs have received considerable attention. Because of their hardness and abrasive nature of reinforcement elements like silicon carbide (SiC) particles these materials are difficult to machine. For experimental investigation of tool wear and surface roughness, homogenised 5% SiC-p aluminium MMC material is used in this study. [1].

By Vortex method, three 2024 Aluminium alloy metal matrix composites (MMCs) of varying sizes are reinforced and weight fractions till 30% of the weight of Al2O3 particles were fabricated. From the results, it was inferred that the hardness and the tensile strength of the composites increased with decrease in size and increase in weight fraction of particles [2].

Al2O3-ZrO2 with a high level of hardness and toughness is known as ceramic steel. Due to its unique properties it can be used as a reinforcement in fabrication of metal matrix composites. In this study, nanoparticles of Al2O3-10% ZrO2 with an average size of 80 nm were used to

IOP Conf. Series: Materials Science and Engineering **923** (2020) 012014 doi:10.1088/1757-899X/923/1/012014

fabricate Al matrix composites containing 0.5, 1, 1.5 and 2 wt.% of the reinforcement. The results revealed that with increasing the reinforcement content, density decreased while yield, ultimate tensile strength and compressive strength increased [3].

One of the most conventional types of metal matrix composites is aluminum matrix composite. This paper deals with the effect of production parameters on wear resistance of Al–Al2O3 composites. Alumina powder with a particle size of 12, 3 and 48 l and pure aluminum powder with particle size of 30 l were used [4].

Al 6061 alloys are used broadly in many engineering applications in the fields of construction and transportation where mechanical properties like hardness and tensile strength are essential. The addition of TiB2 to the metal matrix increases the matrix phase strength [5].

## 3. Methodology

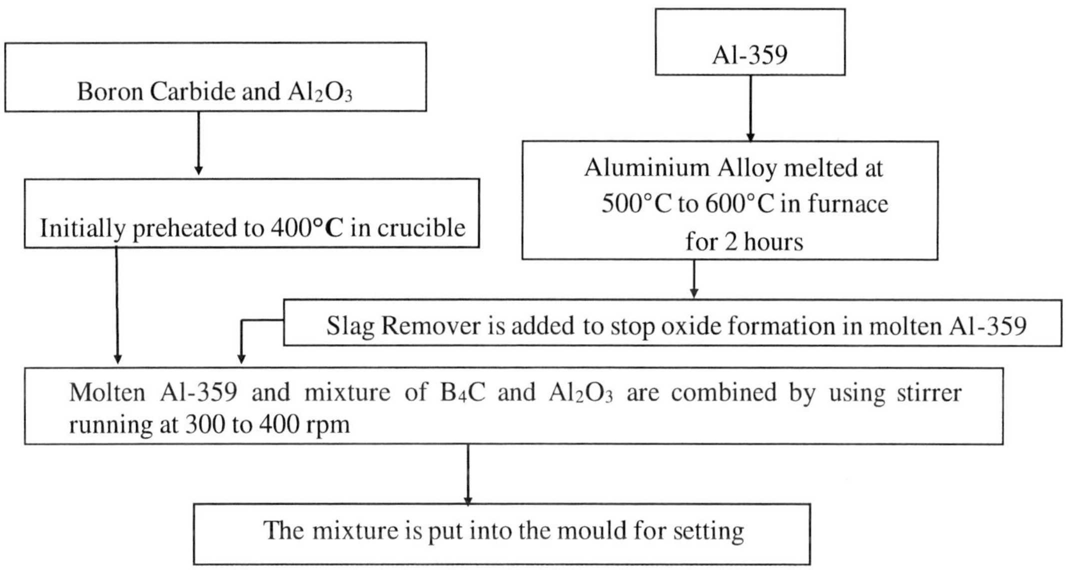


**Figure 1** Methodology

## 4. Composition of the Sample

**Table 1.** Composition of Sample

| | |
|---|---|
| Sample 1 | Al-359 + 5% $B_4C$ |
| Sample 2 | Al-359 + 5% $B_4C$ + 2.5% $Al_2O_3$ |
| Sample 3 | Al-359 + 5% $B_4C$ + 5% $Al_2O_3$ |

## 5. Processing of the composites

In the experiment performed, Al-359 is used as the matrix phase. Using an electric furnace Al-359 is melted at a temperature of 1000°**C** in a graphite crucible of 2kg capacity. Initially boron carbide and aluminium oxide are preheated to 400°**C** in a crucible before mixing with molten Al-359. A mild steel stirrer coated with zirconium is used to avoid contamination of melt. The composite is formed due to the uniform distribution of $B_4C$ and $Al_2O_3$, the reinforcement phase in the Al-359 alloy, the matrix phase by stirring at a speed of 300 to 400rpm. The formation of slag takes place during the chemical reaction taking place in the Al-359 alloy.

## 6. Tests performed on the specimen

There are various tests performed on the specimen. They are Compressive strength test, hardness test and corrosion test.

### 6.1 Compressive Strength Test

Compressive strength is the one where capacity of the materials can be analyzed so that how much load the component can withstand against the tensile strength which can be able to withstand loads which tends to elongate. Various things can be analyzed according to the composition and the compression strength differs according to the material used.

### 6.2 Hardness Test

Hardness test is the type of mechanical test for analyzing material properties. It is generally used in the engineering design, structure analysis and in the materials development. The main purpose of the test is for the determination of suitability of material for the particular application to be performed. The various types of hardness tests are: Macro Hardness test (>1kg) (Rockwell, Brinell and Vickers testing), Micro Hardness Test (<1kg).

### 6.3 Corrosion test

The test in which salt spray is the standardized method for the corrosion test method. This is used to check the corrosion resistance of materials. The appearance of the material corrosiveness is determined by after a period of time that is after a stipulated time for the material.

## 7. Results and Discussions

### 7.1 Compression Test

#### 7.1.1 Sample 1

The composition of sample 1

Al-359 + 5% $B_4C$

IOP Conf. Series: Materials Science and Engineering **923** (2020) 012014 doi:10.1088/1757-899X/923/1/012014

The compression value is 151.96 KN

**Test Report**

**Test Details**

| | | | |
|---|---|---|---|
| Name of Organisation & Address: | OMEGA LAB, Guindy | Date: | 08-03-2019 |
| Material Name: | Id: Al -359 + 5% B4C Aluminium Sample | TOCR No: | 19-7649 |
| Serial No: | C1 | Test Method: | ASTM D695 |

**Input Data**

| | |
|---|---|
| Mode of Test | Compression |
| Sample Type | Flat |
| Thickness | 13.90 mm |
| Width | 50.50 mm |
| Area | 701.95 mm² |
| Gauge Length | 50.00 mm |
| Final Gauge Length | 0.000 mm |
| Final Gauge Length | 0.000 mm |

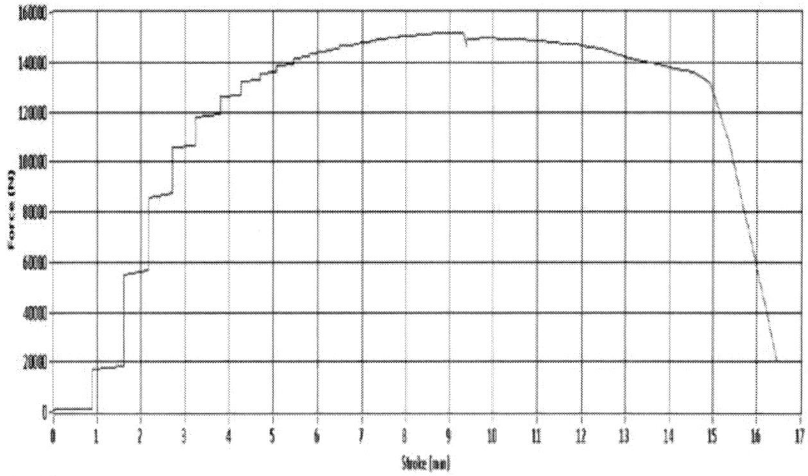


**Results**

Fmax 151.96 KN

T-Rex Analytical Lab.

**Figure 2.** Sample 1

### 7.1.2 Sample 2

The composition of sample 2

Al-359 + 5% $B_4C$ + 2.5 % $Al_2O_3$

The compression value is 172.45 KN

IOP Conf. Series: Materials Science and Engineering **923** (2020) 012014 doi:10.1088/1757-899X/923/1/012014

### 7.1.3 Sample 3

**Test Report**

**Test Details**

| | | | |
|---|---|---|---|
| Name of Organisation & Address: | OMEGA LAB, Guindy | Date: | 08-03-2019 |
| Material Name: | Aluminium Sample | TOCR No: | 19-7649 |
| Serial No: | C2 | Test Method: | ASTM D695 |
| Batch No: | Id: Al -359 + 5% B4C + 2.5% Al2O3 | | |

**Input Data**

| | |
|---|---|
| Mode of Test | Compression |
| Sample Type | Flat |
| Thickness | 13.66 mm |
| Width | 50.45 mm |
| Area | 689.15 mm² |
| Gauge Length | 50.00 mm |
| Final Gauge Length | 0.000 mm |
| Final Gauge Length | 0.000 mm |

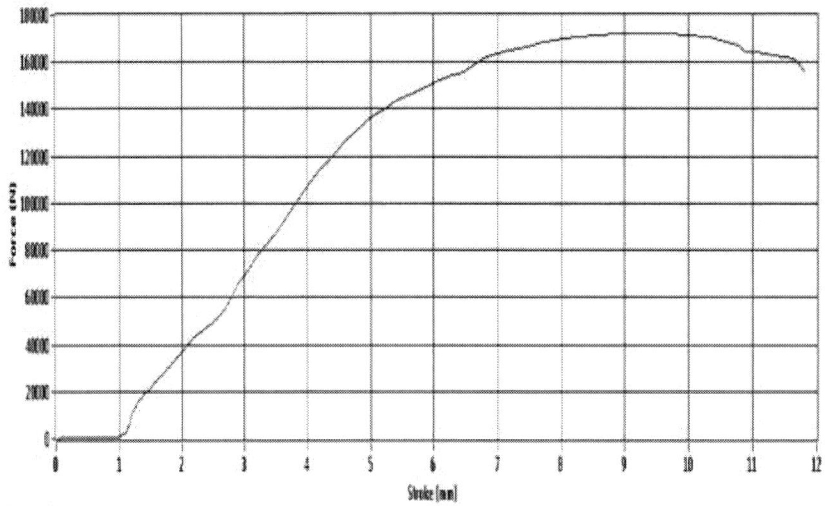


**Results**

| | |
|---|---|
| Fmax | 172.45 KN |

T-Rex Analytical Lab.

**Figure 3**. Sample 2

The composition of sample 3

Al-359 + 5% $B_4C$ + 5% $Al_2O_3$

The compression value is 223.6

IOP Conf. Series: Materials Science and Engineering **923** (2020) 012014 doi:10.1088/1757-899X/923/1/012014

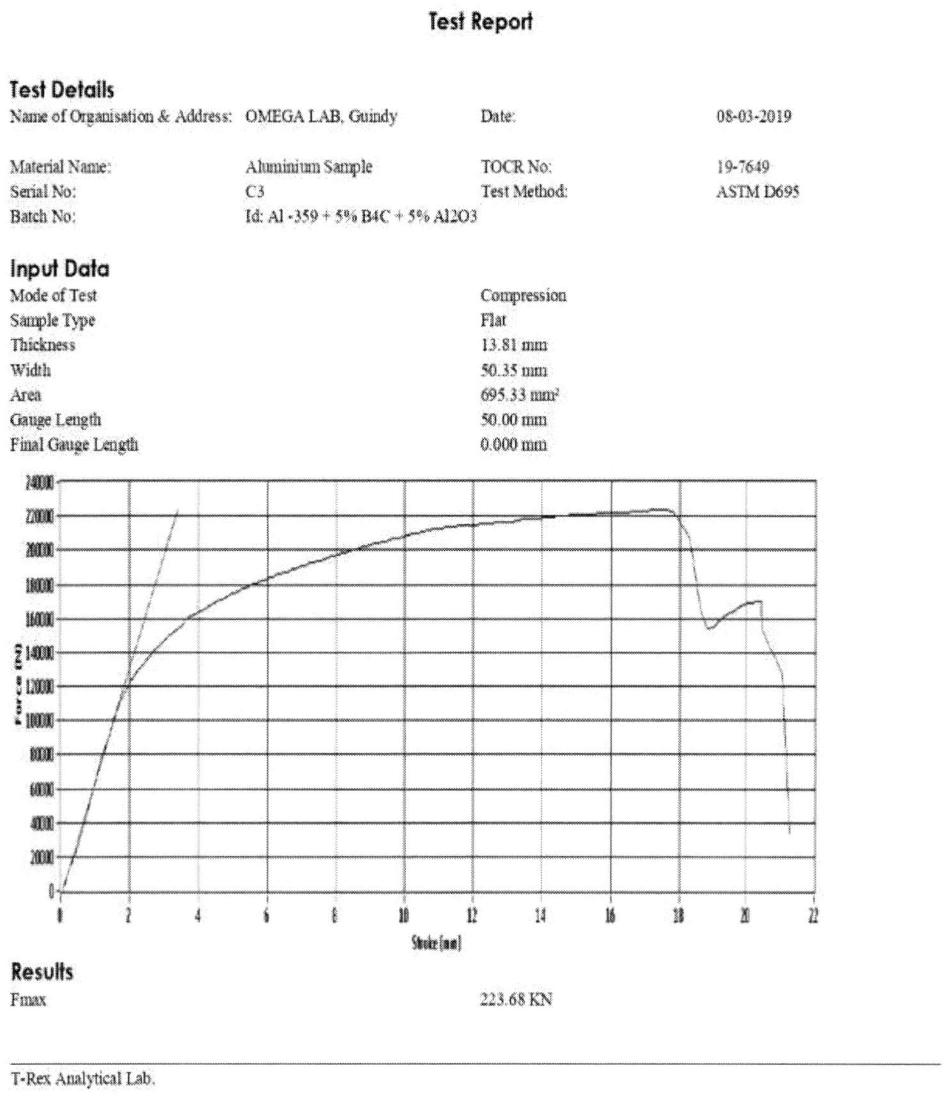
**Test Report**

**Test Details**

| | | | |
|---|---|---|---|
| Name of Organisation & Address: | OMEGA LAB, Guindy | Date: | 08-03-2019 |
| Material Name: | Aluminium Sample | TOCR No: | 19-7649 |
| Serial No: | C3 | Test Method: | ASTM D695 |
| Batch No: | Id: Al -359 + 5% B4C + 5% Al2O3 | | |

**Input Data**

| | |
|---|---|
| Mode of Test | Compression |
| Sample Type | Flat |
| Thickness | 13.81 mm |
| Width | 50.35 mm |
| Area | 695.33 mm² |
| Gauge Length | 50.00 mm |
| Final Gauge Length | 0.000 mm |



**Results**

| | |
|---|---|
| Fmax | 223.68 KN |

T-Rex Analytical Lab.

**Figure 4.** Sample 3

**Table 2.** Compression Test of Al-359, Boron carbide ($B_4C$) and Aluminium Oxide ($Al_2O_3$)

| Sample Number | Sample Composition | Compression Load (KN) |
|---|---|---|
| 1 | Al-359 + 5% $B_4C$ | 151.96 |
| 2 | Al-359 + 5% $B_4C$ + 2.5% $Al_2O_3$ | 172.45 |
| 3 | Al-359 + 5% $B_4C$ + 5% $Al_2O_3$ | 223.68 |

**Table 3.** Hardness Test (using Vickers Hardness machine)

| Sample Number | Sample Composition | Hardness Value (HV) |
|---|---|---|
| 1 | Al-359 + 5% B4C | 99,94,97 |
| 2 | Al-359 + 5% $B_4C$ + 2.5% $Al_2O_3$ | 103,103,104 |
| 3 | Al-359 + 5% $B_4C$ + 5% $Al_2O_3$ | 112,115,104 |

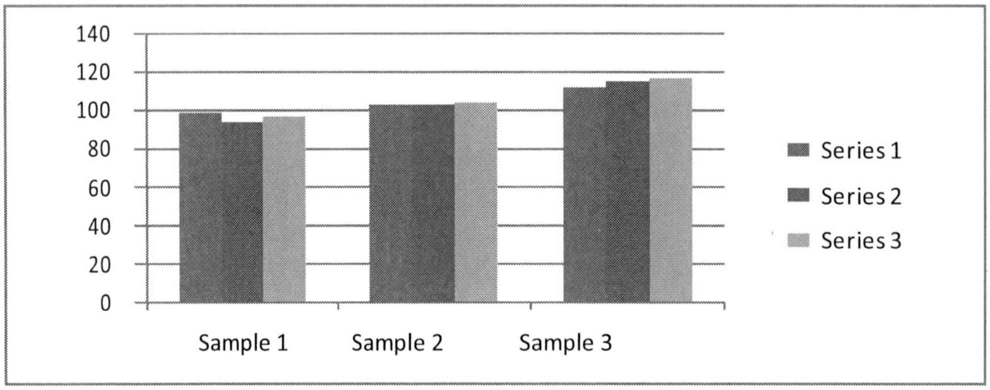


**Figure 5.** Hardness Test Comparison

## 8. Conclusion

After the thorough analysis of the specimens final observations and conclusions have been arrived upon various aspects.

1. In the compression test, when compared with the three-sample produced, the third sample is good when compared to other two samples and the compressive strength is best for the particular composition respectively, that is sample 3 is good.

2. In the hardness test conducted by using Vickers hardness method, the hardness is good compared to other two samples that is sample 3 is good than the others.

Therefore sample 3, having the composition Al-359 + 5% B4C + 5% Al2O3 is the best sample when compared to other two samples.

## References

[1] Emamy M, Abdizadeh H, Lashgari H R, Najimi A A 2011 The effect of Sr and grain refining elements on the microstructure and tensile properties of A356-10%B4C metal matrix composite Mechanics of Advanced Materials and Structures 18:3 210-27 DOI: 10.1080/15376494.2010.516884.

[2] Dipti Kanta Das, Purna Chandra Mishra, Saranjit Singh, Swati Pattanaik 2014 Fabrication and heat treatment of ceramic reinforced aluminium matrix composites – a review International Journal of Mechanical and Materials Engineering 9 Article Number: 6

[3] Girisha K B, Dr. Chittappa H C 2014 Wear performance and hardness property of A356.1 aluminium alloy reinforced with zirconium oxide nano particle International Journal of Engineering Sciences & Research Technolohy (IJSERT) 3(6) ISSN: 2277-9655

[4] Chandramohan V, Arjunraj R 2014 Preparation of Hybrid Aluminium Metal Matrix Composites by Using Stir Process IJESRT ISSN: 2277-9655

[5] Lawrance C A, Suresh Prabhu P 2015 Al 6061 – TiB2 metal matrix composite synthesized with different reaction holding times by In-Situ method International Journal of Composite Materials p-ISSN: 2166479X e-ISSN: 2166-4919, 5(5): 97-101 doi: 10.5923/j.materials.20150505.02

[6] Dr. Nallusamy S 2016 Analysis of wear resistance, cracks and frictional properties of metal matrix composites with SiC additives and Al2O3 as reinforcement Indian Journal of Science and Technology 9(35):01-06

# Design and impact analysis of a jeep bumper made of composite material

**J Jones Mervin*, M Mariyappan, B Ramesh, M Sathish kumar.**
Department of Mechanical Engineering, St.Josephs Institute of Technology,OMR, Chennai, Tamilnadu,India.
Emai:*jones.mervin1@gmail.com,

***Abstract*** While designing the bumper in the automobile considering the impact during the collision is important. The first part that gets involved in the collision is the bumper. The design of the bumper has to be optimized in a way that it absorbs more energy in order to reduce the impact on the car and passengers. Materials play a vital role in the resistance of a bumper, hence the material to be used has to be chosen wisely. Another important aspect of bumper design is that what percent it contributes to the total weight of the car. The usage of composite material can solve both the criteria of weight reduction and equivalent performance to existing bumper materials such as ABS (steel reinforced), and other metal bumpers.
*Keywords:* Front bumper, nylon 6 with Nano clay, ABS, SOLIDWORKS, ANSYS, AFRP.

## 1 Introduction

In automobiles structure it is the front most or the rear most part. It is used to sustain and absorb the energy during the collision. Generally steel and aluminum were used as bumper materials but in recent trends the composite materials are slowly replacing the conventional metals. In another way the front bumpers have more aesthetic value than the rear bumpers. As composite materials can be molded or casted in any shape they can sophisticatedly used as an alternative for the existing materials. Composite materials are also light in weight which makes it a preferable material to the bumper design.

## 2 Problem statement

Nowadays accidents are frequent and cause damage to the bumpers and the vehicle body. The crashworthiness of the bumper has to be ensured in order to reduce the damage caused physically to the vehicle and the passengers. Upon collision the bumper has to be replaced to ensure the safety of the passengers. The replacement costs of bumpers tend to be expensive as claimed by the consumers. It's the prime duty of the automobile manufacturers is to satisfy the need of the customer in an effective manner. Composites materials have been in the lights of research in the past decade. Alternate materials which can withstand the impact of the collision and also add some aesthetic value to the bumper.

## 3 Literatutre survey

Untill now many research and experiment have been done on bumper and materials. Alen John Sanu Alex, have studied on Composite Materials used for Automotive Bumper in Passenger vehicles, in which many plastic based composite materials and their properties are studied. In another study by Sarawut Rimdusit1, Parkpoom Lorjia1, Kuljira Sujirote, and Sunan Tiptipakorn **in** Physical and mechanical characteristics of Kevlar fiber reinforced-PC/ABS Composites, Kevlar reinforced fibers which are generally termed as Aramide fiber reinforced polyesters(AFRP) has similar impact properties as the current material being used in automotive front bumpers. Design and Strength Validation of Front Car Bumper Using Composite Material by Vishwanatha R H, Ajith A, Anand N, Punith Raju K B, Vasantha Kumar concluded that composite materials such as fiber reinforced plastics can be used in bumpers.

## 4 Solid modelling

The solid modelling of the front bumper of a jeep is done using SOLIDWORKS 2013 Edition. A solid model gives more information about the specific features of the component or the part. The geometric and topological information of the part can be easily interpreted and modified if needed. In the modern era of manufacturing more complex shapes are being used which can be only shown in a 3D model.

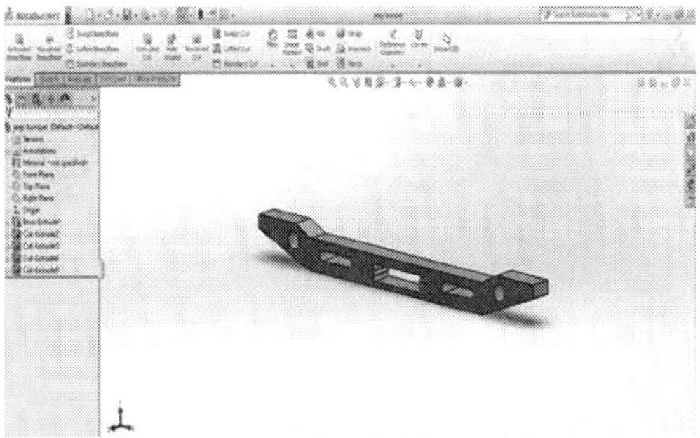

**Figure 1:** modelling of the components in SOLIDWORKS

## 5 Impact analysis of bumper

### *5.1 Organization of the ANSYS software*

For the analysis of bumper, we are using ANSYS WORKBENCH. In ANSYS, we use preprocessors to enter the basic data such as modelling, editing, copying the binary files etc... Then we proceed to the solution processor step to apply the various loads, constraints to the model and to evaluate the various results.

### *5.2 Performing typical ANSYS analysis*

The ANSYS software has a range of finite element analysis capabilities ranging from a simple linear, static to complex nonlinear, dynamic model. In this analysis we have used explicit dynamic analysis of a front bumper. In this project we have done the modelling of the front bumper in the SOLIDWORKS software. The analysis consists of the following steps which can be classified in three categories as shown in table 1.

**Table 1:** Steps in ANSYS analysis

| PREPROCESSOR | SOLUTION PROCESSOR | POST PROCESSOR |
|---|---|---|
| Assigning element type | Analysis definition | Read results |
| Geometry definition | Constant definition | Plot results on graph |
| Material definition | Load definition (velocity) | View animated results |
| Mesh generation | | |
| Model display | | |

The solid model is assigned the material and is meshed to divide the model into number of elements which makes the result to be accurate. The meshing of the model is shown in the figure 2.

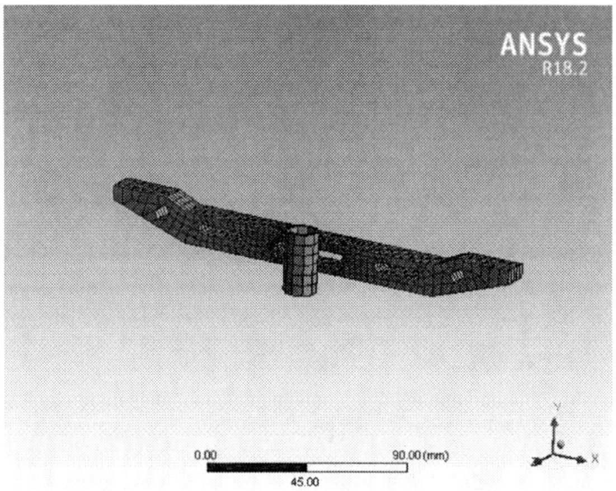


**Figure 2:** Meshing of the components in ANSYS WORKBENCH 18.2

The solid model in workbench is used for impact analysis by applying a velocity of 10 m/s (36 km\hr.). It is impacted against a pole made of structured steel. The pole is a fixed support. It is shown in the figure 3.

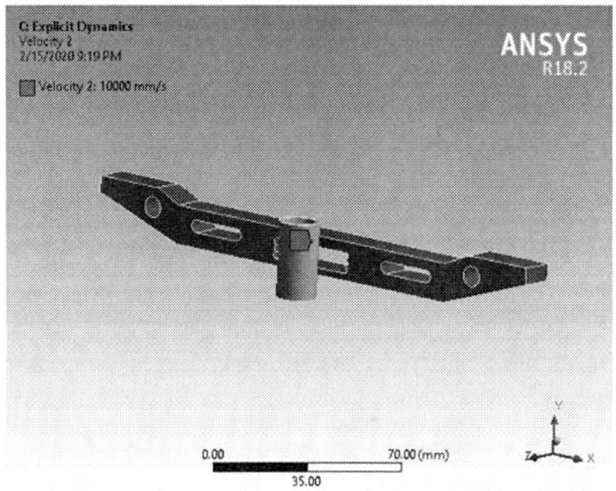


**Figure 3:** Simulation of the Impact phenomena in ANSYS WORKBENCH 18.2

*5.3 Materials properties*

In this way the impact analysis of the bumper solid model is done using the following materials:

*5.3.1 Acrylonitrile butadiene (ABS):*

ABS composite is the currently used bumper material in the automotive industry. It is reinforced with steel.

**Table 2:** properties of ABS (steel reinforced)

| Young's Modulus MPa | Density $Kg/m^3$ | Poisson's Ratio | Bulk Modulus MPa | Shear Modulus MPa |
|---|---|---|---|---|
| 2300 | 1140 | 0.33 | 2254.9 | 864.66 |

IOP Conf. Series: Materials Science and Engineering **923** (2020) 012015 doi:10.1088/1757-899X/923/1/012015

*5.3.2 Aramide fibre reinforced polyester composite*

In Aramide fibers composites the Kevlar-49 fibers are reinforced in polyester resin which can yield resistance towards impact.

**Table 3**: properties of AFRP

| Young's Modulus MPa | Density $Kg/m^3$ | Poisson's Ratio | Bulk Modulus MPa | Shear Modulus MPa |
|---|---|---|---|---|
| 4500 | 1450 | 0.31 | 3947.4 | 1717.6 |

*5.3.3 Nylon 6 with nano clay reinforcement*

Nylon 6 (or polyamide 6) is a polymer possessing good toughness, high tensile strength, elasticity and luster. Nano clays are nanoparticles of layered mineral silicates. The most commonly used Nano clay for material
reinforcement is montmorillonite, a 2-to-1 layered smectite clay mineral with a platy structure.

**Table 4:** properties of nylon 6 (Nano clay)

| Young's Modulus MPa | Density $Kg/m^3$ | Poisson's Ratio | Bulk Modulus MPa | Shear Modulus MPa |
|---|---|---|---|---|
| 5000 | 1130 | 0.39 | 7575.8 | 1798.6 |

*5.4 ANSYS results*

*5.4.1 ANSYS results for ABS composite*

Considering bumper material as acrylonitrile butadiene styrene plastic with steel reinforcements is impacted against pole made of structured steel the results are as follows which are shown in figure 4 and figure 5

*5.4.1.1 Equivalent stress in ABS Composite*

Maximum equivalent stress = 23.895 MPa

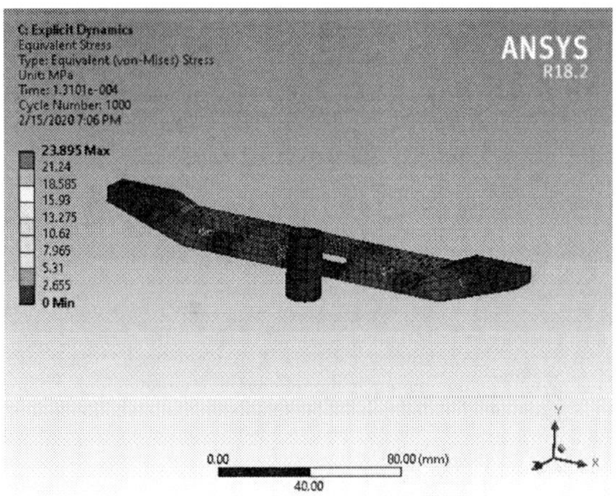


**Figure 4:** Equivalent Stress in ABS composite

*5.4.1.2 Deformation in ABS Composite*

Maximum deformation = 0.4mm

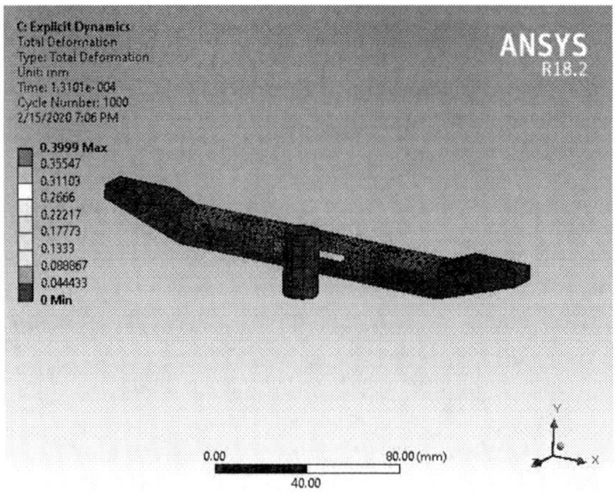


**Figure 5:** Total deformation of ABS composite

*5.4.2 ANSYS results for AFRP composite*

Considering bumper material as Aramide fiber reinforced polyester(kevlar29) is impacted against pole made of structured steel the results are as follows which are shown in fig. 6 and fig. 7

*5.4.2.1 Equivalent stress in AFRP Composite*

Maximum equivalent stress = 23.906MPa

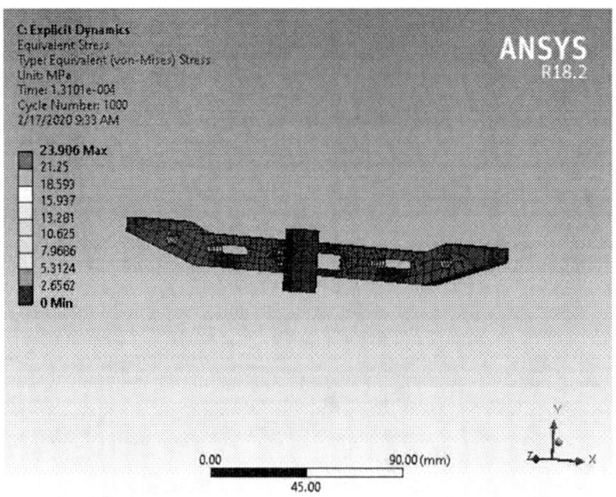


**Figure 6:** Equivalent Stress in AFRP composite

*5.4.2.2 Deformation in AFRP Composite*

Maximum deformation = 0.45132mm

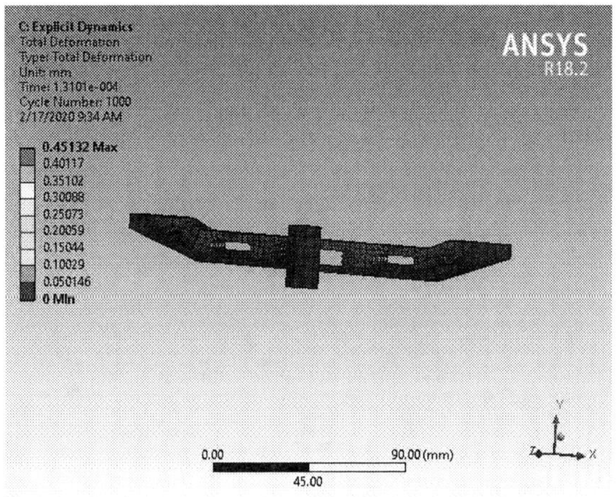


**Figure 7:** Total deformation of AFRP composite

*5.4.3 ANSYS results for nylon-6 with nanoclay composite*

Considering bumper material as Nylon 6 Nano clay is impacted against pole made of structured steel the results are as follows which are shown in fig. 8 and fig. 9

*5.4.3.1 Equivalent stress in Nylon-6 Nano clay Composite*

Maximum equivalent stress = 23.91MPa

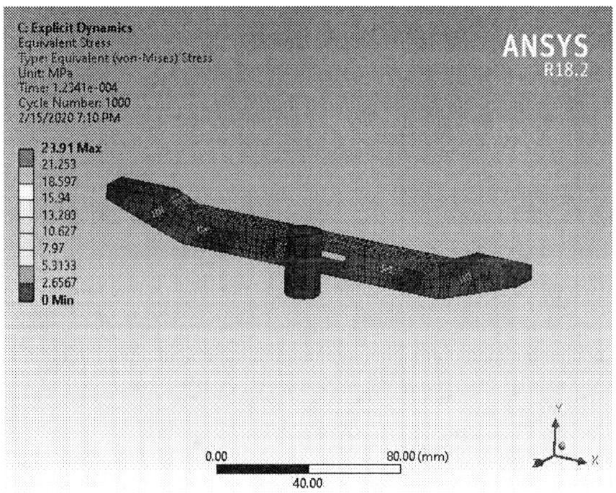


**Figure 8**: Equivalent Stress in Nylon-6 Nano clay Composite

*5.4.3.2 Deformation in Nylon-6 Nano clay Composite*

Maximum deformation = 0.159mm

IOP Conf. Series: Materials Science and Engineering **923** (2020) 012015 doi:10.1088/1757-899X/923/1/012015

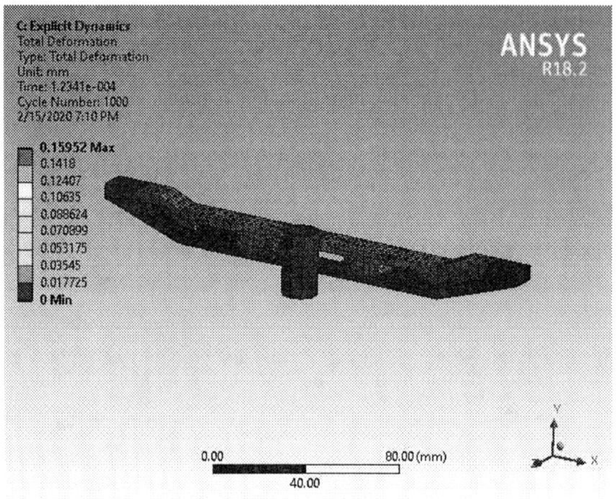


**Figure 9:** Total deformation of Nylon-6 Nano clay composite

### *5.5 Mathematical modelling and results*

Let,

**M** = Mass of the Bumper in Kg
**S** = Velocity of bumper in m/s,
**V** = Volume of the bumper in $m^3$,
**E** = Modulus of elasticity in $N/m^2$,
**σ** = stress generated in bumper in $N/m^2$,
**KE**= change in kinetic energy in bumper in J,
**SE**= change in kinetic energy in bumper in J.

#### *5.5.1 Assumptions made*

- The collision is perfectly elastic so that Momentum is conserved.
- The system is loaded within elastic limit to apply theories of failure.
- The time for collision is very less, that stresses generated rapidly and distribute in fraction of second.
- The whole kinetic energy of Bumper is converted into strain energy of Bumper only.
- The friction between Pole and Bumper is neglected.
- Neglecting effect of bending and shearing of Pole and Bumper for analysis

#### *5.5.2 Formulation*

Kinetic energy of bumper before collision,

$KE = ½*M*S^2$ (Equation 1)

Strain Energy of bumper after collision

$SE = (\sigma^2 * V)/2E$ (Equation 2)

Since the total kinetic energy of Bumper is converted into strain energy of Bumper, so according to law of conservation of energy, total Energy of the system is constant. So, balancing energies of the Bumper.

KE = SE

i.e.,

$\frac{1}{2}*M*S^2 = (\sigma^2 * V)/2E$ (Equation 3)
From equation 3, we can find stress generated in Bumper
So, Stress generated by bumper is given by,

$\sigma^2 = M*S^2*E/ V$
$\sigma = ((M* S^2*E)/ V)^{0.5}$ (Equation 4)
$\sigma = Kf * (M* S^2*E/ V)^{0.5}$................ (Equation 5)

Where, Kf = Ks* Ksz * Kc * Ke.
Kf = Resultant Factors corresponding

**Ks** = Shape Factor

**Ksz**= Size Factor

**Kc** = Stress Concentration Factor

**Ke**= Endurance Limit Factor

**Table 5**: Design factors of the bumper materials

| | Shape Factor (Ks) | Size Factor (Ksz) | Stress Concentration Factor (Kc) | Endurance Limit Factor (Ke) | Resultant Stress Factor (Kf) |
|---|---|---|---|---|---|
| ABS | 1 | 1 | 1.5 | 1 | 1.5 |
| AFRP | 1 | 1 | 1.3 | 1 | 1.3 |
| NYLON 6 NANO CLAY | 1 | 1 | 1 | 1 | 1 |

Now, Applying Material conditions for Bumper

| 1) *For ABS Plastic* | 2) *For AFRP Plastic* | 3) *For Nylon 6 Nano clay* |
|---|---|---|
| Mass, M = 8.954Kg,<br>Velocity, S = 10 m/s,<br>Volume, V = $8.6108*10^{-3}$ $m^3$,<br>Modulus of Elasticity, E = $2300*10^6$ $N/m^2$,<br>So, from equation 5,<br>$\sigma = K_f * (M* S^2*E/ V)^{0.5}$<br>$\sigma$ = 24.28 $N/m^2$<br>$\sigma$ = 24.28 Mpa. | Mass, M = 12.48Kg,<br>Velocity, S = 10 m/s,<br>Volume, V = $8.6108*10^{-3}$ $m^3$,<br>Modulus of Elasticity, E = $1950*10^6$ $N/m^2$,<br>So, from equation 5,<br>$\sigma = K_f * (M* S^2*E/ V)^{0.5}$<br>$\sigma$ = 21.85 $N/m^2$<br>$\sigma$ = 21.85Mpa. | Mass, M = 9.726 Kg,<br>Velocity, S = 10 m/s,<br>Volume, V = $8.6108*10^{-3}$ $m^3$,<br>Modulus of Elasticity, E = $5000*10^6$ $N/m^2$,<br>So, from equation 5,<br>$\sigma = K_f * (M* S^2*E/ V)^{0.5}$<br>$\sigma$ = 23.79 $N/m^2$<br>$\sigma$ = 23.79 MPa. |

**Table 6**: Comparison of results of Impact stresses

| **Sr. No** | **Material** | **Numerical Impact Stress (MPa)** | **ANSYS Impact Stress (MPa)** | **Percentage Error ( % )** |
|---|---|---|---|---|
| 1 | **ABS** | 24.28. | 23.80 | 2.01 |
| 2 | **AFRP** | 21.85 | 23.906 | -5.43 |
| 3 | **NYLON 6 NANO CLAY** | 23.79 | 23.91 | -0.5 |

## 6 Conclusion

From this report it can be concluded that computer model can be used to predict impact analysis of bumper and stress pattern within bumper system has been studied. Also, mathematical model results provide validation for ANSYS simulation model results. Depending on the outcome, composite materials for bumper are selected for impact strength analysis. From, impact analysis of bumper it is suggested that

- Nylon 6 Nano clay is suitable from repairing cost point of view, because it has minimum deformation and
- Aramide fiber reinforced polyester has minimum stress hence it is the best material for impact point of view and can absorb maximum collision energy.

## 7 Acknowledgement

The authors express their deepest gratitude to Department of Mechanical Engineering of St. Josephs Institute of technology for providing their valuable time and full-fledged support to accomplish this research work.

## REFERENCES

[1] Sarawut Rimdusit1, Parkpoom Lorjia, Kuljira Sujirote, and Sunan Tiptipakorn, *Physical and Mechanical Characteristics of Kevlar Fiber-reinforced PC/ABS Composites,*ENGINEERING JOURNAL **Volume 16 Issue 1 Page no 56-66**

[2] Vishwanatha R H, Ajith A, Anand N, Punith Raju K B, Vasantha Kumar, *Design and Strength Validation of Front Car Bumper Using Composite material* **e-ISSN: 2395-0056 p-ISSN: 2395-0072 Page no 3427 - 3434**

[3] Alen John, Sanu Alex, PG Scholar, Department of Mechanical Engineering, Mar Baselios College of Engineering and Technology, Trivandrum, Kerala, INDIA, *A Review on the Composite Materials used for Automotive Bumper in Passenger Vehicles* **Volume-4, Issue-4, August-2014, ISSN No.: 2250-0758 Page Number: 98-101**

*[4]* A.Calienciug ,Gh.N. Radu ,*Design and FEA crash simulation of a composite bumper* **Series I: Engineering Sciences • Vol. 5 (54) No. 1 – 2012 Page no;8-12**
*[5]* Olumide Osokoya,Department of Aeronautical and Automotive Engineering,Loughborough University,*An evaluation of polymer composites for car bumper beam.* **Int. J. Automotive Composites, Vol. 3, No. 1, 2017 Page no 44-59**
[6] Bhavesh A. Bohra, Prof. D. B. Pawar *the comparative analysis of frontal car bumper during impac*t Dept. Of Mechanical Engineering, Gawai Institute of Engineering and Technology, Darapur, Maharashtra, India. **Vol.3, Issue 12, December 2014 page no 48-52**

IOP Conf. Series: Materials Science and Engineering **923** (2020) 012016 doi:10.1088/1757-899X/923/1/012016

# DESIGN AND ANALYSIS OF HELICAL COIL HEAT EXCHANGER

**Kewin Titus. A[1], Khaja Fareedudeen Ahmed. K. S[2], Sabarish Kumar. P[3], Santhosh. D[4], Arun Vasantha Geethan. K[5].**

[1,2,3,4,5]Department of Mechanical Engineering, St. Joseph's Institute of Technology, OMR, Chennai 119.

kewintitus@gmail.com

**Abstract**. Heat exchangers has been widely used in power plants and industries for heating or cooling applications. Heat is transferred from one medium to another medium by using heat exchangers. Helical coil heat exchangers with two different shell configurations are designed. First shell configuration with a central core and second without a central core. Both the configurations has a copper helical coil. Designing of heat exchangers is done with counter flow arrangement by using CATIA V5 R2015. The performance of both the configurations are analysed and compared by using Fluid flow (Fluent) in ANSYS WORKBENCH 18.1for CFD simulations. Further the helical tube materials Copper and Aluminium which are widely used in heat exchangers are also analysed in shell with core configuration and heat transfer rates are also compared using CFD (Fluent) analysis.

## 1. Introduction

The main objective of heat exchangers is to enhance the heat transfer between two fluids. The helical coil heat exchanger is designed with a helix angle of 13° which makes it compact and the heat transfer is greatly improved when compared to straight tubes. Due to the curved geometry of helical coil a centrifugal force is induced on the moving fluid. This results in development of a secondary flow. This secondary flow is the reason for increased heat transfer rates in helical coils.

Two types of shell configurations are designed one with a core and another without a core. Both these configurations has a common material copper as inner tube material. Both the shell and tube fluids is water flowing in a counter flow arrangement. Performance of both the configurations are analysed.

Aluminium is also widely used as a tube material in heat exchangers. So copper is replaced by aluminium as tube material and the configuration used is shell with core. The heat transfer rates of copper and aluminium are analysed and compared.

Amitkumar S Putterwar, A.M.Andhare has designed and thermally evaluated helical coil heat exchanger experimentally in a counter flow arrangement and how the heat transfer is improved due to secondary flow development [1]. Mr. M.D.Rajkamal et al has studied the use of POCO HTC graphite, ASTM SA 179 carbon steel and copper as inner tube materials in a shell with no core configuration using Ansys CFX 15.0 software and the results were compared [2]. Devendra Borse, Jayesh V. Bute has studied the helical coil heat exchanger on the basis of Dean Number, heat transfer coefficient and

IOP Conf. Series: Materials Science and Engineering **923** (2020) 012016 doi:10.1088/1757-899X/923/1/012016

Reynolds number and it infers that helical coiled tube heat exchanger is a promising modification over conventional type of heat exchanger [3]. Sreejith K, Jaivin A Varghese et al experimentally found that the effectiveness of helical coil heat exchanger is found to be higher than that of the straight tube heat exchanger for all the inlet temperatures of water. They have studied the variation of effectiveness with inlet temperature of hot water for both helical coil and straight tube heat exchangers. From the results obtained they have concluded that helical coil heat exchanger is having better effectiveness than straight tube heat exchanger [4]. Subin Michael has studied the changes in temperature profiles for different tube materials and the effectiveness of heat exchangers are calculated. CFD simulations are done for different industrial materials [5]. Shiva Kumara and K.Vasudev Karanth has studied the performance helical coiled tubular heat exchanger used for cooling water under constant wall temperature conditions by CFD simulations. CFD results of helical coiled tubular heat exchanger are compared with the results of straight tubular heat exchanger of the same length under identical operating conditions. Results showed increase in heat transfer and increase in nusselt number when compared to straight tubular heat exchanger [6].

## 2. MATERIAL PROPERTIES

Two commonly used materials in heat exchangers are aluminium and copper. These materials are used as inner tube materials and Mild steel is used as shell material

The properties of materials used are:

| **MATERIAL** | **DENSITY** ($kg/m^3$) | **SPECIFIC HEAT** (J/kg-K) | **THERMAL CONDUCTIVITY** (W/m-K) |
|---|---|---|---|
| Copper | 8978 | 381 | 387.6 |
| Aluminium | 2719 | 871 | 202.4 |

## 3. HEAT EXCHANGER DIMENSIONS

3.1 *Shell dimensions*

| **S. No** | **DIMENSIONAL PARAMETERS** | **DIMENSIONS** (mm) |
|---|---|---|
| 1. | Inner diameter ($D_o$) | 220 |
| 2. | Shell length(L) | 295.625 |
| 3. | Core diameter($D_i$) | 110 |
| 4. | Shell thickness | 4 |

3.2 *Helical coil dimensions*

| **S. No** | **DIMENSIONAL PARAMETERS** | **DIMENSIONS** (mm) |
|---|---|---|
| 1. | Tube outer diameter($d_o$) | 12.5 |
| 2. | Tube inner diameter($d_i$) | 10 |
| 3. | Tube thickness(t) | 1.25 |
| 4. | Mean coil diameter($D_m$) | 165 |
| 5. | Coil outer diameter | 177.5 |
| 6. | Pitch(p) | 18.75 |
| 7. | No of turns(n) | 12 |

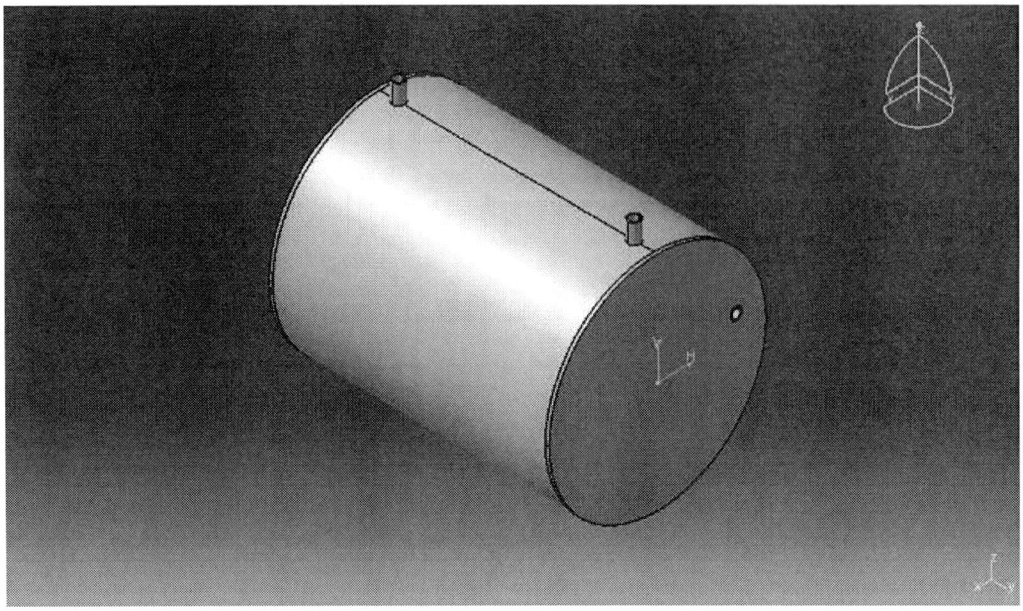

Fig. 1 CATIA 3D model of heat exchanger

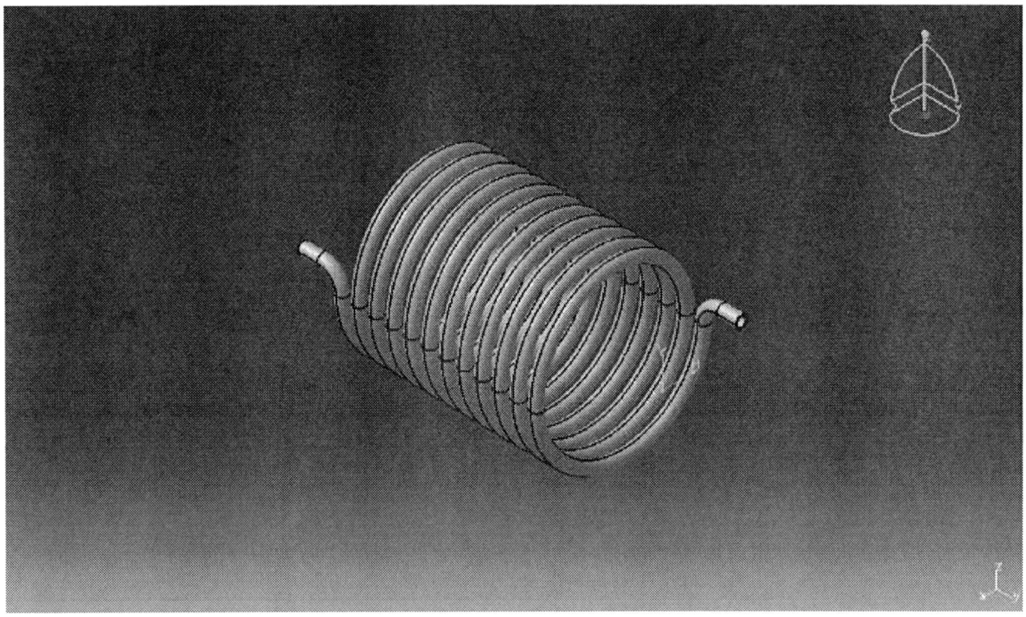

Fig. 2 CATIA 3D model of helical tube

IOP Conf. Series: Materials Science and Engineering **923** (2020) 012016 doi:10.1088/1757-899X/923/1/012016

## 4. MESH DETAILS

3D Model designed and assembled in CATIA is imported to Ansys Workbench and the volume extract feature available in SpaceClaim modeling is used to fill tube fluid volume and shell fluid volume before meshing

Smoothing: High and Fine

| PART NAME | NODES | ELEMENTS |
|---|---|---|
| Shell | 23760 | 73161 |
| Helical tube | 166980 | 91056 |
| Shell fluid | 370661 | 2000359 |
| Tube fluid | 192426 | 888060 |
| All domains | 753827 | 3052636 |

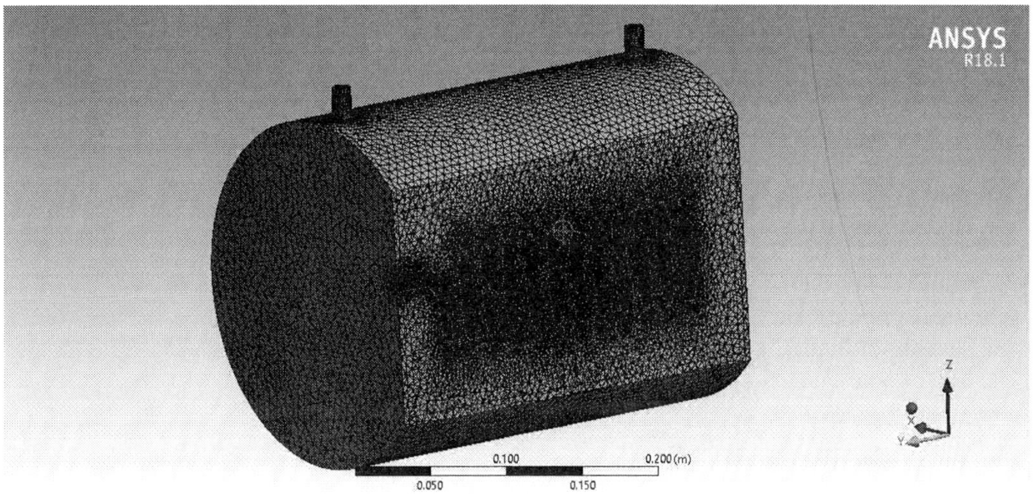


Fig. 3 Mesh cross section of shell without core

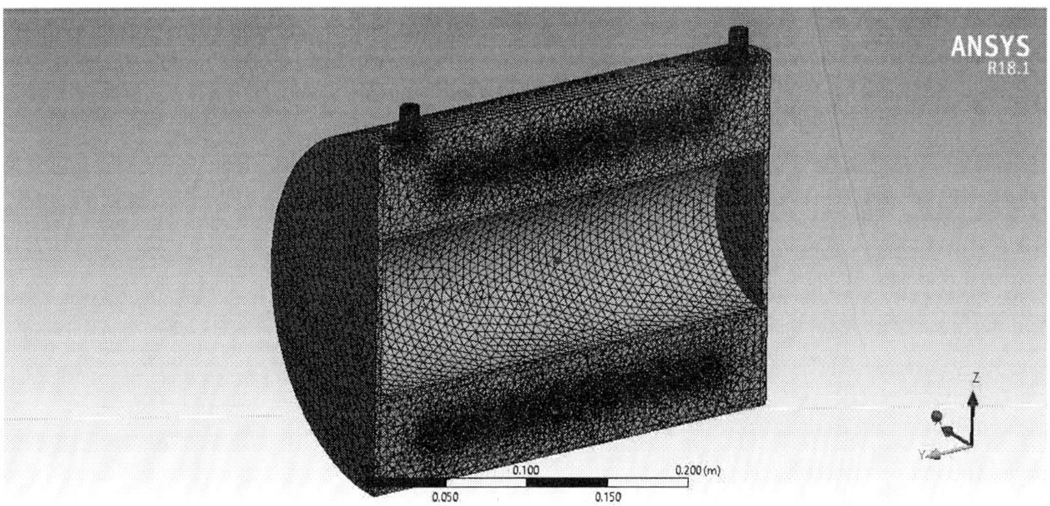


Fig. 4 Mesh cross section of shell with core

IOP Conf. Series: Materials Science and Engineering **923** (2020) 012016 doi:10.1088/1757-899X/923/1/012016

## 5. CALCULATIONS

### *5.1 Fluid flow volume calculations*

Length of coil ($L_{coil}$) = $n \sqrt{((\pi D_o)^2 + p^2)}$
= 8598.34mm
Volume occupied by coil = $(\pi d_o^2/4) * L_{coil}$
= 1067794.2 $mm^3$
Volume of shell (with core) = $(\pi * (D_o^2 - D_i^2) / 4) * L$
= 8428254.95 $mm^3$
Volume of shell fluid (with core)
= 8428254.9-1067794.2
= 7360460.7 $mm^3$
Volume of shell (without core) = $(\pi * D_o^2 / 4) * L$
= 11237673.27 $mm^3$
Volume of shell fluid (without core)
= 10169879.07 $mm^3$

### *5.2 Heat transfer calculations*

$Q = m_h * C_{ph} (t_2 - t_1) = m_c * C_{pc} (T_1 - T_2)$
Mass flow rate of hot fluid ($m_h$) = 0.045 kg/s
Mass flow rate of cold fluid ($m_c$) = 0.050 kg/s
Specific heat of hot fluid ($C_{ph}$) = 4182 J/kg-K
Specific heat of cold fluid ($C_{pc}$) = 4182 J/kg-K
*Inlet and outlet temperatures*:
$T_1$ – Entry temperature of hot fluid (K)
$T_2$ – Exit temperature of hot fluid (K)
$t_1$ – Entry temperature of cold fluid (K)
$t_2$ – Exit temperature of cold fluid (K)

*5.2.1 Shell with core and copper tube:*
$T_1$ = 323 K
$T_2$ = 307.4 K
$t_1$ = 293 K
$t_2$ = 305.3 K
Q = 0.050 * 4182 * (305.3-293)
= 2571.93 J/s

*5.2.2 Shell without core and with copper tube:*
$T_1$ = 323 K
$T_2$ = 309.9 K
$t_1$ = 293 K
$t_2$ = 303.2 K
Q = 0.050 * 4182 * (303.2-293)
= 2132.82 J/s

*5.2.3 Shell with core and aluminium tube:*

$T_1 = 323$ K

$T_2 = 312.8$ K

$t_1 = 293$ K

$t_2 = 301.2$ K

Q = 0.045 * 4182 * (323-312.8)

= 1919.53 J/s

**6. ANALYSIS AND RESULTS**

For viewing the temperature contour plots and velocity profiles at the inlets and outlets two clip planes are used.

Clip plane 1 – To view the shell inlet and outlet

Clip plane 2 – To view the tube inlet and outlet

Volume Rendering is done to both shell fluid and tube fluid with transparency of 0.2.

*6.1 Shell with core and copper tube:*

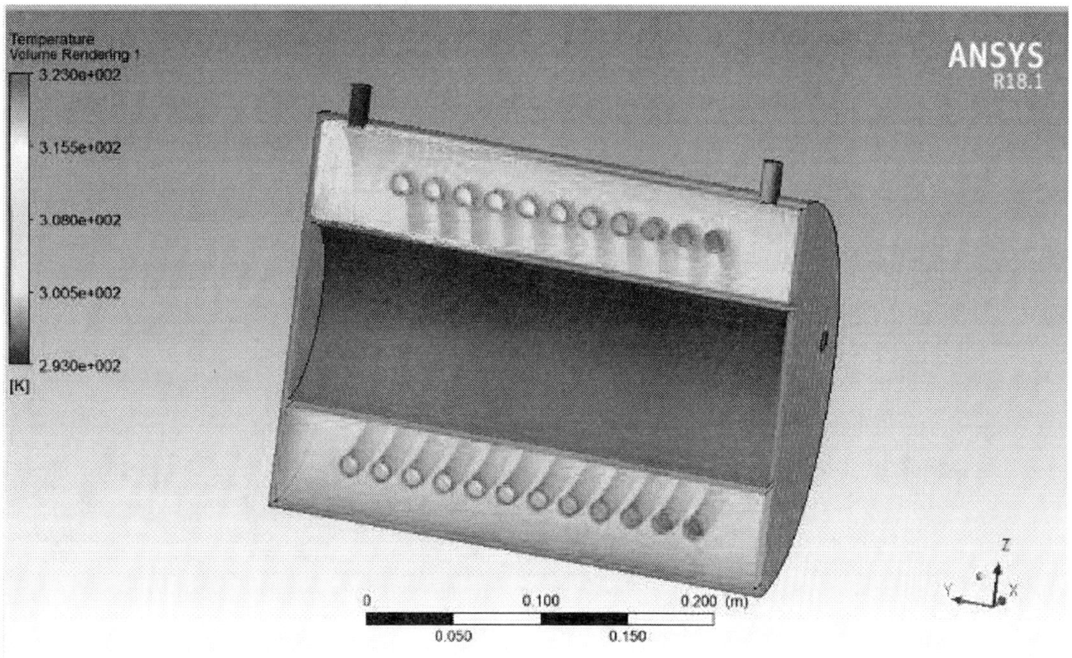


Fig. 5 Temperature contour plot at clip plane 1

IOP Conf. Series: Materials Science and Engineering **923** (2020) 012016 doi:10.1088/1757-899X/923/1/012016

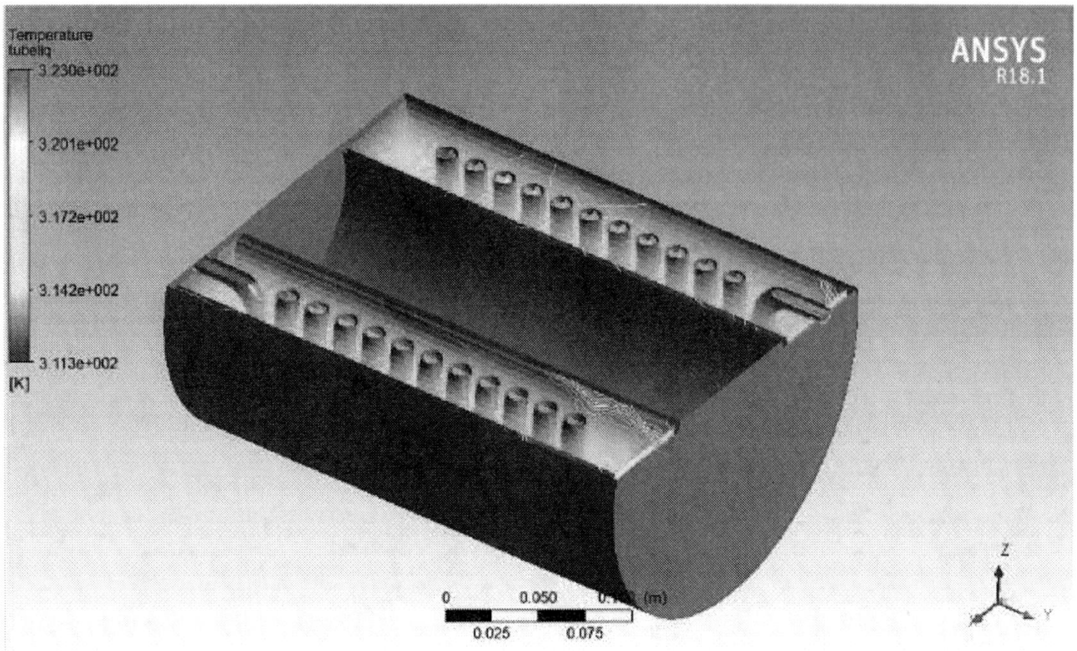


Fig. 6 Temperature contour plot at clip plane 2

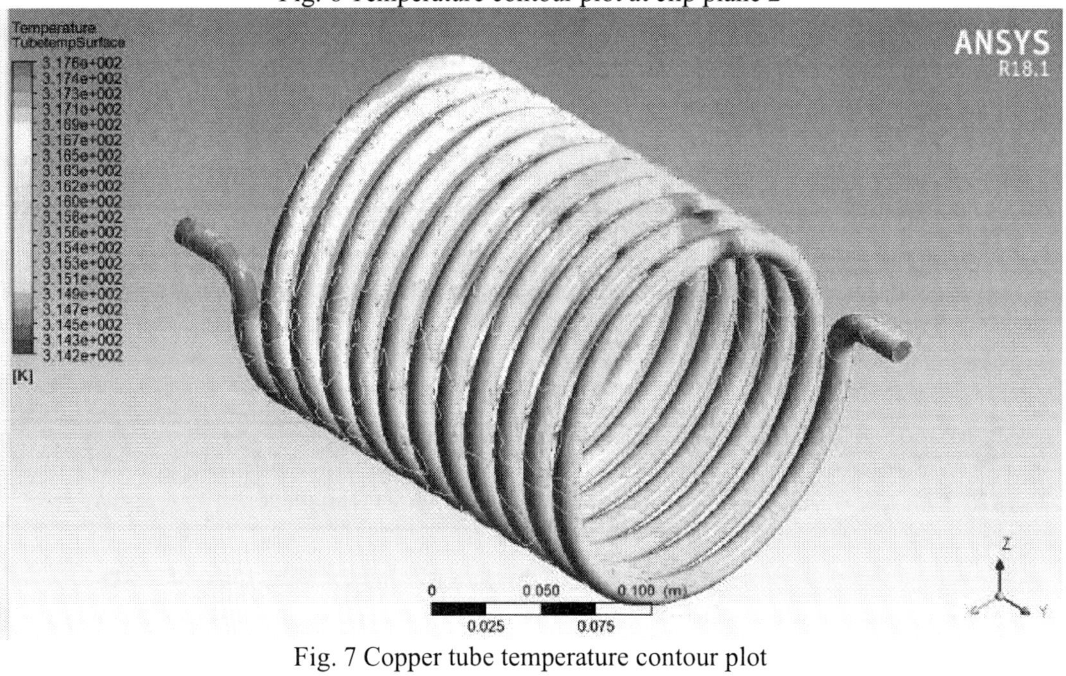


Fig. 7 Copper tube temperature contour plot

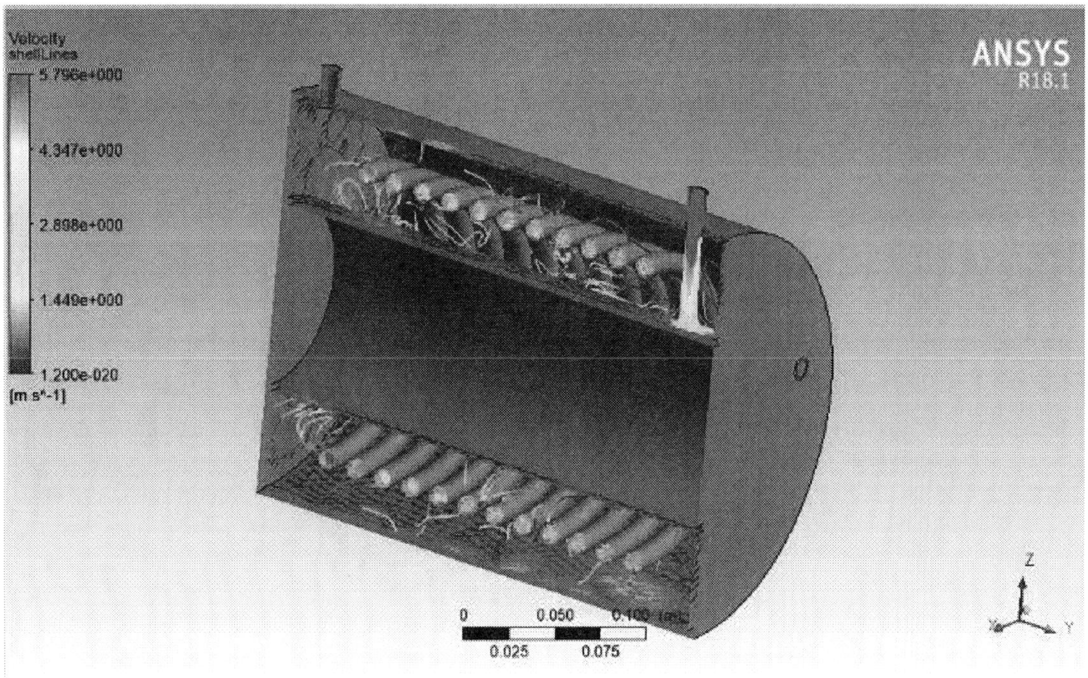


Fig. 8 Shell fluid velocity with streamlines
(no of visible lines is 200)

*6.2 Shell without core and with copper tube:*

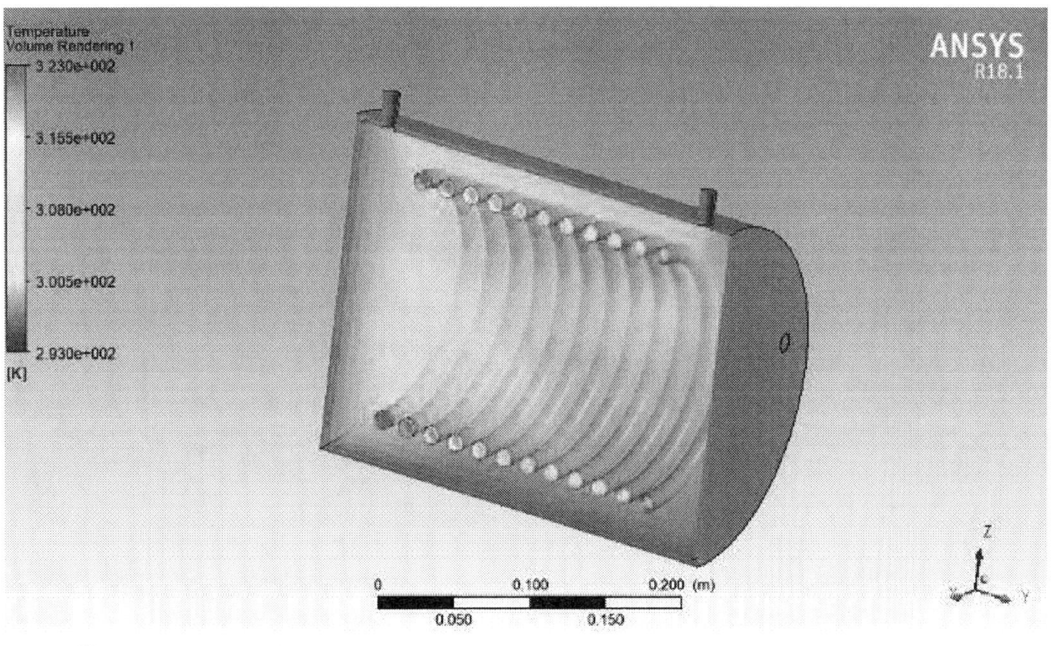


Fig. 9 Temperature contour plot at clip plane 1

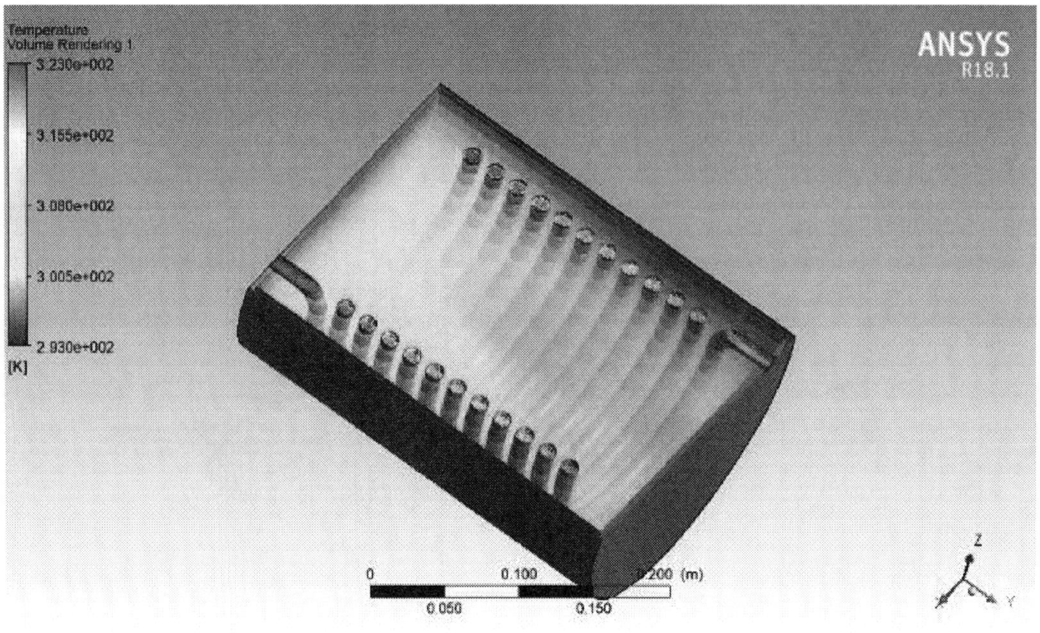


Fig. 10 Temperature contour plot at clip plane 2

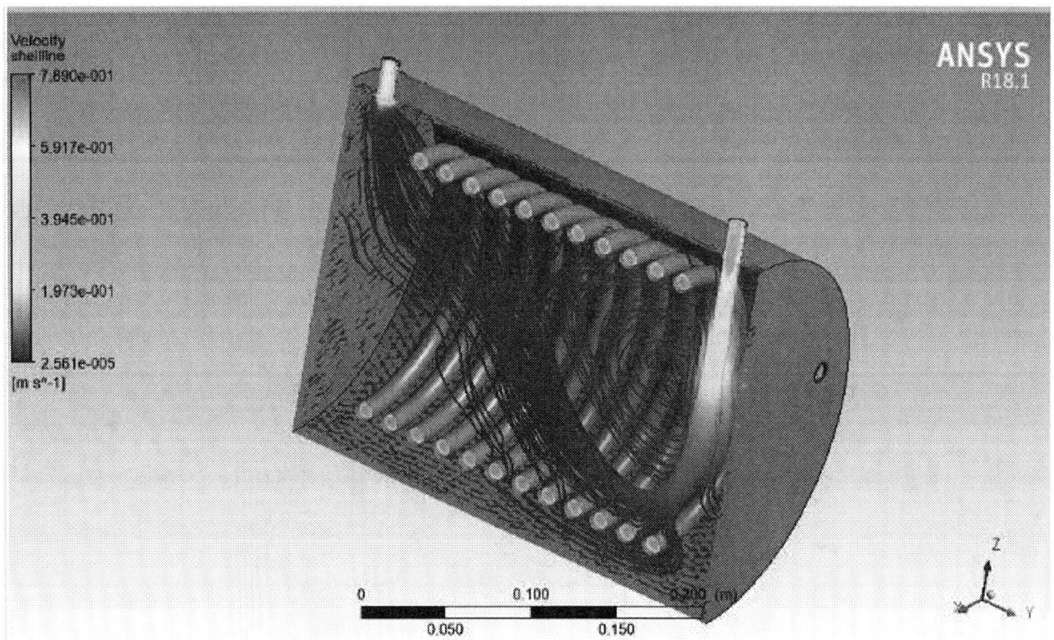


Fig. 11 Shell fluid velocity with streamlines
(no of visible lines is 200)

*6.3 Shell with core and aluminium tube:*

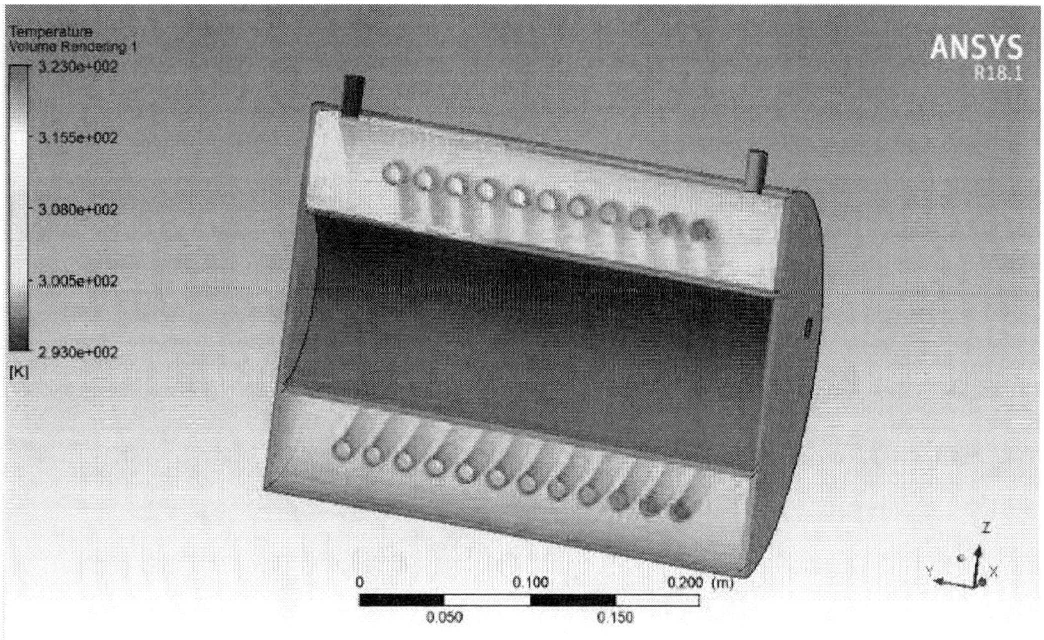


Fig. 12 Temperature contour plot at clip plane 1

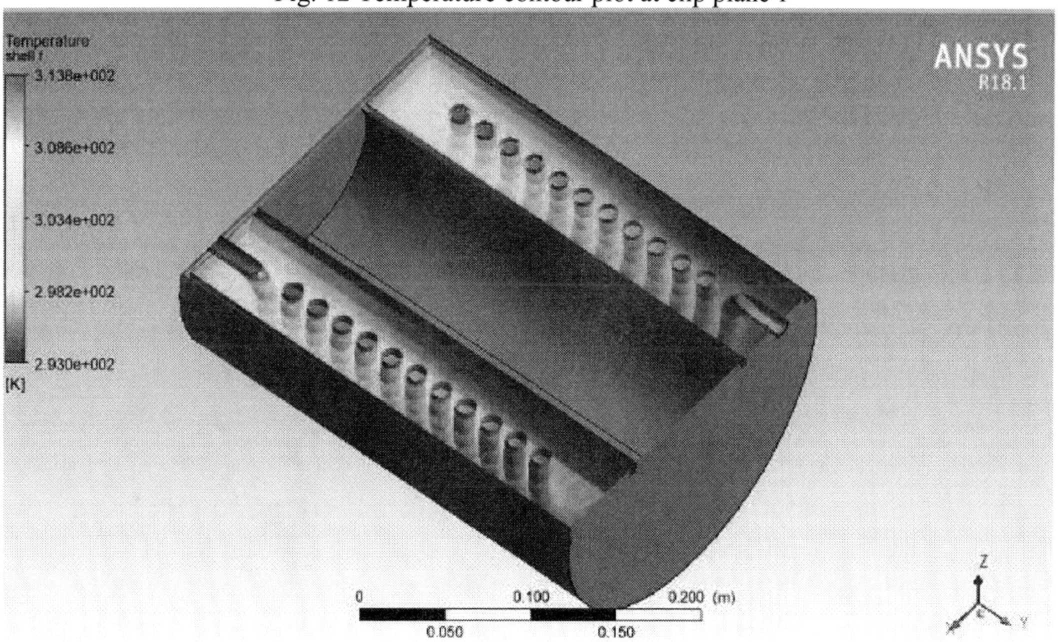


Fig. 13 Temperature contour plot at clip plane 2

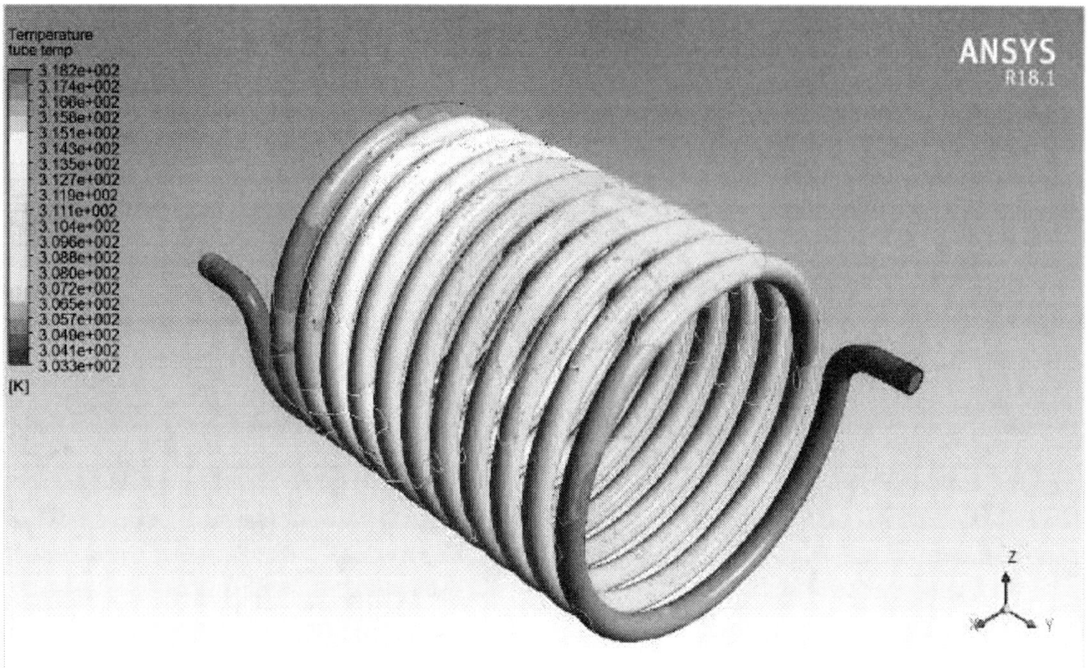


Fig. 14 Temperature contour plot of aluminium tube

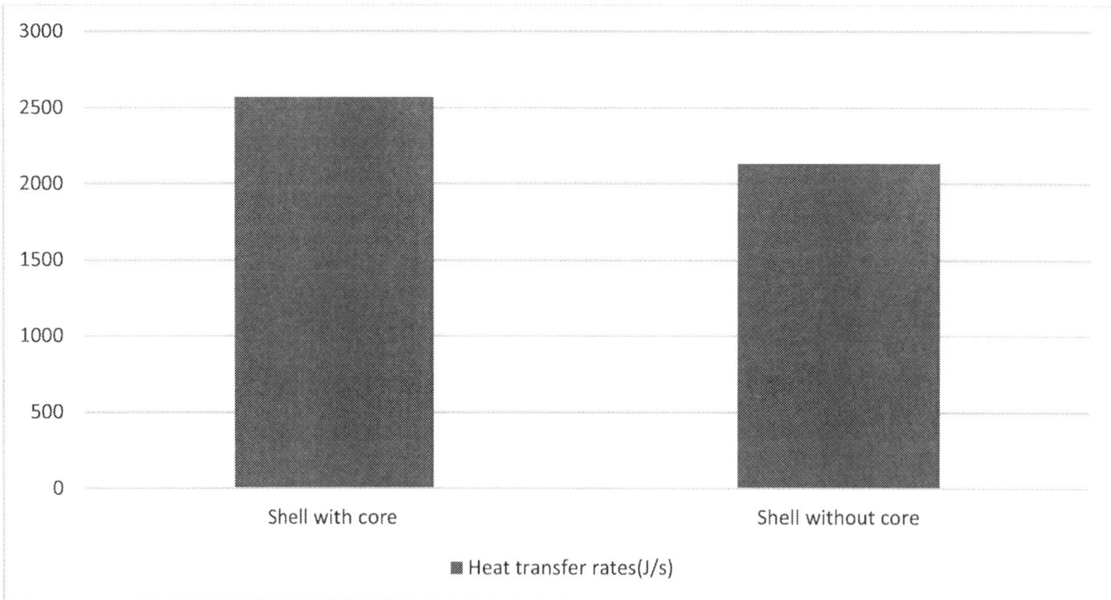


Chart. 1 Heat transfer rates of different shell configurations

IOP Conf. Series: Materials Science and Engineering **923** (2020) 012016 doi:10.1088/1757-899X/923/1/012016

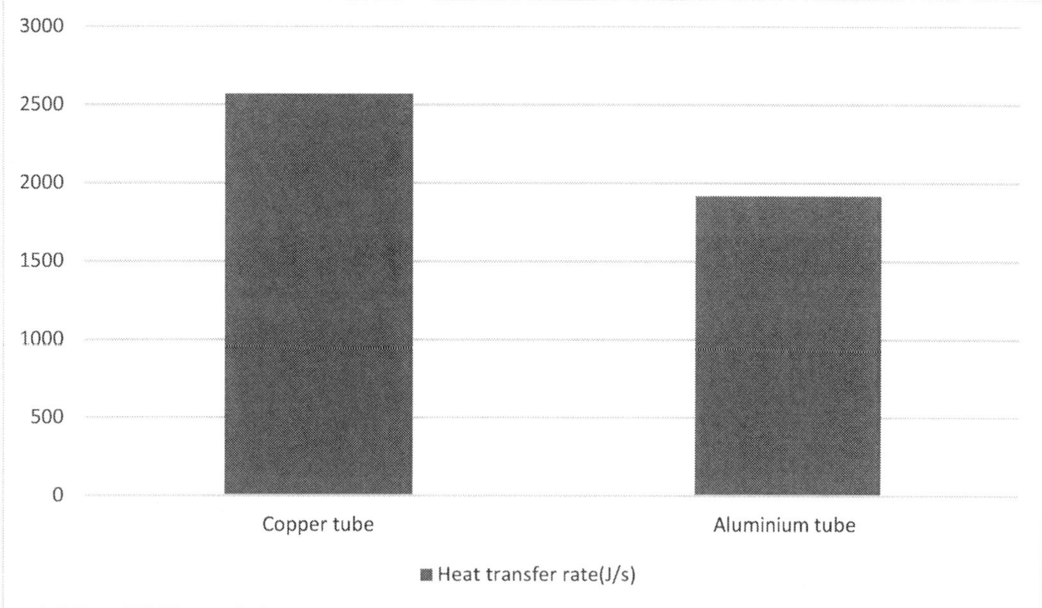

Chart. 2 Heat transfer rates of different tube materials

## 7. CONCLUSION

Thus the design and analysis is helical coil heat exchanger is done and has the following conclusions.

- The performance of both the shell configurations are analysed. The heat transfer to shell fluid in configuration 1(shell with core) is higher than configuration 2(shell without core). This is due to the reduced volume of fluid flow in configuration 1, whereas in configuration 2 the volume through which the fluid flows is 38% higher than configuration 1 and the velocity of flow is much better in configuration 1. This is concluded by comparing the velocity streamline plot in figures 8 and 11. Due to the economic use of cold fluid and increased heat transfer rates observed in configuration 1 it is preferred over configuration 2.
- Even though copper has a better thermal conductivity aluminium has better specific heat and less weight in comparison to copper. Thus aluminium is also preferred in various applications. The increased thermal conductivity of copper can be observed in figure 7 and 14 where the tube temperature distribution is much higher in copper tube than aluminium and the heat transfer rates are also much higher in copper tube than aluminium.

## REFERENCES

[1]Amitkumar S Putterwar, A.M.Andhare-"Design and thermal evaluation of shell and helical coil heat exchanger" ,International journal of research in engineering and technology. eISSN: 2319-1163

[2] Mr. M.D.Rajkamal, M. Mani Bharathi, Shams Hari Prasad M, Santhosh Sivan.M, Karthikeyan.S, R .H.Bahruteen Ali Ahamadu, S.KALIAPPAN, Dr.T.Mothilal – "Thermal analysis of shell and tube heat exchanger", International Journal of Pure and Applied Mathematics Volume 119 No. 12 2018, 14299-14306.

[3] Devendra Borse, Jayesh V. Bute- "A Review on Helical Coil Heat Exchanger", International Journal for Research in Applied Science & Engineering Technology (IJRASET), Volume 6 Issue II, February 2018.

[4] Sreejith K., T.R. Sreesastha Ram, Jaivin A. Varghese, Manoj Francis, Mossas V.J., Nidhin M.J., Nithil E.S., Sushmitha S.- "Experimental Investigation of a Helical Coil Heat Exchanger", : International Journal of Engineering And Science, Vol.5, Issue 8 (August 2015), PP -01-05.

[5]Comparative CFD Analysis of Shell and Serpentine Tube Heat Exchanger by Subin Michael , International Research Journal of Engineering and Technology (IRJET),e-ISSN: 2395 -0056 ,Volume: 04 Issue: 02, Feb -2017.

[6] Shiva Kumara and K.Vasudev Karanth –"Numerical Analysis of a Helical Coiled Heat Exchanger using CFD" ,International Journal of Thermal Technologies ,ISSN 2277 – 4114.

IOP Conf. Series: Materials Science and Engineering **923** (2020) 012017 doi:10.1088/1757-899X/923/1/012017

# Static analysis of different spoke structure of airless and conventional tyre

**T.Prabhuram[1], S. C. Meenakshi Sundaram[2*], S.Jegadeeswer[2], V. Sathish Kannan[2]**

[1] Associate Professor, Department of Mechanical Engineering, St. Joseph's Institute of Technology, Chennai, India
[2] U.G. Students, Department of Mechanical Engineering, St. Joseph's Institute of Technology, Chennai, India

E-mail: * msaigle24@gmail.com

**Abstract**. An airless tyre is known as non-pneumatic tyres which are not advocated by air pressure. The major issues related to pneumatic tyre are puncture and flatness due to wear and tear, which lead to the development of airless tyres. The aim of this work is to conduct static analysis of airless tyre, considering various 3D printing materials with different spoke structures. The spoke structures like honeycomb, triangular and diamond and the 3D printing materials such as Acrylonitrile Butadiene Styrene + Polycarbonate (ABS+PC), Polyethylene Terephthalate (PET) and High Impact Polystyrene (HIPS) are considered for modelling of airless tyre. For conventional tyre, Neoprene rubber is considered. The static analysis is conducted using ANSYS workbench 19.2, and the total deformation, equivalent elastic strain, equivalent von-mises stress and strain energy are compared and evaluated for conventional and airless tyre.

Keywords: Airless tyre, Static analysis, 3D printing materials, ABS+PC, PET, HIPS

## 1. Introduction

Robert William Thomson is the inventor of first pneumatic tyre or conventional tyre. The pneumatic tyre is made up of composition of butadiene rubber, styrene butadiene rubber and natural rubber which encloses air to reduce vibration and improve traction. The main disadvantage of these tyres is that it flattens due to wear and tear or due to sudden deformations consecutively within a short duration and is needed to be replaced. But, an airless tyre doesn't get flattened due to wear and tear and they need not to be replaced often. So, the pneumatic tyres are replaced with airless tyres and these tyres are known as Tweel that is blend of the words tyre and wheel. So, Tweel doesn't have standard wheel hub assembly and was initially developed by a French Company Michelin in 2005. The Structure of Tweel consists of a cable-reinforced band of conventional tyre rubber with molded thread; shear band creates a compliant contact patch with the surface, polyurethane spokes which absorbs energy (i.e.) it provides spring action and wheel hub which connects with axle. The spokes in wheel acts as shock absorbers in airless tyre instead of compressed air in conventional tyre which doesn't use an inner tube filled with compressed air is an significant advantage of airless tyre over conventional tyre and thus, it will not burst, leak pressure wear or tear due to sudden deformations. Currently, most airless tyre on the market are made of solid rubber or plastic. Due to its ability in not flattening it is used in space rovers for space exploration missions. So, it is used for heavy load carrying and transporting applications.

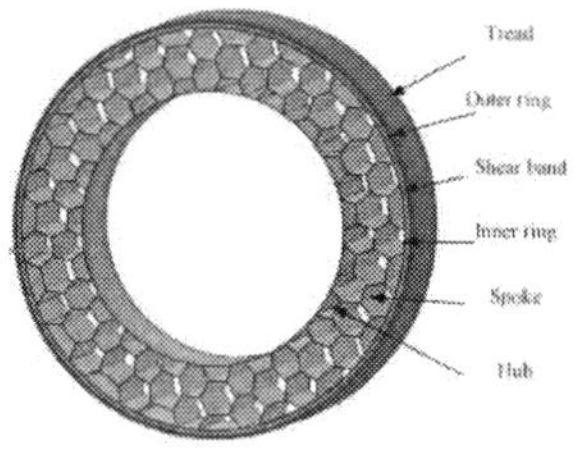


**Figure 1.** Structure of Tweel (i.e.) Airless Tyre

## 2. Literature review

Aramian studied the different additive manufacturing techniques used for cermets [2]. Domingues et al fabricated a composite material by blending the used tyre granules with plastic waste. Also the possible additive manufacturing processes are explored for the utilization of the fabricated composites [3]. Kaufui and Hernandez conducted a detailed review of additive manufacturing process. The need and possible methodologies used in additive manufacturing are explored [4]. Kucewicz et al performed finite element simulation of non-pneumatic tyre. Static radial deflection test was conducted and the different parameter like stress distribution, contact pressure and displacemet are studied at constant load conditions [5]. Aboul yazid et al conducted quasi static two dimensional analysis on Michelin's Tweel. Tyre with and without composite ring are investigated. The outcome shows that the spokes have more impact on tyre's behaviour [1]. Periyasamy and Vijayan generated models of airless and conventional tyre for a car and analysed the parameters like deflection, von-misses stress and strain energy [9]. Prajwal and Seetharama developed the airless and air tyre by considering the rubber as the material and conducted the FEA with static compressive load [10]. Vinay et al developed a model of airless tyre with aluminium alloy for hub, polyurethane for spokes, AISI 4340 for outer ring and rubber for thread. They also considered three different type of honeycomb structures and conducted the analysis to find the deformation and stress developed [11]. Libin Rajan et al studied the airless tyre for a tractor. Three types of tyre namely Michelin Tweel, honeycomb structure and Bridgestone airless tyre are considered and static analysis were conducted. The effects of thickness of spoke and reinforcement layer on total deformation are studied [6].

## 3. Problem Definition

The design and manufacturing related issues of airless tyre are recent areas of the research. Many academicians and industrialist are working on development of new materials and 3D printing methods for airless tyre. The objective of this paper is to conduct finite element analysis of airless tyre with different spoke structures and different 3D printing materials. The spoke structures considered for this study are honeycomb, diamond and triangular. The various 3D printing materials such as ABS + PC, PET and HIPS are selected for analysis. The finite element analysis of conventional tyre with neoprene rubber is also conducted. Finally the total deformation, equivalent stress, equivalent elastic strain and strain energy are compared for different models. The three dimensional models were constructed in CATIA V5 and finite element analysis is carried out in Ansys workbench 19.2.

## 4. Design parameters

The various parameters considered for modelling and analysis of airless tyre are described in this section. Three different materials were considered for model generation. They are Acrylonitrile Butadiene Styrene + Polycarbonate (ABS+PC), Polyethylene Terephthalate (PET) and High Impact Polystyrene (HIPS). The boundary conditions such as load of 3193 N and mesh element size of 50mm are considered. For airless tyre, hub is considered as fixed support. The rim is considered as the fixed support for conventional tyre.

IOP Conf. Series: Materials Science and Engineering **923** (2020) 012017 doi:10.1088/1757-899X/923/1/012017

**Table 1.** Tyre dimensions for modelling.

| Airless tyre dimensions | | Conventional tyre dimensions | |
|---|---|---|---|
| Outer diameter of Tyre | 540 mm | Outer diameter of Tyre | 500 mm |
| Inner diameter of Tyre | 500 mm | Inner diameter of Tyre | 420 mm |
| Outer diameter of Hub | 340 mm | Outer diameter of Rim | 420 mm |
| Inner diameter of Hub | 320 mm | Inner diameter of Rim | 400 mm |

*4.1. CAD Model of tyre*

The CAD model of the airless tyre with different spoke structures are created using CATIA V5 software and shown in the Figure 2.

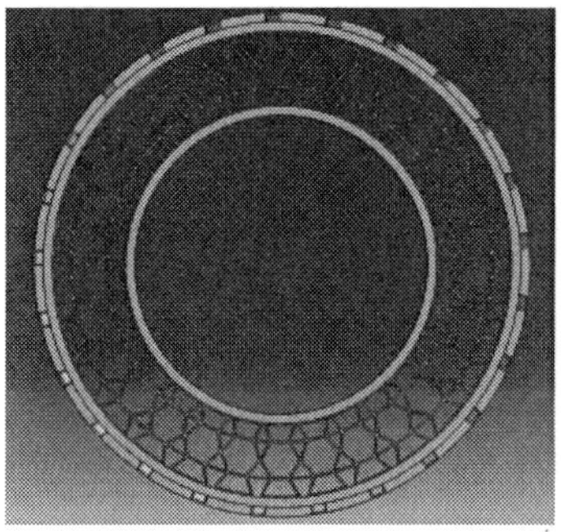

**Figure 2(a).** Honey Comb Structure

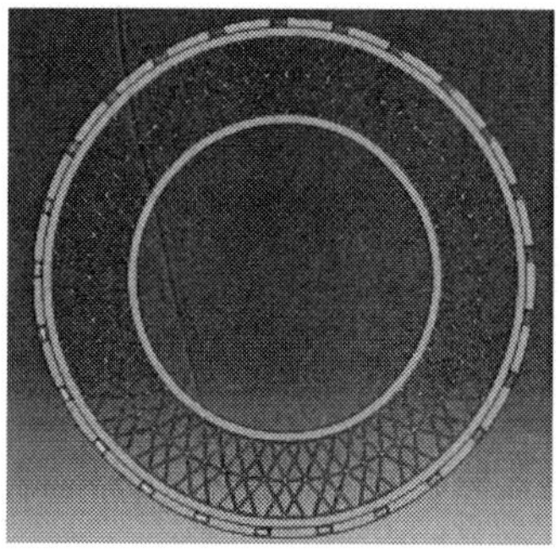

**Figure 2(b).** Diamond Structure

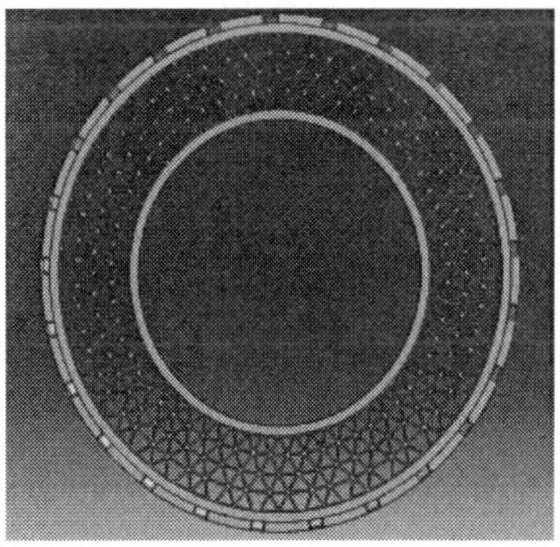

**Figure 2(c).** Triangular Structure

**Figure 2(d).** Conventional Tyre

## 5. Results and discussions

### *5.1. Total Deformation*

*5.1.1. Honeycomb structure.* The honeycomb spoke structure model is taken for statistic structural analysis. The three different material properties namely ABS+PC, PET and HIPS are attached to the model. The following figure shows the total deformation for different materials.

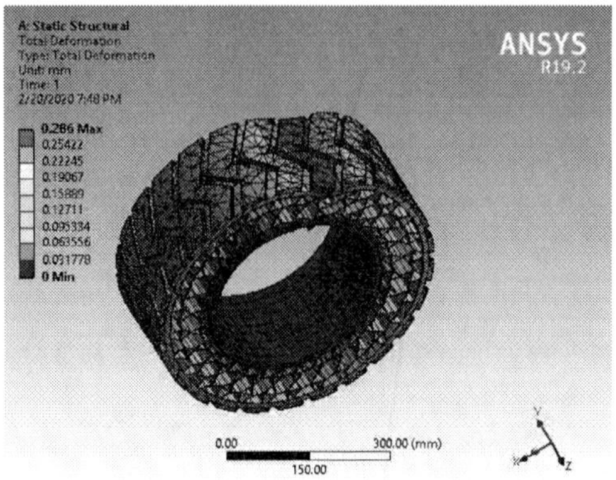


**Figure 3(a). ABS+PC**

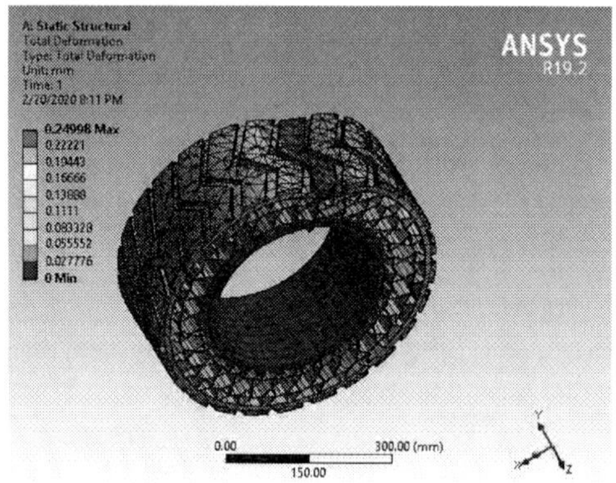


**Figure 3(b). PET**

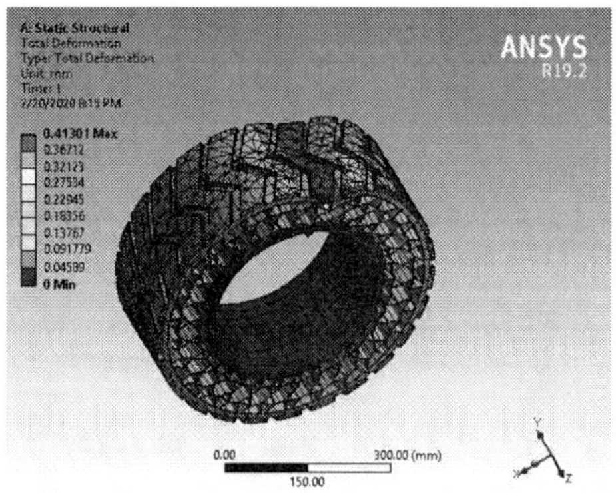


**Figure 3(c). HIPS**

*5.1.2. Diamond structure.* The diamond spoke structure model with three different material properties namely ABS+PC, PET and HIPS are analysed to find the total deformation. The following figure shows the total deformation for different materials.

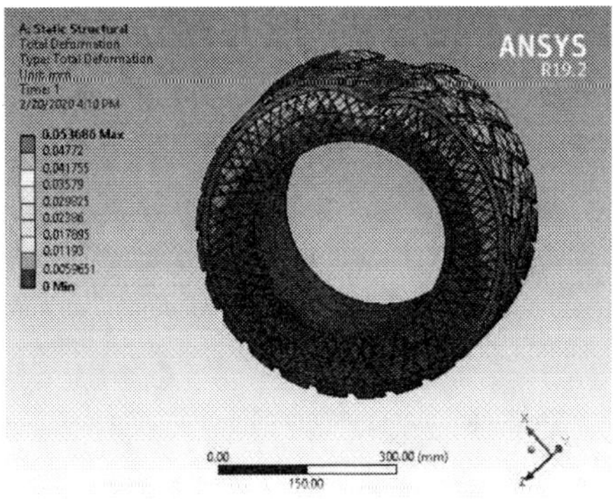


Figure 4(a). ABS+PC

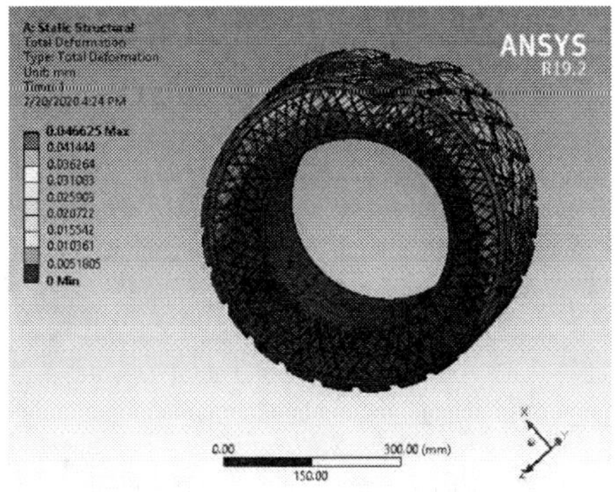


Figure 4(b). PET

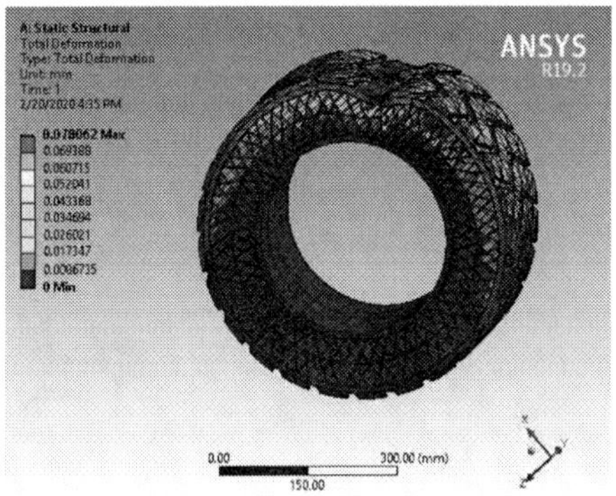


Figure 4(c). HIPS

*5.1.3. Triangular structure.* The triangular spoke structure model with three different material properties namely ABS+PC, PET and HIPS are analysed to find the total deformation. The following figure shows the total deformation for different materials.

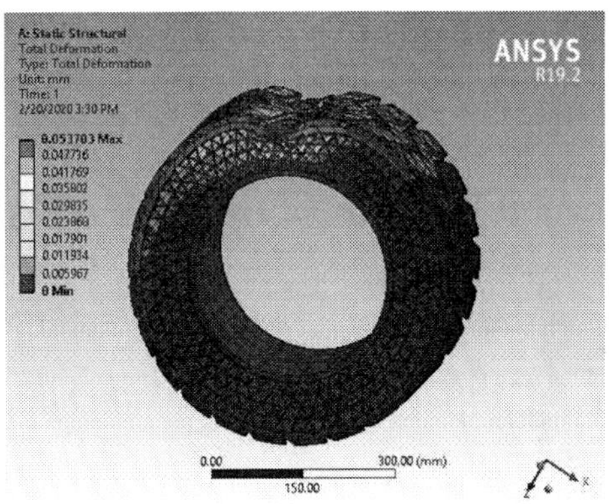


Figure 5(a). ABS+PC

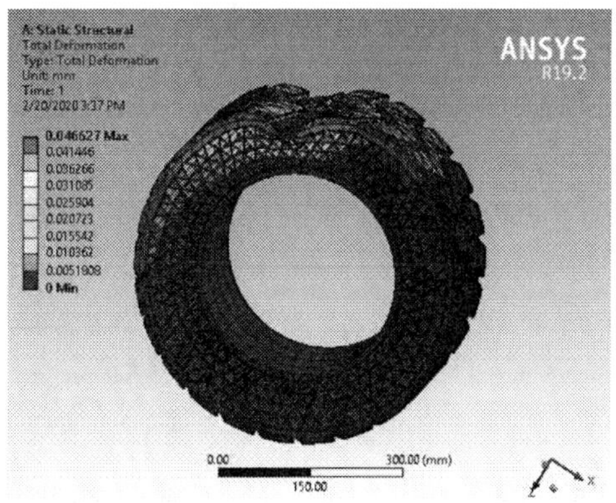


Figure 5(b). PET

IOP Conf. Series: Materials Science and Engineering **923** (2020) 012017 doi:10.1088/1757-899X/923/1/012017

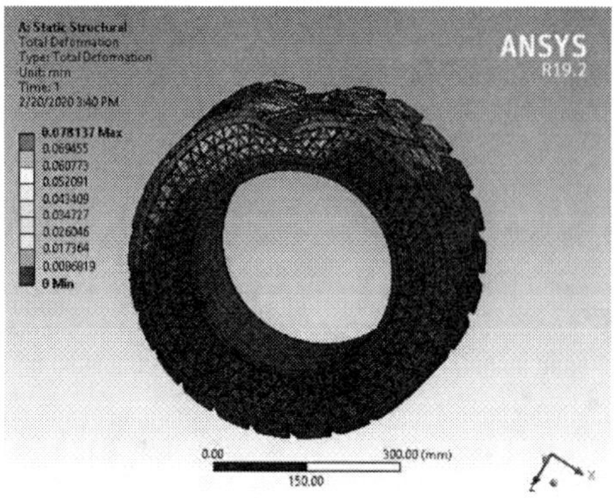


**Figure 5(c). HIPS**

*5.2. Equivalent elastic strain.*

*5.2.1. Honeycomb Structure.* The honeycomb spoke structure model with three different material properties namely ABS+PC, PET and HIPS are analysed to find the equivalent elastic strain. The following figure shows the results of equivalent elastic strain for different materials.

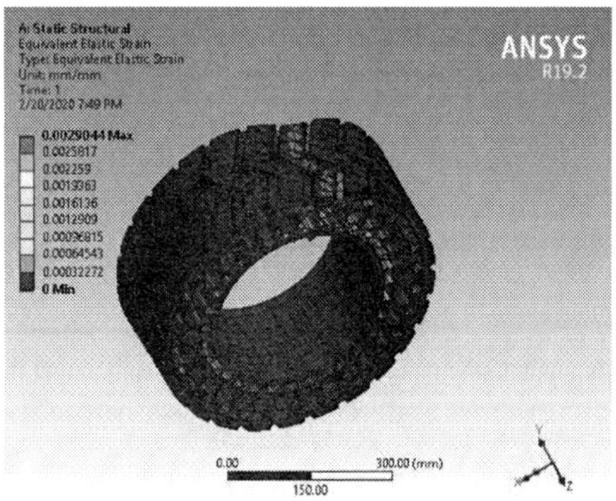


**Figure 6(a). ABS+PC**

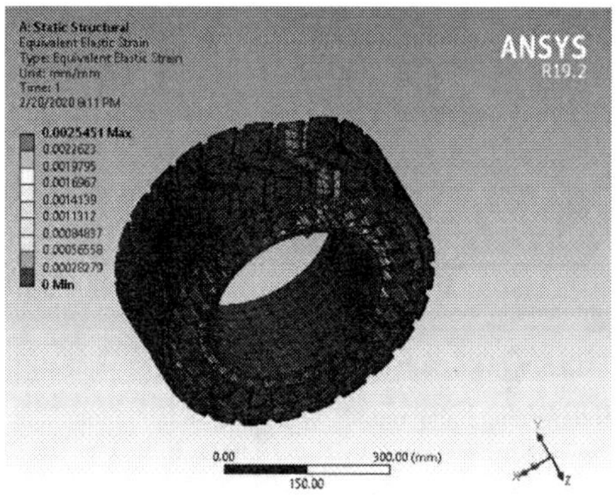


**Figure 6(b). PET**

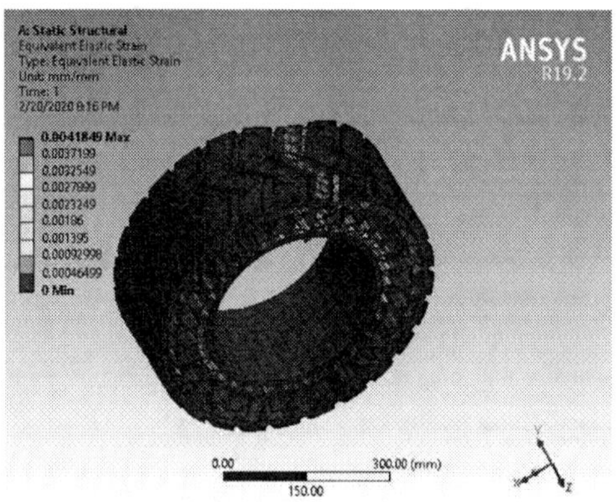


**Figure 6(c). HIPS**

*5.2.2. Diamond structure.* The diamond spoke structure model with three different material properties namely ABS+PC, PET and HIPS are analysed to find the equivalent elastic strain. The following figure shows the results of equivalent elastic strain for different materials.

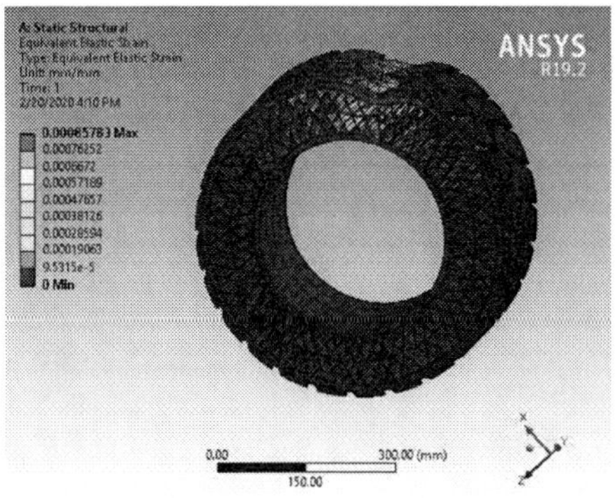


**Figure 7(a). ABS+PC**

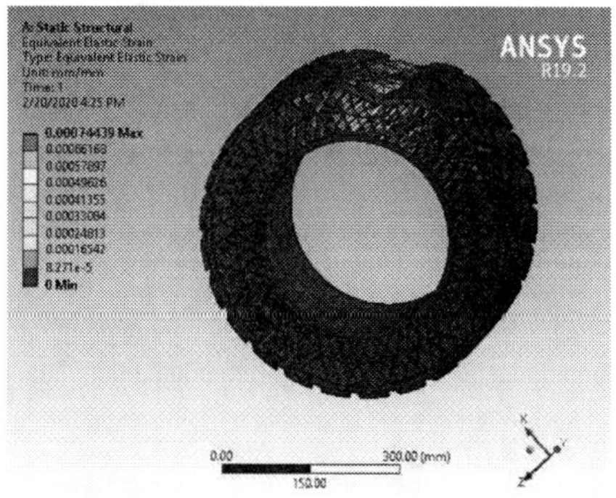


**Figure 7(b). PET**

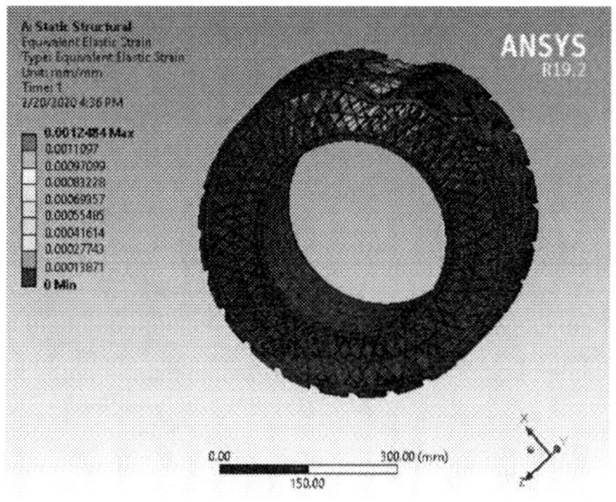


**Figure 7(c). HIPS**

*5.2.3. Triangular structure.* The triangular spoke structure model with three different material properties namely ABS+PC, PET and HIPS are analysed to find the equivalent elastic strain. The following figure shows the results of equivalent elastic strain for different materials.

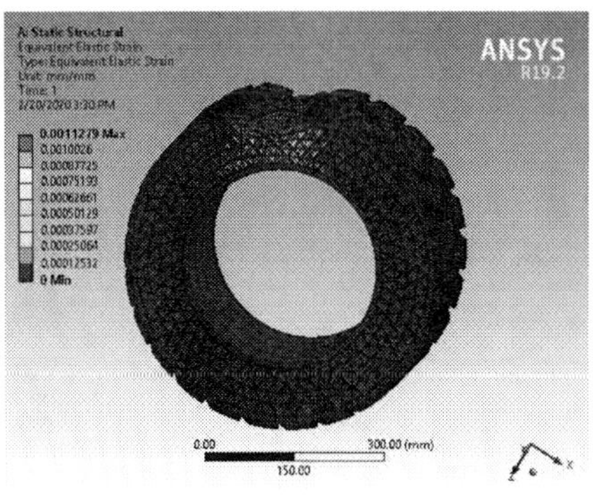


**Figure 8(a). ABS+PC**

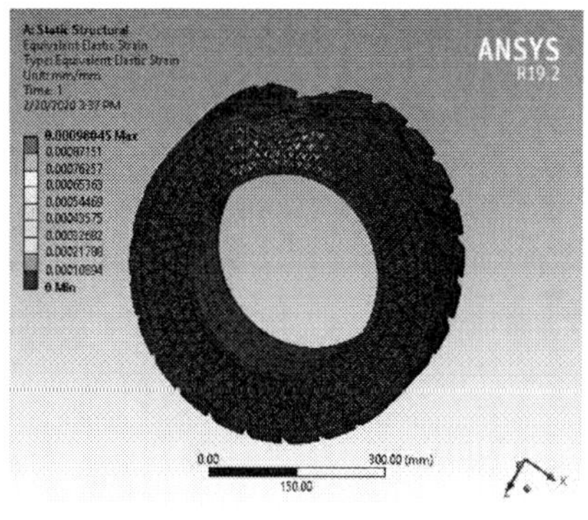


**Figure 8(b). PET**

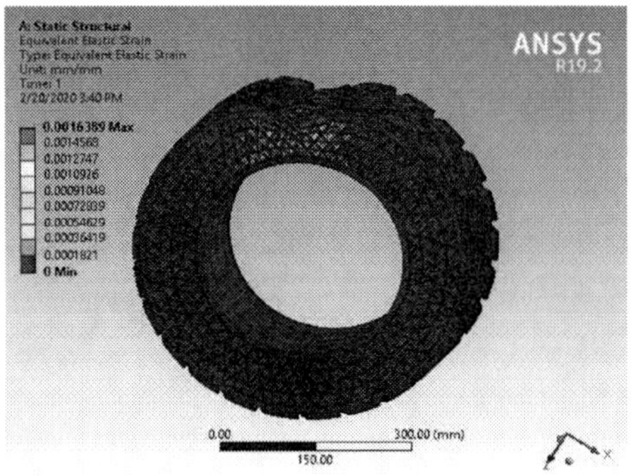


**Figure 8(c). HIPS**

### *5.3. Equivalent (Von-mises) Stress*

*5.3.1. Honeycomb structure.* The honeycomb spoke structure model with three different material properties namely ABS+PC, PET and HIPS are analysed to find the equivalent (von-mises) stress. The following figure shows the result of equivalent (von-mises) stress for different materials.

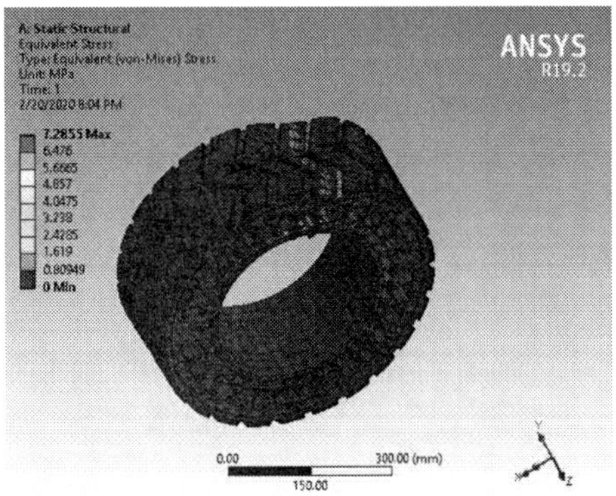


**Figure 9(a). ABS+PC**

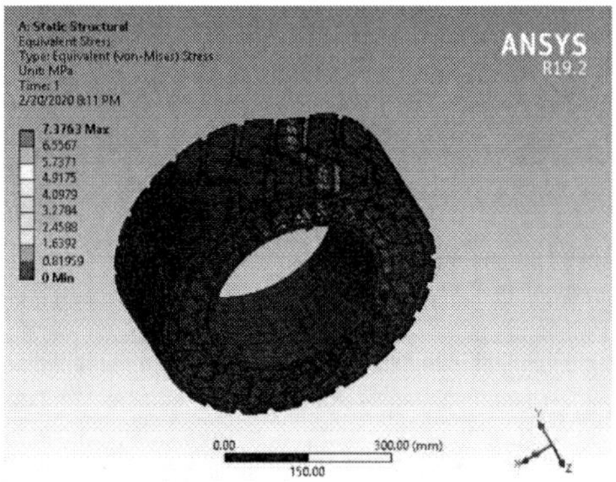


**Figure 9(b). PET**

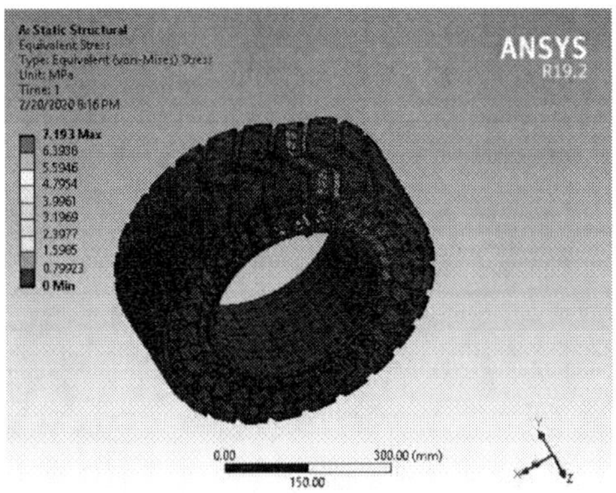


**Figure 9(c). HIPS**

*5.3.2. Diamond structure.* The diamond spoke structure model with three different material properties namely ABS+PC, PET and HIPS are analysed to find the equivalent (von-mises) stress. The following figure shows the results of equivalent (von-mises) stress for different materials.

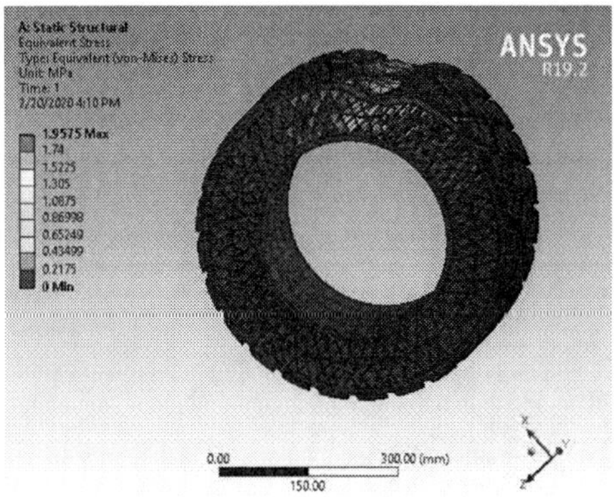


**Figure 10(a). ABS+PC**

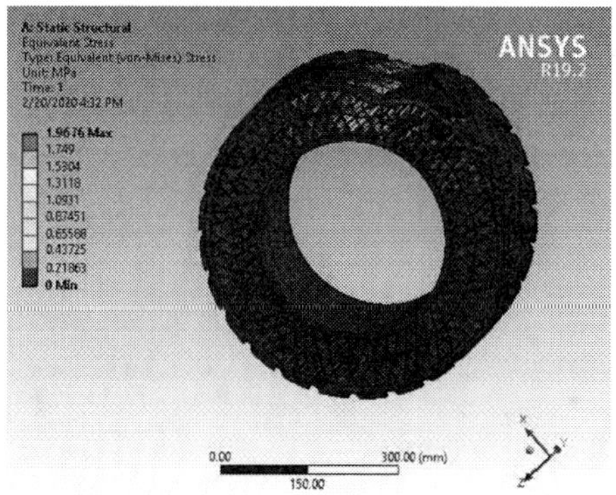


**Figure 10(b). PET**

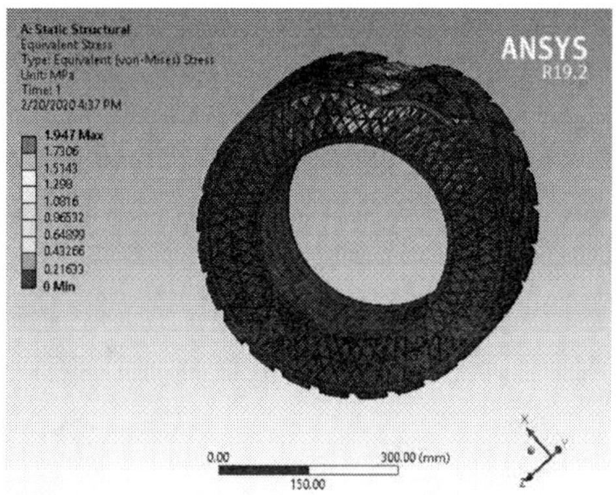


**Figure 10(c). HIPS**

*5.3.3. Triangular structure.* The triangular spoke structure model with three different material properties namely ABS+PC, PET and HIPS are analysed to find the equivalent (von-mises) stress. The following figure shows the results of equivalent (von-mises) stress for different materials.

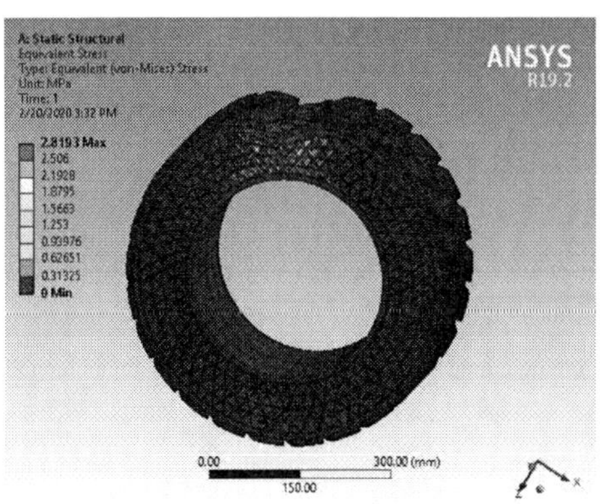


**Figure 11(a). ABS+PC**

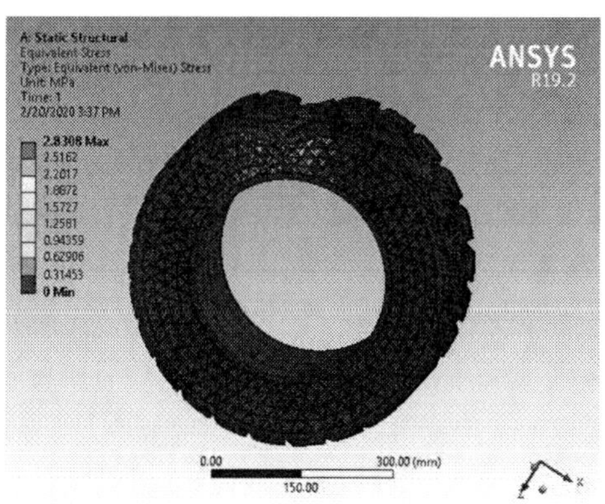


**Figure 11(b). PET**

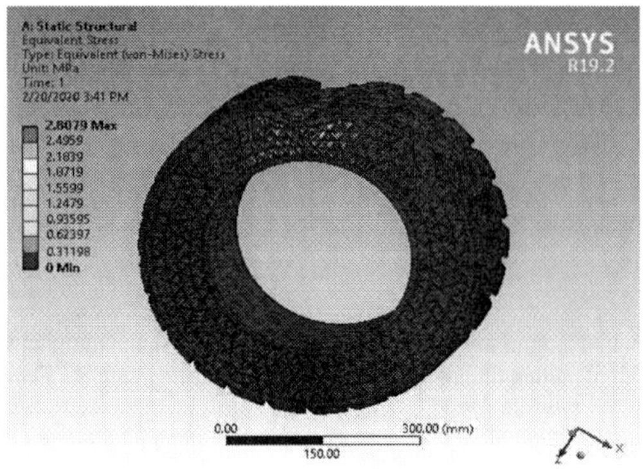


**Figure 11(c). HIPS**

## *5.4. Strain energy*

*5.4.1. Honeycomb structure.* The honeycomb spoke structure model with three different material properties namely ABS+PC, PET and HIPS are analysed to find the strain energy. The following figure shows the results of strain energy for different materials.

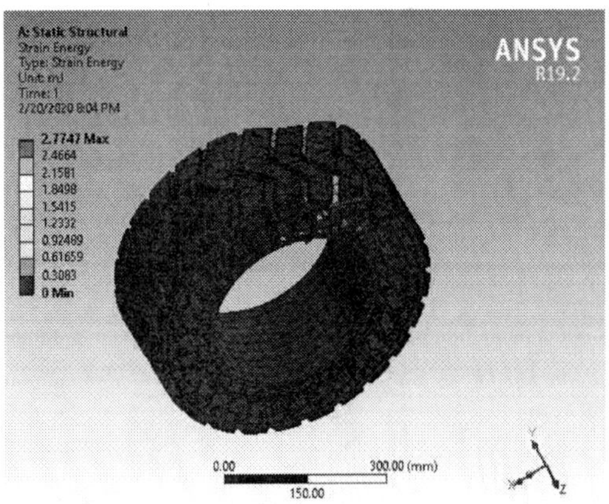


**Figure 12(a). ABS+PC**

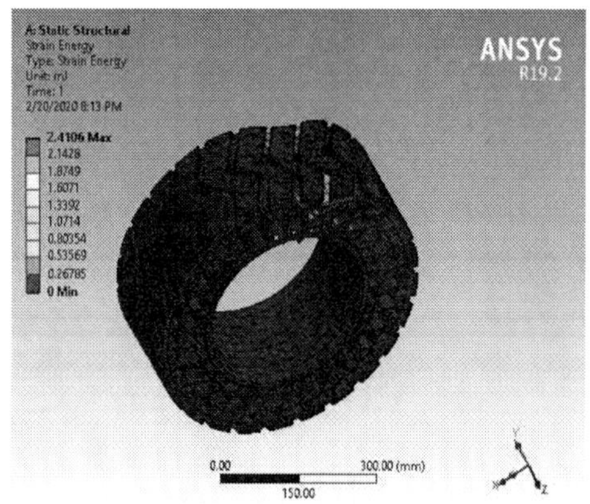


**Figure 12(b). PET**

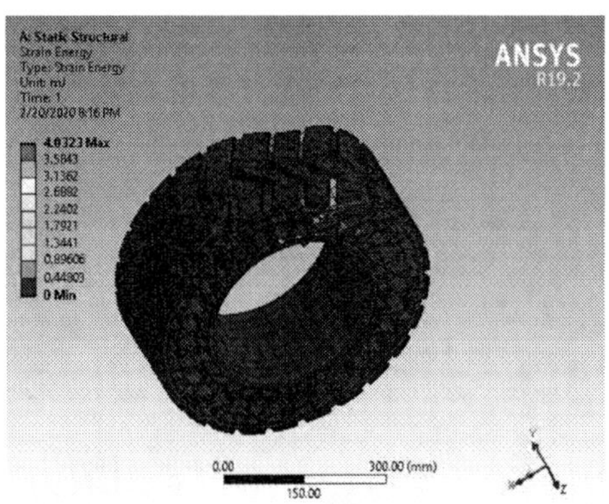


**Figure 12(c). HIPS**

*5.4.2. Diamond structure.* The diamond spoke structure model with three different material properties namely ABS+PC, PET and HIPS are analysed to find the strain energy. The following figure shows the results of strain energy for different materials.

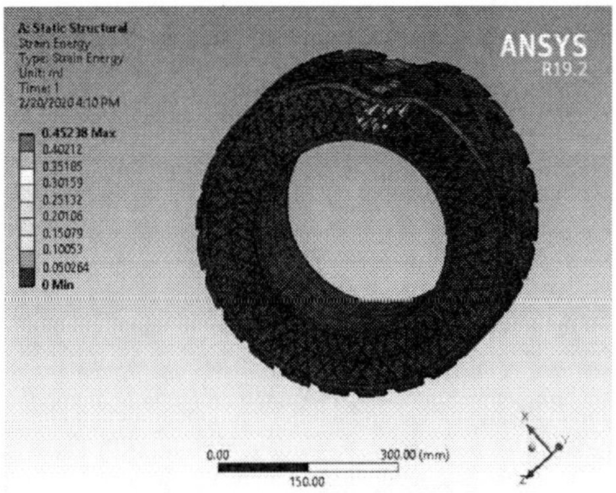


**Figure 13(a). ABS+PC**

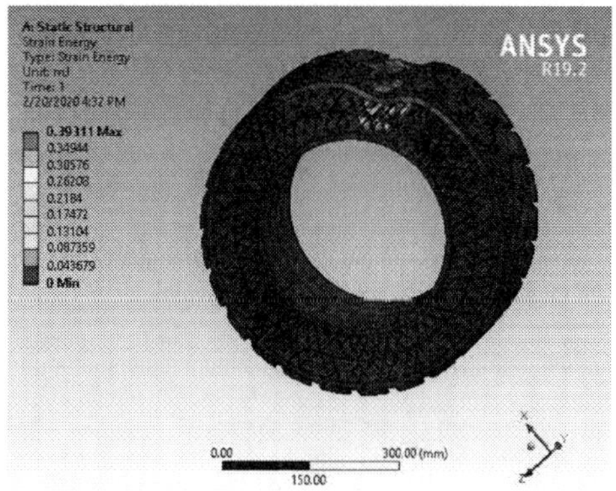


**Figure 13(b). PET**

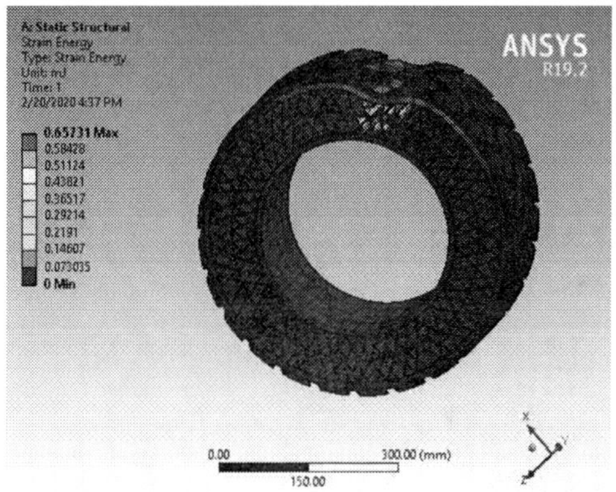


**Figure 13(c). HIPS**

*5.4.3. Triangular structure.* The triangular spoke structure model with three different material properties namely ABS+PC, PET and HIPS are analysed to find the strain energy. The following figure shows the results of strain energy for different materials.

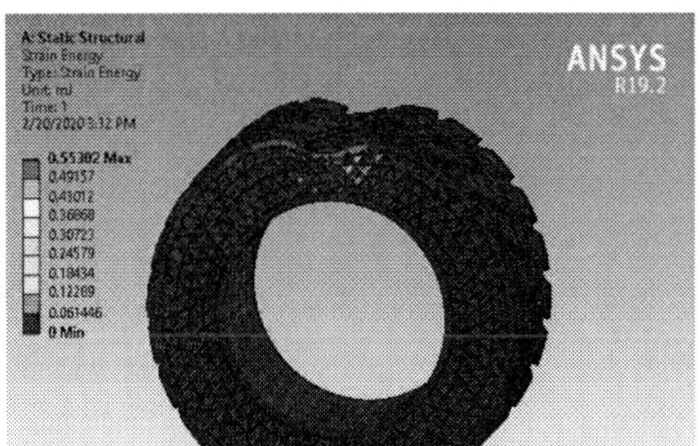


**Figure 14(a). ABS+PC**

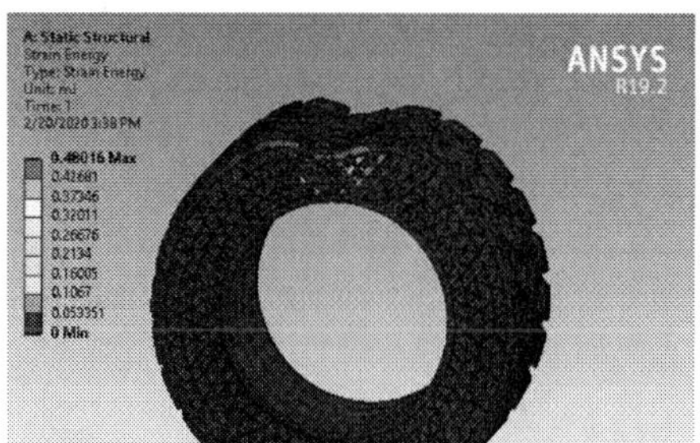


**Figure 14(b). PET**

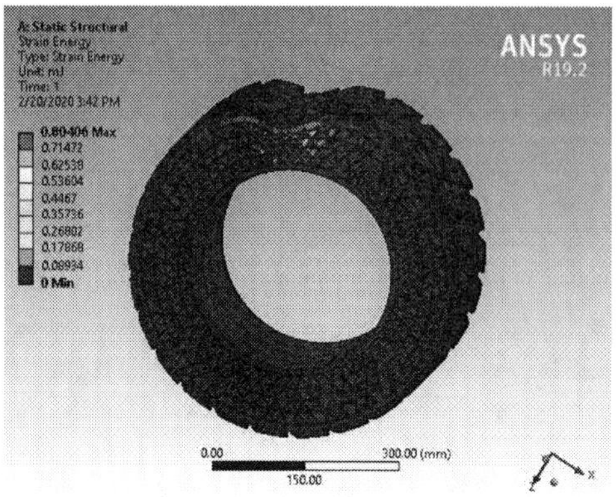


**Figure 14(c). HIPS**

### *5.5. Conventional tyre*

The conventional tyre is modelled with aluminium alloy for rim, neoprene rubber for tyre and air is used in place of inflated tubes. The finite element analyses were conducted and the results for total deformation, equivalent elastic strain, equivalent von-mises stress and strain energy are shown in figure.

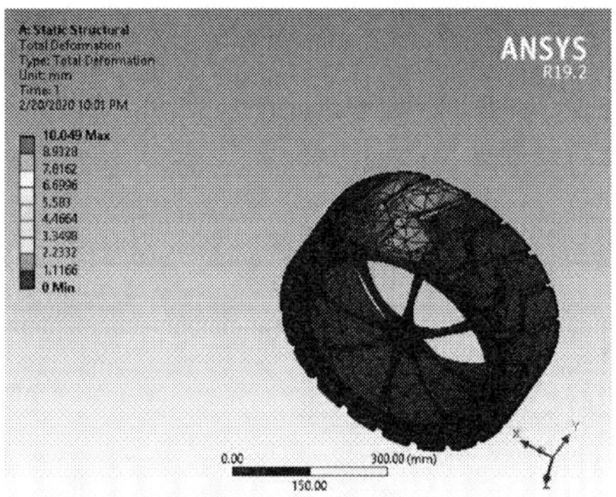


**Figure 15(a).** Total Deformation

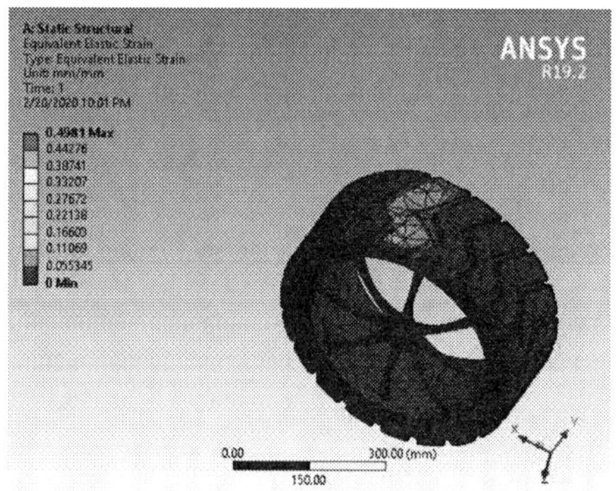


**Figure 15(b).** Equivalent elastic strain

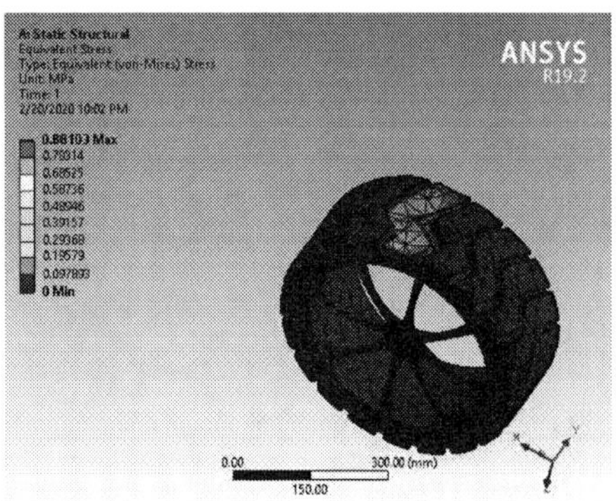


**Figure 15(c).** Equivalent (von-mises) stress

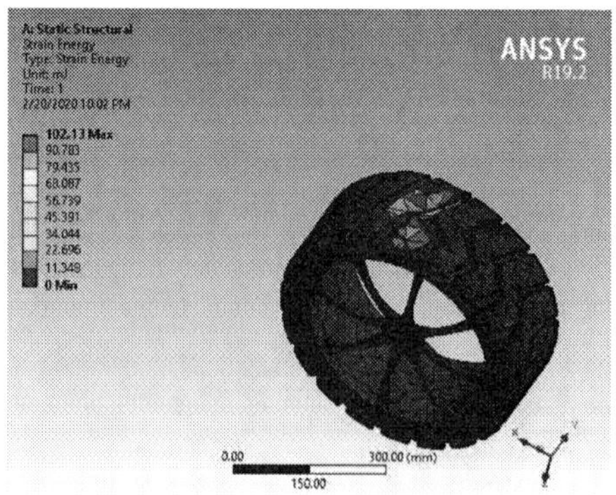


**Figure 15(d).** Strain energy

IOP Conf. Series: Materials Science and Engineering **923** (2020) 012017 doi:10.1088/1757-899X/923/1/012017

*5.6. Consolidated results.*
The following table shows the consolidated results of total deformation, equivalent elastic strain, equivalent von-mises stress and strain energy of airless and conventional tyre. It also gives a comparative analysis of all three materials used for airless tyre.

**Table 2.** Consolidated results of FEA

| S. No | Materials Used | Type of Analysis | Spokes structure | | |
|---|---|---|---|---|---|
| | | | Honeycomb | Diamond | Triangular |
| 1 | ABS +PC | Total Deformation (mm) | 0.286 | 0.053686 | 0.053703 |
| | | Equivalent Elastic Strain | 0.002904 | 0.000857 | 0.001128 |
| | | Equivalent (Von-mises) Stress (MPa) | 7.2855 | 1.9575 | 2.8193 |
| | | Strain energy (mJ) | 2.7747 | 0.45238 | 0.55302 |
| 2 | PET | Total Deformation (mm) | 0.24998 | 0.046625 | 0.046627 |
| | | Equivalent Elastic Strain | 0.002545 | 0.000724 | 0.000980 |
| | | Equivalent (Von-mises) Stress (MPa) | 7.3763 | 1.9676 | 2.8308 |
| | | Strain energy (mJ) | 2.4106 | 0.39311 | 0.48016 |
| 3 | HIPS | Total Deformation (mm) | 0.41301 | 0.078062 | 0.78137 |
| | | Equivalent Elastic Strain | 0.004185 | 0.001248 | 0.001639 |
| | | Equivalent (Von-mises) Stress (MPa) | 7.193 | 1.947 | 2.8079 |
| | | Strain energy (mJ) | 4.0323 | 0.65731 | 0.80406 |
| 4 | Conventional tyre | Total Deformation (mm) | 10.049 | - | - |
| | | Equivalent Elastic Strain | 0.4981 | - | - |
| | | Equivalent (Von-mises) Stress (MPa) | 0.88103 | - | - |
| | | Strain energy (mJ) | 102.13 | - | - |

From the tabulation, we understand the diamond spoke structure model gives better results than the other spoke model. Similarly in different material constraints, for all three materials the diamond spoke structure yields better result. So for the given boundary conditions and applied load, the diamond spoke structure airless tyre is more rigid and durable.

## 6. Conclusion

Use of air-less tyre eliminates disadvantages faced by pneumatic tyres. Air-less tyre provides uniform adhesion and uniform wear while absence of air. The airless tyres need not to be replaced very often and if it wears out and needed to be replaced the outer band of the airless tyre is replaced which saves material in manufacturing point of view. In pneumatic tyres, there was a high possibility of the flatten which can cause more number of accidents. Using airless tyres it's an outsized impact on safety. If the vehicles with airless tyres are used then one needn't worry about safety. Another benefit is that they're going to be reused and earlier tyres go flat and wish to be thrown away so there was the need for the event of airless tyres. Driving these vehicles reduces mind-stress. In this work there dimensional model of airless tyre with different spoke structures are generated and finite element analysis is carried out. From the FEA results of different structures, the airless tyre with diamond structure proves to be more durable than other structures and the conventional tyre.

In future the dynamic analysis of the same model will be performed and the outcomes are to be studied. The composite materials for 3D printing will be developed and by applying these material properties the model will be further studied for static and dynamic load conditions.

## References

[1] A.M. Aboul-Yazid, M.A.A. Emam, S. Shaaban and M.A. El-Nashar 2015, "Effect of spokes structures on characteristics performance of non-pneumatic tires", *International Journal of Automotive and Mechanical Engineering*, Vol.**11**, pp. 2212-23, January-June2015.

[2] Atefeh Aramian, Seyed Mohammad Javad Razav, Zohreh Sadeghian, Filippo Berto 2020, "A review of additive manufacturing of cermets", *Additive Manufacturing* **33,** 101130.

[3] Domingues J, Marques T, Mateus A, Carreira P, Malça C 2017, "An additive manufacturing solution to produce big green parts from tires and recycled plastics", *Procedia Manufacturing* **12**, 242 – 248

[4] Kaufui V.Wong and Aldo Hernandez 2012, "A Review of AdditiveManufacturing", *International Scholarly Research Network ISRN Mechanical Engineering*, Article ID 208760, 10 pages doi:10.5402/2012/208760

[5] Kucewicz, M., Baranowski, P., & Małachowski, J 2017. "Airless Tire Conceptions Modeling and Simulations", *Lecture Notes in Mechanical Engineering*, 293–301. doi:10.1007/978-3-319-50938-9_30

[6] Libin Rajan, ShobithNambiar, C. Ayyanar, Akash Verma,S. Subhash Raj 2019, "Design and Comparative Analysis of Non-Pneumatic Tires for a Tractor", *International Journal of Innovative Technology and Exploring Engineering*, Vol. **8** Issue-7

[7] Manibaalan C, Balamurugan.S, Keshore,Dr.Joshi.C.Haran 2013, "Static Analysis of Airless Tyre", *International Journal of Scientific and Research Publications*, Vol. **3**, Issue 8

[8] Nibin JacobMathew, Dillip Kumar Sahoo, MithunChakravarthy, E 2017, "Design and Static Analysis of Airless tyre to Reduce Deformation", *IOP Conference Series: Materials Science and Engineering*, **197**, 012042

[9] Periasamy K, S.Vijayan 2014, "Design and development of air-less car tire", *International Journal of Advances in Engineering & Technology*, Vol. **7**, Issue 4, pp. 1312-17

[10] Prajwal chatra, Seetharama K S 2017, "Analysis of Airless Tyre for Static Forces", *International Journal of Advance Mechanical and Mechanics Engineering*, Vol.**1** Issue 1 Page 16-21

[11] Vinay T V , Kuriakose J Marattukalam, Sachu Zachariah Varghese, Shibin Samuel 2015, "Modeling and Analysis of Non-Pneumatic Tyres with Hexagonal Honeycomb", *International Journal on Recent Technologies in Mechanical and Electrical Engineering*, Vol. **2**, Issue 3, 019 – 024

# Crash Analysis on Automobile Bumpers

**V. Sathish Kannan[1]*,J. Shenbaga Surendar[2],S C Meenakshi Sundaram[3],S. Jegadeeswer[4],R.Venkatesh[5]**

[1,2,3,4,5]Department of Mechanical Engineering, St. Joseph's Institute of Technology, OMR Chennai-600 119,India

*E-mail: neyvelisathish007@gmail.com

**Abstract**: Bumper is one of the essential parts in an automobile. During collision, bumper is used as a protection shield to safeguard passengers and the body of the vehicle. We know that, death in accident is very common not only in India but also all over the world. Therefore, safety of passengers and automobile body must be considered as primary criteria at the time of vehicle design. ABS is a material which has high impact strength which is used for manufacturing automobile bumper. In this research, we endeavour to analyse three different front bumpers of expensive cars which are highly competitive in the automobile market. CATIA V5R21 is the software used for modelling the bumpers and ANSYS 19.2 is used for analysing each bumper with appropriate velocities. In this research, thickness of all the bumpers is considered as 5 mm and the concrete wall where it is collided to make impact collision is of thickness 10 mm. By using explicit dynamic condition with exact boundary conditions, we determine the best or optimistic design by comparing the results of various factors such as equivalent stress, total deformation, directional deformation and equivalent plastic strain.
***Keywords:*** ABS (Acrylonitrile butadiene styrene), ANSYS19.2, CATIAV5R21, Analyse, Stress, Deformation

## 1. Introduction

In olden days vehicles are mostly used for transportation .As the years pass on vehicle becomes more essential member of a family. They consider vehicle has an additional assert for them so they are willing to buy an expensive vehicle .Nowadays automobile industries are updating themselves based on the customer satisfaction. Hence they improved their vehicle standards to become popular in the market .While buying a vehicle customer gives a high priority to safety features bumpers are one of the their essential safety features in automobile .during impact collision it prevent the damage of front and rear end .A good bumper design will have better aerodynamic property .Different companies make different bumper design to show their uniqueness.

The role of automotive bumpers has changed considerably over the past 70 years. The later performance is achieved by a combination of careful design, material selection to obtain a particular balance of stiffness, strength and energy absorption. Stiffness and Energy absorption are essential criterion. Stiffness is important because vehicle design consideration limits the packaging space for the bumper design to deform under load and Energy absorption is important because bumper must limit the amount of the impact force transmitted to the surrounding rails and vehicle frame. Automotive bumper plays a very important role in absorbing impact energy (original purpose of safety) and styling stand point aesthetic purpose. Now a day, automotive industry concentrates on optimization of weight and safety.

### *1.1. Standards for bumper*
As bumper improves the aesthetic look of an automobile and also its design need to satisfy the roads condition. If bumper ground clearance is good it can travel easily in all road condition

### *1.2. India*
India is the 10$^{th}$ largest producer of automobile in the world .In the year 2016 our country has registered 10 million trucks and 230 million cars . In 1989, the Central Motor Vehicle Rules (CMVR) became effective. Under Rule 126 of the CMVR, each manufacturer must pass the test for their original prototype .All the vehicle to be launched need to satisfy pollution control norms .generally Indian vehicle manufacturer can easily design based on it but foreign vehicle need a slight variation to meet with Indian standard.

### *1.3. Features of best bumper*
- Good aerodynamic course
- Light in weight
- Easy to assemble
- Withstand load during impact collision
- Aesthetic design

### *1.4. Types of bumpers*
- Steel bumper
- Carbon fibre bumper
- plastic bumper

#### *1.4.1. Steel Bumper*
It can easily withstand impact load without complete deformation so the stress is induced inside the vehicle .It is tedious to manufacture it is also heavier in weight .Alloying of steel can improve the properties.

#### *1.4.2. Carbon fibre bumper*
It is highly expensive to manufacture and light in weight .it can easily withstand load compare to other material but it cannot be reformed

#### *1.4.3. Plastic Bumper*
Nowadays plastics are used to make bumper because they are lighter in weight and easy to manufacture .It is also cost efficient .It can be easily recyclable .It can also easily withstand impact load.

#### *1.4.4. ABS Material in Automobile Bumpers*
Early in 1870s bumpers were used as safety features in vehicles and those bumpers were made of steel and aluminium alloys. But those materials were so heavy to be recycled so the industries were searching for a new light weight material which has high strength to withstand the impact load and also capable of being recycled. In 1950s Automobile industries started manufacturing bumpers using polymers and plastics. Later on they discovered a new material called Acrylonitrile Butadiene Styrene (ABS) which has high impact strength to withstand collision.

In this research we have made both the extreme ends of the wall as fixed support and the bottom two layers of the model is also as displacement support and right and left edge of the model is also consider as displacement support.

## 2. Literature Survey

The bumper has been designed with different materials and analysis has been performed on the designed bumper in order to find out which has better durability. Previously crash analysis is performed on the cylindrical block with different velocities and they have also used only gravitational pull with friction coefficient zero for performing the crash test. In this project we have assumed a concrete wall with thickness 30mm and crash test is performed on the front bumper with its bottom layer as the fixed point.

## 3. Design of Bumper

The bumper has been designed using CATIA V5 with two different types with ABS material. The surface thickness of the bumper is 5mm and a concrete wall of 30mm thickness and bumper is made to be in contact with the concrete wall.

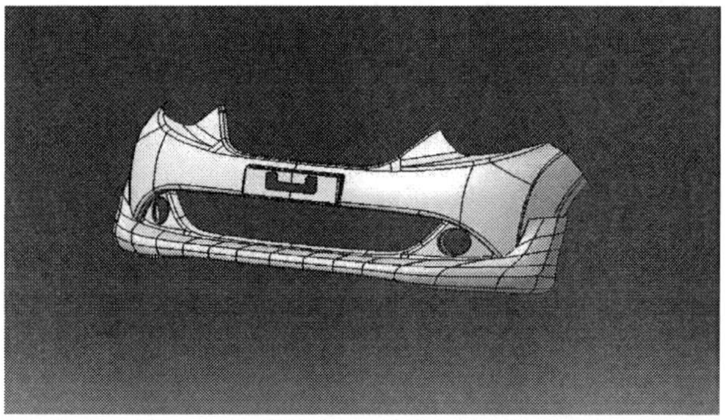

**Figure1:** Design of Bumper type 1

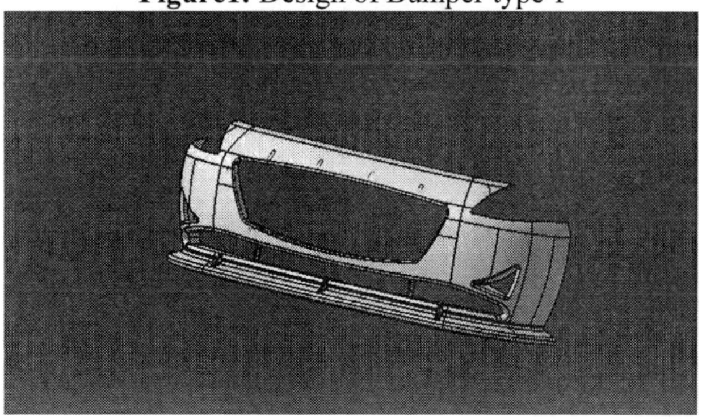

**Figure2:** Design of Bumper type 2

## 4. Boundary Conditions

The two ends of the concrete wall are fixed and the bumper bottom two layers are assumed to be a displacement support

The thickness of concrete wall is 30mm and thickness of bumper is 5mm and the velocity is assumed to be 55 km/hr

## 5. Results

### *5.1. Analysing of Bumper 1*

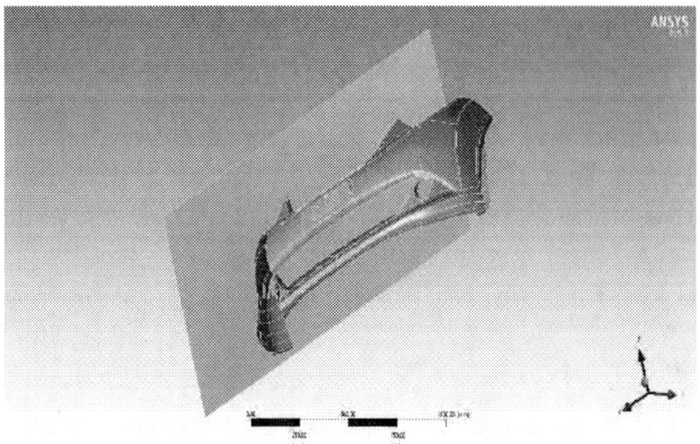


**Figure3:** Geometry of Bumper 1 in ANSYS 19.2

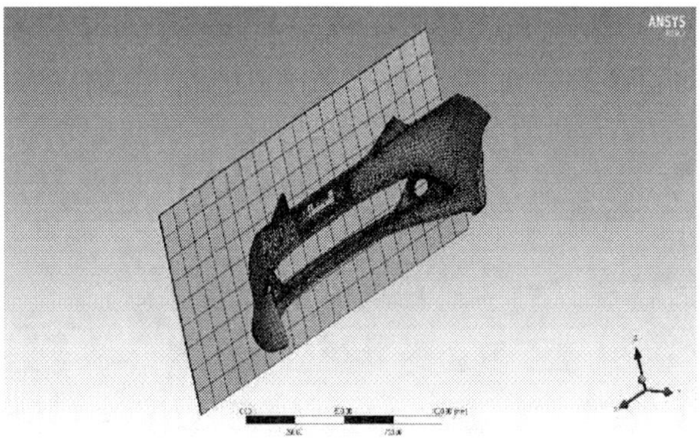


**Figure4:** Meshing of Bumper 1

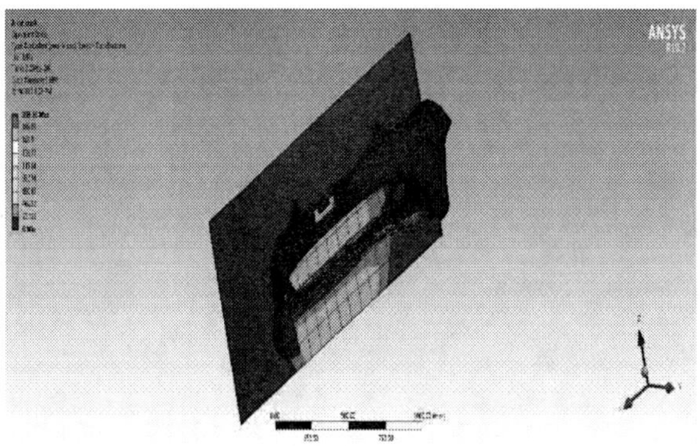


**Figure5:** Equivalent Von mises stress on Bumper 1

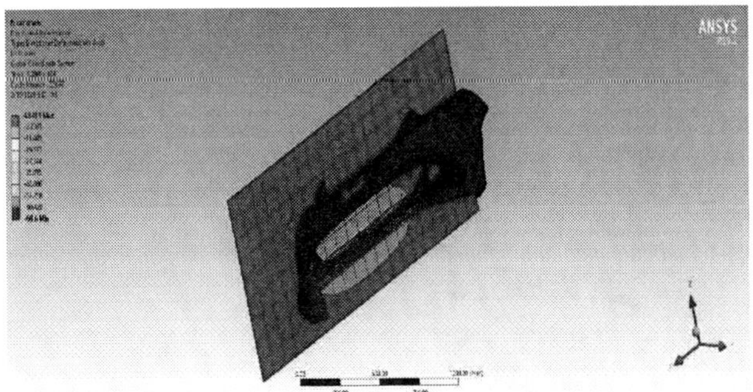


**Figure6:** Directional Deformation of Bumper 1

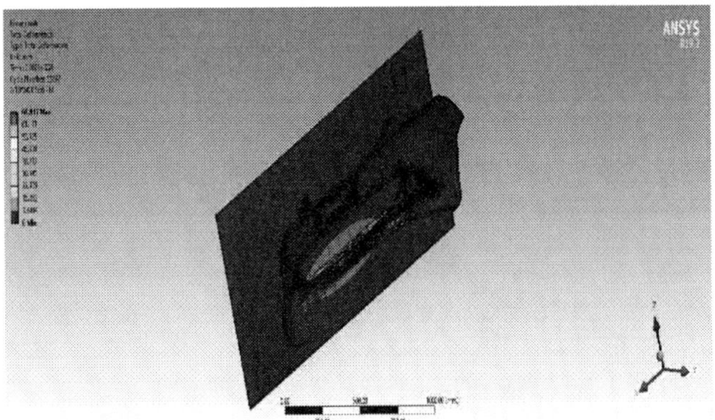


**Figure7:** Total Deformation of Bumper 1

*5.2. Analysing of Bumper 2*

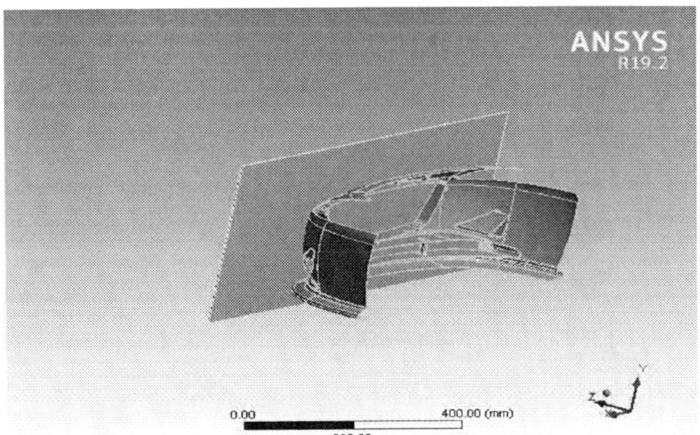


**Figure8:** Geometry of Bumper 2 in ANSYS 19.2

IOP Conf. Series: Materials Science and Engineering **923** (2020) 012018 doi:10.1088/1757-899X/923/1/012018

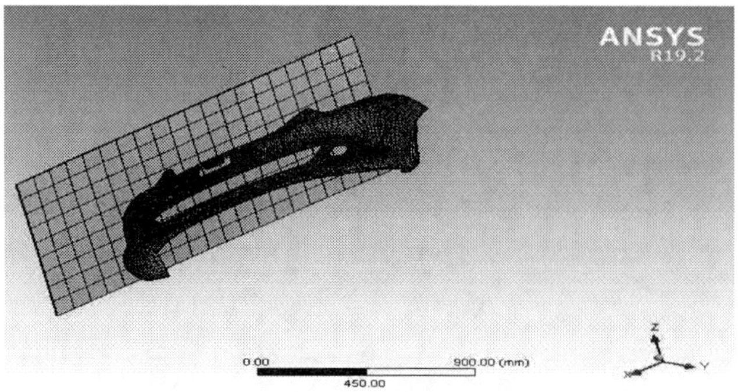


**Figure9:** Meshing of Bumper 2

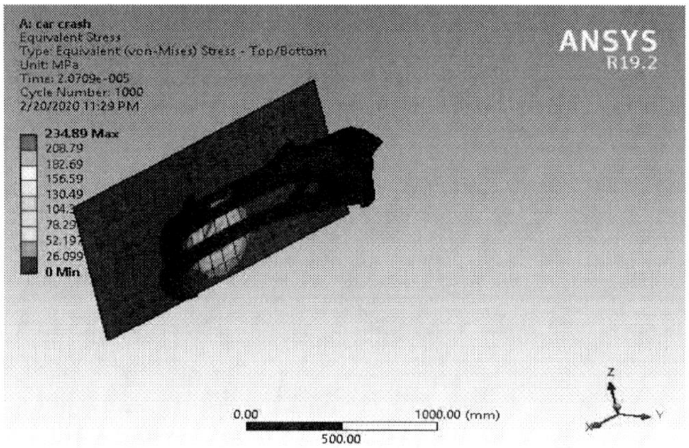


**Figure10:** Equivalent Von mises stress on Bumper 2

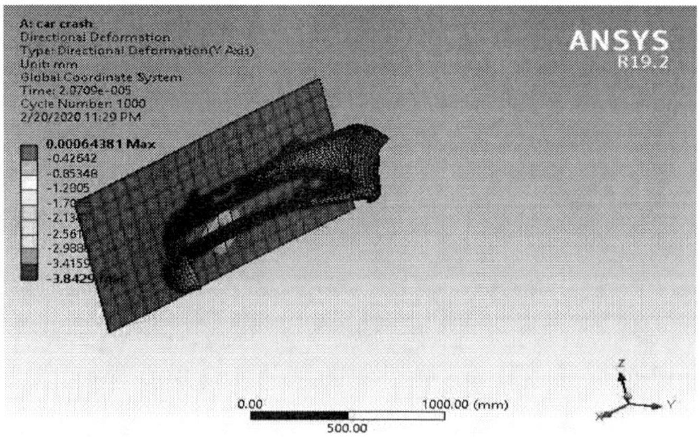


**Figure11:** Directional Deformation of Bumper 2

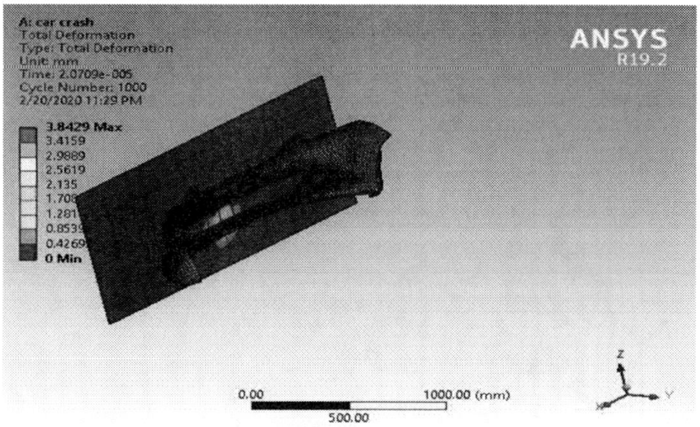


**Figure12:** Total Deformation of Bumper 2

## 6. Bar Chart

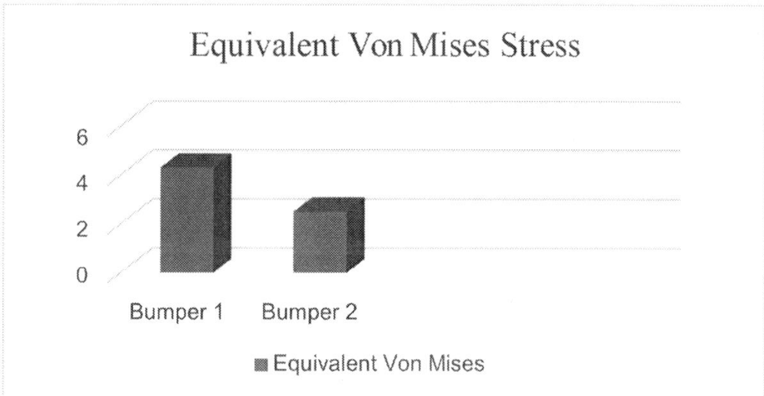


**Figure13:** Comparison of Von-Mises stress between two bumpers(N/mm$^2$)

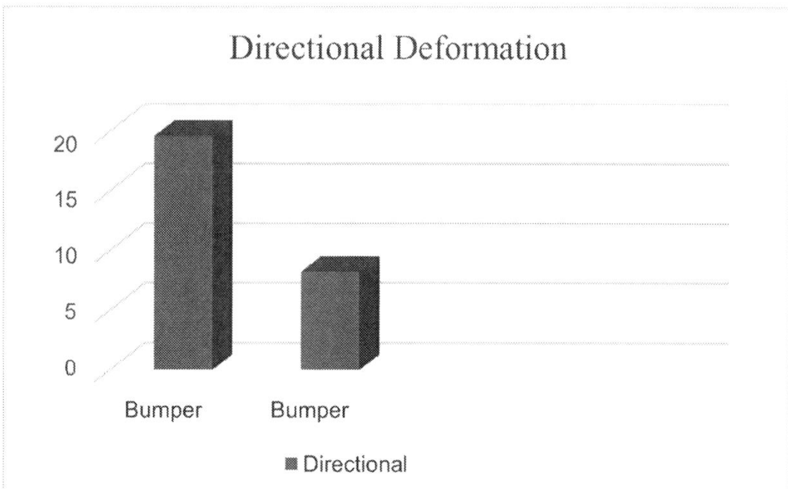


**Figure14:** Comparison of Directional deformation between two bumpers

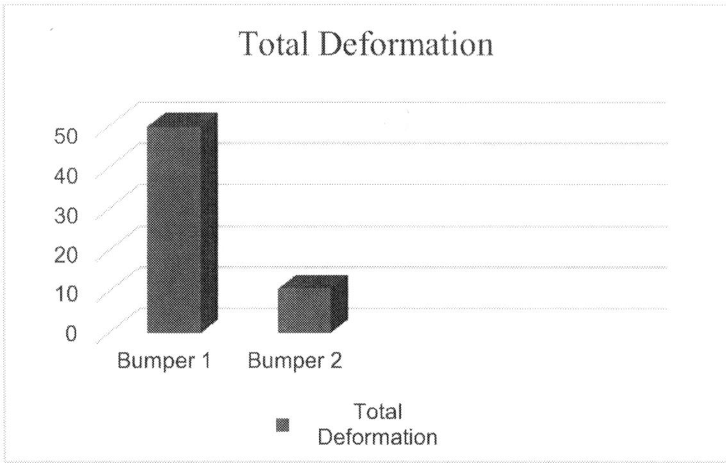


**Figure15:** Comparison of Total deformation between two bumpers

## 7. Conclusion

Thus from the above plotted results hereby we conclude that bumper 1 has more durability and highly safer than bumper 2. From this we have understand that the design is more essential factor for a product effectiveness. In this we have learnt that for a simple variation in design which made a huge difference in product life

## References

[1] M. Anil Kumar N. Phani Raja Rao "Design and Analysis of Car Bumper by Varying Materials and Speeds"- *International Journal on Recent Technologies in Mechanical and Electrical Engineering*,**Vol.2,Issue 3**,pp.56-61

[2] Sumeet Kumar Shukla, Dr.Suman Sharma "Impact Analysis of Car Front Bumper to Enhance Crashworthiness"- *International Journal of Engineering trends and Technology*-**Vol.46,Issue 5**,pp.257-261

[3] K.Suneeta,A.Ramanjaneya Reddy,P.Hussain,B.Sidda Reddy,S.Sudhakar Babu "Design and Analysis of Car bumper"-*International Journal of Mechanical Engineering Research and Technology*-**Vol.2,Issue 2**,pp.30-45

[4] Gandla Pradeep,P.Chandra Kumar "Design and Experimental Analysis on Car Bumper with Composite Materials"-*International Research Journal of Engineering and Technology*-**Vol.05,Issue 10**,pp.54-58

[5] M.Meghana,Ch.Shashikanth,M.Pradeep kumar "Impact Analysis of Bumper and Car Chassis Frame Due to Frontal Collision for Different Materials"-*International Journal of Engineering and Management Research*,**Vol.6,Issue 1**,pp.131-135

[6] Marzbanrad JM, Alijanpour M, and Kiasat MS, "Design and analysis of automotive bumper beam in low speed frontal crashes", Thin Walled Struct., **47** (2009),pp.902- 911.

[7] P.Sampath Roa,S.Saikumar "Design Analysis of Bumper Beam Subjected to Offset Impact Loading for Automotive Applications"-*International Journal of Mechanical Engineering & Technology*-**Vol.6,Issue 5**,pp.64-71.

IOP Conf. Series: Materials Science and Engineering **923** (2020) 012019 doi:10.1088/1757-899X/923/1/012019

# Design and Finite Element Analysis of Prosthetic Hand Controlled by Wireless Gestures for Differently-abled People

**G Shanmugasundar[1], M Dharanidharan[2], D Vishwa[3], A Jayaprakash[4], P Abimanyu[5]**

[1] Associate Professor, Department of Mechanical Engineering, Sri Sai Ram Institute of Technology, Chennai-600 044, India.
[2,3,4,5] UG-Student, Department of Mechanical Engineering, Sri Sai Ram Institute of Technology, Chennai- 600 044, India.

**Abstract.** This paper focuses on the design and finite element analysis of the prosthetic hand for differently-able people. The prosthetic hand consists of two sections: one is the transmitting section and another one is the receiving section. The transmitting section consists of a glove embedded with a series of flex sensors connected with a microcontroller and a wi-fi module. The flex sensors are fixed along with each finger which captures and converts the motion of the fingers into analog signals and is transmitted to the receiving section through the wi-fi module. The receiving section consists of the prosthetic hand in which each finger is actuated by the servo motors. The transmitted signals are received by the wi-fi module in the receiver section through the microcontroller and are converted into digital signals. The fingers of the prosthetic hand are actuated by the servo motors corresponding to the received digital signals. The methodology involved in this research is the designing, meshing, and analyzing of a prosthetic hand using the PTC Creo software and then assembling electronic components. The analysis of the prosthetic hand involves the analysis of equivalent stress and total deformation at various regions of the prosthetic hand by selecting steel as a material to fabricate.

## 1. Introduction

Robotics is a multidisciplinary field that is a combination of mechanical, computer science, electrical and electronics fields. Nowadays, robots have replaced many works which involve more human effort and it also improves our lifestyle. The word 'Prosthetics' refers to the artificial system which is similar to that of the human body part [1]. In recent years, the robotics field makes many changes in the prosthetics area. The hand, one of the most important and complex parts of our human body can be replicated as a prosthetic hand by using the robotic systems. Several kinds of research have been carried to day-to-day to develop an effective prosthetic hand. Our research aims to design a prosthetic hand which is affordable to all customers by using easily available electronics components [1]. The prosthetic hand is designed and then it is meshed and analyzed by using PTC Creo software to find the stress, strain, and total deformation at various regions of the prosthetic hand. Not only for differently-abled persons, but the enlarged model of this design can also be used in industries to handle heavy, harmful, and high-temperature objects. The prosthetic hand may help the differently-abled people to do their day-to-day activities like taking, holding, etc., and helps them live a normal life like other people.

## 2. Literature review

There are many types of research have been carried out on the prosthetic arm around the world till now. The prosthetic hand and arm design researches emerged in the late 1980s. In 1986, S. Jacobsen et al. of Utah university and MIT [2] have developed a simple prosthetic hand consists of three fingers and one thumb and it was used to observe the mobility of machines for construction purposes. Robert Vinet et al. [9] designed a multifunctional prosthetic hand by using the four-bar mechanism for the individual fingers in 1995. They analyzed the angle of flexion for relative motions of the hand and analyzed the mechanism. Hun-Pang Huang and Chun-Ying [8] Chiang used a digital signal processor (DSP) to analyze the EMG signals for distinguishing the pattern recognition from the hand motions in 2000 and it provided a correct value of about 87.5% for the hand motions. In early 2002, M. C. Carrozza et al. [5] have developed a novel prosthetic hand based on bio-mechatronics design. The proposed hand is designed to have enhanced grasping capabilities by involving various mechanisms, actuators, sensors, and embedded systems while maintaining approximately the same dimension and weight.

In 2005, Stefan Schulz et al. [4] designed a hand having 15 degrees of freedom controlled by small flexible fluid actuators. The fluid actuators are placed within the hand and it can handle different objects reliably and naturally. Achintya Das et al. [10] designed a prosthetic hand by using micromotors and Programmable Interface Controller (PIC) in 2010 which enables them to do appropriate linear movements and they used Lagrangian function of the motion of the finger to study their model. Also, Arindam Chatterjee et al. [7] developed a new type of prosthetic hand in 2010 which was controlled wirelessly by using a remote control system. The remote control consists of four switches for doing four different operations by applying different grip forces. In 2017, Punit Kumar Rohilla et al. [3] conducted an experimental analysis of the prosthetic hand by using different materials by using ANSYS. They concluded that Nylon 6 has a suitable material for manufacturing the prototype of the prosthetic hand. Moin Uddin Atique [6] designed a cost-effective myoelectric prosthetic hand in 2018 for wrist amputees instead of manipulating all the fingers which is complex and expensive. The microcontroller interface detects the EMG signals and it converts into motion by using several DC gear motors.

## 3. Design of prosthetic hand

The prosthetic hand has been designed similar to the actual human hand dimensions. The 3D model of the prosthetic hand has been created with PTC Creo. The designed model has 14 degrees of freedom, that is, every 4 fingers will have 3 degrees of freedom and the thumb will have 2 degrees of freedom. The sections of the fingers are connected by a pin rotary joint and the finger sections are also designed accordingly to avoid unusual finger movements. The sections of the finger are connected to the servo motor through a nylon wire rope. The front and rear side of the designed prosthetic hand is shown in figure 1 and figure 2.

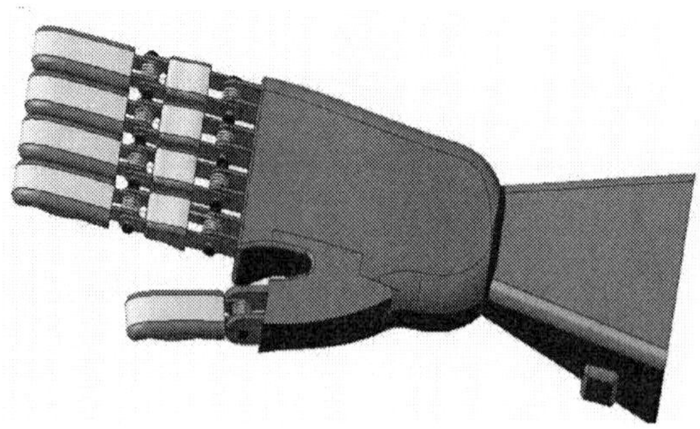

**Figure 1.** The front side of the Prosthetic hand

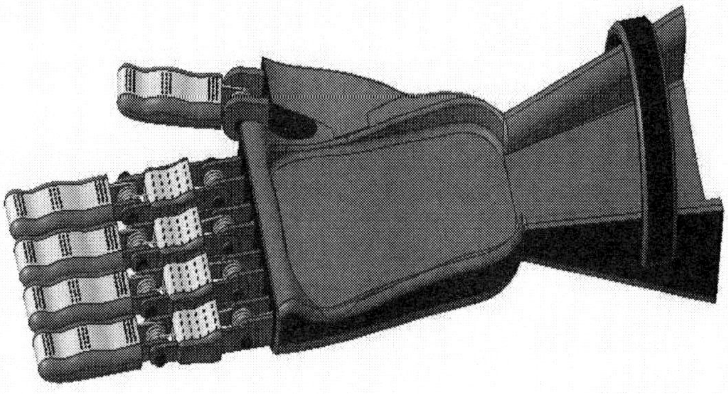

**Figure 2.** The rear side of the Prosthetic hand

## 4. Analysis of prosthetic hand

The above designed prosthetic hand has been done finite element analysis by using the PTC Creo software. The static stress analysis has been done to the individual fingers and the palm section of the prosthetic hand. Static analysis is defined as the analysis of stresses, strains, deformations, and the forces acting in the components due to the applied loads without including the inertia and damping effects. The prosthetic hand is analyzed for the equivalent stress (von-misses stress) and total deformation. The analysis has been done by selecting the material as steel and the results of the analysis are tabulated below. A load of 30 N is applied to the prosthetic hand and the resultant stress and deformation at various points are analyzed.

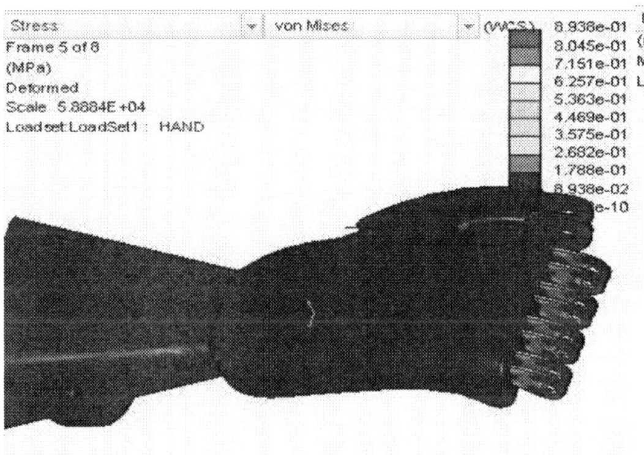


**Figure 3.** Equivalent stress analysis

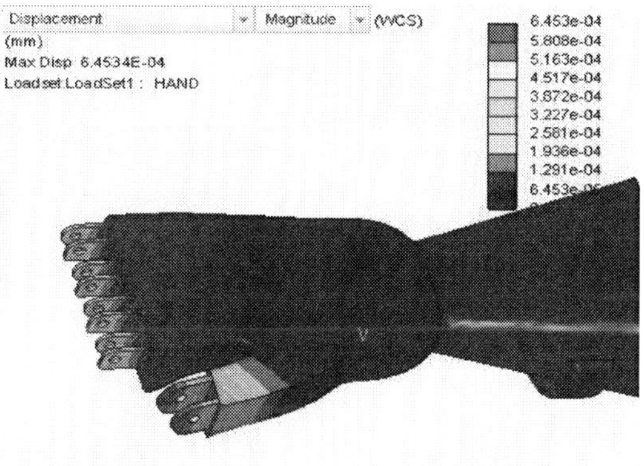


**Figure 4.** Deformation analysis

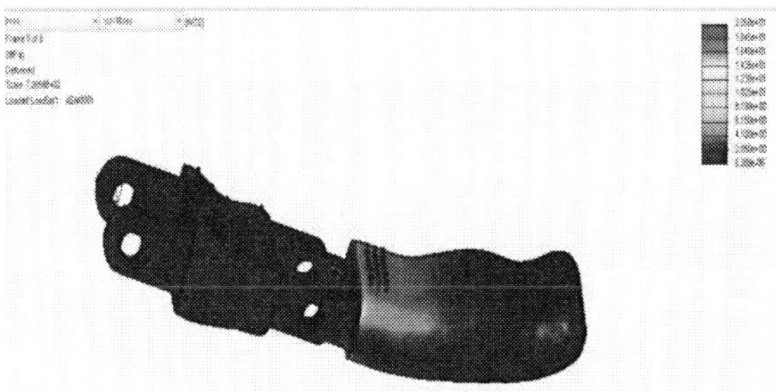

**Figure 5.** Equivalent stress analysis

**Figure 6.** Deformation analysis

Figure 3 shows the equivalent stress analysis or von mises stress analysis of the palm section of the prosthetic hand and figure 4 shows the total deformation analysis of the palm section. Figure 5 and figure 6 show the equivalent stress analysis and deformation analysis of the individual finger. As all the fingers design are the same, the above-analyzed result will apply to all the fingers.

## 5. Results of Analysis

From the above analysis of the prosthetic hand, it is found that the maximum equivalent stress of 0.8938 Mpa is acting on the palm section. The equivalent stress is maximum at the joints which connect the fingers with the palm. The deformation of about $6.453\times10^{-4}$ mm occurs on the palm section at the joint which connects the thumb with the palm section. In the fingers, the maximum equivalent stress of about 0.205 Mpa is acting on the fingers, and the maximum deformation of about 0.023 mm acts on the edges of the fingers. From the analysis, the design is safe to handle the loads which are handled by the human hands. The below table 1 compares the analysis results of the prosthetic hand at various sections.

**Table 1.** Analysis results of the prosthetic hand

| S.no. | Parameter | Value |
|---|---|---|
| 1. | Maximum equivalent stress at the palm section | 0.8938 Mpa |
| 2. | Maximum deformation at the palm section | $6.453\times10^{-4}$ mm |
| 3. | Maximum equivalent stress at the individual fingers | 0.205 Mpa |
| 4. | Maximum deformation at the individual fingers | 0.023 mm |

## 6. Circuit Connection

The prosthetic hand consists of two sections: one is the transmitting section and another one is the receiving section. In the transmitting section, a series of five flex sensors embedded in each finger of the glove is connected to the microcontroller and a wi-fi transmitter is also connected to the microcontroller. In the receiving section, the prosthetic hand which is designed previously is connected to five servo motors to actuate the five fingers by using the nylon braid wires. The servo motors are connected to the microcontroller in which the battery and wi-fi receiver is also connected. Figure 7 and figure 8 shows the circuit connection of the transmitting and receiving section respectively.

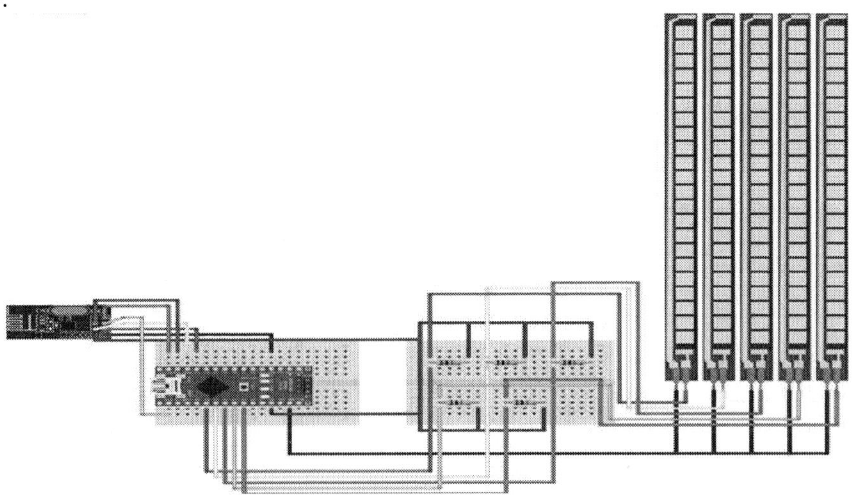

**Figure 7.** Circuit of Transmitting section

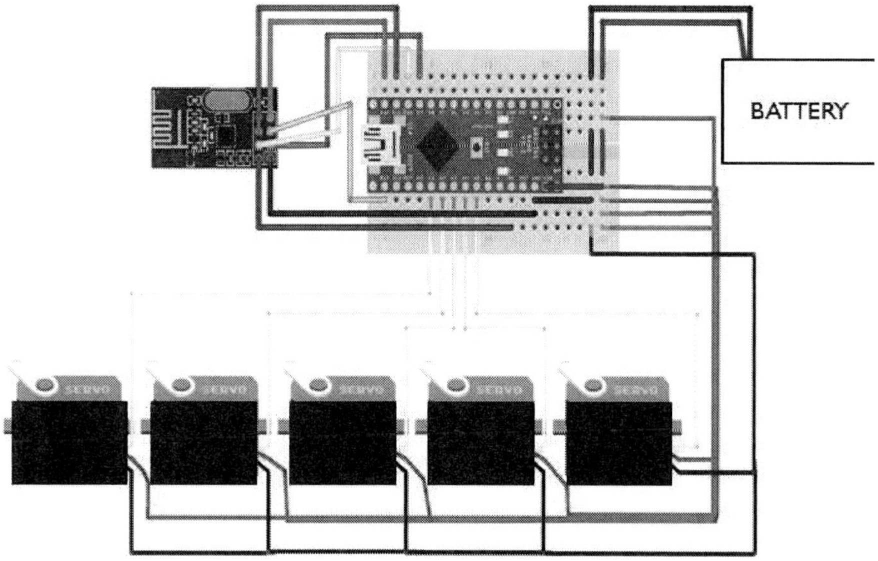


**Figure 8.** Circuit connection of receiving section

## 7. Working Procedure and Result

The person has to wear gloves in one hand and the prosthetic hand should be inserted into another hand. When the person moves the fingers of the hand in which the transmitter section attached, the flex sensor bends which generates an analog signal and it is transmitted to the receiver section through the microcontroller and wi-fi transmitter module. The analog signals are received by the wi-fi receiver module and it is processed and converted into digital signals by the microcontroller. The digital signals are used to actuate the servo motors which are attached to the prosthetic hand by using nylon braid wires. Thus, the gestures of one hand are converted into the prosthetic hand movement by using the flex sensors, servo motors, microcontroller, and wi-fi module. The flow chart of the working procedure is shown in figure 9.

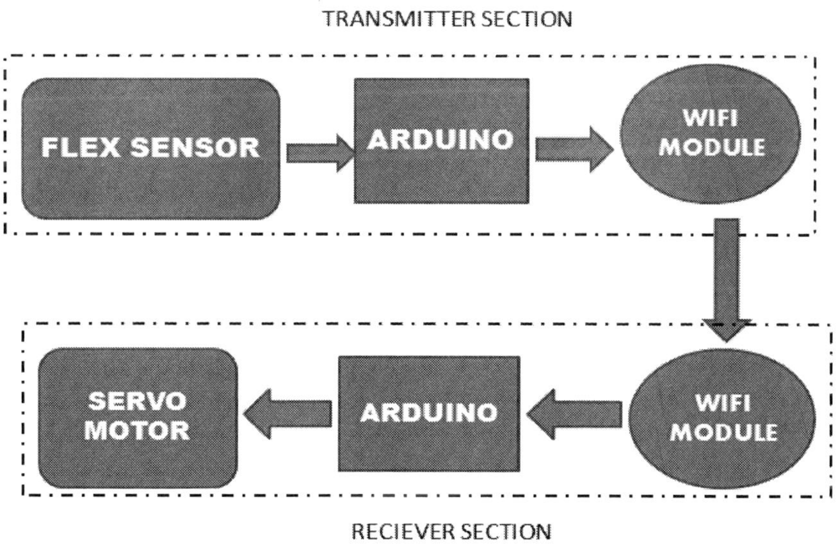


**Figure 9.** Flow chart representing the working procedure of prosthetic hand

## 8. Conclusion

This research work is about the design and finite element analysis of prosthetic hand controlled by wireless gestures for differently-abled people. This work clearly explains the key concepts which are involved in the conversion of hand gestures in one hand into the movements of prosthetic hand by using flex sensors, servo motors, microcontroller, and wi-fi module. From the analysis, the equivalent stress and total deformation of the prosthetic hand have been calculated and it is also found that steel is suitable to fabricate the prosthetic hand. The prosthetic hand is designed for differently-abled persons to do their daily activities. Not only for differently-abled persons, but it can also be used to handle harmful, hazardous objects in industries.

## 9. References

[1] Shruthi K 2018 Low-cost prosthetic hand using flex sensors and servo motors *International Journal of Engineering Research & Technology (IJERT) NCESC – 2018* Vol 6 Issue 13.

[2] S. Jacobsen, E. Iversen, D. Knutti, R. Johnson and K. Biggers 1986 Design of the Utah/M.I.T. Dextrous Hand *IEEE International Conference on Robotics and Automation* pp 1520-1532.

[3] Rohilla P V, Suresh B, Dinesh S, Nitin P, Sudip S and Angana M S 2017 Material selection for the prosthetic hand *International Journal of Biomedical Engineering* Vol. 3 Issue 1.

[4] Schulz, S P, Christian R, Markus M, Jan M and Ralf B 2005 A hydraulically driven multifunctional prosthetic hand *Robotica* Vol 23 pp 293-299.

[5] Carrozza M C, Bruno M, Silvestro L, R Zecca and Massimiliano Dario 2002 The development of a novel prosthetic hand *IEEE/ASME Transactions on Mechatronics* Volume 7 pp 108 - 114.

[6] Moin R, Atique and Khondkar 2018 A cost-effective myoelectric prosthetic hand *Journal of Prosthetics and Orthotics* Vol 30 pp 231-235.

[7] Chatterjee, Arindam B, Amardeep K, Amod G, Kanta A and Ravinder 2010 Design of wireless prosthetic hand *Conference on Emerging Medical Instrumentation.*

[8] Han-Pang H and Chun-Ying C DSP-based controller for a multi-degree prosthetic hand *IEEE International Conference on Robotics and Automation* Vol 2 pp 1378-1383.

[9] Vinet R, Y Beaudry and Nicolas D G 1995 Design methodology for a multifunctional hand prosthesis *Journal of rehabilitation research and development* Vol 32 pp 316-324.

[10] Neogi B, Tibarewala D N and A Das 2010 Design and realization of prosthetic hand and its free movement *International Journal of Systemics, Cybernetics and Informatics* pp 973-4864.

[11] G. Shanmugasundar, P. Jagadeeshwar, S. Adithya, V. Nagappan, and M. Bhaskar, "Design, fabrication and analysis of personal vacuum assisted climber," in *Journal of Physics: Conference Series*, 2019, vol. 1362, no. 1, doi: 10.1088/1742-6596/1362/1/012057.

[12] G. Shanmugasundar, R. Sivaramakrishnan, S. Meganathan, and S. Balasubramani, "Structural optimization of an five degrees of freedom (T-3R-T) robot manipultor using finite element analysis," in *Materials Today: Proceedings*, 2019, vol. 16, doi: 10.1016/j.matpr.2019.05.231.

[13] G. Shanmugasundar, R. Sivaramakrishnan, and S. Venugopal, *Modeling, design and static analysis of seven degree of freedom articulated inspection robot*, vol. 655–657. 2013.

[14] G. Shanmugasundar and R. Sivaramakrishnan, "Computer aided modelling and simulation of a generic robot for inspection," *Int. J. Appl. Eng. Res.*, vol. 10, no. 24, 2015.

IOP Conf. Series: Materials Science and Engineering **923** (2020) 012020 doi:10.1088/1757-899X/923/1/012020

# Pressure Vessel Design by Design by Analysis Route

K M B Karthikeyan[1], T Balasubramanian[2], A Ruskin Bruce[3], P Premkumar[4].
[1, 2, 3, 4] *Assistant Professors, Mechanical Department. St. Joseph's College of Engineering, India*
*kmbkarthekeyavarshan@gmail.com*

**Abstract:** Pressure vessels are conventionally designed based on the Design by Rule (DBR) approach. Standard vessel geometries are designed using simple formulae and charts, formulated as standards and codes based on the rules proposed in the ASME Boiler and Pressure Vessel Code and the European Standard prEN13445-3. The stress categorization problem faced employing DBR methodology was overcome by employing Limit Load Analysis using the DBA approach. A simple unfired vertical cylindrical pressure vessel with torispherical pressure head and Y-forged skirt support was analyzed for its structural integrity using ANSYS software. The results obtained for two different materials, high strength steel P500-QT and the conventional low strength steel P355 employing DBA approach was compared with the calculated results using the DBR approach. The maximum allowable pressure ($P_{MAX}$) for P500-QT material was found to be 2.5X the maximum pressure calculated using the DBR approach as per ASME codes and 3X times as per EN13445-3 standard. For P355 material the maximum allowable pressure ($P_{MAX}$) was found to be 2.32X the maximum pressure calculated using the DBR approach as per ASME codes and 2.6X times as per EN13445-3. The DBR methodology proves to be highly conservative whereas the use of DBA leads to less conservative design parameters.

## 1. Introduction

Pressure vessels are employed as storage vessels, industrial processing chambers, reactor pressure vessels, space tanks, and integrated thermal units. Conventionally pressure vessels are designed using standard codes framed by international agencies like ASME BPV- Section VIII, Div 1 & 2 and European Standard prEN13445-3. DBR is a highly conservative and uneconomical approach, which considers safety as prime design criteria. Nevertheless, design by rule (DBR) approach is still dominant. Stress categorization route also referred to as the DBR approach is employed to calculate the primary, secondary and peak stresses reflecting the nature of their failure mechanisms such as gross plastic deformation (GDP), fatigue failure and ratcheting [1] [7]. The design of pressure vessel warrants for an explicit understanding of the material response, operating stresses and loads, external influence, failure modes, construction details, detection equipment, risk control mechanisms and feedback controls for the structural integrity of the vessel [2].

Elasto-plastic or non-linear analysis, considering geometric discontinuities using commercial FE software such as ANSYS, will be useful in determining the limit load capability of the pressure vessel [3-4]. The limit load is defined as the maximum load beyond which the pressure vessel collapses. In the textbooks, frequent reference is made to ASME Section VIII Division-1 and Division-2. In European countries prEN 13445-3, Annex-B, Direct Route for Design by Analysis (DBA) and EN 13445-3: Unfired pressure vessels - Part 3: Design, Annex C [5] for stress categorization (SCL) route are followed for designing pressure vessels [5]. DBR is based on a set of simple formulae used to determine either the minimum thickness or the maximum allowable working pressure for different pressure load conditions [3]. DBA allows approximation techniques like finite element methods, which can complement [7] DBR in the design of pressure vessels.

## 2. Design by Analysis (DBA) Methodology

The DBA approach is further classified into Stress Categorization (SCL) – indirect route and Design by Analysis (DBA) –direct route using FE codes.

### *2.1. DBA using Commercial FE Software ANSYS*

There are two basic hardening rules used by ANSYS to prescribe the modification of the yield surface. For Kinematic hardening, the yield surface remains constant in size and translates in the direction of yielding as shown in figure.1.(a). For Isotropic hardening, the yield surface expands uniformly in all directions with the plastic flow as shown in figure.1.(b). The Proposed study is confined to rate-independent multilinear kinematic hardening (MKIN).

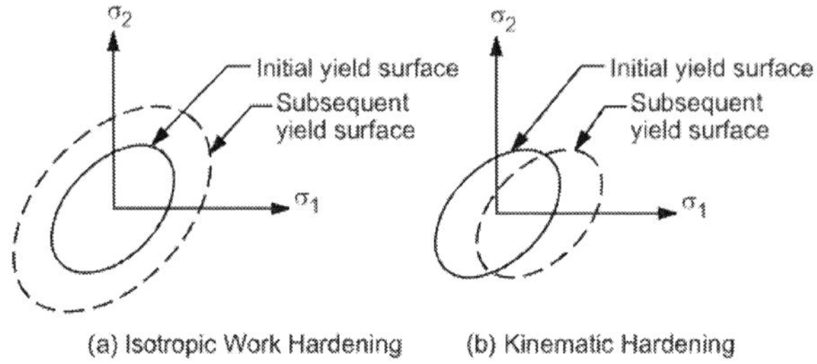


Figure.1. Types of Hardening Rules used in ANSYS [8]

### *2.2. Problem Formulation and Methodology*

The study investigates the simple unfired vertical cylindrical pressure vessel (shell) with Y-forged skirt support and torispherical pressure head, analyzed for structural integrity using ANSYS elastic analysis for stress categorization and in-elastic analysis for limit load capability analysis. The high strength pressure vessel steel P500-QT and a low alloy steel P355 are taken for investigation [7]. The proposed methodology is presented in figure.2. Secondary loads like nozzle cylinder interactions, nozzle moments, discontinuity forces, accidental loads and temperature effects are not considered for the analysis. The design specification is listed in table.1, 2 respectively.

## 3. Design of Pressure Vessel by DBF Approach

### *3.1. Design of Cylindrical Shell*

Allowable design stress (f or S) as per ASME code is denoted by equation-1. Allowable design stress (f or S) as per prEN 13445 -3 European standards is denoted by equation-2. The Thickness of the cylindrical shell is calculated using equation-3. The axial stress ($\sigma_{xy}$) was found using equation-4. Design specifications and results are presented in the table.3, 4.

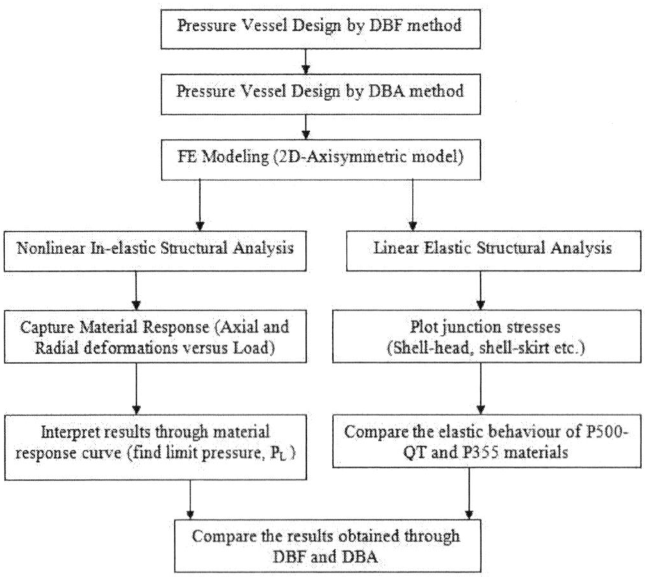


Figure.2. Methodology: Pressure Vessel Design by DBA Route

Table.1 Pressure Vessel Design Specification [7]

| | |
|---|---|
| Component | Cylindrical Shell, Head, Y-Forged Skirt Support |
| Material | P500-QT, P355 Pressure Vessel Steels |
| Design Pressure, P | P500-QT : 8.25 MPa; P355: 8.25 MPa |
| Design Temperature | 120°C |
| Inner Diameter of Vessel / Shell, D | 2900 mm |
| Cylindrical Shell Inner Radius, R | 1450 mm |
| Maximum Allowable design Stress, S or f | P500-QT : 299 MPa; P355: 250 MPa |
| Load Case | Internal Pressure, P, MPa |
| Corrosion Thickness | 2 mm to 4 mm maximum |
| Insulation thickness | Not Required (Unfired Vessel Application) |
| Pressure Head Type | Torispherical Head |
| Skirt type | Y-Section Support (Forged Section) |
| Skirt height | 1000mm |
| Overall height of Vessel | 5800mm |
| Fluid /Application | Hydrocarbon / Storage |
| Joint Efficiency, E | 100% assumed |

Table.2 Material Properties of Commercially Available Pressure Vessel Steels P500-QT and P355

| Material Properties | P500-QT | P355 |
|---|---|---|
| Young's modulus ,E (GPa) | 210 | 200 |
| Poison Ratio ($\gamma$) | 0.29 | 0.28 |
| UTS ,$\sigma_u$ ($R_{M/20}$), MPa | 640 | 600 |
| Yield value ,$\sigma_y$ ($Rp_{0.2/t}$), MPa | 580 | 380 |
| Density ,$\rho$ (Kg/m$^3$) | 7872 | 7850 |

Table.3 Cylindrical Shell Design Specification

| | |
|---|---|
| Design Pressure (P) | 8.25 MPa [7] |
| Corrosion Allowance (t) | 2mm to 4 mm |
| Cylindrical Shell Inner Radius (R) | 1450 mm |
| Joint efficiency (E or $E_J$) | 100% |

$$f = \min\left(\frac{R_{p(0.2/t)}}{1.5};\frac{R_{m/20}}{2.14}\right) \quad (1) \qquad T = \frac{PR}{SE - 0.6P} \quad (3)$$

$$f = \min\left(\frac{R_{p(0.2/t)}}{1.5};\frac{R_{m/20}}{2.4}\right) \quad (2) \qquad \sigma_{xp} = \frac{PD}{4t} \quad (4)$$

Table.4 Design Stress and Thickness of Cylindrical Shell Calculated using DBF Approach

| Material | Calculated Thickness, T mm | Corrosion Allowance 4 mm | Final shell thickness (Commercially available thickness), mm | Design Stress S or f, MPa |
|---|---|---|---|---|
| P500-QT | 40 | 4 | 50 | min (386.67 ; 299) = 299 |
| P355 | 48 | 4 | 50 | min (253.33 ; 250) = 250 |

### *3.2. Design of Torispherical Head*

For thickness and axial stress due to pressure load, equations 5 and 6 are used [3] [5]. Design specifications and results are presented in table.5 and 6.

Table.5 Cylindrical Shell Design Specification

| | |
|---|---|
| Design Pressure(P) | 8.25 MPa [7] |
| Head Inner Radius ($R_1$) | 1450 mm |
| Allowable Stress (f or S) | P500-QT : 299 MPa; P355 : 250 MPa |
| Joint Efficiency (E or $E_J$) | 100t% |
| Thickness of cylindrical shell ($T=t=e_a$) | 50 mm |

$$\sigma_{xy} = \frac{P(R_1 + 0.2t)}{2Et} \quad (5) \qquad P_{MAX} = \frac{2fze_a}{D_M} \quad (6)$$

Table.6. Design Stress and Thickness of Torispherical Head Calculated using DBF Approach

| Material | Calculated Thickness, $T_2$ mm | Corrosion Allowance 2 mm | Final shell thickness (Commercially available thickness), mm | Design Stress S or f, MPa |
|---|---|---|---|---|
| P500-QT | 20 | 2 | 25 | 299 |
| P355 | 23 | 2 | 25 | 250 |

### *3.3. Stresses in Cylindrical Shell and Torispherical Head*

The stresses at or near the juncture of pressure vessels are called as the discontinuity or secondary stresses. They are secondary in extent but can be of very high magnitude. The secondary stresses acting at the cylindrical shell are calculated using equations 7 and 8 to calculate the longitudinal (σy) and hoop stresses (σx) respectively [3] [5]. But in the spherical portion (refer equation 9) both the stresses are equal i.e. σy= σx [3]. The value of't' for the shell portion is 50mm. and for the spherical region is 25mm. Where P=8.25MPa [7] and R=1450mm. Thus from the above design calculations, the thickness of the vessel components like shell, head, and the skirt was found.

$$\sigma_y = [PR / 2t] \quad (7) \qquad \sigma_y = \sigma_x = [PR / 2t] \quad (9) \qquad \sigma_x = [PR / t] \quad (8)$$

| Design Check: P500-QT and P355 Material | |
|---|---|
| Longitudinal stress ($\sigma_y$) in cylindrical shell | 119.625 MPa < S (Deign is valid) |
| Hoop stress ($\sigma_x$) in cylindrical shell | 239.25 MPa < S (Design is valid) |
| Stresses in spherical head, $\sigma_y = \sigma_x$ | 239.25 MPa < S (Design is valid) |
| **Thickness of vessel components like shell head skirt** | |
| Thickness of cylindrical shell (T or t) | 50mm |
| Thickness of torispherical shell ($t_2$) | 25mm |
| Thickness of skirt support ($t_s$) | 20mm |

### *3.4. Maximum Allowable Stresses in Cylindrical Shell and Torispherical Head*

The maximum permissible pressure according to the design by rules approach is calculated as laid out in Section 7.6.6.3 in prEN13445. The general formula used for the calculation of the maximum permissible pressure in EN13445 for cylindrical pressure vessels is represented by Equation 10 [3] [5]. As per the DBR the allowable pressure (Pmax) for the material under study is tabulated in table.7

$$P_{MAX} = \frac{2fze_a}{D_M} \quad (10)$$

Table.7. Design Stress and Thickness of Torispherical Head Calculated using DBR Approach

| | |
|---|---|
| Mean diameter of a cylindrical shell ($D_M$) | 3000 mm |
| Weld efficiency (z) | 100T% |
| Thickness of pressure vessel under study ($e_a$) | 50 mm |
| Allowable design stress (f) | P500-QT: 299 MPa; P355: 250 MPa |

| Material | Maximum allowable pressure ($P_{max, DBR}$) MPa |
|---|---|
| P500-QT | 9.96 |
| P355 | 8.33 |

## 4. Design of Pressure Vessel by DBA Approach

Initially, the elastic analysis was carried out to find the stresses at the junctions of shell and head, shell and skirt and the yield pressure is observed from the elastic analysis. Secondly, nonlinear analysis considering the material nonlinearity was conducted. A 2D axisymmetric analysis was carried out using ANSYS software. The material modeling for nonlinearity is done by selecting the inelastic, rate-independent Mises plasticity [8] and incorporating the multilinear kinematic hardening option by assigning the true stress strain data for P500-QT and P355 steel material as shown in figure.3.

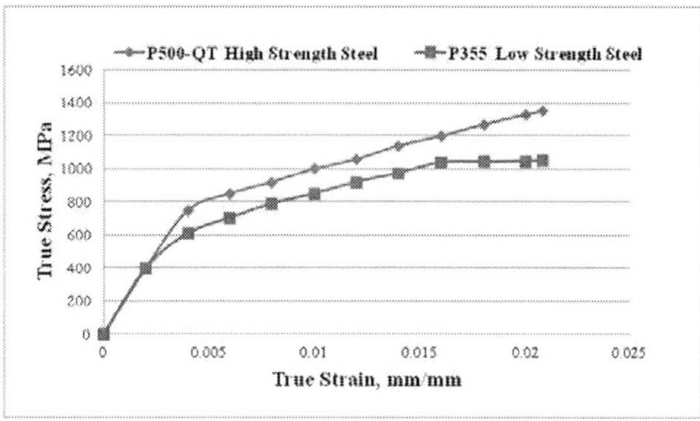


Figure.3. True Stress-Strain Curve for P500-QT and P355 Material

## 5. Elastic Analysis of Pressure Vessel

The stress distribution in the discontinuity regions is marked as SCL (Stress Categorization Lines) as shown in figure.4. Five nodal points equally spaced along with the thickness, designated as O-outer, NO- near outer, M-middle, NI- near inner and I-inner are considered for elastic analysis. The following junctions are considered for stress categorization analysis
a) SCL-1: Shell to pressure head taper start junction; b) SCL-2: Shell to pressure head taper end junction; c) SCL-3: Y-forged section to the head junction; d) SCL-4: Y-forged section to the skirt junction; e) SCL-5: Y-forged section near the curvature and the head junction; f) SCL-6: Y-forged section near the curvature and the skirt junction

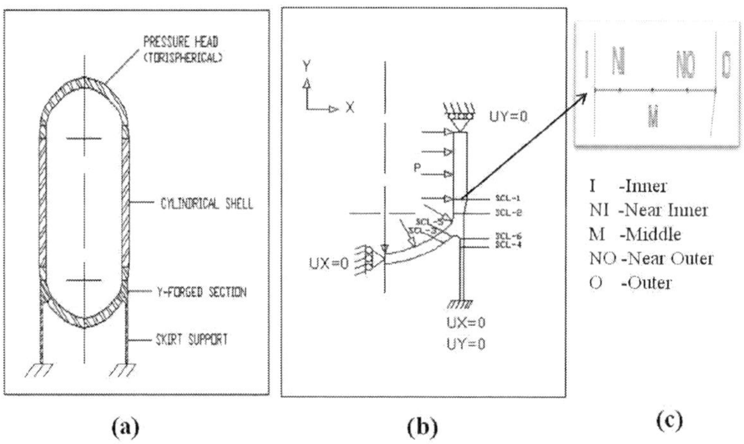


Figure.4. a) Pressure Vessel Schematic b) 2D Finite Element Model with Axisymmetric Boundary Conditions c) Stress Categorization Lines (SCL – Inner, Near Inner, Middle, Near Outer, Outer)

The discontinuity stresses developed in the shell to skirt junctions and the shell to head junctions can be determined using elastic FE analysis, which by analytical method becomes tedious and time-consuming. The stress distributions in the discontinuity region from the inner area to the outer area of the vessel are captured (refer figure 5-8). Also, in-elastic or plastic analysis of the vessel was conducted considering the material nonlinearities, modeled using true stress-strain material data. An increment of 1MPa is given to the yield pressure until we obtain the limit load of the vessel. Thus using ANSYS finite element package we conduct both elastic and elastic-plastic analysis for the pressure vessel taken for study.

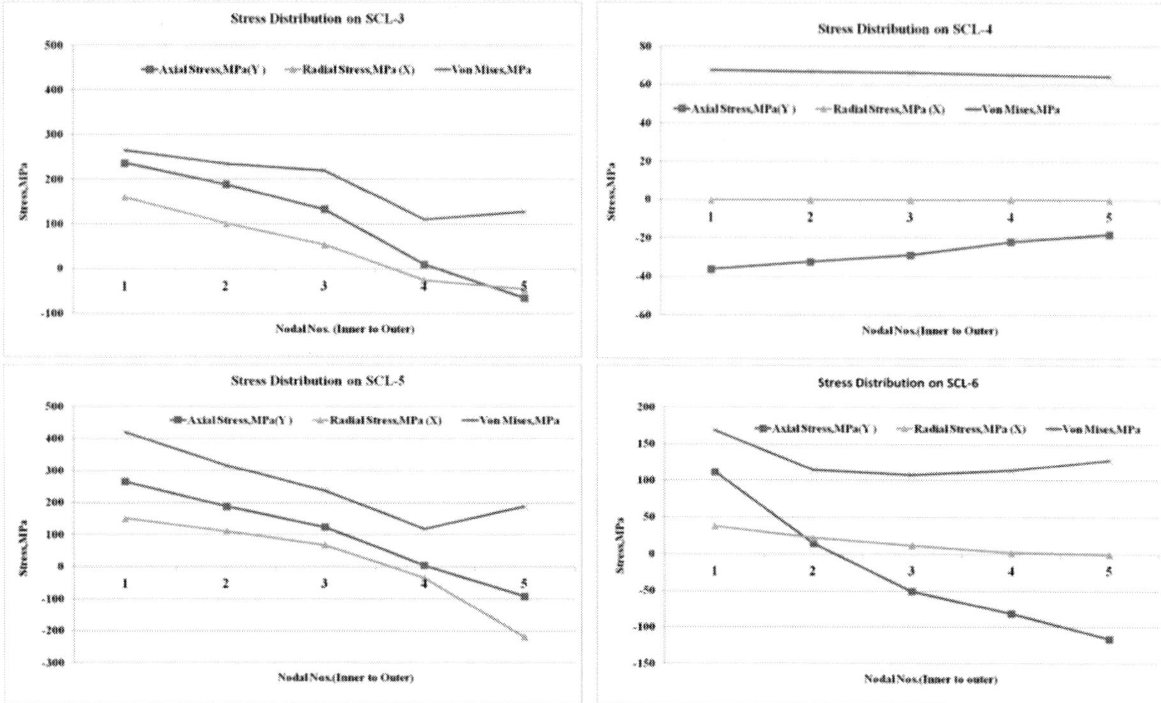


Figure.5. Stress Distribution Curves at Stress Categorization Lines 1 to 6 for P500-QT Material Subjected to Yield Pressure - Elastic Analysis

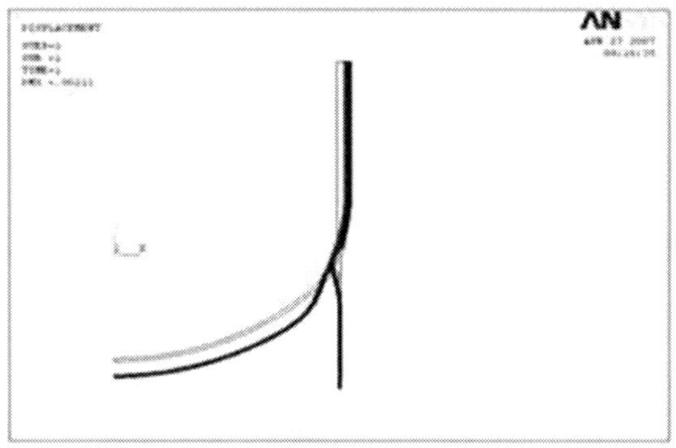

a) Deformation Plot

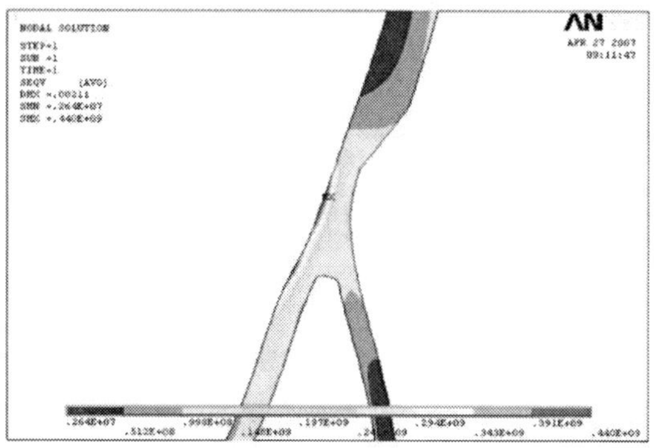


b) Von-Mises Plot

Figure.6. Deformation and Von Mises Plot at Yield Pressure for P500-QT Material- Elastic Analysis

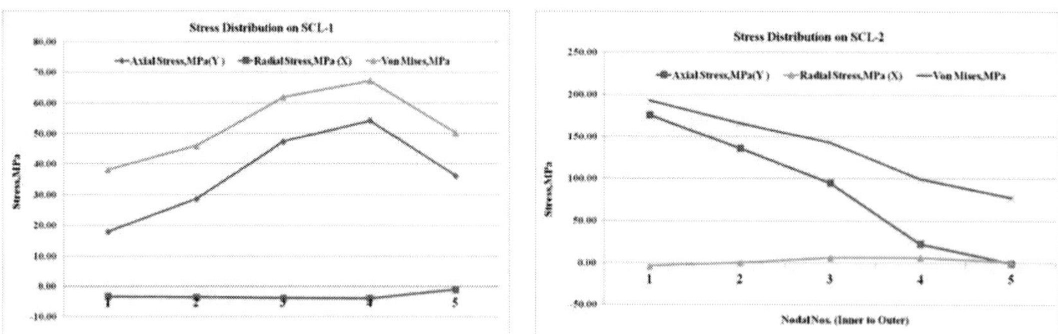

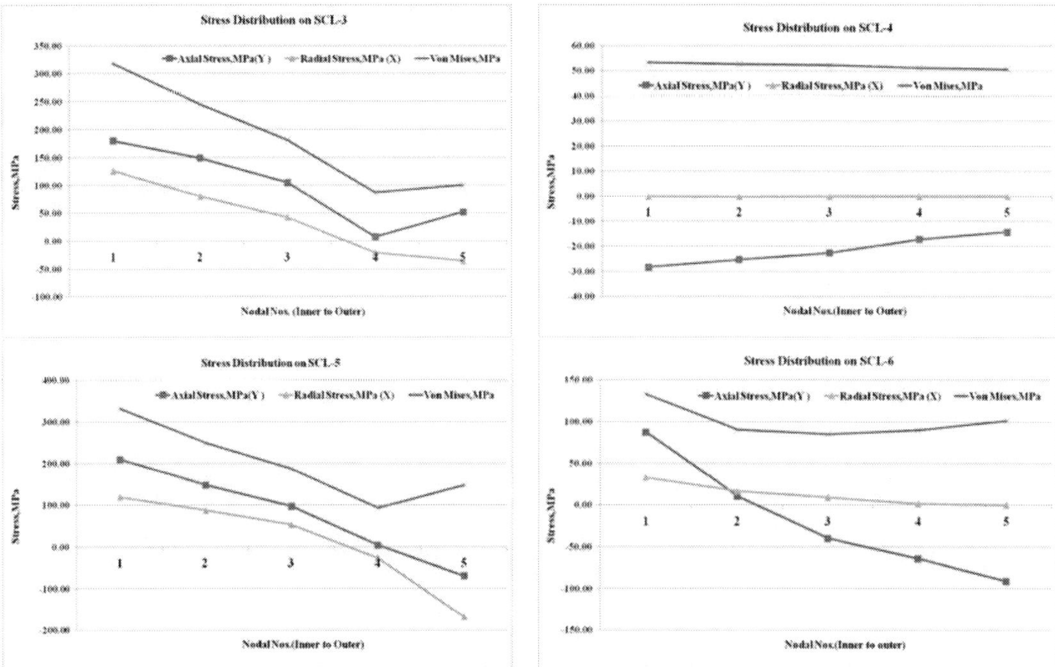


Figure.7. Stress Distribution Curves at Stress Categorization Lines 1 to 6 for P355 Material Subjected to Yield Pressure- Elastic Analysis

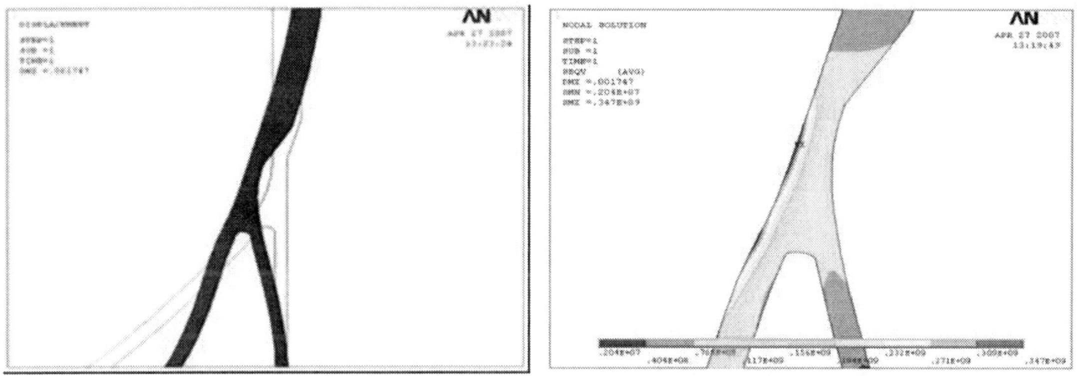


(a) Deformation plot (b) Von Mises stress plot

Figure.8. Deformation and Von Mises Stress Plot at Yield Pressure for P355 Material- Elastic Analysis

## 6. In-Elastic Analysis of Pressure Vessel

The in-elastic analysis was carried out to find the limit or collapse load of the pressure vessel. The deformation along axial (UY) and radial (UX) are captured for different pressure loading. The pressure loading cycle with an increment of 1MPa starting from yield pressure [7] was followed: $P_y$ (pressure at yield) to $P_L$ (limit pressure- to be determined from analysis). The criteria for finding the collapse load in the elastoplastic analysis are presented below;

(a) With a sudden increase in the pressure, the vessel deforms drastically, a condition at which the finite element analysis gives an error stating that the element is highly deformed. This happens in any large strain analysis problems.

(b) Also, the principal plastic strain increment for large strain analysis should be less than or equal to 5 % [7]. Thus the limit pressure value should have a limiting strain increment (refer figure 9).

In our case, this criterion was satisfied to a major extent but produced an error of +1.4% (i.e.6.4% maximum) increase in plastic strain (refer figure.9) for the pressure vessel steel P500-QT. But P355 material falls within the limit. Thus from the investigation, the limit or collapse load ($P_L$) for the pressure vessel and the plastic strain increment (successive strain increment) values obtained from the analysis using ANSYS package is tabulated in table 8.

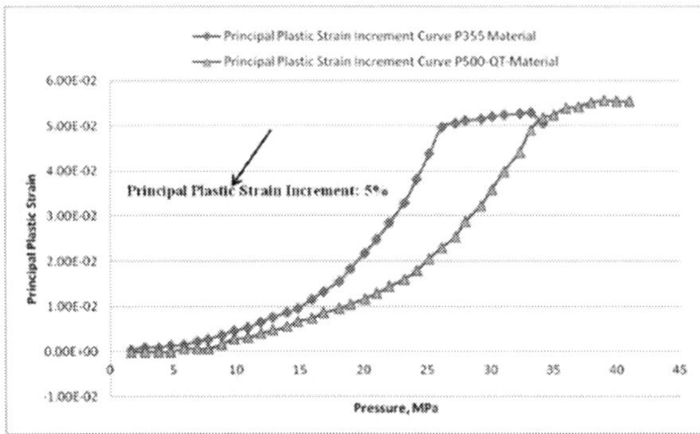


Figure.9.Principal Plastic Strain versus Load for P500-QT and P355 Material

Table.8. Limit Load ($P_L$) for P500-QT and P355 Pressure Vessel Steel: ANSYS Results

| Material | Limit load, $P_L$, MPa From ANSYS | Principal plastic strain Increment (%) | Principal plastic strain Error (%) |
|---|---|---|---|
| P500-QT | 37.25 | 6.4 | 1.4 |
| P355 | 29.05 | - 0.17 (negligible) | NIL |

## 7. Maximum Allowable Pressure as per DBA

From DBA-inelastic analysis, $P_L$ = 29.05 MPa (P355 material); 37.25 MPa (P500-QT material) was observed. As per ASME standard and pr EN13445 standard, the maximum allowable pressure ($P_{max, DBA}$) is calculated using equations 11 and 12 respectively. The maximum allowable pressure using FE analysis following the DBA approach is depicted in table.9.

$$P_{max, DBA} = (2/3)PL \quad (11) \qquad P_{max, DBA} = [P_L / \gamma_p] \quad (12)$$

Where, $P_L$ = Limit pressure obtained through analysis, MPa
$\gamma_p$ = Partial factor of safety for ferritic steel material [7]. ($\gamma_p$ =1.25-P500-QT and $\gamma_p$=1.3383 for P355)

Table.9. Maximum Allowable Pressure from FE analysis following DBA approach

| Material | Maximum Allowable Pressure ($P_{max,DBA}$), MPa, as per ASME standard | Maximum Allowable Pressure ($P_{max,DBA}$), MPa, as per EN13445-3 European standard |
|---|---|---|
| P500-QT | 24.83 | 29.8 |
| P355 | 19.37 | 21.70 |

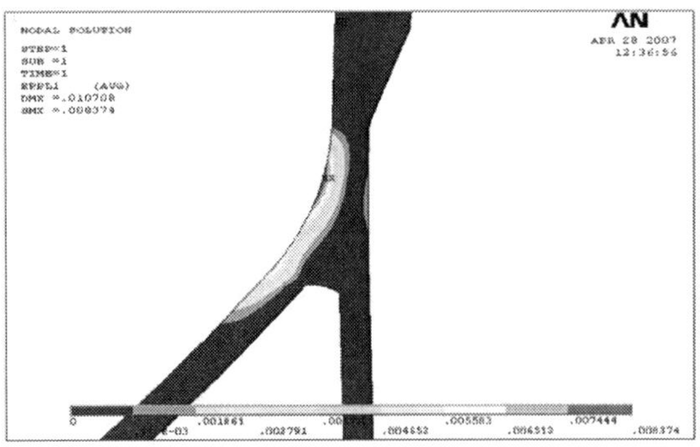

(c) Principal plastic strain at 19.25 MPa

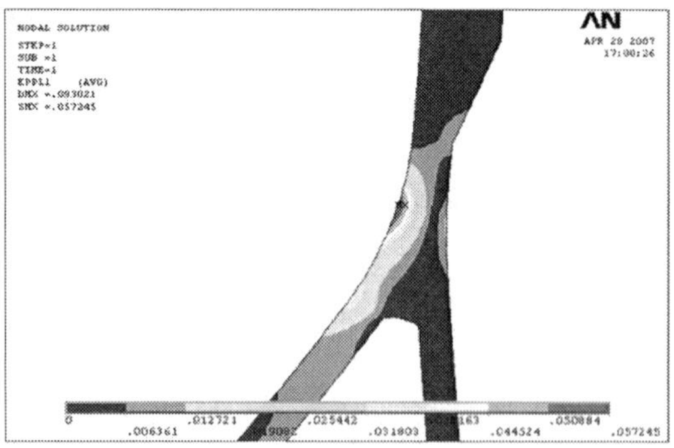

(e) Principal plastic strain at 37.25 MPa

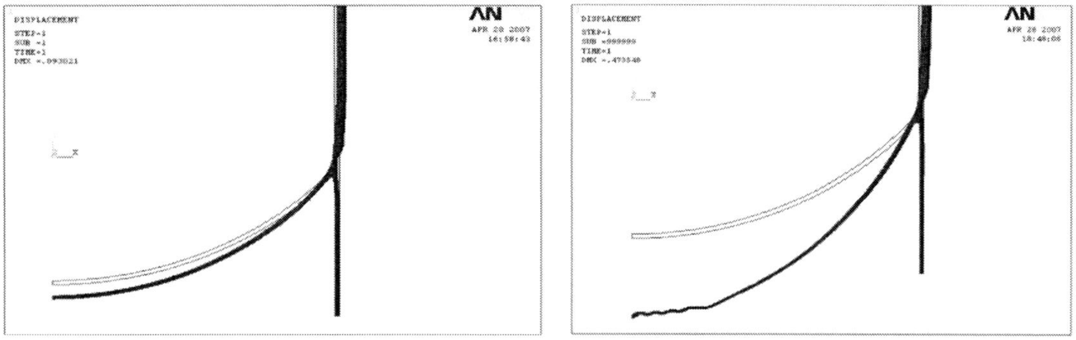

(a) Deformation at P=37.25 MPa (b) Deformation (failure) at P=38.25MPa

Figure.10.Principal Plastic Strain at different Pressure Loads for P500-QT Material: In-Elastic Analysis

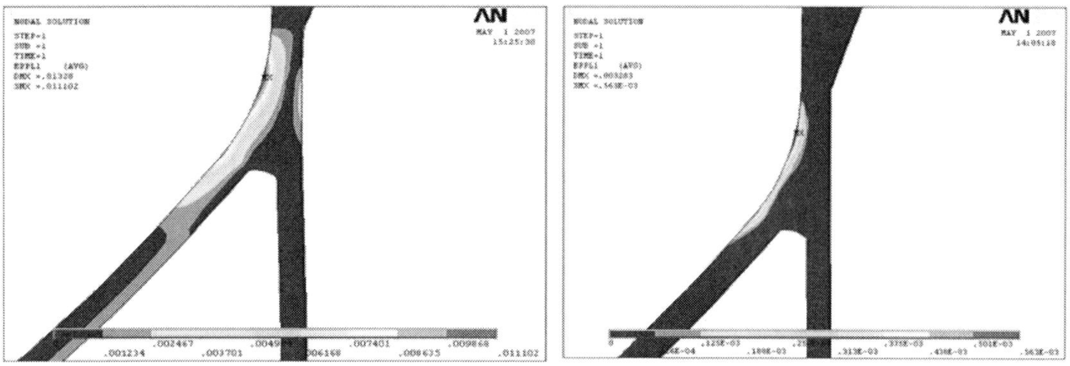

(c) Principal plastic strain at 18.35 MPa (e) Principal plastic strain at 10.35 MPa

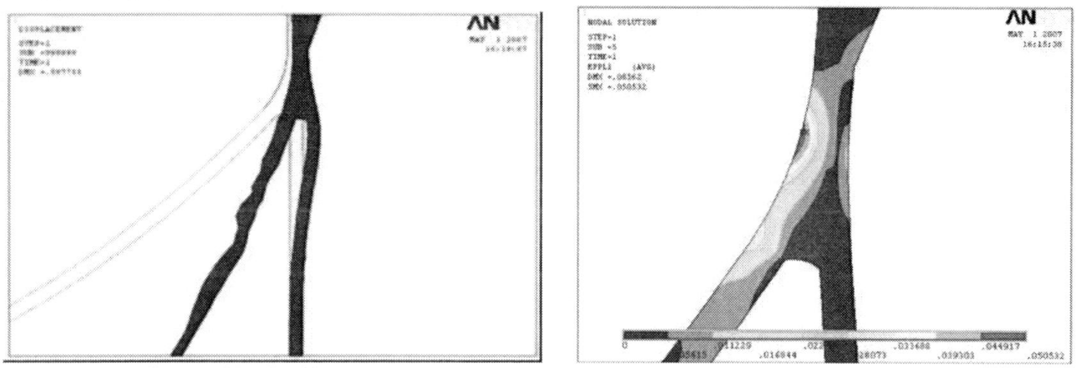

(c) Failure at 29.15 MPa (e) Principal Plastic Strain at 29.05 MPa

Figure.11.Principal Plastic Strain at different Pressure Loads for P355 Material: In-Elastic Analysis

## 8. Comparison of DBR with DBA

The maximum allowable pressure calculated using the design by rule (DBR) and design by analysis (DBA) approach following both ASME and European standards is presented in table.10. The results reveal that steels with high yield to tensile ratio are penalized using the DBR design approach [7].

Table.10.Maximum Allowable Pressure according to DBF and DBA

| Material | $P_{max,DBA}$, MPa | $P_{max,DBA}$, MPa (ASME) | $P_{max,DBA}$, MPa (EN13445-3) |
|---|---|---|---|
| P500-QT | 9.96 | 24.83 | 29.8 |
| P355 | 8.33 | 19.37 | 21.70 |

## 9. Results And Discussion

The maximum allowable pressure for the pressure vessel taken for analysis as per the DBR method is found to be highly conservative. Both the high strength and low strength steel materials are penalized by the DBF methodology [7] in terms of the load carrying capacity of the material (refer table.11). This means the designed pressure vessel can withstand more pressure than the maximum allowable pressure as designated by DBF methodology.

Table.11.Allowable Pressure concerning to DBA and DBF

| Material | Comparison of DBA and DBF as per ASME standard | Comparison of DBA and DBF as per EN13445-3 standard |
|---|---|---|
| P500-QT | $P_{DBA}=2.5\ P_{DBF}$ | $P_{DBA}=3.0\ P_{DBF}$ |
| P355 | $P_{DBA}=2.32\ P_{DBF}$ | $P_{DBA}=2.6\ P_{DBF}$ |

The conclusions drawn from this study can be summarized as follows:

a) The designed pressure vessel can withstand more pressure than the maximum allowable pressure as designated by DBR methodology. The application of DBA leads to much less conservative design parameters [7] for less critical pressure vessel applications.

b) High safety performance is mandatory for pressure vessel applications. Pressure vessels used in nuclear reactors and operating in adverse conditions should necessarily operate under high safety. But this may not be required for vessels used for storage applications (other than harmful liquids and gases).

c) Plastic strain increment error of 1.4% from its limit of maximum 5% for P500-QT material, because of modeling and meshing errors occurred in ANSYS package during the analysis phase. Mesh convergence study could reduce such errors. DBA can be integrated with DBR in designing pressure vessels for normal and less critical storage applications.

***References***

[1]. A. Lietzmann, J. Rudolph, E. Weib, *Chemical Engineering and Processing*, Vol. 35, 1996, pp. 287-293.
[2]. F.A. Simonen, Chapter-16, Vol.1, 2010, UNESCO-Encyclopaedia Life Support Systems (UNESCO-EOLSS)
[3]. ASME (1998), Section VIII, Division 2, the American Society of Mechanical Engineers, New York, NY.
[4]. Jones D. P., and Holliday J. E., *ASME J. Pressure Vessel Technol*. Vol. 122, 2000, pp. 146–151.
[5]. EN 13445-3 : 2014, Unfired pressure vessels - Part 3, European Committee for Standardisation, Technical Committee CEN/TC 54.
[6]. Trevor Seipp, Basics of Design by Analysis in ASME Section VIII, Division 2, www.betch.com, 25 November, 2018
[7]. A.Th. Diamantoudis, Th. Kermanidis, *International Journal of Pressure Vessels and Piping*, Vol82, 2005, pp. 43–50
[8]. https://ansyshelp.ansys.com/

IOP Conf. Series: Materials Science and Engineering **923** (2020) 012021 doi:10.1088/1757-899X/923/1/012021

# Study and evaluation of process parameter on Nimonic 75 alloy by Electrochemical micromachining

**C S Shamli[1*], P Hariharan[2], N Yuvaraj[3] and E Rajkeerthi[4]**

[1,3]Department of Mechanical Engineering, Vel Tech Rangarajan Dr Sagunthala R & D Institute of Science and Technology, Chennai, India

[2,4]Department of Manufacturing Engineering, Anna University, Chennai, India

*E-mail: shamlimit@gmail.com

Abstract. Electrochemical micromachining (EMM) is an effective technique for creating micro features in difficult to cut materials like super alloys. Nimonic 75 a super alloy is used to fabricate the aero combustor parts as it can withstand hot gases during the gas turbine operations. In EMM process, Nimonic 75 alloy of thickness 500 μm has been machined by applying two combination of electrolytes. A solid tungsten carbide with a diameter of 500 μm has been used as the tool in the present examination. The variable process parameters considered are machining voltage, micro tool feed rate and duty ratio and their influence on performance characteristics namely circularity entry, exit and surface roughness were studied. By using multi-response technique TOPSIS, the best combination of process parameter is found. SEM was employed to observe the morphology of the machined surfaces.

*Keyword*: Micromachining, EMM, Superalloy, Optimization

## 1. Introduction

Micromachining can be prescribed as the production of micro product/micro feature in macro component in the range of 1-999 μm. Combustor liner is one such component where number of micro holes are machined in order to achieve thin film cooling. Nimonic 75 is a nickel based super alloy, used for fabricating aero combustor liner as it can withstand high temperature. LBM and EDM are the processes used for machining micro-holes in combustor liner, to ensure the liner is cool enough to maintain its structural integrity. Due to their drawbacks like heat affected zone and thermal stresses, electrochemical micromachining (EMM) can be employed, as it combines benefits like absenteeism of HAZ and tool wear [1]. Electro chemical machining (ECM) is the better alternative process for generating micro features in hard-to-machine materials and EMM is a favored process for creating cooling holes in nickel based super alloys [2] [3].

EMM is effective for generating high precision operations, thus it is widely used in current research and industrial practices [4]. ECM can be defined as the opposite of electroplating, used for machining conductive and hard-to-machine materials. The metal removal is attained by aniodic dissolution of material. In electrochemical machining process the tool is made as the cathode, the workpiece as anode and inter-electrode gap is maintained between them, the whole arrangement is submerged in a electrolytic cell. While electricity is given, the anode starts to dissolve and the mirror image of the tool is obtained, electrolyte is pumped in the inter-electrode gap to remove the debris from the machining zone to achieve good surface finish on the specimen. Sandip et al [5] presented an examination of fitness of electrolyte for machining titanium alloy by EMM process, in order to examine the efficiency of electrolytes.

Dengyong wang et al [1] states that anodic dissolution depends on the workpiece material, the electrolyte, power supply and the significant process parameters in ECM are frequency, voltage, duty cycle. Jerzy et al [6] investigated the outcome of feed rate and voltage on process efficacy. Mohanty et al [7] examined the machinability of inconel 825 by using copper as a tool and studied the significance of machining parameters on response characteristics like surface roughness and MRR. Elevation in EMM performance features can be attained by establishment of short pulse period [8].

Hariharan et al [9] presented a approach in order to optimize the process parameters on drilling nimonic 75 alloy by adopting EMM process. This work states about the augmentation of multiresponse characteristics of AWJ process, accomplished by utilizing TOPSIS method [10]. Literature analysis notify that only minimal investigations were presented on the subject of process parameters and combinations of electrolytes in EMM of Nickel based super alloys. This work objects to recognize the outcome of machining parameters on nimonic 75 alloy under various combination of electrolytes.

## 2. Materials and Methods

### *2.1 Selection of work materials*

A circular shaped Tungsten carbide with a diameter 500μm is used as the cathode tool. Table 1 indicate the chemical composition of nimonic 75 alloy. Nimonic 75 is a nickel based super alloy that find application in aero combustor liner due to its innate properties.

**Table 1.** Nimonic 75 alloy – chemical composition

| Component | Ni | Cr | Cu | Fe | C | Mn | Si | Ti |
|---|---|---|---|---|---|---|---|---|
| Nominal Weight (%) | Remaining | Max 21 | Max 0.5 | Max 5 | Max 0.15 | Max 1.0 | Max 1.0 | Max 0.6 |
| Actual Weight (%) | 75.24 | 18.63 | 0.328 | 4.326 | 0.119 | 0.421 | 0.483 | 0.453 |

### *2.2 Selection of electrolyte*

The first electrolyte (E1) selected for experimentation is a combination of $C_2H_6O_2$ (2.2ml), NaBr (2.2g), HF acid (3.2ml) and consolidation of $NaNO_3$ (5g), NaCl (2.5g) is considered as the second electrolyte (E2) based on trial experiments.

### *2.3 Experimental arrangement*

The ECM arrangement consist of following components as indicated in figure 1. The metal is held in a fixture and placed inside the machining chamber and electrolyte tank aids in recycling the electrolyte by utilizing a filter. When power is switched on, the prescribed feed rate to approach the metal is attained by adopting a stepper motor.

IOP Conf. Series: Materials Science and Engineering **923** (2020) 012021 doi:10.1088/1757-899X/923/1/012021

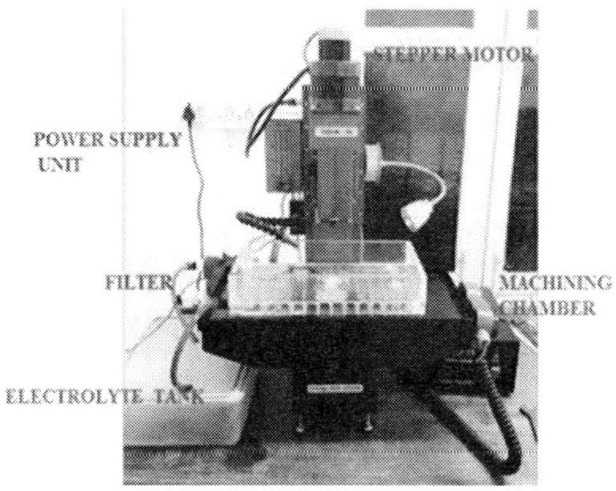


**Figure 1.** ECM arrangement

## 3. Experimental Design

To evaluate the impact of process parameters on the response characteristics like circularity entry, exit and surface roughness, a set of experiments were conducted. The experimental plan has three process parameters namely machining voltage (A), micro tool feed rate (B) and duty ratio (C) as shown in table 2.

**Table 2.** Experimental Design

| Parameters | Levels |
| --- | --- |
| A | 18, 20, 22 |
| B | 0.5, 0.6, 0.7 |
| C | 33, 50, 66 |

### *3.1 Performance characteristics*

Circularity: It can be described as the diametric deviation of the machined hole and denoted by equation 1. Circularity entry ($C_{entry}$) and circularity exit ($C_{exit}$) of the drilled hole is assessed by means of Video measuring system.

$$\text{Circularity} = D_{max}-D_{min} \quad (1)$$

Surface roughness: The irregularity of the machined area is measured by the most common parameter $R_a$, which represents the arithmetic average of the profile. Surface roughness (SR) is assessed by using Talysurf CCI Lite, non contact 3D surface roughness tester.

### *3.2 Experimental observations*

Circularity entry, circularity exit and surface roughness (SR) is calculated for each experiment and displayed in table 3.

**Table 3.** Experimental observation with response

| S. No | Voltage (V) | Tool feed rate (µm/s) | Duty cycle (%) | E1 Circularity entry (mm) | E1 Circularity exit (mm) | E1 SR (µm) | E2 Circularity entry (mm) | E2 Circularity exit (mm) | E2 SR (µm) |
|---|---|---|---|---|---|---|---|---|---|
| 1 | 18 | 0.5 | 33 | 0.003 | 0.005 | 0.458 | 0.002 | 0.004 | 0.345 |
| 2 | 18 | 0.6 | 50 | 0.016 | 0.016 | 0.601 | 0.001 | 0.001 | 0.389 |
| 3 | 18 | 0.7 | 66 | 0.033 | 0.035 | 0.73 | 0.014 | 0.01 | 0.412 |
| 4 | 20 | 0.5 | 50 | 0.039 | 0.038 | 0.504 | 0.026 | 0.008 | 0.647 |
| 5 | 20 | 0.6 | 66 | 0.046 | 0.041 | 0.787 | 0.017 | 0.041 | 0.717 |
| 6 | 20 | 0.7 | 33 | 0.014 | 0.029 | 0.639 | 0.021 | 0.021 | 0.688 |
| 7 | 22 | 0.5 | 66 | 0.114 | 0.12 | 2.281 | 0.021 | 0.015 | 0.847 |
| 8 | 22 | 0.6 | 33 | 0.108 | 0.103 | 1.171 | 0.024 | 0.018 | 0.97 |
| 9 | 22 | 0.7 | 50 | 0.085 | 0.074 | 1.12 | 0.067 | 0.071 | 0.854 |

## 4. Evaluation and Discussion

*4.1 Effect of voltage on response parameters under different electrolytes*

From figure 2 it is noticed that circularity entry and exit are proportional to the machining voltage, as directed by faraday's law of electrolysis. E2 exhibit lower circularity entry, exit and surface roughness values, at the chosen levels of voltage when compared to E1 combination. The presence of sodium chloride in E2 combination has ensured required amount of metal removal, as it contains aggressive anions, whereas the presence of sodium nitrate have resulted in lower surface roughness, due to the presence of oxidizing anions resulting in better surface finish. Mixed electrolytes provide combined advantages effect, thus the perfect blend of non-passive and passive electrolyte resulted in lower circularity entry and exit, which shows that the metal is eliminated in desired region only.

Figure 2 also depicts that stray current rises with voltage, hence incrementing the circularity entry and exit values. Circularity error (entry and exit) is more predominant in E1 electrolytic combination, at the chosen voltages, since E1 combination contains hydrofluoric acid which is a strong etchant, due to the etching factor, the atoms has been removed rapidly, as a result the metal is removed from undesired region and this yield poor geometrical accuracy and surface finish.

Based on the outcomes, it is instituted that circularity entry error was scaled down in the range of 67 to 63.7 %, circularity exit from 73% to 65% and surface roughness increased from 35% to 47.6% by using E2 electrolyte.

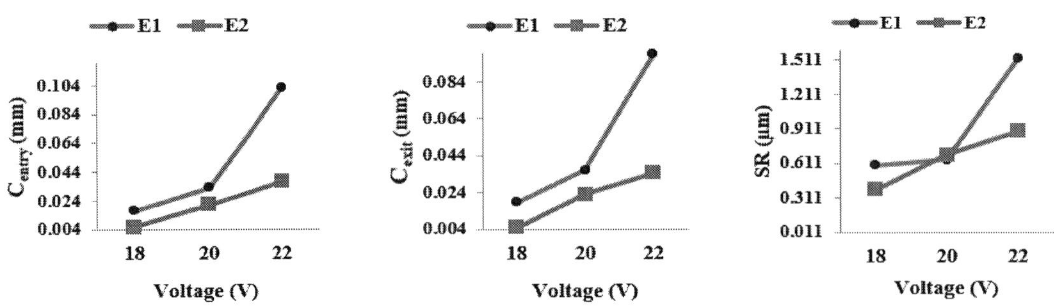


**Figure 2.** Effect of voltage on response parameters

*4.2 Effect of tool feed rate on response parameters under different electrolytes*
At lesser tool feed rate(TFR), tool rest at a specific region of workpiece for elongated duration resulting in desired circular feature. During high tool feed rate, the span of machining declines and therefore the desired circular feature is not obtained. By adopting low feed rate, high dimensional accuracy and finish can be obtained in the machined hole.

By using E2 combination, the hydrogen bubble formation was reduced at the tip of the microtool, because of the combination of nonpassive and passive electrolyte, which result in selective electrochemical dissolution. In relating the E1 electrolytic combination with E2, E2 yields lower values of circularity entry, circularity exit and surface roughness, at the chosen levels of tool feed rate.

From figure 3 it is found that at tool feed rate of 0.7 μm/sec, sufficient exposure to metal under E1 combination is not attained and hence the values of circularity entry and exit is lessened. Circularity entry error was decremented in the range of 69 to 22.7 %, circularity exit from 83% to 26% and surface roughness increased from 26% to 37% by using E2 electrolyte.

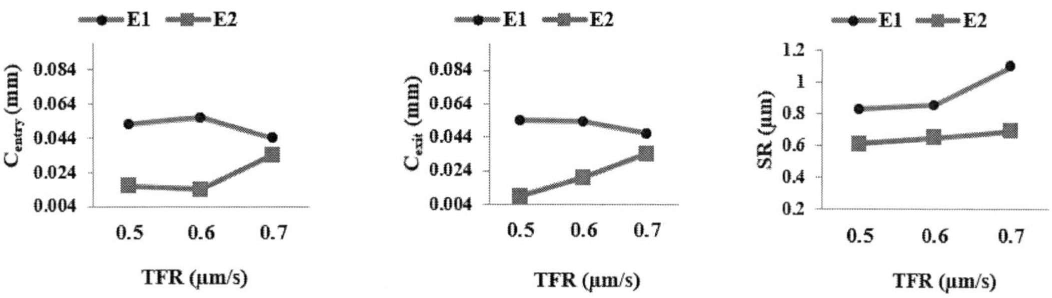


**Figure 3.** Effect of tool feed rate on response parameters

*4.3 Effect of duty ratio on response parameters under different electrolytes*
Figure 4 shows the significance of duty ratio on response parameters of the machined hole. At higher values of duty ratio, pulse on time is more which increase the electrochemical dissolution not only in linear path but also in sideward path which result in added metal removal from the sideward direction of the material triggering circularity entry and exit error as shown in Figure 4.

When pulse off time is more, metal is dissolved in a linear pattern because increased pulse off period ensure to remove heat and sludges from the machining slot and diminish the metal removal from the sideward direction ensuring low circularity(entry and exit). E2 combination is more advantageous over E1 due to the existence of sodium nitrate which has less throwing power ensuing in controlled metal removal whereas sodium bromide in E1 combination has high throwing power.

At 66% of duty ratio it is observed that circularity entry and exit values were scaled down for E2 combination as elevated values of duty ratio result in less metal removal. At higher duty ratio, pulse off period is less than pulse on period resulting in untimely removal of sludges from the machining zone thus affecting the metal removal which influence the values of circularity entry and exit.

From the results, it is observed that circularity entry error was increased in the range of 63 to 73% and circularity exit from 68% to 66% and surface roughness increased from 16% to 47% by using E2 electrolyte.

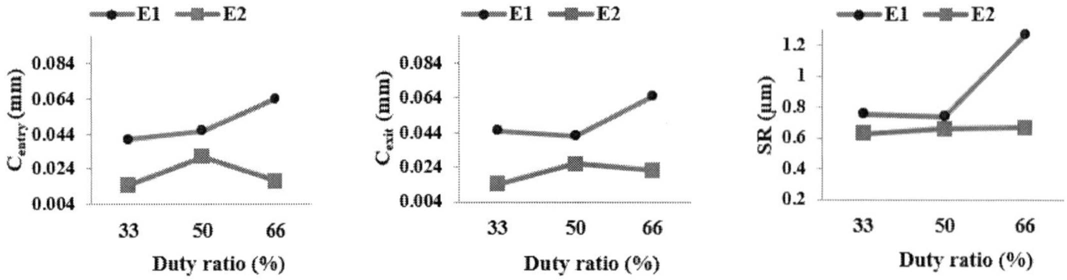


**Figure 4.** Effect of duty ratio on response parameters

*4.4 Multiresponse optimization*

In this study, multiresponse optimization technique namely, TOPSIS was used to examine the performance features of electro chemical micromachining process. The TOPSIS is defined as Technique for order preference by similarity to ideal solution and is reported by Olson [11]. This approach differs from other optimization methods since it uses positive and negative ideal solutions of each response. Among various alternatives, TOPSIS identifies the best solution which is very closer to the ideal solution.The following steps were considered in the multiresponse technique TOPSIS for the optimization of process parameters in the electro chemical micromachining of nimonic 75.

Step 1: Normalized value

In this step, all the observations (responses) were converted to normalized value (dimensionless) for an equal consideration of responses by using equation 2.

$$\text{Normalized values, } r_{ij} = \frac{x_{ij}}{\sqrt{\sum_{i=1}^{n} x_{ij}^2}} \quad (2)$$

where, i = Experimental runs, j = Output responses

Step 2: Weighted Normalized Matrix

Processing of this method requires input data for allocating the weights to the performance measurements. The standard weighing procedure SIMOS technique was used for this purpose. The description has been provided by Yuvaraj and Pradeep Kumar [10]. In the SIMOS technique, responses were arranged from the least to the most important. Details are tabulated in table 4.

**Table 4.** Weighted Normalized Matrix

| Variable | No of variables | Place of position | Non-normalized weight | Sum % |
|---|---|---|---|---|
| Surface roughness | 1 | 1 | (1/6)*100 = 16.66 ~ 17 | 17 |
| Circularity entry and Circularity exit | 2 | 2, 3 | For two variable (5/6)*100 = 83.33~ 83 | 83 |
| Total | 3 | 6 | | 100 |

The weighted normalized values were obtained using equation 3.

$$\text{Weighted Normalized Matrix, } \gamma_{ij} = r_{ij} \times w_j \quad (3)$$

Step 3: Determination of Greatest and Poorest Performance. In this step, the greatest and the poorest performance of each response were identified through its desirability criteria.

$S_j^+$ = greatest performance; $S_j^-$ = poorest performance

Step 4: Measurement of Positive and Negative ideal distance. The positive ($D_i^+$) and the negative ideal distance ($D_i^-$) for all the experimental runs were obtained using equations 4 and 5. These distances were measured from the greatest and the poorest performance of responses which are tabulated in table 5.

$$D_i^+ = \sqrt{\sum_{j=1}^{n}(\gamma_{ij} - S_j^+)^2} \quad i = 1 \text{ to } m \tag{4}$$

$$D_i^- = \sqrt{\sum_{j=1}^{n}(\gamma_{ij} - S_j^-)^2} \quad i = 1 \text{ to } m \tag{5}$$

Step 5: Closeness coefficient values

The closeness coefficient values $C_i$ were obtained using equation 6. $C_i$ values and their ranks are given in Table 5.

$$C_i = \frac{D_{ij}^-}{D_{ij}^+ + D_{ij}^-} \tag{6}$$

**Table 5.** $C_i$ values and their ranks

| S. No. | Closeness co-efficient values | | | | | | Rank | |
|---|---|---|---|---|---|---|---|---|
| | E1 | | | E2 | | | E1 | E2 |
| | $D_{ij}^+$ | $D_{ij}^-$ | $C_{ij}$ | $D_{ij}^+$ | $D_{ij}^-$ | $C_{ij}$ | | |
| 1 | 0.000 | 0.361 | 0.970 | 0.009 | 0.352 | 0.990 | 1 | 1 |
| 2 | 0.025 | 0.323 | 0.927 | 0.010 | 0.364 | 0.973 | 2 | 2 |
| 3 | 0.094 | 0.268 | 0.741 | 0.073 | 0.292 | 0.799 | 4 | 3 |
| 4 | 0.106 | 0.260 | 0.709 | 0.129 | 0.263 | 0.671 | 5 | 4 |
| 5 | 0.123 | 0.227 | 0.648 | 0.201 | 0.176 | 0.467 | 6 | 8 |
| 6 | 0.059 | 0.233 | 0.799 | 0.136 | 0.156 | 0.535 | 3 | 7 |
| 7 | 0.361 | 0.001 | 0.002 | 0.123 | 0.245 | 0.665 | 9 | 5 |
| 8 | 0.314 | 0.071 | 0.184 | 0.146 | 0.223 | 0.604 | 8 | 6 |
| 9 | 0.181 | 0.134 | 0.425 | 0.328 | 0.122 | 0.271 | 7 | 9 |

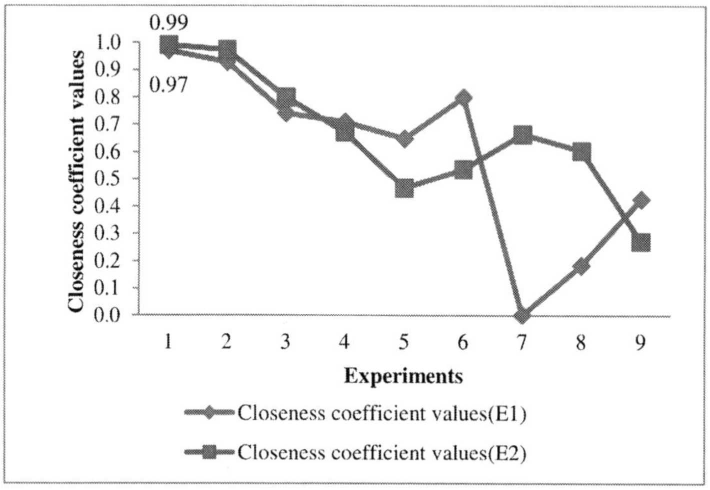


**Figure 5.** Closeness coefficient values for each trial

Figure 5 shows the closeness coefficient values of electrochemical machining process, in which high peak value of 0.99 has been recorded for E2 which is closer to ideal solution at the range of 18V, tool feed rate of 0.5 μm/s and at duty ratio 33%. Table 5 reveals the closeness coefficient value for each experiment. From the outcomes, it is noticed that E2 has higher $c_i$ value over E1. The indicated result had happened due to the impact of E2 on the response characteristics of EMM process. This further confirms that E2 combination can be used for EMM of nimonic 75 alloy over E1, due to the presence of aggressive and oxidizing anions. Enhancement in $C_i$ for E1 and E2 are 78% and 70% respectively, as shown in table 6.

**Table 6.** Comparison of initial and optimal setting results

| Conditions | Setting level | | Response parameters | | | | | | $C_i$ value | |
|---|---|---|---|---|---|---|---|---|---|---|
| | | | E1 | | | E2 | | | | |
| | E1 | E2 | $C_{entry}$ (mm) | $C_{exit}$ (mm) | SR (μm) | $C_{entry}$ (mm) | $C_{exit}$ (mm) | SR (μm) | E1 | E2 |
| Initial setting parameters | A3B2C1 | A3B3C2 | 0.108 | 0.103 | 1.171 | 0.067 | 0.071 | 0.854 | 0.18 | 0.27 |
| Optimal | A1B1C1 | A1B1C1 | 0.003 | 0.005 | 0.458 | 0.002 | 0.004 | 0.345 | 0.97 | 0.99 |

Analysis of variance (ANOVA) is considered as a advantageous method for statistical inference, which was performed by means of MINITAB considering a significant level of 5%. By comparing the various input factors taken in EMM, voltage is the most influencing factor in E1 and E2 combination. As the levels of voltage chosen, influence not only the metal removal but also the error associated in machining features. Table 7 and 8 reveals the voltage contribution of 87% and 47.7% in the presence of E1 and E2 combination.

**Table 7.** ANOVA Result - E1

| Source | Degrees of freedom | Sum of squares | Mean square | F-Value | P-Value | % Contribution |
|---|---|---|---|---|---|---|
| A | 2 | 0.794 | 0.397 | 30.56 | 0.032 | 87.00% |
| B | 2 | 0.0108 | 0.0054 | 0.42 | 0.705 | 1.2% |
| C | 2 | 0.0818 | 0.0409 | 3.15 | 0.241 | 8.96% |
| Error | 2 | 0.0259 | 0.0129 | | | 2.85% |
| Sum | 8 | 0.9126 | | | | 100% |

**Table 8.** ANOVA Result - E2

| Source | Degrees of freedom | Sum of squares | Mean square | F-Value | P-Value | % Contribution |
|---|---|---|---|---|---|---|
| A | 2 | 0.30946 | 0.15473 | 2.29 | 0.304 | 47.74% |
| B | 2 | 0.14128 | 0.07064 | 1.05 | 0.488 | 21.8% |
| C | 2 | 0.06262 | 0.03131 | 0.46 | 0.683 | 9.66% |
| Error | 2 | 0.1349 | 0.06745 | | | 20.81% |
| Sum | 8 | 0.64826 | | | | 100% |

## 5. Hole Morphology

Figure 6 and 7 depicts the scanning electron microscopic images of machined profile at the range of 18 volt, 0.5 μm/sec tool feed rate and duty ratio at 33% for E1 and E2, conforming to the maximum closeness coefficient. Comparing the SEM images of optimal parametric setting of E1 and E2 conditions, the micro-hole machined by E2 combination possess reduced circularity and surface roughness. Improved accuracy and better surface finish were achieved in E2 combination when compared to E1 combination. As even metal dissolution have occurred in the machined region, due to the effect of combination of electrolytes used. Whereas in E1 combination uneven material removal and surface irregularities have occurred due to the absence of oxidizing anions which in turn expose the metal surface resulting in increased metal removal and falling of material owing to the presence of etchant in E1.

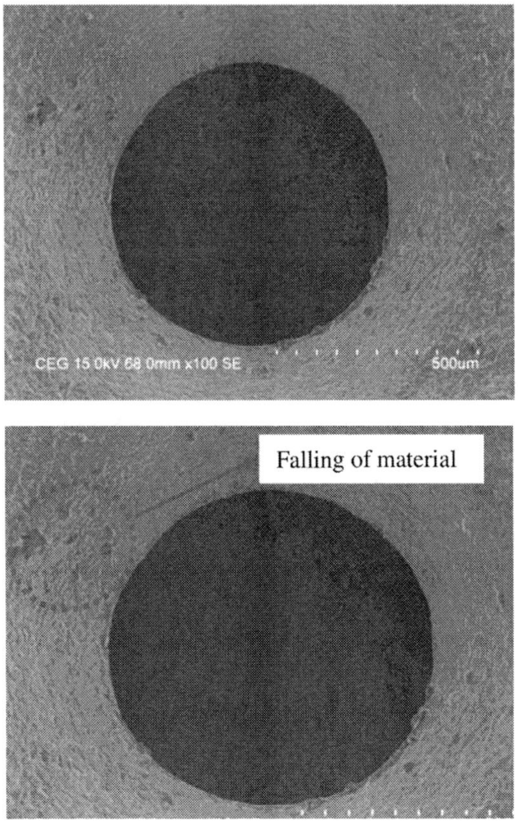


**Figure 6.** SEM image of drilled hole at 18 volt, 0.5 µm/sec tool feed rate and duty cycle at 33% for E1

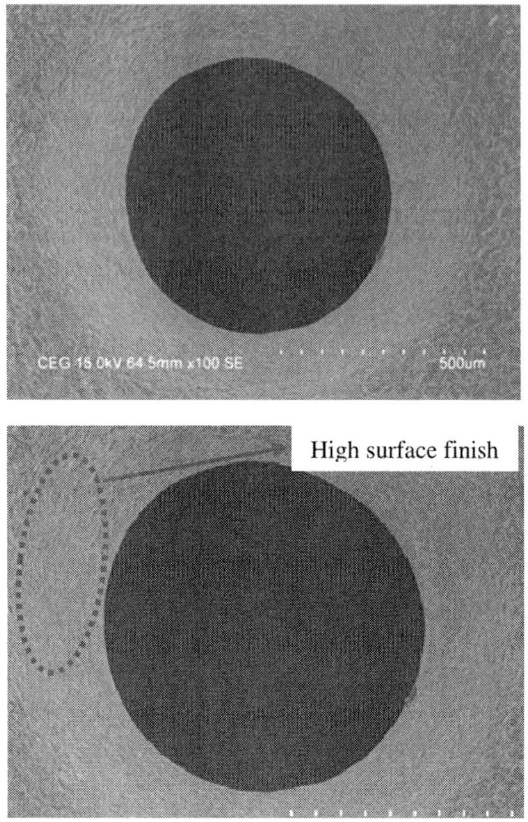


**Figure 7.** SEM image of drilled hole at 18 volt, 0.5 µm/sec tool feed rate and duty cycle at 33% for E2

## 6. Conclusion

EMM was instigated on nimonic 75 alloy to examine the effect of process parameters on different features like circularity entry, circularity exit and surface roughness. Utilizing multi-response technique TOPSIS optimization was performed. From this investigation, the succeeding results have been consolidated.

- Experimental outcome expose that E2 produce lesser values of surface roughness, circularity entry and circularity exit when compared to E1.
- For E1 and E2 combination, lesser range of micro tool feed rate, duty ratio and machining voltage produce preferred surface roughness and lesser circularity entry and circularity exit values.
- Based on ANOVA, voltage is the utmost influencing parameter in E1 and E2 investigation.
- The TOPSIS outcomes showed that the overall enhancement of $c_i$ is 78% and 70% in E1 and E2.

## REFERENCES

[1] Wang D, Zhu Z, Wang N, Zhu D and Wang H 2015 Investigation of the electrochemical dissolution behavior of Inconel 718 and 304 stainless steel at low current density in NaNO3 solution *Electrochimica Acta* **156** 301-307

[2] Sharma S, Jain V K and Shekhar R 2002 Electrochemical drilling of inconel superap;plloy with acidified sodium chloride electrolyte *The International Journal of Advanced Manufacturing Technology* **19** 492-500

[3] Rajurkar K P, Sundaram M M and Malshe A P 2013 Review of electrochemical and electrodischarge machining *Procedia Cirp* **6** 13-26

[4] Bhattacharyya B, Munda J and Malapati M Advancement in electrochemical micro- machining *International Journal of Machine Tools and Manufacture* **44** 1577-1589

[5] Anasane S S and Bhattacharyya B 2016 Experimental investigation on suitability of electrolytes for electrochemical micromachining of titanium *The International Journal of Advanced Manufacturing Technology* **86** 2147-2160

[6] Mohanty A, Talla G, Dewangan S and Gangopadhyay S 2014 Experimental study of material Removal rate, surface roughness & Microstructure in electrochemical Machining of inconel 825 *In 5th International & 26th All India manufacturing technology, Design and Research Conference AIMTDR* 174(1-6)

[7] Kozak J, Rajurkar K P and Makkar Y 2004 Selected problems of micro-electrochemical machining *Journal of Materials Processing Technology* **149** 426-431

[8] Reddy M M K 2013 Influence of pulse period and duty ratio on electrochemical micro machining (emm) characteristics *International Journal of Mechanical Engineering and Applications* **1** 78-86

[9] Pillai H P, Sampath S C, Elumalai R, Hariharan S and Natarajan Y 2017 Influence of Process Parameters on Electrochemical Micromachining of Nimonic 75 Alloy *International Mechanical Engineering Congress and Exposition, American Society of Mechanical Engineers Digital Collection* **2** V002T02A007

[10] Yuvaraj N and Pradeep Kumar M 2015 Multiresponse optimization of abrasive water jet cutting process parameters using TOPSIS approach *Materials and Manufacturing Processes* **30** 882-889

[11] Olson D L 2004 Comparison of weights in TOPSIS models *Mathematical and Computer Modelling* **40** 721-727

# Processing and Testing of Aluminum based Hybrid Surface Composites by Friction Stir Processing

**C.Devanathan[1]*, E.Shankar[2], A.Sivanand[3], B.Haribabu[4]**

[1,2,3] Assistant Professor, Department of Mechanical Engineering, Rajalakshmi Engineering College, Thandalam, Chennai 602 105.
[4] Training officer, L& E operation, NTTF, Chennai.

devanathan.c@rajalakshmi.edu.in

**Abstract.** The ever increasing service necessities in industries like automobile and aerospace demands the engineers to use the aluminium based composites which are reinforced by either single or mixture of ceramic particles. When hybrid reinforcements are used, the resulting composites are having superior mechanical properties. In the present experimental work, the Al 5083 was reinforced by Titanium oxide and aluminium oxide to the depth of 2mm and processed by solid state friction stir processing technique. To prepare the surface composites, three non-consumable rotating tools were used with different tool pin profile such as square, straight cylinder and taper cylinder. The other parameters like tool rotational speed, tool traverse speed, number of passes were maintained uniform. Totally three workpieces were processed in order to inspect the ultimate tensile strength, yield strength and micro hardness. The test results shown that the square tool pin has delivered the highest strength and hardness than the other tool pin profile. The micro hardness test results showed that the hardness of the reinforced zone is higher than unreinforced region.

**Keywords**: Hybrid Composites, Friction Stir Processing, Tool pin profile, Ultimate tensile strength, Micro hardness

## 1. Introduction

Over the last few decades, in the area of metal joining techniques, friction stir welding was created a revolutionized development. Friction stir processing is a technique invented by Mishra and his team as a modification of friction stir welding and was initially used for developing ultra-fine grained super plastic aluminum alloys. The process was emerged as an efficient tool for processing and developing the materials, for modifying the microstructure of the materials.

Hybrid surface composites are the group of materials which contains more than one type of reinforcement particles dispersed in the matrix. Hybrid composite takes the benefits of all the reinforcements and therefore exhibits the improved properties than the other materials [1,2]. In general hybrid composites are prepared by various methods like powder metallurgy, plasma spraying technique, pressure infiltration method, and squeeze casting. In this list friction stir process is added as a unique technique for manufacturing surface composites by means of modifying the grain structure.

The major approaches found in literature for developing surface composites are groove method, drilled hole method, cover plate method. In all three methods non consumable rotating tool is used to ensure the proper stirring of reinforcement in to the matrix material. The main function of the tool is heating and deforming the workpiece material. Friction stir processing technique has different process parameters which are grouped as machine related variables, tool related variables, and material properties based variables as shown in the Figure1.

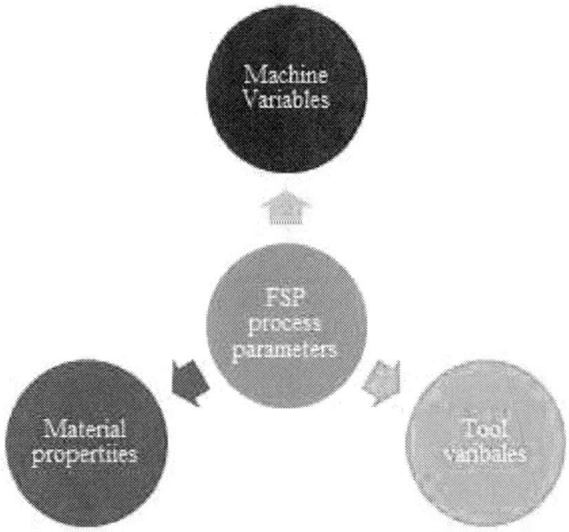


**Figure 1:** FSP process parameters

The variables controlled in machine are rotational speed, traverse speed, plunge force and tilt angle. As far as tool concern, dimensions of the shoulder and pin, profile of the pin is significant factors in developing the heat in the process. Hosseini et al [3] used CeO2 and CNT reinforcements to fill the groove machined in Al5083 base material and found that tensile strength increases with increasing the volume fraction of the carbon nano tubes. Ramnath et al [4] studied the mechanical properties of aluminum based hybrid composites which were reinforced using Al2O3 and B4C. The prepared composites exhibited the superior impact strength and hardness than the unreinforced alloy. On the other hand tensile strength and flexural strength of the prepared composites were lower than base materials due to improper stirring and uneven distribution of reinforcement particles in the alloy.

Few authors [5 – 8] have carried out a detailed study on the mechanical behavior of materials and tribological properties of aluminium based hybrid composites reinforced with combination of synthetic ceramic particles like alumina ($Al_2O_3$), silicon carbide (SiC), boron carbide ($B_4C$), tungsten carbide (WC), graphite (Gr), silica ($SiO_2$), Carbon Nano Tube (CNT). The researchers concluded that the reinforcement particles have a noticeable contribution in improving the mechanical properties of the composites prepared.

## 2. Experimental methods:

The aluminum 5083 was selected as base material with 4mm thickness. The small groove of size 1mm width and 2mm depth was cut on the base material to fill the reinforcement particles. Titanium oxide and aluminum oxide were taken equal amount as reinforcement particles, mixed thoroughly and filled in the machined groove.

The tool is considered as a heart of the process, because it produces the heat and ensures the proper mixing of the reinforcement particles in to the matrix aluminum. In the present work, three different tools were machined with different pin profiles namely taper cylinder, square of side 4mm, and straight cylinder with 6mm diameter. The tool is made from H13 steel and heat treated to improve the hardness. The other process parameter like tool rotational speed, traverse speed and number of passes were fixed constant. The details of the process parameters were given below in Table 1.

**Table 1**. Details of Process Parameters

| S.No | Process Parameter | Details |
|---|---|---|
| 01 | Tool rotational speed | 1000 rpm |
| 02 | Tool traverse feed | 100 mm/min |
| 03 | Tool pin profile | Square, Straight cylinder, Taper cylinder |

Totally three experiments were conducted on vertical milling machine for analyzing the mechanical properties like ultimate tensile strength, yield strength, and micro hardness. The machine used, tool and sample specimen prepared were shown in the Figure 2 as [A], [B], [C] respectively.

**Figure** 2 [A] Machine Used, [B] FSP tool, [C] workpiece with groove

The friction stir processed workpiece using all the three tools are shown in the Figure 3.

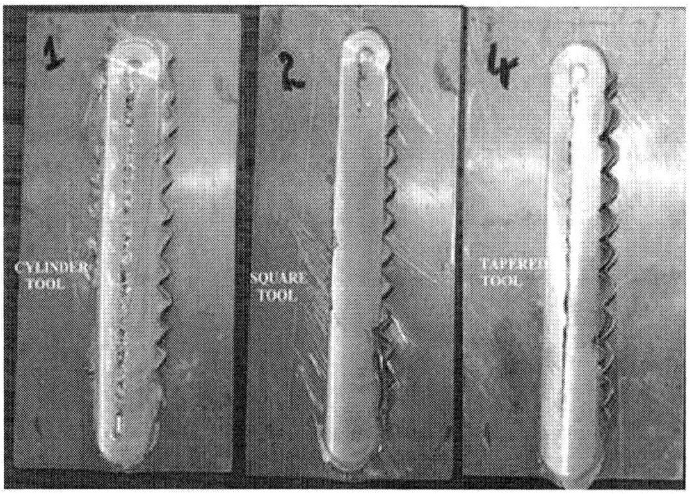


**Figure 3**: Processed Specimen

## 3. Results and discussion

In order to analysis the tensile strength and yield strength of the friction stir processed workpieces, the dog bone like sample was prepared as per ASTM E8 standard. The shape and dimensions to be followed for preparing the specimen is given in the Figure 4.

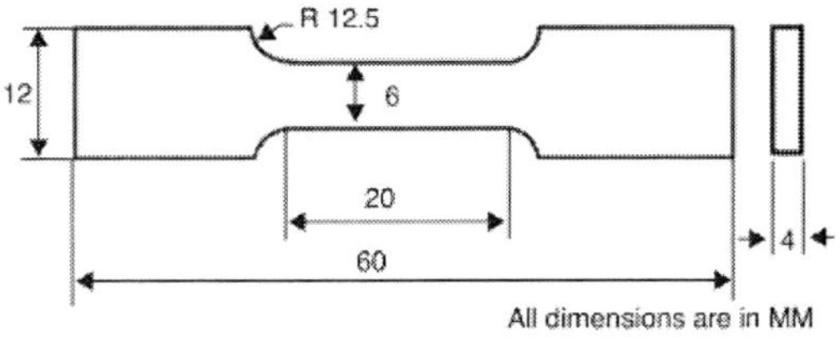


**Figure 4:** ASTM E8 standard for tensile test

From each workpiece three samples were cut using Wire Electrical Discharge Machining method. The samples were tested for its tensile strength, yield strength on Computerized Universal Testing Machine. Form the samples it was observed that the specimens were failed at processed zone. The specimen after the tensile test was shown in the Figure 5.

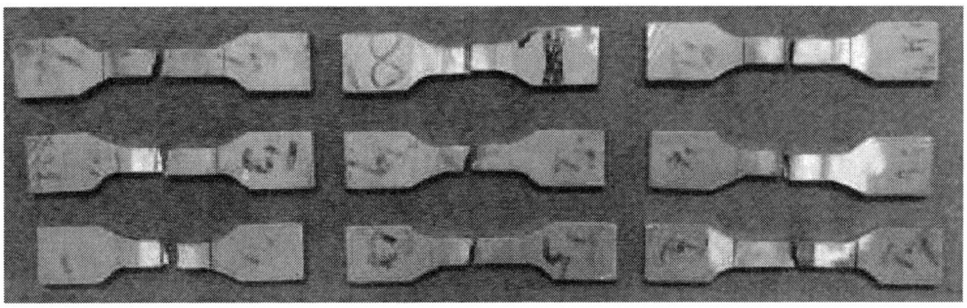

**Figure 5**: Tensile specimen after test

The computer controlled UTM was used to plot the curve between stress and strain. The sample result produced for one of the square pin profiled tool is shown in the Figure 6. All the details like dimensions of the specimen prepared, the standard followed was given above the curve and the results obtained after the test were mentioned bottom of the report.

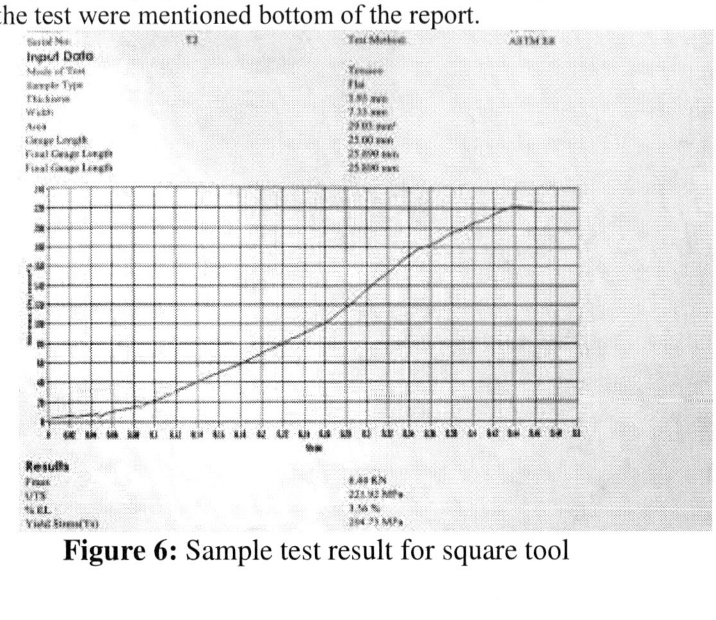


**Figure 6:** Sample test result for square tool

IOP Conf. Series: Materials Science and Engineering **923** (2020) 012022 doi:10.1088/1757-899X/923/1/012022

The ultimate tensile strength and yield strength of the specimens were given in the Figure 7. From the sample prepared it was observed that cylindrical and tapered pin were had a tunnel like defect throughout the length. The specimen prepared using square pin had slightly better processed surface. It was proven by the test results that when compared with the other two profiles the square pin tool had better tensile strength and yield strength.

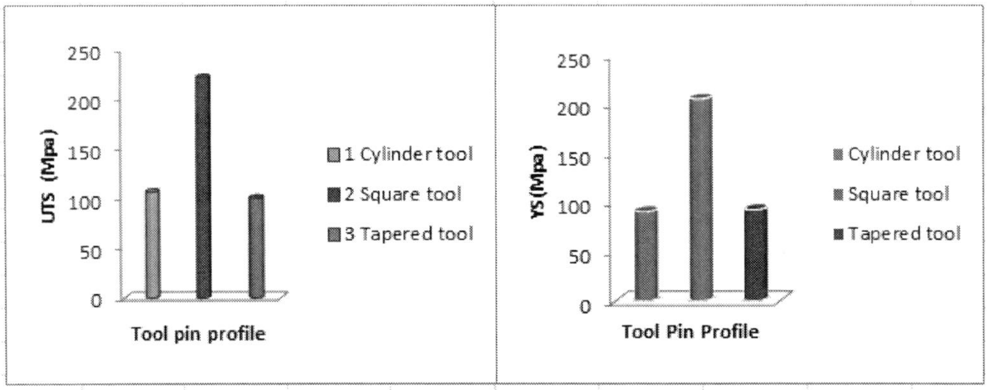


**Figure 7:** UTS and YS of FSP specimens

Micro hardness testing is a type of finding a material hardness or resistance to penetration when the samples were thin foil or small region of the composite materials. The samples were prepared and tested as per ASTM E384 standard. The prepared samples from the friction stir processed specimen are shown in the Figure 8.

**Figure 8**: Micro hardness sample specimen

The hardness values were measured at base material and processed zone at different locations. The average hardness of the materials are considered and given in the Figure 9. For all the three tools, the processed zone had higher hardness than the base material. This is due to inclusion of reinforcement particles in to the base materials.

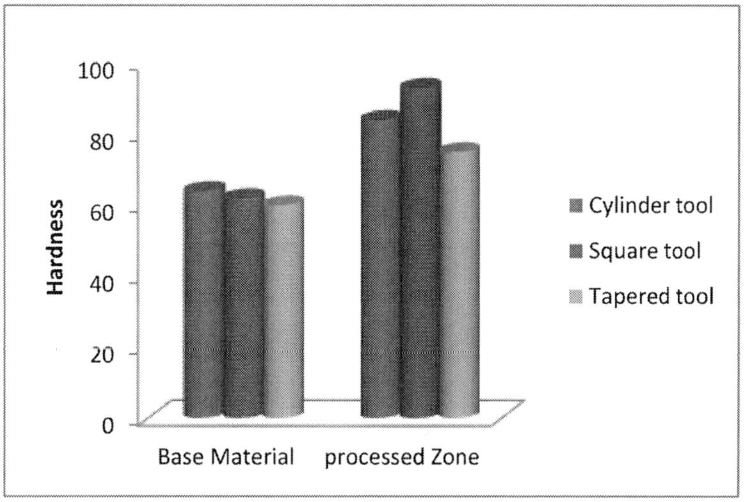


**Figure 9**: Micro hardness of FSP specimens

Based on the available literature, this is because in the stir zone the material recrystallization occurs due to the stir of the material by rotating tool. Among the three tool considered for study, square pin tool had produced better hardness values in all three zones than the other two tools by means of pulsating stirring action.

## 4. Conclusion

As a unique material processing method, Friction stir processing has great advantages in preparation of surface composites and modification of materials due to solid state processing method. This article presents the details of the aluminium based hybrid surface composites prepared by three different tools via square, cylinder and tapered. The composites were tested for its mechanical properties such as ultimate tensile strength, yield strength and micro hardness.

The test results revealed the ultimate tensile strength reached maximum of 221.92 MPa for square pin tool profile followed by cylindrical tool and tapered tool of 106.64 Mpa and 100.27Mpa respectively. The yield strength was obtained maximum as 204.73 Mpa for square tool. The results of the micro hardness test showed that at stir zone the hardness was high for all three different tool pin. This was due to stirring action of the tool ensures the proper distribution of reinforcement particles in to the base material.

## References

[1] Zhang, X.X., 2007, *Mater. Letters 61*, 3504-3506.
[2] Fei, W. D., Wang, L. D. 2004, *Thermochimica Acta* 455,7-10.
[3] Hosseini, S.A.; Ranjbar, K.; Dehmolaei, R.; Amirani, A.R. 2015, *J. Alloy. Compd.* 622, 725–733.
[4] VijayaRamnath B, Elanchezhian C, Jaivignesh M, Rajesh S,Parswajinan C, Siddique Ahmed Ghias A, 2014, *Materials Design*, 58:332–8.
[5] Poovazhagan L, Kalaichelvan K, Rajadurai A, Senthilvelan V, 2013, *Procedia Engineering* 64:681–9.
[6] Suresha S, Sridhara BK, 2010, *Materials Design*, 31(4):1804–12.
[7] Garg HK, Verma K, Manna A, Kumar R,2012, *Int J Latest Res Sci Technol* 1(1):36–44.
[8] Pandi G, Muthusamy S, 2012, *Procedia Engineering* 38:1399–408.

# Theoretical Investigation of Friction Welding on Magnesium and Aluminium Alloy

**R. Praveen Kumar[1]*, R. Rajavel[1], G.Navaneethakrishnan[2]**
[1]Academy of Maritime Education and Training / Assistant Professor, Department Of Mechanical Engineering, Chennai, India
[2]K.Ramakrishnan College of Technology / Assistant Professor, Department of Mechanical Engineering, Trichy, India
*Corresponding Author: praveen71989@gmail.com

**Abstract:**

In this study, the weld ability of Aluminium Alloys 5083, 6061, 7075 and Magnesium AZ31 was investigated using Friction Welding. Magnesium AZ31 is the base alloy used and other Aluminium alloys were welded with Magnesium Alloy. These Alloys are welded by changing the parameters. Parameters which were used are speed of the spindle, pressure applied during friction, time for which the materials are kept under friction time, Forging Pressure and Forging time. Once the materials are welded by the process of friction welding the weld specimen joints are examined for its changes in its microstructure by the Scanning Electron Microscope (SEM). In order to evaluate the welded specimens mechanical properties the tests such as tensile test and micro hardness tests are carried out. It is found that a specific strength by friction welding method is achieved by aa combination of Magnesium alloy AZ31 along with Alumunium alloys such as 5083, 6061, 7075 by the experimental results.

**Key Words:** Friction welding, Microstructure, Tensile Strength, Forging Pressure, Friction Pressure, Friction Time.

## 1. Introduction

In the all available engineering metals, Magnesium which is the 8$^{th}$ most common element is found to be the lightest metal possessing a density value of 1.74 g/cm$^3$ and also more than four times lighter than steel whose density value is 7.86 1.74 g/cm$^3$ and lighter than alumunium of density 2.7 1.74 g/cm$^3$ by 35%. For its explicit properties of magnesium like high specific strength and very less weight makes its alloy the one of the lightest metals that could be employed for defense industry, aerospace industry and automobile industry. While these alloys are likening to plastics in terms of their density and evaporation properties, they also have the mechanical properties of a metal. It is a material sought in terms of density, strength, and rigidity advantage for fuel efficiency and performance enhancement. In addition, magnesium is much more rigid than engineering plastics and is a much more recyclable material.[4] Recyclability, as important as light weight, has become a key factor in the selection of materials for the protection of raw materials and energy resources of the world. Unalloyed magnesium has low strength and toughness values. Therefore, it is used by alloying. Al, Zn, and Mn from alloying elements increase strength, toughness, and corrosion resistance, respectively. However, more than 2% Zn causes red shortness [2].

Due to the extensive properties of aluminum alloys like the good formability and its high specific strength they have found its applications in many fields like communication equipment, aerospace industries and electronic industries. [3].

Table 1 Properties of Magnesium and Aluminium [4]

| Property | Magnesium | Aluminium |
|---|---|---|
| Crystal structure | hcp | FCC |
| Density at 20°C ($g/cm^3$) | 1.74 | 2.70 |
| Coefficient of thermal expansion 20–100°C ($\times 10^6$/C) | 25.2 | 23.6 |
| Elastic modulus [Young's modulus of elasticity] ($10^6$ Mpa) | 44.126 | 68.947 |
| Tensile strength (Mpa) | 240 (for AZ91D) | 320 (for A380) |
| Melting point (°C) | 650 | 660 |

The design flexibility and weight reduction capacity of the AL-Mg composite structure makes the combination to expand its application and also to join them easily by the usage of the traditional methodology called fusion welding. [1]. But along the fusion zone a very large number of continuously distributed brittle intermetallic compounds (IMCs) were available which in turn makes the weld material to undergo severe thermal cracking thus making the Al-Mg fusin weld unfit for industrial applications. Due to the excellent mechanical properties and very good solid state character, this solid state welding technique has found its way for the joining of both similar and dissimilar materials. Due to the limitation in the welding process of Linear Friction Welding (LFW) it is not feasible to join few structural components such as the bars.[3].

By decreasing the welding time and lowering the heat input in the joints of Al and Mg bars by friction welding the formation of IMC's could be controlled effectively. For its improved material savings and reduced production times the friction welding process has fond its applications in automobile industry and aerospace industry.[3]. Mainly for joining Al alloy for its usage in the aerospace industry and marine automotive industry and other commercial applications the solid state welding method of friction welding is being used. [5]. Due to the wide applications over the other conventional welding methodology such as zero fuming, absence of spatter or porosity or blowholes, very less distortion, non-requirement of special surface treatment methods, nil consumable requirement like filler wire and non-requirement of shielding gas Friction Welded joints have an improved mechanical properties compared to the conventionally welded materials. Despite of its unmatchable advantages this friction welding process has its disadvantage too as it leaves an exit hole when the tool is withdrawn at the end of the welding process which is not preferable in most of its applications. [6]. In order to overcome this disadvantage, for continuous trajectory an offset is being introduces in its path, or for non-continuous trajectory it has been continued into a dummy plate, or the unwanted part with the hole has been machined off. The friction welded joints with certain geometries faces limitation in its application because of rigid clamping of the plates which are to be welded and very large down forces.[5]. Due to the formation of cracks, void in weld zone and expulsion it is very difficult to join magnesium alloys by conventional methods, whereas friction welding has the capacity of joining magnesium alloy as it do not melt the alloy which removes the solidification problem. A good quality weld is obtained in friction welding since it doesn't require any filler material which removes the metallurgical problems in the weld zone.[7].

The magnesium alloy named AZ31 having very good mechanical properties is commercially made available in various forms such as cylindrical rods and possess brittleness and minimum ductility at room temperature. It is been able to achieve super plastic behavior by increasing the temperatures under certain constrained conditions from the AZ31 magnesium alloy which is in cylindrical rod from. By the process of refining and homogenizing of the grain structure of the cylindrical rod of AZ31 magnesium alloy we are able to improvise its formability and ductility. [8]. For various combinations of aluminium alloys and for joining many dissimilar alloys and metals the process of friction stir welding has shown wide joining possibilities as it doesn't involve any form of consumables or melting of the materials. Here we study about the magnesium alloy AZ31 by the friction welding method, its optical metallography and the base metal and weld zone microstructures comparison is provided by transmission electron microscopy. [9]. Due to formations of IMC's is limited because of very less processing temperature and the IMC's distribution is also dispersed by high strain rate and adverse plastic deformation during friction welding we get high quality Al-Mg alloy welded joints. [10].

## 2. Methodology

### 2.1 Model description

To begin with the alloys such as alumunium 5083, alumunium 6061, alumunium 7075 and magnesium AZ31 are taken and investigated for their mechanical properties. These alloys are being welded together by varying five different process parameters like the speed of the spindle, pressure applied during friction, time for which the materials are kept under friction time, Forging Pressure and Forging time and the welded joints are tested for the mechanical properties like its tensile strength, its hardness and its microstructure which were displayed in Table 1.

In order to fuse the materials with one another by plastically displacing them with the mechanical friction, this generates heat between the two workpieces which are rotating in relative motion along with a lateral force known as 'upset' is known as friction welding process that is a solid-state welding process.[11] that is used for welding of various metals and other thermoplastics which find their applications in the aviation industry and automotive industry. [12]. As there is no melting occurring this friction welding is not the traditional fusion welding process but it involves more of forge welding technique.

### 2.2. Parameters Description

The spindle speeds and feeds are two different machining parameters which refers the two separate velocities that are in use with the machine tool practice, which are called as the cutting speed and feed rate. Due to their combined effect in the cutting process they are considered as a pair but both of them could be taken individually and analyzed on their own. [13]

Friction pressure is the pressure applied at the joining of two metals during the process of the friction welding. The pressure is applied at the joining end of both metals. The duration of friction pressure depends on the friction time. Friction acts on materials that or in content with one another. Plays vital role in friction welding. In this process more amount pressure applied at one practices area, so that the welding process in very effective. [14]

Due the impact of the normal force during the friction time, the work piece will break away and initiate itself to move. The kinetic frictional force is almost the same value of the number of times the coefficient of kinetic friction to that of the normal force. As it is mentioned earlier, this force is always opposite to the direction of motion.

IOP Conf. Series: Materials Science and Engineering **923** (2020) 012023 doi:10.1088/1757-899X/923/1/012023

The shaping of the metal by the utilization of local compressive forces is one of the manufacturing processes known as forging in which a hammer or a power hammer or a die is often used to deliver a blow. Based on the temperature the forging operation is performed this forging is classified into three different ways,

- cold forging (a cold working method)
- warm forging
- hot forging (a hot working method)

In the last two methods the metal will be heated usually in the forge. [15].

Forging time is the time taken for the forging pressure applied during the process of friction welding which is often known as the time interval of forging process. Forging time is mostly same as the frication time in friction welding process. [16]

### 2.3. Chemical Compositions of Alloys

The magnesium alloy AZ31 which are commercially purchased are of extruded 19 mm rods used for the investigation whose chemical composition is given in Table 2. [17].

Table 2 Chemical compositions of magnesium AZ31

| Alloy | Al | Zn | Mn | Fe | Si | Cu | Ni | Mg |
|---|---|---|---|---|---|---|---|---|
| AZ31 | 3.26 | 0.92 | 0.43 | 0.004 | 0.001 | 0.000 | 0.001 | bal. |

The chemical analysis which gave the clear chemical composition of alumunium alloy along with its tensile strength is given in Table 3. [18].

Table 3 Chemical composition of aluminium and its tensile strength

| Aluminium | % Sn | % Pb | % Zn | % Mn | % Fe | % Ni | % Si | % Mg | % Sb | % Cr | % Ti | % Cu | % Al | Tensile strength (MPa) |
|---|---|---|---|---|---|---|---|---|---|---|---|---|---|---|
| | 0.00500 | 0.03360 | 1.14000 | 0.11800 | 0.57400 | 0.01220 | 0.55400 | 0.17100 | 0.00300 | 0.02420 | 0.01340 | 0.59300 | 96.76000 | 200 |

## 3 Result and Discussion

### 3.1 Forging characteristics

The pectoral representation of the friction and the forge load characteristics are displayed in Figure 1. The duration of welding process, the speed of rotation, the friction and forge load are the factors which are to be explored. Before the commencement of the welding operation the ends of the samples are cleaned in order to minimize the organic contamination along the welding zone. the thermal phenomenon changes which occur during the welding is determined by the welding parameters such as speed of rotation, the duration of the weld, the load applied for creating friction. A small amount of temperature is achieved at the weld-zone by increased friction load which in turn produces very minimum intermetallic compound at the interface, while considering the interface consolidation the forging load becomes more predominant. When the resultant weld properties are examined for the effect of the parameters the findings state that the amount of friction load during the heating period and the rotation speed gives the conditions of deformation which activates the surfaces in contact that limits the time conditions and temperature for joint heating that minimizes the formation of intermetallic compounds. [19].

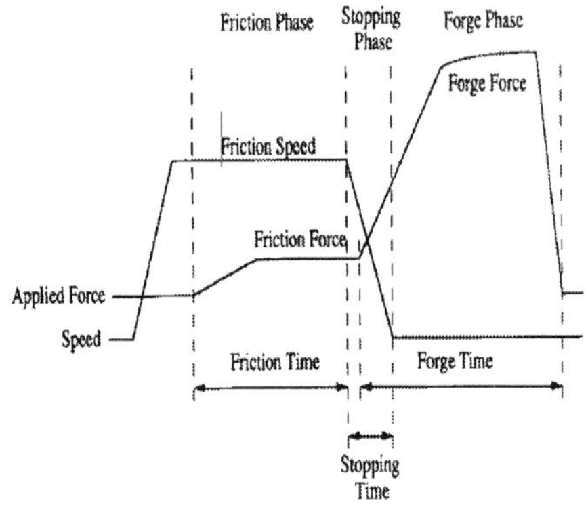


Figure.1 characteristics of the friction and the forge load
5

### 3.2. Distribution of microstructure:

When investigating the micro hardness it resulted in an uneven distribution and also the stir zone hardness was much higher than that of the base material. [20]. In the middle of the stir zone the maximum hardness value was recorded which twice the hardness value of the base material was. The hardness value was lower when it was measured along the 1.5 mm line that the 3 mm and 4.5 mm line because of the absence of intermixing which occurred between the 5083 Al alloy and AZ31 Mg alloy in top region which is clearly shown in Figure 2. The reason for the sharp variation in the weld zone was the intercalated structure and the onion ring structure. [21].

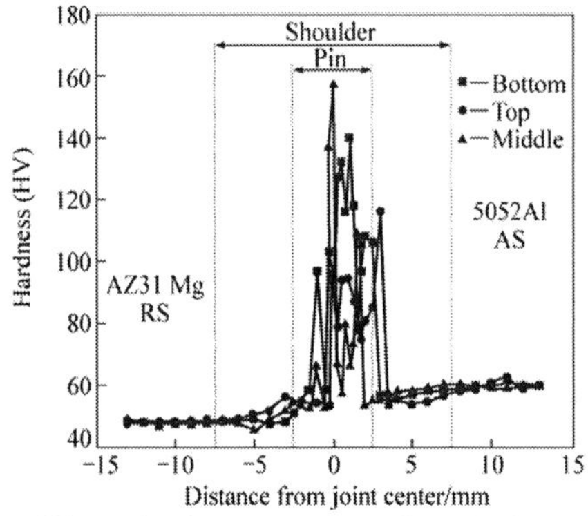


Figure 2.Microhardness profiles of microstructure from Mg to Al with different locations

### 3.3 Tensile strength of friction welding:

For minimizing the heating the workpieces are machined with coolant. The samples were tested for the tensile test at the room temperature with the traverse of the weld to its same weld direction during which along the heat affected zone (HAZ) the majority of specimens occurred. The AZ31 magnesium alloy tensile test results are shown in Figure 4 which are welded at the same speed and pressure. It is concluded that, in certain extension the tensile strength is higher when the travel speed is also higher. At 90 mm/min travel speed the joint strength was same as the base material strength, but when the travel speed is increased above 90 mm/min the joint strength started to decrease. [22].

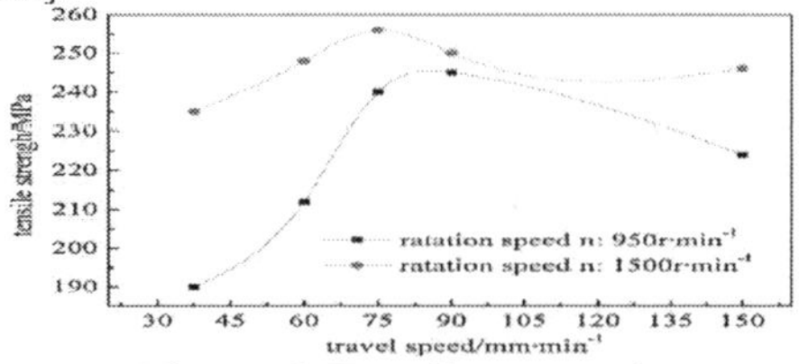


Figure 3 Tensile strength

The normally rolled AZ31 sheets under H24 condition which are partially annealed whose grain size ranges from 10–20 μm and the asymmetrically rolled sheets mechanical properties are given in Table 3. For enhancing the room temperature ductility it is suggested for reverse rolling as it is effective in AZ31 magnesium alloy. For clarifying the higher ductility of the material which is being possessed by reverse rolling is compared to unidirectional rolling, and the fractured specimens optical microstructures were observed. [23].

Table 4. Tensile mechanical properties of the materials which are processed by normal rolling and differential speed rolling

| Rolling method | YS, MPa | UTS, MPa | $e_f$, % | Ref. |
|---|---|---|---|---|
| Normal rolling | – | 314 | 9 | [10] |
| Differential speed rolling | | | | |
| Reverse | 271 | 311 | 14.6 | This work |
| Unidirectional | 258 | 300 | 7.9 | This work |

### 3.4. Microstructure and properties of FW of similar Metals

The effect of Friction Welding conditions and aging treatment on the mechanical properties of type 7075-T6 aluminium alloy (A7075) friction welded joints. By giving the heat input lower than the conventional method the work pieces are welded by using LHI method. The braking forces are not able to instantly stop the rotation of the work piece in the conventional method of continuous drive friction welding. [24]. It is impossible to decrease the speed to instant zero by the relative speed that is in between both the specimens and there is a deformation in the joint that is welded during braking occurs which is shown in Figure 4a. In the LHI method the joining of the specimens is achieved by an electromagnetic clutch in order to prevent the deformation which occurs during braking while there is a decrease in the rotational speed. There is an instant decrease in the relative between the specimens and is brought to zero when the clutch is being released which is shown in Figure 4b. Here the friction pressure is being maintained such that the influence of the braking time over the deformation becomes negligible. [25]. Therefore the base metal strength was equal to the strength of the LHI method welded joint of low carbon steel (LCS). There were very minimum axial shortening and flash in LHI method joints than conventional method joints. Hence it is inferred that the LHI method heat input was very less than the conventional method heat input. [26].

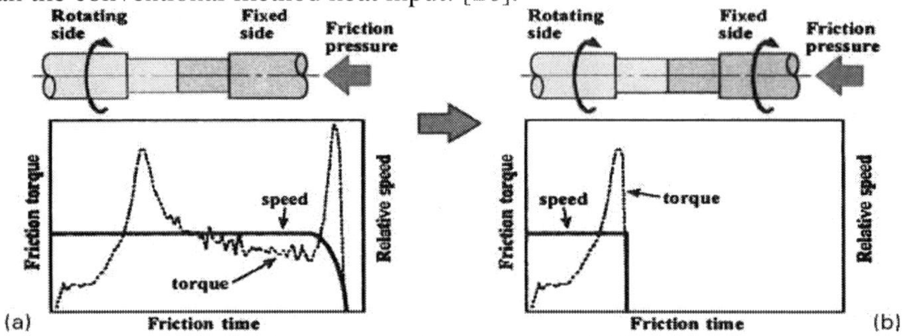


Figure 4 (a) Conventional Method, 4 (b) LHI Method

### 3.5. Effect of friction pressure and friction time on the joint efficiency:

The efficiency of joint and the weld parameters are shown in Figure 5. There could be a result of unbounded regions as a result of irregular heating effect when we decrease the heating times. The strength of the friction welding may decrease due to slow cooling rate and elongated heating times. As there is an increase in the friction time there is a decrease in the joint strength for the specimens which are tested are shown in Figure 5. The clusters and defects that are preset in a composite material may decrease its ductility when it is subjected to loading and those areas may fail permanently. These are the attributed factors for the material failure at the interface which is known as the weld zone (WZ). [27].

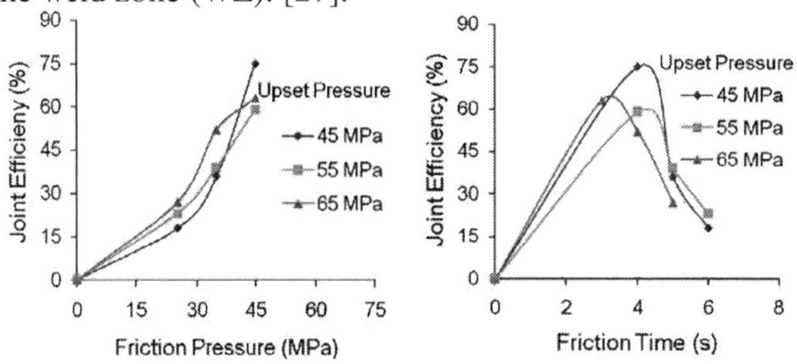


Figure 5 (a) Joint efficiency vs. friction pressure 5 (b) Joint efficiency vs. friction time

Table 5. Welding parameters [27]

| Test no. | Friction pressure (MPa) | Friction time (s) | Upsetting pressure (MPa) | Upsetting time (s) |
|---|---|---|---|---|
| T1 | 25 | 6 | 45 | 5 |
| T2 | 35 | 5 | 45 | 5 |
| T3 | 45 | 4 | 45 | 5 |
| T4 | 25 | 6 | 55 | 5 |
| T5 | 35 | 5 | 55 | 5 |
| T6 | 45 | 4 | 55 | 5 |
| T7 | 25 | 5 | 65 | 5 |
| T8 | 35 | 4 | 65 | 5 |
| T9 | 45 | 3 | 65 | 5 |

### 3.6 Tensile tests

The strength was 97% of the original material in the case of 5083-5083 joint, while it was 67% with the case of 6061-6061 joint, whereas the strength of the 5083-6061 and 6061-6061 is the same. The tensile strength is dependent on the type of the similar or dissimilar alloys being welded. The friction welding of similar material like 5083 and 6061 alumunium alloys and dissimilar materials like 5083 and 6061 alumunium alloys were done and the combinations were joined successfully.[28]. Thus friction welding has joined the dissimilar alloys and the properties like tensile strength and hardness distribution were based on the alloy combination. [29].

Table 6. Tensile strength

| | Strength (MPa) | Elongation (%) |
|---|---|---|
| Alloy | | |
| 5083 | 328 ± 2 | 24 ± 1 |
| 6061 | 320 ± 2 | 16 ± 1 |
| Joint | | |
| 5083-5083 | 318 ± 2 | 21 ± 3 |
| 6061-6061 | 199 ± 6 | 11 ± 1 |
| 6061-5083 | 202 ± 3 | 7 ± 1 |

**3.7 Mechanical data**

The samples tested for the initial strain rates from 3 _ 10_5 to 10_2 s_1 for their variation in the nominal true stress with elongation which is shown in Figure 6. All the samples which were taken for the test failed except the sample which was tested at 3 _ 10_5 s_1 as the tensile pull was terminated while it was in the elongation of 50% anticipating the long test duration. [30]. Assuming uniform deformation and constant volume under the basis of the displacement of the load and the cross load the nominal true stress was calculated. [13]. following a flow softening the flow stresses reached the peak at relatively low elongations are observed by the mechanical test results which indicate the strain hardening at very high strain rates. Before gradual flow softening the hardening values are going up to elongations of 50-100% at very low strain rates. A maximum elongation of 475% is observed at a strain rate of 10_4 s_1and at a high strain rate of 10_2 s_1 the elongation value decreased to 160%. The conditions that correspond to superplastic and non-superplastic flow the experiments were made on GBS and texture focused on strain rates of 10_4 and 10_2 s_1 are based on ductility value. The corresponding strain rate sensitivity and ductility are consistent with the observations which are made. [31].

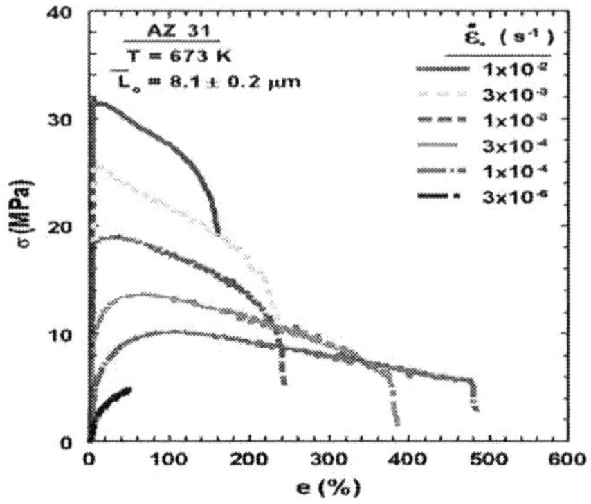


Figure 6 True stress–elongation behavior of AZ31

**3.8 Results and discussion:**

Based on its excellent electrical conductivity than copper and 100% recyclable nature without losing its original characteristics and their very high strength to weight ratio aluminium alloys are considered best than magnesium alloys are inferred from the works.[32]. Aluminium oxide quickly resists corrosion and the resulting surface coat of aluminium oxides resists further corrosion, by air, water and Chemicals this protective coating is clear, colorless and non-staining. [33]. The strong alumunium alloys or the alloy steels are comparable with the strength-to-weight ratio of the precipitation hardened magnesium alloys. [34-36]. Low density, specific modulus, stand greater column loading per unit weight, high thermal conductivity that allows rapid heat dissipation are the properties of the magnesium alloys. When magnesium alloy compared with alumunium alloys they are very difficult to be deformed by cold working and they are also very costly but their efficiency do not match with the other materials of the range. [35-37]. Some aluminium alloys especially those in the 5xxx and 7xxx series (example 5083 and 7075) are so strong that they could be used in armour structures. Aluminium 5083 alloy cannot be heat treated and Aluminium 7075 alloy can be heat treated. [38].

Aluminium alloy majorly used in ship building vehicle bodies and pressure vessels etc. It is preferred majorly because of its lightweight and resistant to corrosion aluminium alloys maintains strength at high temperatures. Magnesium alloys also lightweight material and has high tensile yield strength. Both the materials all lightest of the commonly used alloys. [39-40]

## 4. Conclusion

This research paper investigates the process of collection of materials and completion of research through literature paper. The systematic studies that have been so for carried out and they are said to optimize parameters of the friction welding in order to attain better Mechanical properties in Magnesium and Aluminium dissimilar joints. With the knowledge of various research papers, we will further proceed to the welding process magnesium AZ31 is to be joined with aluminium 5083, aluminium 6061 and Aluminium 7075 after that various test will be carried out to evaluate their Mechanical properties like tensile strength, hardness and microstructure. Analyzation of the test results will take place after that all these processes

1. Keeping in mind the prime target of environmental conservation these magnesium alloys when implemented in automobile industry can be a key factor for weight reduction which will effect in the decrease of fuel consumption and $CO_2$ emissions.

2. Taking the property of ductility the Al-condition weld by friction welding increases the strength of weld joint than the O-condition weld of parent material of all trials.

3. With the help of the spindle speed and the x-axis torque we may calculate the heat input in the Al-Mg friction weld and by placing the Al on advancing side and by offsetting the tool to Al, the heat input is increased and then decreased with the increase of traverse speed and rotation speed.

4. The reaction layer that was made with the IMCs layer and Mg solid solution layer. The thickness of IMCs layer in Al/Mg joints increased and the Mg solid solution has transformed the IMCs gradually as the friction time increased.

5. The Mg base material micro hardness was lower than the Al/Mg friction interface. As the friction time increased the thickness of the hardened layer on Mg side also increased, whereas on the Al alloy side, the micro hardness in the zone close to the friction interface was lower than that of the Al base material.

**References:**

1. L.E. Svesson , L. Karlsson, H. Larsson, B. Karlsson, M.Fazzini, and J. Karlsson, Microstructure and Mechanical Properities of Friction Stir Welded Aluminium alloys with Special reference to AA 5083 and AA 6082.
2. Uğur Çalıgülü, Mustafa Türkmen, Ali Kaya Gür, Remzi Kallı, Friction Welding Of AZ63 Magnesium Alloys.
3. Zhida Liang, Guoliang Qin*, Liyuan Wang, Xiangmeng Meng, Fei Li, Microstructural characterization and mechanical properties of dissimilar frictionwelding of 1060 aluminum to AZ31B magnesium alloy, Key Laboratory for Liquid-Solid Structural Evolution and Processing of Materials Ministry ofEducation, Shandong University, Jinan 250061, PR China
4. Mustafa Kemal Kulekci, Magnesium and its alloys applications in automotive industry. Received: 21 May 2007 / Accepted: 9 October 2007 / Published online: 28 November 2007 # Springer-Verlag London Limited 2007
5. N. T. Kumbhar and K. Bhanumurthy, Materials Science Division Bhabha Atomic Research Centre, Friction Stir Welding of Al 6061 Alloy
6. Emel Taban a,*, Jerry E. Gould b, John C. Lippold, Dissimilar friction welding of 6061-T6 aluminum and AISI 1018 steel:Properties and microstructural characterization
7. N. Afrin a, D.L. Chena,∗, X. Caob, M. Jahazi b, Microstructure and tensile properties of friction stir welded AZ31B magnesium alloy
8. B.M. Darras, M.K. Khraisheh ∗, F.K. Abu-Farha, M.A. Omar, Friction stir processing of commercial AZ31magnesium alloy
9. J. A. ESPARZA, W. C. DAVIS, E. A. TRILLO, L. E. MURR, Friction-stir welding of magnesium alloy AZ31B, Department of Metallurgical and Materials Engineering, The University of Texas at El Paso, El Paso, TX 79968-0520, USA.
10. Banglong Fu, Guoliang Qin∗, Fei Li, Xiangmeng Meng, Jianzhong Zhang, Chuansong Wu, Friction stir welding process of dissimilar metals of 6061-T6aluminum alloy to AZ31B magnesium alloy
11. A.Vairis*, M. Frost , High Frequency Linear Friction Welding of a titanium Alloy
12. A. Fuji, M. Kimura, T. H. North, K. Ameyama, and M. Aki, Mechanical properties of titanium-5083 aluminium alloy friction joints
13. S. Fukumoto, H. Tsubakino, K. Okita, M. Aritoshi, and T. Tomita, Friction welding process of 5052 aluminium alloy to 304 stainless steel

14. W H Jiang and R Kovacevic*, Feasibility study of friction stir welding of 6061-T aluminium alloy with AISI 1018 steel. Research Centre for Advanced Manufacturing, Southern Methodist University, Richardson, Texas, USA
15. M. Kimura a,*, A. Fuji b, S. Shibata c,1, Joint properties of friction welded joint between pure magnesium and pure aluminium with post-weld heat treatment
16. A. Gerlich*1, M. Yamamoto2 and T. H. North3, Local melting and cracking in Al 7075-T6 and Al 2024-T3 friction stir spot Welds
17. Yu Yoshida*2, Lawrence Cisar*2, Shigeharu Kamado and Yo Kojima, Effect of Microstructural Factors on Tensil Properties of an ECAE-Processed AZ31 Magnesium Alloy
18. Mumin Sahin, Joining of stainless-steel and aluminium materials by friction welding
19. Bekir S, Yilbas , Ahmed Z. Sahin , Nafiz Kahraman, Ahmed Z. Al- Garni .Friction welding of St- Al and Al –Cu materials
20. Material Flow Behavior during Friction Stir Welding of Aluminum, BY K. COLLIGAN
21. School of Mechanical and Automotive Engineering, South China University of Technology Guangzhou 510640, China, Dissimilar friction stir welding between 5052 aluminum alloy and AZ31 magnesium alloy
22. Wang Xunhonga,*, Wang Kuaishe a,b, Microstructure and properties of friction stir butt-welded AZ31 magnesium alloy
23. H. Watanabe, T. Mukai, K. Ishikawa, Differential speed rolling of an AZ31 magnesium alloy and the resulting mechanical poperties
24. S.D. Meshram*, T. Mohandas, G. Madhusudhan Reddy. Friction welding of dissimilar pure metals
25. A. A. McLean, G. L. F. Powell, I. H. Brown and V. M. Linton , Friction stir welding of magnesium alloy AZ31B to aluminium alloy 5083
26. M. B. Uday*, M. N. Ahmad Fauzi, H. Zuhailawati and A. B. Ismail, Advances in friction welding process: a review
27. M. Srinivasan a,⇑, C. Loganathan b, V. Balasubramanian c, Q.B. Nguyen d, M. Gupta d, R. Narayanasamy e, Feasibility of joining AZ31B magnesium metal matrix composite by friction welding
28. J. A. Esparza, W. C. Davis, E. A. Trillo, L. E. Murr, Department of Metallurgical and Materials Engineering, The University of Texas at El Paso, El Paso,, Friction-stir welding of magnesium alloy AZ31B
29. G. Liu, L.E. Murr, C-S. Niou, J.C. McClure, and F.R. Vega Department of Metallurgical and Materials Engineering, Microstructural Aspects Of The Friction-Stir Welding Of 6061-T6 Aluminum
30. M.W. Mahoney, C.G. Rhodes, J.G. Flintoff, R.A. Spurling, And W.H. Bingel, Properties of Friction-Stir Welded 7075 T651 Aluminum
31. R. Panicker a, A.H. Chokshi a,*, R.K. Mishra b, R. Verma b, P.E. Krajewski, Microstructural evolution and grain boundary sliding in a superplastic magnesium AZ31 alloy
32. I . Shigematsu, Y.-J. Kwon, K. Suzuki, T. Imai, N. Saito, Joining of 5083 and 6061 aluminum alloys by friction stir Welding

33. A.C. Somasekharan, L.E. Murr*, Microstructures in friction-stir welded dissimilar magnesium alloys and magnesium alloys to 6061-T6 aluminum alloy
34. S. Rajakumar, C. Muralidharan, V. Balasubramanian. Establishing empirical relationships to predict grain size and tensile strength of friction stir welded AA 6061-T6 aluminium alloy joints
35. M. Peel ∗, A. Steuwer, M. Preuss, P.J. Withers, Microstructure, mechanical properties and residual stresses as a function of welding speed in aluminium AA5083 friction stir welds
36. Zhikang Shen, Xinqi Yang ⇑, Zhaohua Zhang, Lei Cui, Tielong Li, Microstructure and failure mechanisms of refill friction stir spot welded 7075-T6 aluminum alloy joints
37. Jiuchun Yan, Zhiwu Xu *, Zhiyuan Li, Lei Li, Shiqin Yang, Microstructure characteristics and performance of dissimilar welds between magnesium alloy and aluminum formed by friction stirring
38. Sori Wona,1, Bosung Seoa,1, Jong Min Parka, Hyo Kyu Kima, Kuk Hyun Songb, Seok-Hong Minc, Tae Kwon Hac, Kwangsuk Parka, Corrosion behaviors of friction welded dissimilar aluminum alloys
39. Sunggon Lim, Sangshik Kim, Chang-Gil Lee, And Sungjoon Kim, Tensile Behavior of Friction-Stir-Welded Al 6061-T651
40. A. Kostka,a,* R.S. Coelho,a J. dos Santosb and A.R. Pyzall, Microstructure of friction stir welding of aluminium alloy to magnesium alloy.

IOP Conf. Series: Materials Science and Engineering **923** (2020) 012024 doi:10.1088/1757-899X/923/1/012024

# Tribological and thermogravimetric analysis of pineapple fibre reinforced epoxy composite

**J Lilly Mercy[a#], Daniel Samuel Parmar[a] and Shlok Srivastava[a]**

[a]School of Mechanical Engineering, Sathyabama Institute of Science and Technology, Chennai, Tamil Nadu 600119, INDIA.

[#]Corresponding author, email: lillymercy.j@gmail.com

**Abstract** Natural fibre composites are preferred for various static and dynamic applications owing to its biodegradable capacity and strength. Pineapple fibres obtained by the retting process from pineapple leaf were successfully reinforced with epoxy resin with different fibre volume percentage. The wear rate and the thermogravimetric analysis of the polymer composite were studied. It was discovered that 30% reinforcement of pineapple fibre resulted in better wear and higher glass transition temperature due to the optimal fibre volume fraction and enough wetting of the resin on the fibre.

## 1. Introduction

Natural resources are God's gift to mankind. Natural fibres have gathered the attention of researchers due to its biodegradable nature after its usage lifetime, which is an essential lookout in today's polluted world. However, the usage of natural fibres gets limited due to its less strength when compared to synthetic tailored fibre, attack by microorganisms particularly when it is exposed to moist conditions, the non-uniformity in the thickness, length and composition of each fibre as it is not man made. Hence the configuration of the fibre can be expressed as a wider range rather than a numeric value with precision unlike synthetic fibres.

However, the alarming threat of pollution and the call of the hour to reinvent man made materials have pushed the researchers to optimise these natural fibres tailoring to suit the needs of the society, without harming its natural integrity. Natural fibres extracted from jute, coir, hemp, cotton, banana etc., have been in use for more than a decade in the domain of composite materials. The search for new natural fibres has not exhausted and particularly in natural fibres in research is the pineapple fibre extracted from the leaves of the pineapple plant.

Pineapple leaf fibre is glossy, smooth and has a subtle surface than other natural fibres [1]. They have high specific strength, flexural and torsional rigidity equal to jute fibres [2]. Pineapple fibres have proved to surpass the previous mechanical properties of polymer-based composite [3]. However, the water absorbing nature of the fibres need to be addressed with suitable surface modification process [2,3]. Ahmad et al [4] have performed tensile, bending and shear tests on pineapple fibre reinforced polyester matrix and concludes that different concentrations in fibre gives rise to different mechanical properties. Pineapple fibres were reinforced in polypropylene matrix and studied for its mechanical properties. It was observed that the tensile strength is higher than the pure resin while the flexural strength was lower due to fibre-fibre interactions, voids and less dispersal [5]. Composites of pineapple fibres mixed in High density polyethylene were fabricated and tested for its water absorption properties.

The advancement in the fibre volume fraction led to the increase in moisture absorption which decreased the durability of the composite [6].

To manage the water absorption property of the pineapple fibres, it is mandatory to perform at least two successive chemical treatments, to deliver high mechanical properties. Additives and binders can be added to enhance adhesion [7]. Higher cellulose content in pineapple fibres increase the tensile strength of the fibre. A homogenous distribution of compatibilizers reduce the hydrophilic effect of cellulose and hinders fracture propagation and thereby maintains the tensile strength [8].

This paper approaches towards the study of tribological and thermogravimetric analysis of the pineapple fibre reinforced epoxy samples.

## 2. Materials and Methods

### *2.1. Fabrication of the composite material*

Pineapple fibres were refined with NaOH solution and dried were used for the study. The fibres were procured from Go green products, Chennai. The fibres were checked for uniformity in length and the experimental procedure was done.

**Table 1.** Attributes of Resin and Hardener.

| Property | Specification | Unit | Araldite LY556 | Aradur HY951 |
|---|---|---|---|---|
| Viscosity at 25°C | ISO 12058 | Mpa.s | 10,000-12,000 | 10-20 |
| Density at 25°C | ISO 1675 | Gm/cc | 1.15-1.20 | 0.97-0.99 |
| Flash point | ISO 2719 | °C | >200 | >180 |

For preparation of the specimen, wet hand lay-up method was used. The mylar sheet was kept in the mould followed by the application of epoxy resin in the mould. The epoxy LY556 resin having the density of 1.15-1.20 with hardener HY951 of the density 0.97-.99, was mixed in the ratio of 1:10. Initially the resin is applied in the mould, followed by keeping the pineapple leaf fibre having the specific gravity of 1.3 gm/cc in horizontal direction. Again, the epoxy resin is applied and with the assistance of roller the epoxy resin is impregnated in the fibres. In all three layers of pineapple was sandwiched into one to make a firm layer. Later the specimen was kept at 25°C for 24 hrs. The preparation of the composite sample is shown in Figure 1.

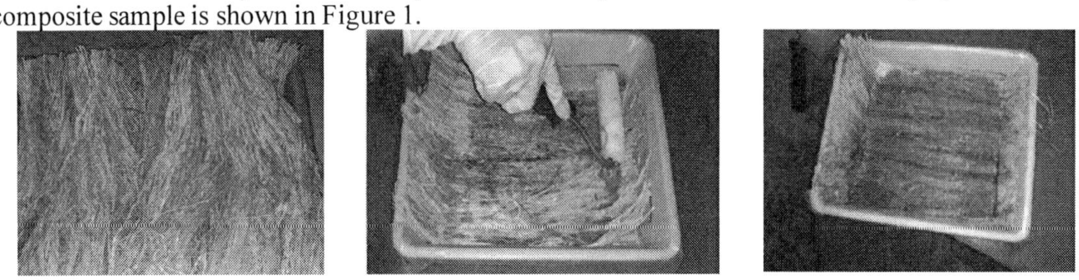

(a) Pineapple fibres (b) Layup of pineapple fibre and epoxy resin (c) One layer of fibre and resin done

**Figure 1.** Composite specimen preparation.

Two samples were produced where Sample A consisted of 50% of pineapple leaf fibre and 50% epoxy resin and the Sample B consisted of 30% of pineapple leaf fibre and 70% epoxy resin. After hardening the specimen was cut by the means of water jet machining into 25mm diameter and 10mm height. The specimens used for wear test is shown in Figure 2.

(a) Specimen A with 50% pineapple fiber and 50% Epoxy resin.

(b) Specimen B with 30% pineapple fibre and 70% epoxy resin

**Figure 2.** Samples for wear testing.

*2.2 Wear test*

The dry sliding wear studies were conducted using pin on disc machine shown in Figure 3. The composite material was taken as the disc and the pin made up of carbon steel.

**Figure 3.** Pin on disc wear testing machine.

The settings on the pin on disc wear testing machine is given in Table 2.

**Table 2.** Wear testing parameters.

| Parameter | Setting |
|---|---|
| Applied load (N) | 10 |
| Sliding velocity (m/sec) | 1.5 |
| Sliding distance (m) | 500 |
| Sliding diameter (mm) | 46 |
| Speed (rpm) | 623 |
| Time (Secs) | 333.33 |

## 3. Results and Discussions

### *3.1. Wear testing*

It was concluded that sample A (50% pineapple leaf fibre and 50% epoxy resin) whose initial weight was 14.181 gm and after the test the final weight was 13.502 gm which shows about 4.7880% wear loss. Whereas the sample B (30% pineapple leaf fibre and 70% epoxy resin) the initial mass was 17.263gm which after the test it was found 17.122gm that is 0.816775% wear loss. Sample B was denser than Sample A as the resin volume was higher.

With the advancement in the time, the volume of wear of sample A increases from 0 to 700, whereas it increases from 0 to 200 for sample B. The coefficient of friction of sample A was found increasing from 0 to 0.99 but it was 0 to 0.2 for sample B. The frictional force of sample A was 15N whereas it was found 2N for sample B. This describes that sample B (30% pineapple and 70% epoxy resin) is much durable. Figures 4-6 shows the results obtained from the Wear test. On an average the coefficient of friction of Sample A was found to be 0.83 and for Sample B it is 0.17. The lesser the fibre content, less the coefficient of friction leads to more durability.

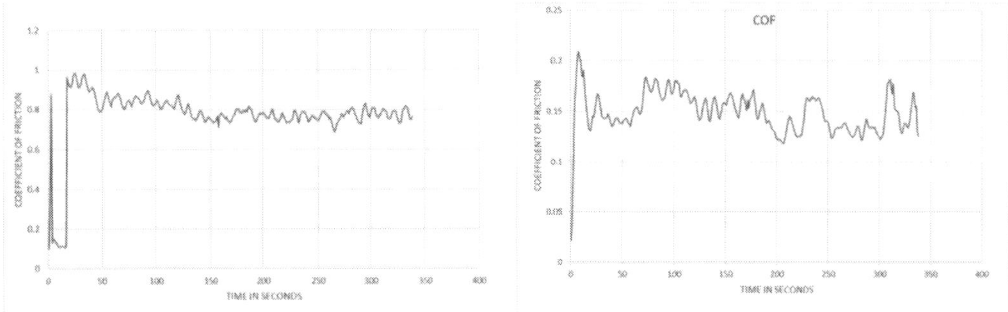


Sample A Sample B

**Figure 4.** Coefficient of friction of the samples.

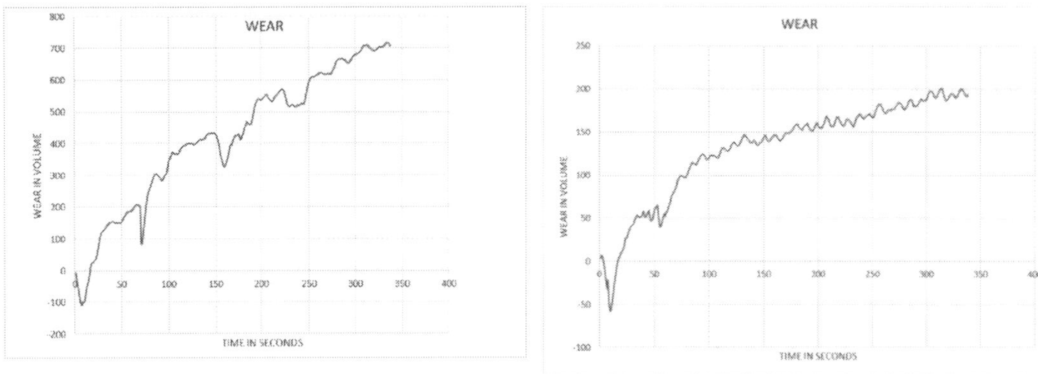


Sample A Sample B

**Figure 5.** Wear volume of the samples.

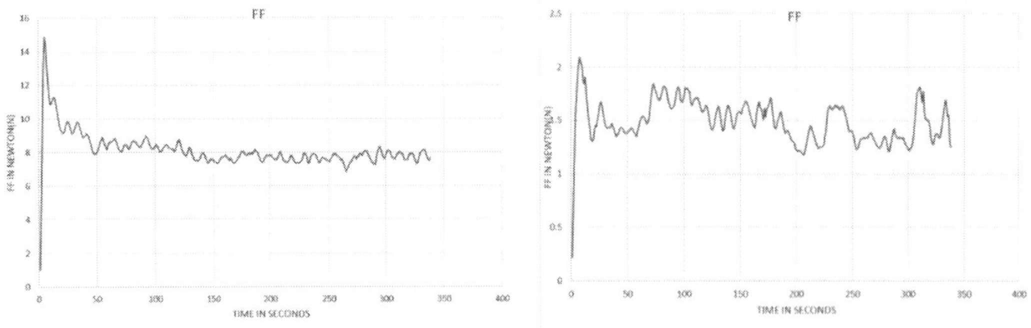


Sample A Sample B

**Figure 6.** Frictional force in the samples.

*3.2 Thermogravimetric Analysis:*

Thermogravimetric analysis was conducted using ASTM D-3418 standard with an increase of 2°C/min rate of heating. Figure 7 shows the heat flow Q across various temperatures for Sample A and Sample B. It could be observed that the Sample A yields around 52°C. Epoxy has been proved to take up between 105 to 135 °C (9). The reason of drastic reduction is the nonbonding or non-adhesiveness of the pineapple fibre with the epoxy resin. And also the higher weight percentage of fibre leaves the sample starved of the resin leading to lesser adhesiveness. Whereas Sample B yields at 2 points- one at 67.15 °C and another at 108.27 °C. The presence of two $T_g$ points show that the two materials yields at two different points independently. The fibre yields at lesser temperature well before epoxy. However, fibres in Sample B shows better performance than Sample A because of better boding due to the enhancement in the resin content.

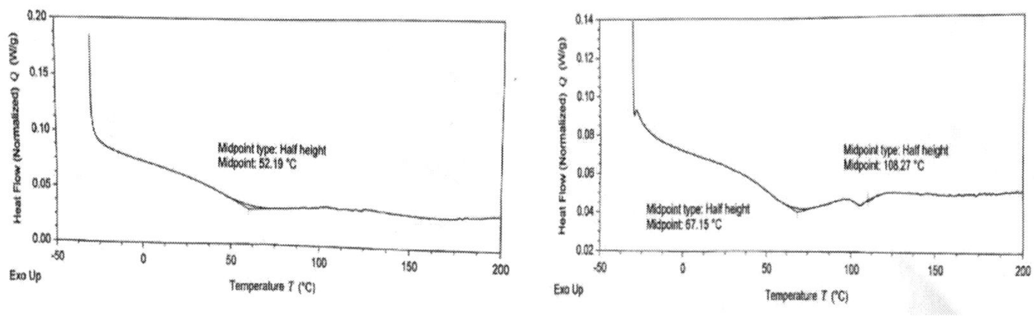


Sample A Sample B

**Figure 7.** TGA graphs for the samples.

## 4. Conclusion

Samples of pineapple reinforced epoxy composite were prepared using hand lay-up process and tested for wear and thermal properties. It was confirmed that the sample with 30% of fibre reinforced had better performance and results compared to 50% of reinforcement. However, the optimal amount of reinforcement can be decided only after testing few more samples with different reinforcement percentages.

## 5. References

[1] C. T. C. Py and J. J. Lacoeuilhe, The pineapple—cultivation and uses, *Maisonneuve & Larose*, (Quae, Paris, France), 1987.

[2] M.Asim, Khalina Abdan, M.Jawaid, M.Nazir, Zahra Dashtizadeh, M.R.Ishak and M.Enamul Hoque, A review on pineapple leaves fibre and its composites, *International Journal of Polymer Science*, 2015, http://dx.doi.org/10.1155/2015/950567.

[3] N. Lopattananon, Y. Payae and M. Seadan, Influence of Fiber Modification on Interfacial Adhesion and Mechanical Properties of Pineapple Leaf Fiber-Epoxy Composites, (*Wiley-Interscience)*, 2008.

[4] Ahmad Pauzi, Aris Doyan and Susilawati, Pineapple fibers leaves mechanical properties, *IOSR Journal of Mechanical and Civil Engineering*, **Vol.15**, Issue 2, pp.56-64.

[5] R.M.N.Arib, S.M.Sapuan, M.M.H.M. Ahamad, M.T.Paridah and H.M.D.Khairul Zaman, Mechanical properties of pineapple leaf fibre reinforced polypropylene composites, **Vol.27**, Issue 5, 2006, pp.391-396.

[6] A.Danladi and J.Shuaib, Fabrication and properties of pineapple fibre/high density polyethylene composites, *American Journal of materials science*, **Vo.4**, Issue 3, 2014, pp. 139-143.

IOP Conf. Series: Materials Science and Engineering **923** (2020) 012024 doi:10.1088/1757-899X/923/1/012024

[7] Santosh Sadasiv Todkar and Suresh A. Patil, Review on mechanical properties evaluation of pineapple leaf fibre reinforced polymer composites, (Composites Part B), **Vol.174**, 2019, 106927.
[8] Das PK, Nag D, Debnath S and Nayak LK. Machinery for extraction and traditional spinning of plant fibres. *Indian Journal of Traditional Knowledge* 2010;**9(2)**:386–93.
[9] J. Lilly Mercy and S. Prakash, Investigation on the dynamic mechanical properties of Self healing glass fibre reinforced plastics, *International Journal of Vehicle Structures and Systems*, **Vol. 9(2)**, pp.113-116.

IOP Conf. Series: Materials Science and Engineering **923** (2020) 012025 doi:10.1088/1757-899X/923/1/012025

# ENVIRONMENTAL SCANNING ELECTRON MICROSCOPY AND SURFACE ROUGHNESS ON AL -8011 HYBRID METAL MATRIX WITH CHICKEN BONE FLY ASH

**Prabhu M.K[1], sasikumar sai nishanth[2], RamPrasath k.m[2], Sayant Pradeep[2] and Ram shuriya I[2]**

[1]Assistant professor [M.E], sri krishna college of technology
[2]B.E Mechanical, sri krishna college of technology

E-mail: sayeenishanth@gmail.com (sasikumar sai nishanth)

**Abstract.** Aluminum Metal Hybrid matrix composites, which have made changes to pure material with respect to natural environmental properties like rusting, wear and tear, light, weight etc. In this study, Aluminum 8011 hybrid metal matrix composite fabricated by stir casting methods using L 9 orthogonal array. The environmental pr operties has been investigated, Environmental scanning electron microscopy, surface roughness were investigated with the casted aluminum hybrid composite. The aluminum hybrid composite was casted with 2%mg with variations of 2%, 3%, 4% of chicken bone ash and 3%, 6%, 9% of SiC as obtained from the microstructure and Environmental scanning electron microscopy. Pin on disc in which wear test shows improved material properties of hybrid matrix composite. Surface Roughness of different parameter been analyzed with taguchi method using design of experiments.

*Keywords:* Al-8011, Chicken bone fly ash, scanning electron microscopy, surface roughness, stir casting

## Introduction

In account of research development in the nanotechnology over decades as the need for hybrid, metal nano composites have improved the performance of the material. The hybrid metal matrix composites has a wide range of factors accessed with the proper distribution of nanoparticles is the major work in the hybrid composites. As the improved results of the alloy Al-8011 casted with combinations of chicken bone ash.

In addition, Sic 2%, 3%, and 4% and 2% of magnesium. In which the chicken bone fly ash has increased the hardness and reduced wear properties of the casted material. The fly ash act as a low cost nanocomposite material has helped to change the properties of the aluminum. The microstructure shows that material has been tightly packed, as reinforcements arehighly magnetic to bond the material. This hybrid composite can change the aviation industry as our material shows results of increased hardness ,reduced wear and tear , and ductile . Investigation of Mechanical properties.

Formulated Hybrid Composite Aluminum 8011 Reinforced with B4C and Al2O3 by Stir Casting Method,R.badrinath, has reviewed that hard ceramic particles increase the strength of the aluminum hybrid composites [1] . The Comparison of the Microstructure and Hardness of Al-B and Al- Mg-B Composites by Mohammad Reza Derakhshesh has studied the various bonding properties of the nanocrystals by the microstructure and scanned electron microscopy [8] .Experimental analysis of the microstructure and properties of Al-8011 by variation of copper content, Tauheed Khursheed has reviewed that hardness and tensile test. It shows that test results informed in phrase that different amount of salts added through the scanned electron microscopy has increased the hardness and wear properties [6] . Preparation and Characterization of an Aluminum 6061 Alloy based Metal Matrix Composite G. Jims John Wessleyhas studied that good bonding between the reinforcements and hybrid matrix composite [4] .Wear Behavior of Al 8011-Chicken Bone Ash Metal Matrix Composites Arun Prakash, has reviewed that abrasive wear and the influence of load, surface hardness,sliding speed, wear reinforcement fracture and morphology are the parameters that look into the wear properties [3] . Hardness and Tensile Properties of Knee Brace Produced from Cow Bone and Periwinkle Shell Composites, Abdul KareemSuleiman has studied those properties ofadding bone fly ash has increased the hardness of the hybrid metal matrix[2] .

## 2. Experimental setup

MATERIALS

An Aluminum 8011 alloy with silicon carbide and magnesium as the primary alloying element selected. Chicken Bone fly ash selected as a reinforcement nanomaterial of 300 microns .The study is concentrated on hardness[5] , wear, surface roughness and scanned electron microscopy properties of the composite materials.

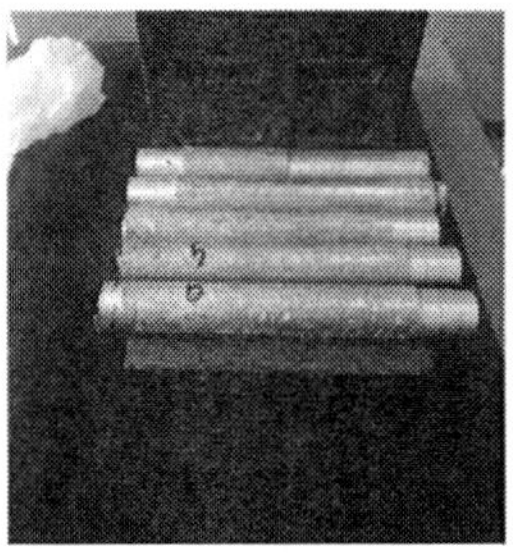

Fig 1- stir cast material

STIR CASTING

The Reinforcements of chicken bone fly ash, silicon carbide and magnesium has been distributed into matrix by Stir casting of hybrid metal matrix composites when alumina particles into aluminum melt by stirring. Molten aluminum 8011 matrix alloys containing the silicon carbide powders[9]. Motorized stirrer in the furnace was used to stir the nano particles. The formulated molten metal, with silicon carbide particles, can then be die casted .Stir casting advisable for hybrid metal matrix composites up to 20% volume of reinforcement material. during the melting and casting processes. The final distribution of reinforcements of casted material depends on properties changed

.

IOP Conf. Series: Materials Science and Engineering **923** (2020) 012025 doi:10.1088/1757-899X/923/1/012025

The experimental boundary varies with the nature of the materials with the liquedification process, physique of stirrer, compactness, and rate of hardness[7] .The sharing of the strength depends on the molten matrix depends on the area of the motorized stirrer, fixing of the motorized stirrer in the melt, with decreasing in temperature, of 720c and then attribute of the reinforcements were added

## 3. Experimental results

ENVIRONMENTAL SCANNING ELECTRON MICROSCOPY [SEM]

A scanning electron microscope (SEM) test scans a concentrated electron beam over a polished surface to generate an image. The electrons in the beam interact with the specimen, manufacturing different signals that can be nearly used to obtain information using vacuum chamber about the surface topography and composition. A beam of electrons is focused onto the surface of a specimen-using microscope, camera .Then electromyography is similar with respect to both X, and Y directions, the electrons that are returned back on the specimens surface was collected and then multiplied in magnifications.

The microscope can project highly magnified images because the area scanned on the specimen surface can be varied but the displayed image, so the area of the hybrid matrix specimen scanned to the fixed area of the display is a function of the microscopes magnification. Electrons, which was created by the source, were created up to 40 kV. The size in the Condenser lenses and then images are pointed towards the specimen through the Objective lens reaches the specimen .The wavelength is scanned across the hybrid matrix sample creating election-solid interactions. It gives different types of results given by –the Secondary electrons give information about the hybrid matrix topography, the scattered electrons which give results about al-8011 hybrid matrix composition.

**Figure 2.**

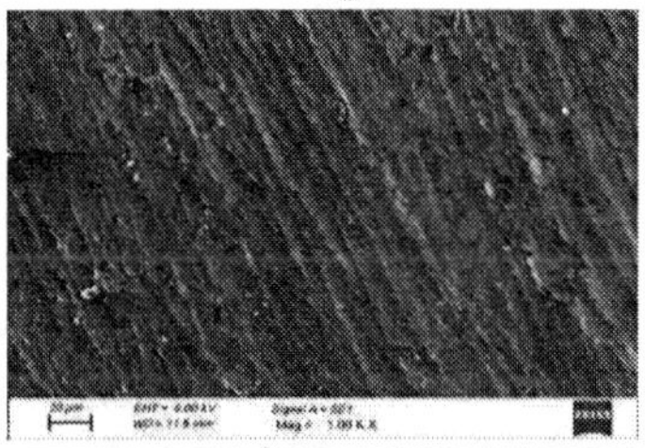

**Figure 3.**

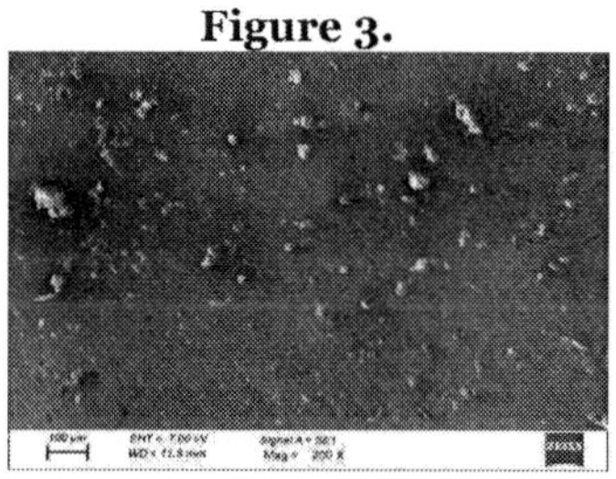

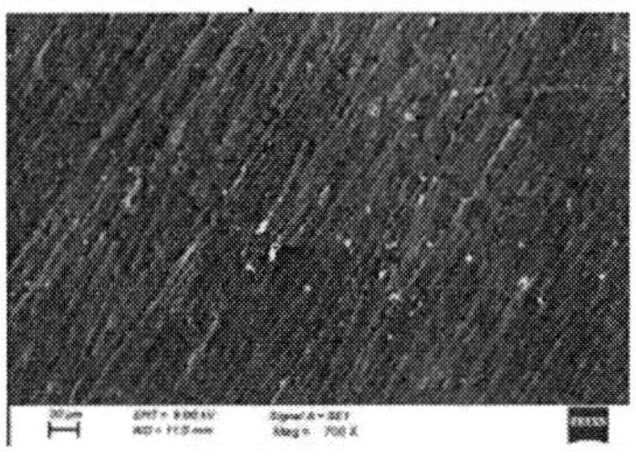

**Figure 4.**

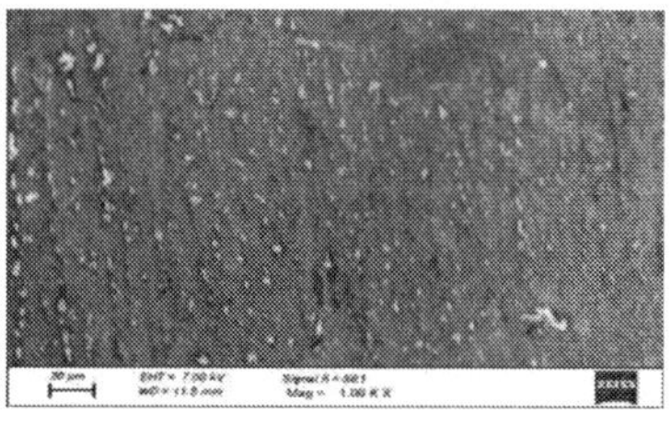

**Figure 5.**

**Figure 6.**

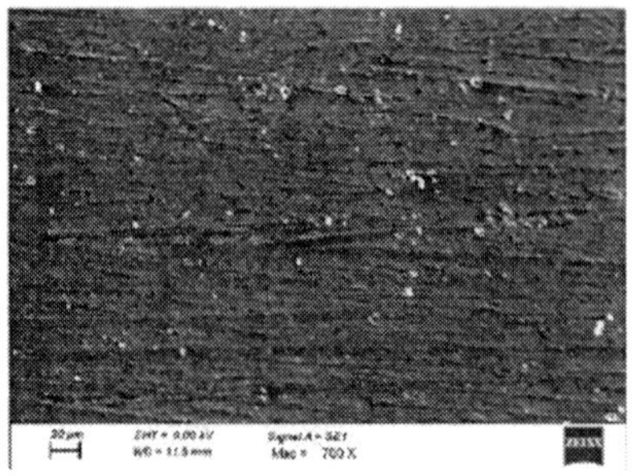

**Figure 7**

**As samples shown from fig 2 – 7 samples . as the 6 th sample shows that reinforced material is tightly bonded with al-8011**

SURAFCE ROUGHNESS

As It is not advisable to machine a specimen with perfectly smooth surface using cutting tools .the disortion and fluctuation are bound to occur. Factors affecting Vibration, material quality , type of machining, rigidity of the machine center, type& material of cutting tool , cutting condition like feed, speed, depth of cut. A portable stylus which is drawn across the surface at a constant speed for a fixed distance across the material. An wavelength signal was obtained and magnify to produce a enlarged vertical magnification as the test report shows that samples from 1 to 6 effectively as it shows that the surface is smooth at sample 6.

| SPEED | FEED 1 | DEPTH OF | Ra | Rq | Rz | Ra1 | Rq1 | Rz1 | Ra2 | Rq2 | Rz2 | Ra3 | Rq 3 | Rz 3 | Ra 4 | Rq 4 | Rz4 | Ra5 | Rq 5 | Rz 5 |
|---|---|---|---|---|---|---|---|---|---|---|---|---|---|---|---|---|---|---|---|---|
| 600 | 1 | 0.5 | 1.051 | 1.312 | 7.063 | 0.768 | 0.971 | 4.599 | 0.608 | 0.769 | 4.253 | 1.46 | 1.894 | 8.785 | 2.345 | 2.887 | 13.04 | 3.339 | 4.283 | 19.349 |
| 600 | 1.5 | 1 | 4.652 | 5.618 | 23.054 | 1.398 | 1.622 | 7.434 | 1.274 | 1.492 | 6.054 | 1.926 | 2.416 | 11.741 | 2.059 | 3.231 | 14.529 | 4.479 | 5.523 | 24.423 |
| 600 | 2 | 1.5 | 4.923 | 5.939 | 23.06 | 4.688 | 5.55 | 24.266 | 1.171 | 1.368 | 5.562 | 3.147 | 3.88 | 16.338 | 3.888 | 4.686 | 18.354 | 6.084 | 7.407 | 30.904 |
| 1200 | 1 | 1 | 1.354 | 3.286 | 9.321 | 1.458 | 1.784 | 6.018 | 0.794 | 0.941 | 7.332 | 2.346 | 2.033 | 9.011 | 4.074 | 3.468 | 16.487 | 4.667 | 5.337 | 21.334 |
| 1200 | 1.5 | 1.5 | 6.793 | 7.339 | 26.12 | 1.786 | 3.145 | 9.773 | 1.39 | 1.672 | 8.543 | 3.045 | 2.794 | 14.324 | 4.874 | 5.12 | 16.994 | 5.604 | 7.042 | 17.127 |
| 1200 | 2 | 0.5 | 7.049 | 7.329 | 26.974 | 5.374 | 7.567 | 27.024 | 1.278 | 1.694 | 9.359 | 3.997 | 4.069 | 16.974 | 5.324 | 5.667 | 17.634 | 7.741 | 7.823 | 28.607 |
| 1800 | 1 | 1.5 | 2.697 | 5.326 | 14.654 | 2.369 | 3.697 | 9.334 | 1.367 | 0.963 | 11.365 | 6.304 | 5.347 | 11.324 | 7.069 | 9.334 | 19.427 | 9.485 | 12.041 | 16.954 |
| 1800 | 1.5 | 0.5 | 3.049 | 8.012 | 27.649 | 4.021 | 7.612 | 14.347 | 1.697 | 1.687 | 13.784 | 6.774 | 7.924 | 15.627 | 7.847 | 12.687 | 19.847 | 9.991 | 10.664 | 19.654 |
| 1800 | 2 | 1 | 3.557 | 8.706 | 30.24 | 6.147 | 12.701 | 29.342 | 1.984 | 1.934 | 19.375 | 8.947 | 12.667 | 19.427 | 9.039 | 8.648 | 21.378 | 11.024 | 6.774 | 32.014 |

Table 1: surface roughness data

IOP Conf. Series: Materials Science and Engineering **923** (2020) 012025 doi:10.1088/1757-899X/923/1/012025

| Level | Ra | Rq | speed |
|---|---|---|---|
| 1 | - 16.28 | - 16.28 | -19.34 |
| 2 | - 17.79 | - 17.79 | -20.14 |
| 3 | - 19.73 | - 19.73 | -21.97 |
| 4 | - 21.94 | - 19.64 | |
| 5 | - 24.25 | - 22.11 | |
| 6 | - 19.64 | - 22.60 | |
| 7 | - 22.11 | - 20.02 | |
| 8 | - 20.02 | - 21.94 | |
| 9 | - 22.60 | - 24.25 | |
| Delta | 7.97 | 7.97 | 2.63 |
| Rank | 1.5 | 1.5 | 3 |

Table 2: surface roughness result

| Level | Ra | Rq | speed |
|---|---|---|---|
| 1 | 4.329 | 4.329 | 6.610 |
| 2 | 5.467 | 5.467 | 7.599 |
| 3 | 7.974 | 7.974 | 10.407 |
| 4 | 10.212 | 6.397 | |
| 5 | 13.036 | 9.103 | |
| 6 | 6.397 | 9.978 | |
| 7 | 9.103 | 7.353 | |
| 8 | 7.353 | 10.212 | |
| 9 | 9.978 | 13.036 | |
| Delta | 8.707 | 8.707 | 3.798 |

Table 3: surface roughness result

The graph shows that the surface roughness weaken with elevate in speed.

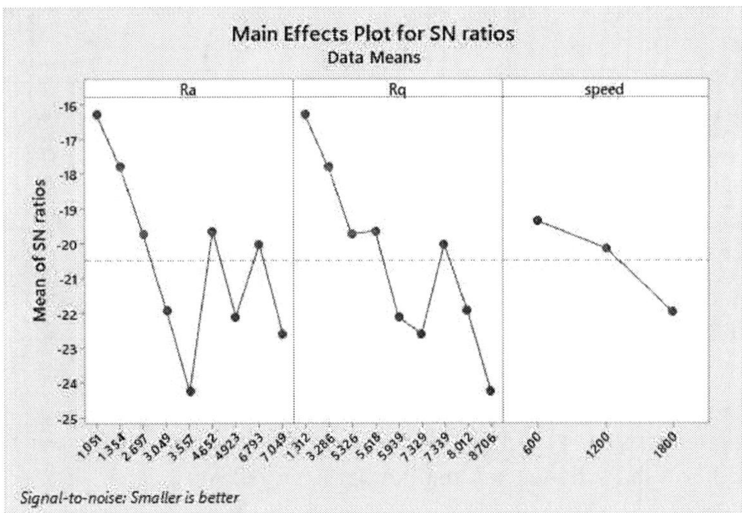


Figure 8: surface roughness graph

Conclution

1.ENVIRONMENTAL SCANNING ELECTRON MICROSCOPY [SEM]

As the test results show that in addition of the chicken one fly ash the results show that materials are tightly bonded as the nano composite bonds with al-8011.

2.SURFACE ROUGHNESS

As the test results show that the material has undergone various parameters on turning process .with speed of 600, 1200, 1800 respectively and variations of feed and depth of cut .on analyzing the casted specimens the Ra ,Rq and Rz values was studied and analysied with respect to design of engineering using taguchi method and found that surface roughness is less at higher speed

FUTURE WORK

As the material shows that surface roughness is high at lower speed an addition of graphite will lower the surface roughness at lower speeds.

REFRENCES

[1] Investigation of Mechanical Properties on Newly Formulated Hybrid Composite Aluminium 8011 Reinforced With B4C and Al2O3 by Stir Casting Method R. Kandan 2017

[2] Hardness and Tensile Properties of Prophylactic Knee Brace Produced from Cow Bone and Periwinkle Shell Composites Abdulkareem Sulaiman 2019

[3] Wear Behaviour of Al-Chicken Bone Ash Metal Matrix Composites Arun Prakash. J, 2016

[4] Preparation and Characterization of an Aluminium 6061 Alloy based Metal Matrix Composite
G. Jims John Wessley 2019

[5] Composite Materials of Al-7075,TiB2,TiC Tensile strength and Rockwell Hardness Test
Pullareddy.M 2016

[6] Experimental analysis of microstructure and mechanical properties of Al-8011 by variation of copper content Tauheed Khursheed 2019

[7] Processing, microstructure and hardness of TiN/(Ti, Al)N multilayer coatings S.J. Suresha 2006

[8] The Comparison of the Microstructure and Hardness of Al-B and Al-Mg-B Composites
Mohammad Reza Derakhshesh 2006

[9] Wear Characteristics of Hybrid Al 6063 Matrix Composites Reinforced with Graphitand Fly Ash Particulates Manojkumar M. 2012

# Laser parameter optimization and unidirectional sliding wear performance of surface textured Ti6Al4V alloy

**A Karthikeyan, R Mithun Krishnan, M Prem Ananth*, S Ponnuvel and V Sridharan**
Department of Mechanical Engineering, Sri Venkateswara College of Engineering, Chennai- 602 117, India

*Email: premananth@svce.ac.in

Abstract. In this research work, a study on the coefficient of friction observed in samples made from Ti6Al4V titanium alloy textured with different types of patterns was made. Analysis of Variance (ANOVA) based on design of experiments was done to understand the contribution of type of pattern, laser power and number of passes used for texturing in affecting the coefficient of friction. The analysis indicated that the samples textured with hatch pattern produced minimum coefficient of friction while grid pattern produced maximum coefficient of friction. Along with hatch pattern single pass and 5W power setting in the laser engraving machine produced the optimum condition. The interaction between the pattern type and number of passes also contributes for the variations in the coefficient of friction.
*Keyword:* Titanium Alloy, Texturing, coefficient of friction and Anova.

## 1. Introduction

The exceptional properties such as lesser modulus, higher corrosive resistance, superior biocompatibility and higher strength to weight ratio possessed by titanium alloys leads to its higher application in the field of human joint replacement, automobile and aerospace applications [1-2]. The wear resistances of titanium alloys are poor and poor wear resisting property limited its applications in sliding contact. Hence, a suitable attempt to improve the surface properties of titanium alloys is necessary [3]. The titanium alloy surface properties can be enhanced by creating dimples through laser texturing. The regularly patterned surface creates dimples which will perform as a reservoir and also capture the wear debris formed during sliding test [4]. Laboratory wear test analysis such as pin on disc is normally carried out to reproduce the practical applications and also to study the useful information on friction and wear characteristics [5]. Many researchers have undertaken statistical analysis of wear behavior of metals and reported the outcome based on it. The experiments were planned according of Design of Experiment. The wear experiments for aluminum-based metal matrix composites were designed based on L27 orthogonal array. The influence of reinforcement, load, sliding distance and sliding speed on wear were studied statistically and reported [6]. Analysis of Variance (ANOVA) was performed to understand the contribution of various input process parameters on wear of hybrid metal matrix composites. Statistical analysis was instrumental in analysing the role of various input parameters on output characteristic studied [7]. Baradeswaran et.al used orthogonal array using Minitab software to design their experiment and statistically analyzed and optimized the wear test parameters [8]. Sachit et.al reported that the applied load was the most contributing factor for wear on tantalum niobium reinforced composite based on their statistical analysis results [9]. Virkunwar et al used orthogonal array design with three factors at three levels to conduct dry sliding wear test on AL6061 reinforced with rice husk ash particles. The input factor that contributed the most for wear was found out by conducting ANOVA. The statistical results were validated with confirmation experiment [10].

The paper presents the statistical analysis of coefficient of friction in various surface modified titanium alloy during wear tests under dry sliding condition. The experiments were carried out as per full factorial design and statistical analysis by performing ANOVA. The influence of individual design factor on the coefficient of friction and the most influential factor on the wear characteristic has been identified. From the data means curve the testing condition that gave most preferred output was found out. The interaction of various process parameters has also been discussed.

## 2. Experimental Procedure

Grade 5, Ti6Al4V was procured as a cylindrical rod and machined as a cylindrical pin of 8 mm diameter and 28 mm length. This pin is used as specimen in the pin on disc experiment and the experimental setup is shown in Figure 1. The rectangular patterns different shapes were designed by EZCAD software and engraved on top of the pin specimen by fiber laser marking machine. The following geometries were engraved on the pin specimen: circle, ellipse, hatch and grid. The laser parameters such as Wavelength: 1064 nm; Power: 5, 10 and 15W; Number of Pass: one, two and three; Pulse: 2 kHz; Speed: 200 rpm were used for texturing on the pin surfaces. The circular dimple diameter is 100 μm and its pitch is 160 μm. Major and minor axis length is 125 μm and 80 μm respectively for elliptical pattern machining. The dimple density range of is about 25-35%.

**Figure 1.** Pin on Disc Tribometer experimental setup

ASTM standard G99-04 was used for pin-on-disc testing procedures. The pin on disc tribometer consists of a rotating disk in contact with preloaded stationary cylindrical pin. Force sensor is provided to measure the frictional force during sliding and Linear variable differential transducer is also provided to measure the wear parameters continuously during sliding. Heat treated EN-31 with 63 HRC steel disc of thickness 8 mm and diameter 165 mm was used as a counter-body. The wear test was carried-out under 50 N normal load at 2 m/s velocity and the test duration is 30 minutes. The friction data was continuously monitored and recorded during the wear test and it is reported in Table 2.

The experiments were designed using full factorial design of experiments with three factors and multiple levels. The factors and levels are given in Table 1. A total of 36 experiments were conducted (Table2). Coefficient of friction was the output taken for further analysis.

IOP Conf. Series: Materials Science and Engineering **923** (2020) 012026 doi:10.1088/1757-899X/923/1/012026

**Table1.** Factors and Levels.

| **Factors** | **Levels** | | | |
|---|---|---|---|---|
| Pattern | Circle | Ellipse | Hatch | Grid |
| Power (W) | 5 | 10 | 15 | - |
| Pass | 1 | 2 | 3 | - |

**Table 2.** Experimental Details.

| Sl. No | Pattern | Power | Pass | Coefficient of friction |
|---|---|---|---|---|
| 1 | Circle | 5 | 1 | 0.549 |
| 2 | Circle | 5 | 2 | 0.574 |
| 3 | Circle | 5 | 3 | 0.592 |
| 4 | Circle | 10 | 1 | 0.550 |
| 5 | Circle | 10 | 2 | 0.550 |
| 6 | Circle | 10 | 3 | 0.664 |
| 7 | Circle | 15 | 1 | 0.627 |
| 8 | Circle | 15 | 2 | 0.666 |
| 9 | Circle | 15 | 3 | 0.588 |
| 10 | Ellipse | 5 | 1 | 0.562 |
| 11 | Ellipse | 5 | 2 | 0.695 |
| 12 | Ellipse | 5 | 3 | 0.590 |
| 13 | Ellipse | 10 | 1 | 0.620 |
| 14 | Ellipse | 10 | 2 | 0.565 |
| 15 | Ellipse | 10 | 3 | 0.603 |
| 16 | Ellipse | 15 | 1 | 0.610 |
| 17 | Ellipse | 15 | 2 | 0.573 |
| 18 | Ellipse | 15 | 3 | 0.604 |
| 19 | Hatch | 5 | 1 | 0.564 |
| 20 | Hatch | 5 | 2 | 0.598 |
| 21 | Hatch | 5 | 3 | 0.575 |
| 22 | Hatch | 10 | 1 | 0.642 |
| 23 | Hatch | 10 | 2 | 0.562 |
| 24 | Hatch | 10 | 3 | 0.541 |
| 25 | Hatch | 15 | 1 | 0.564 |
| 26 | Hatch | 15 | 2 | 0.553 |
| 27 | Hatch | 15 | 3 | 0.538 |
| 28 | Grid | 5 | 1 | 0.458 |
| 29 | Grid | 5 | 2 | 0.697 |
| 30 | Grid | 5 | 3 | 0.647 |
| 31 | Grid | 10 | 1 | 0.686 |
| 32 | Grid | 10 | 2 | 0.639 |
| 33 | Grid | 10 | 3 | 0.623 |
| 34 | Grid | 15 | 1 | 0.690 |
| 35 | Grid | 15 | 2 | 0.689 |
| 36 | Grid | 15 | 3 | 0.645 |

## 3. Results and Discussion

The frictional characteristics of different shape of the dimples are compared in Figure 2. The friction coefficient values shown in figure for different shape patterns tested at 50 N normal load and patterns are engraved at 5 W power with single pass. The higher value of friction coefficient is observed in grid pattern and large fluctuation in friction coefficient is observed for ellipse and circle pattern. The friction coefficient for hatch pattern is observed as steady, comparatively low and less fluctuating as in Figure 2. The average coefficient of friction values was calculated from the pin on disc experimental values for the specimens engraved with different powers and passes and it is shown in Table 2.

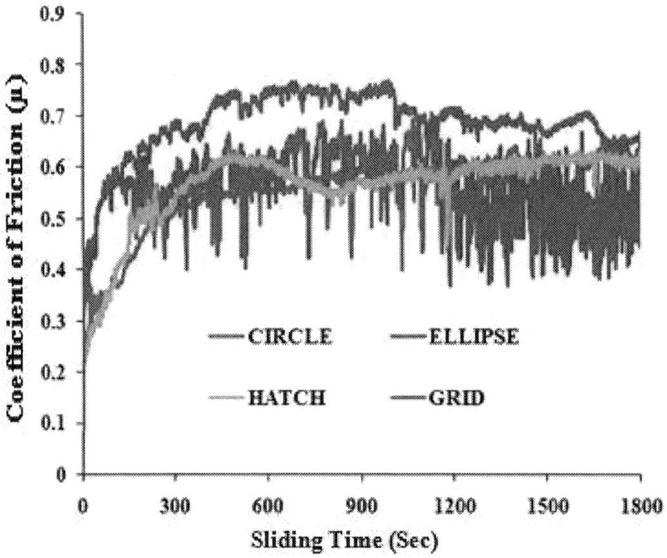


**Figure 2.** Coefficient of Friction Vs Sliding Distance for the different dimple shapes

ANOVA was performed to understand the effect of input variables on the coefficient of friction. The details are Tabulated in Table.2. From the analysis of variance, it is observed that the type of pattern made in the specimens has a contribution of 21.96%. Interaction between power and pass found to be contributing for 26.11% of total variation.

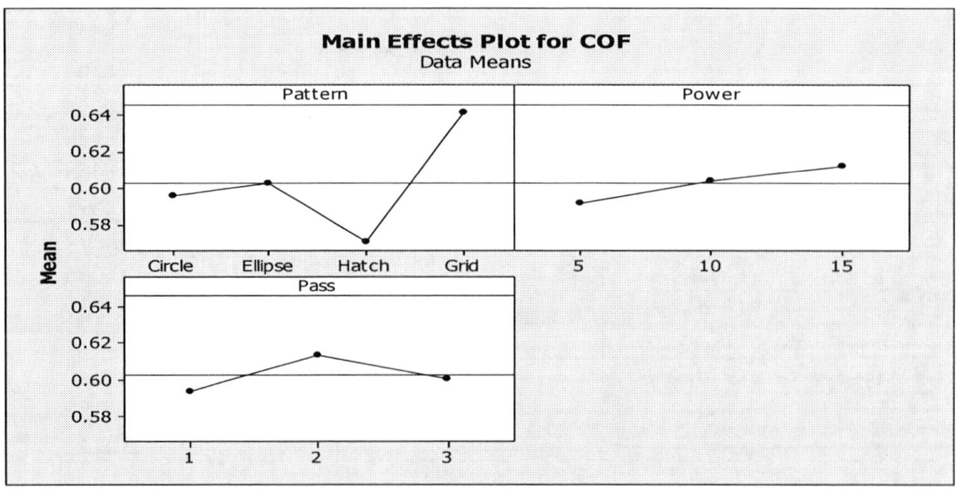


**Figure 3.** Main effect plot

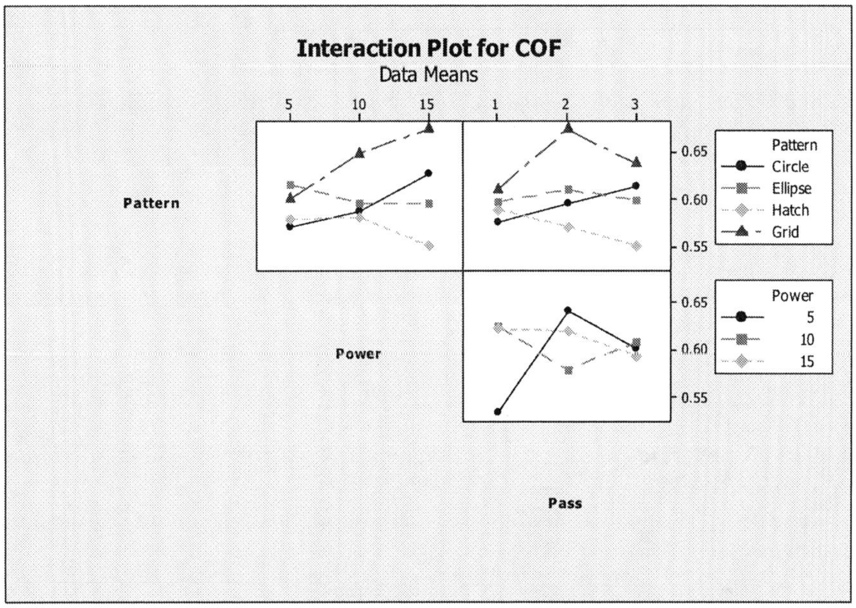


**Figure 4.** Interaction effect plot

From the main effect plot shown in Figure 3 and from the values in Table 3, the optimum combination for minimum coefficient of friction was found to be A3B1C1. This suggests that the coefficient of friction between the disc material and the specimen was at its minimum value when hatch pattern was used. Other optimum input conditions for laser engraving were observed to be 5W power and single pass. Also, from the Figure 3, it is clear that the grid pattern produces maximum coefficient of friction, hence is not recommended. From the interaction plot shown in Figure 4, it is observed that the range of variations in the coefficient of friction values with different patterns were minimum with single pass and 5W power setting in the laser engraving machine.

**Table 3.** ANOVA details.

| Source | DF | SS | MS | Ftest | Ftable at 95% a | P (%) |
|---|---|---|---|---|---|---|
| Pattern | 3 | 0.023219 | 0.007740 | 3.31 | 3.49 | 21.96 |
| Power | 2 | 0.002546 | 0.001273 | 0.54 | 3.89 | 2.41 |
| Pass | 2 | 0.002435 | 0.001218 | 0.52 | 3.89 | 2.30 |
| Pattern*Power | 6 | 0.013231 | 0.002205 | 0.94 | 3.00 | 12.52 |
| Pattern*Pass | 6 | 0.008594 | 0.001432 | 0.61 | 3.00 | 8.13 |
| Power*Pass | 4 | 0.027601 | 0.006900 | 2.95 | 3.26 | 26.11 |
| Residual Error | 12 | 0.028085 | 0.002340 | | | 26.57 |
| Total | 35 | 0.105711 | | | | 100 |

## 4. Conclusions

In this study the following conclusions are derived.

1. The friction coefficient for hatch pattern is observed as steady, low and less fluctuating as compared with other patterns.
2. Pattern type has a contribution of 21.96%. Minimum coefficient of friction was observed with hatch pattern.

3. Single pass and 5W power setting in the laser engraving machine produced optimum engraving conditions.
4. Grid pattern produces maximum coefficient of friction, hence not recommended.
5. Interaction between pattern and pass was observed to be 26.11%.

**Acknowledgement**
This work was supported by the All India Council for Technical Education (AICTE) under the Scheme of Research Promotion Scheme (RPS) with a reference number of File No. 8-50/RFID/RPS/POLICY-1/2016-17. We would like to acknowledge them for the financial and technical support.

**References**

[1] Christoph Leyens and Manfred Peters 2003 “Titanium and Titanium Alloys” *Wiley-VCH Verlag GmbH& Cos*, Germany.
[2] Boyer 1996 “An overview on the use of titanium in the aerospace industry”, *Materials Science and Engineering*, **213**: 103-114.
[3] Kathy Wang 1996 “The use of titanium for medical applications in the USA. *Materials Science and Engineering:* A. Aug 15;**213**(1-2):134-7.
[4] Etsion I. 2005 Jan 1 State of the art in laser surface texturing. *J. Trib.*.**127**(1):248-53.
[5] Psyllaki PP, Jeandin M, Pantelis DI, and Allouard M 2000 Aug 21, Pin-on-disc testing of PE-CVD diamond-like carbon coatings on tool steel substrates. *Surface and coatings Technology*.**130**(2-3):297-303.
*[6]* Patil, Shanawaz, and Mohamed Haneef 2019 "A statistical analysis of optimization of wear behavior of Al7075-Beryl-Graphene hybrid composites using Taguchi technique." *Materials Today: Proceedings.*
[7] Natrayan, L., and M. Senthil Kumar 2019 "Optimization of wear behaviour on AA6061/Al2O3/SiC metal matrix composite using squeeze casting technique–Statistical analysis." *Materials Today: Proceedings.*
[8] Baradeswaran, A., A. Elayaperumal, and R. Franklin Issac 2013 "A statistical analysis of optimization of wear behaviour of Al-Al2O3 composites using Taguchi technique." *Procedia Engineering* **64**: 973-982.
[9] Sachit, T. S., et al. 2019 "Optimization of dry sliding wear behavior of aluminum LM4-Ta/NbCnano composite using Taguchi technique." *Materials Today: Proceedings.*
[10] Virkunwar, Anwesh K., Shouvik Ghosh, and RanjanBasak 2019 "Wear characteristics optimization of Al6061-Rice husk ash metal matrix composite using Taguchi method." *Materials Today: Proceedings.*

# LUBRICATION PERFORMANCE OF CASTOR OIL BLENDED WITH OTHER VEGETABLE OILS IN GRINDING OF INCONEL 625

**Suresh Kannan I[1], Karthikeyan S[2], Ravikumar K[3]**

[1,2,3] Faculty in Department of Mechanical Engineering,
SRM Institute of Science and Technology, Vadapalani, Chennai – 600 026
E-mail address: iskannancim@gmail.com

**Abstract.** The present work is aimed to investigate the effect of mixed vegetable oils in surface quality during surface grinding of Inconel 625 by vitrified bonded alumina grinding wheel. Three different blends of vegetable oils were considered for the study in which castor oil was base oil and blending oils were soya bean oil, corn oil and rice bran oil. Castor oil has good lubricating properties but its high viscous and poor rheological properties make it directly unviable for lubrication applications. In this study, the rheological properties of the castor oil were modified to cater to the specified needs for grinding of Inconel 625 by mixing it with some other vegetable oils. The blends were prepared in 1:1 proportion. Experiments were conducted using the blends of vegetable oils in grinding of Inconel 625 under identical work conditions. The finished work piece was measured for its surface quality with respect to its average surface roughness, surface morphology and temperature generated on the work piece. The castor oil and soybean blend produced the lowest surface roughness, Ra = 0.0738 μm & Rz = 0.4299μm, with near absence of deep grooving and smearing of chip material on the surface, in comparison with other blends due to enhanced rheological properties.

**Keywords:** Grinding, Inconel 625, Vegetable based cutting fluid, Surface roughness, blended castor oil

## 1. INTRODUCTION

Grinding is a finishing process, which generates high temperature in the workpiece due to grit-work material interaction. The friction between grinding wheel and workpiece is the major source for heat generation, where some part of heat is lost through chips. High temperature involved in grinding process leads to large power consumption, poor surface quality and surface integrity on the workpiece [1]. Grinding process requires large amount of metal working fluids (MWF) to reduce the amount of heat generation by reducing friction with the aid of applying appropriate MWF in the grinding zone [2]. Traditionally copious amount of mineral oil-based cutting fluids is being used as MWF in grinding difficult to cut materials like Inconel 625. However, mineral oil exhibits very poor impact on environment. About 80% of occupational diseases to machine operators are due to the exposure of MWF in the working environment [3]. Dry machining is the target to eliminate the cutting fluid in the machining industry thus reduces the machining cost and ecological impact. In contrary absence of MWF increase the grinding wheel- workpiece interaction thus diminish the surface quality, tool life and increase cutting forces. Minimum Quantity Lubrication is another approach to reduce the cutting fluid in grinding, where minute droplets of lubricants are directed to the grinding zone. Although MQL reduces the MWF consumption, it still uses them in the form of mist or droplets which increases health hazards for the workers like respiratory issues [4]. Another approach to grind hard metals by sustainable manner is cryogenic grinding. Irrespective of its eco-friendly nature, it has produced some negative effects in grinding of hard to cut materials. PP Reddy and Ghosh (2016) explored the scope of cryogenic coolant (LN2) application in surface grinding of AISI52100 using vitrified Alumina Grinding wheel [5]. In their work a high power consumption, excessive wheel wear and more tensile residual stress on the ground surface were observed for LN2 coolant when compared with soluble oils. Vegetable oils have proved to exhibit better lubricity due to its long fatty acid chains in their molecules [6]. Asadauskas et al. (1997) exhaustively investigated the castor oil, super-refined mineral oil, and high-oleic corrosive sunflower oil for understanding the thickness, oxidation security, statement age tendency, unpredictability, lubricity, and similarity with added substances [7]. The outcomes of literature study show that the castor oil displays to be potential biodegradable base oil. Although the high viscosity of castor oil provides the better lubrication in the grinding zone, it has poor rheological attributes. This restricts their usage in high speed machining operations. Further these poor properties can be balanced by mixing another vegetable oil with low viscosity such as soybean, maize, peanut, sunflower, palm, and rapeseed oils with castor oil [8]. However, its application in grinding and other metal cutting operations is not explored in detail in the past. The main aim of the present work is to investigate the role of mixed vegetable oils in surface finish of ground Inconel 625 workpiece.

## 2. EXPERIMENTAL PROCEDURE

The work material selected for grinding in this investigation is Inconel 625 which is used where high strength is required at elevated temperature such as aero engine. Its chemical composition is listed in Table.1 The grinding experiments were done on surface grinding machine having maximum rotation speed of 2400 rpm and Power of 1.5kW. The grinding conditions are wheel speed: 24 m/s; Feed: 3m/min and depth of cut: 7, 14 μm under dry and lubricated environment. All grinding experiments were done with Alumina grinding wheel having fine alumina grits with vitrified bond. The grinding temperature was measured with the aid of non-contact Laser Infrared Thermometer (Make: MCP, Model: Ir-20). Average surface roughness ($R_a$) and ten point average roughness ($R_z$) of the ground surface were measured using computerized surface roughness tester (Make: Surfcom, Model: 1400G). Further the selected ground region was observed under Scanning Electron Microscope (SEM) to reveal the grinding marks and surface finish features.

Table.1: Chemical composition of Inconel 625

| Elements | Ni | Cr | Fe | Mo | Nb | C | Mn | Si | Ti | S | Al |
|---|---|---|---|---|---|---|---|---|---|---|---|
| **Percentage** | 63 | 20.3 | 4.31 | 8.81 | 3.33 | 0.004 | 0.03 | 0.018 | 0.017 | 0.006 | 0.17 |

Castor oil was selected as base oil for all grinding experiments. Other vegetable-based oils such as soya bean, corn oil and rice bran oil were mixed individually with base oil (castor oil) in the proportion of 1:1 to alter the rheological property. The flash point, fire point and viscosity of the mixture were tested using Cleveland open cup apparatus and Redwood viscometer respectively and listed in Table: 2.

Table: 2 Flash point, Fire point and Kinematic viscosity of castor oil and other blends

| Oil Blends | **Flash Point (°C)** | **Fire point (°C)** | **Kinematic Viscosity (cSt) at 60 °C** |
|---|---|---|---|
| **Castor oil** | 230 | 250 | 198.6 |
| **Castor oil + Soyabean oil** | 260 | 310 | 25.50 |
| **Castor oil + Corn oil** | 300 | 325 | 28.81 |
| **Castor oil + Rice Bran oil** | 310 | 330 | 30.45 |

## 3. RESULT AND DISCUSSION

### *3.1 Grinding Temperature*

Grinding temperature is one of the most important factors that affects the surface quality of ground workpiece. During grinding, an abrasive wheel is in contact with the workpiece causing an abrupt rise in temperature. The detrimental effects of surface cracking, burn marks and tensile residual stress are often directly related to high grinding temperature. The variations of the temperature on the ground surface of the workpiece with the different vegetable oil at similar grinding conditions are shown in Fig.1. It is clearly evidenced that dry grinding developed a temperature to a minimum of 1.5 times higher than that of produced by other blended oil lubricating conditions. Use of blended vegetable oil based cutting fluids reduce the impact of rubbing abrasive grits with workpiece by forming a thin film of lubrication layer and reduce the wear of micro abrasives by dulling. The castor oil + rice bran oil lubricating conditions produced least grinding temperature of 350°C when compared with other combinations, which was mainly due to good thermal properties such as high flash point (310°C) and fire point (330°C) and high viscosity (30.45 cSt at 60°C). However, the impact of such low temperature of castor oil + rice bran oil lubrication was not reflected on the surface roughness and surface morphology.

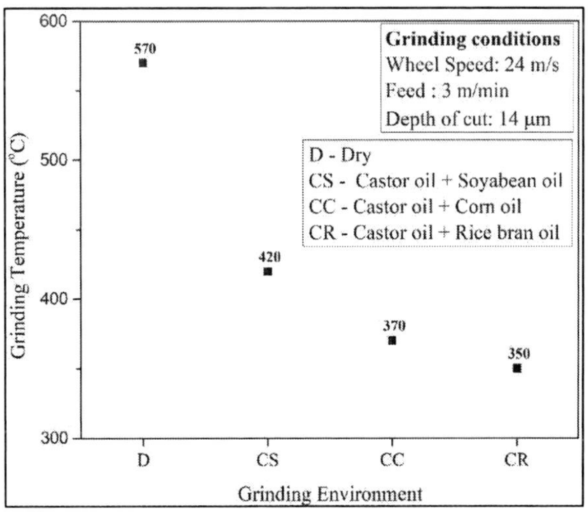


Figure.1 Variation of grinding temperature during grinding of Inconel 625 at cutting speed: 24 m/s; feed: 3 m/min and depth of cut: 14 µm with different blended vegetable-based metal working fluids

*3.2 Surface Roughness*

Figure 2 and 3 depicts the variations of surface roughness (Ra and Rz) of ground surface at cutting speed= 24 m/s, Feed = 3 m/min and depth of cut = 7, 14 μm. The average surface roughness (Ra) increased with increasing depth of cut in the case of grinding with dry and all other lubricated conditions. It is thus obvious that the surface roughness produced under lubricated conditions is superior to that of finish produced under dry environment. Among three vegetable blend, castor oil + Soya bean oil produced smoother surface (Ra = 0.0738) compared with its counterparts. The lowest kinematic viscosity of the blend (25.5 cSt at 60°C) allows it to reach the grit workpiece interface without any difficulty which in turn reduce the friction and average surface roughness. The poor surface quality produced under castor oil + rice bran oil is probably due to its insufficient rheological behavior even after mixing. The ten-point average roughness, Rz also followed the same trend as Ra.

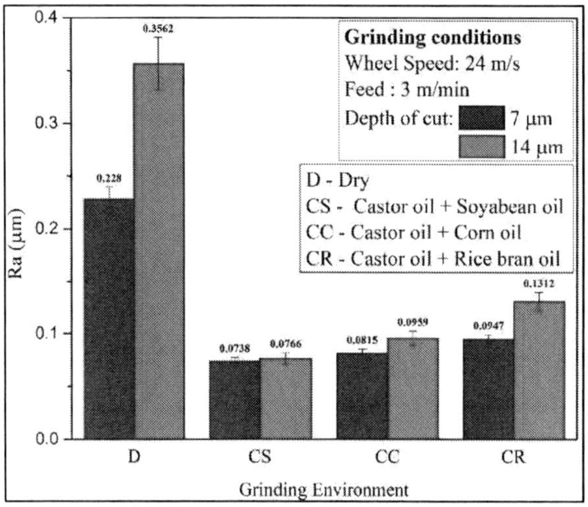


Figure.2 Variation of Ra after grinding of Inconel 625 at cutting speed: 24 m/s under different blended oil lubrication environments

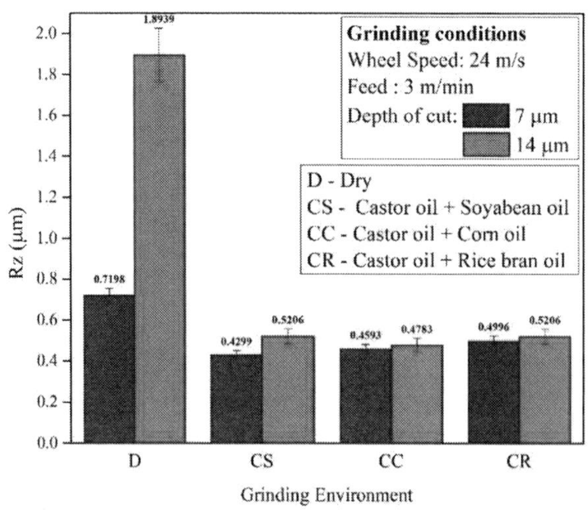


Figure. 3 Variation of ten point mean surface roughness ($R_z$) after grinding of Inconel 625 at cutting speed: 24 m/s under different lubrication conditions

In contrary variation of Rz observed under castor oil + rice bran oil is almost nearer to the values that are produced by other lubrication conditions.

*3.3 Surface Morphology of ground surface*

The surface of the ground specimens was observed under SEM to reveal the grinding marks and other surface features at higher magnification. Figure 4. depicts the magnified view (2000X) of ground surface after grinding at speed = 24 m/s, Feed = 3 m/min and Depth of cut: 7, 14 μm. It is observed that in all lubrication conditions grinding marks were visible. But the severity of the marks and other features determine the quality of ground surface. Deep grooving and smearing were observed on the surface under dry environment (Figure. 4(a)) which may be due to the increased rubbing action of abrasive grits with work material. Castor oil + soya bean oil blend produced almost flaw free surface which was mainly due to very smooth interaction of grit over workpiece during grinding.

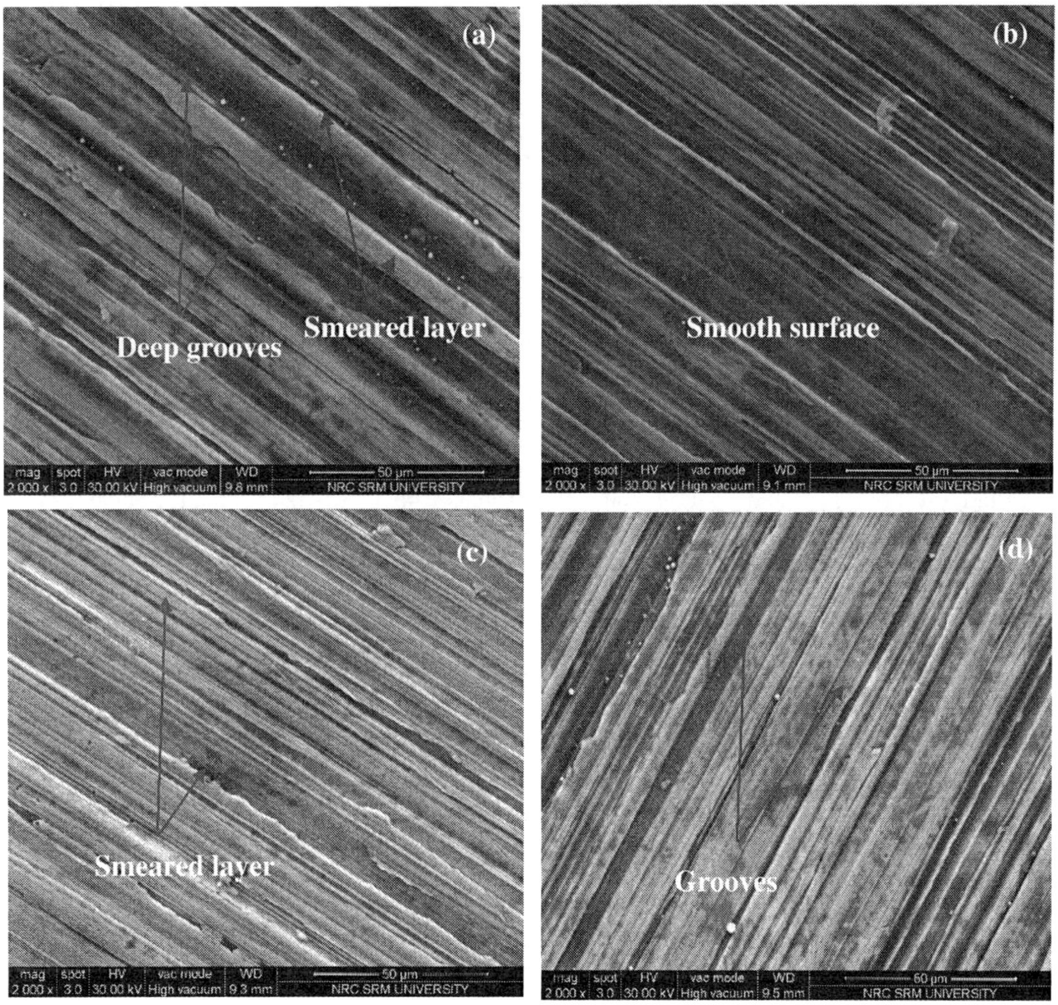


Figure. 4 SEM images of ground surface at cutting speed: 24 m/s table feed: 3 m/min and depth of cut: 14 μm under (a) dry (b) castor oil + soya bean oil (c) castor oil + corn oil (d) castor oil + rice bran oil environments

There is a redeposited layer on the surface produced by castor oil + corn oil blend. In contrast castor oil + rice bran oil produced irregular grooving on the work surface. It is evidenced that there is a lack of penetration of high viscous castor oil + rice bran oil combination particularly at depth of cut = 14 μm.

## 4. CONCLUSION

Three different blends of Castor oil mixed with other vegetable oils (i.e., soybean, corn and rice bran oil) were prepared, each at a ratio of 1:1. Castor oil was taken as point of comparison and nickel based super alloy Inconel 625 was chosen as workpiece material. The lubrication performance of blended oils during grinding was experimentally investigated. From the investigated results obtained, the following conclusions were construed.

On comparison, the flash and fire point of castor and soya bean oil blend was least at 260°C and 310°C respectively. But rice bran and castor mixture show the highest flash and fire point which are 310°C and 330°C. As the rice bran shows the highest value in flash and fire point, it is best suitable for grinding at high temperature.

The viscosity of rice bran and castor oil blend shows higher value (30.5 cST) than castor and corn oil mixture (28.81 cSt) or castor and soya bean oil blend (25.5 cSt). So higher viscosity oils will have flow problems and lesser viscosity is good for lubricant in flood lubrication.

Among three blends, castor oil + soya bean oil produced lower average roughness (Ra = 0.0766 μm) when compared with the results of blends, castor oil + corn oil (0.0959 μm) and castor oil + Rice bran oil (0.1312 μm), but the variation is very marginal.

Many deep grooves, rough surfaces were observed in dry grinding environment. Meanwhile, the surface morphology of the workpiece under the blended oil environment was found enhanced. Therefore, the blend of soybean and castor oil produced the best surface quality with nearly free from deep grooving and smearing of chip material on the surface, when compared with other blends.

**References**


[1] Brinksmeier, E., Meyer, D., Huesmann-Cordes, A. G., & Herrmann, C. (2015). Metalworking fluids—Mechanisms and performance. CIRP Annals, 64(2), 605-628.Katna, R., Suhaib, M., & Agrawal, N. (2020).

[2] Nonedible vegetable oil-based cutting fluids for machining processes–a review. Materials and Manufacturing Processes, 35(1), 1-32.

[3] Shashidhara, Y. M., & Jayaram, S. R. (2010). Vegetable oils as a potential cutting fluid—an evolution. Tribology international, 43(5-6), 1073-1081.

[4] P. Sreejith, B. Ngoi, Dry machining: machining of the future, Journal of Materials Processing Technology 101 (2000) 287–291.

[5] Reddy, P. P., & Ghosh, A. (2016). Some critical issues in cryo-grinding by a vitrified bonded alumina wheel using liquid nitrogen jet. Journal of Materials Processing Technology, 229, 329-337.

[6] Iyappan, S. K., & Ghosh, A. (2019). Small quantity lubrication assisted end milling of aluminium using sunflower oil. International Journal of Precision Engineering and Manufacturing-Green Technology, 1-9.

[7] Asadauskas, S., & Erhan, S. Z. (1999). Depression of pour points of vegetable oils by blending with diluents used for biodegradable lubricants. Journal of the American Oil Chemists' Society, 76(3), 313-316.

[8] Guo, S., Li, C., Zhang, Y., Wang, Y., Li, B., Yang, M., Z Xianpeng & Liu, G. (2017). Experimental evaluation of the lubrication performance of mixtures of castor oil with other vegetable oils in MQL grinding of nickel-based alloy. Journal of Cleaner Production, 140, 1060-1076.

# EFFECT OF LOAD AND SLIDING DISTANCE ON TRIBOLOGICAL PROPERTIES OF AISI 1040 STEEL BY MAGNETRON SPUTTERED TI-6AL-4V COATING

R. Malkiya Rasalin Prince[1], D. Arulkirubakaran[2], V. Naveen[3], J. Allan Paulraj[4], S. Arokkia Julius Raja[5]

[1,2,3,4,5]Department of Mechanical Engineering, Karunya Institute of Technology and Sciences, Coimbatore - 641114, Tamilnadu, India.

russelmecher@gmail.com, arul.kirubakaranbe@gmail.com, vnaveen2204@gmail.com, allanpaulraj@gmail.com, arojuliusraja472@gmail.com

**ABSTRACT:** In the present study, Ti-6Al-4V-$2B_4C$ coatings are deposited using a DC magnetron sputtering system on AISI 1040 steel substrates. The composition and elemental confirmation of the coating was explored by SEM, XRD and EDS. Wear and friction tests were trained to study the influence of sliding distance and load on tribological behaviour of Ti-6Al-4V-$2B_4C$ coated AISI 1040 steel for machine elements. Wear tests were performed on coated AISI 1040 steel substrates counter to E-52100 steel balls by using a ball-on-disc wear testing apparatus under 2*N* and 3*N* normal loads at room temperature. The different sliding distances (63, 78, 94 and 110 *m*) has taken for wear test under room temperature and low coefficient of friction obtained for higher sliding distance and load. Ti-6Al-4V-$2B_4C$ coated AISI 1040 steel reduces the frictional coefficient from 0.65 to 0.23 and improves the wear resistance. Specific wear rate of the coating reaches the maximum value with higher sliding distance. On the outset, it is clearly showing that the formation of oxide layers of Ti-6Al-4V-$2B_4C$ coatings mainly affects its tribological characteristics.

**KEY WORDS:** Magnetron sputtering, sliding distances, wear, friction, Ti-6Al-4V-$2B_4C$.

## 1. INTRODUCTION

Aeronautical and medical fields are focusing on Ti-6Al-4V alloy (titanium grade-5) in line for its solitary strength-to-weight proportion [1]. Industrial machinery and components are habitually failed during working due to surface damage [2]. It is also has been rectified through surface modifications techniques especially by wear-resistive coatings. High-quality films with a good vary of industrially necessary have been developed through magnetron sputtering deposition technique [3, 4]. One of the familiar materials in the engineering field is carbon steel AISI 1040 and properties enhancement is achieved by various hard coatings [5]. Nitrides and carbides with higher hardness take on a vital role in the surface designing field.

Carbide coatings deliver higher hardness than nitride coatings which exhibit low wear nature on steel [6]. Applications of boron carbide coating are boundless; such as fusion reactors and coatings on nozzles etc. Higher hardness has been achieved through $B_4C$ coatings on steel and it is often used as fusion reactors [7, 8]. In this study, wear characteristics of Ti-6Al-4V-2$B_4C$ coated AISI 1040 steel is studied by a ball-on-disc wear test machine. The sliding distance and load were selected with a suitable range of working environments for genuine machine components.

## 2. MATERIALS AND METHODS

**Figure 1:** SEM of (a) Pure AISI 1040 steel (b) Ti-6Al-4V-2$B_4C$ Green compact

Fig. 1(a) shows the surface morphology of the pure AISI 1040 steel. The target is the prepared forms of pure Ti, Al, V and $B_4C$ powders. Pre-compacted Ti, Al, V and $B_4C$ raw powders exhibit an irregular shape is shown in Figure 1 (b). AISI 1040 steel is coated through the magnetron sputtering process with prepared Ti-6Al-4V-2$B_4C$ target. Substrates of 3 *mm* thicknesses with 60 mm diameter AISI 1040 steel plates are taken for coating. Before sputtering the chamber was vacuumed to $1x10^{-5}$*bar* for a free flow of atoms in the chamber. The temperature of 100 ± 20 °*C* was kept up during the coating with 0.5*h* and 1*h* coating time duration has been fixed for all coatings. Gas pressure range of (Ar) 0.001 to 0.1 *bar* has been maintained as during sputtering coating. Wear test carried out by ball-on-disc wear tester on AISI 1040 coated steel with loads (2 and 3*N) and* different sliding distances (63, 78, 94 and 110 *m*). AISI 52100 steel balls with 10 *mm* diameter and hardness of 450 *VHN* were used as the counter body. The loss of mass was weighed with high accuracy of ±0.001 *mg* by way of a sensitive electronic balance and weight loss has been converted into volume loss. Further corresponding wear parameters such as SWR and CoF were calculated according to the reference [9].

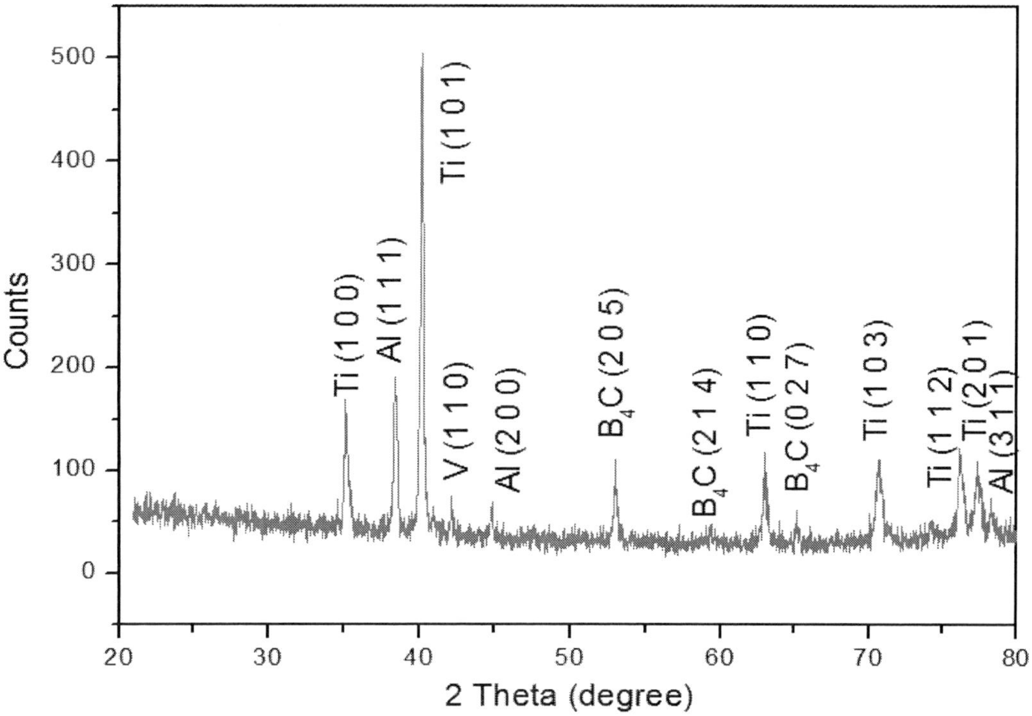


**Figure 2:** XRD pattern of coating

XRD peaks of $B_4C$ reinforced Ti-6Al-4V is shown in Figure 2. Elemental presence is confirmed by elements peaks; Ti peaks represented by high-intensity peaks using 2θ angle of 35.2°, 40.1°, 62.8°, 70.5°, 76.2° and 77.2° are connected with (1 0 0), (1 0 1), (1 1 0), (1 0 3), (1 1 2) and (2 0 1) crystal planes are authenticate by JCPDS 89-2762. Also, another element of Aluminium was acknowledged by different peaks with 2θ values of 38.4°, 44.8° and 78.3° through (1 1 1), (2 0 0) and (3 1 1) planes authenticate by JCPDS file no 89-4037. Vanadium peak is identified at (1 1 0) plane with a 2θ angle of 41.2° and authenticate by JCPDS file no 65-6689. The 2θ values of 53.4°, 58.8° and 65.1° through (2 0 0), (2 1 4) and (0 2 7) planes have confirmed the presence of Boron carbide and authenticate by JCPDS file No 75-0424. Based on the presence of different peaks present in the coating it is confirmed that no oxide peak is identified in the coating.

# 3. RESULTS

## 3.1. Friction coefficients

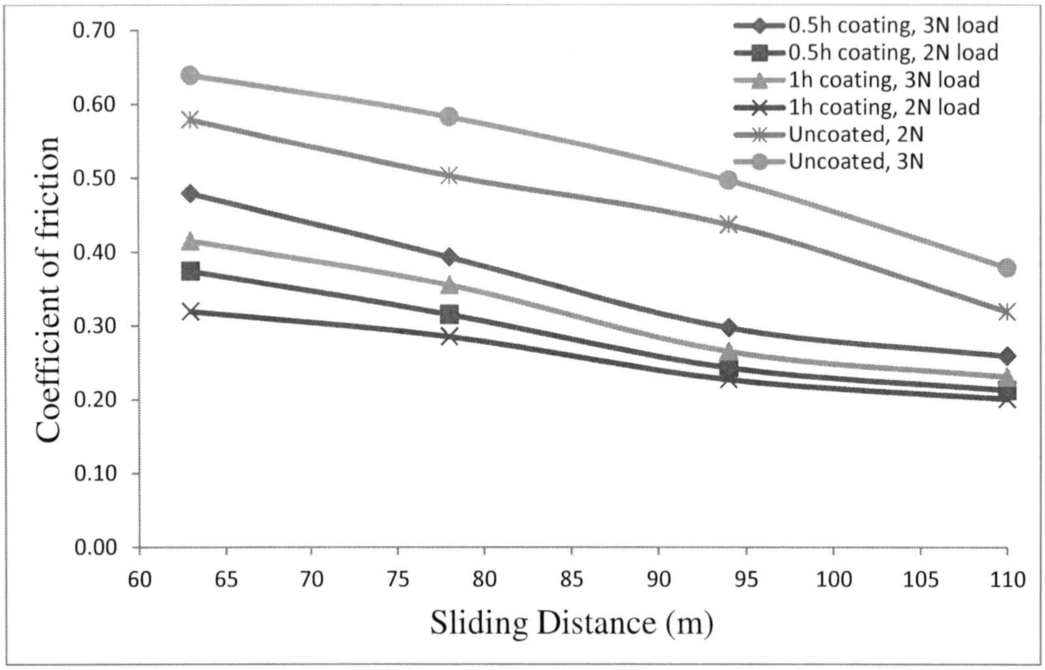


**Figure 3:** Coefficient of friction within sliding distance under 2N and 3N load

Fig. 3 deals with the frictional parameter of coatings against various sliding distances. In the earlier stage higher coefficient of friction has been noticed for all sliding distances then decreased gradually and reaches almost steady state value. The testing sliding distances of 63–110 *m* has noticed with some coefficient of friction fluctuations also noticed probably at higher loads and sliding distances. The various range of coefficient of friction is recorded in Fig.4 with different load and sliding velocity. The results were obtained with the loads of (2N, 3N) at normal working atmosphere and (0.52, 0.65, 0.79 and 0.92 *m/s*) sliding velocities. The formation of protective layer maybe happened and protects the sliding surfaces from frictional behaviour; hence decreased level of coefficient of friction is noticed for all working samples and sliding distances. Once the sliding velocity rises from (0.52 *m/s* to 0.92 *m/s*) subsequently coefficient of friction decreases by the effect of protective layer. For all sliding distances lower value of coefficient of friction is obtained and maximum value is obtained as 0.65 for of 3 N against 110 m sliding distance of and minimum of 0.31 for 2 N against 63 m sliding distance is obtained. Initially, the coefficient of friction value is higher and it decreases bit by bit due to the formation of oxide between the ball and coated surface. The developed oxide layer acted as a protective layer between the ball and coated surface for all loads with various sliding velocity. The above results were compared with other findings achieved under matching test circumstances [10]; frictional coefficients of this Ti-6Al-4V-2$B_4$C coating is lower.

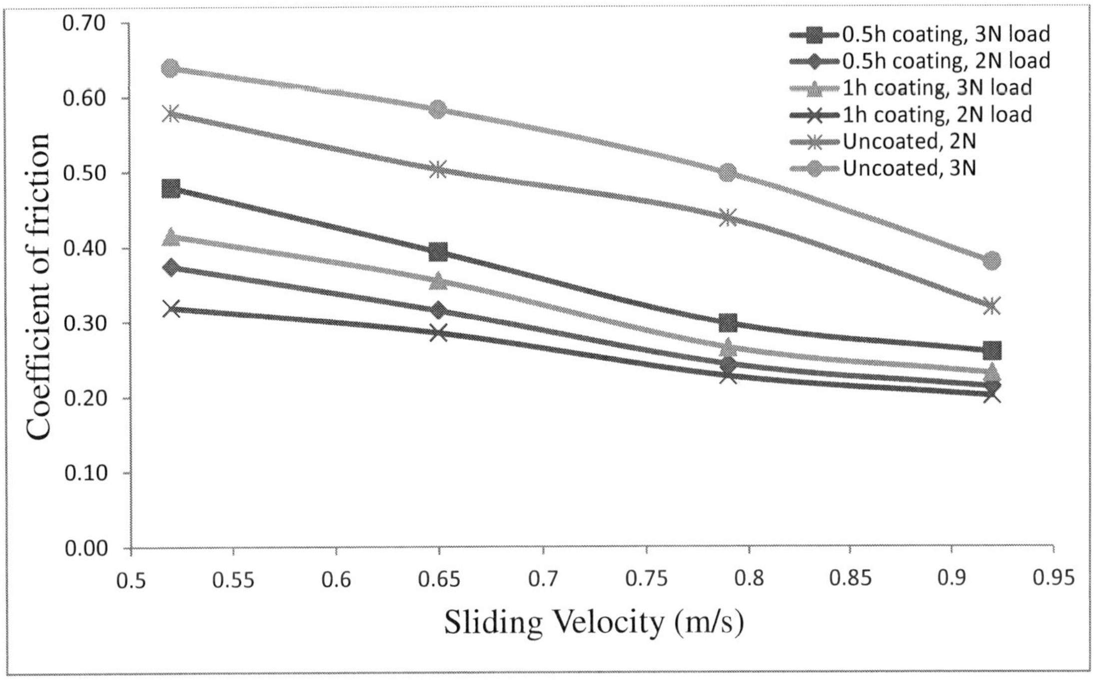


**Figure 4:** Coefficient of friction for different sliding velocities and normal loads

### 3.2. Analysis of wear rates of counter body coating

Failure of coating is expressed for all testing cycles that express the increased life of all coatings which confirms through opposition of ruptured nature. Fig.5 shows wear scar images of AISI 52100 steel counter body. Based on the SEM results it is identified that broadened wear marks and wear track width for higher loads under similar sliding distance circumstance. Moreover, sliding distances varies from 63–110 *m* and wear track width incremental is noticed accordingly under similar loads. Also, during wear test Ti-6Al-6V-2$B_4$C coatings are not ruptured and fully stable; hence not identified any ejected particles form coatings. The wear scar areas are mainly influenced by counter body and sliding distance. The 110 *m* sliding distances produces the higher wear areas with a sliding velocity of 0.92 *m/s* and 3 *N* load.

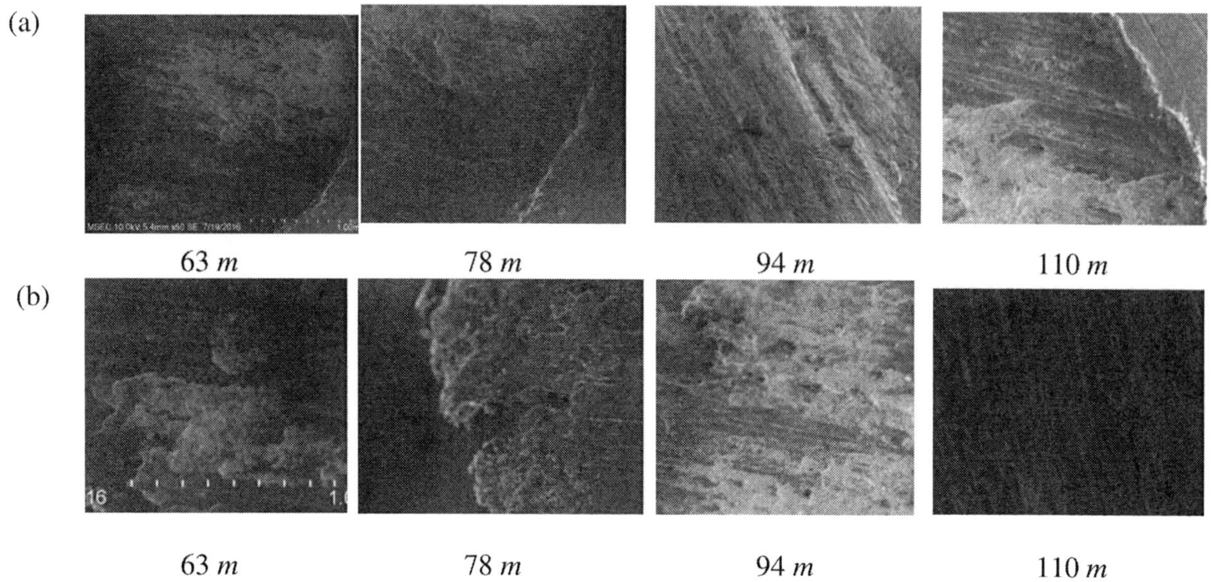


**Figure 5:** Worn surfaces of counter surface with various sliding distances: (a) 2*N* load, (b) 3*N* load

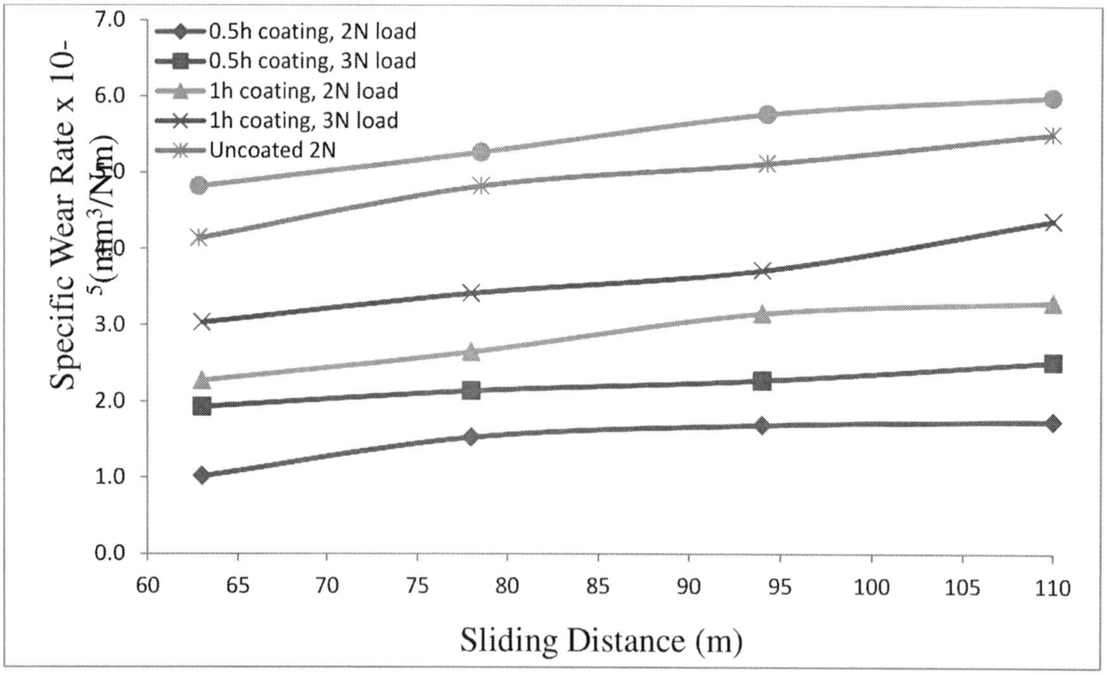


**Figure 6:** Specific wear rates at different sliding distances and loads

Figs. 6 illustrate the specific wear rates of E-52100 steel ball and Ti-6Al-6V-2$B_4C$ coatings. It is observed that for all sliding distances with applied load the specific wear rate is increased gradually. The maximum specific wear rate of 4.4726x$10^{-5}$ $mm^3/Nm$ and normal wear volume of 0.00881 $mm^3$ was observed during sliding. According to the derived results, a higher wear rate is observed with 110 *m* sliding distance and obtained higher and lower specific wear rates of 4.4726x$10^{-5}$ $mm^3/Nm$ and 1.0132x$10^{-5}$ $mm^3/Nm$ individually. The weight loss during sliding is denoted in Figs. 7 that shows the overall weight loss with sliding distance and load. The higher weight loss of 0.008 g is noted for 3N load and weight losses of all others are lower. Weight loss of the coating is also influenced by surface hardness of the counter body as well as the hardness of the coated surface. The weight loss and specific wear rates of this Ti-6Al-6V-2$B_4$C coating are produced significant results with similar load and sliding distance.

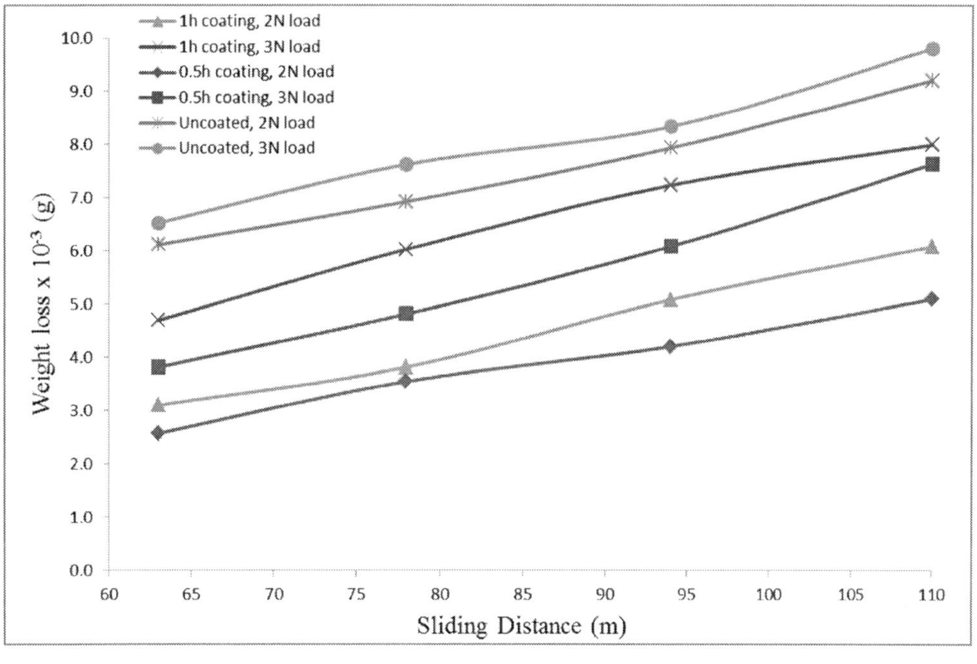


**Figure 7:** Wear loss at different sliding distances and loads

## 4. DISCUSSION

### 4.1. Frictional behavior

The present discussion focused on the effect of $B_4$C reinforcement on frictional behavior coatings. In sliding contacts, CoF varies with sliding distance and loads; the reinforcement plays a major role in reducing the CoF. At the initial stage, the contact between the surfaces of a specimen in the ball and disc led to low frictional force, so that CoF decreases due to the ploughing effect. The reinforcement from the coating is protruding out at the normal load to reduce the CoF. Under heavy load, there might be higher wear and CoF which leads to poor tribological behaviour. The addition of hard reinforcement material ($B_4C$) results in the dropping of CoF. It shows that CoF decreases gradually due to the development of an oxide layer in between the disc and ball during sliding. During sliding, heat is generated by friction between contact surfaces in addition to the adhesion of flat and spherical surfaces of sliding. Due to the relative motion of the counter surface, the continuous generation of frictional heat and low heat dissipation rate occur due to time lacking. Again, as a result of the superior flow ability of the coating surface and more sliding action; frictional heating is reduced and subsequently, the coefficient of friction is also reduced.

### 4.2. Wear behavior

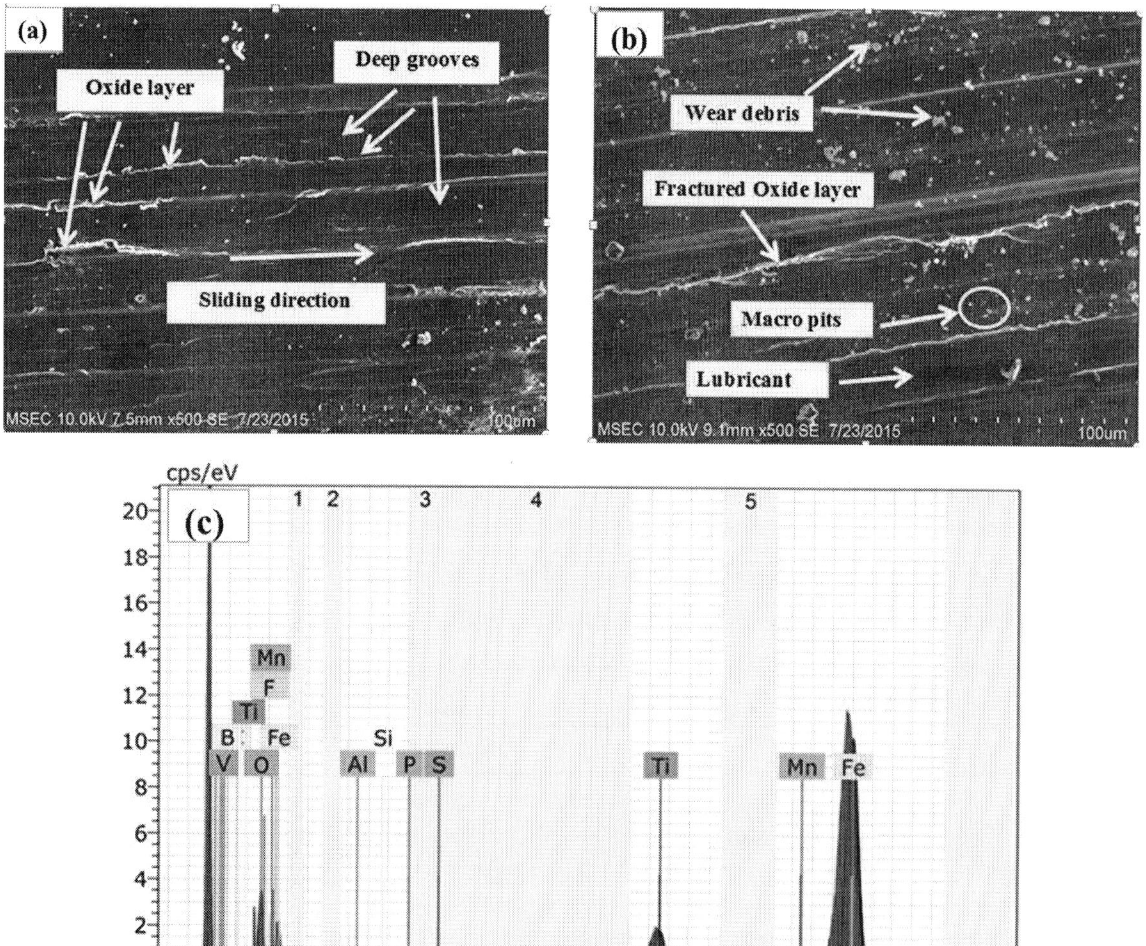


**Figure 8:** (a-b) SEM image of the wear track (c) EDS of worn surface

Based on load and sliding distance increment; the formation of the transfer layer increases gradually. The transfer layers were found in between the sliding surfaces for sliding distance higher than 63 *m*. Besides, when the sliding distance increases more than 63 *m* wear rate also increases accordingly. The formation of the transfer layer controls the wear behaviour of sliding surfaces due to its lubricating nature. Figure 8 (a) illustrates the worn surface analysis of Ti-6Al-4V-2$B_4C$ coated AISI 1040 steel by SEM which includes transferred oxide layers and Figure. 8 (b) confirms the presence of oxide layer by Energy Dispersive Spectrum (EDS) analysis. In addition, Based on obtained results, the presence of icosahedron-based borides in the $B_4C$ increases the nanohardness of the coating as well as low wear rate. The presence of hard $B_4C$ ceramic particles controls the wear rate of the contact surfaces during sliding even for higher loads and sliding distances. Additionally transferred layer and hardness of the counter body (E-52100 steel ball) are the key factors connected to the wear of the coating.

## 5. CONCLUSIONS

Wear behaviour of Ti-6Al-6V-2$B_4C$ coating was investigated and results are drawn.

- The Ti-6Al-6V-2$B_4C$ coatings are not ruptured up to 110 *m* sliding distance under the normal atmospheric test conditions.
- The lubricating film is created between coating and E-52100 ball during sliding by developed oxide layer which supports to improve the wear resistance also gradually decrease the CoF for higher loads and sliding distances.
- The Ti-6Al-6V-2$B_4C$ coatings delivers better wear performance even for higher loads and sliding distances; thus E-52100 steel ball shows higher wear rates.
- The higher and lower value of coefficient of friction of Ti-6Al-6V-2$B_4C$ coatings is 0.49 (under 3 *N* and 110 *m*) and 0.23 (under 2 *N* and 63 *m*) respectively.
- As an outset, Ti-6Al-6V-2$B_4C$ coated AISI 1040 steel is a good replacement for engineering applications such as bearings, tappets, gudgeon pins, etc.

## REFERENCES

[1] Bai L L, Li J, Chen J L, Song R, Shao J Z and Qu C C 2016, *"Effect of the content of $B_4C$ on microstructural evolution and wear behaviors of the laser-clad coatings fabricated on Ti6Al4V"*, Optics & Laser Technology, **Vol. 76,** pp. 33–45.

[2] Huiying Zhu, Yaran Niu, Chucheng Lin, Liping Huang, Heng Ji and Xuebin Zheng 2013, *"Microstructures and tribological properties of vacuum plasma sprayed B4C–Ni composite coatings"*, Ceramics International, **Vol. 39,** pp. 101-110.

[3] Kelly P J and Arnell R D 2000, *"Magnetron sputtering: a review of recent developments and applications"*, Vacuum, **Vol. 56**, pp. 159-172.

[4] Bejarano G, Caicedo J M, Baca E, Prieto P, Balogh A G and Enders S 2006, *"Deposition of $B_4C$/BCN/c-BN multilayered thin films by r.f. magnetron sputtering"*, Thin Solid Films, **Vol. 494,** pp. 53 – 57.

[5] Sanchez-Hernandez Z E, Dominguez-Crespo M A, Torres-Huerta A M, Onofre-Bustamante E, Andraca Adame J and Dorantes-Rosales H 2014, *"Improvement of adhesion and barrier properties of biomedical stainless steel by deposition of YSZ coatings using RF magnetron sputtering"*, Materials Characterization, **Vol. 91,** pp. 50–57.

[6] Miletic A, Panjan P, Skoric B, Cekada M, Drazic G and Kovac J 2014, *"Microstructure and mechanical properties of nanostructured Ti–Al–Si–N coatings deposited by magnetron sputtering"*, Surface and Coatings Technology, **Vol. 241,** pp. 105–111.

[7] Qing Xiang, Daixiong Zhang and Jirao Qin 2016, *"Catholic electrophoretic deposition of nano-$B_4C$ coating"*, Materials Letters, **Vol. 176,** pp. 127-130.

[8] Abou Gharam A, Lukitsch M J, Balogh M P and Alpas A T 2010, *"High temperature tribological behaviour of carbon based ($B_4C$ and DLC) coatings in sliding contact with aluminum"*, Thin Solid Films, **Vol. 519,** pp. 1611–1617.

[9] Vettivel S C, Selvakumar N and Leema 2013, *"Experimental and prediction of sintered Cu–W composite by using artificial neural networks"*, Materials and Design, **Vol. 45,** pp. 323–335.

[10] Saduman Sen and Ugur Sen 2008, *"Sliding wear behavior of niobium carbide coated AISI 1040 steel"*, Wear, **Vol. 264,** pp. 219–225.

IOP Conf. Series: Materials Science and Engineering **923** (2020) 012029 doi:10.1088/1757-899X/923/1/012029

# Process Capability & Fascinating applications of Friction Stud welding

**D.S.Samuvel Prem Kumar**

Asst.Professor, Department of Mechanical Engineering, JP College of Engineering, Tenkasi

**R.Pravin**

Under Graduate Student, Department of Mechanical Engineering, Mepco Schlenk Engineering College (Autonomous), Sivakasi.

**S.Kavin Raj**

Under Graduate Student, Department of Mechanical Engineering, Mepco Schlenk Engineering College (Autonomous), Sivakasi.

**Ms. B.Benita**

Assistant Professor, Department of Computer Science & Engineering, Francis Xavier Engineering College, Tirunelveli-627002

**N.Samuel Dinesh Hynes**

Independent Researcher, & Former Part Time Lecturer, Department of Computer Science & Engineering, Government College of Engineering, Tirunelveli-627007.

**R.Nagarajan**

Under Graduate Student, Department of Mechanical Engineering,Mepco Schlenk Engineering College(Autonomous),Sivakasi.

**Abstract**. In earlier days, welding only involves joining two metals of same kind and dissimilar metals were usually joined by fasteners. Friction Stud welding is a true engineering marvel of making welds between dissimilar metals without any melting or fillers involved. The friction stud welding doesn't involve heat generation from electricity and it is purely based on friction. Besides these facts the friction stud welding has numerous advantage and applications. This paper focuses on the process of stud welding, application and its limitations.

**Keywords**. Friction; Temperature Distribution; Stud Welding; Impact strength; Dissimilar Joints.

IOP Conf. Series: Materials Science and Engineering **923** (2020) 012029 doi:10.1088/1757-899X/923/1/012029

## 1. Introduction

Friction stud welding [1-23], Friction Plug Welding [24-28], Diffusion bonding [29, 30], Friction Welding [31-44] Friction Drilling [45-63], and Friction Riveting [64-67] are few processes employed for joining of dissimilar metals. Friction stud welding, a solid phase welding technique that uses a stud rotating at a high speed, being forced against a substrate and generating heat by friction [2,5]. The metal surfaces reach the temperature of plasticity and they flow plastically under pressure, surface impurities are expelled and a forged weld is formed [7]. The melting point of metals are much higher than the maximum temperatures attained. It is a unique type of friction welding process. Portable equipment for friction stud welding is available for the use on construction work sites, offshore, underwater and in workshops [12,46]. These units are portable and are lighter and smaller than the large static friction welding machines. The weld time being short, around only 4 seconds for a 10mm diameter stud. Weld quality is consistently high, and when destructively tested, failure invariably occurs in the weaker parent material and well away from the weld.

### 1.1 Significance of Friction Stud Welding

The friction stud welding process is well suited for deep water naval applications, short-term emergency repairs and for submarine rescue. Ocean engineers pursued the application commercially for offshore platform repairs. Moreover, Aluminum/stainless steel friction stud welded joints are inevitable in pipes of tanks of liquid propellants in Satellite Launch vehicles and rocket fuel tanks [50]. Recently, there is a need of joining a steel ear thing pin, which is in the shape of a stud to an aluminum car body. Production of burner assembly of fossil boilers and smart valve blowing system requires an irresistible need of friction stud welding of M12 steel studs [22].

## 2. Process Capability of Friction Stud Welding

- The relatively low temperature of weld formation indicates that the process could be adapted for applications such as welding on live pipelines and in explosive environments. Some of the potential problems encountered with arc welding such as contamination of the weld with hydrogen, nitrogen and oxygen, in the presence of electric arc and liquid phase in the metal, are avoided.
- The rapid weld cycle time is typically around 5 to 10 seconds and this method of weld formation results in a fine grain structure. In the "as welded" condition the residual stresses are compressive which tend to result in good fatigue life [9].

### 2.1 Limitations

The process can only be used for the welding of relatively small components such as studs or plugs, which can be rotated at high speed, onto a work piece.

- The systems used are limited typically 25 mm diameter studs and plugs for filling holes up to typically 25 mm diameter (plug welding).The system requires a rigid clamp for holding the welding tool on the work piece and withstand the force applied to the stud during welding.

### 2.2 Applications

- Joining a steel ear thing pin to an aluminum car body
- Production of burner assembly of fossil boilers and smart valve blowing system
- Joining of anodes in an off-shore drill rig
- Friction Stud Welding through paint in seawater discharge pipes
- Friction Stud Welding was used to weld 150mm long M12 steel studs during hull repair.

## 3. Experimentation

The schematic of the experimental set-up is given in Fig. 1. In the present work, friction stud welding is carried out with 12 mm diameter aluminium stud and 12 mm diameter mild steel workpiece with processing conditions mentioned in Table 1.
The lathe machine has been modified into the friction stud welding machine. The tool post has been replaced by the pneumatic cylinder and tool holder. The tool inserts for the tool holder has been prepared to hold the stud of diameter 6, 10,12mm. Two pneumatic cylinders are used to generate axial pressure cylinder and to apply dynamic braking respectively. The brake used is band brake system. The hand braking system has been modified to pneumatic braking system to reduce the braking time because the braking time as an effect on the quality of the weld. The pneumatic circuit has been modified. The 2/2 solenoid valve has been replaced by 3/2 solenoid valve which would cut the power to the electric motor and actuates both axial pressure cylinder and pneumatic braking cylinder. The quick exhaust valve has been connected at one of the opening of pneumatic braking cylinder. The axial pressure cylinder produces the both friction pressure and forging pressure required for the process.

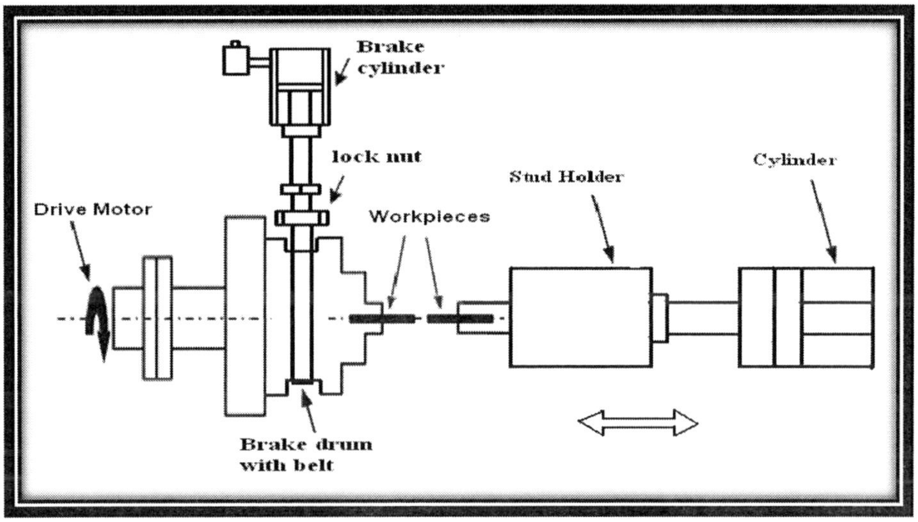


Figure 1: Experimental arrangement

Table 1: Processing conditions

| SI.NO | PARAMETER | RANGE |
|---|---|---|
| 1. | FRICTION PRESSURE | 2.0 bar |
| 2. | FORGING PRESSURE | 4.0 bar |
| 3. | FRICTION TIME | 6 sec |
| 4. | FORGING TIME | 3 sec |
| 5. | SPEED | 1600 rpm |

IOP Conf. Series: Materials Science and Engineering **923** (2020) 012029 doi:10.1088/1757-899X/923/1/012029

3.1 Geometry of parts

The experiment specimens were machined from mild steel and aluminium on geometry in below. Geometry of parts is given in Fig. 2 and Fig. 3. Friction stud welding of aluminium with mild steel is characterised by extremely short welding times and high frictional and compressive forces. As a result, the joining plane is kept extremely narrow, to prevent the biuld up of intermetallic compounds and phases.

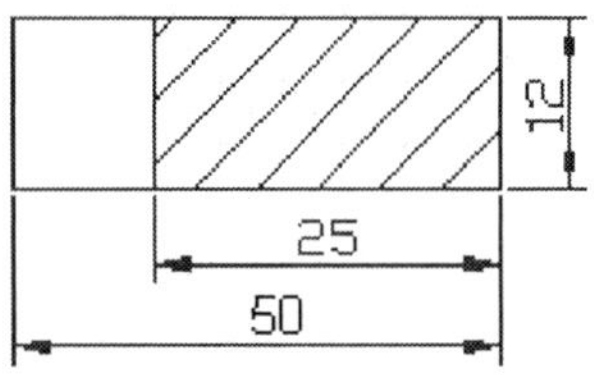


Figure 2 Stud dimensions

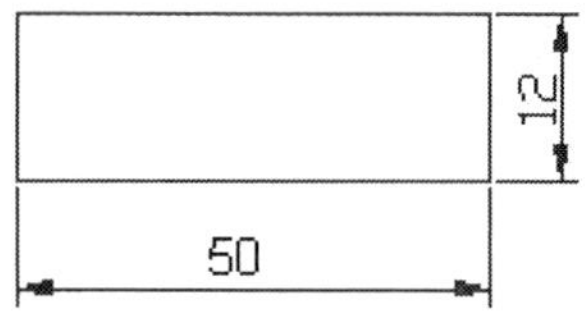


Figure 3 workpiece dimensions

## 4. Results

4.1 Impact Strength

Impact strength is the energy need to break a specimen by an impact load. The dissimilar joints are subjected to Charpy testing to find the impact strength and their results are given in Table 2.

Table 2: Impact strength

| MATERIAL | IMPACT STRENGTH (Joules) |
|---|---|
| Pure Aluminium Vs Pure Aluminium | 15 |
| Mild steel Vs Mild steel | 13 |
| Pure Aluminium Vs Mild steel | 15 |
| AA6o63 Vs AISI 1030 | 15 |

4.2 Microstructure

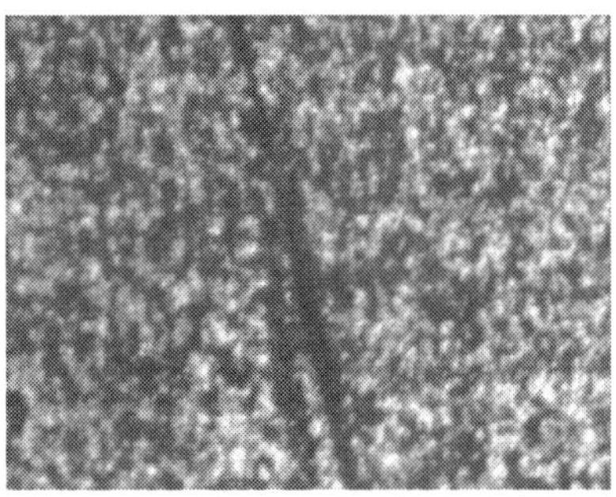

Figure 4: Pure Aluminium Vs Pure Aluminium

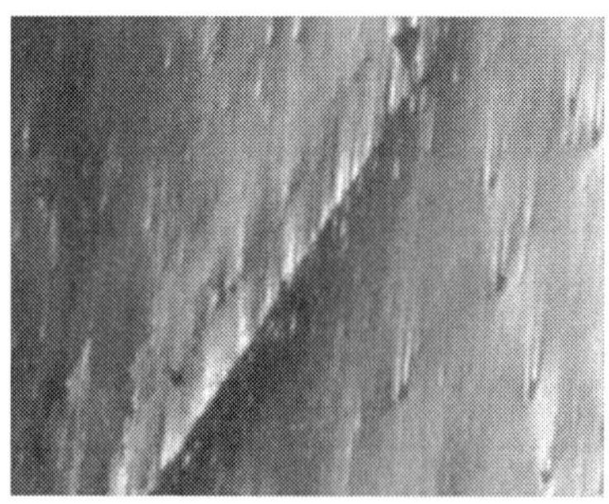

Figure 5: Mild steel Vs Mild steel

Microstructure shows the mechanism of bonding during friction stud welding. Figure 4 shows the micro structure at the interface of friction stud welded pure aluminium-aluminium joints.

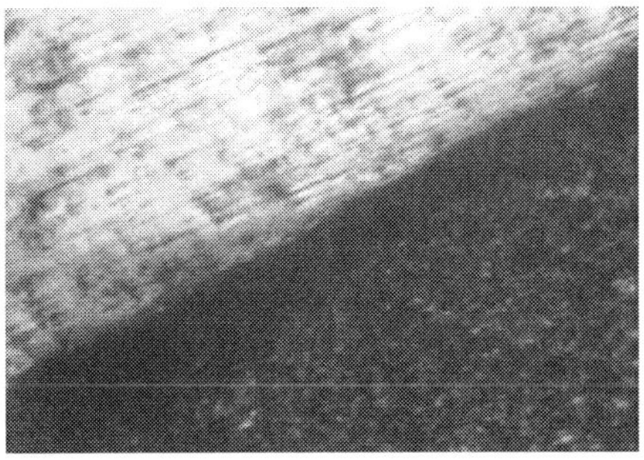

Figure 6: Aluminium - Mild steel

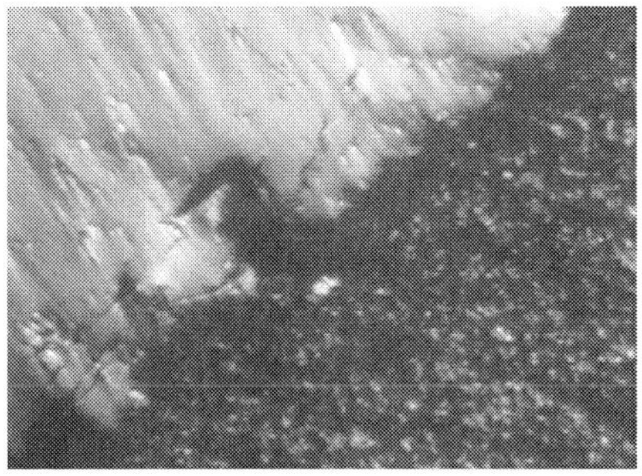

Figure 7: AA6063 Vs AISI 1030

Figure 5 shows the micro structure at the interface of friction stud welded mild steel-mild steel joints. Figure 6 shows the micro structure at the interface of friction stud welded pure aluminium-mild steel joints. Figure 7 shows the micro structure at the interface of friction stud welded AA 6063 – AISI 1030 joints.

## 5. Conclusion

Friction stud welding finds more applications in repair works and hole fillings. This is also suitable for welding small components. The friction stud welding can be probably exploited for particular application as the mentioned above and utilized especially in case of welding of dissimilar metals. Friction stud welding of different dissimilar materials such as Aluminium-Copper, Aluminium-Brass, and Aluminium-Stainless steel can be done by varying the critical process parameters. Online monitoring of the entire process could be carried out with a dedicated computer system.

## References

[1] M.A.Arafin, M.Medraj, D.P.Turner, P.Bocher "Transient liquid phase bonding of Inconel 718 and Inconel 625 with BNi-2: Modeling and experimental investigations" Journal of Materials Science and Engineering A 447(2007) 125-133.

[2] Rajesh N et al (2014) "Ultrasonic evaluation of friction stud welded AA6063/AISI1030 steel joints" Mater Des 62:118–123.

[3] Rajesh Jesudoss Hynes N, Shenbaga Velu P, "Simulation of friction welding of alumina and steel with an aluminium interlayer", Int J Adv Manuf Technol, Springer, DOI 10.1007/s00170-015-7874-8.

[4] Grant O.Cook.III.CarlD.Sorensen "Overview of transient liquid phase and partial transient liquid phase bonding" Journal of Material Science (2011) 46:5305-5323 DOI 10.1007/s10853-011-5561-1.

[5] HongshengChen, Chongsheng Long, Tianguo Wei Gao, HongxingXiao, Le Chen "Effect of Ni interlayer on partial transient liquid phase bonding of Zr-Sn-Nb alloy and stainless steel" Materials and Design 60 (2014) 358-362.

[6] ZhihongZhong,ZhangjianZhou,ChangchunGe "Brazing of doped graphite to Cu using stress relief interlayers" Journal of Materials Processing Technology 209 (2009) 2662-2670.

[7] Rajesh Jesudoss Hynes N et al," Numerical Simulation of Heat Flow in Friction Stud Welding of Dissimilar Metals", Arabian Journal for Science and Engineering, 2014, 39, 3217-3224.

[8] Rajesh N et al (2014) "Ultrasonic evaluation of friction stud welded AA6063/AISI1030 steel joints" Mater Des 62:118–123.

[9] Rajesh Jesudoss Hynes N, Shenbaga Velu P, "Simulation of friction welding of alumina and steel with an aluminium interlayer", Int J Adv Manuf Technol, Springer, DOI 10.1007/s00170-015-7874-8.

[10] Rajesh Jesudoss Hynes N et al," Numerical Simulation of Heat Flow in Friction Stud Welding of Dissimilar Metals", Arabian Journal for Science and Engineering, 2014, 39, 3217-3224.

[11] V.Jalilvand, H.Omidvar, M.R.Rahimipour, H.R.Shakeri "Influence of bonding variables on transient liquid phase bonding behavior of nickel based superalloy IN-738LC" Materials and Design 52 (2013) 36-46.

[12] Rajesh N et al (2014) "Investigation on joining of aluminium & mild steel by friction stud welding" Mater Manuf Process 27:1409–1413, DOI: 10.1080/10426914.2012.667894

[13] Rajesh et al (2014) "Numerical simulation on joining of ceramics with metal by friction welding technique" Int J Mod Phys Conf Ser 22:190–195. DOI:10.1142/S2010194513010118

[14] N. Rajesh Jesudoss Hynes, P. Nagaraj, J.A.J.Sujana, "Ultrasonic Evaluation of Friction stud welded AA6063/AISI1030 steel joints," Materials & Design, Elsevier publications, Volume 62, pp 118–123, October 2014

[15] Rajesh Jesudoss Hynes N, Nagaraj P, Palanichamy R, Arumugham CAK & Angela Jennifa Sujana J, "Numerical Simulation of heat flow of Friction Stud Welding of Dissimilar metals," The Arabian Journal for Science and Engineering, Springer Publications, DOI 10.1007/s13369-013-0932-3, 2014, Volume 39, Issue 4, pp 3217–3224
[16] Rajesh Jesudoss Hynes N, Nagaraj P & Angela Jennifa Sujana J , "Mechanical Evaluation and Microstructure of Friction welded Aluminium-Mild steel joints," The Arabian Journal for Science and Engineering Springer Publications, DOI 10.1007/s13369-014-1082-y, June 2014, Volume 39, Issue 6, pp 5017–5023.
[17] Rajesh Jesudoss Hynes N, Nagaraj P & Meby Selvaraj R (2013) , "Finite Element based Thermal Modelling of Friction Welding of Dissimilar Materials," International Journal of Modern Physics Conference Series,ISSN: 2010-1945 DOI: 10.1142/S201019451301012X, vol. 22, pp.196–202.
[18] Rajesh Jesudoss Hynes N, Nagaraj P &Vivek Prabhu M (2013), "Evaluation of Bending Strength in Friction Welded Alumina/Mild Steel Joints by Applying Factorial Technique," International Journal of Modern Physics Conference Series,ISSN: 2010-1945, DOI: 10.1142/S201019451301012X, vol. 22, pp.184-189.
[19] Rajesh Jesudoss Hynes N, Nagaraj P & Joshua Basil S (2013), "Numerical Simulation on Joining of Ceramics with Metal by Friction Welding Technique," International Journal of Modern Physics Conference Series,ISSN: 2010-1945, DOI: 10.1142/S2010194513010118, vol. 22, pp.190-195.
[20] Rajesh Jesudoss Hynes N, Nagaraj P and Thanga Kumar P, "Thermal Modeling of Friction Plug Welding," International Journal of Applied Engineering Research, ISSN 0973-4562 Vol. 9 No.26 (2014) pp. 9031-9033.
[21] Rajesh Jesudoss Hynes N, Nagaraj P and Tharmaraj R, "Prediction of Thermal Profile During Friction Stud Welding of Aluminium - Mild Steel Joints," International Journal of Applied Engineering Research ISSN 0973-4562 Volume 10, Number 8 (2015), pp. 6107-6110.
[22] Rajesh Jesudoss Hynes N, Nagaraj P & Prakash P, "Numerical Modeling of Temperature Distribution in Friction Stud Welding of Dissimilar Metals," International Journal of Applied Engineering Research, ISSN 0973-4562 Vol. 9 No.26 (2014) pp. 9023-9027.
[23] Rajesh Jesudoss Hynes N, Nagaraj P & Prakash P, "Mathematical Model to Predict Heat Flow in Underwater Friction Stud Welding," Advanced Materials Research, doi:10.4028/www.scientific.net/AMR.984-985.596, Vols.984-985 (2014), pp.596-599.
[24] Rajesh Jesudoss Hynes N, Nagaraj P &Thanga Kumar P, "Mathematical Modeling of Friction Plug Welding with Preheating effect," Advanced Materials Research, doi:10.4028/www.scientific.net/AMR. 984-985.600, Vols. 984-985 (2014), pp.600-603.
[25] Rajesh Jesudoss Hynes N, Nagaraj P & Prakash P, "Thermal Analysis on Joining of Dissimilar Metals by Friction Stud Welding," Advanced Materials Research, doi:10.4028/www.scientific.net/AMR.984-985.592, Vols. 984-985 (2014), pp.592-595.
[26] Rajesh Jesudoss Hynes N, Tharmaraj R, "One Dimensional Thermal Model for Heat Flow in Friction Stud Welding," Journal of Thermal Engineering and Applications. 2015; 2(2), pp. 22–27.
[27] N Rajesh Jesudoss Hynes and P Shenbaga Velu,"Thermo Mechanical Modelling of Friction Welding using Finite Element Method," European Journal of Advances in Engineering and Technology, 2015, 2(7), pp. 29-32.
[28] N. Rajesh Jesudoss Hynes, P. Shenbaga Velu, "Simulation on Friction Welding Of MgAZ31 / AA 6061 T6 Joints," Recent Advances in Computer Science, Recent Advances in Computer Engineering Series, ISSN: 1790-5109, Volume 32, 2015, pp. 524-528.
[29] N. Rajesh Jesudoss Hynes, R. Tharmaraj, P. ShenbagaVelu, and R. Kumar, "Finite element based simulation on friction stud welding of metal matrix composites to steel,"American Institute of Physics (AIP) Conference Proceedings, doi: 10.1063/1.4946607, ISSN: 0094243X, Volume 1728 (2016); pp.020556-1 to 020556-4.
[30] R. Meby Selvaraj, and N. Rajesh Jesudoss Hynes, Finite element approach in thermal modelling of friction stud welding, AIP Conference Proceedings 2142, 110006 (2019); https://doi.org/10.1063/1.5122466
[31] R. Sankaranarayanan, and N. Rajesh Jesudoss Hynes, Friction riveting for joining of wide range of dissimilar materials, AIP Conference Proceedings 2142, 150004 (2019); https://doi.org/10.1063/1.5122553
[32] N. Rajesh Jesudoss Hynes, P. Shenbaga Velu, R. Tharmaraj, R. Kumar, "Numerical investigation on friction welding of alumina / AA 6063 T6 joints," American Institute of Physics (AIP), doi: 10.1063/1.4946601, ISSN: 0094243X, Conference Proceedings, Volume 1728, (2016); pp 020550-1 to 020550-5.
[33] N. Rajesh Jesudoss Hynes, M. Abeyram Nithin, "Simulation on friction taper plug welding of AA6063-20Gr metal matrix composite", American Institute of Physics (AIP) Conference Proceedings, doi: 10.1063/1.4946607, ISSN: 0094243X, Volume 1728, (2016); pp 020556-1 to 020556-4.
[34] N. Rajesh Jesudoss Hynes, R.Kumar, J. Angela Jennifa Sujana, "Modelling of Process Parameters of Friction Stud Welding Using Fuzzy Logic System", International Journal of Advanced Engineering Technology-IJAET, E-ISSN 0976-3945, Volume 8, Issue 1, 2016; pp. 413-417.
[35] N. Rajesh Jesudoss Hynes, P. ShenbagaVelu, "Simulation of friction welding of alumina and steel with aluminum interlayer," The International Journal of Advanced Manufacturing Technology, DOI:10.1007/s00170-015-7874-8, ISSN 0268-3768, October 2017, Volume 93, Issue 1–4, pp 121–127.
[36] Rajesh Jesudoss Hynes N, Angela Jennifa Sujana J & Karuppasamy P, "Simulation of Friction Stud Welding Process with an inter-metallic layer," International Journal of Applied Engineering Research, ISSN 0973-4562 Vol. 9 No.26 (2014) pp. 9028-9030.
[37] N. Rajesh Jesudoss Hynes, P. Shenbaga Velu, R. Kumar, M. Karthick Raja, "Investigate the influence of bonding temperature in transient liquid phase bonding of SiC and copper", Ceramics International, Volume 43, Issue 10, July 2017, Pages 7762–7767.
[38] N. Rajesh Jesudoss Hynes, M. Karthick Raja, "Numerical simulation on influence of bonding temperature in transient liquid phase bonding," American Institute of Physics (AIP) Conference Proceedings, doi: 10.1063/1.4946594, ISSN: 0094243X, Volume 1728, (2016); pp 020543-1 to 020543-4.
[39] P. Shenbaga Velu, N. Rajesh Jesudoss Hynes, "Numerical Analysis of Friction Welded Titanium Joints," Journal of Achievements in Materials and Manufacturing Engineering, Vol. 76, Issue 1 (2016), pp. 26-29.
[40] Dr.N.Rajesh Jesudoss Hynes, Mr.Shenbaga Velu, "Effect of rotational speed on Ti-6Al-4V-AA 6061 friction welded joints," Elsevier, Journal of Manufacturing Processes, DOI: 10.1016/j.jmapro.2018.02.014; Volume No: 30, Issue No: -1; pp. 288-297.

[41] Rajendran, T.P., Hynes, N.R.J., Nikolova, M.P. et al. Influence of heat treatment on friction-welded joints made of high-carbon high-chromium tool steel/low-carbon steel for tooling applications. J Braz. Soc. Mech. Sci. Eng. 42, 87 (2020).

[42] Packiaraj Rajendran, T., Rajesh Jesudoss Hynes, N. & Christopher, T. Characterization of high-carbon high-chromium tool steel/low-carbon steel friction-welded joints for industrial tooling applications. J Braz. Soc. Mech. Sci. Eng. 40, 316 (2018).

[43] Dr.N.Rajesh Jesudoss Hynes, Mr.Shenbaga Velu, "Friction push plug welding in airframe structures using Ti-6Al-4V plug," Springer, Journal of the Brazilian Society of Mechanical Sciences and Engineering, DOI: 10.1007/s40430-018-1088-6; Volume No: 40, Issue No: 3; pp. 158-165.

[44] Mr.Shenbaga Velu, Dr.N.Rajesh Jesudoss Hynes, "Microstructural and Mechanical properties on Friction Welding of dissimilar metals used in motor vehicles," Materials Research Express, DOI: 10.1088/2053-1591/aaabe6; Volume No: 5, Issue No: 2; pp. 26521-26532.

[45] N. Rajesh Jesudoss Hynes, R. Kumar, R. Tharmaraj, P. Shenbaga Velu, "Production of aluminium metal matrix composites by liquid processing methods," American Institute of Physics (AIP) Conference Proceedings, doi: 10.1063/1.4946609, ISSN: 0094243X, Volume 1728, (2016); pp 020558-1 to 020558-5.

[46] Dr.N.Rajesh Jesudoss Hynes, Mr.Vivek Prabhu, Dr.P.Nagaraj, "Joining of hybrid AA6063-6SiCp-3Grp composite and AISI 1030 steel by friction welding," Elsevier, Defence Technology, DOI: dx.doi.org/10.1016/j; Volume No: 13, Issue No: 1; pp. 338-345.

[47] Meby Selvaraj R, Rajesh Jesudoss Hynes N, "Assessment of Influencing Factors on Mechanical and Electrical Properties of Al/Cu Joints," DOI: https://doi.org/10.1063/1.5033163; AIP Conference Proceedings, Volume 1953 D, Issue No: 1; pp. 1300191-1300195.

[48] Rajesh Jesudoss Hynes N, Raja S, "Experimental Study on Joining of AA6063 and AISI 1040 steel," DOI: https://doi.org/10.1063/1.5033164; AIP Conference Proceedings, Volume 1953 D, Issue No: 1; pp. 1300201-1300204.

[49] Shenbaga VeluP, Rajesh Jesudoss Hynes N, "Numerical Modeling of Friction welding of Bi-metal joints for Electrical applications," DOI: https://doi.org/10.1063/1.5033272; AIP Conference Proceedings, Volume 1953 D, Issue No: 1; pp. 1400971-1400975.

[50] Rajesh Jesudoss Hynes N, Nagaraj P & Angela Jennifa Sujana J (2013),"Controller for friction stud welding machine," IEEE Conference proceedings on Energy Efficient Technologies for sustainability, pp.879-882.

[51] Rajesh Jesudoss Hynes N, Nagaraj P & Angela Jennifa Sujana J (2013), "Analytical Modeling of Temperature profile in friction stud welding of dissimilar metals," IEEE Conference proceedings on Energy Efficient Technologies for sustainability, pp.827-829.

[52] N. Rajesh Jesudoss Hynes, M. V. Maheshwaran, "Numerical analysis on thermal drilling of aluminum metal matrix composite,"American Institute of Physics (AIP) Conference Proceedings, doi: 10.1063/1.4946597, ISSN: 0094243X, Volume 1728 (2016); pp.020546-1 to 020556-4.

[53] N.Rajesh Jesudoss Hynes, M. Muthukumaran, N. Rakesh and C.K.Gurubaran, "Numerical Analysis in Friction Drilling Of AISI 1020 Steel and AA 6061 T6 Alloy," Recent Advances in Environmental and Earth Sciences and Economics, Energy, Environmental and Structural Engineering Series, ISSN: 2227-4359, Volume 39, 2015, pp. 145-149.

[54] Mr.R.Kumar, Dr.N.Rajesh Jesudoss Hynes, "Finite-element simulation and validation of material flow in thermal drilling process," Springer, Journal of the Brazilian Society of Mechanical Sciences and Engineering, DOI: 10.1007/s40430-018-1091-y; Volume No: 40, Issue No: 3; PageNo: 162-172.

[55] N. Rajesh Jesudoss Hynes et al, "Optimum Bushing Length in Thermal Drilling of Galvanized Steel Using Artificial Neural Network Coupled with Genetic Algorithm,"Materiali in tehnologije / Materials and technology, ISSN 1580-2949, doi:10.17222/mit.2016.290, 51 (2017) 5, pp. 813–822.

[56] Dr.N.Rajesh Jesudoss Hynes, R.Kumar, "Simulation and Experimental Validation of Al7075-T651 Flow Drilling Process," Journal of the Chinese Society of Mechanical Engineers, 2017, Vol.38, No.4, pp. 413-420.

[57] Mr.R.Kumar, Dr.N.Rajesh Jesudoss Hynes, "Process optimization for maximizing bushing length in thermal drilling using integrated ANN-SA approach," Journal of the Brazilian Society of Mechanical Sciences and Engineering, DOI: 10.1007/s40430-017-0820-y; Volume No: 39, Issue No: 12; pp. 5097-5108.

[58] Kumar R, Rajesh Jesudoss Hynes N, "Numerical Analysis of Thermal Drilling Technique on Titanium sheet metal," DOI: https://doi.org/10.1063/1.5033158; AIP Conference Proceedings, Volume 1953 D, Issue No: 1; pp. 1300141-1300144.

[59] Vijayabaskar P, Rajesh Jesudoss Hynes N, "Simulation of Friction Stir Drilling Process," DOI: https://doi.org/10.1063/1.5033284; AIP Conference Proceedings, Volume 1953 D, Issue No: 1; pp. 1401091-1401094.

[60] Vignesh N J, Rajesh Jesudoss Hynes N, "Thermal Analysis of Friction Riveting of Dissimilar Materials," DOI: https://doi.org/10.1063/1.5033285; AIP Conference Proceedings, Volume 1953 D, Issue No: 1; pp. 1401101-1401104.

[61] R Kumar, N Rajesh Jesudoss Hynes, J Angela Jennifa Sujana, "Multi-objective optimization of green technology thermal drilling process using grey-fuzzy logic method", Journal of Cleaner Production (SCI Journal, IF 7.100), Volume 236, Available online 21 July 2019.

[62] R. Kumar, N. Rajesh Jesudoss Hynes, Anish Khan, Simulation of thermo-mechanical behaviour of friction drilling process, Int. J. Computational Materials Science and Surface Engineering, Vol. 9, No. 1, 2020, pp. 70-84.

[63] R.Kumar, N. RajeshJesudoss Hynes, "Thermal drilling processing on sheet metals: A Review", Elsevier, International Journal of Lightweight Materials and Manufacture, Available online 19 August 2019.

[64] Vignesh N J, Rajesh Jesudoss Hynes N, "Thermal Analysis of Friction Riveting of Dissimilar Materials," DOI: https://doi.org/10.1063/1.5033285; AIP Conference Proceedings, Volume 1953 D, Issue No: 1; pp. 1401101-1401104.

[65] Sankaranarayanan R, Rajesh Jesudoss Hynes N, "Prospects of Joining Multi-Material Structures," DOI: https://doi.org/10.1063/1.5033165; AIP Conference Proceedings, Volume 1953 D, Issue No: 1; pp. 130021-130025.

[66] R. Sankaranarayanan, and N. Rajesh Jesudoss Hynes, Friction riveting for joining of wide range of dissimilar materials, AIP Conference Proceedings 2142, 150004 (2019); https://doi.org/10.1063/1.5122553

[67] Hynes, N.R.J., Vignesh, N.J. & Velu, P.S. Low-speed friction riveting: a new method for joining polymer/metal hybrid structures for aerospace applications. J Braz. Soc. Mech. Sci. Eng. 42, 434 (2020). https://doi.org/10.1007/s40430-020-02519-8

# A Review On Friction Stir Welding And It's Processes

B Vijaya Ramnath*, C Elanchezhain , C Vignesh, V Sudharsan, C Selva Ganesh

Department of Mechanical Engineering, Sri Sairam Engineering College, West Tambaram, Chennai-600 044

*E-mail: **vijayaramnathb@gmail.com**

Abstract. Friction stir Welding (FSW) is a solid-state joining process which removes the common problems encountered in conventional welding processes. FSW is normally suitable for alloys of aluminium, magnesium and steel. Sectors like automotive, aerospace, ship building and electronics industries are facing problems with fusion welding, since there is a huge difference in chemical and physical properties of components to be joined. This paper reviews the basic principle and methodology of FSW and covers all the technical aspects like mechanical characterization, similar and dissimilar combinations, micro structural characterization.

*Keyword:* Friction stir welding, mechanical characteristics, process parameters, FEA modeling

## 1. Introduction

Friction Stir Welding (FSW) is a method of solid-state metal joining in which non-consumable rotating device used for joining two plates is a specially designed pin. Instead, the device is traversed along weld axis. The tool surface typically has dual behaviour namely heat generation and mechanical sweeping of softened metal such as aluminium alloy. FSW is an ideal process for producing low cost with high performance joints.

## 2. Review on Mechanical Properties of FSW

Shayan Eslami et al. (2017) researched during high frequency, the fatigue behaviour of the dissimilar polymers welded in lap-joints have been observed to be failed due to fatigue stresses and specimens failed by thermal stresses. Martin Reimann et al. (2017) performed research in AA7075-T651, using refill friction stir spot welding, on the microstructure and mechanical properties of keyhole repair welds. The measurements of the thermal cycle showed that high heating levels and peak temperatures in the weld center are up to 540 deg Celsius. It was found that a loss occurs in two types of fracture. The microstructure, mechanical properties, and shape memory behaviour of FSW nitinol were investigated by Mani Prabu et al.(2017). Owing to grain refinement, the yield strength of the specimen weld joint improved by 17 % compared to the base material but the initial phase was maintained using the SMA. Sharma et al.(2017) studied the mechanical properties of Al-Zn-Mg joints type FSW steel. The joint formed was subjected to bend test, stiffness test, optical examination and SEM analysis. It is derived from the regression model that the scale of the aluminium grains increases as the welding speed rises and the rotational speed decreases. Chao Zhang et al.(2017) studied the 9 % Cr microstructure and mechanical properties of reduced FSW joint ferritic / martensitic steel activation. The plant refining and the martensitic transition are the two main characteristics of the weld region's microstructure. The SiC nano-powder results of dissimilar Al-Mg alloy to A316L Stainless Steel on the FSW were studied by Fallahi et al.(2017). For in-situ production of the hybrid composite in the stir zone of the bi-metallic joint, the effects of Pass numbers and traverse velocity on the second phase distribution, micro-structural features, and mechanical properties of dissimilar joints were analysed. Hemath Kumar et al.(2016) examined the mechanical behaviour of Al-based composite material for friction stir welding and the results showed that the tension of the welded joint improved and the tensile properties were adequate compared to the base metal. Ramesh et al.(2016) studied the characterization of the microstructure and toughness of the friction stir welded brass plate and found that rotational speed had a significant effect on crack forming and the frequency of dynamic re-crystallization in the stir zone was shown by the inclusion of fine grains and higher hardness. Shayan Eslami et al. used a newly developed method to examine the impact of friction stir welding parameters for lap joint of dissimilar polymers. It was noticed that the welds developed with this tool significantly improved the surface quality and strength of the welds and the rotational and traverse speeds play the most effective role with respect to the overall tensile strength of the specimen. Xueqi Lv et al.(2017) studied the microstructure and mechanical properties of Al alloy and Mg alloy. They found that difference in the welding mechanism ,weld properties and improvement in process temperature, direction material flow and mechanical interlocking of the Al-Mg joint .

## 3. Review on Effect of Process Parameters of FSW

Shanavas et al., used reaction surface technique to achieve parametric optimization of friction stir welding parameters of marine grade aluminium alloy. A defect-free weld is found in and around the stir region from the macro-and micro-structure analysis. As a consequence of complex re-crystallisation, grain refining occurred in the stir zone. Farzad et al.(2017) studied the effects of processing parameters of micro-structure and mechanical behaviour's of underwater FSW of Al5083 alloy. The water-cooling cycle at the time of welding defined very high cooling rate so that the grains do not have enough time to grow and shape a zone in the welded region. Suresh et al.(2017) proposed an substitution to fusion welding to increase stress corrosion cracking resistance maraging steel. The separation of alloying elements and formation of reverted austenite pool where suppressed by FSW. The result shows that the improved SCC resistance of FSW maraging steel weldments as compared to base metal. Shayan Eslami et al.(2017) researched friction stir welded dissimilar polymers for the assessment of fatigue existence. He conclude that in cyclic loading , the friction stir welded joints with brittle behaviour exhibited good resistance. It was found that the experiment fell and necked due to high level of examination and resulted in retreated side of weld. Wen-hui Liu et al.(2017) tracked optical microscopy (OM) and X-ray diffraction (XRD) to study the microstructure and texture evolution. The evolutions of structure of α-Mg and β-Li have also been studied. The results showed an elongation in the phases α-Mg and β-Li, and thin laminar sandwich structure is established. Paoletti et al.(2014) performed friction swirl spot welding to 3 mm thick polycarbonate sheets by creating an Artificial Neural Network to anticipate the welded area's mechanical behaviour. They concluded that Artificial neural network was used to model the impact of friction stir spot welding parameters on the maximum plunging force, maximum torque and joint strength of polycarbonate sheet welds.

## 4. Review on FEA Modeling on FSW

El-Sayed et al.(2018) used the finite element simulation of aluminium alloy AA5083-O FSW method to estimate the temperature distribution and residual thermal stresses by experimentally validating it using thermal image camera. They found that the peak temperature obtained through the use of threaded device pin profile is greater than that obtained through the use of tapered ones. Haidong et al. (2018) developed a FSW-induced simulation analysis of welding stress. The characteristics of stress are shown in the analytical results of the temperature distribution and stress evolution and it is stated that the radial stress in the welded structure decreases with a rise in the depth of the impression and the tensile stress decreases in the welded structure in the higher advancing speed. Noor Zaman Khan et al. (2017) examined the flaws in the FSW clean manufacturing process. The study revealed that the faults are induced by insufficient consolidation of material behind the pin, ineffective heat generation and excessive movement of the material. The multiple faults such as tunnelling faults, hooking faults, inadequate root penetration and voids are listed for their reasons and suggestions.

## 5. Review on Corrosion analysis and other metal welding using FSW

Multi-criteria decision making approach for the selection of optimum process parameters for FSW of Aluminium 2024 alloy and the optimum process parameter conditions that give maximum response are rotational speed of 1000 rpm, welding speed of 0.08 m/min and tool offset of 0 mm were been conducted by Sudhagar et al. (2017) . Nidhi Sharma et al.(2017) analysed Aluminium to Copper's FSW. Work mentions the faults that occurred during the dissimilar Al-Cu FSW, and describes the potential remedies. The thrust force and the force component as well as the torque empowered for FSW are reported. The temperature distribution followed by the flow stress is also studied. Kamal Jayaraj et al.(2017), investigated the electro-chemical corrosion behavior os stir-zone of AA6061 Aluminum-AZ31B magnesium alloy dissimilar joints. The potentio-dynamic polarization test was performed to compare the empirical relationship of the FSW stir zone corrosion rate of dissimilar joints. Ranvijay Kumar et al.(2018) investigated the weldability of FSW thermo-plastic materials. It was observed that the selection of pin profile shoulder diameter, reasonable speed, travel speed, tilt angle, plunge depth are the input parameters to be selected with optimum material flow and heat generation. Hou Longgang et al.(2017) studied the corrosion behavior of FSW Al-Mg-Zn alloys and the impact of pre-welded temperature conditions on the FSW alloy's corrosion behaviour. It was found that with the increase in the zinc content, both maximum corrosion depth and dominated corrosion mode differs in different welding regions after heat treatment and post-welding.

## 6. Conclusion

This paper discusses various aspects of FSW such as Tool Geometry, different materials, gripping device (Fixture), optimization technique, temperature distribution, process impact on the environment, fracture mode, micro-structure and various mechanical properties of different materials. This research article also discusses Numerical analysis of FSW as well as the latest trends. But there are still there are some issues left to debate in Numerical analysis. For industrial applications, therefore, it can be assumed that FSW is an economical and environmentally friendly welding process.

## REFERENCES

[1] Shayan Eslami, Poulo J Taveres, P MG P Moreira, “Fatigue Life Assessment of Friction Stir Welded Dissimilar Polymers”, 2nd International Conference on Structural Integrity,(4-7), pp. 1433-38, 2017.M.B. Karamis , A. Tasdemirci, F. Nair // Failure and tribological behaviour of the AA5083 and AA6063 composites reinforced by SiC particles under ballistic impact, 34 (2003) 217–226

[2] Martin Reimann, Jannik Goebel, Jorge F. dos Santos, "Microstructure and Mechanical Properties of keyhole repair welds in AA7075-T651 using refill Friction Stir Welding", Materials and Design, (132), pp.283-294, 2017.
[3] S.S.Mani Prabu, H.C.Madhu, Chandra S.Perugu, K.Akash, P.Ajay Kumar, Satish, V.Kailas, Manivannan, Anbarasu, I.A.Palani, "Microstructure, Mechanical Properties and shape memory behaviour of Friction Stir Welding nitinol", Materials Science and Engineering A, (693), pp.233-236, 2017.
[4] C.Sharma, C.Upadhyay, D.K.Dwivedi, P.Kumar, " Mechanical Properties of Friction Stir Welding armor grade Al-Zn-Mg alloy joints", Transactions of Non-Ferrous Metals Society Of China, (27), pp.493-506, 2017.
[5] Chao Zhang, Lei Cui, Yongchang Liu, Chenxi Liu, Huijun Li, " Microstructure and Mechanical Properties of Friction Stir Welding on 9% Cr reduced activation ferritic/martensitic steel", Journals of Materials Science and Technology, 2017.
[6] A.A.Fallahi, A.Shokuhfar, A.Ostovari Moghaddam, A.Abdolahzadeh, " Analysis of SiC nano-powder effects on Friction Stir Welding of dissimilar aluminium-magnesium alloy to A316L stainless steel", Journal of Manufacturing Processes, (30), pp.418- 430, 2017.
[7] G. Hemath Kumar1, Babu Vishwanath1, Rajesh Purohit, R. S. Rana and Saurabh Singh Rajpurohit, "Mechanical Behaviour Of Friction Stir Welding On Aluminium Based Composite Material", 6th International Conference of Materials Processing and Characterization, (4), pp.5336–5343, 2017.
[8] Ramesh Ga, Srimuthunath Ib, Santha Kumar Nb, Sugandipriya Sb, Aravindhan V, "Characterization of microstructure and hardness of friction stir welded brass plate", 6th International Conference of Materials Processing and Characterization, (5), pp.2721–2725, 2018.
[9] Shayan Eslami, Paulo J Tavares, P M G P Moreira, "Fatigue life assessment of friction stir welded dissimilar polymers", 2nd International Conference on Procedia Structural Integrity, (5), pp.1433-1438, 2017.
[10] Xueqi Lv, ChuanSong Wu, Chunliang Yang, G.K. Padhy, "Weld microstructure and mechanical properties in ultrasonic enhanced friction stir welding of Al alloy to Mg alloy", Journal of Materials Processing Tech. (254), pp.145–157, 2018
[11] S.Shanavas, J.Edwin Raja Dhas, " Parametric optimization of Friction Stir Welding parameters of marine grade Aluminium alloy using response surface methodology", Transactions of Non-Ferrous Metals Society Of China, (27), pp.2334-44, 2017.
[12] Farzad Heirani, Alireza Abasi, Mohammad Ardestani, "Effects of processing parameters on Microstructure and Mechanical Properties of underwater Friction Stir Welding of Al5083 alloy", Journal of Manufacturing Processes, (25), pp.77-84, 2017.
[13] Suresh.D.Meshram, Archana G.Paradkar, G.,Madhusudhan Reddy, Sunil Pandey, "Friction Stir Welding: An alternative to fusion welding for better stress corrosion cracking resistance of maraging steel", Journal of Manufacturing Processes, (25), pp.94-103, 2017.
[14] Shayan Eslamia , Tiago Ramosa, Paulo J. Tavaresa, P. M. G. P. Moreira "Effect of friction stir welding parameters with newly developed tool for lap joint of dissimilar polymers", Procedia Engineering, (114),pp. 199 – 207, 2015.
[15] Wen-hui Liu , Xiao Liu , Chang-ping Tang , Wei Yao , Yang Xiao , Xu-he Liu, "Microstructure and texture evolution in LZ91 magnesium alloy during cold rolling", Journal of Magnesium and Alloys, 2017.
[16] A. Paoletti, F. Lambiase, A. Di Ilio, "Optimization of friction stir welding of thermoplastics", Procedia CIRP, (33),pp.562 – 567, 2015.

[17] M.M.El-Sayed, A.Y.Shash, M.Abd-Rabou, "Finite element modeling of aluminium alloy AA5083-O Friction Stir Welding process", Journal of Materials Processing Tech, (252), pp.13-24, 2018.
[18] Haidong Yu, Bin Zheng, Xinmin Lai, " A modeling study of welding stress induced by Friction Stir Welding", Journal of Materials Processing Tech, (254), pp.213-220, 2018.
[19] Noor Zaman Khan, Zahid A.Khan, Arshad Noor Siddiquee, Abdulrahman M.Al-Ahmari, Mustafa H.Abidi, "Analysis of faults in clean fabrication process of Friction Stir Welding", Transactions of Non-Ferrous Metals Society Of China, (27), pp.1507-16, 2017.
[20] S.Sudhagar, M.Sakthivel, Prince.J.Mathew, S.Ajith Arul Daniel, "A multi-criteria decision making approach for process improvement in Friction Stir Welding of aluminium alloy", Measurement, (108), pp.1-8, 2017.
[21] Nidhi Sharma, Zahid A.Khan, Arshad Noor Siddiquee, "Friction Stir Welding of aluminium to copper-An overview", Transactions of Non-Ferrous Metals Society Of China, (27), pp.2113-36, 2017.
[22] R.kamal Jayaraj, S.Malarvizhi, V.Balasubramanian, " Electro-Chemical corrosion behavior of stir zone of Friction Stir Welding dissimilar joints of AA6061 aluminium-AZ31B magnesium alloys", Transactions of Non-Ferrous Metals Society Of China, (27), pp.2181-92, 2017.
[23] Ranvijay Kumar, Rupinder Singh, I.P.S.Ahuja, Rosa Penna, Luciano Feo, " Weldability of thermo-plastic materials for Friction Stir Welding-A state of art review and future applications", Composites Part B, (137), pp.1-15, 2018.
[24] Hou Longgang, Yu Jiajia, Zhang Di, Zhuang Linzhong, Zhou Li, Zhang Jishan, " Corrosion Behaviour of Friction Stir Welding Al-Mg-Zn alloys", Rare Metal Materials and Engineering, (46), pp.2437-44, 2017.

# 3D Printing of Poly Lactic Acid Bone scaffold combined with Poly Ethylene Glycol for tissue engineering applications

**Dinesh Kumar S[1]*, Yogesh P[2]**
1* Professor, Department of Mechanical Engineering, Faculty of Engineering and Technology, Jain University, Bangalore, Karnataka, India.

2 UG Scholar, Department of Mechanical Engineering, Faculty of Engineering and Technology, Jain University, Bangalore, Karnataka, India.

* E-mail: softdin@gmail.com

**Abstract**. Additive Manufacturing (AM) is a new trend in manufacturing in which computer aided design (CAD) 3D model file is used to make a solid physical model of the designed part. Fused deposition modelling (FDM) a technique in additive manufacturing has revolutionized engineering in various applications. One of the significant impacts of this technology is on the medical industries. The most substantial reason is because of this technology's facility to convert 3D medical imaging data into solid objects. The productional cost of an implant and other instruments reduced because of the ease of this technology which significantly reduced the cost charged from the patients. Polylactic acid (PLA) is conceivably one of the best-known polymers shaped from renewable raw materials such as sugar cane and corn starch. Poly(ethylene glycol) (PEG) is an FDA approved low toxicity material which is in re water-soluble liquid state or waxy solid-state. It is used in preparations of cosmetics and pharmaceuticals. Most of the emulsifying or wetting agents and lubricants are manufactured using PEG. Also, PEGs are suitable plasticizers owing to their low glass transition temperature. PLA/PEG has a hopeful probable for large-scale manufacture of green composites towards various applications in biomedical fields. In this research work, PEG reinforced polylactic acid composite were produced in 2, 5 and 10 wt. % PEG content by a single-screw extruder in filament form. The results obtained during the experiment indicated that PEG concentration enhances the processing of PLA, but changes in various features like relevant surface, geometry ad structure were observed. The addition of PEG in PLA had significantly changed the rate of degradation of the FDM fabricated scaffolds.

**Keywords**: Additive Manufacturing, Poly Lactic Acid, Fused Deposition Modelling, Poly Ethylene Glycol, Bone Scaffolds

## 1. Introduction

Additive Manufacturing (AM) is a new trend in manufacturing in which computer aided design (CAD) 3D model file is used to make a solid physical model of the designed part [1]. The growing opportunities and innovative challenges in 3D printing has attracted industrialists towards rapid prototyping for various components and academics to learn the fabricating of 3D models with this technology, as this uses trifling detrimental chemicals during manufacturing a component [2–4]. The AM has paved a new way of manufacturing with optimized use of materials, FDM is widely used because of ease of printing models by extruding a thermoplastic filament and printing it layer by layer.

Recent research on 3D printed scaffolds exhibits the scope of various degradable polymers such as polylactic acid (PLA), polyglycolic acid, polycaprolactone, chitosan and their copolymers being used in AM fabricated scaffolds. [5]. Currently, most biomedical applications use PLA as a biodegradable polymer which is approved by the FDA [6-7]. Though numerous researches have used this polymer and extensively studied, the usage of this polymer for 3D printed scaffolds in tissue engineering applications through FDM is scarcely reported. All the recent reports of PLA -based scaffold fabricated through FDM use the control of temperature during the print process along with freeze-drying as a post-processing technique to set in molecular modifications in the PLA matrix. [10]. Poly (ethylene glycol) (PEG) is an FDA approved low toxicity material which is in re water-soluble liquid state or waxy solid-state. It is used in preparations of cosmetics and pharmaceuticals. Most of the emulsifying or wetting agents and lubricants are manufactured using PEG. Also, PEGs are suitable plasticizers due to their low glass transition temperature [8-9].

In this work the blending of PLA and PEG for various composition ratios and 3D printing of porous scaffold with the PLA/PEG composite through FDM with greater structural and degradable properties is discussed.

## 2. Experimental

### *2.1. Materials*

Extrusion grade Poly Lactic Acid (PLA) polymer (trade-name Ingeo 2003D) was purchased. The material has a specific gravity of 1.24 and yield Modulus of 3.5GPa (as provided by the manufacturer) Pure Pharma grade Poly ethylene Glycol (PEG 400) was purchased. The specifications and properties as per the manufacturer report are with a density of 1.127 kg/litre, Molecular Weight 380-420 and the viscosity is 6.8 -8.0 cSt.

### *2.2. Materials processing and sample preparation*

#### *2.2.1. Compounding*

The Polylactic acid and PEG were solvated separately in chloroform (5%w/v) with addition of PEG in concentrations (0, 2, 5, 10% w/w) and the composites were named PLA-2, PLA-5 and PLA-10 respect to the PEG concentrations. A homogeneous polymer blend solution was achieved by placing the prepare composites for 48hours on an orbital mixer.

#### *2.2.2. Filament extrusion*

The blended composites were fed into single screw extruder (screw diameter = 3.175mm, L/D ratio = 20 and rod die diameter =3mm). The rotation speed of the screw and collection rate were controlled to acquire a final diameter of the extruded filament of 1.75mm with a screw speed at 20 rpm. The Table 1 depicts the details of process control parameters for different samples.

**Table 1.** Conditions used to produce PEG-filled PLA

| Composite Name | PEG wt % | Extruder Temperature (°C) Hopper | Screw | Die | *Motor Speed (rpm)* |
|---|---|---|---|---|---|
| **PLA** | 0 | 148 | 175 | 175 | 20 |
| **PLA-2** | 2 | 148 | 170 | 170 | 20 |
| **PLA-5** | 5 | 142 | 170 | 170 | 20 |
| **PLA-10** | 10 | 145 | 170 | 170 | 20 |

*2.2.3. Scaffold Design and Fabrication*

The PLA/PEG scaffolds were fabricated using a Flashforge Desktop 3D Printer. The feed to the 3D printer was the filaments prepared using the extruder. However, since there is fluctuation in filament diameter during extrusion, not all prepared filaments were printable. As shown in figure [1] The doubled layer scaffolds, which are Orthogonal displaced with the distance between struts D=400μm and each layer of height $\Phi$=200μm.The samples of neat PLA, PLA-2, PLA-5 and PLA-10 were successfully printed with the following printing parameters: 100% object infill; no raft; 0.35 mm nozzle diameter; 0.20 mm layer height; 220$^0$C nozzle temperature and 55$^0$C bed temperature.

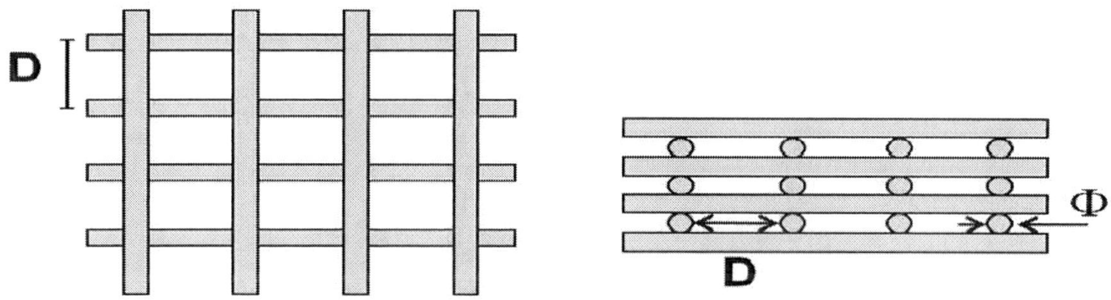


**Figure 1.** Axial and cross-section view of designed 3D scaffold of orthogonal scaffolds: D = 400 μm; Φ = 200 μm

*2.3. Scaffold Characterization*

*2.3.1. Differential Scanning Calorimetry*

Differential Scanning Calorimetry tests were carried out to measure the glass transition temperature and degree of crystallinity. The samples were initially heated from 15$^0$C to 220$^0$C at a rate of 100C/min and then it was kept for an isothermal stay at 220$^0$C for 5minutes. Then the temperature of the samples was reduced from 220$^0$C to -15$^0$C at a rate of -10$^0$C/min and reheated at the same temperature from -15$^0$C to 220$^0$C. Then the glass transition temperature (Tg) was measured.

*2.3.2. Morphological SEM study*

The morphological study of the fabricated scaffolds and the prepared samples were observed under a scanning electron microscope (SEM). The prepared polymer film samples were kept in simulated body fluid (SBF), and the SEM was inspected after t= 0, 2, 4, 6 and 8 weeks for possible surface modifications during the degradation. This analysis was primarily done to evaluate the scaffold's key parameters like structural solidity of the intermediate layers and struts, surface morphology and design of the 3D scaffold. The morphology analysis plays a vital role in the quality assessment of the porosity in the theoretical geometry and the fabricated scaffold.

### *2.3.3. Mechanical Properties*

The mechanical properties were evaluated using a Universal Testing Machine (Instron 5567) was used to study and compare the mechanical properties of all the prepared PLA/PEG bio-composites at room temperature. The tensile specimen and the tests were conducted according to the ASTM D638 standards.

**Figure 2.** Universal Testing Machine (Instron 5567)

### *2.3.4. Porosity*

The porosity evaluation for the fabricated scaffolds of varying composition of PEG in the PLA matrix was performed by calculating the theoretical volume porosity percentage (%Vol theoretical) with the considered design (Fig.1). An assumption that there is no touching due to the blend between struts from end-to-end was made so that the values of layer height and strut diameter was the same.

$$\%\text{Vol theoretical} = (1 - (Ø^2/4)\ \pi N_c N_l /(wh)) \times 100\%$$

Where Ø, w and h refer to the strut diameter, scaffold width and scaffold height in millimetres as per the scaffold design. Moreover, Nc denotes the number of cylinders (struts) per layer, while Nl denotes the number of layers per scaffold.

## 3. Results and Discussions

### *3.1. Morphological SEM Study and Porosity Evaluation*

From the SEM analysis it was observed that there was a striking variation between the three scaffolds of varying PEG concentration. Upon increasing the PEG concentration, there was a shift from the designed initially squared pores into slightly rounded ones.

The Neat PLA scaffold had a even flat exterior whereas the PLA-10 had a slightly irregular coarse spherical structure. During the degradation process, i.e. after 8 weeks of absorption of the scaffolds in Simulated Body Fluid, Neat PLA did not show any degradation, and the surface remained the same. In the other case, PLA-10 shows great contrasts in terms of their exterior morphology and intense degradation with several cracks at the surface.

Figure 3 shows the SEM micrograph of PLA-5 which indicates the scaffolds surface showing the micro porosity due to the evaporation of the solvent. Figure 4 presents the theoretical volume porosity percentages of the PLA/PEG scaffolds fabricated with the varying concentration of PEG in the PLA matrix. From the graph, we can interpret that the porosity of the scaffold decreases from 98.5% (Neat PLA) to 86.5% (PLA-10) as PEG concentration increases from 0 to 10% (w/w).

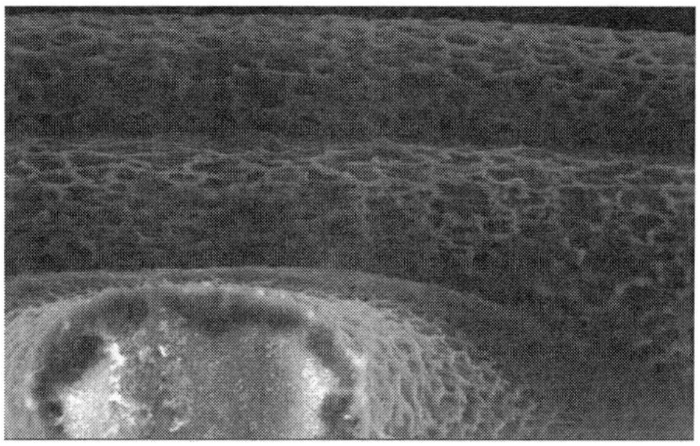

**Fig 3.** SEM image of PLA-5 showing micro porosity

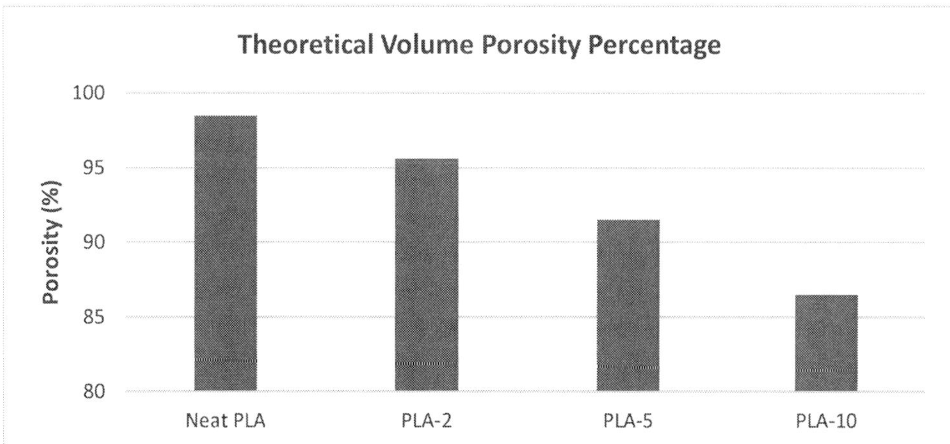


**Fig 4.** Theoretical Porosity % for 3D Printed scaffolds

### *3.2. Differential Scanning Calorimetry Analysis*

The thermal characteristics of Neat PLA and PLA/PEG composite samples are shown in Table 2.Here the results obtained from the DSC analysis exhibit that increase in the concentration of PEG in the PLA matrix increases the percentage of crystallinity (Xc), i.e. increase in the hardness and density of the composite but decreases the Glass transition temperature (Tg). For the increase in the PEG concentration from 0% to 10% (w/w), the value of Tg decreases from 57.59°C (Neat PLA) to 29.25°C (PLA-10), and Xc substantially increases from 2.1% (Neat PLA) to 22.13% (PLA-10).

**Table 2.** Glass transition temperature (Tg) and Crystalline fraction (%Xc) of PLA/PEG blends

| Samples | Tg ( °C) | Xc (%) |
|---|---|---|
| **Neat PLA** | 57.59 | 2.10 |
| **PLA-2** | 50.78 | 7.62 |
| **PLA-5** | 41.6 | 14.40 |
| **PLA-10** | 29.25 | 22.13 |

A gradual rise of Tg values along the degradation rate was reported, particularly for the materials with a higher concentration of PEG. During the degradation period in SBF, at t=4 weeks Neat PLA showed a gradual increase whereas the PLA-5 exhibited a steeper increase which depicted an improvement in crystallinity. Overall,the addition of PEG has a strong influence on the degreation rate of the fabricated scaffold.

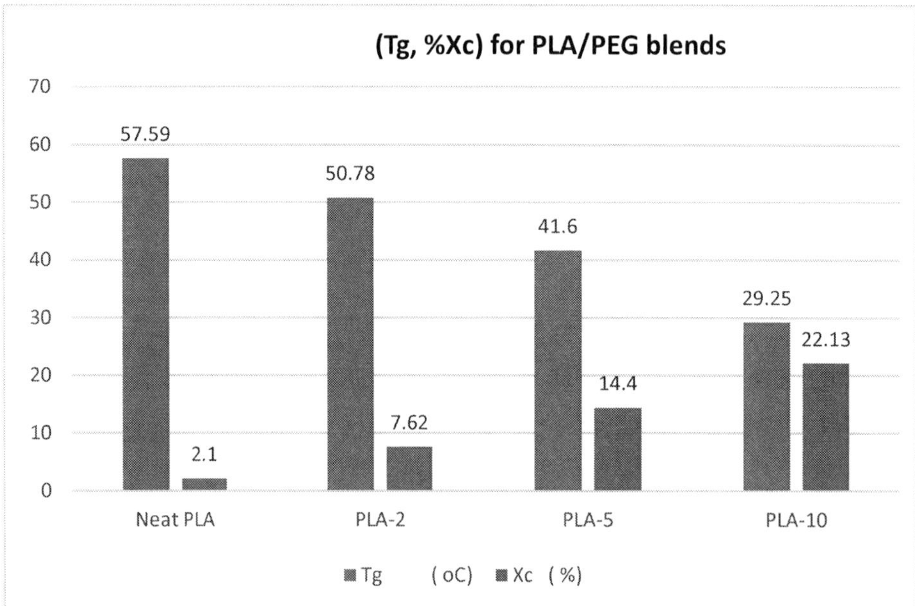


**Figure 5.** Initial thermal properties (Tg, %Xc) for different PLA/PEG blends

### *3.3. Mechanical Properties Evaluation*

The comparison of the mechanical properties of the neat PLA and PLA/PEG composite samples are shown in Table 3. The measured mechanical properties of the neat PLA are Young's Modulus of 3306 MPa, elongation at the breakpoint is 3.8% and the yield strength of 61.0MPa. As previously stated, 'increase in the concentration of the PEG increases the hardness', leading to the decrease of yield strength and Young's Modulus but with least change in the values of elongation at the breakpoint. This least change occurs because of the variation in the chemical structure of PLA and PEG. Thus it can be understood that stiffness and strength of the PLA/PEG composite, gradually reduced during the variation of PEG from 2-10% wt.

**Table 3.** Mechanical Properties of PLA/PEG composite

| Samples | Stress at Break (MPa) | Elongation at Break ( %) | Young's Modulus (Mpa) |
|---|---|---|---|
| **Neat PLA** | 61.0 | 3.8 | 3306 |
| **PLA-2** | 53.6 | 4.1 | 3751 |
| **PLA-5** | 48.9 | 3.2 | 3568 |
| **PLA-10** | 35.3 | 3.5 | 3042 |

## 4. Conclusions

Biopolymer composites consisting of a polylactic acid matrix reinforced with Poly ethylene glycol was synthesized successfully and filaments were extruded to make the 3D scaffolds. The preliminary results indicate that polymer composites can be achieved with direct mechanical mixing of PLA and PEG. An extensive work on the addition of the PEG in the PLA matrix was done to achieve a novel bio-composite and has been investigated to develop a 3D printed Bone scaffold through Fused Deposition Modelling (FDM) for tissue engineering applications. The modernity of the research was by optimizing some basic parameters like material printability, thermal stability, printing conditions and design. This supported the fabrication of the 3D printed PLA/PEG scaffold with the variation of PEG concentration (2, 5, 10% w/w). The morphological, structural, thermal and mechanical characterization was conducted on both the film samples (2D) and the fabricated scaffolds (3D).

The results achieved through the work make apparent that the addition of various weight percentage of PEG gives rise to notable changes in the Structure and degradation rate of the PLA/PEG composite. With more weight percentage of PEG, there is a significant increase in the hardness, stiffness, surface roughness and wettability which reduces the mechanical properties like yield strength, Young's modulus of the fabricated 3D scaffold. The blending of 5% (w/w) PEG with PLA is an efficient method for the realistic approach for 3D printing of Bone scaffolds as it weaves the structural, thermal properties and degradation rate with fair stability.

## References

[1] Kumar, S.D., Kannan, V.N. and Sankaranarayanan, G. (2014), Parameter optimization of ABS-M30i parts produced by fused deposition modeling for minimum surface roughness, International Journal of Current Engineering and Technology, Vol. 1 No. 3, pp. 93-97.
[2] Dinesh Kumar, S.; Ekanthamoorthy, J.B.; Senthil Kumar, K. (2015), Study of development and applications of bioactive materials and methods in bone tissue engineering. Biomed. Res., 26, S55–S61.
[3] Scaffaro, R.; Botta, L.; Lopresti, F.; Maio, A.; Sutera, F. (2017) Polysaccharide nanocrystals as fillers for PLA based nanocomposites. Cellulose, 24, 447–478.
[4] Kumar, D., Babu, G., & Krishnan, S. (2019). Study on mechanical & thermal properties of PCL blended graphene biocomposites. Polímeros: Ciência e Tecnologia, Vol. 29 No. 2.
[5] Frone, A. N., Panaitescu, D. M., Chiulan, I., Nicolae, C. A., Vuluga, Z., Vitelaru, C., & Damian, C. M. (2016). The effect of cellulose nanofibers on the crystallinity and nanostructure of poly (lactic acid) composites. Journal of materials science, 51(21), 9771- 9791.

IOP Conf. Series: Materials Science and Engineering **923** (2020) 012031 doi:10.1088/1757-899X/923/1/012031

[6] Garlotta, D. (2001). A literature review of poly (lactic acid). Journal of Polymers and the Environment, 9(2), 63-84.
[7] A.N. Frone, S. Berlioz, J.F. Chailan, D.M. Panaitescu.(2013). Morphology and thermal properties of PLA-cellulose nanofibers composites. Carbohydr Polym. 91(1):377-84.
[8] Annunziata O, Asherie N, Lomakin A, Pande J, Ogun O, Benedek GB. (2002) Effect of polyethylene glycol on the liquid–liquid phase transition in aqueous protein solutions. Proc Natl Acad Sci USA .99:14165–70
[9] Chieng, B.W.; Ibrahim, N.A.; Wan Yunus, W.M.Z.; Hussein, M.Z. (2013) Plasticized poly(lactic acid) with low molecular weight poly(ethylene glycol): Mechanical, thermal, and morphology properties. J. Appl. Polym. Sci., 130, 4576–4580.
[10] Tsuji, H. (2011). Poly (lactic acid): synthesis, structures, properties, processing, and applications. Wiley.

IOP Conf. Series: Materials Science and Engineering **923** (2020) 012032 doi:10.1088/1757-899X/923/1/012032

# An experimental study of drilling parameters and determination of ovality during the drilling of grade 5 titanium alloy with TiN coated carbide drill tools

**Nivin Joy[1*], S.Prakash[2], Guggilapu Sri Harsha[3], A.Krishnamoorthy[4]**
[1,2,3,4]School of Mechanical Engineering, Sathyabama Institute of Science and Technology, Chennai, Tamil Nadu, India

[1*]Email: nivinjoy@gmail.com

**Abstract.** Titanium is particularly a hard to cut element. Alloys of Titanium are also regarded as hard-to-machine materials. Out of many machining operations on Titanium, drilling is treated as very important operation as it is crucial in many applications. Drilling is normally considered among the final steps of fabrication of parts and has critical economic importance. Hence it is desirable to develop an economical and feasible drilling process for Titanium alloy and improve the effectiveness of current methods since even a small mistake in the drilling operation can lead to throwing away the entire part and having to replace the entire part. On account of above facts, an experimental trial has been made to study the machining aspects during drilling of the Ti-6Al-4V alloy. The primary aim of this research paper is to study various issues in drilling holes on grade 5 Titanium alloy with Titanium Nitride (TiN) coated carbide tool. Drill bits are selected in varying diameters and drilling procedure is carried out under different feed rates and speed. The output responses measured are temperature, thrust force and torque which are analyzed and studied. Also, the hole quality is studied by measuring the Hole Ovality on a profile projector. FESEM of drill bits is carried out to analyze the pattern of wear occurred on drill bits while drilling.

***Keywords:*** Drilling, Ti-6Al-4V, Carbide drill, Temperature, Force, Torque, Coating, Ovality, FESEM.

## 1. Introduction

Titanium is one of the very hard metal that has been found on the Earth. It has some excellent material properties making it a good choice of material suitable for many critical applications. Titanium is bio compatible and more importantly corrosion resistant which makes it a suitable material choice for medical implants [1] and aerospace applications [2] respectively. Previous researches point out that Titanium machining is a tedious task [3][4][5] resulting in a time and money wasting scenario. This is still an ongoing research where researchers try to develop an optimized approach in machining of Titanium and its alloys [6]. One main drawback in Titanium machining is the development of enormous temperature at the drill zone. This temperature will negatively affect the drilling procedure by gradual wear of drilling tool [7]. So, it is really a necessity to develop an optimized model of drilling of Titanium alloy by suitably selecting the right drilling parameters, right material of drill bit so as to minimize the negative effects that may occur during the drilling operation.

## 2. Experiment - Materials and Methods

### *2.1. Base Materials*
The base material Ti-6Al-4V alloy required for the experiment was procured in plate form. Table 1 displays the chemical composition of the plate used.

**Table 1.** Chemical Composition of Ti-6Al-4V Alloy

| TITANIUM % | ALUMINIUM % | VANADIUM % | CARBON % | Fe % |
|---|---|---|---|---|
| 90.100 | 5.7100 | 3.200 | 0.590 | 0.2660 |

The drilling experiment is carried out on a conventional vertical machining center [VMC] available at Karunya University, Coimbatore with table size of 1280 x 240 mm. The thrust force values and torque values are measured using a three-component Kistler dynamometer. Coated Carbide tool coated with TiN (Titanium Carbide) is selected as the tool. The drilling tools are selected with different diameters of 6 mm, 8 mm and 10 mm.

### *2.2. Input Parameters*
The input parameters chosen for the experiment is shown in Table 2. The experiments are planned based on the design of experiments using Taguchi's approach (Taguchi and Konishi, 1987). Three factors have been identified as process parameters, viz. Spindle Speed, Drill Diameter and Feed Rate each at three levels that affect the quality of holes produced in drilling as per Taguchi's orthogonal array of experiments.

**Table 2.** Drilling Experiment Parameters

| Drilling Parameters | A | B | C |
|---|---|---|---|
| **Speed (rpm)** | 500 | 1000 | 1500 |
| **Diameter of Tool (mm)** | 6 | 8 | 10 |
| **Feed Rate (mm/min)** | 20 | 40 | 60 |

### *2.3. Drilling Operation*
As the experiment pattern has been devised based on L9 orthogonal array, 9 holes out of which 3 each hole for each drill bits were drilled and the corresponding output parameters like Force, Torque and Temperature were recorded. A Ti-6Al-4V plate of 150 mm x 80 mm x 5 mm size was drilled using carbide drill bits coated over with Titanium Nitride and diameters of 6 mm, 8mm, 10 mm diameters. The Ti-6Al-4V plate and rill bits used in the experiment is shown in Figure 1.

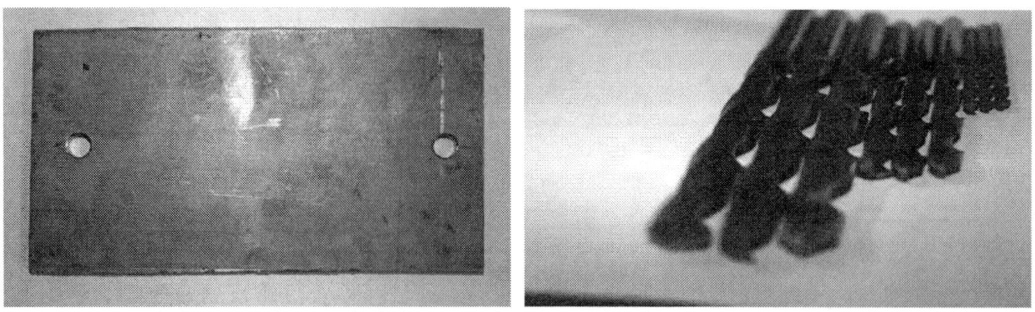

**Figure 1.** Ti-6Al-4V plate and TiN coated Carbide drill bits used in the drilling experiment

## 3. Result and Discussion

### *3.1. Measurement of Thrust Force and Torque*

Thrust force values and torque values measured while carrying out the drilling experiment are recorded with Kistler dynamometer interface. The plate is mounted on the dynamometer which send the pulses to the computer interfaced with it and plots the graphs for Force and Torque. In this experiment drilling of a Ti-6Al-4V plate of 150 x 80 x 5 mm dimensions is carried out. Carbide drill bits coated with Titanium Nitride (TiN) of different diameters 6 mm, 8 mm, 10 mm are used for drilling titanium plate. The force graphs and torque graphs obtained from the interfaced computer helps in studying the variations of force and torque developed during drilling of Titanium. Sample graphs obtained during the drilling experiment are shown in Figure 2.

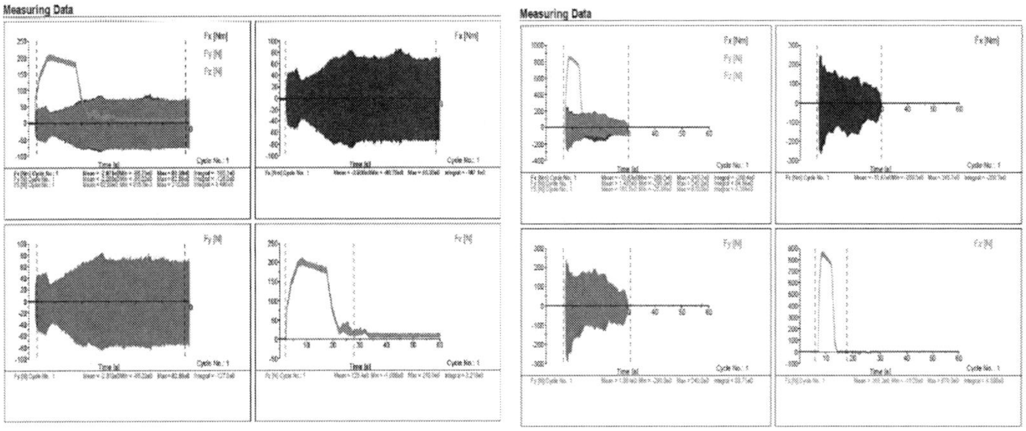

**Figure 2.** Sample graphs obtained during drilling

The values obtained for Force, Torque while drilling operation has been devised according to a L9 array and is shown in the Table 3 below:

**Table 3.** Output Readings of Force and Torque recorded during drilling experiment

| SI.No | SPEED (RPM) | FEED (mm/min) | DRILL DIA. (mm) | THRUST FORCE (N) | TORQUE (Nm) |
|---|---|---|---|---|---|
| 1 | 500 | 20 | 6 | 210 | 86.36 |
| 2 | 750 | 60 | 6 | 258.9 | 92.09 |
| 3 | 1000 | 40 | 6 | 179.2 | 88.88 |
| 4 | 500 | 40 | 8 | 382.5 | 139.80 |
| 5 | 750 | 20 | 8 | 246 | 120.2 |
| 6 | 1000 | 60 | 8 | 329.10 | 137.6 |
| 7 | 500 | 60 | 10 | 810 | 245.7 |
| 8 | 750 | 40 | 10 | 581.5 | 168.1 |
| 9 | 1000 | 20 | 10 | 381.7 | 241.1 |

IOP Conf. Series: Materials Science and Engineering **923** (2020) 012032 doi:10.1088/1757-899X/923/1/012032

*3.2. Temperature Measurement*

During the machining of hard to cut elements like Titanium, temperature plays a crucial role. While machining enormous quantity of temperature is produced. This temperature will affect the tool and result in tool failure. Hence developing a model to survive this amount of temperature is an important task inorder to protect the tool and have an extended tool life with less wear. Coating the drill bits with Titanium Nitride (TiN) helps to an extend in protecting the drill bits to withstand the wear and erosion of drill bits as a result of machining.

Temperature values measured during the drilling experiment is shown in Table 4.

**Table 4.** Temperature values measured during drilling experiment

| SI.No | SPEED (RPM) | FEED (mm/min) | DRILL DIA. (mm) | TEMPERATURE (ºC) |
|---|---|---|---|---|
| 1 | 500 | 20 | 6 | 40 |
| 2 | 750 | 60 | 6 | 39.8 |
| 3 | 1000 | 40 | 6 | 44.3 |
| 4 | 500 | 40 | 8 | 72 |
| 5 | 750 | 20 | 8 | 61 |
| 6 | 1000 | 60 | 8 | 44 |
| 7 | 500 | 60 | 10 | 36 |
| 8 | 750 | 40 | 10 | 77 |
| 9 | 1000 | 20 | 10 | 72 |

*3.3 Measurement of Hole Ovality*

Hole Ovality may be described as the deviation from perfect circularity. In short it shows how far the drilled hole is circular in shape. Measurement of Ovality is carried out after the drilling experiment on a profile projector. Profile projector was calibrated as per DOLG - NLQM/001/07 with metric traceability. Hole ovality is measured at Hole entry and hole exit.

The profile projector in which the experiment was carried out is shown in Figure 3.

**Figure 3:** Profile Projector

The following parameters were set on the profile projector:

- Least count =0.1
- Range=0-25mm
- Magnification=10x

The following readings are taken by using profile projector at Hole Entry:

**Table 5:** Experimental Readings of Ovality Test in Hole Entry

| SL.NO | DRILL DIA (mm) | INITIAL READING (mm) | FINAL READING (mm) | ACTUAL READING (mm) |
|---|---|---|---|---|
| 1 | 6 | 3.36 | 9.89 | 6.53 |
| 2 | 6 | 4.51 | 11.16 | 6.65 |
| 3 | 6 | 9.45 | 15.55 | 6.10 |
| 4 | 8 | 15.25 | 23.80 | 8.65 |
| 5 | 8 | 22.70 | 14.25 | 8.55 |
| 6 | 8 | 14.64 | 6.70 | 8.06 |
| 7 | 10 | 0.90 | 11.22 | 10.32 |
| 8 | 10 | 10.45 | 20.99 | 10.54 |
| 9 | 10 | 14.40 | 25 | 10.60 |

The following readings are taken by using profile projector at Hole Exit:

**Table 6:** Experimental Readings of Ovality Test in Hole Exit

| SL.NO | DRILL DIA (mm) | INITIAL READING (mm) | FINAL READING (mm) | ACTUAL READING (mm) |
|---|---|---|---|---|
| 1 | 6 | 24.90 | 18.38 | 6.52 |
| 2 | 6 | 18.34 | 24.99 | 6.64 |
| 3 | 6 | 24.09 | 18.12 | 6.03 |
| 4 | 8 | 17.24 | 9.06 | 8.19 |
| 5 | 8 | 17.06 | 9.15 | 8.09 |
| 6 | 8 | 8.22 | 16.75 | 8.530 |
| 7 | 10 | 11.35 | 21.77 | 10.35 |
| 8 | 10 | 21.15 | 10.60 | 10.55 |
| 9 | 10 | 11.36 | 21.24 | 10.12 |

### *3.4 FESEM Analysis of Drill Bits Before and After Drilling*

Inorder to analyze the wear that has happened on the drill bits during the drilling experiment FESEM analysis was carried out on the drill bits before and after the drilling experiment. This helps in comparing the drill bit morphology that happens during the machining process.

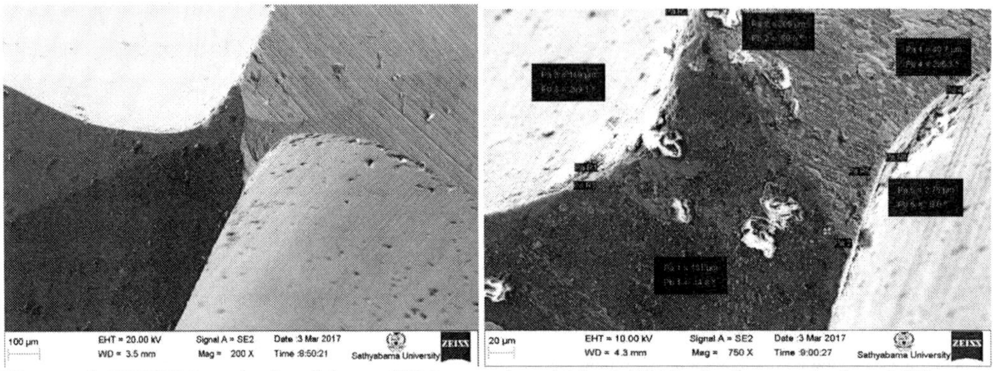

**Figure 4.** FESEM analysis of 6 mm TiN coated carbide drill bits before and after drilling

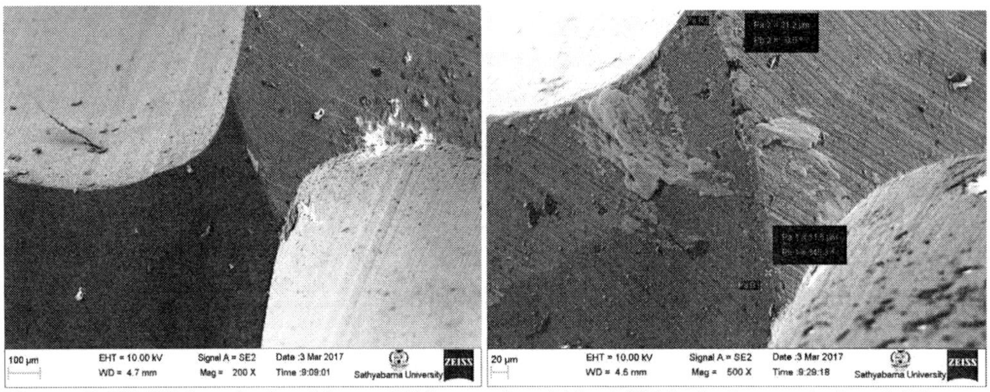

**Figure 5.** FESEM analysis of 8 mm TiN coated carbide drill bits before and after drilling

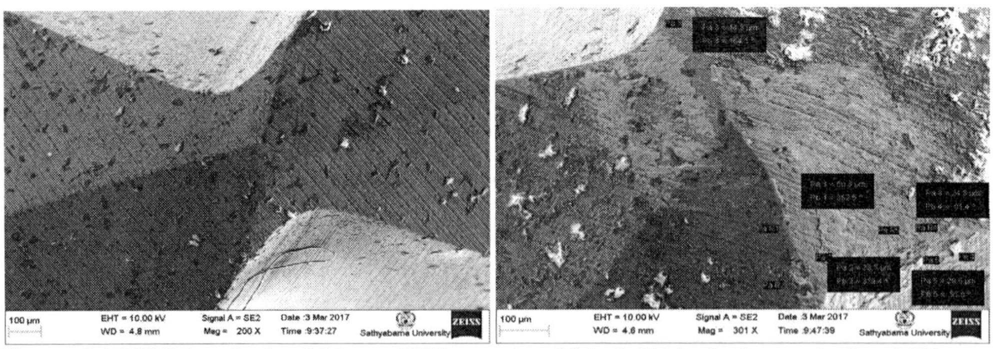

**Figure 6.** FESEM analysis of 10 mm TiN coated carbide drill bits before and after drilling

### *3.5 Chip Formation*

As a result of Titanium drilling chips are formed and those were collected. These chips enable the study of drilling parameters during the progression of drilling experiment.

The chips collected during the drilling experiments is shown in Figure 7.

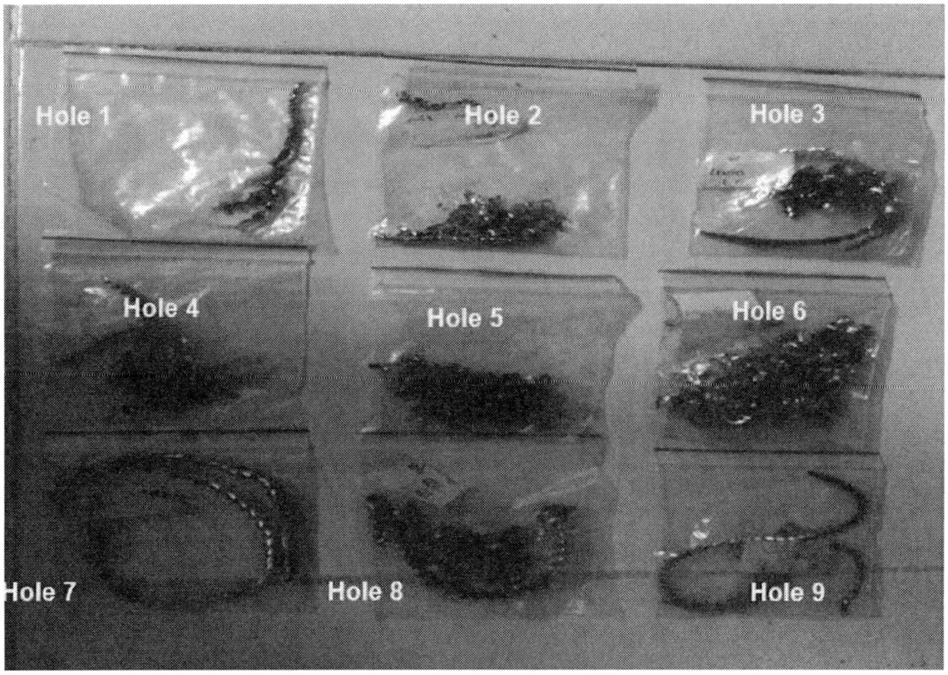


**Figure 7:** Chips formed during the drilling experiment

In comparison to other metals the formation of chips while drilling of Titanium alloys is fairly complex. Generally, there are three types of chips formation like continuous, continuous with build-up edge and discontinuous chips. In our experiment at the end of drilling of holes we were able to get only two types of chip types - continuous chips and discontinuous chips. After closer examination it was found that when the feed rate was low long and continuous chips were formed. And in contrast, when there was high feed rate the chips formed were non continuous chips as seen in the above figure. It was also observed that with the increase in speed and feed the chips were fragile and broken.

## 4. Conclusion

In the present work we have tried to make a small contribution to the area of drilling of Titanium alloy. We have carried out drilling in Grade 5 alloy of Titanium with carbide tools which is coated with Titanium Nitride (TiN). During drilling experiment the output responses like temperature, thrust force and torque were measured and analyzed. Measurement of Ovality was also carried out to determine the hole quality. FESEM analysis was carried out to study the wear pattern of the drill tool. Chips formed were collected.

The following observations were made from the experiment:

- There was considerably high heat production throughout the drilling process which eventually affected the tool. There was increase in drilling temperature at higher speed and feed rates.
- Titanium Nitride (TiN) coating on the drill bit slowed down the tool wear to some extent.
- Feed rate is the key parameter affecting the output responses in the drilling experiment. Also, with respect to increment in feed rate, there is gradual increment in tool wear.
- At higher speed the holes were found to be more circular which was found through the Ovality test.

- The chips formed were collected and at higher feed rates the chips were fragile and broken.
- From the FESEM analysis carried out on the drill bits before and after drilling, it is seen that there is considerable parent material deposition happening on the tool.

## 5. Acknowledgement

The authors would like to express gratitude to staff of Centre for Research in Design and Manufacturing Engineering (CRDM), Karunya Deemed University, Coimbatore.

## References

[1] Brian E Hurless, Sam Froes F H, 2002, Advanced Materials and Processes, p.37
[2] Yamashita Y, Takayama I, Fuji H, Yamazaki T, 2002, Applications and features of titanium for automotive industry. Nippon Steel Technical Report, 85:11–14.
[3] Siekmann H J, 1955, Tool Eng. 34(1):78-82
[4] Komanduri R, Reed W R, 1983, Evaluation of Carbide Grades and a new cutting geometry for Machining Titanium Alloys, Wear, Vol. 92 pp 113-123
[5] Ezugwu E O, Wang Z M, 1997, Titanium alloys and their machinability—a review, Journal of materials processing technology 68, no. 3, pp: 262-274
[6] Joy N, Prakash S, Krishnamoorthy A et al., 2019, Experimental investigation and analysis of drilling in Grade 5 Titanium alloy (Ti-6Al4V), Materials Today: Proceedings, Volume 21, Part 1, 2020, Pages 335-339. DOI: 10.1016/j.matpr.2019.05.458
[7] Rahman M, Seah W K H, Teo T T, 1997, The Machinability of Inconel 718, J. Mater. Proc. Tech., 63: 199–204

IOP Conf. Series: Materials Science and Engineering **923** (2020) 012033 doi:10.1088/1757-899X/923/1/012033

# DISPLACEMENT STUDY OF GLASS FILLED UNSATURATED POLYESTER COMPOSITE PLATE UNDER TRANSVERSE IMPACT FORCE

N.JAYAKUMAR[1]* and S.SINTHUJA[2]

[1] DrawingI Incorporated, Baltimore, Maryland, United States of America.

[2] Department of Marine Engineering, AMET Deemed to be University, Chennai, India

*E-mail: jainanda2008@gmail.com

Abstract : A finite element analysis is offered for analyzing the response of glass fiber filled reinforced unsaturated polyester composite plate due to transverse drop weight impact. The analysis could be used to estimate the displacements of the composite plate during impact and the stress distribution throughout the plate thickness. A thin SHELL 163 ANSYS-LS DYNA explicit element has been used for modeling the composite plate and SOLID 168 ANSYS-LS DYNA explicit element has been used for modeling the drop weight impactor.
*Keywords*: Modeling, Explicit Element, Drop Weight Impact

## 1. INTRODUCTION

Impact resistance of composite materials, especially fiber reinforced plastics (FRP) is a very important area in modern day research. Fiber reinforced plastics structures usually react very poorly to transverse impact trials. In spite of the reservations about the mechanisms that lead to damage, it is difficult to establish the effect of an impact event on the performance of a composite structure after impact. In certain cases, the damage won't be visible which could lead the materials with significant damage being allowed to continue in service.

Fiber Reinforced Plastics composites are one of the most preferred materials of choice over the past couple of decades for use in areas like automobiles, aerospace, piping, sports and construction. These preferences are attributed mainly due to their strength-to-weight ratio, superior mechanical properties and their environmental resistance. As most of the applications are subjected to impacts of some nature, ranging from dropping of tools to bomb blasts, it is imperative to study the impact behavior of these materials in order to design them comprehensively . We should also remember that conventional composite materials perform very poorly when subjected to transverse impacts.

S.N.A.Safri et al. experimented impact tests using drop weights of less intensity and observed the damages that occurred on various materials with different specimen thickness due to these impact loads. Eventually, they came out with a finding that E-800 had more strength in comparison with other types of similar materials. [1]

Rakesh Reghunath et al. conducted an investigation to evaluate the behavior of bidirectionally oriented glass filled composite material. They prepared specimens and carried out failure mode analysis and found out certain key factors that could affect the potency and firmness in the constitution of the materials. [2]

Muhammed Tijani ISA et al. carried out experiments with lesser intensity impacts on dissimilar composite materials. The outcome showed that strength of materials were largely influenced by the factors such as fracture in substances and rupture in glass fibers. They also examined alternate options which could render better results and concluded that using hybrid glass strands for reinforcing would lead to curtailing the above mentioned harms in materials. [3]

Sunith Babu.L and H. K. Shivanand investigated the properties of laminates that were filled with both glass strands and carbon. The test was conducted based on ASTM standards D3029 and D7137. They analyzed the performance of specimen with 2 mm thickness with a centrally acting impact load. [4]

In this research endeavor, we are examining the drop weight impact of glass fiber reinforced unsaturated polyester composite plate by finite element analysis.

## 2. EXPERIMENTAL SETUP

### 2.1 Analysis by Finite Element Method

Finite element method is perhaps the most popular numerical technique to resolve complex engineering problems. The concept of FEA is the discretization of a given model into finite elements, applying boundary conditions and loads as per the problem to solve it numerically.

In our work, we have adopted iterative methods to analyze the performance of the glass filled unsaturated polyester plate under a transverse impact load.

### 2.2 Modeling of Composite Plate

Orientation of [0°/90°] was considered for the composite plate. Thin SHELL 163 element was used for meshing the composite plate. It was a 4-noded element with twelve degrees of freedom at each node. SHELL 163 is an element with 4 nodes with the capabilities of membrane as well as bending. We made sure that both normal loads and in-plane loads are allowed to be used. The dimension of composite plate was 200 X 200 X 2.75 mm and properties for the glass filled unsaturated composite plate were supposed as given in table 2.1 and the boundary conditions were kept fixed on all sides of the plate.

| | |
|---|---|
| $E_x$ | 122 GPa |
| $E_y$ | 7.95 GPa |
| $\nu_{12}$ | 0.32 |
| $G_{12} = G_{23} = G_{13}$ | 5.52 GPa |
| P | 1584 kg/mm$^3$ |

*Table 2.1 Material properties of composite plate*

### 2.3 Modeling of Drop Weight Impactor

Solid 168 element was used for meshing the impactor. It is a 10-Node Tetrahedral Structural Solid component with each node having three degrees of freedom. The impactor was modeled as a sphere ball with radius of 12.5 mm. The Linear-Elastic model with material properties are given in table 2.2.

| | |
|---|---|
| E | 215 GPs |
| N | 0.30 |
| P | 7860 kg/mm$^3$ |

*Table 2.2 Material properties of drop weight impactor*

## 3. RESULTS AND DISCUSSION

The impact analysis by numerical method using the commercially available analysis software is presented here. The assembly of the model is given in figure 3.1. And the deflection of the centre of the plate Vs time is given in figure 3.2.

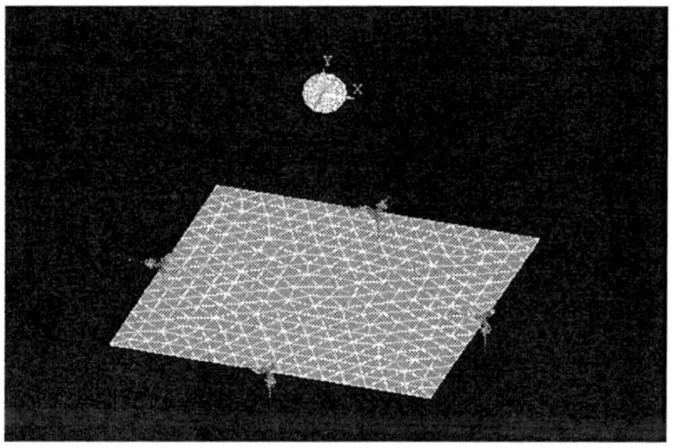


*Fig. 3.1. Assembly of the impact model*

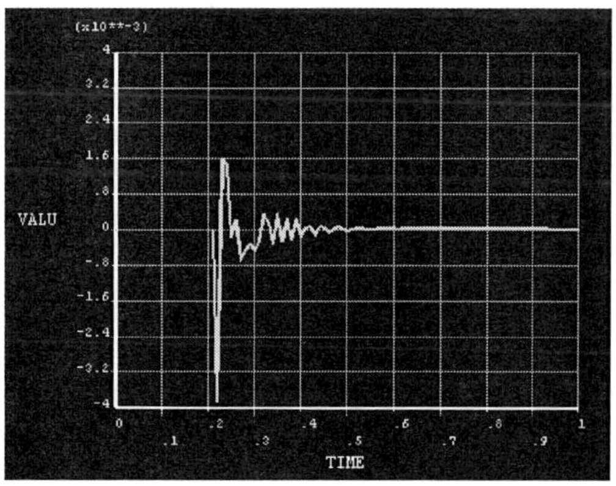


*Fig. 3.2. Deflection of the composite plate with respect to time.*

In figure 3.2, Y axis represents deflection of the plate and X axis represents time. It is obvious from the above graph that the magnitude of the deflection reaches its maximum in 0.02 seconds from the start of impact and the maximum deflection is about 3.8 mm. It was programmed so as to avoid the repeated impact on the plate. The plate deflects to a maximum value and due to its elastic property the impactor rebounds. The plate deflects back to a value of

1.6 mm and after some small vibrations the plate regains its original position as the kinetic energy disappears.

The corresponding stress distribution at the centre of the plate along the fiber direction with respect to time is given in figure 3.3.

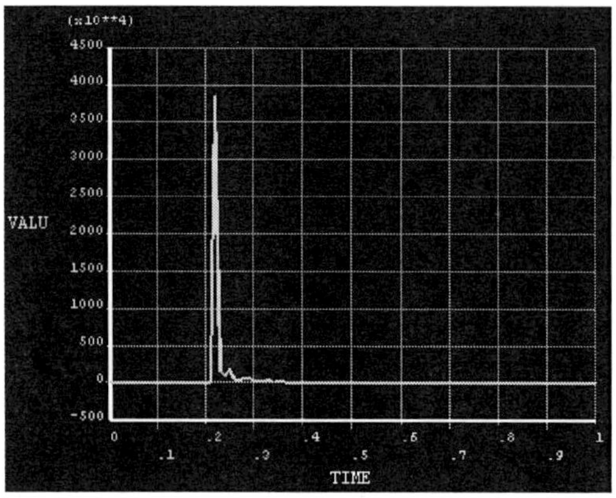


*Fig. 3.3. Stress Vs Time*

The stress in Pascal is plotted in Y axis of the graph and time in X axis. The value of the maximum stress is about 38 MPa.

## 4. CONCLUSION

Numerical analysis was conducted to learn the impact behavior of the glass filled unsaturated polyester composite plate due to dropping of an impactor from the given altitude. During the examination, a finite element procedure was developed to study the response of the composite plate to a foreign object impact.

The plate was modeled using a thin SHELL 163 ANSYS-LS DYNA explicit element and the drop weight impactor was modeled using SOLID 168 ANSYS-LS DYNA explicit element.

It was observed that the maximum deflection of the plate was 3.8 mm and it occurred in .02 seconds. The total impact duration was about 0.16 seconds. Part of the kinetic energy was used in deflecting the plate to a maximum and because of the kinetic energy the impactor bounced back.

## REFERENCES

1. S. N. A. Safri, M. T. H. Sultan and F. Cardona, "Impact damage evaluation of glass-fiber reinforced Polymer (GFRP) using the drop test rig – an experimental Based approach", *ARPN Journal of Engineering and Applied Sciences, vol. 10, no. 20, November 2015.*

2. Rakesh Reghunath, Mahadevan Lakshmanan, and K M Mini , "Low velocity impact analysis on glass fiber reinforced composites with varied volume fractions", *International Conference on Materials Science and Technology (ICMST 2012) IOP Publishing IOP Conf. Series: Materials Science and Engineering 73 (2015)*

3.

Muhammed Tijani ISA, Abdulkarim Salaw AHMED, Benjamine Olufemi ADEREMI, Razaina Mat TAIB, Hazizan Md AKIL, Ibrahim Ali MOHAMMED-DABO, "Drop Weight Impact St udies of Woven Fibers Reinforced Modified Polyester Composites", *Leonardo Electronic Journal of Practices and Technologies, Issue 24, January-June 2014, p. 97-112.*

4. Sunith Babu L, H. K. Shivanand, "Impact Analysis of Laminated Composite on Glass Fiber and Carbon Fiber", *International Journal of Emerging Technology and Advanced Engineering, Volume 4, Issue 6, June 2014.*

5. N Rajesh Mathivan, J Jerald, "Experimental Investigation of low velocity impact Characteristics of woven glass fiber epoxy matrix composite laminate", *Journal of materials and design,Vol 31, 2010, pp.4553-4560.*

IOP Conf. Series: Materials Science and Engineering **923** (2020) 012034 doi:10.1088/1757-899X/923/1/012034

# Quality Transformation to Improve Customer Satisfaction: Using Product, Process, System and Behaviour Model

V Poornachandrika* and M Venkatasudhakar
Department of Mechanical Engineering, Vel Tech Rangarajan Dr. Sagunthala R&D Institute of Science and Technology, Chennai, Tamilnadu, India.

*Email : Poornachandrika.vasu@yahoo.com

**Abstract.** In recent years automotive manufacturing industries systematically improved its quality parameters principally through strategies with in supply chain. However the current performance not enough, industry dynamics and new technology development require a major quality transformation. Implementation of various quality improvements is the aim of achieving customer delight in many companies. Despite the marked improvements that have come from quality management, quality process and information technology system, without a transformation within an organizations culture, quality management will continue to suffer. This paper presents a case study on application of behaviour model of quality blue box with Plan, Control, Assure concept through Veto right approach, Process model of robust process automation in customer claim handling and system model of mature audit tool. This work is carried out in a leading automotive commercial vehicle control system manufacturing company. This case study illustrating to get zero defect by applying behaviour, process and system model. After implementation there is significant change in reduction of customer claims.
Key words: Veto right, Quality blue box, Robust process automation, Quality transformation

## 1. Introduction

Any transformational strategy is about making drastic and significant changes within a business to change the course of its short and long term viability. It is a process of developing a strategic plan for modifying an organization business process through the focus of quality by fixing various rules, policies, procedures and processes to move an organization from current state to future state.

Current role of quality environment, where quality is not always appreciated. Quality not invited into decision making as quality is seen as blocking point. Moderator of the failure resolution process, reactive and high share of trouble shooting (fire fighting) not heard when problems are raised or perceived as undermining. As for quality improvement it is well understood and described in literature. With reference from Julia Madrid,(2017) delivering value to the customer through superior quality is one of the greatest challenges for manufacturing enterprises today. The success of the product on the global market can be achieved only with the fulfilment of customer requirements. In the premium and luxury segment of the automotive industry, these challenges take another dimension with the constant battle against cost pressure and shortening the production lifecycle. Thus successful premium vehicle design characterized by excellence as a combination of both manufacturing and perceived quality. The proposed integrated quality framework incorporates manufacturing quality with the perceived quality. It can serve as a basis for quality department by combining the sub quality function for seamless communication with in quality to break the silos within department for better feedback throughout the

IOP Conf. Series: Materials Science and Engineering **923** (2020) 012034 doi:10.1088/1757-899X/923/1/012034

flow of future methods development that incorporates customer's holistic quality perception.

Om Prakash Yadhav,et.al,(2006), proposed a comprehensive framework for target planning for customer satisfaction driven quality improvement efforts in the product development process. This framework facilitates a link between corporate decision making and engineering decision making by integrating best practices and structuring technical activities. Potential vehicle attributes are classified and prioritized for further improvement using Kano model and quality function deployment. With that it is concluded in general approach quality function deployment is used to convert the customer needs or requirements and translating them into specific plans to produce products to meet those needs. The "voice of the customer" is the term to describe these stated and unstated customer needs or requirements. Here it is linked into Kano model to bring the metric for satisfaction to achieve them. Here it is limited to only capturing the customer satisfaction on new product stage.Therefore the main focus on any automotive manufacturing industry in a dynamic environment is to sustain their performance in all aspect. Not only on component level or any other particular functional level also.

Daniel Tuic,(2016) states that, a new method of using TRIZ methodology for the validation of the corrective permanent action during complaints in completion of the 8D [8 dimension] report in the automotive industry. It focuses on complaints which appear in the development phase. During the production, specific methodologies regarding the complaints management for control and analysis are defined, then in the development phase these become more sensitive. Therefore a new method of using TRIZ in validation of corrective permanent actions, before being implemented on the product.

According to Anderson et al. (1994 articulated integrative theory of quality management to describe and explain the process and effectiveness of adopting the Deming management method. In their theory development, he identified and defined seven constructs as forming the theoretical blocks for describing, explaining, and predicting the organizational performance impact of quality management adoption. The results of this paper helped the organization to streamline the cross cultural quality management. Eight hypotheses were identified and tested, framed questionnaire where measured to get the real insight of gaps for adoption of quality management. Finally it is concluded organizational culture to be more supportive of quality management.

Transformational Change and Leader Character with reference from Gerard H Seijts, et.al,(2017) shows, ccollaboration and accountability is the key leadership skill for cultural transformation. This research talks about the clear top down demonstration of leadership quality to bring transformation across organization.

Deming out of crisis which was published in 1982 during post second world war. The crux of the book was identifying and explaining Deming fourteen points for quality management. But the current scenario was not effectively followed in any of the industries. There are three major insights take ways from the book are, first one is "Leadership should be focused on the system which is exactly part of this transformation journey", second one is "statistical control required to have any hope of achieving improvement" and third one is "focus on quality first, throughput will follow".

Edosomwan, J.A. (1996), stated transformation journey is the organizational transformation process creates pains, change, risk, uncertainty and, most importantly rewards that are beneficial to individual and organizations. Transformation is a wakeup call in a competitive environment. This is an exciting time in a world with fewer economic boundaries and increased competition in both the public and private sectors. Trends show that organizations are waking up to the challenges. Many organization realizing that the old bureaucratic business structures can no longer respond to the new customer driven market economy.

Teng, et.al (1996), briefs that organizational transformation was being done by changing business process and organizational change frame work. Reengineering and organizational change, lessons from a comparative analysis of company experiences by Alessio Ascari, et,al (1995) demonstrates that culture, people, process and structure change will bring the result to achieve the breakthrough results in the dynamic environment.

Total quality management and socio technical system approaches organizational change, A.B.Rami Shani, et.al (1996), states that unique integration of management principle brings the radical change in the organization which helps them to achieve the desired results and maintain their performance in competitive business environment.

Process reengineering and the dynamic balance of the organization, as per John Hendry (1995), three dimensions was brought the change they are co-ordination, motivation and control and efficiency and learning.

After analysing the related papers it is understood that, the quality improvements discussed neither process nor product nor system nor behaviour. Transformation is not just improving any of the aspect which is stated above. This case study has uniqueness of clustering all process, system and more specific towards behaviour which marked as important for transformation.

## 2.Problem Description

Why Quality transformation is required:

Due to continuous escalation from leading commercial vehicle Original Equipment Manufacturer, the organization has urgency to raise the company standard to automotive quality standards. This transformational change journey started at April 2018. Step back and reflections has been done along with management team to take the organization to next level. The following are the current gap which is addressed during brain storming with management.

Independent and empowered quality organization formulated to encounter the issues which affecting the current performance. Veto right, approval, bringing pot risk to table, true customer focus, mindset and behavior of process owners, healthy friction and collaboration between quality function and out of quality function, frontloading of projects, risk mitigation during developments, proactive approach rather than reactive it needs more than just structural change to lift quality performance in all functions.

The industry bench mark for commercial vehicle are "zero customer complaint in 'zero km", claims per million is less than 30, Production Part Approval Process (PPAP) right first time on time greater than 95%, and internal audit non conformance on time closure rate is greater than 90%. Considering this metrics customer started escalating about the performance of the company. Hence it is recommended to go for complete transformation to improve the current scenario. As shown in Table 1, the data were collected to measure the current level of the organization.

**Table 1**: Leading key performance indicator of the organization

| Key performance indicator | April 2018 – March 2019 |
|---|---|
| Number of customer complaint | 800 |
| Launch CPM [Claims per million] | 200 |
| PPAP right first time on time | 70% |
| Internal audit non conformance on time closure rate | 30% |

### *2.1 Customer claim handling process*

Figure 1 explains the current process for customer claim handling. Gap in receiving the claim information on time and appropriate. Once the information is received there is no distribution channel to create the complaint awareness from top management to operator level. Initial investigation requires manual follow up with respective production team. No trigger or status about the report to customer. Once the initial response is done, the team has to work to analyze the complaint for permanent solution. In the existing process no defined timeline for each reporting like containment and final report.

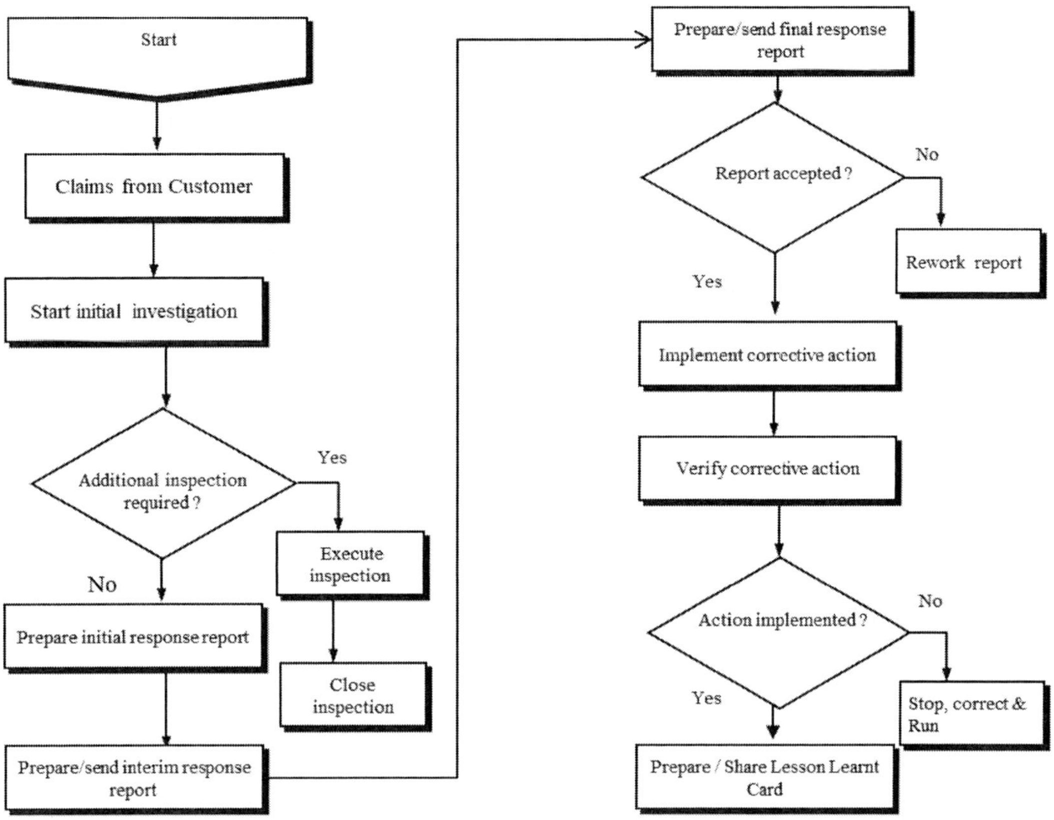


**Figure .1**. Process flow diagram of claim handling process before implementation of new methods

This leads to delay in response and action implementation. Due to delayed implementation complaints are continue to receive before cut off failures. Each stage of activities being done manually. This also pulled the satisfaction level of customer down.

### *2.2 New product quality*

The role of new product quality in project management is to check the sample before it is submitting to customer. The mindset of project management is that quality is seen as blocking point into decisions. Project quality is not involved in important decision like finalizing the customer requirements, new supplier selection and so on. The project management is not having the structured approach of cross functional mapping for involving quality team. Less involvement of quality team in project management leads the organization to receive more escalations while submitting the new products.

### *2.3 Audit Management*

Audit are performed only because "Must Do" to satisfy customer. Importance and opportunity of internal audits is not understood by the organization which affects the image of the organization during external audit and customer audits. Also this audit score is nowadays a part of customer score card and the organization is in the tremendous pressure to establish the internal audit process in a right context. The challenges are struggles with too many open or delayed post audit activities. Some gaps are being closed without deep root cause analysis which results in recurring issues. This is not complaint with industry standards that may cause negative customer feedback. Figure 2 gives the information of on time audit non conformance closure rate from 2018 quarter 1 ( Jan to Mar 2018) to quarter 1 of 2019 (Jan to Mar 2019).

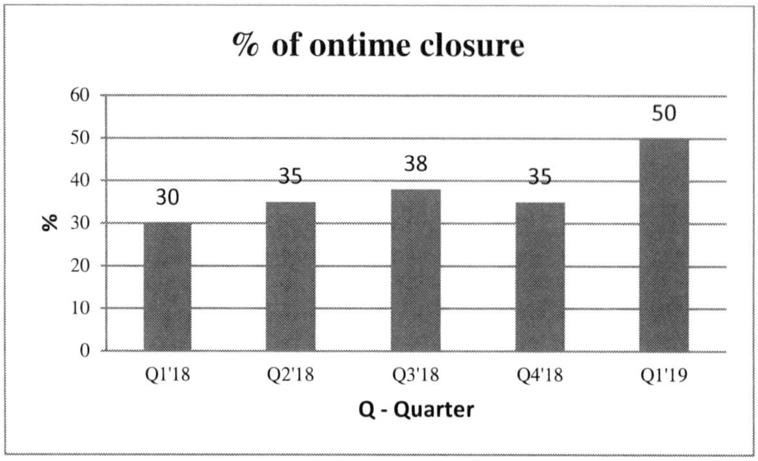


**Figure 2**. On time non conformance closure rate before implementation

Current problems are differentiated in three aspects like behavior, process and system with respect to quality transformation. The identified gaps will be addressed in new methods which will be outlined in detail below. The use of technology for handling customer complaints especially to receive notification from customer on time. Change the behavior of people using veto right concept and survey method to improve on time audit gap closure rate.

## 3. Methodology

To improve the quality performance in view of customers the organizations have done product related initiatives. By instituting additional inspection which caused extra investment to management. However the performance in front of customer is not improved. Every month the organization tend to receive different sort of issues which is clearly indicating that, systemic gaps in the process and behavioural aspect of the people in core supply chain. Hence it is suggested to use new approach to bring the cultural transformation instead of only adding inspection controls and product related poka yokes to eliminate human error

### *3.1 Transformation in customer claims handling*

Customer claims handling process completely reworked and new methods are amended, the current challenges are ineffective management of claims which leads customer dissatisfaction. The flow of handling process from start to end updated with clear responsibilities which are introduction of RPA [Robotic Process Automation]. RPA is the technology that allows anyone today to configure computer software or a "ROBOT" to emulate and integrate the actions of a human interacting within digital systems to execute a business process. RPA robots utilize the user interface to capture data and manipulate applications like humans do. They interpret, trigger responses and communicate with other systems in order to perform on vast variety repetitive task only substantially better. RPA software robots never sleeps. Makes zero mistakes and costs a lot than an employee with perfection.

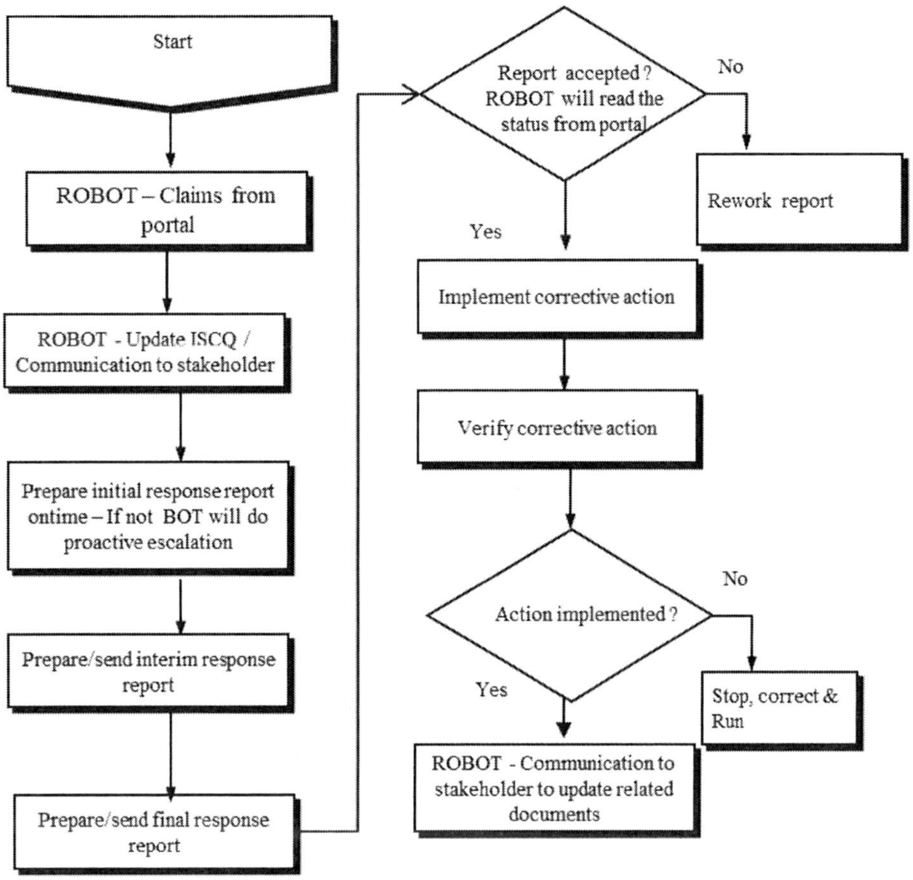


**Figure 3**. Claims handling process after introduction of RPA [ROBOT]

Figure 3 explains the use of RPA for customer claim handling process, become agile and brought lot of transparency between customer and supplier. The RPA will read the claim information automatically from customer portal the moment customer updated the data in respect of vendor code then the ROBOT picks information and will update them in a Integrated Supply Chain Quality portal (ISCQ) which is exclusive for the organization. ISCQ will send the communication to respective stakeholders and wait for immediate response if the immediate response will not be updated in the ISCQ the RPA will send the escalation to management also it will send warning notification. Incase of repeat claims similarly it will work for each steps of problem solving which is described in Figure 3 and also this will not be limited to sending the feedback to customer. It also go back to the product directory and search the similar product and process and give the trigger to process owner for horizontal deployment. If the same failure mode was not captured in PFMEA [Process Failure Mode Effect Analysis] it will also send the trigger to process design engineer to update the same. The help of RPA, response time to customer drastically improved and number of claims reduced by effective horizontal deployment also updation of failure mode in PFMEA act as a proactive tool for quality to eliminate defect outflow with appropriate controls.

### *3.2 Transformation in project management*

The project management function faced lot of issues while submitting new products to customer especially PPAP RFToT (Right First Time on Time) and sample submission, this leads to loss of business, so management decided to have robust new development process to eliminate the discussed issues. PCA concept with veto right approach shows in Figure 4 which supports the project management organization. In existing approach during planning quality interface was missing. Gap in requirement capturing and cascading requirements which leads to launch claims.

To ensures the interface to the customer for quality to drive the right quality mindset in the project teams and constantly radars and manages risk through all project phases. In order to ensure customer requirements are rightly understood and validated in the proposed product and process design and to achieve flawless launch. Redefined project management organization structure mentioned below. Veto right is a concept of empowering quality team by changing the behavior of the individual in an organization. Transformation result cannot be achieved because the behavioral management challenges are frequently underestimated. Especially resistance to accept change results failure in transformation process. Employee needs to recognize and understand their roles in order to satisfy the customer requirements. Earlier quality is ignored now it is empowered that is called veto right given to quality.

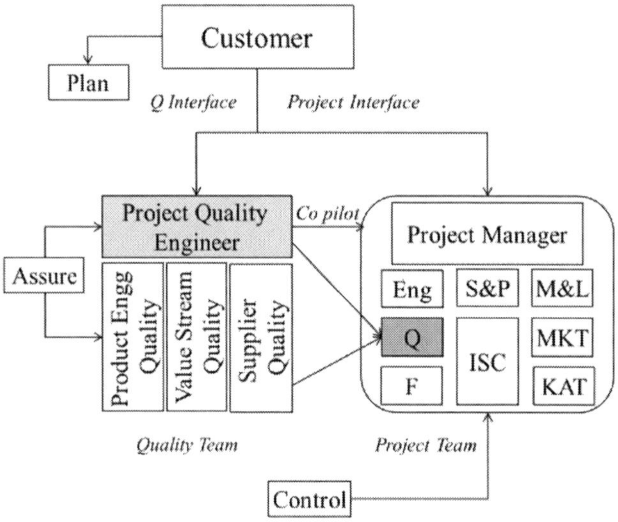


**Figure 4** : Project management revamped structure - Quality blue box with PCA concept through Veto right approach

Assure phase explains the synergy with in quality team. Product, supplier and value stream quality is working together to achieve the common goal of the organization. In control phase the role of project quality is to act as a co-pilot of project manager and interact with customer for quality aspects and connecting the quality dots in the projects also tracking of quality deliverables in the project. Gate keeper for quality and has veto right in project reviews, this all brings the filtration of quality issues in early stage and control the defect outflow to customer. In Figure 4, Q interface is quality interface between customer and project quality. Project staff comprises all functions. Eng means engineering, S&P denoted as sourcing and purchasing, M&L states that manufacturing and logistics, Q for quality. ISC stands for integrated supply chain, MKT means marketing, KAT is key account team, F for finance.

### *3.3 Transformation in audit process*

Quality transformation will not reach to the next level without mature audit culture. However the organization decided to change the perception of audit. Make employees to understand the purpose and terms of engagement, auditors to understand their roles and mission. The questionnaire has been formulated to conduct a campaign to understand the real voice of auditor and auditee in the organization. The questionnaire contains ten aspects. Table 2 gives an overview of the ten aspects of the questionnaire in a logical order. Seventy five auditors, seven process owners with six hundred and fifty auditees in five manufacturing locations of the organizations were participated in the campaign. It is measured in a 1 to 5 scale. 1 is poor response, 3 moderate response and 5 is good response. A rating scale provides more than two options, in which the respondent can answer in neutrality over a questions being asked. The criteria of the questionnaire defined based on the issues faced in audit process. The criteria were finalized using brainstorming session of core team members in the quality and key production functions. The overall score is calculated according to the number of responses under each ratings. Responses under rating 5 eliminated due to prioritization. 80% of the rating falls under 1 and 3 scale and 20% for scale.

**Table 2** : Survey questionnaire and result

| Serial Number | Criteria | Description | Rating – 1 | Rating -3 | Rating -5 | Overall score |
|---|---|---|---|---|---|---|
| 1 | General | Internal, external business environment | | | | No response |
| 2 | Reason | Real reasons/motives for conducting an audit and closing the gaps timely | | | | No response |
| 3 | Audit process | The process of practically executing the audit and the importance of certain issues (time, conviction, involvement of auditor, auditee) | x | | | 30% |
| 4 | Team aspect | The extent to which team culture existed | | x | | 20% |
| 5 | Training | Scope of the training existing for audit | | x | | 15% |
| 6 | Resource | Resource adequacy to complete an audit on time | | x | | 15% |
| 7 | Benefits / effectiveness | The benefits and effectiveness of audits measured in terms of organizations leading key performance indicator | | | x | 7% |
| 8 | Influences | External factor which is affecting the audit | | | x | 6% |
| 9 | Concern | Major concern for conducting an audit and closing non conformities on time | | | x | 7% |
| 10 | Help | How global organization supports for audit process | | | | No response |

According to the survey poor and moderate response are compiled and overall score calculated. The questionnaire returns generated a considerable amount of data which identified the root concerns of the implementation of effective audit process. Based on the overall score forty five actions were identified and implemented. The above questionnaire helped to address the systemic gap in the process.

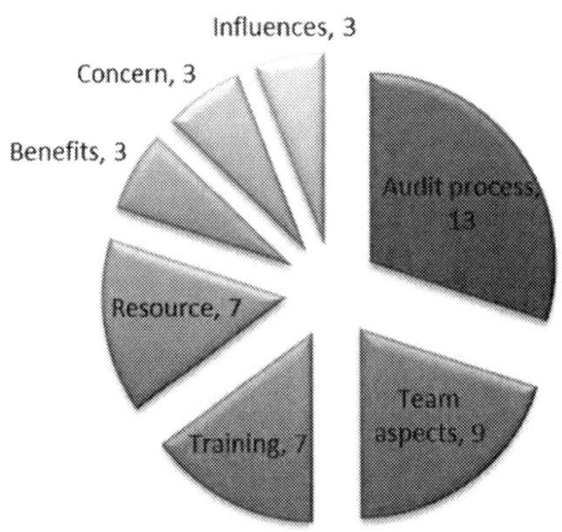


**Figure 5**: Action implementation break up

Figure 5 outlines the breakup of the actions. Thirteen actions were implemented in audit process they are audit planning, execution and monitoring using automation and dedicated person for managing the tool for all five manufacturing locations. Nine actions related to coordination between auditee and auditors. Seven actions for training of entire stake holders. Seven actions related to resource to implement the gaps. Finally nine actions related to benefits, concerns and influences. After the campaign the breakthrough changes are observed in the auditing process.

## 4. Results and Discussions

By implementing the customer claims handling process using RPA, human interaction is eliminated for documentation which was resulted in quick response to customer. A wide gap in updating the process documents was missing during the manual process which is made robust. Without consolidated view of data, analysis gap was observed which is eliminated. BOT will give the repeat claim notification for better focus to rectify the ineffective corrective action. Use of RPA system people inefficiency identified and resolved. The real benefit of RPA is to update the horizontal deployment data base which is the input for designer while making new developments.

Empowering project quality by changing the project management staffing structure and given all authority to quality to work collaborates with in project team. This early involvement of quality, understanding, implementing customer requirements brings RFToT. Veto right approach resulted in drastic reduction of launch claim. Involvement of project quality from business award to start of production is measured and monitored in a systematic way. During every phase review project quality have Veto right to claim the right for quality. These discipline in the process made all new development launch without flaw.

**Table 3**: Result after implementation

| Key performance indicator | Before implementation Average / month | After implementation | | | Average/ Month (Second Quarter) |
|---|---|---|---|---|---|
| | | July 2019 | August 2019 | September 2019 | |
| Number of customer claims | 66 | 5 | 4 | 4 | 4 |
| Launch CPM | 30 | 0 | 0 | 0 | 0 |
| PPAP Right first time on time | 70% | 95% | 100% | 100% | 98% |
| Internal audit non conformance on time closure rate | 30% | 85% | 91% | 100% | 92% |

The response to the questionnaire in audit implementation helped the organization to streamline the process. Each section in the questionnaire have found to be important and action initiated to the responses with immediate effect which gives the motivation to those who participating in audit. Employees feel accountable for closing the gaps identified during the audit, auditors feel accountable for audit advocacy and transparency.

In table 3, number of customer complaint average 66 per month after implementation of RPA model this is drastically reduced to single digit. Launch claims per million is zero and PPAP RFToT 98, this is achieved because of addressing issues proactively in phase reviews by empowered quality team. Addressing internal audit non conformance improved to 92% due to right system control which was streamlined through campaign.

## 5. Conclusion:

Quality transformation is an integral part of business world. Accelerate the rate of quality improvement in core supply chain. It is focusing on excelled quality mindset. Implementation of process, system and behavior model improved the organization leading key performance with respect to quality. Comparing the result obtained in the case study between the traditional practice of quality management and transformation models shows the greater results. It can be concluded that the elimination of human intervention in claims handling process gives better result. Early involvement of project quality in project management brings significant improvement for flawless launch of projects which helped the organization to increase the market out performance and zero recalls. Implementation of audit process has limitations, caused by issues such as understanding of audit requirements and practical aspects of managing the resources and keeping it up to date. It was indicated that the employees found it difficult to quantify the true benefit of the audit result in terms of customer claim reduction and problem prevention. Hence recommendation were made on the opportunities existing to improve the audit process.

IOP Conf. Series: Materials Science and Engineering **923** (2020) 012034 doi:10.1088/1757-899X/923/1/012034

## REFERENCES

[1] Om Prakash Yadhav, Praveen and S.Goel, (2006), 'customer satisfaction driven quality improvement target planning', *International journal of production economics*, Vol.113; PP 997-1011.

[2] Daniel Tuic,(2016), 'TRIZ model used for complaint management in the automotive product development process', *Procidia social and behavioral sciences*, Vol.221; PP 424-422.

[3] Julia Madrid,(2017), 'Towards overcoming the boundaries between manufacturing and perceived quality', *Procidia CIRP*, Vol.63; PP 733-738.

[4] Anderson, J., Rungtsanatham, F. and Schroeder (2015), ' A theory of quality management underlying the Deming management method', *Academy of Management Review*, Vol.19, PP: 472-509.

[5] Gerard H. Seijts, JeffreyGandz (2017), 'Transformational Change and Leader Character', *Kelley School of Business, Western University*,Vol.1442, PP: 11.

[6] Deming, W.E. (1982), 'Out of Crisis' *Boston. Massachusetts*: MIT/CAES.

[7] Edosomwan, J.A. (1996), 'Organizational transformation and process re-engineering', *Library cataloging in pub. Data.*

[8] JTC Teng, V Grover, KD Fedler, (1996) 'Developing Strategic Perspectives on Business Process Reengineering: From Process Reconfiguration to Organizational Change', *University of south Carolina*, Vol.24, No.3, PP : 271-294.

[9] ALESSIO ASCARI, Associate, M&&my and Company, Milan; MELINDA ROCK, Manager, Andersen Consulting, London; SOUMITRA DUTTA, Associate Professor of Technology Management, INSEA D, France, (1995), 'Reengineering and Organizational Change: Lessons from a Comparative Analysis of Company Experience', *European management journal*, Vol.13, No.1, PP : 1-30.

[10] A.B.Rami Shani, Yoram Mitki, (1996), 'Reengineering Total Quality Management and Sociotechnical Systems Approaches to Organizational Change: Towards an Eclectic Approach, *Journal of quality management*, Vol.1 No.1, PP: 131-145.

[11] John Hendry, (1995), 'Process Reengineering and the Dynamic Balance of the Organisation' *European management journal*, Vol.13. No.1, PP: 52-57.

IOP Conf. Series: Materials Science and Engineering **923** (2020) 012035 doi:10.1088/1757-899X/923/1/012035

# Influence of Dimple Geometry on Coating Performance Under Uni-Directional Sliding

**M Prem Ananth, V Vijay, M Dhinesh kumar*, G Kishore Kanna and M Manoj.**

Department of Mechanical Engineering, Sri Venkateswara College of Engineering, Chennai 602117, India

*Email:selectdhineshkumar@gmail.com

**Abstract.** The 316L stainless steel as very low carbon content than 316 steel alloys and 316L is material generally used for medical implant purpose and other industrial applications because of its excellent corrosion properties. The only disadvantage with 316L is it has poor wear resistance. By surface texturing the tribological properties can be improved. DLC (Diamond Like Carbon) Coating were made over the surface texturing. DLC Coating is a nano-composite coating which has unique properties of natural diamond low friction, high hardness, and also high corrosion resistance. Various dimple structures were made on the pin to study about its performance and the ideal dimple structure is selected by using pin on disc experiment. D2 Steel disc is taken as the counter-part for this experiment.

## 1. Introduction

Grade 316 as carbon content in the range 0.03% to 0.08% and Grade 316 is a general molybdenum-bearing grade which is second in importance to 304 amongst the austenitic stainless steels and the molybdenum gives 316 better overall corrosion resistant properties than Grade 304, specific higher resistance to pitting and crevice corrosion in chloride environments. Grade 316L Stainless Steel offers higher creep, stress to rupture and tensile strength at elevated temperatures. Typical Applications of stainless steel 316L includes Food preparation equipment particularly in chloride environments, Pharmaceuticals, Marine applications, Architectural applications, Medical implants, including pins, screws and orthopedic implants like total hip and knee replacements, fasteners.

## 2. Fabrication of pin

316L Stainless Steel Rod with 540mm length and 8mm diameter is cut into 45mm length using lathe machine, so that 12 number of pin pieces is available for experimental work. 316L Stainless Steel Rod of length 45mm of diameter 8mm is step turned to diameter 4mm of length 5mm from any one side of the job as shown in Figure 1. The surface of 4mm diameter is mirror finished for surface texturing as shown in Figure 2.

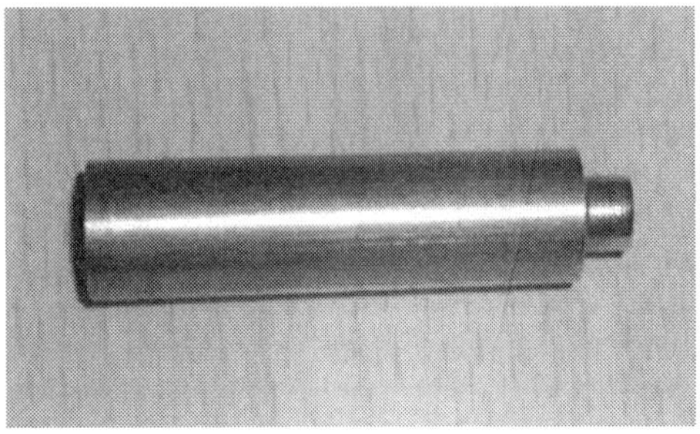

**Figure 1.** Fabrication of pin

**Figure 2.** Mirror finish surface

## 3. Fabrication of disc

D2 steel is an flexible high-carbon (1.40% to 1.60%), air hardening tool steel and also it has high chromium (11.0% to 13.0%) content with high wear and abrasion resistant properties. It is also heat treatable and offers a hardness in the range of 55-62 HRC, and D2 steel is machinable in the annealed condition. D2 steel shows little deformity on correct hardening and due to its high chromium content it has mild corrosion resisting properties in the hardened condition. D2 steel rod is machined to 55mm diameter and 10mm thickness to form as disc and the machined disc is hardened at a temperature range of 900-1000°C as shown in Figure3. The hardness gets increased from 20 HRC to 60 HRC after the heat treatment.Both sides of the surface of disc is made smooth by surface grinding as shown in Figure4.

**Figure 3.** Fabrication of disc

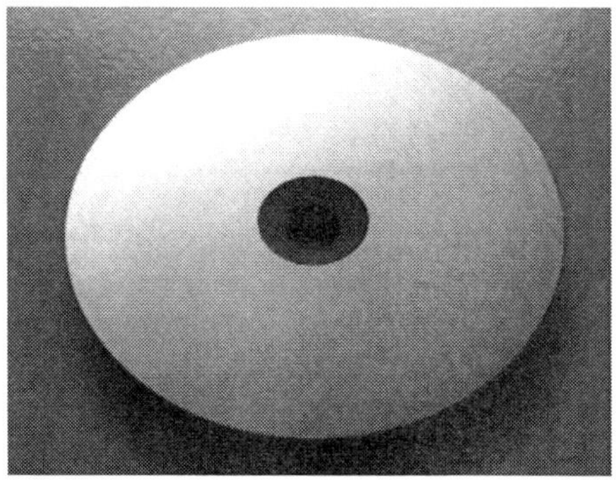

**Figure 4.** Surface grinded disc

## 4. Surface texturing

Surface Texturing of Circular dimple, Horizontal Ellipse (major axis in horizontal direction), Vertical Ellipse (major axis in vertical direction) and Inclined Ellipse (major axis at 45° to x-axis) were made over the mirror finished surface over the pin. The laser surface texturing was effectuated on the stainless steel 316L pin (test specimen) with a Nd:YAG laser beam[1] and the dimple depth is of 2.5micron. Dimples will lower the coefficient of friction[2]. The dimensions of various shapes of surface texturing are shown in Figure 5 and Figure 6.

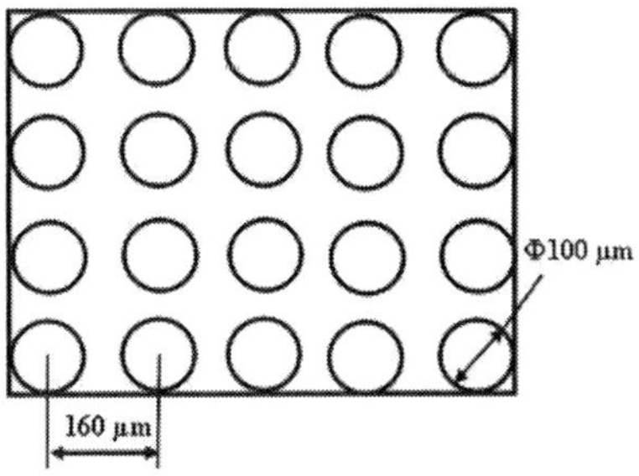


**Figure 5.** Schematic of the rectangular array circular texturing pattern

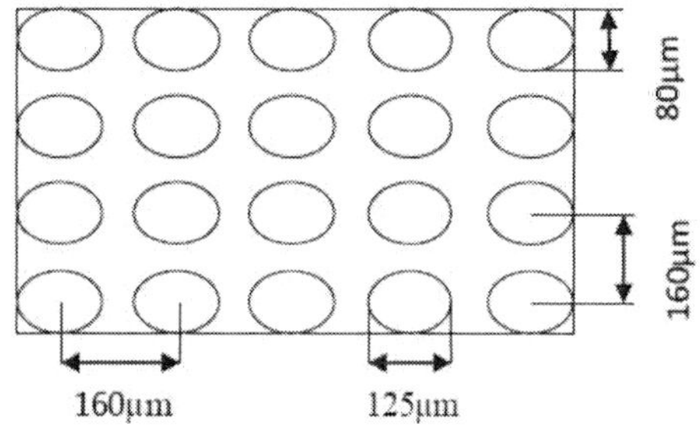


**Figure 6.** Schematic of the rectangular array elliptical texturing pattern

The SEM (Scanning Electron Microscope) were taken on the surface textured area on the pin[3]. The dimples made on the pin were clearly visible as shown in Figure 7 and Figure 8.

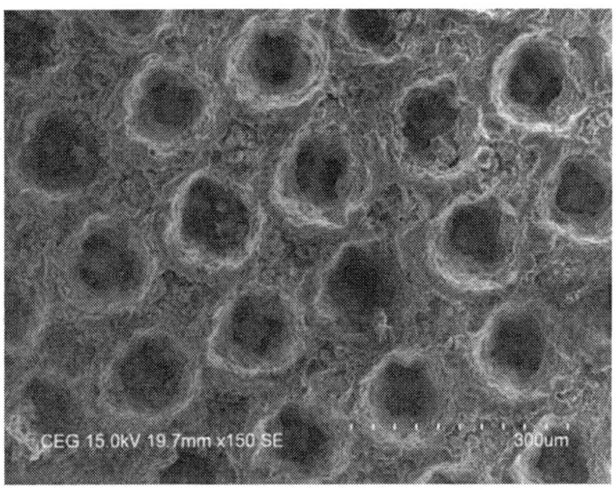


**Figure 7.** Circular dimples

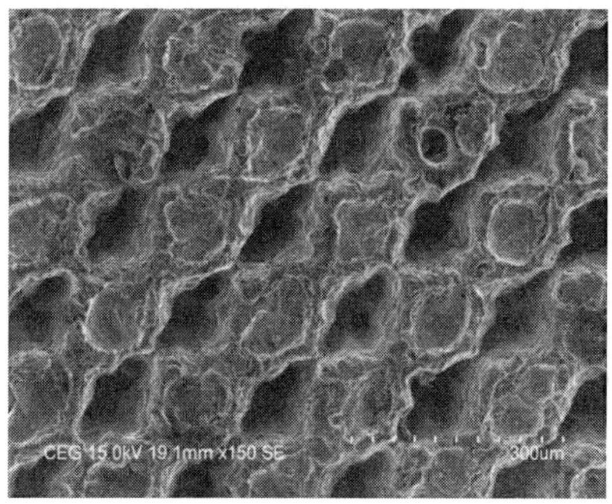


**Figure 8.** Elliptical dimples

## 5. DLC coating

Diamond-like carbon (DLC) is amorphous carbon material which shows some properties of diamond and the characteristics of DLC are excellent abrasion and corrosion protection, low friction coefficient and excellent protection against cold welding. DLC coating finds its application in aerospace (aerospace fasteners and components), medical instruments, cutting and plastic injection moulding tools. It has an outstanding gliding property [4].DLC coating were applied over the surface textured area of4micron thickness through PVD method (Physical Vapor Deposition) [5,6] as shown in Figure 9.

**Figure 9.** DLC coated pin

## 6. Micro hardness

The Vickers test can be used for all metals and has one of the widest scales among hardness tests since the required calculations are independent of the size of the specimen. Vickers Pyramid Number (HV) or Diamond Pyramid Hardness (DPH) is the unit of hardness. The micro hardness for 316L SS is 272.5HV and for DLC coating is 343.9HV, where HV indicates hardness scale (Vickers).

## 7. Experimental setup pin on disc

The Pin on Disc is to determine the wear of the material during sliding[7]. Here 316L SS serves as pin, and the D2 steel acts as disc. The experiment is conducted with three different loads 10N, 20N, 30N as shown in Figure 10,11 and 12 on four different surface texture DLC coated pin. The Pin on disc experiment has sliding distance of 2000m with a velocity of 2m/s. The coefficient of friction for each surface textured pin with DLC coating is determined and compared.

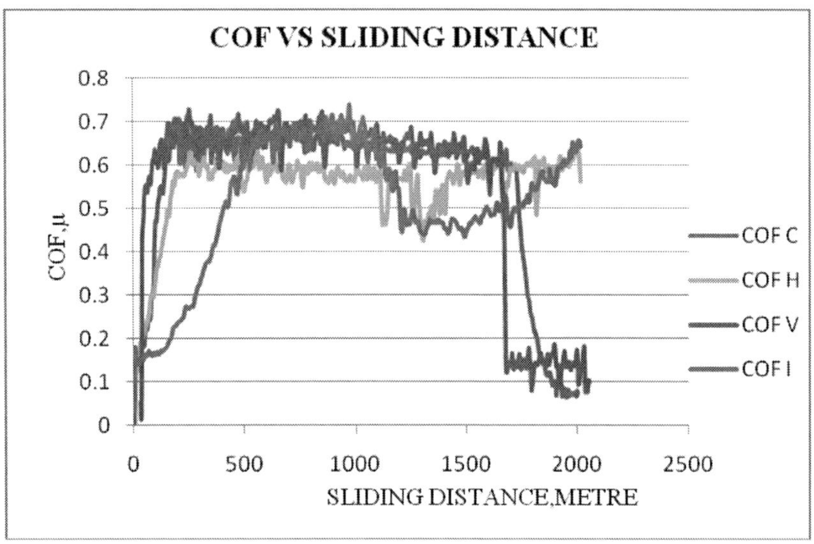


**Figure 10.** Coefficient of friction vs sliding distance for load 10N

In the Figure 10, graph has been made between COF versus Sliding Distance for the load of 10N. It can be observed that at lower normal load, the COF drops from an initial high value to very low value. It is clear that coulomb friction law is no longer valid when self-lubrication occurs[8].

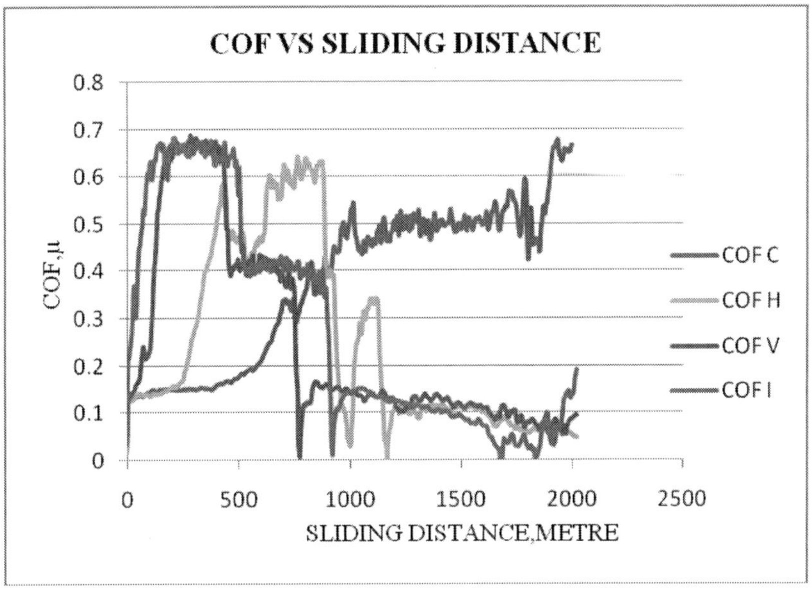


**Figure 11.** Coefficient of friction vs sliding distance for load 20N

In the Figure 11, graph has been made between COF versus Sliding Distance for the load of 20N. The COF increases up to a maximum value & then decreases before the end of the test. This behaviour is probably due to the formation of a tribo-film which builds up during sliding & reaches maximum effectiveness in decreasing COF after a given sliding distance[9].

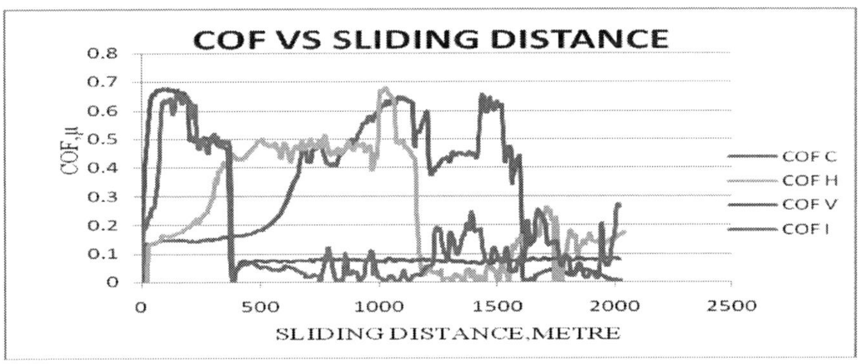


**Figure 12.** Coefficient of friction vs sliding distance for load 30N

In the Figure.12, graph has been made between COF versus Sliding Distance for the load of 30N. The DLC coated samples attain the lowest value of COF under most applied loads. This is probable due to the high hardness (H) of the coating which is effective in supporting the low shear strength graphitised tribo-film. At higher normal loads the COF starts to oscillate and this oscillation in the COF with sliding distance could be because of stick slip phenomenon.

## 8. Average coefficient of friction

The Average coefficient of friction for the above graphs for various loads and various dimples shapes has been made. It clearly defines the coefficient of friction for various DLC coated surface texturing.

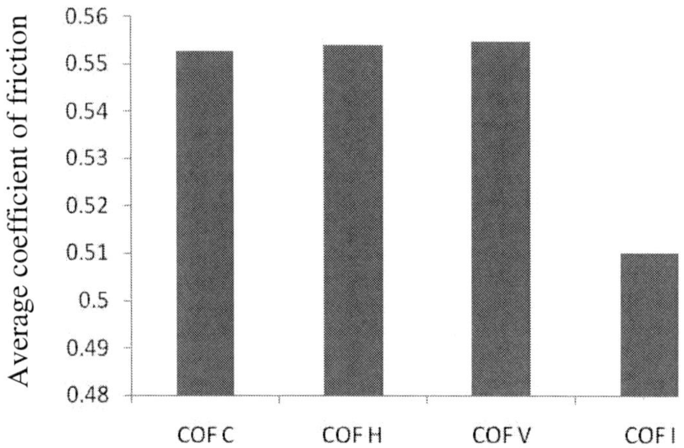


**Figure 13.** Average COF for load 10N

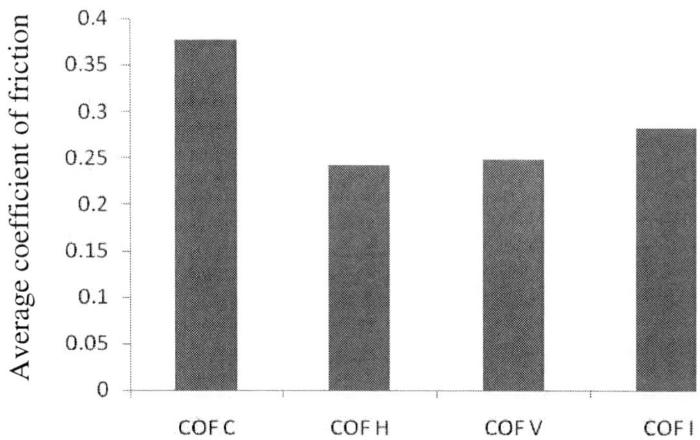


**Figure 14.** Average COF for load 20N

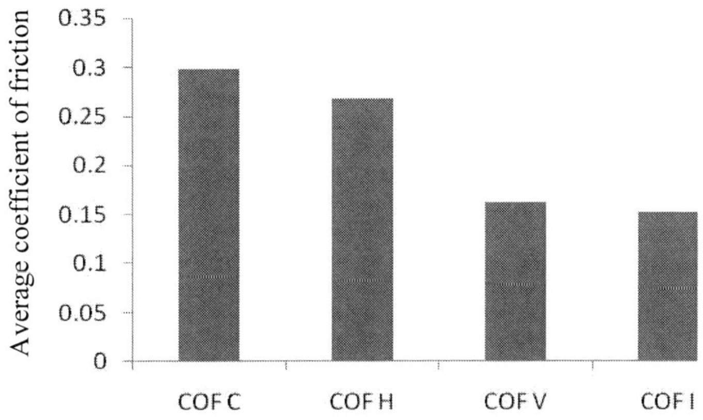


**Figure 15.** Average COF for load 30N

## 9. Specific wear rate

The specific wear rate can be calculated from the formula :

Specific wear rate = (Initial mass – Final mass)/(ρxFxL) Where ρ = density of the pin g/mm$^3$,F=Normal Load N, L = Sliding distance m.

Unit of Specific wear rate is mm$^3$/Nm.

The Specific wear rates for various DLC coated dimples were compared for corresponding loads. The specific wear rate for various dimples have been compared and analysed.

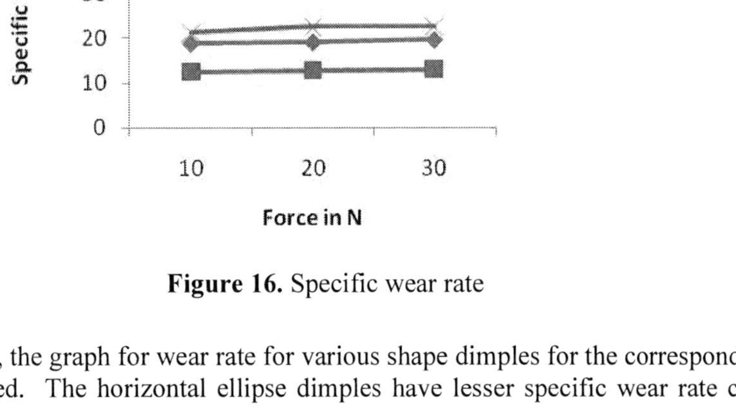


**Figure 16.** Specific wear rate

In the Figure 16, the graph for wear rate for various shape dimples for the corresponding load has been made& compared. The horizontal ellipse dimples have lesser specific wear rate compared to other dimples. This is probably due to the higher temperature & duration. The reason is that ellipse major axis is perpendicular to sliding condition results in a greater converging wedge & also possesses optimum load carrying capacity [11].

## 10. Results and discussion

- Experimental investigation of DLC coated with various surface textured shapes were analysed using pin on disc apparatus with loading conditions of 10N, 20N and 30N with sliding speed of 2m/s and sliding distance of 2000m.
- The average coefficient of friction for various surface textured DLC coated pin were analysed.
- For 20N horizontal ellipse surface texturing has better coefficient of friction compared to other surface texturing.
- For 30N inclined ellipse surface texturing has better coefficient of friction compared to other surface texturing.
- The wear rate has been calculated and has been analysed for various surface texturing.
- The horizontal ellipse surface texturing has lower wear rate compared to other surface texturing.

**Acknowledgement**
This work was supported by the All India Council for Technical Education (AICTE) under the Scheme of Research Promotion Scheme (RPS) with a reference number of File No. 8-50/RFID/RPS/POLICY-1/2016-17. We would like to acknowledge them for the financial and technical support.

## References

[1] Yuki Kuroiwa, Auezhan Amanov. "Effectiveness of surface texturing for improving the anti-seizure property of copper alloy"; Malaysian International Conference 2013; Pg:600-606.
[2] Kristian Tonder. "Dimpled Pivoted Plane Bearings: Modified Coefficient" Published in Tribology International Pg:2303-2307.
[3] M. Prem Ananth, R.Ramesh. "Tribological improvement of titanium alloy surfaces through texturing and TiAlN coating"; Surface Engineering, Vol30; Pg:758-762.
[4] Dipankar Choudhury. "A novel tribological study on DLC-coated micro-dimpled orthopedics implant interface"; Biomedical materials 2015; Pg:121-131.
[5] J. Gerth and U. Wiklund: "The influence of metallic interlayers on the adhesion of PVD TiN coatings on high-speed steel", Wear,2008, Pg:885–892.
[6] J. L. Mo, M. H. Zhu, B. Lei, Y. X. Leng and N. Huang: "Comparison of tribological behaviours of AlCrN and TiAlNcoatings – deposited by physical vapor deposition", Wear, 2007, Pg:1423–1429
[7] L. Rapoport, A. Moshkovich and V. Perfilyev: "Wear life and adhesion of solid lubricant films on laser-textured steel surfaces",Wear, 2009, Pg:1203–1207.
[8] Pei Yutao T, de Jong: "Breakdown of the Coulomb friction law inTi-C/a-C", JournalofAppliedPhysics,2006.
[9] Y. T. Pei, D. Galvan & J. Th. M. De Hosson: "Super-low Friction Behavior Of Nanostructured DLC Composite Coatings", WIT Transactions on Engineering Sciences, Vol 49.Pg:185-194.
[10] Pradeep L.Menezes, Kishore: "Influence of surface texture on coefficient of friction and transfer layer formation during sliding of pure magnesium pin on 080 M40 (EN8) steel plate", Wear, Vol 261 20 September 2006, Pg:578-591.
[11] Haiwu Yu, Xiaolei Wang: "Geometric Shape Effects of Surface Texture on the Generation of Hydrodynamic Pressure Between Conformal Contacting Surfaces", Tribol Lett (2010) 37:123–130.

IOP Conf. Series: Materials Science and Engineering **923** (2020) 012036 doi:10.1088/1757-899X/923/1/012036

# INFLUENCE OF TOOL PIN PROFILE IN UNDERWATER FRICTION STIR WELDING OF ALUMINIUM 6061 ALLOY

**Premkumar.P, Ruskin Bruce.A , YogeshKrishnan P.R**

Department of Mechanical Engineering, St.Joseph's College of Engineering,

Chennai 600119, Tamilnadu, India.

E-mail : prem1125@gmail.com

Abstract- Light metals like Aluminium and Magnesium are usually joined by welding processes like Friction Stir Welding (FSW). Aluminium being light weighted is used in various aeronautical applications. During welding using FSW, the Heat affected zone (HAZ) obtained is very much high due to direct exposure of air. In order to minimise the heat affected zone and also to improve the Weld strength, an attempt has been made to weld Aluminium Al 6061 alloy using Underwater Friction stir welding (UFSW). By carrying out the FSW process underwater, the hardness of the welded structures is also increased due to rapid cooling. In this present study, the effect of different pin profiles like cylindrical, square and conical profiles on the welded specimen is studied. Consequently, the mechanical properties of these welded specimens are also compared and analysed.

*Keywords*- Underwater Friction stir welding, Heat affected Zone, Hardness

## 1.INTRODUCTION

Friction Stir Welding is the obvious choice to weld Aluminium Alloys which are difficult to weld like the 5xxx series or the high strength alloys of 2xxx and 7xxx series (1). FSW involves rotation of non-consumable tool over the material specimens. The shoulder contact between tool and work piece provides frictional heat for the work pieces to be welded. The heat generated during the plunging, travelling and dwelling phases are sufficient to soften and join the work pieces (2). FSW reduces the porosity, surface cracking, material shrinkage of the welded joints (3). However, the thermal cycles experienced during FSW leads to reduction of mechanical properties due to coarsening and dissolution of strengthening precipitates which resulted in poor joint properties (4,5). In most of the Aluminium Welded specimens the Thermo-Mechanically Affected Zone (TMAZ) and the Heat Affected Zone (HAZ) are weak regions (6-8). The coarsening and dissolution of strengthening precipitates is prevented by rapid cooling (9). To avoid defects like Hydrogen Embrittlement, Oxidation, Porosity, Underwater Friction Stir Welding is used. The underwater conditions increase the weld properties because of lower heat input (10). However, there are only limited studies to prove the cooling effects of FSW on mechanical and micro structural properties. This urges researchers to implore more on the cooling effects of the specimen (11). The optimization results of UFSW indicated that the mechanical properties of the specimens showed a 6% increase than the FSW samples (12). High hardness and tensile strength of the welded specimen attributes to their use in the aerospace and submarine industry (13). The micro hardness distribution of UFSW is greater at the TMAZ and HAZ than the FSW Joint (14). Abbas et al. concluded that the optimum parameters to obtain a strong weld were 1300 rpm and a tool tilt angle of $4^0$ (15). In this study, the influence of tool pin profile in Underwater Friction Stir welding of Aluminium 6061-T6 Alloys are studied. Different pin profiles like Cylindrical, Conical, Square profiles are taken into consideration. Hardness, tensile and micro structural properties of the samples was studied. Optimisation of these results was done using RSM optimization software.

## 2. METHODOLOGY OF THE STUDY

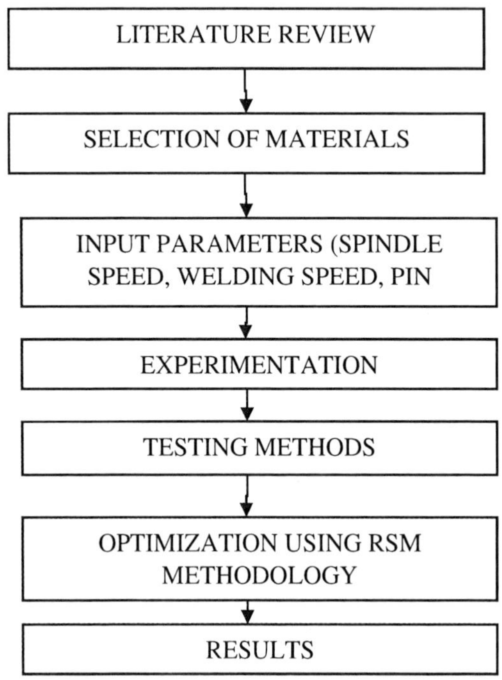


**Fig 1.** Methodology of study

## 3. EXPERIMENTATION

### Selection of Materials

Aluminium 6061- T6 alloys are chosen for the study for their good machinability and workability. Aluminium 6061 alloys have excellent joining characteristics. The chemical composition of the aluminium alloy is listed in table 1.

**Table 1.** Chemical composition of Al 6061

| Component | Weight % |
|---|---|
| Aluminium | 95.8-98.6 |
| Chromium | 0.04-0.35 |
| Copper | 0.15-0.4 |
| Iron | 0.7 Max |
| Magnesium | 0.8-1.2 |
| Silicon | 0.4-0.8 |
| Tin | 0.15 Max |

| | |
|---|---|
| Zinc | 0.25 Max |

These alloys exhibit excellent corrosion resistance. Owing to their wide availability, these alloys are used for a variety of applications including couplings, brake pistons, bike frames, etc. Moreover, the industrial applications of Al 6061-T6 are highly desirable and economical. All these characteristics adds to the selection of Aluminium Al 6061- T6 alloy.

The dimensions of the work piece were selected according to the standard dimensions followed in various experiments. The dimensions of Al 6061- T6 alloys include the length of 75mm, width of 100mm and thickness of 5mm.

**Fig 2.** Prepared samples of UWFSW

**Tool Design**

Among various tools available to carry out FSW process, the best suited HSS tool steel (HSS) is selected. The HSS tool steel is used in the Vertical Milling Centre (VMC). The three different tool pin profiles selected are displayed here.

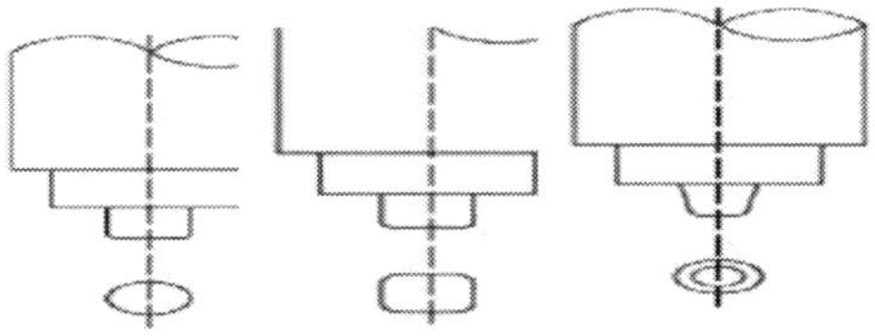


**Fig 3** Different tool profiles

The dimensions of the HSS tool were determined based on the optimized parameters through literature survey. Moreover, an ideal fit for the collect is also needed for strong machining. The dimensions are listed below.

**Table 2.** Tool Dimensions

| PIN PROFILE | SHOULDER DIAMETER (mm) | PINDIAMETER (mm) |
|---|---|---|
| Cylindrical | 20 | 5 |
| Conical | 20 | 5 |
| Square | 20 | 5 |

The HSS tools prepared are heat treated to withstand high temperatures and to produce a sound weld.

**Parameters selection and L12 orthogonal array**

After a rigorous literature study, the parameters that have an impact on the properties of the welded specimen are selected. The process parameters selected for the study include Spindle speed, Welding speed and the tool pin profile. The range of these parameters was also identified beforehand. The L12 orthogonal array showing different combination of parameters is listed in table 3.

**Table 3.** L12 Orthogonal Array

| **WELDING SEQUENCE** | **SPINDLE SPEED (rpm)** | **WELDING SPEED (mm/min)** | **TOOL PIN PROFILE** |
|---|---|---|---|
| 1 | 1400 | 20 | SQUARE |
| 2 | 1600 | 24 | SQUARE |
| 3 | 1500 | 26 | SQUARE |
| 4 | 1300 | 22 | SQUARE |
| 5 | 1400 | 24 | CONICAL |
| 6 | 1300 | 26 | CONICAL |

| | | | |
|---|---|---|---|
| 7 | 1500 | 22 | CONICAL |
| 8 | 1600 | 20 | CONICAL |
| 9 | 1400 | 26 | CYLINDRICAL |
| 10 | 1300 | 24 | CYLINDRICAL |
| 11 | 1500 | 20 | CYLINDRICAL |
| 12 | 1600 | 22 | CYLINDRICAL |

**Underwater Friction Stir Welding**

Underwater Friction Stir Welding is carried under Vertical Milling Centre (VMC). The Jig setup to withhold water during welding is shown in figure 4. The vertical Milling Centre used for experimentation is shown in figure 5. The aluminium plates were clamped to the bed of the VMC Machine. Different combinations of process parameters including spindle speed, welding speed, tool pin profiles are used to conduct the experiments. The experimentation was carried out using HSS tool steel by joining Al 6061- T6 plates of dimensions 75mm length, 100mm width, 5mm thickness.

**Figure 4-** Jig to clamp Aluminium plates

**Figure 5.** Vertical Milling Centre used for UFSW

**Fig 6** Welding of Aluminium specimens

*Square* *Conical* *Cylindrical*

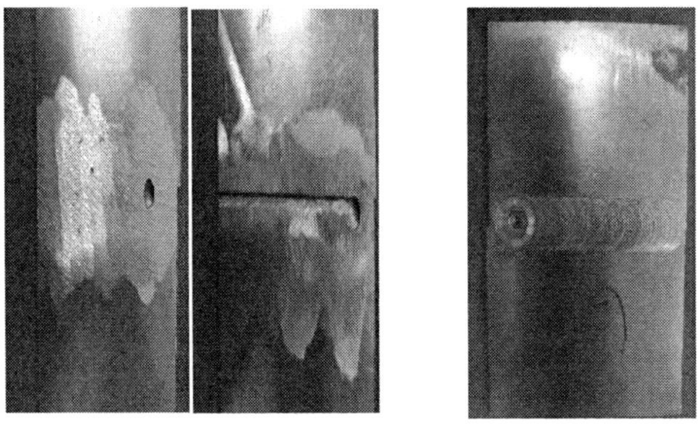

**Figure 7**- Weld samples obtained from different pin profiles

## 4. Tensile Test

Tensile test was conducted according to ASTM B557 standards to non-ferrous metals. The workpieces are machined according to the specified standards. An Universal testing machine is used to conduct the tensile test. Tensile test results provide us insights about the ultimate tensile strength, yield strength, % elongation, etc. A maximum load of 100 KN was applied on the specimen. The ultimate tensile testing machine is shown in figure 8. The samples obtained after tensile fracture is shown in figure 9.

**Figure 8.** Universal Testing machine **Figure 9.** Samples after Tensile test

The tensile test results obtained after testing is listed in table 4. The table shows results of Tensile Strength obtained from different welded samples

**Table 4.**Tensile Test results

| Welding sequence | Spindle speed (rpm) | Welding speed (mm/min) | Tool Pin profile | Tensile Strength (MPa) |
|---|---|---|---|---|
| 1 | 1400 | 20 | Square | 79 |
| 2 | 1600 | 24 | Square | 77 |
| 3 | 1500 | 26 | Square | 72 |
| 4 | 1300 | 22 | Square | 72 |
| 5 | 1400 | 24 | Conical | 62 |
| 6 | 1300 | 26 | Conical | 65 |
| 7 | 1500 | 22 | Conical | 57 |
| 8 | 1600 | 20 | Conical | 63 |
| 9 | 1400 | 26 | Cylindrical | 44 |
| 10 | 1300 | 24 | Cylindrical | 49 |
| 11 | 1500 | 20 | Cylindrical | 42 |
| 12 | 1600 | 22 | Cylindrical | 45 |

## 5.Hardness Test

Hardness is defined as resistance to indentation or cracks. Hardness test was carried on welding specimens using the Rockwell Hardness Testing Machine. The Scale B in Rockwell Testing machine is used for determining hardness in the welded samples. Load of 100 KN is applied on the welded samples to obtain indentation. The setup of Rockwell Hardness machine is shown in figure 10. Three values of hardness were obtained on each sample to obtain precision results. Hence a total of 36 tests were conducted on welded samples. Sample subjected to hardness testing is shown in figure 11. The results obtained as a result of the hardness test is tabulated in Table 5.

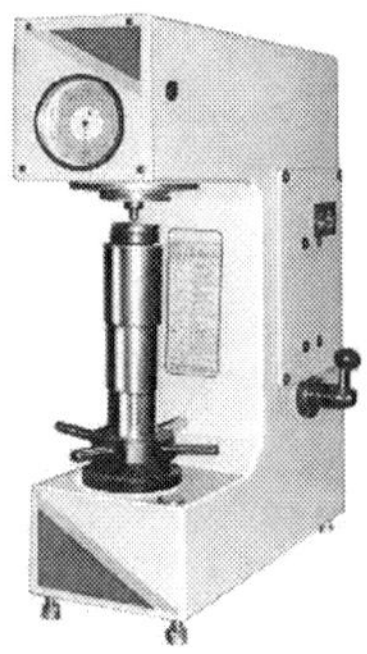

**Figure 10** – Rockwell Hardness Testing Machine

**Figure 11-** Sample subjected to Hardness test

**Table 5.** Hardness Test Results

| Welding sequence | Spindle speed (rpm) | Welding speed (mm/min) | Tool Pin profile | Hardness (RHN) |
|---|---|---|---|---|
| 1 | 1400 | 20 | Square | 80 |
| 2 | 1600 | 24 | Square | 78 |
| 3 | 1500 | 26 | Square | 75 |
| 4 | 1300 | 22 | Square | 75 |
| 5 | 1400 | 24 | Conical | 70 |
| 6 | 1300 | 26 | Conical | 62 |
| 7 | 1500 | 22 | Conical | 53 |
| 8 | 1600 | 20 | Conical | 58 |
| 9 | 1400 | 26 | Cylindrical | 48 |
| 10 | 1300 | 24 | Cylindrical | 52 |
| 11 | 1500 | 20 | Cylindrical | 49 |
| 12 | 1600 | 22 | Cylindrical | 54 |

## 6. Optimization

IOP Conf. Series: Materials Science and Engineering **923** (2020) 012036 doi:10.1088/1757-899X/923/1/012036

The optimization results were obtained using RSM methodology. The L12 orthogonal arrays as well as the results obtained from various tests were utilized to obtain optimized results. Various optimization results are discussed below,

**Response Surface Methodology**

The response surface methodology was carried out on L12 orthogonal array. The box- Behnken model and the coded coefficient values were tabulated in the tables.

**Table 6.**Design of response surface model

| Factor | Name | Units | Type | Low Actual | High Actual |
|---|---|---|---|---|---|
| A | SPINDEL SPEED | rpm | Numeric | 1300 | 1600 |
| B | WELDING SPEED | mm/min | Numeric | 20 | 26 |
| C | PIN PROFILE | | Numeric | 1 | 3 |

| Response | Name | Units | Obs | Analysis | Min. | Max. | Mean | Std. Dev. |
|---|---|---|---|---|---|---|---|---|
| Y1 | Hardness | RHN | 12 | Polynomial | 48 | 75 | 59.28 | 8.46349 |
| Y2 | Tensile strength | Mpa | 12 | Polynomial | 42 | 79 | 60.554 | 12.6769 |
| Y3 | Yield Strength | Mpa | 12 | Polynomial | 37 | 72 | 53.683 | 11.9639 |

regression equation for tensile strength, yield stress, hardness is obtained using the methodology. They are explained as follows,

$$\text{Hardness} = 59.15372 - 2.84845*A - 0.57905*B - 9.7448*C - 0.6819*A*B + 1.721067*B*C + 0.104433*C*A$$

Hardness = 69.8933 RHN

Tensile Strength = 59.79- 0.68*A-0.58*B -16.42*C -4.13*A*B- 2.87*A*C + 2.36*B*C

Tensile strength = 78.63706 MPa

Yield Stress = 52.83-0.047*A -2*B-15.67 *C -4.62*A*B -3.07*A*C + 2.13*B*C

Yield stress = 72.00603 MPa

### Nomenclature

A- Spindle Speed
B- Welding speed
C- Tool pin profile

## 7. Results and Discussion

Based on the regression equation, various plot of residuals for different parameters is obtained. They are shown in figures 12, 13, 14.

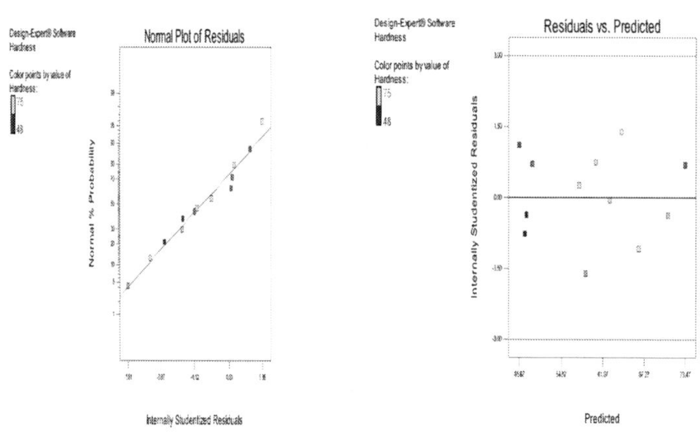


**Figure 12-** Normal plot, Residuals vs Predicted plot for Hardness

IOP Conf. Series: Materials Science and Engineering **923** (2020) 012036 doi:10.1088/1757-899X/923/1/012036

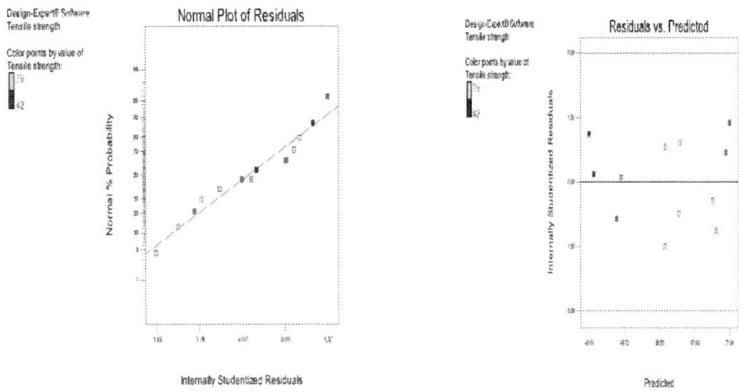


**Figure 13-** Normal plot, Residuals vs Predicted plot for Tensile Strength

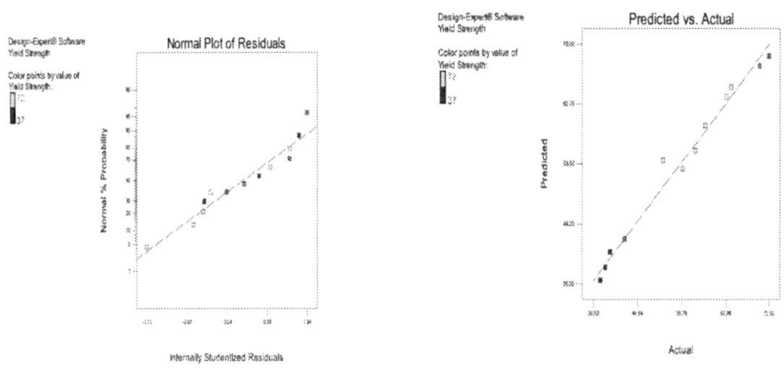


**Figure 14-** Normal plot, Actual vs Predicted plot for yield stress

## Contour and 3D surface plot

The contour plot and 3D surface plots of different parameters were obtained using the Design Expert software. The values of maximum tensile strength, yield stress and hardness are compared to obtain optimized results. They are illustrated in the below images.

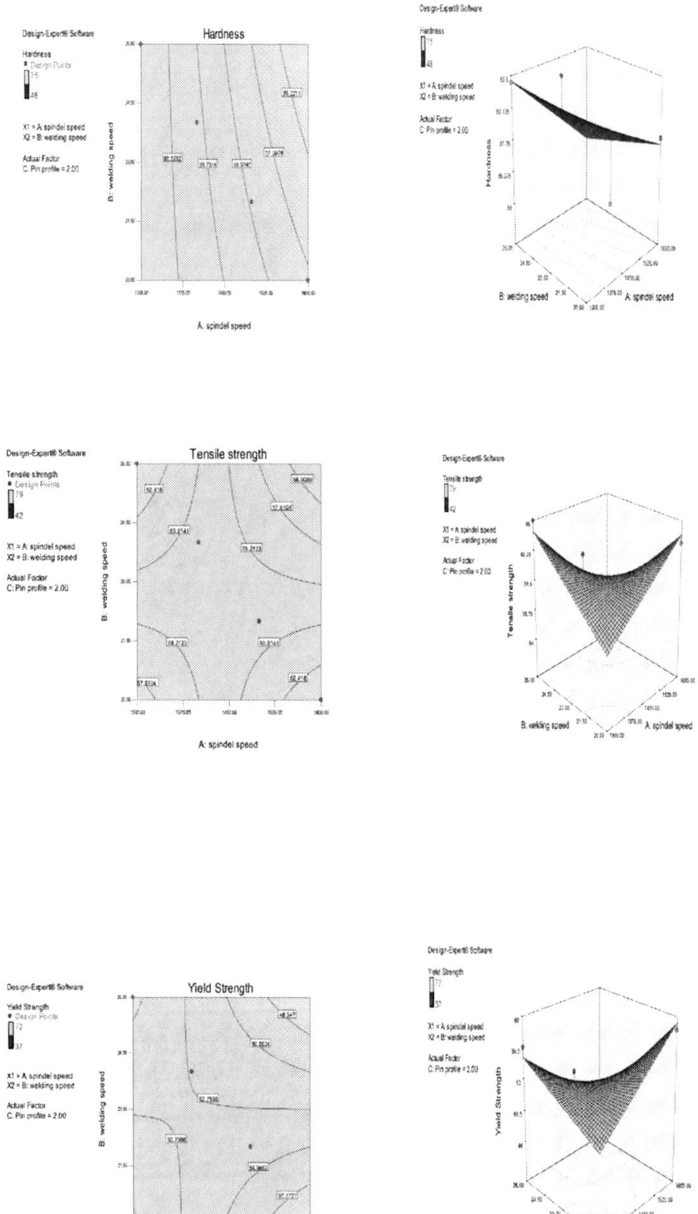


**Figure 15.** Contour and 3D surface plot for Spindle speed vs Welding speed

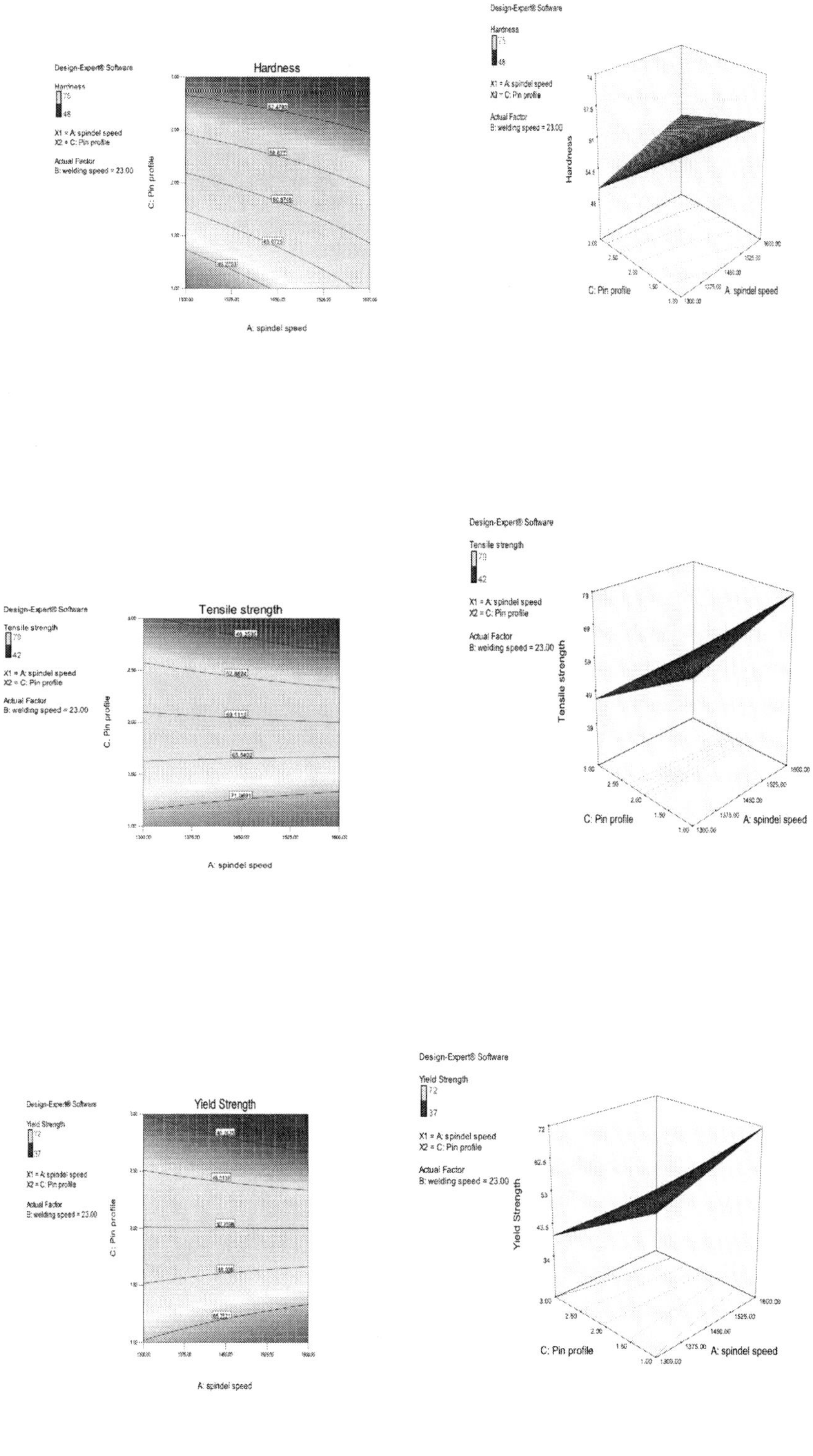


**Figure 16**. Contour and 3D surface plot for Spindle speed vs Pin Profile

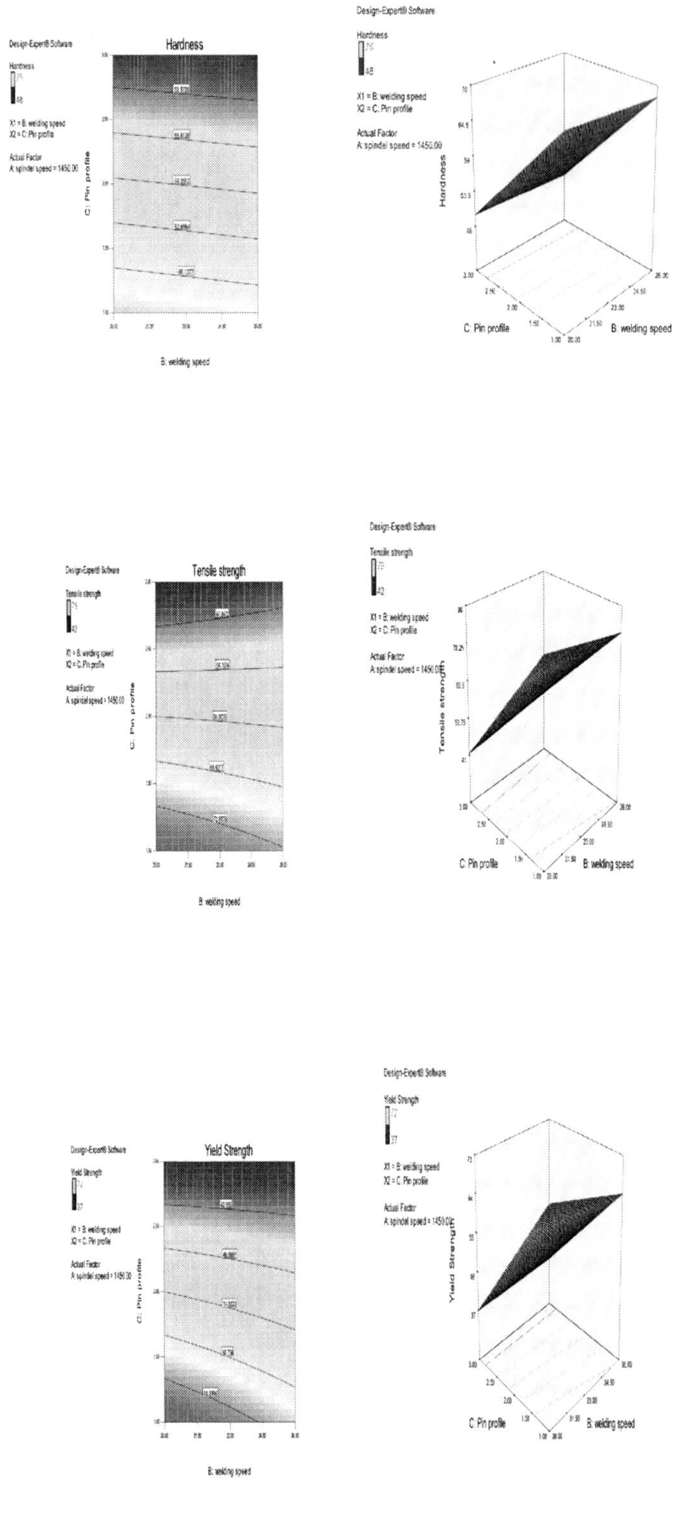


**Figure 17.** Contour and 3D surface plot for Welding Speed vs Pin Profile

From the plots, it is clear that the tensile strength, welding speed and spindle speed has greater impact than the tool pin profiles used to weld the samples. The surface plot clearly visualizes that maximum tensile strength is obtained at moderate range of parameters and not the maximum or minimum.

**Response Optimization- Tensile Strength**

Optimization of the obtained values provides us the best optimized results for carrying out the weld. The maximum strength was found to be 79Mpa for square profile at a tool traverse speed of 20 mm/min and a spindle speed of 1400 rpm. The results obtained after multi-response table is tabulated here.

**Table 7.** Response Optimization

| Spindle speed | Welding speed mm/min | Pin profile | Hardnes s RHN | Tensile strength MPa | Yield Strength MPa | Desirability |
|---|---|---|---|---|---|---|
| 1437.87 | 20 | Square | 69.893 | 78.63706 | 72.00603 | 0.929442 |

**Conformity Test**

To confirm the correctness of obtained values, conformity test was conducted. The results obtained from the conformity test is tabulated below,

**Table 8.**Conformity Test

| Pin Profile | Spindle Speed (rpm) | Welding Speed (mm/min) | T1 MPa | T2 MPa | T3 MPa | Mean Tensile Strength (MPa) |
|---|---|---|---|---|---|---|
| Square | 1400 | 20 | 78.4 | 80 | 77 | 78.46 |

| Pin Profile | Spindle Speed (rpm) | Welding Speed (mm/min) | H1 RHN | H2 RHN | H3 RHN | Mean Hardness (RHN) |
|---|---|---|---|---|---|---|
| Square | 1400 | 20 | 80 | 78 | 75 | 77.66 |

**Effect of Tool pin profile**

Tool pin profiles have a varying impact on the strength of the weldments. Of the three profiles used, Square pin profile yielded the best results. This is because of the ratio between swept volume and static volume. The ratio between swept volume and static volume decides the path of flow of plasticized material from the leading edge of the tool to the trailing edge of the tool. The ratio is 1.56 for Square profiles which is the maximum among different profiles. Moreover, at 1200 rpm, square pin produces as much as 80 pulses/ sec while a triangular pin generates 60 pulses/sec. As a result, square pin produces better results.

## 8. Conclusion

This study involving Underwater Friction Stir Welding of Al 6061- T6 alloys provides quality insights about the optimized parameters and the influence of tool pin profiles to obtain a sound weld.

The investigation provided great scope to study about the welding of Aluminium alloys, the strength of the welds, etc. We have inferred the following from the study,

- RSM methodology was used to obtain optimized parameters in this Underwater Friction Stir Welding of Aluminum Al6061-T6 alloys.
- A maximum strength of 79 MPa was obtained by the sample welded with square profile with a spindle speed of 1400 rpm and tool traverse speed of 20mm/min.
- The minimum Tensile strength of 48 MPa was obtained by the aluminum sample welded with a cylindrical profile with a spindle speed of 1400 rpm and tool traverse speed of 26mm/min.
- The optimum machining condition to obtain a sound weld was found to be with a spindle speed of 1437 rpm, tool traverse speed of 20mm/min with a square pin profile.
- Square pin profile yielded better results due to high ratio of swept volume to static volume.
- Besides, the square profile generates a greater stirring action, leading to the soundness of the weld.
- Hardness is uniformly distributed throught the weldment and increase in hardness is attributed to the rapid cooling of the welded samples.

## REFERENCES

[1] C. Y. Lee, W. B. Lee, J. W. Kim, D. H. Choi, Y. M. Yeon, and S. B. Jung, "Lap joint properties of FSW dissimilar formed 5052 Al and 6061 Al alloys with different thickness," *Journal ofMaterials Science*, vol. 43, no. 9, pp. 3296–3304, 2008.

[2] M. B. Durdanovic, M. M. Mijajlovic, D. S. Milcic, D. S. Stamenkovic. 2009. Heat Generation during Friction Stir Welding Process. Journal of the Serbian Tribology Society.

[3] Esther T. Akinlabi, Stephen A. Akinlabi. 2012. Friction Stir Welding Process: A Green Technology. International Journal of Mechanical, Aerospace, Industrial and Mechatronics Engineering. Vol:6, No:11.

[4] Threadgill PL, Leonard AJ, Shercliff HR, Withers PJ, Friction stir welding of aluminium alloys, Int Mater Rev 2009; 54:49-93

[5] Liang XP, Li HZ, Li Z, Hong t, Ma B, Liu SD, Study on the microstructure in a friction stir welded 2519-T87 Al Alloy, Mater Des 2012;35:603-8

[6] Simar A. Bechet Y, De Meester B, Denquin A, Pardoen T. Microstructure, local and global mechanical properties of FSW of Al6005A-T6. Master Sci Engg A 2008;486: 85-95

[7] Sivaraj P, Kanagarajan D, Balasubramanian V. Effect of post weld heat treatment on tensile properties and mechanical properties of FSW armour grade AA7075- T651 Aluminium alloy. Def Technol 2014; 10:1-8

[8] Dixit V, Mishra RS, Lederich RJ, Talwar R. Influence of process parameters on microstructural evolution and mechanical properties in friction stirred Al-2024(T3) Alloy. Sci Tech Weld Join 2009; 14:346-55

[9] TulikaGarg,Priyank Mathur, Varun Singhal, Underwater Friction Stir welding:An Overview (2014)

[10] Zhao Y, Lu Z, Yan K, et al. Microstructural characterizations and mechanical properties in underwater friction stir welding of aluminum and magnesium dissimilar alloys. Mater Des 2015; 65: 675–681.

[11] Guo JF, Chen HC, Sun CN, et al. Friction stir welding of dissimilar materials between AA6061 and AA7075 Al alloys effects of process parameters. Mater Des 2014;56: 185–192.

[12] Jie LH, Jie ZH, Lei Y. Homogeneity of mechanical properties of underwater friction stir welded 2219-T6 aluminium alloy. J Mater Engg perform 2011:20; 1419-22

[13] Huseyin Tarik Serindag, Binnur Goren Kiral, Dokuz. 2017. "Friction Stir Welding of AZ31 Magnesium Alloys – A Numerical and Experimental Study". *Latin American Journal of Solids and Structures*. 14, pp. 113-130.

[14] L.Hui-Jie; Z. Hui-Jie; H. Yong-Xian. 2009. Mechanical properties of underwater friction stir welded 2291 Aluminium alloy. Trans. Nonferrous Met. Soc. China 20 (2010). 1387-1391.

[15] Mohd Abbas. Neha Mehani. Atishey Mittal. 2014. Feasibility of Underwater Friction Stir Welding and Its Optimization Using Taguchi Method. International Journal of Engineering Sciences & Research Technology.

# Analysis and Optimization of Abrasive water Jet Machiningprocesses on the hybrid nano particle reinforced Aluminum alloy matrix composite material

**R.Venkatesh[1], Vaddi Seshagiri Rao[2], S.Ashwin kannan[3], S.M.Mullaikodi[4]**

[1]Research Scholar, Department of Mechanical Engineering, Sathyabama Institute of Science and Technology, Chennai, India.
[2&4]Department of Mechanical Engineering, St.Joseph's College of Engineering, Chennai, India.
[3]Department of Mechanical Engineering, St.Joseph's Institute of Technology, Chennai, India.

[1]venkateshrj73@gmail.com,
[2]principal@stjosephs.ac.in,[3]ashwinkannan.mech@gmail.com,
[4]mullaikodism@gmail.com

**Abstract**. The Nano hybrid particles reinforced aluminium based MMC has been prepared and characterized. The materials chosen for the fabrication of the composites include Aluminium Al6061 alloy(70%) as the base or matrix material in the composite reinforced with hybrid Nano materials as the particulate reinforcements (Cu-9%,Gr-9% and Al2O3-12%) were prepared by following the stir casting route. The developed material has been characterized and subjected to AWJM process. The input parameters were carefully chosen. The Experimental design was formulated for the orthogonal array of L30. Observations were made and the results were taken to the Design Expert software so as to do the optimization process. Reinforcing a hybrid mixture comprising of an alloy, a metal, a ceramic and allotropy of carbon in the matrix and analysing their properties are in itself unique and such works are scanty. The material thus fabricated showed few novel characteristics. The conduciveness of processing the material, aimed at providing the good quality surface and the ease of reducing the bulk material were investigated. Unique tool for optimization proposed by RSM "the Central composite Design" was put in to use in this work. Bearings, Bushes etc. known for their structural supports could be manufactured using this material exhibiting good machinability aspects.

**Keywords.** Hybrid nano composites, Design of experiments, ANOVA, RSM, Central composite Design, WP, AFR, TS,SOD

## 1. Introduction

The manufacturing sector relentlessly works to find a novel combination of materials that reduces the weight considerably, without compromising on their intended performance. The investigations so far carried out by erstwhile researchers have found material combinations that enhanced the performances as well. The support structures of machineries, automobile spares and even in house hold appliances, the composites have taken a prominent role. Copper alloys are extensively used as a link device that supports a fixed member and a rotary member of machineries, owing to its load carrying capacity and wear resistance. The advent of materials under nano regime and becoming a particulate reinforcement in composite materials, show a positive role in related applications. Fretting wear is a phenomenon under heavy load conditions that can better be defeated with these MMCs comprising of nano reinforcements.

Hybridization of the composites, with the constituent reinforcements in form of a Ceramic, Carbon and another metal has been attracting the researchers for the past couple of previous decade. Characterizations pertaining to Wear, Corrosion and the Mechanical properties have been comprehensively done by the researchers like Michael Oluwatosin Bodunrin et al. [1].T.Rajmohan et al. [2] utilized the power metallurgical path of developing the Aluminum based MMC and did an extensive work on the characterization. More over the present day research aspirants take the direction that leads to Aluminum matrix composites on a large scale. It has become obvious from the research reports, which strongly advocate, that the presence of particulate reinforcements increases the resistance of the work material towards Abrasion and wear at very high loads. Wear resistance at high loads increases with the increased volume of particulate reinforcements, ascertains Mohan Hossein et al. [3]. Copper alloys could expect a potential threat from Aluminum composites, as the better replacements in structural support systems, in the days to come. The difference in their melting point places Aluminum at the top, ahead of copper, when it comes to stir casting of Metal based hybrid composites.

Michael Oluwatosin Bodunrin et al. also reviewed the proportions and combinations of materials which are used as reinforcements, their impacts and roles were broadly debated alongside thereinforcements' influences. A predominant portion was spent to analyse the factors lie in the reinforcements that influence the characteristics of the composites as a whole. The techniques adopted in fabricating a composite material were also meticulously studied. The subtleties involved in the tribological characterization were stringently investigated. The investigators followed die compaction technique for the copper alloys, since copper exceeds the 1000°C range to turn in to a molten metal and the toxic inheritances of the copper alloys largely confine them to technique of powder metallurgy.

Ceramic reinforcements like Alumina, SiC and metal oxides were preferred widely. Present day applications find the presence of carbides of tungsten, silicon and even tantalum for supporting the structural rigidity and the thermal stability they provide to the structures is an additional feature. Hussain I Alalkawi et al. [4] developed aluminum based MMC using the stir casting technique with nano alumina taking the role of reinforcement enforcing a betterment in the mechanical properties. To improve the vigor of the hybrid composites, the numbers of reinforcements with appropriate compatibilities have been increased by the investigators. The allotropic form of the element carbon in Graphite plays a vital role in MMCs, where the lubricating property and wear resistances are the key. CNT in different forms have instigated the insight of the investigators in metallurgical and research laboratories. Yahya Altunpak et al. [5] investigated for, the ease with which an aluminium matrix hybrid composites could be drilled effectively, a better drilled surface at reasonable time. Suryanarayana et al. [6] made a deliberate and a detailed discussion on the Synthesizing methods on the variety of metal and ceramic matrix using the physical mode called as Ball milling techniques. Otherwise termed Mechanical alloying. Material compacting and sintering are the conventional techniques which aid the researchers.Thoughthe hindrances due to porosity are more, the researchers stilllean towards stir casting techniques.

It has become essential for the investigators to understand the nature of the developed novel materials and to get them machined. Thyagarajan Rajmohan et al. [7] had preferred the central composite design for optimizing the drilling process Baradeswaran et al. [8] and Ke chu et al. [9] the researchers attempted to reinforce the carbon contents in form of Graphite and CNT in Al-$Al_2O_3$ and Cu-Cr composites respectively to study the wear resistive capability and the betterment achieved in the mechanical properties of the composites, and their predictions came true through the positive results. The modern day investigators know the inevitability to choose an appropriate machining process and the parameters need to be optimized, says Bhaskar Chandra Kandpal et al. [10] and Conducted a detailed review analysis of thespark erosion on Al based MMC. Preetkanwal Singh Bains et al.[11], comprehended the methods of fabricating MMCs and did Optimizations have a maximized MRR without compromising on the surface quality. Gnanavel Babu et al.[12] initiated a refinement of the AJM process of a AA6061-B4C,HBN MMC, hybrid in nature, and reported it.

## 2. Experimental method

**Table 1.** Details of Material compositions and Techniques Adopted

| S.No | Base Material | Reinforcents | % Composition | Technique adopted | Furnace & Temperatur e | Speed of the stirrer and duration |
|---|---|---|---|---|---|---|
| 1 | Aluminum Alloy 6061 (70%) | Alumina<br>Copper<br>Graphite | 12<br>9<br>9 | Planetary Ball mill ( Mechanical alloying at powder to ball ratio 10:1) 12 hrs of milling for individual elements and 8hrs for the hybrid mixture | 1000 $^0$C Graphite Crucible Preheating 500$^0$C | 250 rpm at 5 min |

Al, $Al_2O_3$, Cu, and Gr, were preheated at 500$^0$ C in a graphite crucible, for a uniform dispersion of reinforcements in to the aluminum alloy base [13]. Lighter Graphite particles have a buoyancy effect on the alumina particles. The materials were subjected to SEM analysis. The peaks values derived through XRD are similar and matches with the revelations by José Manuel Mendoza-Duarte et al. [14], to ascertain their entry in to the Nano scale regime. Table 1 provides the details of Material compositions and Techniques adopted. The plate like structures were cast with the dimensions of 100x100x10 in millimeters by adopting the Stir cast technique.

## 3. Mechanical Property of the material developed

**Table 2.**Summary of Mechanical Properties and Characterization of the composites

| Mechanical Property | Hybrid Composite Developed | Stress Vs Strain Image obtained from Instron 5900 series | SEM Image |
|---|---|---|---|
| Vickers hardness, (MPa) | 431.4 | 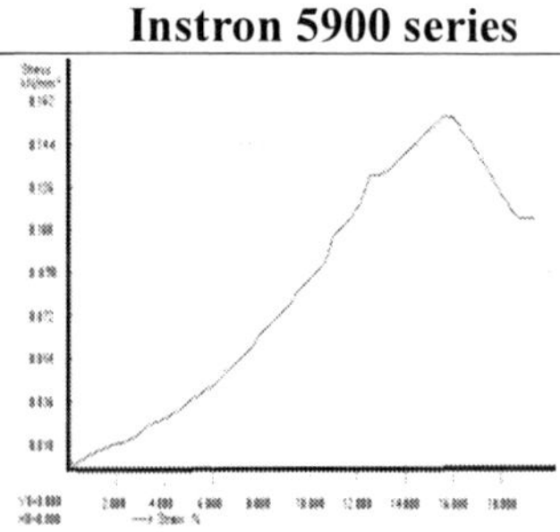 | 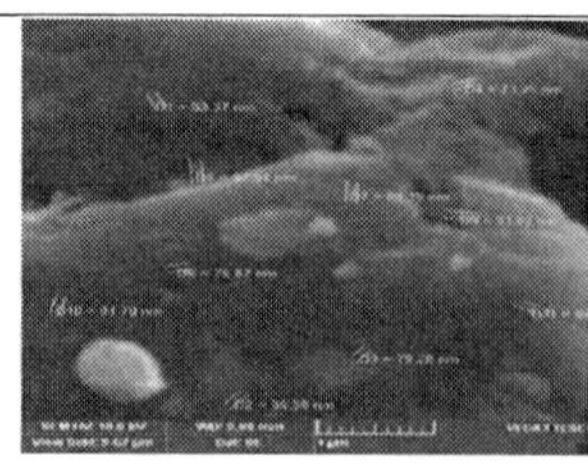 |
| Yield strength(KN/MM$^2$) | 67 | | |
| Ultimate Tensile strength(MPa) | 156 | | |

The tensile testing machine a capacity of 600kN was used to measure the ultimate tensile strength as per the ASTM standards. Gauge length = 60mm and Diameter of the gauge length zone= 12.5mm are the dimensions of the specimen prepared for the tests.

## 4. Machining

The Specification of the "Aqua cut" abrasive water jet machine that has been used for this work is shown in the Table 3.

**Table 3.**Specificationsof the Abrasive Water Jet Machining

| | | |
|---|---|---|
| X-Y cutting travel | Mm | 1575 x 1575 |
| Table Size | Mm | 2337 x 1752 |
| Material support slats | m/min | 102 x 2 |
| Max Traverse speed | m/min | 9 |
| Max Power | Hp | 30 |
| Delivery of Water | Litres/min | 3.2 |
| Max Pressure | Bar | 3800 |
| Repeatability | Mm | 0.05 |
| Orifice Diameter | Mm | 0.25 or lesser |
| Focusing Nozzle Size | Mm | 0.5 or lesser |

The input parameters and their levels are shown in Table V. Minitab software was used to develop the Experimental design to conduct the various experiments. Table IV gives the information pertaining to the experimental observations and the values of the output variables. Method of calculating the outcomes from the observations are given below [15].

MRR = Mean cutting speed × Thickness of the material in mm × Width of cut (mm)

$$MRR = V_c \times t \times b \text{ mm}^3/\text{min} \quad (1)$$

The MRR could be evaluated in Kg/min or in grams of weight, by measuring the weight before and after the machining and subsequently dividing it by the time to machine. The aforesaid formula provides the investigator the value of MRR in Volume. INSPEX, the measuring device with vision sensors was used to test the surface roughness [16] make were used to measure Kerf Width and surface quality respectively.

**Table 4.**Input parameters and their levels with ranges

| Input parameters | Level 1 | Level 2 | Level 3 |
|---|---|---|---|
| Water cutting pressure in MPa | 300 | 325 | 350 |
| Traverse Speed in mm/min | 250 | 300 | 350 |
| Abrasive flow rate, gm/min | 300 | 350 | 400 |
| Standoff distance, mm | 2 | 4 | 6 |

**Table 5.**Experimental Observations

| STD | RUN | WATER CUTTING PRESSURE | TRAVERSE SPEED | ABRASIVE FLOW RATE | SOD (Stand Of Distance) | KERF WIDTH | SURFACE ROUGHNESS | MRR (Material Removal Rate) |
|---|---|---|---|---|---|---|---|---|
| | | MPa | mm/min | grams /min | mm | mm | µm | $mm^3$/min |
| 1 | 12 | 300.00 | 250.00 | 300.00 | 2.00 | 0.669 | 5.78 | 1672.5 |
| 2 | 19 | 350.00 | 250.00 | 300.00 | 2.00 | 0.668 | 5.64 | 1670 |
| 3 | 1 | 300.00 | 350.00 | 300.00 | 2.00 | 0.668 | 5.64 | 2338 |
| 4 | 2 | 350.00 | 350.00 | 300.00 | 2.00 | 0.71 | 5.98 | 2485 |
| 5 | 8 | 300.00 | 250.00 | 400.00 | 2.00 | 0.698 | 5.98 | 1745 |
| 6 | 15 | 350.00 | 250.00 | 400.00 | 2.00 | 0.657 | 5.703 | 1642.5 |
| 7 | 29 | 300.00 | 350.00 | 400.00 | 2.00 | 0.65 | 5.53 | 2275 |
| 8 | 14 | 350.00 | 350.00 | 400.00 | 2.00 | 0.656 | 5.569 | 2296 |
| 9 | 25 | 300.00 | 250.00 | 300.00 | 6.00 | 0.712 | 5.93 | 1780 |
| 10 | 24 | 350.00 | 250.00 | 300.00 | 6.00 | 0.695 | 5.85 | 1737.5 |
| 11 | 4 | 300.00 | 350.00 | 300.00 | 6.00 | 0.657 | 6.137 | 2299.5 |
| 12 | 6 | 350.00 | 350.00 | 300.00 | 6.00 | 0.666 | 6.055 | 2331 |
| 13 | 5 | 300.00 | 250.00 | 400.00 | 6.00 | 0.699 | 5.857 | 1747.5 |
| 14 | 18 | 350.00 | 250.00 | 400.00 | 6.00 | 0.712 | 5.93 | 1780 |
| 15 | 27 | 300.00 | 350.00 | 400.00 | 6.00 | 0.692 | 6.24 | 2422 |
| 16 | 22 | 350.00 | 350.00 | 400.00 | 6.00 | 0.678 | 6.198 | 2373 |
| 17 | 21 | 300.00 | 300.00 | 350.00 | 4.00 | 0.634 | 5.23 | 1902 |
| 18 | 7 | 350.00 | 300.00 | 350.00 | 4.00 | 0.67 | 5.377 | 2010 |
| 19 | 17 | 325.00 | 250.00 | 350.00 | 4.00 | 0.621 | 5.361 | 1552.5 |
| 20 | 28 | 325.00 | 350.00 | 350.00 | 4.00 | 0.598 | 4.99 | 2093 |
| 21 | 10 | 325.00 | 300.00 | 300.00 | 4.00 | 0.611 | 4.89 | 1833 |
| 22 | 26 | 325.00 | 300.00 | 400.00 | 4.00 | 0.621 | 5.639 | 1863 |
| 23 | 9 | 325.00 | 300.00 | 350.00 | 2.00 | 0.659 | 5.676 | 1977 |
| 24 | 13 | 325.00 | 300.00 | 350.00 | 6.00 | 0.721 | 6.07 | 2163 |
| 25 | 3 | 325.00 | 300.00 | 350.00 | 4.00 | 0.611 | 4.867 | 1833 |
| 26 | 11 | 325.00 | 300.00 | 350.00 | 4.00 | 0.611 | 4.89 | 1833 |
| 27 | 30 | 325.00 | 300.00 | 350.00 | 4.00 | 0.605 | 4.87 | 1815 |

| | | | | | | | | |
|---|---|---|---|---|---|---|---|---|
| 28 | 20 | 325.00 | 300.00 | 350.00 | 4.00 | 0.611 | 4.89 | 1833 |
| 29 | 23 | 325.00 | 300.00 | 350.00 | 4.00 | 0.611 | 4.85 | 1833 |
| 30 | 16 | 325.00 | 300.00 | 350.00 | 4.00 | 0.611 | 4.89 | 1833 |

## 5. Result and Discussion

Design Expert Software has been used to find out the relationship between the parametric combinations and their influences in the output variables. The relationships were obtained and shown in the form of graphs. The graphical representations gives a concrete idea. The analysis of Variance have been done to ascertain the role and significance of certain input variables and they have been discussed in detail in this chapter [17]. The abbreviations used in this paper are: WP-Water Cutting Pressure, AFR- Abrasive flow rate, SOD- Stand Off Distance, TS-Traverse Speed.[18].

### *5.1. KERF WIDTH*

**Table 6.**ANOVA for KERF width

| **Source** | **Sum of Squares** | **Df** | **Mean Square** | **F Value** | **p-value Prob > F** | |
|---|---|---|---|---|---|---|
| Model | 0.037 | 14 | 2.662E-003 | 8.13 | 0.0001 | Significant |
| A-WCP | 6.050E-0051 | 1 | 6.050E-005 | 0.18 | 0.6734 | |
| B-TS | 1.352E-003 | 1 | 1.352E-003 | 4.13 | 0.0602 | |
| C-AFR | 2.722E-006 | 1 | 2.722E-006 | 8.317E-003 | 0.9285 | |
| D-SOD | 2.156E-003 | 1 | 2.156E-003 | 6.59 | 0.0215 | |
| AB | 4.951E-004 | 1 | 4.951E-004 | 1.51 | 0.2377 | |
| AC | 2.976E-004 | 1 | 2.976E-004 | 0.91 | 0.3555 | |
| AD | 1.406E-005 | 1 | 1.406E-005 | 0.043 | 0.8386 | |
| BC | 1.381E-004 | 1 | 1.381E-004 | 0.42 | 0.5259 | |
| BD | 8.556E-004 | 1 | 8.556E-004 | 2.61 | 0.1268 | |
| CD | 6.891E-004 | 1 | 6.891E-004 | 2.11 | 0.1674 | |
| $A^2$ | 1.749E-003 | 1 | 1.749E-003 | 0.3869 | 0.0354 | |
| $B^2$ | 7.069E-004 | | 7.069E-004 | 2.16 | 0.1623 | |
| $C^2$ | 2.600E-004 | | 2.600E-004 | 0.79 | 0.3869 | |
| $D^2$ | 0.011 | 0.011 | 32.40 | < 0.0001 | | |
| Residual | 4.910E-003 | 15 | 3.273E-004 | | | |
| Lack of Fit | 4.880E-003 | 10 | 4.880E-004 | 81.33< 0.0001 | significant | |
| Pure Error | 3.000E-005 | 5 | 6.000E-006 | | | |
| Cor Total | 0.042 | 29 | | | | |

The Model F-value is significant because the value of 8.13 justifies it and a 0.01% chance prevails for the F-Value" could occur tobe larger due to noise. The model terms are significant due to Values of "Prob > F" goes below 0.005. In this case D, A2, D2 are significant model terms, which is being supported by the value of 0.1000.

IOP Conf. Series: Materials Science and Engineering **923** (2020) 012037 doi:10.1088/1757-899X/923/1/012037

*5.1.1.Final Equation in Terms of Actual Factors*

KERFWIDTH=+4.08085-0.027001*WATERCUTTINGRESSURE+3.04838E-003*TRAVERSESPEED+4.02394E-003 * ABRASIVE FLOW RATE-0.11743 *STANDOFF DISTANCE+4.45000E-006 * WATER CUTTING PRESSURETRAVERSESPEED-3.45000E-006 * WATER CUTTING PRESSURE * ABRASIVE FLOW RATE-1.87500E-005 * WATER CUTTING PRESSURE * STAND OFF DISTANCE-1.17500E-006 * TRAVERSE SPEEDABRASIVEFLOWRATE-7.31250E-005 *TRAVERSESPEED*STANDOFFDISTANCE+6.56250E-005 * ABRASIVE FLOW RATE * STAND OFF DISTANCE+4.15719E-005 * WATER CUTTING PRESSURE$^2$-6.60702E-006*TRAVERSESPEED$^2$-4.00702E-006*ABRASIVEFLOWRATE$^2$+0.015996*STANDOFFDISTANCE$^2$ (2)

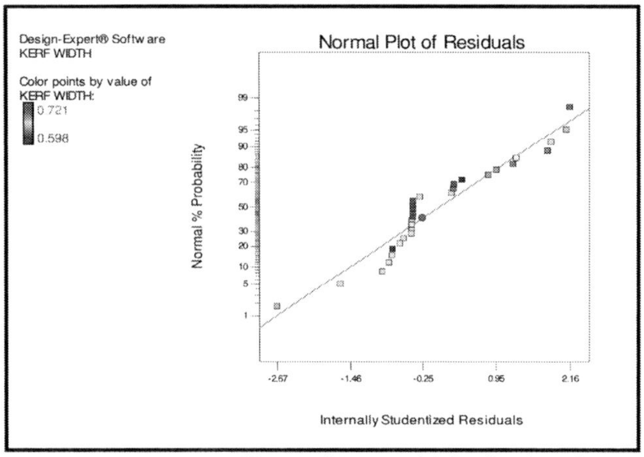


**Figure 1.**Normal Distribution Curve for KERF WIDTH

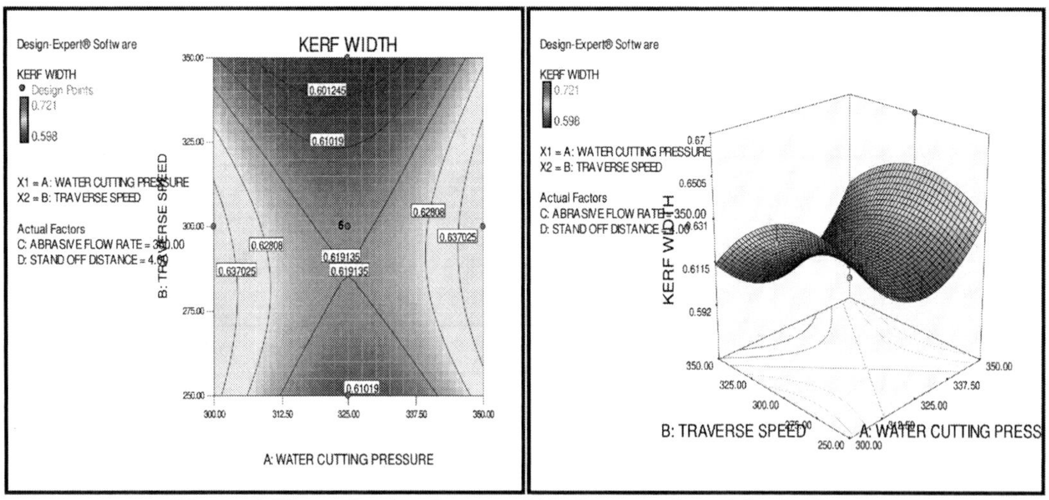


**Figure 2.**Water Cutting Pressure VS Traverse Speed for KERF WIDTH

The water cutting pressure and Traverse speed have no influence on the KERF WIDTH. The width becomes minimum at moderate pressure and traverse speed. The abrasive flow rate goes up with the pressure and water pressure. The higher speed and pressure erodes the material at a uniform rate and maintains a uniform width. Higher pressure and velocity provides the needed kinetic energy and the thrust force to reduce the material a unifrom measure. This is also due to the diameter of the jet which maintains a constant and a uniform cross sectional area. The turbulance of the jet erodes the material more at moderate traverse speed and water pressure where as the low pressure and speed reduces the required force to remove the material. The distribution curve shown in Figure 1. shows the values lie along the mean line and Figure 2. shows the relation ship between WP and TS.

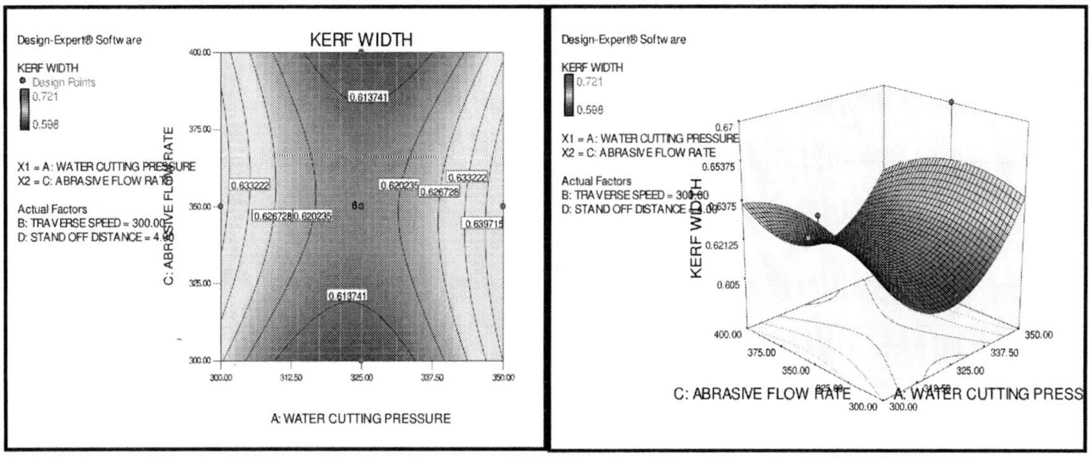


**Figure 3.**Water Cutting Pressure VS Abrasive Flow Rate for KERF WIDTH

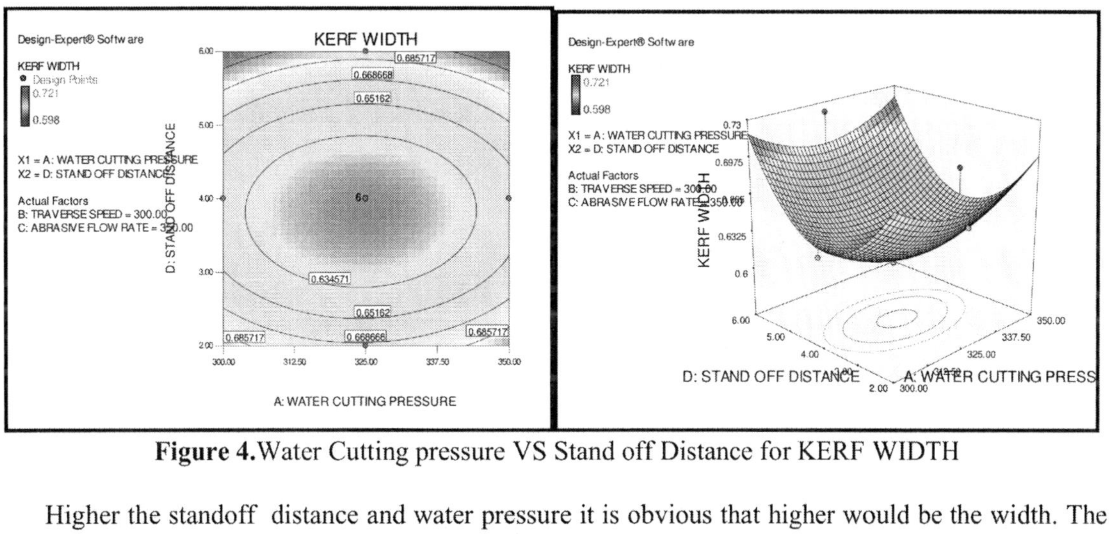


**Figure 4.**Water Cutting pressure VS Stand off Distance for KERF WIDTH

Higher the standoff distance and water pressure it is obvious that higher would be the width. The moderate standoff distances reduces the kerfwidth. Lower pressure and lower stand off distance also keeps the kerfwidth at a maximum. The pressure and kinetic energy plays an important role at higher input process variables whereas the potential energy takes its stake at lower values to maintain the width to be maximum. The moderate values at the middle level may be due to the deviation of the variables from the mean line as shown in the Normal plot. The Figure 3., Figure 4., and Figure 5. explains the role palyed by the parameters WP, AFR and SOD. The graphs clearly explain the parametric roles.

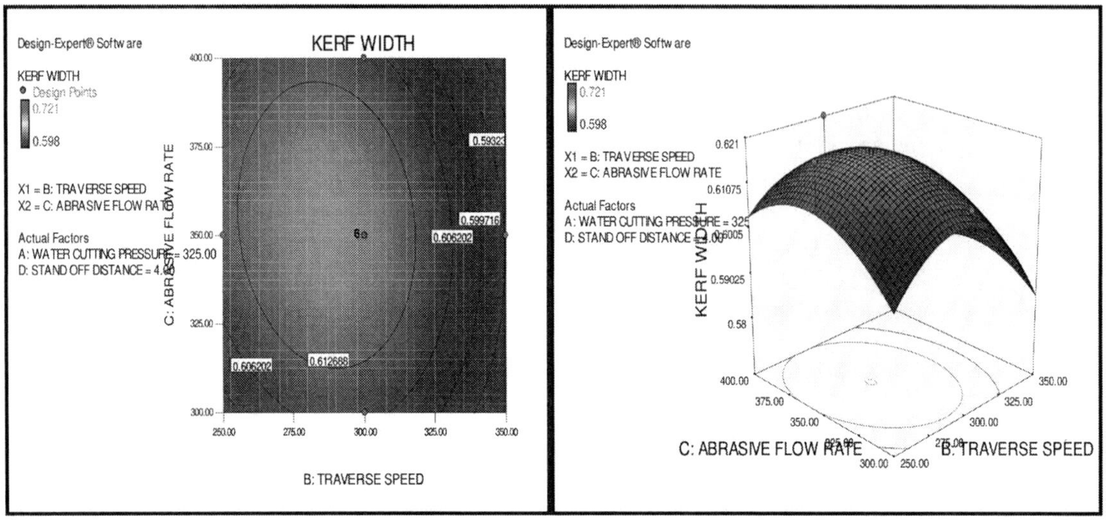


**Figure 5.**Traverse Speed VS Abrasive Flow Rate for KERF WIDTH

The combination of Abrasive flow rate at minimum Traverse speed gives a maximum kerfwidth. More the amount of abrasives that flow in and hit the work piece at lower transverse speeds give ample time for the abrasive materials to reduce the material and maintain the higher width of cut. The lower abrasive flow rates at lower traverse speeds fail to increase the kerf width since the volume of abrasive cutting fluid dips. The interactive effects of water pressure and Standoff distance play a significant role and influence the change in kerfwidth.

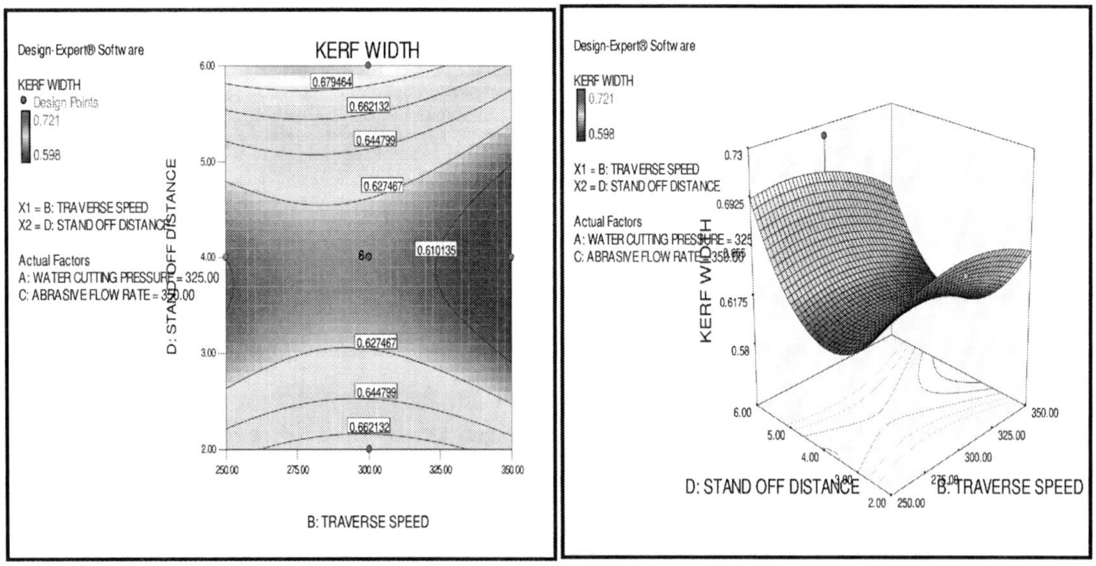


**Figure 6.**Traverse Speed VS Standoff Distance for KERF WIDTH

The higher standoff distance helps to have a greater width irrespective of traverse speed. The value of kerfwidth comes down with the decrease in standoff distance. The Figure 6 and the Figure 7 explain the SOD Vs TS and SOD Vs AFR respectively pertaining to Kerfwidth.

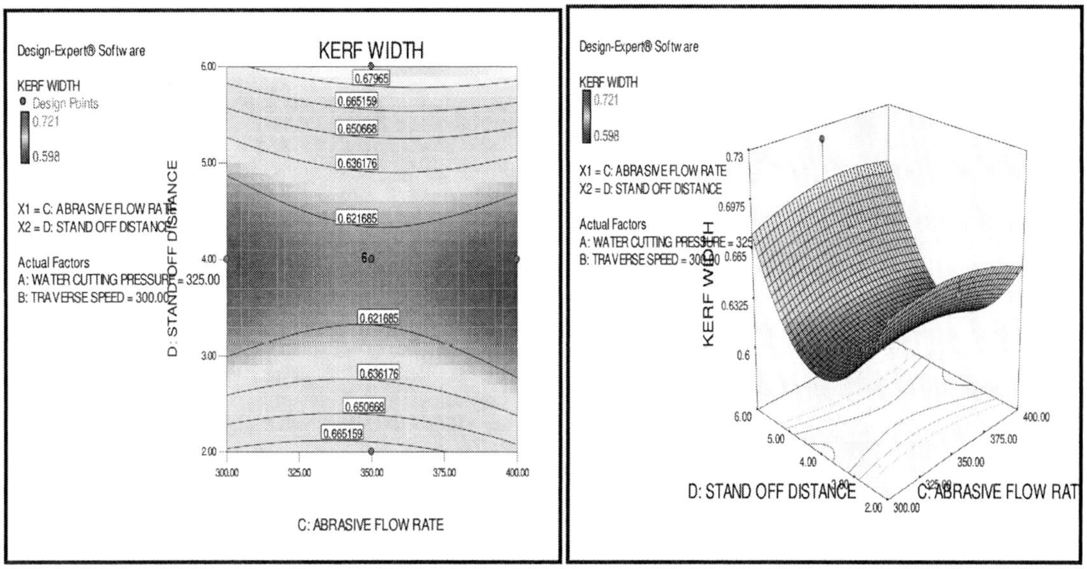


**Figure 7.**Abrasive Flow Rate VS Standoff Distance for KERF WIDTH

### *5.2.SURFACE ROUGHNESS*

**Table 7.**ANOVA for Surface roughness

| Source | Sum of Squares | df | Mean Square | F Value | p-value Prob > F | |
|---|---|---|---|---|---|---|
| Model | 5.51 | 14 | 0.39 | 6.91 | 0.0003 | Significant |
| A-WCP | 2.689E-005 | 1 | 2.689E-005 | 4.720E-004 | 0.9830 | |
| B-TS | 5.270E-003 | 1 | 5.270E-003 | 0.093 | 0.7652 | |
| C-AFR | 0.031 | 1 | 0.031 | 0.54 | 0.4739 | |
| D-SOD | 0.43 | 1 | 0.43 | 7.48 | 0.0154 | |
| AB | 0.029 | 1 | 0.029 | 0.51 | 0.4879 | |
| AC | 3.752E-003 | 1 | 3.752E-003 | 0.066 | 0.8010 | |
| AD | 5.406E-004 | 1 | 5.406E-004 | 9.488E-003 | 0.9237 | |
| BC | 0.019 | 0.019 | 0.33 | 0.5766 | | |
| BD | 0.13 | 1 | 0.13 | 2.30 | 0.1504 | |
| CD | 0.016 | 1 | 0.016 | 0.29 | 0.6003 | |
| $A^2$ | 0.034 | 1 | 0.034 | 0.60 | 0.4491 | |
| $B^2$ | 4.215E-004 | 1 | 7.398E-003 | | 0.9326 | |
| $C^2$ | 0.015 | 1 | 0.015 | 0.26 | 0.6146 | |
| $D^2$ | | 1 | 1.21 | 21.32 | 0.0003 | |
| Residual | 0.85 | 15 | 0.057 | | | |
| Lack of Fit | 0.85 | 10 | 0.085 | 308.94 | < 0.0001 | Significant |
| Pure Error | 1.381E-003 | 5 | 2.762E-004 | | | |
| Cor Total | 6.37 | 29 | | | | |

The Model F-value is significant because the value of 6.91 justifies it and a 0.03% chance prevails for the F-Value" could occur tobe larger due to noise. The model terms are significant due to Values of "Prob > F" goes below 0.005. In this case D, $D^2$ are significant model terms, which is being supported by the value of 0.1000.

*5.2.1.Final Equation in Terms of Actual factors*

SURFACE ROUGHNESS =+31.70811-0.12534* WATER CUTTING PRESSURE -6.47923E-003* TRAVERSE SPEED-0.013731 * ABRASIVE FLOW RATE-1.63789* STAND OFF DISTANCE+3.39500E-005 * WATER CUTTING PRESSURE * TRAVERSE SPEED-1.22500E-005 * WATER CUTTING PRESSURE * ABRASIVE FLOWRATE-1.16250E-004 * WATER CUTTING PRESSURE * STAND OFF DISTANCE-1.36250E-005 * TRAVERSE SPEED * ABRASIVE FLOW RATE+9.04375E-004 * TRAVERSE SPEED * STAND OFF DISTANCE+3.19375E-004* ABRASIVE FLOW RATE * STAND OFF DISTANCE+1.84393E-004*WATERCUTTINGPRESSURE$^2$-5.10175E-006*TRAVERSESPEED$^2$+3.04982E-005*ABRASIVEFLOWRATE$^2$+0.17119*STANDOFFDISTANCE$^2$ (3)

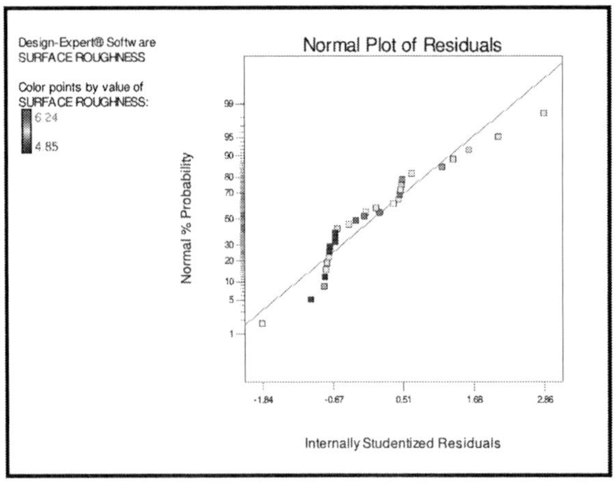


**Figure 8.**Normal Distribution Curve for Surface Roughness

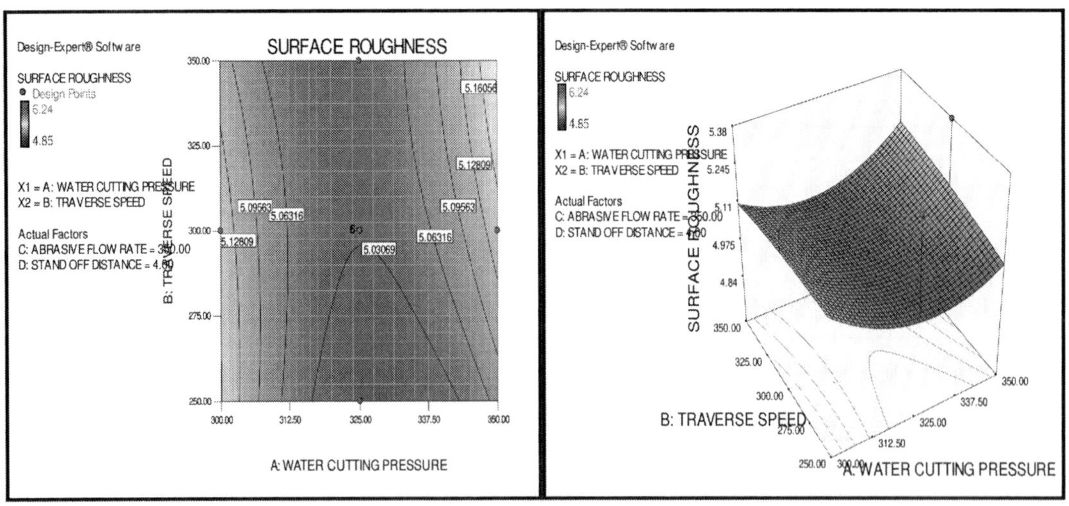


**Figure 9.**Water Cutting Pressure VS Traverse Speed for Surface Roughness

The water cutting pressure and Traverse speed maintains the value of surface roughness constant and uniform, The roughness value takes an edge upwards at higher values of speed and pressure. The combinatorial effects of traverse speed and water pressure works in tandem to make the interaction with the surface at a uniform fashion. The pressure and speed joins together to maintain a very good surface quality. The abrasive particles erode the workpiece at a proportional variation in the joint effort of kinetic and pressure energies.

The abrasive flow rate and water pressure also provide a proportional variation with respect to surface roughness. The intensity of abrasive particles at a proportionately intensified water pressure maintains a smooth exit of materials from the bulk so as to ensure a better and uniform surface finish. This parametric combination has no distictive difference or variation in achieving the surface finish. The sharp edges of the abrasive particles at a given pressure removes the material that ascertains the waviness to be uniform and stable.

Figure 8 gives the normal plot where the values of the control variables lie within the predictable lines and follow the mean line. Figure 13, Figure 14 and Figure 15 explain the interactions between them as well as the role played by the individual parameter in establishing a better surface roughness.

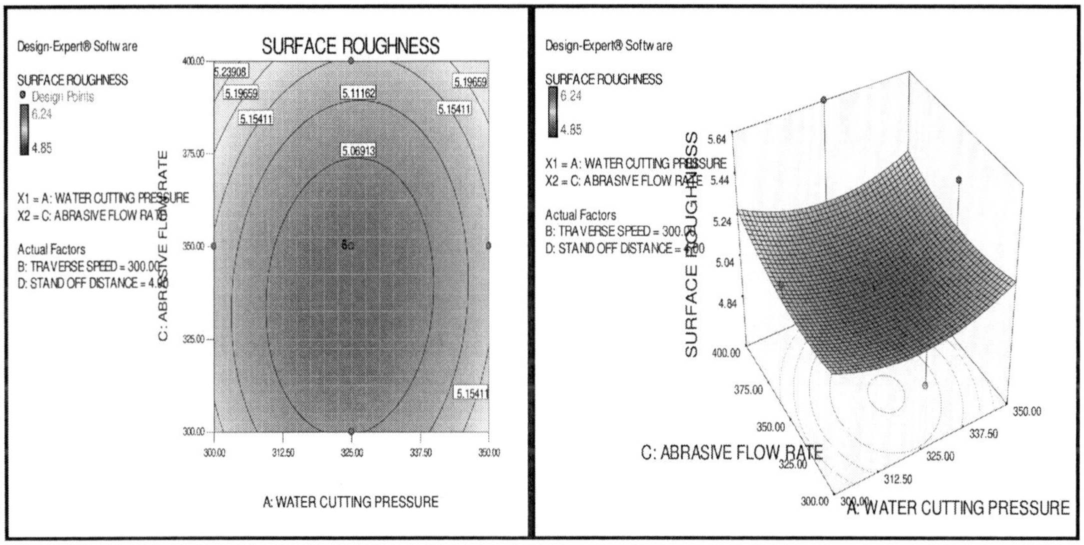


**Figure 10.**Water Cutting Pressure VS Abrasive Flow Rate for Surface Roughness

The water cutting pressure and stanoff distances have an explicit effects on surface roughness. Higher stand off distances increases the surface roughness at higher pressures. The thudding force with which the abrasive particles penetrate the surface and waekens the cohesive forces that bind the base material and the reinforcements. The imbalance created by the hard abrasive particles at higher pressure disrupts the uniformity of the materials layer at the surface and afects the surface. The lower pressure and stand off distance lag the needed impact force of penetration. The surface enegy that helps to deplete the outer layer goes down at the lower stand off distance and during the lower water cutting pressures.

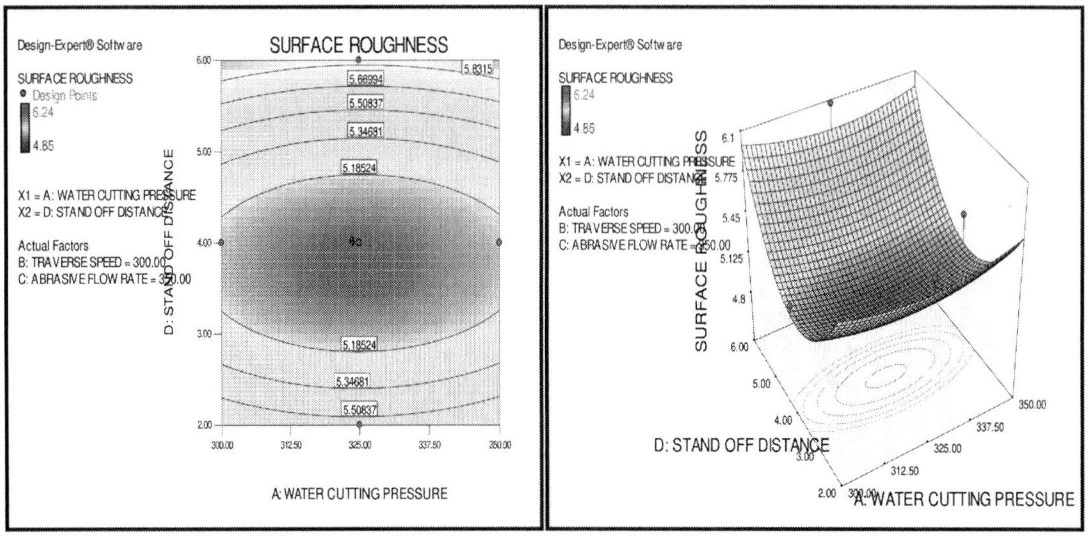


**Figure 11.**Water Cutting pressure VS Stand off Distance for Surface Roughness

Traverse speeds and abrasive flow rates jointly make no appreciable impact in the surface roughness. Whatever be their values they maintain a uniform surface finish. The traverse speed pushes the water jet along with it at almost at the similar and equivalent flow rate and the surface roughness remains unaltered. The variation adopted by these parameters simultaneous have no tellingimpact on the surface quality.

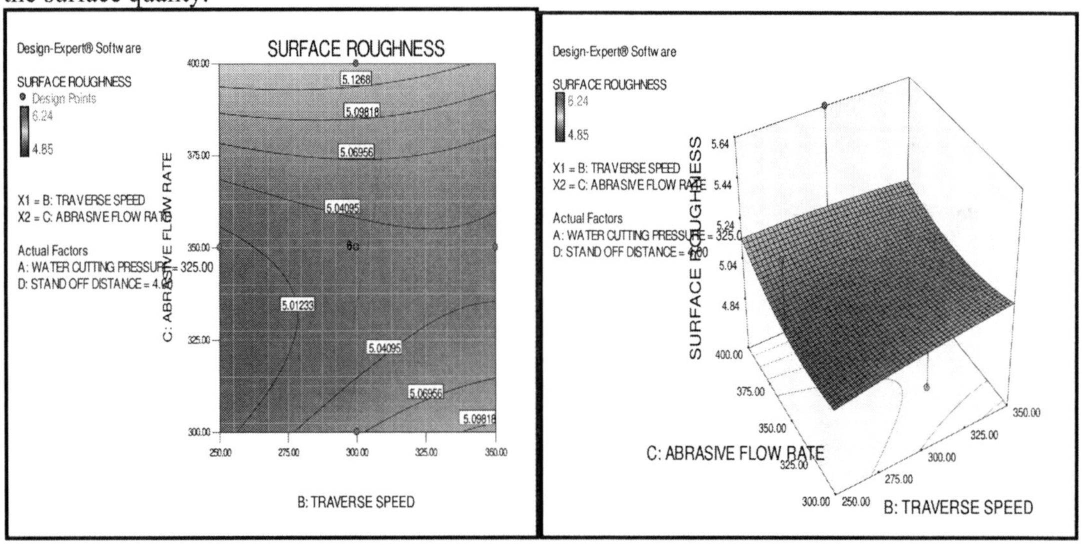


**Figure 12.**Traverse Speed VS Abrasive Flow Rate for Surface Roughness

The effects of traverse speed and standoff distance also give us an interesting understanding, more the stand off distance and traverse speed the surface roughness shoots up. The propellent abrasive jet from the nozzle if takes alonger distances to reach the target material and at the same time if the table traverse is also faster, ther is a greater posibility of material scatter along the surface. There would be a lack of cutting force developed along the surfaces, the speed with which the nozzle or the table is moved, if they fail to take up the impact then the uniformity of removal comes down and it would have a telling impact on the surface quality.

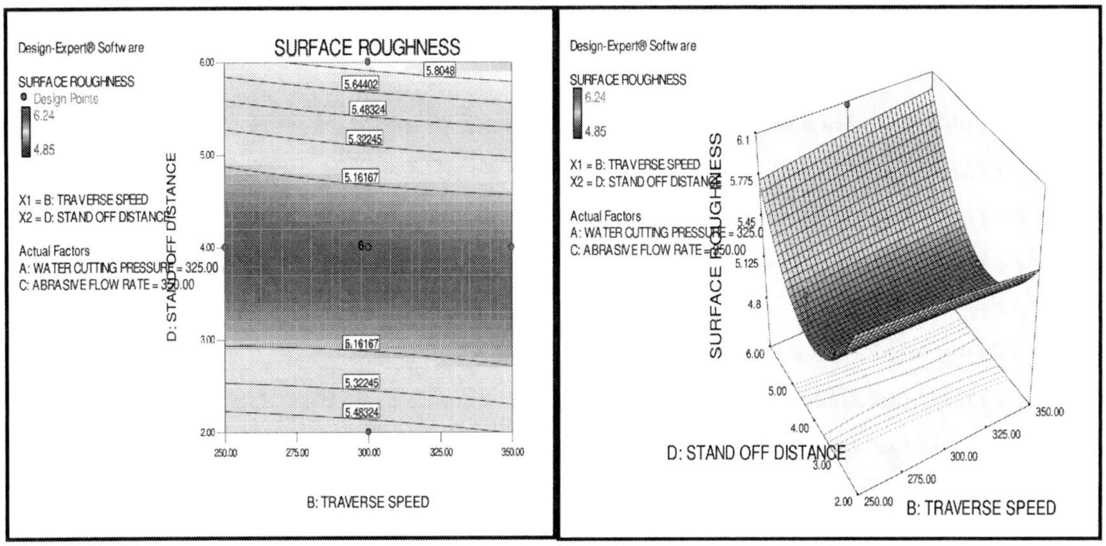


**Figure 13.**Traverse Speed VS Standoff Distance for Surface Roughness

The aforesaid effect is quite similar to the relationship between the traverse speed and standoff distance. The surface roughness values are greater at higher values of these two parameters. More the rate of abrasive particles from a greater height, the volume of materials with needed hardness falling on the work piece material is more. The no of pointed edges of the particles try to remove or damage the already machined surface. The frequency of the abrasive particle coming in to contact with the surface also increases abruptly there by affecting the surface quality. In a way these two combinations may have a reasonable impact in material removal rate as well.

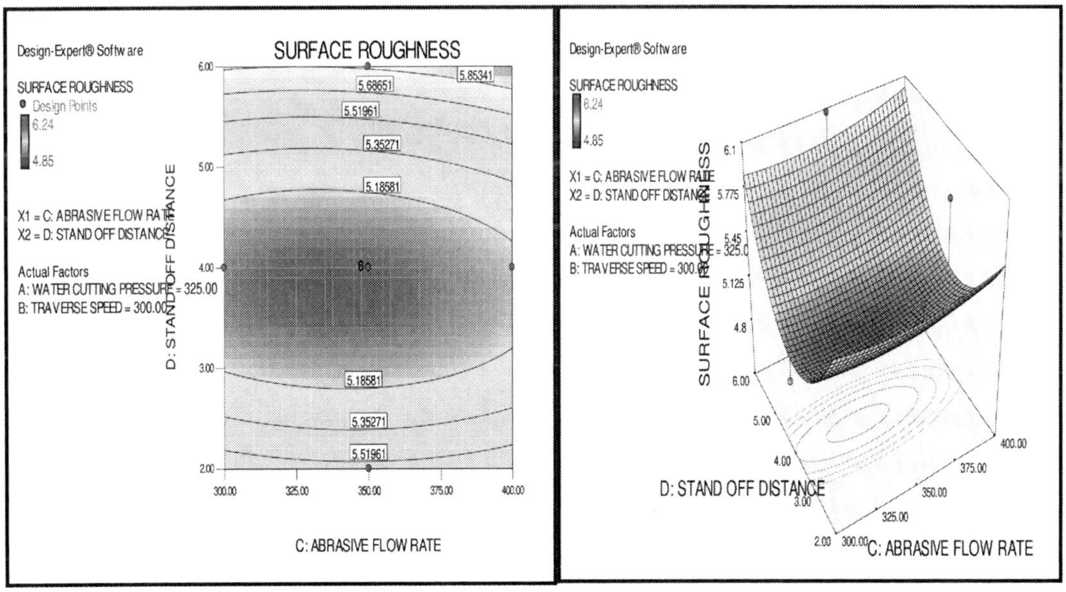


**Figure 14.** Abrasive Flow Rate VS Standoff Distance for Surface Roughness

IOP Conf. Series: Materials Science and Engineering **923** (2020) 012037 doi:10.1088/1757-899X/923/1/012037

*5.3.MATERIAL REMOVAL RATE*

**Table 8.** ANOVA for Material Removal Rate

| Source | Sum of Squares | df | Mean Square | F Value | p-value Prob > F |
|---|---|---|---|---|---|
| Model | 2.048E+006 | 14 | 1.463E+00543.81 | <0.0001 | Significant |
| A-WCP | 1144.01 | 1 | 114.01 | 0.34 | 0.5670 |
| B-TS | 1.733E+006 | 1 | 0.35 | 1.040E-004 | 0.9920 |
| C-ABR | 0.35 | 1 | 0.35 | 1.04 E-004 | 0.9920 |
| D-SOD | 15753.13 | 1 | 15753.13 | 4.72 | 0.0463 |
| AB | 4405.64 | 1 | 4405.64 | 1.32 | 0.2687 |
| AC | 3349.52 | 1 | 3349.52 | 1.00 | 0.3324 |
| AD | 511.89 | 1 | 511.89 | 0.15 | 0.7009 |
| BC | 1269.14 | 1 | 1269.14 | 0.38 | 0.5468 |
| BD | 5023.27 | 1 | 5023.27 | 1.50 | 0.2389 |
| CD | 9096.39 | 1 | 9096.39 | 2.72 | 0.1196 |
| $A^2$ | *15923.04* | 1 | 15923.04 | 4.77 | 0.0453 |
| $B^2$ | *7796.30* | 1 | 7796.30 | 2.33 | 0.1473 |
| $C^2$ | *2270.86* | 1 | 2270.86 | 0.68 | 0.4225 |
| $D^2$ | *95904.40* | 1 | 95904.40 | 28.72 | <0.0001 |
| *Residual* | *50083.82* | 15 | 3338.92 | | |
| *Lack of Fit* | *49813.82* | 10 | 4981.3892.25 | <0.000z | Significant |
| *Pure error* | *270.00* | 5 | 54.00 | | |
| *Cor Total* | *2.098E+006* | 29 | | | |

The Model F-value is significant because the value of 43.81 justifies it and a 0.01% chance prevails for the F-Value" could occur tobe larger due to noise. The model terms are significant due to Values of "Prob > F" goes below 0.005. In this case B, D, $A^2$, $D^2$ are significant model terms, which is being supported by the value of 0.1000.

*5.3.1.Final Equation in Terms of Actual Factor*

MATERIAL REMOVAL RATE =+10000.86349-80.69039 * WATER CUTTING PRESSURE+17.01207 * TRAVERSE SPEED+12.16357 * ABRASIVE FLOW RATE-363.52906* STAND OFF DISTANCE+0.013275 * WATER CUTTING PRESSURE* TRAVERSESPEED-0.011575 * WATER CUTTING PRESSURE * ABRASIVE FLOW RATE-0.11312 * WATER CUTTING PRESSURE * STAND OFF DISTANCE-3.56250E-003* TRAVERSE SPEED * ABRASIVE FLOW RATE-0.17719 *TRAVERSESPEED* STAND OFF DISTANCE+0.23844 * ABRASIVE FLOW RATE *STAND OFF DISTANCE +0.12543 * WATER CUTTING PRESSURE$^2$-0.021942 * TRAVERSE SPEED$^2$-0.011842* ABRASIVE FLOWRATE$^2$+48.09868*STANDOFFDISTANCE$^2$ (4)

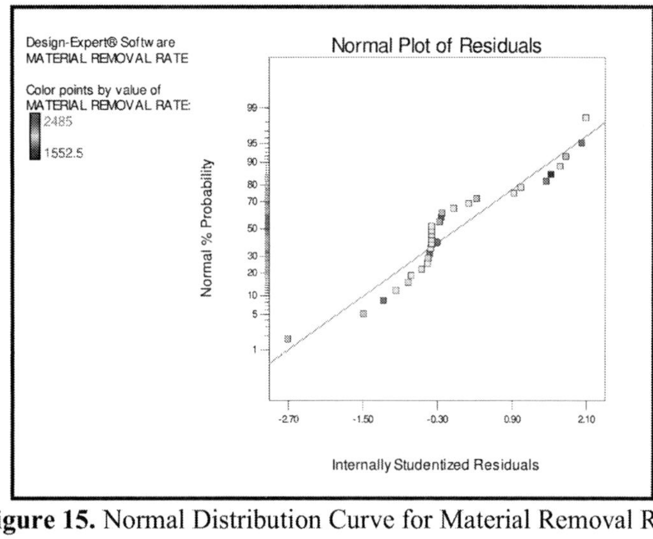


**Figure 15.** Normal Distribution Curve for Material Removal Rate

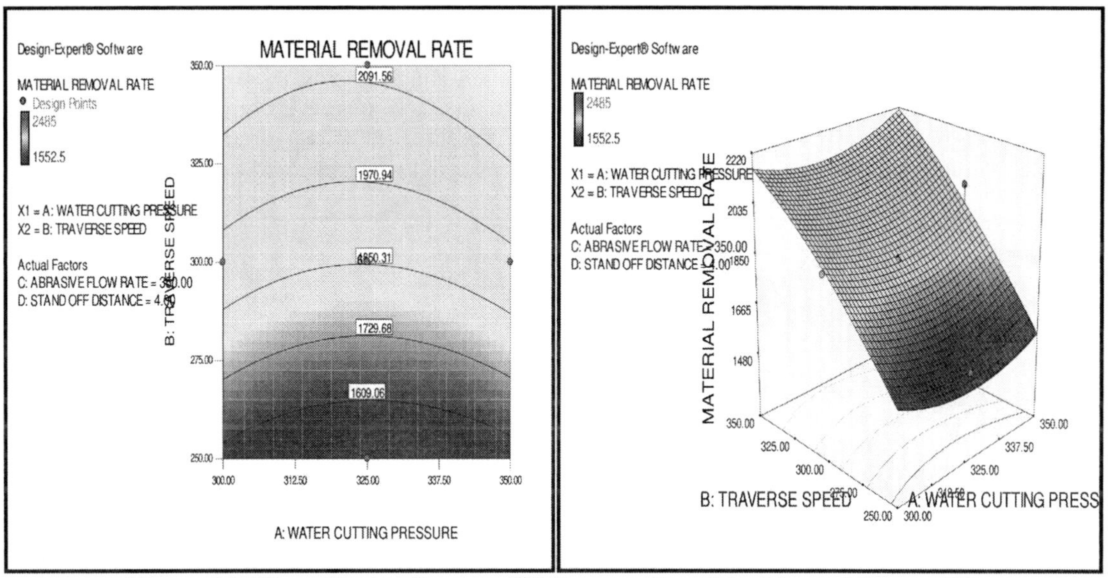


**Figure 16.** Water Cutting Pressure VS Traverse Speed for Material Removal Rate

From the diagram it is clear that the higher cutting pressure and higher traverse speed removes material in large quantitites. The pressure energy removes the material, the abrasive particles with a greater pressure hits the material's surface they distrub the surface tension prevails in the materials. The more the disuption of the surface with pressurized fluid carrying the abrasive particles, the more material would be dislodged from the surface and ultimate the MRR takes a steep ride. The phenomenon mentioned above is vice-versa at the lower value combinations of the control parameters.

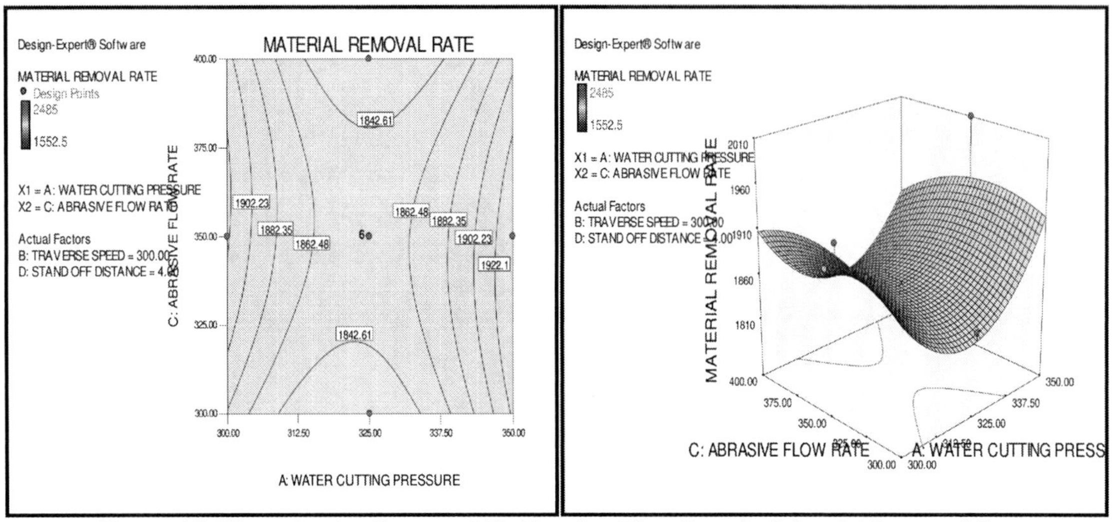


**Figure 17.** Water Cutting Pressure VS Abrasive Flow Rate for Material Removal Rate

The water cutting pressure and abrasive flow rate if increases the material removal rate would also increase, this combination is effective even at the lowervalues of these control parameters. At higher values the pressurized abrasive slurry removes more amount of material, the high pressure and discharge of cutting particles ease the material reduction and increases the volume of material removal. Where as even at lower dischare and at lower pressure, the MRR is more since the abrasives keeps with it the inherent forces and aids the process of material removal.

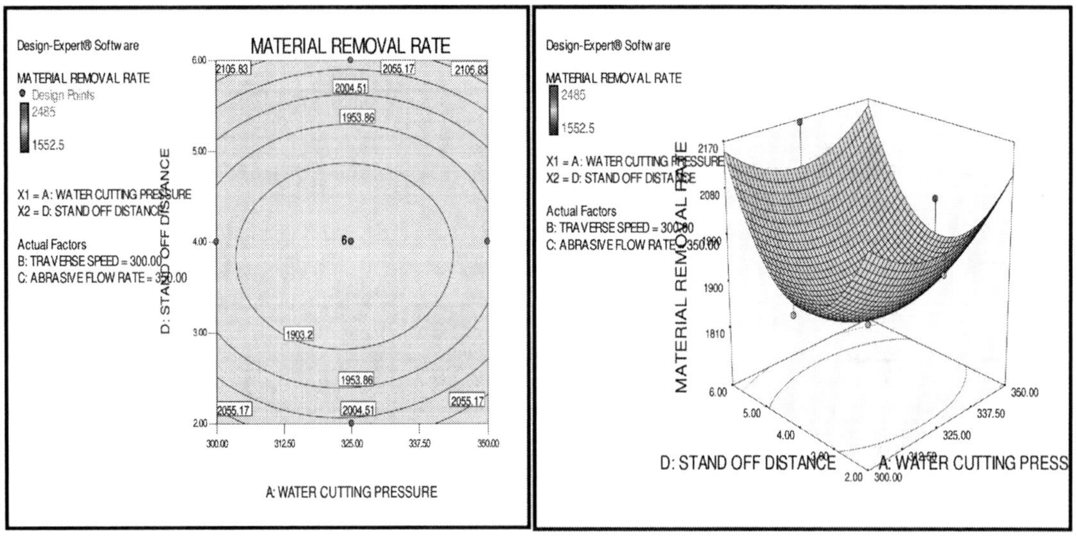


**Figure 18.** Water Cutting pressure VS Stand off Distance for Material Removal Rate

The standoff distance and pressure removes the material at a faster rate irrespective of their levels of performance. The reason for the high material removal at the higher values is a valid guess of any machinists, whereas even at lower values the MRR is high as well. The lower pressure and low standoff distance would give a better MRR only when the traverse speed is low. At lower tranverse speeds, there is a more than sufficient time for the cuting material to disrupt the workpiece material. Lower stand off distances moreover require a higher pressure to remove the material, since the

IOP Conf. Series: Materials Science and Engineering **923** (2020) 012037 doi:10.1088/1757-899X/923/1/012037

quantity of abrasives present in the abrasive water jet would be more and cohesive so as to create an imact. If the abrasive slurry falls from greater heights the cohesiveness loosenes up and the impact would be not of a consequential one.

Figure 15 shows the deviations of the process variables from the mean line whereas the Figure 16, Figure 17, Figure 18 and Figure 19 shows the role WP,TS, AFR and SOD in the pursuit of a better MRR.

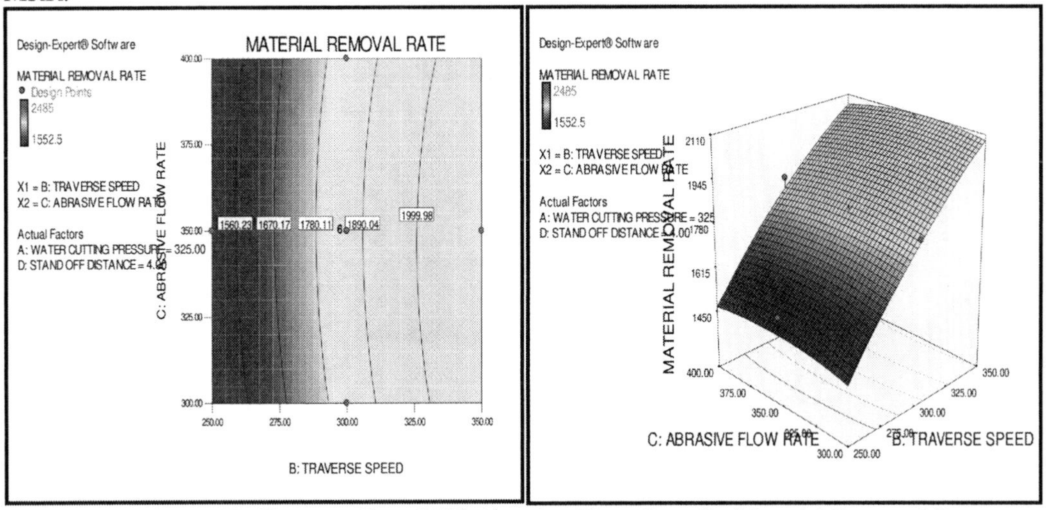


**Figure 19.** Traverse Speed VS Abrasive Flow Rate for Material Removal Rate

The relationship with the Traverse speed and Abrasive flow rate has similar consequences to traverse speed and cutting pressure. The jet with high discharge abrasive slurry acts like a solid cutting tool; if it is traversed across the material then the resulting material removal would be high. The velocity of the table or the nozzle and discharge creates a greater rate of wear and erosion along the edges of the work piece there by provides a higher MRR. At lower speed and lower abrasive flow rate the MRR becomes low because there is an interaction with other parameters like low water pressure and lower standoff distance.

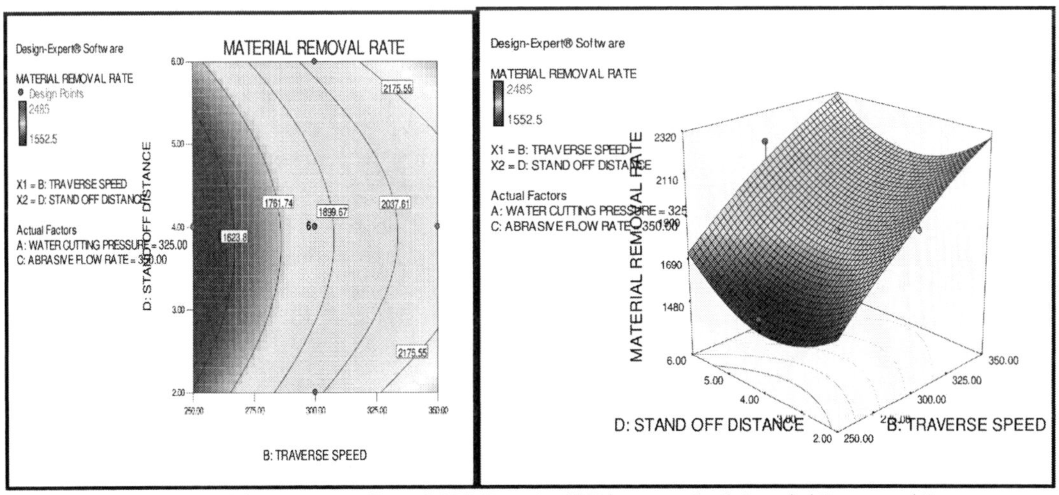


**Figure 20.** Traverse Speed VS Stand off Distance for Material Removal Rate

The traverse speed and standoff distance increases the MRR with their respective incremental values. Because of the virtual position of the nozzle, the energy woud be higher and at higher traverse speeds if the abrasive slurry impacts the workpiece, it creates the phenomenon quite similar to that of cutting the workpieces with a sharp edged tool. The high velocity jet with loads of energy impinges on the material traversing with substantial velocity would providegreater MRR.

The increased abrasive flow rate and higher stand off distance increase the MRR.If the much energized abrasive material in sufficient volume falls over the workpiece material for a period of time,then it creates a much needed disruption and disloges the material in higher volumes, by weakening up the bonds already formed within the bulk material.

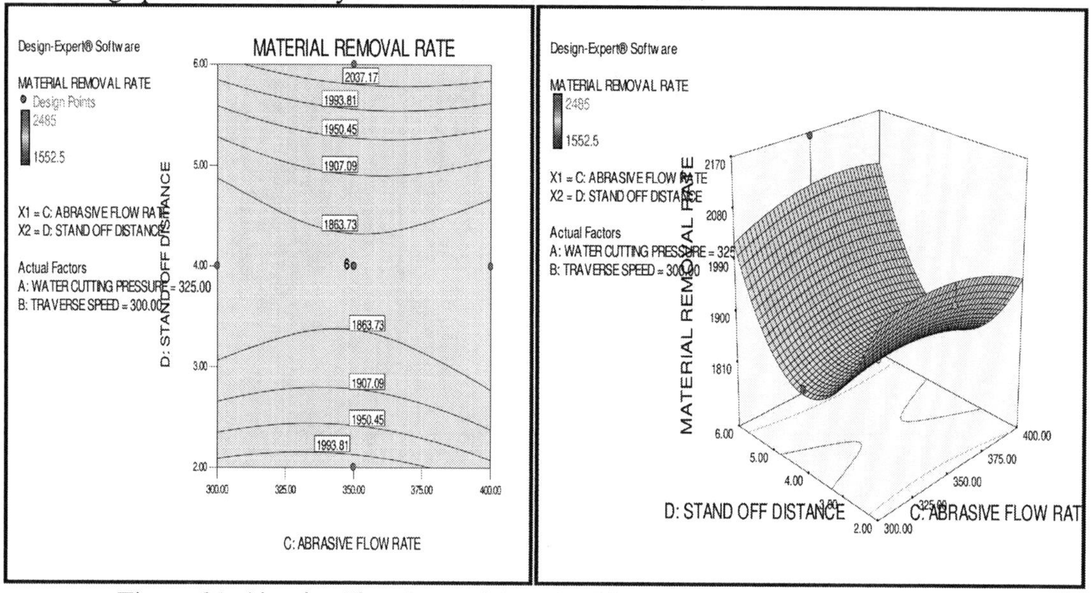


**Figure 21.** Abrasive Flow Rate VS Stand off Distance for Material Removal Rate

## 6. CONCLUSION

The hybrid material made of Aluminium Al6061 alloy(70%) as the base or matrix material in the composite reinforced with hybrid nano materials as the particulate reinforcements (Cu-9%,Gr-9% and $Al_2O_3$-12%) were prepared by following the stir casting route, sidelining the rule of mix. Good mechanical characteristics and resistance to the adverse rubbing consequence are part of material cast as MMC. Central composite design of RSM comprehends the effects and impacts of control parameters meticulously over the machining process. The interaction between individual process parameters and their combinatorial effects have been extensively scrutinized and put in a nutshell in this report. Lower the control variables viz., standoff distance, Abrasive flow rate, Traverse speed and water pressure, better is the surface quality. The impact created by the lower process control variable over the surface roughness is appreciable. Higher the input variables viz., traverse speed, water pressure, standoff distance and Abrasive flow rate, higher is the Material Removal Rate.

## 7. REFERENCES

[1] Michael Oluwatosin 2015, Aluminium matrix hybrid composites: a review of reinforcement philosophies;Mechanical, corrosion and tribological characteristics, jmaterrestechnology.2015; 4(4):434–445,DOI: 10.1016/j.jmrt.2015.05.003

[2] T.Rajmohan 2014, Synthesis and characterization of sintered hybrid aluminium matrix compositesreinforced with nanocopper oxide particles and microsilicon carbide particles, Composites: Part B 59(2014) 43–49 DOI: 10.1016/j.compositesb.2013.10.060

[3] Mohsen Hossein-Zadeh, Characterization of properties of Al–Al2O3 nano-composite synthesized via milling and subsequent casting, Journal of King Saud University – Engineering Sciences (2013) 25, 75–80,DOI:10.1016/j.jksues.2012.03.001
[4] Hussain I Alalkawi 2018, Mechanical Properties of Al 6061/Al2O3 Nanoco*mposite by Stir Casting,Engineering and Technology Journal, Vol. 36, Part A, No. 3, 2018, DOI: 10.30684/etj.36.3A.12
[5] Yahya Altunpak 2012, Drilling of a hybrid Al/SiC/Gr metal matrix composites, Int J Adv Manuf Technology,(2012) 60:513–517 DOI: 10.1007/s00170-011-3644-4
[6] C.Suryanaraya 2013, Mechanically alloyed composites, Progress in Materials Science 58 (2013) 383–502,DOI: 10.1016/j.pmatsci.2012.10.001
[7] Thyagarajan Rajmohan, 2013, Application of the central composite design in optimization of machiningparameters in drilling hybrid metal matrix composites, Measurement 46 (2013) 1470–1481,DOI: 10.1016/j.measurement.2012.11.034
[8] A.Baradeswar, 2013, Study on mechanical and wear properties of Al 7075/Al2O3/graphite hybrid Composites, Composites: Part B 56 (2014) 464–471 DOI:10.1016/j.compositesb.2013.08.013
[9] Khe Chu, 2013, Improvement of interface and mechanical properties in carbon nanotube reinforced Cu–Crmatrix composites, Materials and Design 45 (2013) 407–411 DOI: 10.1016/j.matdes.2012.09.027.
[10] Bhaskar Chandra Kandpal, 2015, Machining of aluminium metal matrix composites with Electricaldischarge machining- A Review, Materials Today: Proceedings 2 ( 2015 ) 1665 – 1671, DOI:10.1016/j.matpr.2015.07.094
[11] Preetkanwal Singh Bains, 2016, Fabrication and Machining of Metal Matrix Composites: A Review,Materials and Manufacturing Processes, 31:5, 553-573, DOI: 10.1080/10426914.2015.1025976
[12] A Gnanavel Babu 2018, Investigation on the cutting quality characteristics of abrasive water jet machiningof AA6061-B4ChBN hybrid metal matrix composites, Materials and Manufacturing Processes,DOI:10.1080/10426914.2018.1453146
[13] S. Gopalakrishnan 2012, Production and wear characterisation of AA 6061 matrix titanium carbideParticulate reinforced composite by enhanced stir casting, Composites: Part B 43 (2012) 302–308,Doi:10.1016/j.compositesb.2011.08.049
[14] José Manuel Mendoza-Duarte 2016, Mechanical and Micro structural Response of an Aluminum Nanocomposite Reinforced with Carbon-Based Particles, Mat. Res. vol.19 supl.1 São Carlos Dec. 2016, Epub Aug 04, 2016. DOI: 10.1590/1980-5373-mr-2015-0625.
[15] N. E. Arun Kumar & A. Suresh Babu. "Influence of input parameters on the near-dry WEDM of Monelalloy," Materials and Manufacturing Processes 33:1, 85-92, 2017. DOI:10.1080/10426914.2017.1279297.
[16] N. E. Arun kumar, A. Suresh Babu D. Murali,,"A Study on Parametric Optimization of wire ElectricalDischarge machining using response surface methodology," Int. J. Chemical Science 14(2), 2018.
[17] B. SATYANARAYANA, G. SRIKAR, "optimization of abrasive water jet machining process parametersusing taguchi grey relational analysis" International Journal of Mechanical And Production Engineering ISSN: 2320-2092, Volume- 2, Issue-9, Sept.-2014.
[18] K.Siva Prasad " A review on current research trends in abrasive water jet machining" International ResearchJournal of Engineering and Technology, Volume: 04 Issue: 06 | June -2017.

IOP Conf. Series: Materials Science and Engineering **923** (2020) 012038 doi:10.1088/1757-899X/923/1/012038

# Investigation on Pineapple and Pirambu Natural Fibres Reinforced with Polyester Hybrid Composite

**Daniel Lawrence I[a], Parameswara Raja G [b], Anbarasan A[c], Gnanaraj J[d]**

[a)] Assistant Professor, Dept of Mech Engg, Loyola Institute of Technology, Chennai, INDIA.
[b)] Assistant Professor, Dept of MechEngg, Mount Zion College of Engg and Technology, Karaikudi, INDIA.
[c)] Assistant Professor,Dept of Mech Engg, Loyola Institute of Technology, Chennai, INDIA.
[d)] *Associate Professor,Dept of Mech Engg, Loyola Institute of Technology, Chennai, INDIA*

*E-Mail: danilawernce@gmail.com

**Abstract**. Composite materials have gained their presence in several industrial sectors. New materials, advanced material treatments and manufacturing methods have been introduced in research and developments today. Experimentation in composite materials has changed the people to come up with new innovations and alternative ideas in day today applications. This work focuses on obtaining the mechanical characterization and morphological study on composite materials which are prepared by compression moulding. The prefabrication tests were conducted to know the specific nature of each fibre material by conducting single fibre and FTIR test. This experimental measurement was observed the better mechanical properties from Tensile, Flexural and Impact test with the specimens of the prepared composite lamina of composite. The morphological study in fractured specimens was observed by using Scanning Electron Microscope and Energy Dispersive X-ray Analysis. This research work were represented the mechanical characteristics and morphological behaviour of pirambu pineapple hybrid polyester composite and also will help us for implementation of new manufacturing applications with composite materials.

Keywords: Composite lamina, Tensile strength, Flexural strength, Impact test, Compression Moulding

## 1. Introduction

Due to the environment awareness and sustainability issues, remarkable achievements were attained in last decade, particularly in the area of material science. The two different natural fibres are used in the composite system forming a hybrid composition of reinforcement. The natural fibres are obtained from the raw material through mechanical and water retting methods. Single polymer is used as matrix over here. Fibres are cut into small segments and packed into the composite system with random orientation. A catalyst with octane solution is used as an interface between the matrix and the reinforcement in the composite system to speed up the rate of the reaction. Sabbie A. Miller et al. [1] has presented an experimental report with bio-based composite material which are used as an alternative for unconventional materials to study its mechanical properties, environmental impacts and creep. MohitSood et al. [2] conducted a review on flexural properties and its effect on fibre treatment with natural fibre reinforced composite. B.K. Kandola et al. [3] conducted mechanical and fire performance study on natural fibre reinforced thermoplastic composites and found out that woven and nonwoven technology can be used to prepare synthetic and natural fibre composite which has good mechanical properties. Swagata Dutta et al. [4] presented a numerical and experimental analysis on fire reaction properties and made a comparative analysis with flax and wool based composites. A. Quintana et al. [5] conducted an comparative analysis between traditional plastic board and bio based composite board based on life cycle assessment and the results converged have showed that the environmental impact is approximately same in both types of boards. Yingji Wu et al. [6] have made an experiment by introducing magnesium hydroxide into the natural fibre reinforced composite for improving the mechanical properties which can be used in automotive applications. Anna Bernstad Saraiva et al. [7] conducted life cycle assessment by comparing cardboard material with natural fibre composite and the results showed that the use of cardboard material has less environmental impact than the composite material. William T. Kern et al. [8] made microscopy and finite element analysis on natural fibre composites to get a view of microcellular voids since voids would cause stress intensity and it would change the properties of non-brittle material to brittle material. N.K. Kim et al. [9] have performed review on thermal behaviour and flammability of natural fibres. Bartosz Zukowski et al. [10] conducted alkali treatment in caraua fibre to check the durability of strain hardening cement based composite by exposing it to natural weathering.

WassamonSujaritjun et al [11] conducted experimental analysis tensile properties of bamboo fibre, vetiver grass fibre and coconut fibre reinforced PLA composites were compared with each other. It is found that Bamboo fibre has most effective improvement in reinforcement of tensile properties of natural fibre reinforced PLA composite. Vivek Mishra et al [12]. Found that the hardness, tensile and impact strength of jute-epoxy composites increases with the increase in fibre loading and they also found that the void content decreases with increase in fibre loading.K Deepak et al [13]. Has made comparative analysis with Biocomposite material using coir as fibres and general polyester resign as matrix with the addition of nano clay material. They found out that the composites which are prepared with and without clay mixture have produced similar results. They also found that use of clay material has improved the interaction between the clay and the resin. They suggested that the use of composite do not meet the strength parameters of conventional composites, and so they can be used for architectural equipments. AshwinSailesh et al [14]. Preformed an experimental work, depending upon the result obtained they found out that the stalking sequence and the interaction between the fibre and the reinforcement plays a vital role in determining the properties of the composite material. G Bujjibabu et al [15]. Conducted experimental analysis on mechanical behavior of banana/coir reinforced polypropylene hybrid composites and found out that the tensile and flexural strength decreases with increasing coir fibre percentage, where the impact strength increases upto certain extent and then it decreases. Ashik K P et al [16]. Conducted experimental analysis and found out that increase in cocknut coir in the glass/coir composite specimen increases water absorption, flexural strength increases at higher proportion of glass fibre and impact strength increases when percentage of epoxy increases.Vijay R et al [17]. Conducted comparative analysis between the stalking sequence with 2 symmetric material layer and one intermediate layer. Here the replaced the intermediate layer with kenaf, sisal and jute fibre for undergoing the comparative experiment. From the results obtained they found out that the stalking sequence which is having the kenaffibre as its intermediate layer has good performance in mechanical properties and characteristics. Sabbie A. Miller et al [18]. Conducted experiment on natural fibre textile reinforced bio-based composites, they found out that jute burlap composites displayed high off-axis and on-axis stiffness and the creep deformation is dependent on the stress level and the orientation of textiles. B.K. Kandola et al [19]. Have conducted experiment from the result they found out that woolen and non-woolen technology can be used to prepare composites with good mechanical properties and this technology can be used for preparing natural-synthetic fibrous performs.Yingji Wu et al [20]. Conducted experimental analysis and found out that the comparable mechanical properties, reduced energy consumption and environmental impacts have the potential to NFRC by replacing it with GFSMC in automobile applications. Bakare et al. [21] studied the mechanical properties of the sisal fibre rubber seed oil polyurethane composite, with and with-out water treatment process. Gonzalez and Ansell [22] suggested that the mechanical properties of chemically treated henequen fibre epoxy composite produces similar results as like the untreated fibre composite. Silva et al. [23] investigated the tensile properties of the sisal fibre for the different fibre gauge length. Herrera-Franco and Valadez-Gonzalez [24] concluded that the stress distribution between the fibres and matrix for a short discontinuous fibre were better than the continuous fibres. Igor et al. [25] investigated the importance of phormium (flax fibre)/epoxy laminated composite with short fibre and long fibre. Various chemical compositions of the fibres were compared with the other natural fibres.

The focus on this work would gain the attention of researchers in implementation of research and several manufacturing applications. The reinforced composite materials are low cost composite materials. The major perspective of this work is to focus on preparing a highly economical material with enhanced mechanical properties for gaining durable manufacturing applications with good safety factor.

## 2. Methodology

### *2.1 Materials and Fabrication*

Two different fibres were used. The materials used for fabrication are pineapple [ananascomosus] and pirambu [arundinaria] fibres. The fibre of pineapple is obtained from the leaves of the pineapple plant by mechanical retting with rolling crushing machine and the fibre of pirambu sticks by water retting method through hand stepping technique. The polyester GP51 matrix was used to fabricate natural fibre reinforced composites to evaluate the fibre performance under different percentage of compositions. These are prepared as small fibres which have a length between them 15mm to 20mm. The selected fibres are different from region and they exhibit variability in their physical and mechanical properties. The natural fibre composite with 20% pineapple [Ficusreligiosa], 10% pirambu [arundinaria], 69.5% polyester GP51 resin, 0.25% of MERP catalyst and 0.25% of octane solution is kept as composition. These two fibres are mixed with polyester matrix with hardener solution.

**Figure 1.** Pineapple-10% and Pirambu-20% compsite lamina

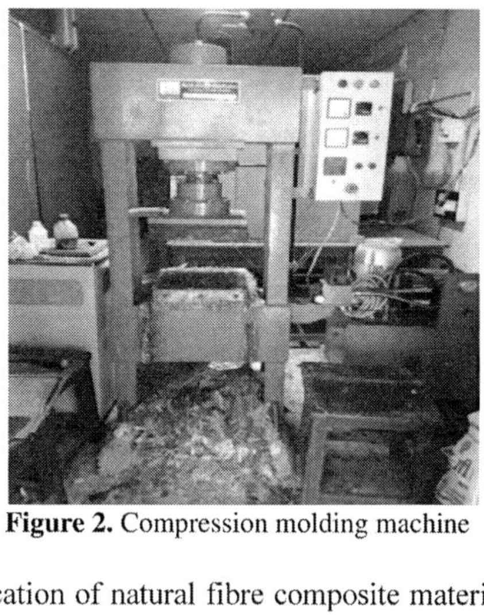

**Figure 2.** Compression molding machine

Compression moulding machine were used for the fabrication of natural fibre composite material. The machine consists of a fixed lower die and the upper die attached to the hydraulic actuation system. A digital control panel with upper and lower die settings and timer settings are used to control the moulding machine. The fabrication processes is executed by starting the machine followed by setting the layup plate for preparing the lamina. The composition for preparing the composite lamina includes 10% pineapple fibre and 20% pirambu fibre as reinforcements, 69.5% polyester GP51 resin, 0.25% of octane solution and 0.25% of MEKP catalyst as matrix. A mixture of matrix and reinforcement is embedded over the layup plate and introduced into the moulding machine. The machine set up is checked for its position before the compression task is executed. Once the position of the layup plate over the lower die is checked the upper die is moved slowly at 2bar pressure until the layup plate is completely compressed. The hot composite lamina is removed from the layup plate and placer under some load for natural cooling in atmosphere.

## 3. Result and discussion

The composite lamina fabricated with the help of compression molding machine. The single fibre test and FTIR were observed to determine the mechanical ability of the individual fibre. The composite lamina was prepared as per ASTM standard to identify the mechanical properties and morphological behavior. Tensile, Flexural and Impact tests were conducted to check the mechanical properties and Scanning electron microscopy and Energy Dispersive X-ray Analysis were studied to point out the morphological behaviour of the composite.

### *3.1 Single fibre strength for Preprocessing*

Pre-processing studies were conducted to identify the material characteristics and properties before implementing them into the fabrication work. The individual examination of these materials would help us in preceding the fabrication processes with effective methods. Single fibre test was conducted to check the strength of individual fibre. The single fibre is loaded gradually at 100N at a displacement of 10mm/min with the help of universal testing machine. The fibre is tested within the closed boundary of a paper frame by sticking the single fibre with adhesive. The specimen is tested under 100mm grip to grip separation from its starting position. The results are obtained in graphical format through computer integrated universal testing machine.

The results obtained by testing single pineapple fibre are plotted in the graphical representation. It shows that the maximum force attained by testing the single pineapple fibre is 2.467766N with the corresponding elongation of 0.945653%. The same force and elongation is obtained at the breaking point.

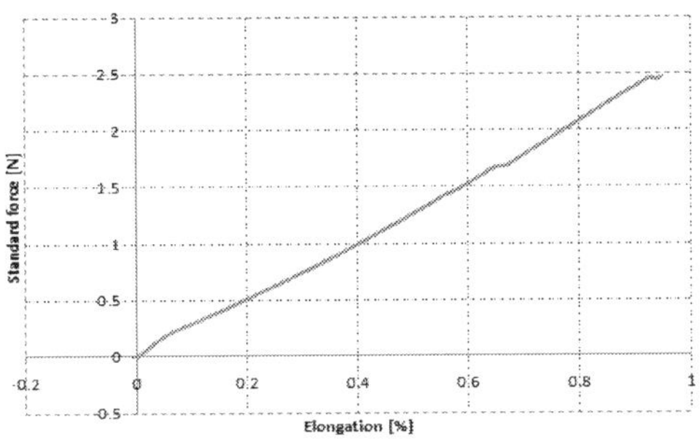


**Figure 3.** Single fibre test for pineapple fibre

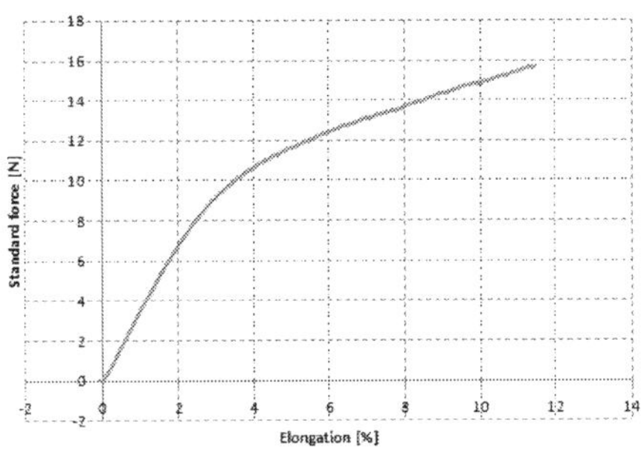


**Figure 4.** Single fibre test for Pirambu fibre

The results obtained by testing single pirambu fibre are plotted in the graphical representation for visualization. It shows that the maximum force attained by testing the single pirambu fibre is 15.697418N with the corresponding elongation of 11.493111%. The same force and elongation is obtained at the breaking point.

### 3.2 *FTIR Test*

Fourier transform infrared spectroscopy was studied to obtain the molecules and chemical compound of fibre element. It helps to identify the chemical substance present in the material. The thin beam of infrared light were made to pass through the specimen, if the wavelength of infrared light matches the wavelength of the specimen then the light rays will be observed by the specimen otherwise the light rays would fall on detector behind the specimen. The results obtained through plots of graphical representation.

IOP Conf. Series: Materials Science and Engineering **923** (2020) 012038 doi:10.1088/1757-899X/923/1/012038

*3.2.1 FTIR test with Pirambu fibre*

The obtained results of pirambu fibre shows that, the specimen has obtained about 95% of transmittance approximately in the functional group region which shows that only 5% of light beam coincides with the wave length of the molecules present in the specimen. The wave numbers 3338.09 and 2918.67 shows the presence of strong ,broad alcohol (O-H) group which is exhibiting stretching vibration at a transmittance of almost 89% and strong alkane (C-H) group which is exhibiting stretching vibration at a transmittance of almost 92% respectively. The wave number of 2116.33 and 2035.28 shows the presence of alkyne (-C≡C-) group with variable intensity at weak or medium level, exhibiting stretching vibration at a transmittance of 95% and the presence of weak nitrile (-C≡N-) group which is exhibiting stretching vibration at a transmittance of 96%. From the observed results that noted 1-Hydroxy-4-Nitrile-2-Butyne is the maximum chemical compound present in pirambu fibre.

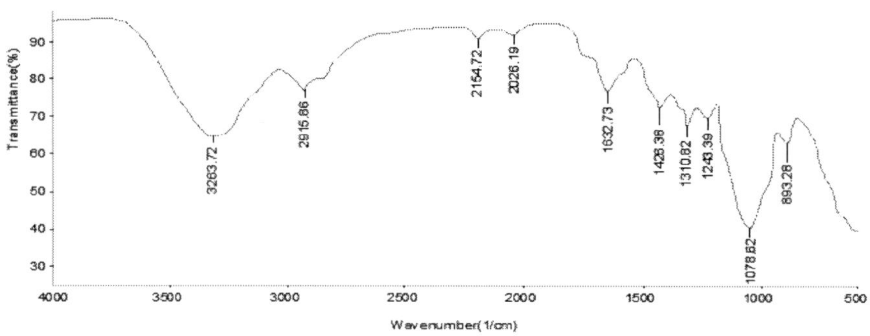


**Figure 5.** FTIR graph for Pirambu fibre

*3.2.2 FTIR test with pineapple fibre*

The obtained results of pineapple fibre shows that, the specimen has obtained about 80% of transmittance approximately in the functional group region which shows that only 20% of light beam coincides with the wave length of the molecules present in the specimen. The wave numbers 3299.17 and 2915.33 shows the presence of strong ,broad alcohol (O-H) group which is exhibiting stretching vibration at a transmittance of almost 65% and strong alkane (C-H) group which is exhibiting stretching vibration at a transmittance of almost 88% respectively. From the observed results that noted formaldehyde is the maximum chemical compound present in pineapple fibre.

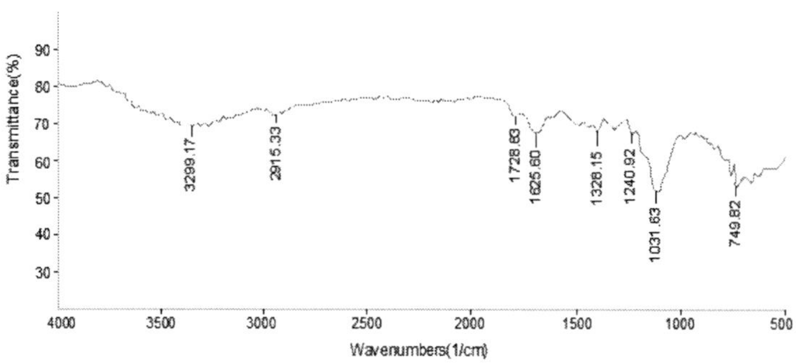


**FIGURE 6.** FTIR graph for Pineapple fibre

IOP Conf. Series: Materials Science and Engineering **923** (2020) 012038 doi:10.1088/1757-899X/923/1/012038

*3.3 Tensile Test*

The tensile test was conducted to check the mechanical response of the composite material by applying tensile load through universal testing machine. The results are obtained in graphical representation of stress strain curve and force length curve as per ASTM D638 standard. From the observations of stress strain and force length graph it is noted that the specimen has obtained ultimate stress at 10.5MPa with 54% of strain, the ultimate force of 394N is obtained at 0.546mm of length, the specimen breaks at 0.561mm of length at a force of 199N.

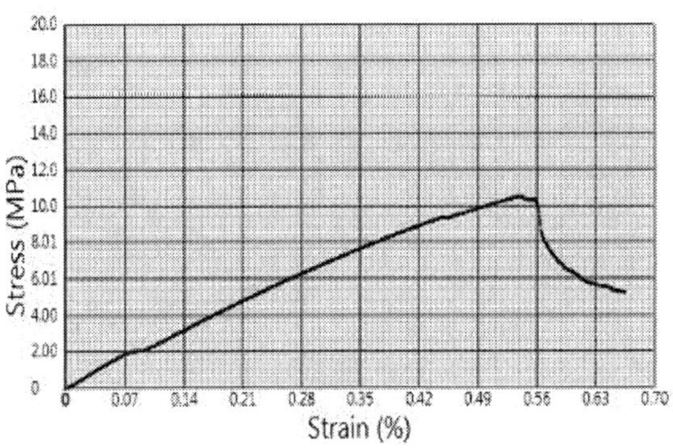


Figure 7. Stress Strain graph for Tensile test

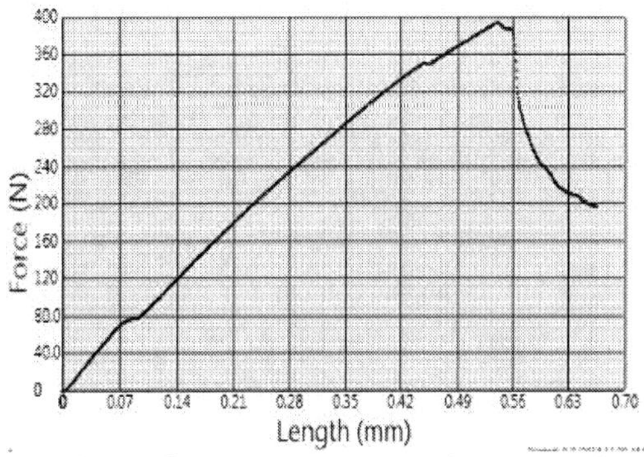


**Figure 8.** Force Length graph for Tensile test

*3.4 Flexural Test*

The Flexural test was conducted to find the ability of the flexural resistance on the fabricated material by applying bending load. This test was conducted with three point bending set up through universal testing machine. The results are obtained in graphical diagram through stress strain as per ASTM D790 standard. The results obtained in the graphical form shows that the specimen were obtained flexural strength of 40.89MPa at 1.08% of strain. The ultimate force is obtained at 30.67N. The specimen breaks at a load of 32.5MPa at 1.215% of strain respectively.

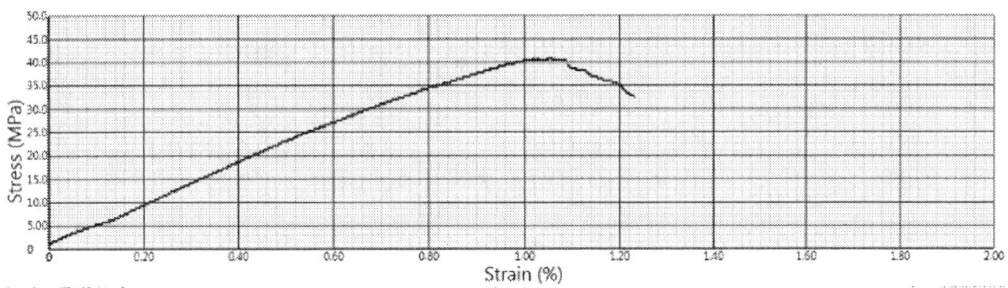


**Figure 9.** Stress Strain graph for Flexural test

*3.5 Impact Test*

Izod impact test was conducted as per ASTM D256 standard. The specimen which is to be tested is placed vertically with one end fixed and the other end facing the striking edge. The specimen obtained that the internal energy of 27.449J, external energy of 2.2292J and total energy of 25.2198J. The maximum strength of 178.341J/m were obtained from the tested specimen.

*3.6 Analysis of SEM Images*

The fractured surface of the Pirambu and Pineapple fibre reinforced composite was analyzed using scanning electron microscope. The observed SEM images shows that the fibre and resin compatibility in natural fibre polymer composites. The SEM image of tensile fractured specimen shows uniform

IOP Conf. Series: Materials Science and Engineering **923** (2020) 012038 doi:10.1088/1757-899X/923/1/012038

crack propagation due to better mixing of the fibre and matrix. The observed images present the random orientation of fibre packed with resin, having void content over it. The porous spot over the merged fibre shows the presence of optimal level of impurities over the composite material. Fibre segregation was found to be good in the fractured region due to the improvement in compatibility between the fibre and resin matrix.

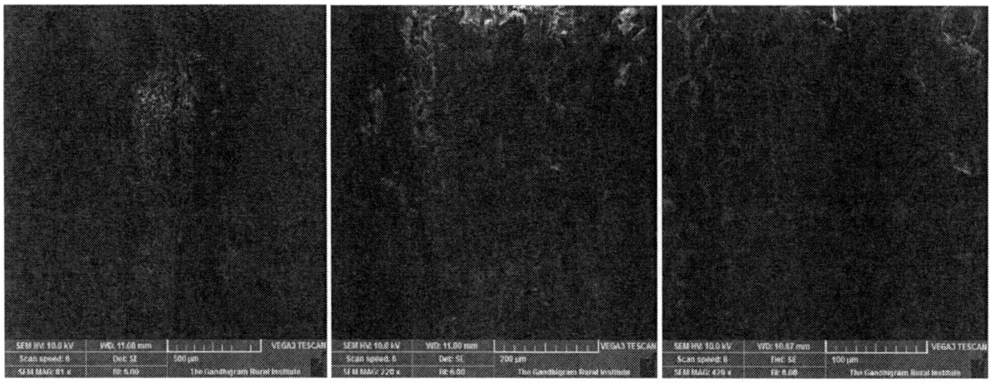

**Figure 10.** SEM image of Pirambu and Pineapple fibre composition for 500 μm, 200 μm, 100μm

*3.7 Energy Dispersive X-ray Analysis*

From the obtained results of the Energy Dispersive X-ray analysis, Carbon element contains the maximum level of un-normalised concentration in weight percent of the element, normalised concentration in weight percent of the element, atomic weight percentage and the error in the weight percent concentration at the 1 sigma level and also the observed results showed that, Zink element contains the minimum level of un-normalised concentration in weight percent of the element, normalised concentration in weight percent of the element, atomic weight percentage and the error in the weight percent concentration at the 1 sigma level.

**Table 1.** EDX chemical composition for Pirambu and Pineapple fibre

| Sl.no | El | AN | Series | Unn.C (wt . %) | Norm.C (wt . %) | Atom.C (at . %) | Error (1 sigma) |
|---|---|---|---|---|---|---|---|
| 1 | C | 6 | K-series | 54.07 | 54.07 | 61.86 | 7.26 |
| 2 | O | 8 | K-series | 42.74 | 42.74 | 36.71 | 6.29 |
| 3 | Si | 14 | K-series | 0.86 | 0.86 | 0.42 | 0.07 |
| 4 | Ca | 20 | K-series | 0.65 | 0.65 | 0.22 | 0.05 |
| 5 | Na | 11 | K-series | 0.34 | 0.34 | 0.21 | 0.06 |
| 6 | Mg | 12 | K-series | 0.36 | 0.36 | 0.20 | 0.06 |
| 7 | Al | 13 | K-series | 0.38 | 0.38 | 0.20 | 0.05 |
| 8 | Fe | 26 | K-series | 0.26 | 0.26 | 0.06 | 0.04 |
| 9 | Cl | 17 | K-series | 0.15 | 0.15 | 0.06 | 0.03 |
| 10 | K | 19 | K-series | 0.12 | 0.12 | 0.04 | 0.03 |
| 11 | P | 15 | K-series | 0.03 | 0.03 | 0.01 | 0.03 |
| 12 | Zn | 30 | K-series | 0.04 | 0.04 | 0.01 | 0.03 |

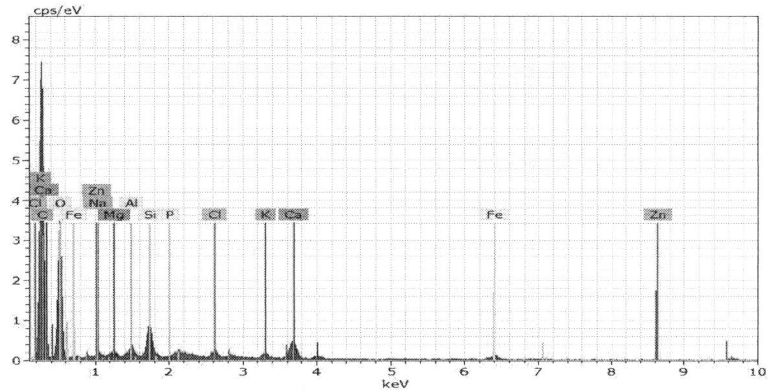


**Figure 11.** EDAX image for Pirambu and Pineapple fibre composition

## 4. Conclusion

The mechanical properties and morphological behavior of pirambu and pineapple polymer matrix composite were studied by preprocessing and post processing evaluation. The better tensile, flexural and impact behaviors of 10.5 MPa, 40.89MPa, and 178.341J/m were obtained pirambu (20%) and Pineapple (10%) reinforced polymer composites. The microstructure and chemical composition where studied by scanning electron microscope through SEM and Energy Dispersive X-ray analysis. The presence of cellulose, hemi-cellulose and lenin contents are observed in detail with some fractures and crack propagations. The mechanical behaviors studied through real time experiments witnessed that both composition of Pirambu and Pineapple reinforced polymer composites can be utilized as the replacement of plastics because of its ease of decomposability and low cost. The obtained result shows that, the specimen is malleable with high bending stress with good elastic and plastic behavior.

## References

1. Miller S A 2018 Natural fibre textile reinforced bio-based composites: Mechanical properties, creep, and environmental impacts *Journal of Cleaner Production* **198** 612-623.
2. Kandola B K, Mistik S I, Pornwannachai W and Anand S C 2018 Natural fibre-reinforced thermoplastic composites from woven-nonwoven textile preforms:Mechanical and fire performance study *Composites Part B: Engineering* **153** 456-464.
3. Dutta S, Kim N K, Das R and Bhattacharyya D 2019 Effects of sample orientation on the fire reaction properties of natural fibre composites *Composites Part B: Engineering* **157** 195-206.
4. Wu Y, Xia C, Cai L, Garcia A C and Shi S Q 2018 Development of natural fibre-reinforced composite with comparable mechanical properties and reduced energy consumption and environmental impacts for replacing automotive glass-fibre sheet molding compound. *Journal of Cleaner Production **184*** 92-100.
5. Saraiva A B, Pacheco E B, Gomes G M, Visconte L L, Bernardo C A and Soares A G 2016 Comparative lifecycle assessment of mango packaging made from a polyethylene/natural fibre-composite and from cardboard material. *Journal of cleaner production* **139** 1168-1180.
6. Kern W T, Kim W, Argento A, Lee E C and Mielewski D F 2016 Finite element analysis and microscopy of natural fibre composites containing microcellular voids *Materials & Design* **106** 285-294.
7. Zukowski B, Santos Mendonça Y G, Andrade Silva F and Toledo Filho R D 2018 The durability of SHCC with alkali treated curaua fibre exposed to natural weathering. *Cement and Concrete Composites* **94** 116-125.
8. Daniel Lawrence I and Kaviprakash G 2014 Design and Analysis of Composite Drive Shaft for Automotive Application *International journal of engineering research and technology* **3** 429-436.

9. Sujaritjun W Uawongsuwan P and Hamada H 2013 Mechanical property of surface modified natural fibre reinforced PLA biocomposites *Energy Procedia* **34** 664-672.
10. Mishra V and Biswas S 2013 Physical and mechanical properties of bidirectional jute fibre epoxy composites *Procedia engineering* **51** 561-566.
11. Deepak K Reddy N S and Naidu T S 2015 Thermosetting polymer and nano clay based natural fibre bio-composites *Procedia Materials Science* **10** 626-631.
12. Sailesh A Arunkumar R and Saravanan S 2018 Mechanical Properties and Wear Properties of Kenaf-Aloe Vera-Jute Fibre Reinforced Natural Fibre Composites *Materials Today: Proceedings* **5** 7184-7190.
13. Bujjibabu G Das V C Ramakrishna M and Nagarjuna K 2018 Mechanical and Water Absorption Behavior of Natural Fibres Reinforced Polypropylene Hybrid Composites *Materials Today: Proceedings* **5** 12249-12256.
14. Ashik K P Sharma R S and Guptha V J 2018 Investigation of moisture absorption and mechanical properties of natural/glass fibre reinforced polymer hybrid composites *Materials Today: Proceedings* **5** 3000-3007.
15. Vijay R and Singaravelu D L 2018 Influence of stacking sequence on mechanical characteristics of Cyperuspangoreifibres based natural fibre composites *Materials Today: Proceedings* **5** 8504-8513.
16. Miller S A 2018 Natural fibre textile reinforced bio-based composites: Mechanical properties, creep, and environmental impacts *Journal of Cleaner Production* **198** 612-623.
17. Kandola B K Mistik S I Pornwannachai W and Anand S C 2018 Natural fibre-reinforced thermoplastic composites from woven-nonwoven textile preforms: Mechanical and fire performance study *Composites Part B: Engineering* **153** 456-464.
18. Wu Y Xia C Cai L Garcia A C and Shi S Q 2018 Development of natural fibre-reinforced composite with comparable mechanical properties and reduced energy consumption and environmental impacts for replacing automotive glass-fibre sheet molding compound *Journal of Cleaner Production* **184** 92-100.
19. Bakare I O Okieimen F E Pavithran C Khalil H A and Brahmakumar M 2010 Mechanical and thermal properties of sisal fibre-reinforced rubber seed oil-based polyurethane composites *Materials & Design* **31** 4274-4280.
20. Gonzalez Murillo C and Ansell M P 2009 Mechanical properties of henequen fibre/epoxy resin composites *Mechanics of Composite Materials* **45** 435-442.
21. Herrera Franco P and Valadez Gonzalez A 2005 A study of the mechanical properties of short natural-fibre reinforced composites *Composites Part B: Engineering* **36** 597-608.

IOP Conf. Series: Materials Science and Engineering **923** (2020) 012039 doi:10.1088/1757-899X/923/1/012039

# Development of Hybrid Friction Material for Brake Pad Application

**Mullaikodi S.M[1], Sethuvelappan.P[2], Kamaraj.M[3], Naveen.E[4], Ramanan.N[5]**

[1]Associate Professor, Department of Mechanical Engineering, St Joseph's College of Engineering, Chennai
[2]UG Scholar , Department of Mechanical Engineering, St Joseph's College of Engineering, Chennai
[3]Associate Professor, Department of Mechanical Engineering, SRM Institute of Science and Technology, Chennai
[4]Assistant Professor, Department of Mechanical Engineering,Sri Sairam Engineering College, Chennai
[5]Engineer, Synce Engineering Service, Chennai

E-mail: ramananinjs2020@gmail.com

**Abstract:** As per the report of world health organization and environmental protection agencies, the existing brake pad materials produce toxic gases. Toxicity leads to create a bad effect on human health. So, there is a need to reduce the percentage of toxic elements at source and develop novel material compositions. It is known that the natural fiber-reinforced material reduces the toxicity. In this direction, an attempt made to fabricate hybrid natural fiber-reinforced composite material. The hand layup method was used to fabricate the hybrid fiber-reinforced composites. The mechanical test namely tensile, flexural, impact, shear, and interlaminar were performed to characterize the composites. The results revealed that hybrid fiber-reinforced composite materials exhibit superior mechanical properties compared to non-hybrid composites. Scanning Electron Microscope was used to analyze the fracture surface of the tested samples.

*Key words:* Friction materials, Brake pad, Hybrid fibre reinforced composites, Mechanical properties.

## 1. Introduction

Composites are formed by combining materials to form an overall structure that is better than the sum of the individual components. When two or more constituent materials having a significant change in the properties combine gives rise to new composite material, with characteristics different from the individual components. The individual components remain completely distinct within the finished structure in terms of their properties and other characteristics. The composite material thus formed may be preferred for many reasons due to their extensive characteristics when compared to that of traditional materials. The growth of the composite materials, plastics, and ceramics have been tremendous over the last thirty years. The volume and number of applications of composite materials have grown steadily, penetrating and conquering new markets relentlessly. Modern composite materials constitute a significant proportion of the engineered materials market ranging from everyday products to sophisticated niche applications. While exhibiting greater advantage over these composites as weight-saving materials, the present challenge is to make them cost-effective. To provide economically viable and stable composite components several new manufacturing techniques are widely used in the composite industries.

It is obvious, especially for composites, that the improvement in manufacturing technology alone is not enough to overcome the cost hurdle. There must be an integrated effort in designing, and other factors which include material processing, tooling, quality assurance, manufacturing, and even programmed management for composites to become competitive in the industry. The composites industry has already begun to recognize that the commercial applications of composites promise tend to offer a wide range of business opportunities than the other sectors. Earlier the applications of the composite materials were mainly focused on aircraft industry, but now the shift has occurred in the applications of other commercial uses, increasingly enabled by the introduction of newer polymer resin matrix materials and high-performance reinforcement fibers of glass, carbon, and aramid, the penetration of these advanced materials has witnessed a steady expansion in uses and volume. The new composite materials even tend to be cost-efficient. The applications of the high-performance FRP widely vary to a greater extent such as composite armoring designed to resist explosive impacts, fuel cylinders for natural gas vehicles, windmill blade industrial drive shafts, support beams of highway bridges and even paper making rollers.

Concerning certain applications, the use of composites has resulted in savings of both cost and weight. Some examples are cascades for engines, curved fairing, and fillets, replacements for welded metallic parts, cylinders, tubes, ducts, blade containment bands, etc. Further, the need for composites for lighter construction materials and more seismic-resistant structures has placed high emphasis on the use of new and advanced materials that not only decreases dead weight but also absorbs the shock and vibration through tailored microstructures. Composites are now extensively being used for rehabilitation strengthening of pre-existing structures that have to be retrofitted to make them seismic-resistant or to repair damage caused by seismic activity.

## 2. Literature Review

On low-velocity impact behavior, the number of the carbon and aramid layers and fiber orientation were reinforced on the fabric hybrid composites with a stratified filled epoxy matrix. The impact tests were done with the drop weight impact system at 90.629 J of energy level. The damage areas were analyzed by the visual inspection and tomography images. The results indicate that the matrix properties have a great influence over the fracture mode of the hybrid composites, while the fault degree of the damaged areas depending on the fiber orientation. The high impact resistance was obtained in the case of hybrid composites with 0° ply orientation 1 The carbon fiber reinforced composites perform better in fatigue loading, in comparison to glass fiber reinforced composites; The effect of polymer aging was also evaluated through thermal aging of neat resin specimens.2. The flexural performance, under both static and fatigue loading, of a glass fiber/carbon fiber hybrid polymer matrix composite material It was found that exposure to 180 _C in the air for 3 months resulted in improved fatigue performance of the composite, with mildly diminished static flexure strength, after six months the performance resulted in moderately degraded fatigue performance and a further reduction in flexure strength, while severely reduced performance (both static and fatigue) for 12 months aging time was found.

The mild improvement in fatigue performance for 3-month aged specimens attributed to the post-cure of the composite3.They have the potential to compete with traditional fiber in all thermal and mechanical properties. The thermal degradation of vetiver fiber composite had more thermal decomposition if they undergo acetylation and silane chemical treatment and could treatment and could have still modified the surface of the fiber. [4].In this work, Jute, Pineapple leaf fiber, and Glass fiber reinforced hybrid composites were prepared. The tensile and flexural properties of the composites with these fibers were found to be increased with fiber content, conforming to the reinforcing action of the fibers. Thus the composites of Jute, Pineapple leaf fiber, and Glass fiber epoxy composites were found to be light in weight, possessed better mechanical properties. Hence these composite materials can be used for applications such as automobile parts, electronic packages, building construction.[5].This paper characterized the impact response of a new hybrid composite material made using Type A (woven carbon fibers / UD flax fibers) and Type B (woven carbon fibers / cross-ply flax fibers) configurations in an epoxy matrix. The Type B composite had better overall impact performance as indicated by lower absorbed energy, higher penetration energy, smaller crack lengths, smaller indentation depths, smaller damage areas, lower temperature rise, and higher impact strength vs. the Type A version of the material. Both the Type A and B hybrid composite improved various impact properties compared to neat flax fiber composites reported in prior literature.

The impact properties of the current composite may potentially have various research and industrial uses depending on the particular needs of the application.[6]. In a new hybrid polymer composite with banana, a natural fiber, as reinforcement and also to study the machining characteristics of the composite. Response surface methodology was used to analyze the influence of machining process parameters on the material removal rate and surface roughness. The effect of machining parameters on the material removal rate and surface roughness were obtained through regression equations and different plots obtained from ANOVA analysis, employed through MINITAB and ANFIS. It is concluded that feed rate has maximum influence on the material removal rate and surface roughness followed by speed and depth of cut. It is observed that cardanol and Cashew nutshell liquid-based composites have similar material removal rates but they vary significantly in surface roughness. Further scope of this investigation is to find the optimum values of feed rate, speed, and depth of cut for material removal rate and surface roughness.it is also planned to investigate the various mechanical properties like tensile behavior, flexural behavior, dynamic behavior, etc. to suggest suitable applications.7 In recent days, technology needs a high demand for the growth of future generations. Hence, to produce a good product which needs to satisfy the customers, it should be developed in an economical way, high quality with minimum possible time challenging among the competitors.

Natural fiber-based polymer composites are the most shows the potential option for synthetic fiber and powerful polymer composites. It enhances the awareness of eco-friendly material makes it crucial to utilize natural fiber as potential reinforcement in polymer composites. Natural fibers are plentifully existing, biodegradable, and recyclable, which makes them conventional in the automobiles industries. The major negative aspect of testing is stretched time, the manufacturing cost is high, mistakenness, machine error, today, most bio composite researchers are moving toward computational methods to model the NFRPC to simulate the mechanical and thermal properties. Studied the tensile properties of flax/basalt hybrid composites with loading conditions.8 They reported that the tensile properties of composites decreased with increasing aging conditions.11 The rigidity and modulus diminished because of the embrittlement of the matrix materials. Comparative attributes were observed for sisal fiber reinforced PP composites [9,10] The affiliation between the cross-section of the natural fiber and the tensile properties.

They modeled and evaluated the tensile performance of hemp fiber with dissimilar elliptical cross-sections. The results appeared that the tensile properties have a physically powerful manipulate on the degree of elasticity. The micro fibril angle and the viscoelastic properties played a very important role in the geometry of the natural fiber 12. Presently, we have used natural fiber material as used in a wind turbine material. Because It is used less weight and observed high wind turbine energy efficiency.

## 3. Materials and Methods

The prime ingredient of the composite material is Carbon fiber known for its toughness, high tensile strength, having good resistance to abrasion and resistance to organic solvents being Non-conductive, Low flammability, Good fabric integrity at elevated temperatures elasticity, good luster and thermal insulators because of their high ratio of surface area to weight. However, the increased surface area makes them much more susceptible to chemical attack with thermal conductivity of the order of 0.05 W/(mK). By trapping air within them, blocks of Carbon fiber make good insulation. The addition of Carbon fibers improves the bending load-carrying capacity of both symmetric and asymmetric Wind Turbine blades. Carbon polymer matrix composites material (90:10) 1kg is prepared by taking 900 grams of Carbon fiber in the shape sheet form (90%) and10 grams of Epoxy Resin. In addition to coconut and vetiver fiber, to add strength, the hybrid Composite Material has been made by the Hand layup Method.

## 4. Hand lay-up method

In the hand layup method, resins are impregnated into fabrics by manual feed which is mainly of roving form. The fibers and resins impinge on the surface with the help of the rollers which has been fed into the mould. Then the fibers are made free under standard atmospheric conditions. This method is very simple and easy but it requires high skill to fabricate the composite. The fibers used are Carbon, in addition to that Coconut and Vetiver fiber are used. Carbon fibers are laid at the top and bottom in most of the composite laminate for better finishing purposes. Three different categories with two samples have been fabricated using hand layup method. All the fibers are laid in a normal direction and kept in a dry condition before it is fed to the laminate. Initially, the releasing agent is applied over the surface to remove the laminate easily. After applied, a thin layer of resin is applied and then the Carbon fiber is laid on the surface. A weight of 5 kg is placed over the Carbon fiber to remove air bubbles if any and it is kept undisturbed for about 3hours. Fibers are also dried in the normal condition to make the fibers moisture-free. After 3 hours the laminate removed from the die material and used various test for engineering applications.

## 5. Testing

### *5.1 Tensile testing*

Strength of the specimen can be found when the material is subjected to tension or compression. Standard dog bone shaped specimens were handlay up according to the ASTM D638 specifications for tensile testing .Each specimen having 30 mm width and 280 mm gauge length, as shown in Figure. 1. The specimen is loaded in computer controlled Universal Testing Machine (ASE – UTN 10) until the failure of the specimen occurs.

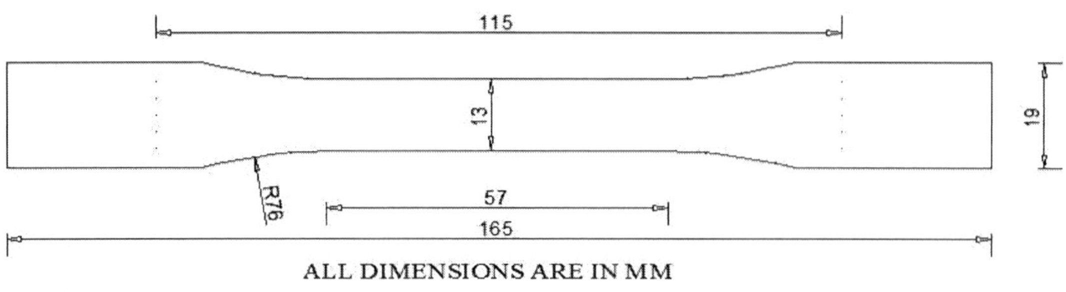


Figure 1. Tensile test specimen

IOP Conf. Series: Materials Science and Engineering **923** (2020) 012039 doi:10.1088/1757-899X/923/1/012039

Tests are conducted on composites of different combinations of reinforcing materials and ultimate tensile strength and ductility are measured. Simultaneous readings of load and elongation are taken at uniform intervals of a load. A tensile test is carried out at room temperature. Uniaxial tensile test is conducted on the constructed the specimen to obtain information regarding the behavior of a given material under gradually increasing stress-strain conditions. The Tensile stress test results the elasticity limit of the Polymer Matrix composite.

### *5.2. Flexural Testing*

The Flexural stress is the quantitative measure that defines the variations of Load concerning the displacement. The rigidness of the Polymer composite fiber can be studied through the flexural testing method. The flexural specimens are equipped as per the ASTM D790 standard (fig2). The 3-point flexure test is the most common flexural test for composite materials. Specimen Deflection is considered by the crosshead position. Test results take account of flexural strength and displacement. The testing process involves placing the test specimen in the universal testing machine and applying force to it until it fractures and breaks. The specimen used for carrying out the flexural test. The tests are regulating at a condition of an average relative humidity of 50%. A graph is drawn for each sample between the force and the displacement in the flexural test.

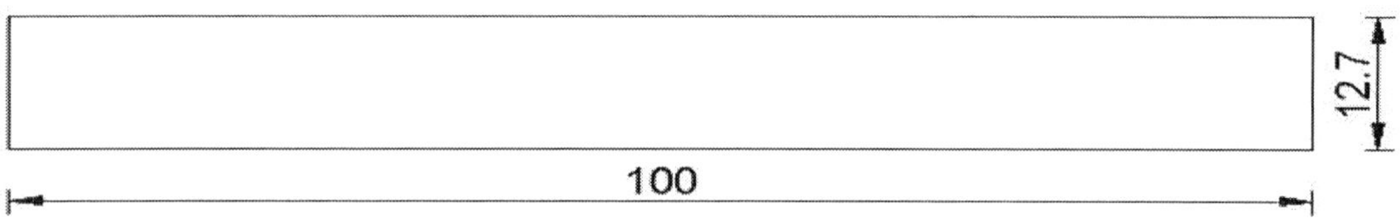


Figure 2. Flexural test specimen

### *5.3. Impact Test*

Impact Test tests the condition for absorbing the energy

Case (i) - When the striker impacts the composite specimen, the specimen absorbs energy until the yield point, then the specimen begins to undergo plastic deformation by absorbing energy, and work hardening occur at the plastic zone.

Case (ii) - When the composite can absorb no more energy, eventually failure occurs.

The Izod impact test is a standardized high strain-rate test which determines the amount of energy takes up by a material. during fracture. This test is carried out on the samples that are made as per the ASTM D256 standard. The absorbed energy by the specimen is noted until fracture takes place during testing.

## 6. Result and discussion

### *6.1. Tensile Test*

Concerning the figure1 standard dog bone-shaped specimens were hand lay up according to ASTM D638 specifications. Each specimen has a 30 mm width and 280 mm gauge length and the following observations were taken.

Table 1. Ultimate tensile strength of composites

| ***Orientation*** | ***Fibers*** | | ***Matrix*** | ***UTS1*** | ***UTS2*** | ***UTS3*** |
|---|---|---|---|---|---|---|
| *type 1* | *Carbon 100%* | *Vetiver0* | *Epoxy* | *12.27* | *12.29* | *13.02* |
| *type 2* | *Carbon 90%* | *Vetiver10* | *Epoxy* | *12.24* | *12.27* | *12.99* |
| *type 3* | *Carbon 80%* | *Vetiver20* | *Epoxy* | *12.23* | *12.36* | *12.78* |
| *type 4* | *Carbon 70%* | *Vetiver30* | *Epoxy* | *12.25* | *12.45* | *12.64* |
| *type 5* | *Carbon 60%* | *Vetiver40* | *Epoxy* | *12.26* | *12.35* | *12.56* |
| *type 6* | *Carbon 50%* | *Vetiver50* | *Epoxy* | *12.1* | *12.02* | *12.45* |
| *type 7* | *Carbon 40%* | *Vetiver60* | *Epoxy* | *12.02* | *12.35* | *12.05* |
| *type 8* | *Carbon 30%* | *Vetiver70* | *Epoxy* | *11.56* | *12.36* | *11.95* |
| *type 9* | *Carbon 20%* | *Vetiver80* | *Epoxy* | *11.28* | *12.09* | *11.64* |
| *type 10* | *Carbon 10%* | *Vetiver90* | *Epoxy* | *11.02* | *11.23* | *11.54* |
| *type 11* | *Carbon* | *Vetiver100* | *Epoxy* | *10.98* | *11.05* | *11.27* |

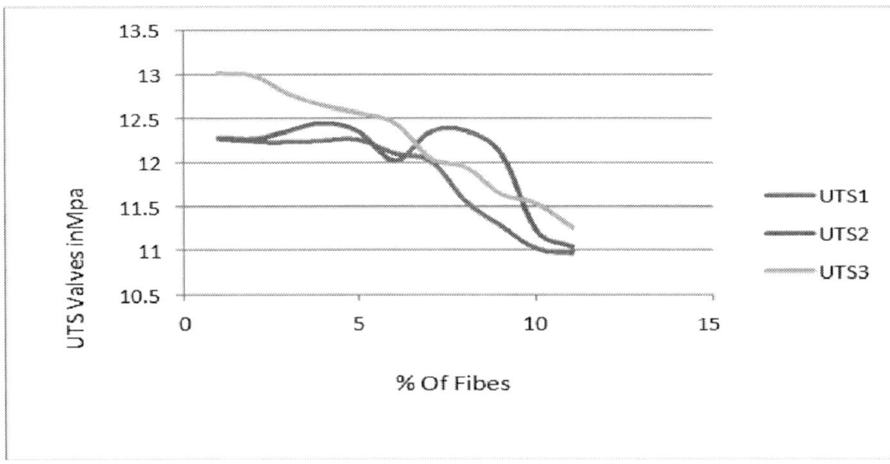


Figure 3. Ultimate tensile strength with % of fibers

The Figure 3 shows the variation between the UTS Values in Mpa to that of the % of fibers. The peak values are recorded at 13, 12.25, 12.25, for UTS 3, UTS 2, UTS 1 curves respectively, and the trough values are recorded at 11, 11.25, 11.25 for UTS 1, UTS 2, UTS 3 curves. This indicates that with the % of the increase in fibers the UTS value decreases.

IOP Conf. Series: Materials Science and Engineering **923** (2020) 012039 doi:10.1088/1757-899X/923/1/012039

*6.2. Flexural Test*

For the figure 2, ***the*** flexural specimens are equipped as per the ASTM D790 standard . Each specimen having 100 mm length and 12.7 mm width and the following observations were recorded

Table 2. Flexural strength of composites

| ***Orientation*** | ***Fibers*** | | ***Matrix*** | ***Flexural Strength N/mm²*** | ***Flexural Strength N/mm²*** | ***Flexural Strength N/mm²*** |
|---|---|---|---|---|---|---|
| *type 1* | *Carbon 100%* | *Vetiver0* | *Epoxy* | *21.5* | *21.9* | *22.23* |
| *type 2* | *Carbon 90%* | *Vetiver10* | *Epoxy* | *21.2* | *21.7* | *22.28* |
| *type 3* | *Carbon 80%* | *Vetiver20* | *Epoxy* | *21.8* | *21.6* | *21.98* |
| *type 4* | *Carbon 70%* | *Vetiver30* | *Epoxy* | *21.4* | *21.4* | *21.78* |
| *type 5* | *Carbon 60%* | *Vetiver40* | *Epoxy* | *21.6* | *21.8* | *21.64* |
| *type 6* | *Carbon 50%* | *Vetiver50* | *Epoxy* | *21.2* | *21.3* | *21.23* |
| *type 7* | *Carbon 40%* | *Vetiver60* | *Epoxy* | *20.8* | *21.1* | *21.89* |
| *type 8* | *Carbon 30%* | *Vetiver70* | *Epoxy* | *20.2* | *20.9* | *20.05* |
| *type 9* | *Carbon 20%* | *Vetiver80* | *Epoxy* | *20.4* | *20.7* | *19.9* |
| *type 10* | *Carbon 10%* | *Vetiver90* | *Epoxy* | *19.58* | *20.4* | *19.92* |
| *type 11* | *Carbon* | *Vetiver100* | *Epoxy* | *19.23* | *20.7* | *19.85* |

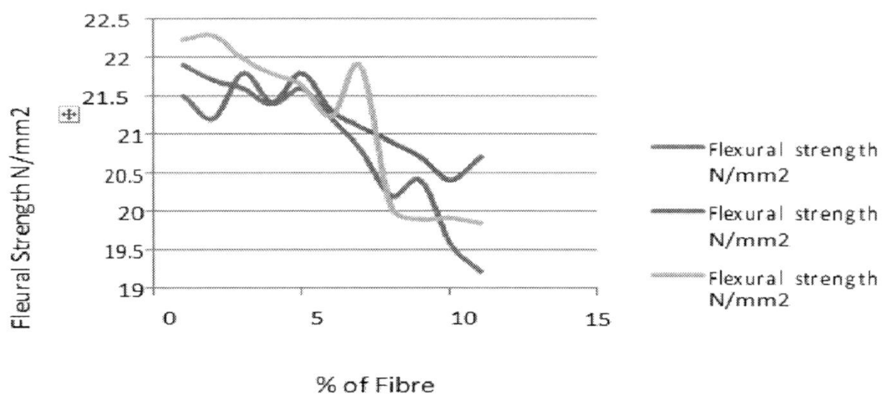


Figure 4. Flexural strength with % of fibers

The result shown that the various combinations in mould method and are discussed with the graph in Flexural testing. The Figure 4 shows the variations between the Flexural strength and the % of fibre.

IOP Conf. Series: Materials Science and Engineering **923** (2020) 012039 doi:10.1088/1757-899X/923/1/012039

the peak values are recorded at 22.25 , 22 , 21.5 corresponding to the flexural strength values 3 , 2, 1 respectively and the trough values are observed at 19.25 , 20 and 20.5 for the flexural strengths 1 , 3 and 2 respectively.

### *6.3. Impact Test*

This test is carried out on the samples are made as per the ASTM D256 standard. The absorbed energy by the specimen is noted until fracture takes place during testing.

Table 3. Impact strength of composites

| *Orientation* | *Fibers* | | *Matrix* | *impact valve in j* | *impact valve in j* | *impact valve in j* |
|---|---|---|---|---|---|---|
| *type 1* | *Carbon 100%* | *Vetiver0* | *Epoxy* | *9.8* | *9.5* | *9.7* |
| *type 2* | *Carbon 90%* | *Vetiver10* | *Epoxy* | *9.7* | *9.2* | *9.9* |
| *type 3* | *Carbon 80%* | *Vetiver20* | *Epoxy* | *9.6* | *9.3* | *9.5* |
| *type 4* | *Carbon 70%* | *Vetiver30* | *Epoxy* | *9.4* | *8.7* | *9.8* |
| *type 5* | *Carbon 60%* | *Vetiver40* | *Epoxy* | *9.3* | *8.5* | *9.7* |
| *type 6* | *Carbon 50%* | *Vetiver50* | *Epoxy* | *9.2* | *8.2* | *9.6* |
| *type 7* | *Carbon 40%* | *Vetiver60* | *Epoxy* | *8.8* | *8.4* | *9.2* |
| *type 8* | *Carbon 30%* | *Vetiver70* | *Epoxy* | *8.6* | *8.3* | *9.1* |
| *type 9* | *Carbon 20%* | *Vetiver80* | *Epoxy* | *8.5* | *8.2* | *8.8* |
| *type 10* | *Carbon 10%* | *Vetiver90* | *Epoxy* | *8.4* | *8.1* | *8.2* |
| *type 11* | *Carbon* | *Vetiver100* | *Epoxy* | *8.3* | *8.5* | *8.1* |

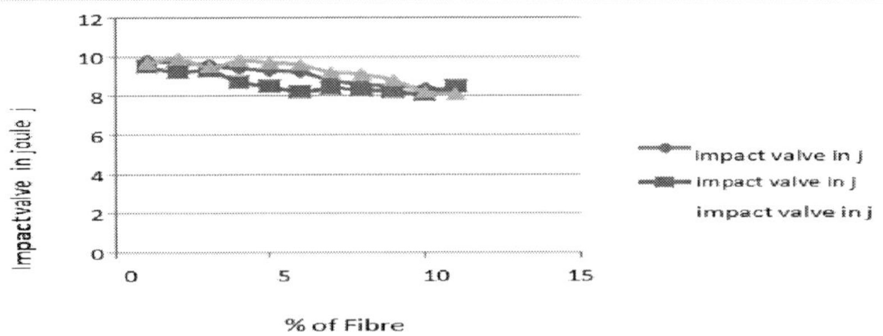


Figure 5. Impact strength with % of fibers

The Figure 5 plots the variation of Impact value to that of the % of Fibre , the peak values are observed at 10 for all the impact value trials and the trough values are observed at 8 . This shows that the Impact value decreases with the increase in % of fibres

*Hardness*

Barcoll hardness test is carried out to find out the hardness of the specimen. Taking major load as 150 grams and minor load as 50 grams this test is carried. After applying the minor load for a minute then the major load should be applied. This test is specifically suited for polymer samples. Hardness scale C should be noted having its pin type indenter. Size of the specimen used for testing is 10*10 cm having a thickness of 4 cm.

Table 4. Hardness of composites

| *Orientation* | *Fibers* | | | *Matrix* | *Hardness1* | *Hardness2* | *Hardness3* |
|---|---|---|---|---|---|---|---|
| *type 1* | *Carbon* | *100%* | *Vetiver0* | *Epoxy* | *33* | *11* | *35* |
| *type 2* | *Carbon* | *90%* | *Vetiver10* | *Epoxy* | *35* | *36* | *32* |
| *type 3* | *Carbon* | *80%* | *Vetiver20* | *Epoxy* | *36* | *35* | *31* |
| *type 4* | *Carbon* | *70%* | *Vetiver30* | *Epoxy* | *32* | *32* | *38* |
| *type 5* | *Carbon* | *60%* | *Vetiver40* | *Epoxy* | *31* | *33* | *32* |
| *type 6* | *Carbon* | *50%* | *Vetiver50* | *Epoxy* | *33* | *35* | *34* |
| *type 7* | *Carbon* | *40%* | *Vetiver60* | *Epoxy* | *32* | *36* | *39* |
| *type 8* | *Carbon* | *30%* | *Vetiver70* | *Epoxy* | *29* | *35* | *37* |
| *type 9* | *Carbon* | *20%* | *Vetiver80* | *Epoxy* | *27* | *32* | *34* |
| *type 10* | *Carbon* | *10%* | *Vetiver90* | *Epoxy* | *29* | *31* | *34* |
| *type 11* | *Carbon* | | *Vetiver100* | *Epoxy* | *30* | *30* | *32* |

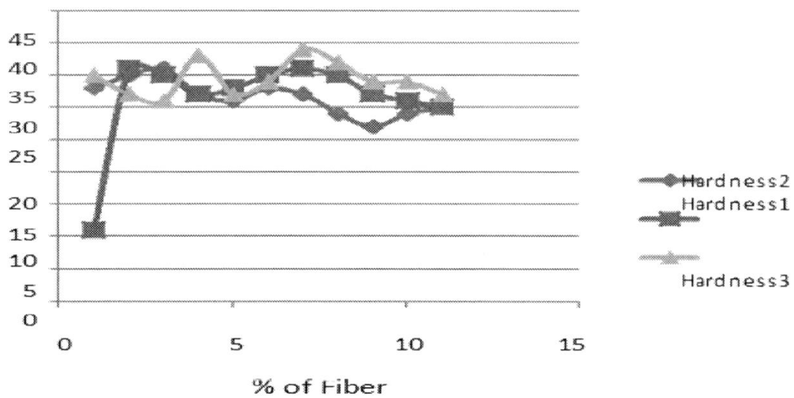


Figure 6. Hardness with % of fibers

IOP Conf. Series: Materials Science and Engineering **923** (2020) 012039 doi:10.1088/1757-899X/923/1/012039

The Figure shows that the various of the combination in mould method and discussed with the graph in Hardness. The highest Hardness values are observed at 35-40 for all the the hardness tests respectively and the through values are recorded at 30 , 35, 37 for the hardness tests 2 , 1 , 3 respectively.

## 7. Fracture surface using SEM

A scanning electron microscope (SEM) is a group of electron microscope that generates images of a sample by scanning it with a focused beam of electrons which are used to observe the topography and morphology of a specimen. The electrons collaborate with atoms in the sample, generating various signals that can be detected and that contain information about the sample's surface topography and composition. The analysis of fractured surfaces of impact test specimens is analyzed using SEM. The samples are taken for the SEM Analysis as they have the large variations in the mechanical test results as well variation in the composition and melting temperature.

*7.1 SEM Image of vetiver*

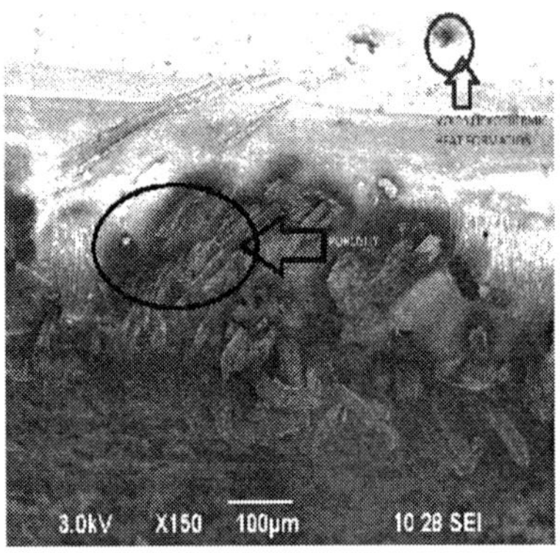

Fig 7: Tensile Fracture surface

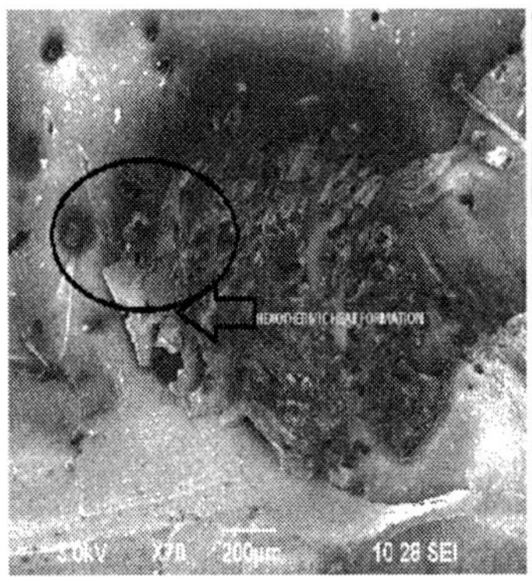

Fig 8 : Flexural fracture

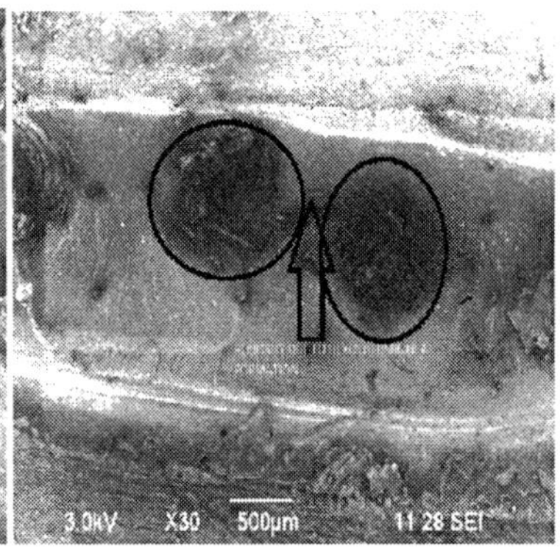

Fig 9: Impact Fracture

The following fracture observations are obtained through the SEM and correspondingly the Fig.7 indicates the tensile fracture surface and Fig.8 indicates the flexural fracture of the specimen and Fig.9 depicts the impact fracture.

## 8. Conclusion

The vetiver and coconut fiber are made by hand layup method. The test result is shown in figures and graphs. The tensile test value results in higher UTS value in Coconut fiber. Hence that material would withstand in the above elastic region. UTS improved when the hardness value is increased hence the material withstand in a loading condition. Corresponding wear resistance is improved. The wear resistance is improved when the material withstands in corrosion wear and flexural and impact value also show the higher flexural strength and joules. The Flexural is improved when the corresponding bending strength is improved. And the impact value is improved when the ballistic load condition also improved. The fracture surface shown in SEM images. The SEM images have shown that the Voids occur in a vetiver fiber. But the coconut fiber results in low voids and porosity. Hence It can be concluded that Coconut Fibre can be used in a Break Pad material because it has no exothermic heat, no voids, porosity, and excessive tensile strength. The wear, corrosion TGA tests will be taken for wind turbine material is an added advantage.

## Reference

[1]. M. Buneaa,*, A. Cîrciumarua, etal, fiber orientation on low velocity impact behavior of the fabric reinforced hybrid compositesdoi.org/10.1016/j.compscitech.2018.11.024.

[2]. Brian Burks, James Middleton, Maciej Kumosa, the compare.ison of carbon fiber glass fiber reinforced composites, August 2012, Composites Science and Technology

[3] Brian Burks, Maciej Kumosa, the flexural performance, dx.doi.org/10.1016/j.compscitech.2012.07.018,

[4] saravanaBavan D. and Mohan Kumar G.C.A Simple hand lay-up method. Tensile and flextural test creativecommons.org/licenses/by-nc-nd/3.0/). Procedia Materials Science 5 ( 2014 ) 605–611

[5] M. Indra Reddy a*, M. Anil Kumar a, Ch. Rama Bahadri Raju a, the tensile and flexural characterization of polymer hybrid composites, (July 29th – 30th) 2016, Materials Today: Proceedings 5 (2018) 458–462

*[6]* Al-Hajaj, Z., Sy, B.L., Bougherara, H., Zdero, R., Impact Properties of a New Hybrid Composite Material, *Composite Structures* (2018), doi:https://doi.org/10.1016/j.compstruct.2018.10.033

[7] Sivakiran.G.a, Yash Gangwaletal., Investigations on Hybrid Polymer Matrix Composite Materials, Materials Today: Proceedings 5 (2018) 7908–7914

[8] [1].Finite element analysis of fiber-reinforced polymer composites J. Naveen Mohammad Jawaid Modeling of
Damage Processes in Biocomposites, Fibre-Reinforced Composites and Hybrid Composites https://doi.org/10.1016/B978-0-08-102289-4.00009-6

[9]. Azwa ZN, Yousif BF, Manalo AC, Karunasena W. A review on the degradability ofpolymeric composites based on natural fibres. Mater Des 2013;47:424e42.

[10]. Mahzan S, et al. UV radiation effect towards mechanical properties of natural fibre reinforced composite material: a review. IOP Conf Ser Mater Sci Eng2017;165:1e9.

[11]. Bakare FO, Ramamoorthy SK, Åkesson D, Skrifvars M. Thermo mechanical properties of bio-based composites made from a lactic acid thermoset resin and flax and flax/basalt fibre reinforcements. Composites Part A 2016.

[12]. Del Masto A, Trivaudey F, Guicheret-Retel V, Placet V, Boubakar L. Nonlinear tensile behaviour of elementary hemp fibres: a numerical investigation of the relationships between3D geometry and tensile behaviour. J Mater Sci 2017;52(11):6591e610

IOP Conf. Series: Materials Science and Engineering **923** (2020) 012040 doi:10.1088/1757-899X/923/1/012040

# Exploration of Properties of Meat Waste Incinerated Fly ash / Cement / Sand Dust associated Fly ash Bricks

**J Thanikachalam*[*] and N Ganesh**

Department of Mechanical Engineering, Mepco Schlenk Engineering College, Sivakasi, Tamil Nadu, India

[*]E-Mail: jthanik@gmail.com

**Abstract**. In this modern technology the generation of meat waste causes the considerable impact to the environment and it creates higher level of air pollution, water pollution and land pollution. These pollution causes ill effects to human beings and other living organisms by the creation of harmful micro-organisms which accounts to spreading of diseases. In order to addressing this issue and to utilize the waste as the part of waste management. The meat waste that are incinerated in the specially designed incinerator which does not create any pollution. The fly ash obtained from the incineration of meat waste was incorporated in the manufacturing of fly ash bricks along with the fly ash obtained from the thermal power plant and these mixture are mixed with the sand / stone dust by using cement as the binding agent. Under various proportions the brick was manufactured according to the NTPC guidelines for manufacturing Quality fly ash bricks. After manufacturing the bricks, Brick's property such as Compression strength test, Water absorption test and the Efflorescence property test were conducted in order to identify the best sample out of all manufactured fly ash brick samples. On manufacturing these types of bricks the environmental caused by the meat waste was reduced drastically.

*Keywords* – Incineration, Meat waste, NTPC, Compression, Efflorescence, Water absorption.

## 1. Introduction

In the present system the incinerators that are handling the wastes are burnt inside the setup but through the exhaust manifold, the burnt gases travels are set free and released to the atmosphere. Because of that the harmful exhaust gases contains $NO_X$, $SO_X$, $CO_2$, CO etc. are released to the surrounding atmosphere and thus creates the environmental pollution. This research primarily focus to reduce the environmental air pollution by treating the exhaust gases as harm free by sending the burnt gases through the smoke collector and particle collector setup and the remaining carbon content gases are made to pass through the tank that contains some amount of water for absorbing the gases. The remaining meat waste burnt fly ash was collected in the incinerator vessel itself and that can be used for manufacturing fly ash bricks under various proportions.

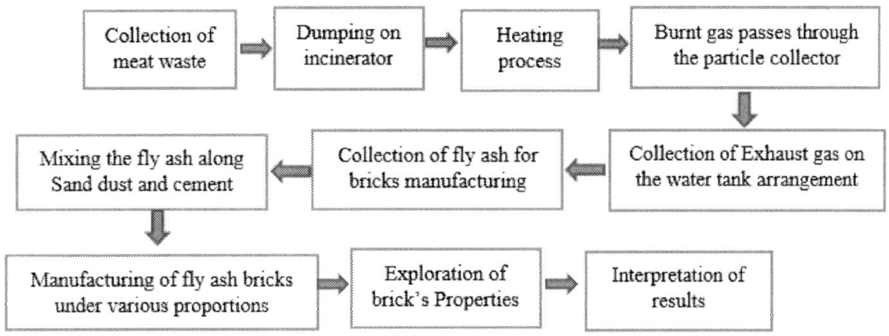


**Figure 1.** Research Flow

Pooja G. Nidoni "Incineration process for solid waste management and effective utilization of byproducts" - MSWI (Municipal Solid Waste incineration) was heat treated and the by-products obtained was incorporated to the plant growth was suggested in an Eco friendly manner with Air pollution control devices. [1]

Ankur tayal et al "Compressive Strength of Fly Ash Bricks" - Here the Fly ash imposed bricks are manufactured and also stated that the fly ash that are used in manufacturing of bricks has the great deal with that of mixing with the concrete, this literature also states the compression strength of various proportions of fly ash bricks. [2]

A. Sumathi et al "Compressive Strength of Fly Ash Brick with Addition of Lime, Gypsum and Quarry Dust" From this literature the compression strength of fly ash brick with addition of lime, gypsum and quarry dust. Here the authors made the fly ash bricks at various proportions and the test was taken.[3]

Tabin Rushad S et al "Experimental Studies on Lime – soil Fly ash Bricks" From this Literature It is Known that MDD – Maximum Dry Density then the standards and denstity of soil, Fly ash are understood then the manufacturing Methodology of Flyash Bricks and the test to be taken are understood.[4]

## 2. Design Modelling of the Incinerator

The 3D Model of incinerator was drawn using Solid works software by considering all the parameters. The Isometric view of the incinerator and the cut section view of the incinerator was shown in the figure 2. Here the meat waste placed on the arrangement as shown in the cut sectional view and subjected to burn. The meat waste on burning, leaves the gases outside with the help of exhaust gases arrangements and it is collected on the water tank arrangement through the exhaust manifold. So no gases are released outside. The harmful gas particles are stick on the particle collector arrangement. So that design confirms the minimised pollution. The prototype was made of mild steel materials.

The electrical connections are done separately for two exhaust fans, Electric heating coil arrangements.

IOP Conf. Series: Materials Science and Engineering **923** (2020) 012040 doi:10.1088/1757-899X/923/1/012040

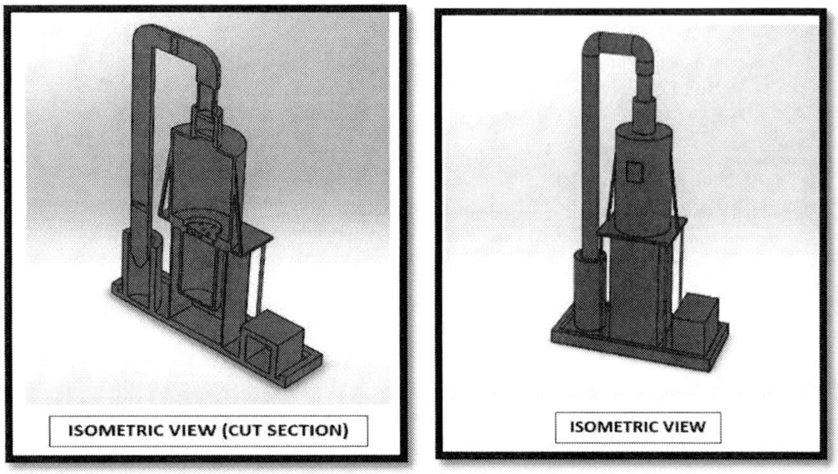


**Figure 2.** Design Modelling

## 3. ANSYS Modelling

The ANSYS modelling of incinerator was done for understanding the gas flow path, temperature, velocity analysis. The results shows the maximised value that are within the permissible limits and in the accepted criteria.

There was no pressure build up inside the incinerator vessel because the burnt gases are released to the surroundings and made to settle on the water tank which was set open to atmosphere. The heavier force from the exhaust fans in the pipeline made the burnt carbon content gases to dissolve in the water, only the remaining gases are send outside which were harmless gases. This ensures that there were no pressure build up in the vessels.

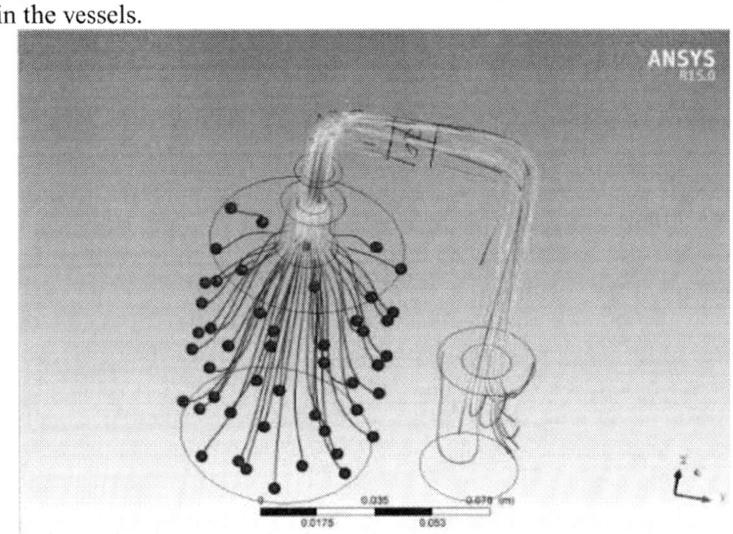


**Figure 3.** Flow Path Analysis

The gas flow path analysis indicates the flow of gases inside the incinerator and also it describes the behaviours of fluids on reaching the pipe and exhaust fans. The fluids (burnt gases) on reaching the narrow area, the velocity increases to a greater value. This was the observation from the above mentioned analysis.

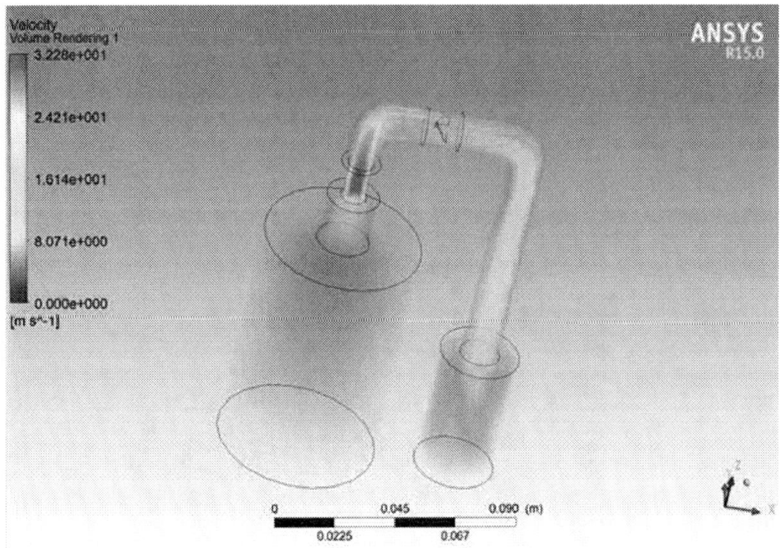


**Figure 4.** Velocity Analysis

The velocity analysis indicates the velocity at which the burnt gases are travelled inside the vessel. Here the velocity is low (8 m/s) in the area of water tank and also in the meat waste placing zone and the velocity is higher (32 m/s) in the area where the fluids reaches the plates.

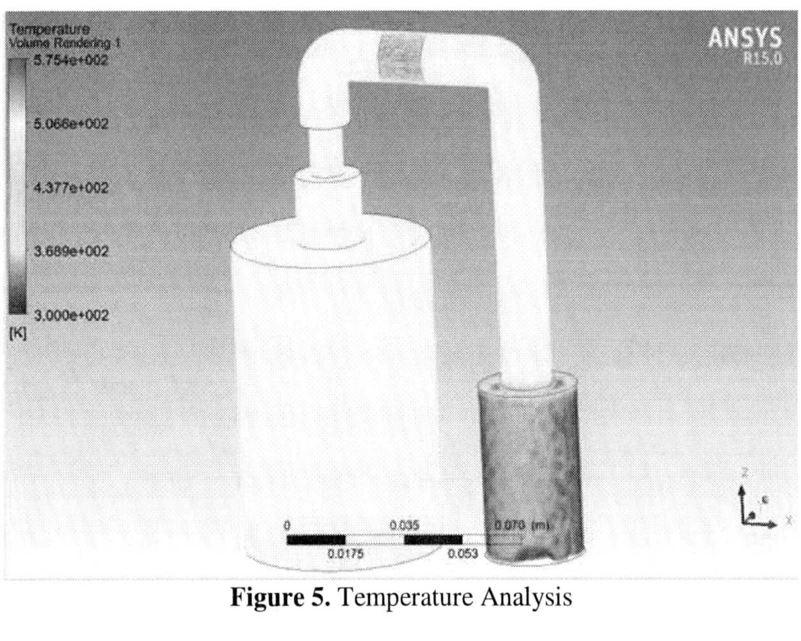


**Figure 5.** Temperature Analysis

The temperature analysis indicates the temperature level of the incinerator while burning, the temperature is of minimised level at the water tank arrangements (368 K) and the temperature is of medium level in the incinerator meat waste placing zone (506 K) and the temperature is of higher (570 K) on the place where the fluids passes the exhaust fan setup.

## 4. Experimental Setup

**Figure 6.** Incinerator Prototype

The entire experimental setup consists of Electric heater, smoke and particle collector setup, Exhaust fans and the ash depositing tank. The meat waste was dried and incinerated by using the prototype as shown above. The entire prototype is manufactured by using Mild Steel material.

## 5. Fabrication of Fly ash Bricks

Using the Fly ash obtained by above method. The meat waste incinerated fly ash and the fly ash obtained from the thermal power plant whose density [4] is of 800 $kg/m^3$ are mixed with proportion of (25:75). Then the sand dust with the density [5] of 1788.07 $kg/m^3$ and the cement with the density [6] of 1500 $kg/m^3$ are mixed. The material requirement for the manufacturing of fly ash was calculated as follows.

### *5.1 Material Allocation*

Weight of the brick Mixture = L x B x H x [{Percentage of Fly ash x Density of Fly ash} + {Percentage of Sand x Density of Sand} + {Percentage of Cement x Density of Cement)]

(L, B, H are the Mould Specifications in m)

Bricks with Five samples of different proportions are made. The proportions allocation is given as follows. Proportion (a: b: c) indicates (Fly ash %: sand dust %: Cement %).

The proportions are made under the guidelines of NTPC guide for manufacturing quality Fly ash bricks.

**Table 1.** Samples and Proportions

| Sample | Proportion (a:b:c) |
|---|---|
| 1 | (50:40:10) |
| 2 | (52:38:10) |
| 3 | (54:36:10) |
| 4 | (56:34:10) |
| 5 | (58:32:10) |

The material allocation is done by the following relations. The model calculation for sample 1 proportion is given as follows.

*5.2 Weight of the brick Mixture*

= 0.23 x 0.1 x 0.09 x [{0.5 x 800} + (0.4 x1788.07) + (0.1 x 1500)]
= 0.00207 x [400 + 715.228 + 150]
= 2.618 kg

Similarly, for the samples 2,3,4,5 the material allocation was done and it is tabulated as follows.

**Table 2.** Table for Material Allocation

| Sample | Proportion (a:b:c) | Weight of Fly ash (kg) | Weight of Sand / Stone Dust (kg) | Weight of Cement (kg) | Total Material Weight (kg) |
|---|---|---|---|---|---|
| 1 | (50:40:10) | 1.309 | 1.047 | 0.261 | 2.618 |
| 2 | (52:38:10) | 1.340 | 0.979 | 0.257 | 2.577 |
| 3 | (54:36:10) | 1.377 | 0.918 | 0.255 | 2.551 |
| 4 | (56:34:10) | 1.397 | 0.848 | 0.249 | 2.496 |
| 5 | (58:32:10) | 1.423 | 0.785 | 0.245 | 2.455 |
| Total Material Weight (kg) | | 6.846 | 4.577 | 1.267 | |
| Total Material Weight for Making 4 Bricks in each Sample in (kg) | | 27.384 | 18.308 | 5.068 | |

*5.3 Manufacturing of Fly ash Bricks*

The Fly ash bricks are manufactured by mixing three products under 5 different proportions with the favour of brick mould. The bricks that are manufactured that are air cured and then water cured and subjected to test conditions.

IOP Conf. Series: Materials Science and Engineering **923** (2020) 012040 doi:10.1088/1757-899X/923/1/012040

**Figure 7.** Manufacturing of Bricks, Air Curing, Water Curing of Bricks

## 6. Experimental Results

For the above brick samples made under various proportions, Water Absorption, Compression Strength and the Efflorescence test were done. The results of such tests are discussed as follows.

### *6.1 Water Absorption Test*

For all the samples the water absorption test was conducted as per NTPC Guidelines and also as per IS 3495 (Part 2): 1976[7]. The brick was soaked in the cold water for 24 hours and the water absorption rate must not be greater than 20%.

$$\text{Water Absorption Test (\%)} = \frac{(W_2 - W_1) * 100}{W_1}$$

Where $W_1$ and $W_2$ are the initial and final Weight of the brick samples in kg.

**Table 3. Water Absorption Test Observation**

| Sample | Proportion (a:b:c) | $W_1$ (kg) | $W_2$ (kg) | $W_2 - W_1$ (kg) | $\frac{(W_2 - W_1)}{W_1}$ | $\frac{(W_2 - W_1) * 100}{W_1}$ |
|---|---|---|---|---|---|---|
| 1 | (50:40:10) | 3.290 | 3.481 | 0.191 | 0.05805 | 5.805 |
| 2 | (52:38:10) | 3.281 | 3.486 | 0.205 | 0.06248 | 6.248 |
| 3 | (54:36:10) | 3.303 | 3.511 | 0.208 | 0.06297 | 6.297 |
| 4 | (56:34:10) | 3.284 | 3.494 | 0.213 | 0.06394 | 6.394 |
| 5 | (58:32:10) | 3.279 | 4.484 | 0.208 | 0.06251 | 6.251 |

Water absorption test was conducted for samples of all the Proportions. All the samples are within the limit as stated in NTPC Guide lines for manufacturing quality fly ash bricks.

From table it was observed that the "Greater the Sand dust proportion will increase the weight of the bricks".Out of these samples, Sample 1 can be used for the construction of water tank applications and other similar applications.

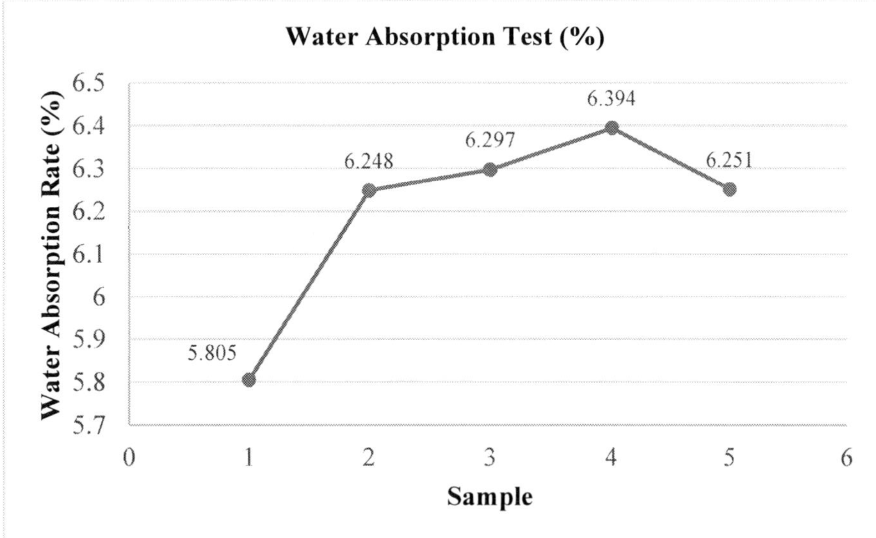


**Figure 8.** Plot of Water Absorption Test

*6.2 Compression Strength Test*

For all the samples the Compression Strength test was conducted as per NTPC Guidelines and also as per IS 3495 (Part 1): 1976[8]. The brick was subjected to Compression Strength Test and whose value shall not be less than 7.5 N/sq.mm.

Area of the Bed = Length x Height of the brick

**Table 4.** Compression Strength Test Observation

| Sample | Proportion (a:b:c) | Dimensions | | | Area of the bed ($mm^2$) | Maximum Load at Failure (KN) | Compression Strength | |
|---|---|---|---|---|---|---|---|---|
| | | Length (mm) | Breadth (mm) | Height (mm) | | | ($N/mm^2$) | ($kg/cm^2$) |
| 1 | (50:40:10) | 230 | 100 | 90 | 20700 | 263.87 | 12.748 | 130 |
| 2 | (52:38:10) | 230 | 100 | 90 | 20700 | 253.74 | 12.258 | 125 |
| 3 | (54:36:10) | 230 | 100 | 90 | 20700 | 223.29 | 10.787 | 110 |
| 4 | (56:34:10) | 230 | 100 | 90 | 20700 | 202.98 | 9.806 | 100 |
| 5 | (58:32:10) | 230 | 100 | 90 | 20700 | 172.14 | 8.316 | 85 |

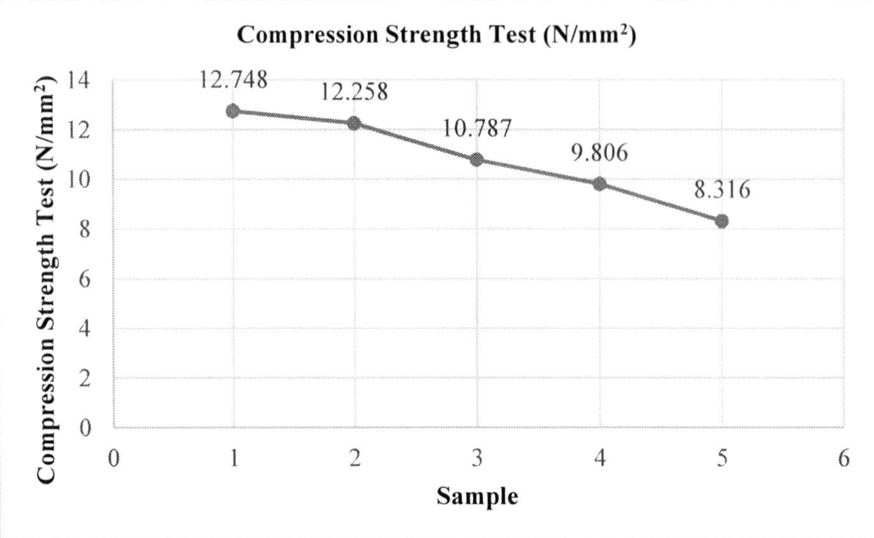


**Figure 9.** Plot of Compression Strength Test

Compression Strength test was conducted for samples of all the Proportions.All the samples are within the limit as stated in NTPC Guide lines for manufacturing quality fly ash bricks[9].

From chart, it was observed "Greater the Fly ash content reduces the compression strength of the bricks". Out of these samples, Sample 1 can be used for the construction of Tall Rise Buildings.

*6.3 Efflorescence Test*

For all the samples the Efflorescence test was conducted as per NTPC guidelines and also as per IS 3495 (Part 3) : 1976[9]. As per the standards the Effloescence rate shall not be more than that of Moderate.

Here none of the samples shows the white salty deposits on the bricks or the Efflorescence rate. All the samples are with in the limit as stated by the Indian standard.

**Table 5.** Efflorescence Property Observation

| Sample | Proportion (a:b:c) | Efflorescence Rating |
|---|---|---|
| 1 | (50:40:10) | Nil |
| 2 | (52:38:10) | Nil |
| 3 | (54:36:10) | Nil |
| 4 | (56:34:10) | Nil |
| 5 | (58:32:10) | Nil |

Efflorescence test was conducted for samples of all the Proportions.All the samples shows no efflorescense i.e., there is no white / salt deposit on the bricks. So that the bricks conforms the NTPC guidelines for quality bricks.

All these samples can be used for construction purposes.

IOP Conf. Series: Materials Science and Engineering **923** (2020) 012040 doi:10.1088/1757-899X/923/1/012040

## 7. Results and Discussion

From the design modelling, none of the gases were settled inside the vessel because of the implementation of Exhaust fan for forcing the gases outside the vessel and the good quantity of Fly ash was collected from the vessel and it was utilized for the useful application of making fly ash.

From the ANSYS modelling of incinerator it was understood the gas flow path analysis, the way by which the fluid behaves within the vessel was understood by means of animated model.

The velocity of the Fluid path was understood by means of analysis model for velocity it was understood there was no sudden hitting or velocity drop within the incinerator here the maximum velocity reaches to 32m/s by providing the inlet velocity to a minimal value.

The temperature withstanding capacity of incinerator while burning the meat waste the temperature that was raised to the maximum of 570 K. Entire incinerator setup was manufactured by means of Mild Steel material which has the melting point around (1535 K).

As per the Experimental results obtained from the above conducted tests, the Efflorescence property of the fly ash bricks was found good and all samples shows the same results. Then the compression strength and the water absorption test was analyzed and out of all the samples best sample was chosen as per analyses done below.

Water absorption strength need to be as low as possible for a quality Fly ash brick and the compression strength needs to be as high as possible for the construction of buildings.

**Table 6.** Final results of all the Observations

| Sample | Proportion (a:b:c) | Efflorescence Rating | Compression Strength ($N/mm^2$) | Water Absorption Rate (%) |
|---|---|---|---|---|
| 1 | (50:40:10) | Nil | 12.748 | 5.805 |
| 2 | (52:38:10) | Nil | 12.258 | 6.248 |
| 3 | (54:36:10) | Nil | 10.787 | 6.297 |
| 4 | (56:34:10) | Nil | 9.806 | 6.394 |
| 5 | (58:32:10) | Nil | 8.316 | 6.251 |

Thus the research work concludes from graph. It was observed that the Sample1 which holds the property of having "High Compression Strength and Low Water absorption Rate". So out of all the samples on analysis basis Sample 1 with the proportion of (50:40:10) – (Fly ash: Sand / Stone Dust / Cement) was the best proportion of all the samples that were manufactured.

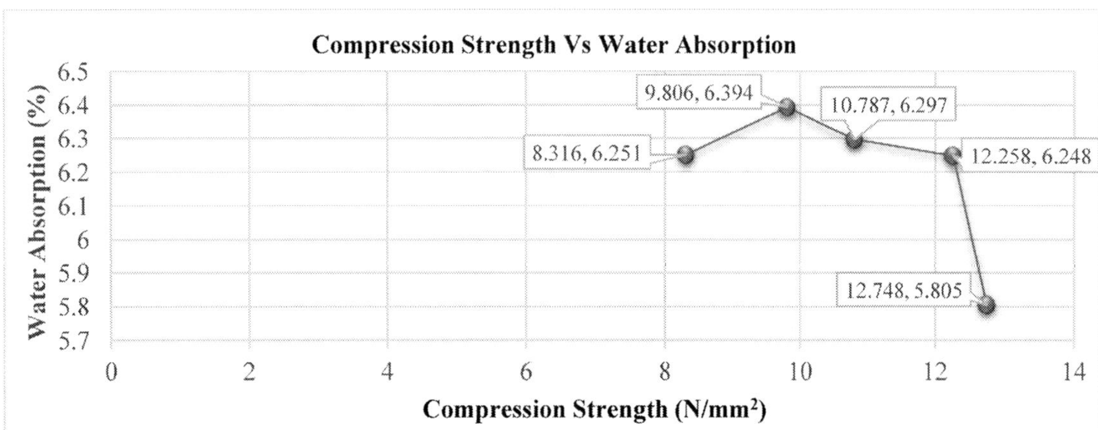


**Figure 10.** Plot of Compression Strength vs Water Absorption

**References**

[1] Pooja G. Nidoni "Incineration process for solid waste management and effective utilization of byproducts" - International Research Journal of Engineering and Technology (IRJET) - Volume: 04 Issue: 12, Dec-2017 - e-ISSN: 2395-0056.

[2] Ankur tayal et al "Compressive Strength of Fly Ash Bricks"- International Research Journal of Engineering and Technology (IRJET) e-ISSN: 2395-0056 Volume: 06 Issue: 05 | May 2019

[3] A. Sumathi et al "Compressive Strength of Fly Ash Brick with Addition of Lime, Gypsum and Quarry Dust" - International Journal of ChemTech Research - Vol.7, No.01, pp 28-36, ISSN: 0974-4290.

[4] Tabin Rushad S et al "Experimental Studies on Lime – soil Fly ash Bricks" - International journal of civil and structural Engineering Volume 1, No 4, 2011- ISSN 0976 – 4399.

[5] Meghashree. M et al "Comparison of Physical Properties between Natural Sand and Manufactured Sand" - IJIRST –International Journal for Innovative Research in Science & Technology| Volume 3 | Issue 07 | December 2016 ISSN (online): 2349-6010.

[6] Density of cement IS 2185.1.2005 Concrete masonry Unit.

[7] Water absorption test – "Water absorption test IS 3495 (Part 2) : 1976".

[8] Compression Strength Test – "Compression strength test IS 3495 (Part 1) : 1976".

[9] Effloresence Test - "Effloresence Test IS 3495 (Part 3) : 1976".

[10] Jeong Hoon Byeon et al "Collection of submicron particles by an electrostatic precipitator using a dielectric barrier discharge" – Journal of Aerosol Science.: 2006 - doi:10.1016/ j.jaerosci.2006.05.003.

[11] Zhang Xiangrong et al "An analysis ofa wire-plate electrostatic precipitator" – Journal of Aerosol Science - Aerosol Science 33 (2002) 1595–1600.

[12] M.J. Cuetos et al "Anaerobic digestion and co-digestion of slaughterhouse waste (SHW): Influence of heat and pressure pre-treatment in biogas yield" – waste Management, doi:10.1016/j.wasman.2010.01.034.

[13] U. Sonesson et al "Environmental and economic analysis of Management systems for biodegradable waste" – Elsevier Resources, Conservation and Recycling 28 (2000) 29–53 - Resources, Conservation and Recycling 28 (2000) 29–53.

[14] Michael L. Murphy - "Fluidized Bed Technology Solution to Animal Waste Disposal" – Seventeenth Annual International Pittsburgh Coal Conference .September, 2000.

[15] M. Adama et al "Heavy Metal Contamination of Soils around a Hospital Waste Incinerator Bottom Ash Dumps Site" – Hinwadi Journal - http://dx.doi.org/10.1155/2016/8926453.

[16] D. R. Edwards et al - "Environmental Impacts of On-Farm Poultry Waste Disposal - A Review" Journal of Bio resource Technology - Bioresource Technology 41 (1992) 9-33.

IOP Conf. Series: Materials Science and Engineering **923** (2020) 012041 doi:10.1088/1757-899X/923/1/012041

# Dilatometric Methods: Insights of other researchers

**E Pavithra***

Associate Professor, Department of Mechanical Engineering, Vel Tech Rangarajan Dr Sagunthala R & D Institute of Science and Technology, Avadi, Chennai, India

*E-mail: epavithrasenthil@gmail.com

Abstract. Measurement of thermal expansion and contraction of a solid material during heating and cooling is done using dilatometer. To obtain best results dilatometers are operated under vacuum or inert atmosphere. The physical and chemical characteristics are inferred from the dilation of the material. It helps to study the properties of the sintered material and the phase transition of the solid material. Few authors also developed a customized dilatometer for their research applications. This paper deals with those different designs, the various material characteristics providing adequate data to make researchers identify the new area of research for the dissimilar materials. The thermo physical chemical property studies based on the dilatometer are reviewed in this paper.

*Keyword: Dilatometer, solid material, thermo physical chemical property*

## 1. Introduction

A well known method for thermal analysis of any solid material is by dilatometry. The change in length and the volume is measured when there is a change in temperature. Also the dilatometer provides the observation on the phase transformations using the microstructure Characterization. As per the ASTM E228-06, dilatometer is measured. The coefficients of thermal expansion are determined between 25°C and 1000°C. It is estimated between ±5% below 500°C and ±3% above 500°C at a heating rate of 3°C/minute [1].The single phase quasi crystalline is formed while heating in the isotherm annealing, prolonged heating results synthesized powder quasicrystals. The same is observed in the investigation of dilatometer and DTA [2]. Mostly Dilatometer is categorized into Linear, volume, capacitive, deformation and non-contact types. Among the above all, capacitive methods are sensitive to small changes in length during measurement of thermal expansion and magnetostrsiction measurements within solids. For various experiments different types of dilatometers were developed namely small dilatometers for magnetic systems, absolute-expansion dilatometers and open-architecture dilatometers for sample mounting arrangement [3]. When compared with X-ray method which estimates correct dimensional changes induced due to thermal variation. Better services are provided when transducing techniques are equipped with sophisticated tools namely high speed photographic techniques, optical interferometer and capacitance change measurement for calibration purpose [4]. Modern dilatometric techniques find wide variety of applications ranging from metals, alloys and engineering; powder metallurgy to polymers; ceramics and glass industries. Soil properties and stratigraphy were evaluated using dilatometer test (DMT) which is a fast and economic tool [5].

IOP Conf. Series: Materials Science and Engineering **923** (2020) 012041 doi:10.1088/1757-899X/923/1/012041

## 2. Insights of Dilatometer

Density measurement was precisely measured by using both dilatometer and traveling microscope with a least count of 0.001cm. For temperature control, dilatometer is covered with water and kept inside a cylindrical glass [6]. The dilatometer provides the data for the study of the volume change. But it is necessary to consider the kinetics of $\gamma$ to $\alpha$ of the sample. So, calculated length change can be deduced as a 'virtual' dilatometer signal. But the unmatched results will not provide the sufficient data. Hence plasticity of the material should be studied to obtain the pure phases [7].

To measure the volumetric growth and determine ASR feature, unlike the presence of fly ash motor and concrete samples containing New Mexico Rhyolite (NMR) and Platte River Gravel (PRG) were examined using dilatometer at different alkalinities [8].

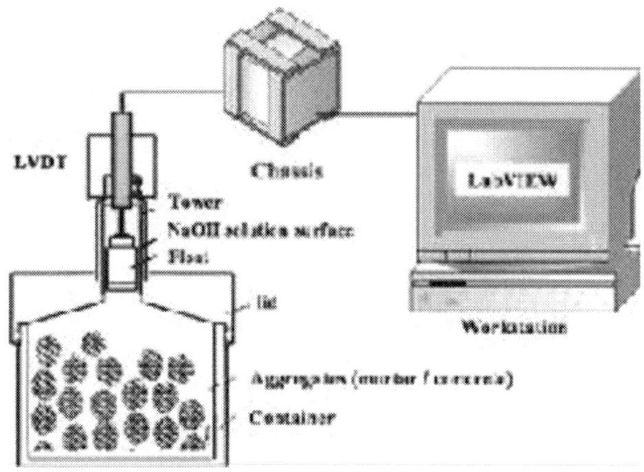


**Figure 1.** Ghanem (2012)

For processing, dilatometer is kept at a distance of 14cm from a spectrum lamp of PRK-2 which works under stabilized working conditions. Thermostat baths were used to carry out the process with temperature control of ±0.05° [9]. With solid state reaction, compositions in series $\{Pb(Fe_{0.5}Nb_{0.5})O_3\}_{1-x} - \{PbTiO_3\}_x$ (x = 0 ->1) were developed. Differential thermal analysis has been studied as a function of x where the phase shift changes from tetragonal/rhombohedra to cubic structure, where the x-ray dilatometry and bending high temperature ranges from 298-773K. Thermal expansion and dielectric co-efficient of 0.2 indicates relax or type behavior of lead iron niobate upto 453K [10]. The effect of isothermal transformation of cold-rolled high strength TRIP steel was investigated using dilatometry. Transformation of temperatures like Ac1, Ac3 and Ms were also studied by dilatometry curves [11].

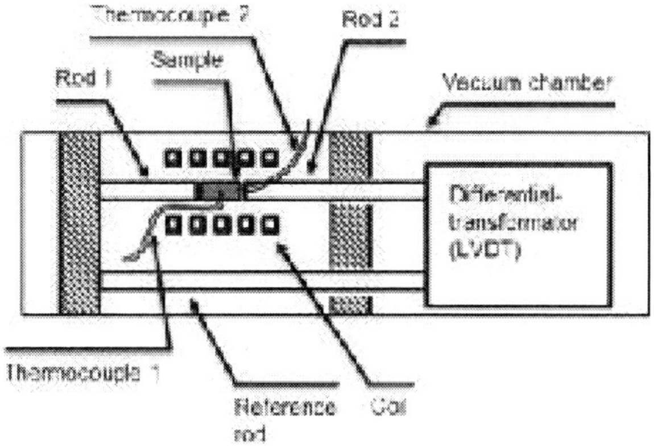


**Figure 2**. Kaiser (2016)

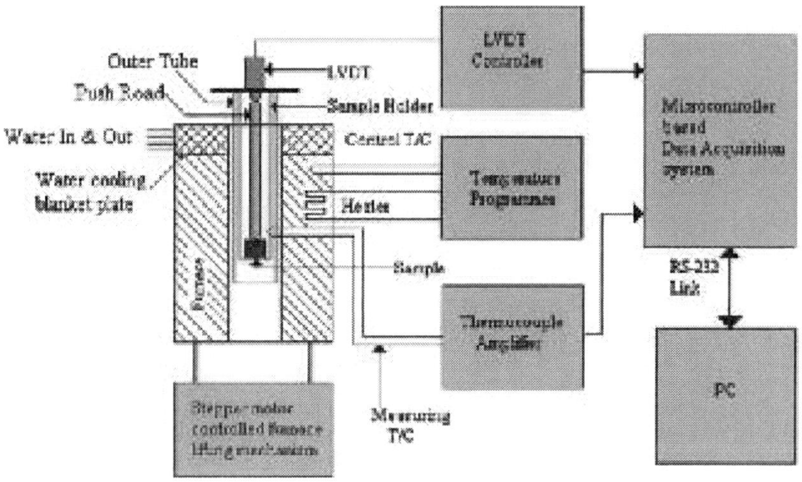


**Figure 3.** Manoj (2016)

In DIL 805, the longitudinal temperature gradient increases as heating rate increases. To describe the gradient of temperature inside the sample, 2D electromagnetic FEM simulations are deployed to develop a numerical model which consists of transfer of heat from the sample to dilatometer roads as well as heat conduction within the sample [12].

The dilatometer developed by Manoj et al was fabricated by the available sources at the low price and its ease in maintenance. It has got high temperature resolution (1 K, ±1 mm, 10nm). It is verified by NIST sapphire and high purity metals with nickel and aluminum. Model will be helpful for the thermo analytical analysis, all over the world [13].

An advanced dilatometer for variety of adhesives is designed domestically for the study of thermal expansion, polymerization shrinkage, change in volume and cooling contraction. By this new dilatometer, the effects of the cure parameters are widely investigated [14].

Through spray forming Zircon sand reinforced LM13 is fabricated for a length of 25mm. For a temperature range of 29–470 °C the thermal coefficient expansion for the base alloy and its corresponding composite is measured using Netzsch DIL 402 PC dilatometer [15]. Test coupons of Φ 5 × 25 mm were precisely machined and polished for dilatometer test. The specimens were heated up to 1273 K, with rate of heating rate at 1 K/min under argon atmosphere with a purity level of ≥99.999% during the measurement on Netzsch DIL-402C. The change in length as a function of temperature was plotted to understand the phase change [16].

Laser typed dilatometer is used to investigate the effect of distortion of forged steel which is homogenized at 1553K from the ingot for the current study. During phase transformation at the non-recrystallized region consists of coarse austenitic grain with the presence of low carbon and high Nb within the steel matrix. The microstructure observation also reveals the same [17]. In Netzsch dilatometer DIL 402C, a push rod dilatometer, Thermal growth of CVD single crystal silicon was measured up to 1100ºC. The resolution was ±0.125nm of the normal growth. It is functioned in the argon atmosphere, at a temperature of 5 K/min with continuous heating and cooling. The repeated experiments provide the average results and thus optimal output is recorded [18].

To determine the temperature of sintering for the pre-alloyed powder 718 under Ar atmosphere, Differential Scanning Calorimetry (DSC) and dilatometric studies were performed. The samples were sintered at 1260–1300ºC for a period of 1-3 hrs based on the results obtained from DSC and dilatometric tests [19]. Based on Vickers hardness, metallographic checks and dilatometry trials confirms the Continuous Cooling Transformation (CCT) diagram for the material. To measure the ferrite-to-austenitic change start ($Ac_1$) and finish ($Ac_3$) temperature a Formastor-Digital deformation dilatometer is used, similarly martensite start temperature and bainite start temperatures are also measured using dilatometer [20].

Thermal expansion using dilatometer is used to determine the expansion properties of lead zirconatetitanate–lead nickel niobate ((1−x)PZT–xPNN, x=0.1–0.5) ceramics for a temperature range of −100 to+450 °C [21]. In Bähr DIL 805 deformation dilatometer, direct strip cast, AZ31 magnesium alloy was examined by the flow curves. Between 300ºC and 400ºC dynamic recrystallization takes place [22].

Slurry which consists of 36.85wt% kaolin/clay, 33wt% water, 17.42wt%, 12.73wt% feldspar along with an industrial commercial deflocculant (Formsil D) was prepared. Using the dilatometer (Netzsch DIL 402 PC), the comparative linear change in length of the slurry was measured up to 1300 ◦C with 15 min dwell time and at 5 K/min heating and cooling rate [23].

Rockwell hardness, X-ray diffraction, impact toughness and dilatometer was used to investigate the IF (internal friction) of high alloyed martensitic carbon steel. The stage could be distinguished during the measurement. Change of material is dependent on the internal friction peak [24]. Classification of $LaAl_{0.3}Ni_{4.7}$ films performed using a capacitance dilatometer revealed that the lattice growth was relative to the partial pressure of $H_2$ over the range 0.01–1.3 atm used in this testing [25]. The transformation of the δ -› α't at or above 150 ºC is studied for the naturally aged sample or doping with 238Pu by using Netzsch 402C dilatometer with a low-temperature furnace and silica sample tube and pushrod [26]. To find the precise measurements of change in length on co efficient of thermal expansion of low thermal expansion materials is found using an optical heterodyne interferometric dilatometer for extreme ultraviolet lithography [27].

## 3. Dilatometry – a Holistic view

Dilatometry thermal analysis is studied on linear and volume matters. Thereby can be used for wide applications. Other researchers have also emphasized on dilatometry. We practically study on varied terms and could be done related to frequency modulation. At the same time, a newer material has to be identified so that we could use dilatometry in all range of sectors and will be used for inventing newer products which will be greatly beneficial. In seismic analysis for construction based work or earthquake work, flat dilatometry can be used to study. Soil pressure, soil properties, mechanical properties like sheer strength and sheer modulus can also be studied using flat dilatometry. Both flat and seismic dilatometry is used for soil based geology study.

## 4. Summary

Newer materials can be developed based on the thermal properties by using dilatometer. The study on the two or more dissimilar metals or alloys can be made with the newly developed thermal expansion characteristics. Moderate level of research has been done using existing studies. However, new researchers could use this as reference in future.

## REFERENCES

[1] Rabin B, Swank W and Wright R 2013 Thermophysical Properties of Alloy 617 from 25 C to 1000 C *Nuclear Engineering and Design* **262** 72-80

[2] Salimon A, Stepashkin A, Tcherdyntsev V, Olifirov L, Klyueva M and Kaloshkin S 2017 Towards the growth of single quasicrystalline grains in Al-Cu-(Fe, Cr) alloys after mechanical alloying and subsequent high temperature heating *Journal of Alloys and Compounds* **720** 95-104

[3] Abe S, Sasaki F, Oonishi T, Inoue D, Yoshida J, Takahashi D, Tsujii H, Suzuki H and Matsumoto K 2012 A compact capacitive dilatometer for thermal expansion and magnetostriction measurements at millikelvin temperatures *Cryogenics* **52** 452-6

[4] Raju S, Sivasubramanian K, Divakar R, Panneerselvam G, Banerjee A, Mohandas E and Antony M 2004 Thermal expansion studies on Inconel-600® by high temperature X-ray diffraction *Journal of nuclear materials* **325** 18-25

[5] Choo H, Lee W, Hong S-J and Lee C 2016 Application of the dilatometer test for estimating undrained shear strength of Busan New Port clay *Ocean Engineering* **115** 39-47

[6] Manohar R, Gupta M and Shukla J 2000 Phase transition studies of some cholesteric liquid crystals and their mixtures using dielectric, optical transmittance and density measurement techniques *Journal of Physics and Chemistry of Solids* **61** 1465-73

[7] Apel M, Benke S and Steinbach I 2009 Virtual dilatometer curves and effective Young's modulus of a 3D multiphase structure calculated by the phase-field method *Computational materials science* **45** 589-92

[8] Ghanem H, Zollinger D, Lytton R and Ghanem N 2012 Determining ASR characteristics using dilatometer method *Construction and building materials* **36** 1008-15

[9] Rafikov S, Leplyanin G, Varisova E, Naletova G and Kazakov V 1979 Control by complex formation of the inhibiting effect of anthracene in free radical polymerization *Polymer Science USSR* **21** 2170-6

[10] Sunder V S and Umarji A 1995 Thermal expansion studies in the lead iron niobiate—Lead titanate system *Materials research bulletin* **30** 427-34

[11] Srivastava A K, Bhattacharjee D, Jha G, Gope N and Singh S 2007 Microstructural and mechanical characterization of C–Mn–Al–Si cold-rolled TRIP-aided steel *Materials Science and Engineering: A* **445** 549-57

[12] Kaiser D, de Graaff B, Dietrich S and Schulze V 2016 A novel procedure to account for high temperature gradients in an induction dilatometer sample during rapid heating *Thermochimica Acta* **646** 8-15

[13] Manoj N, Jain D, Gautam J, Thomas K, Sudarsan V, Pillai C, Vatsa R and Tyagi A 2016 A simple, reliable and cost-effective, high temperature dilatometer for bulk thermal expansion studies on solids *Measurement* **92** 318-25

[14] Yu H, Adams R and Da Silva L F 2013 Development of a dilatometer and measurement of the shrinkage behaviour of adhesives during cure *International Journal of Adhesion and Adhesives* **47** 26-34

[15] Kaur K and Pandey O 2010 Microstructural characteristics of spray formed zircon sand reinforced LM13 composite *Journal of alloys and compounds* **503** 410-5

[16] Chen F-W, Xu G, Zhang X-Y, Zhou K-C and Cui Y 2017 Effect of α morphology on the diffusional β↔ α transformation in Ti–55531 during continuous heating: Dissection by dilatometer test, microstructure observation and calculation *Journal of Alloys and Compounds* **702** 352-65

[17] Yin S-b, Sun X-j, Liu Q-y and Zhang Z-b 2010 Influence of deformation on transformation of low-carbon and high Nb-containing steel during continuous cooling *Journal of Iron and Steel Research International* **17** 43-7

[18] Mazur A V and Gasik M M 2009 Thermal expansion of silicon at temperatures up to 1100 C *Journal of materials processing technology* **209** 723-7

[19] Özgün Ö, Gülsoy H Ö, Yılmaz R and Fındık F 2013 Microstructural and mechanical characterization of injection molded 718 superalloy powders *Journal of alloys and compounds* **576** 140-53

[20] Han W, Li Y, Chen G and Yang Q 2017 Effect of sintering additive composition on microstructure and mechanical properties of silicon nitride *Materials Science and Engineering: A* **700** 19-24

[21] Unruan M, Prasatkhetragarn A, Laosiritaworn Y, Ananta S, Khamman O, Yimnirun R, Guo R and Bhalla A 2010 Thermal expansion behavior and estimated total polarizations of lead zirconate titanate–lead nickel niobate ceramics *Materials Letters* **64** 1960-3

[22] Zi A and Palkowski H 2010 Direct strip casting and hot rolling of an AZ31 magnesium alloy *Materials Science and Engineering: A* **528** 559-65

[23] Kivitz E, Palm B, Heinrich J, Blumm J and Kolb G 2009 Reduction of the porcelain firing temperature by preparation of the raw materials *Journal of the European Ceramic Society* **29** 2691-6

[24] Li S, Deng L, Wu X and Wang H 2010 Low-frequency internal friction investigating of the carbide precipitation in solid solution during tempering in high alloyed martensitic steel *Materials Science and Engineering: A* **527** 6899-903

[25] Kirby D, Chang D, Stratton F and Zinck J 2009 A differential capacitive thin film hydrogen sensor *Sensors and Actuators B: Chemical* **141** 424-30

[26] Mitchell J, Freibert F, Schwartz D and Bange M 2009 Unconventional δ-phase stabilization in Pu–Ga alloys *Journal of nuclear materials* **385** 95-8
[27] Takeichi Y, Nishiyama I and Yamada N 2006 Performance of dilatometer for determining absolute CTE of EUVL LTEMs *Microelectronic engineering* **83** 1617-20

IOP Conf. Series: Materials Science and Engineering **923** (2020) 012042 doi:10.1088/1757-899X/923/1/012042

# Influence of eggshell particles on mechanical and water absorption properties of hemp-glass fibres reinforced hybrid composites

**R Bhoopathi[1] and M Ramesh[2*]**
[1]Mechanical Department, Sri Sairam Engineering College, Chennai-600044, India.
[2]Mechanical Department, KIT-Kalaignarkarunanidhi Institute of Technology, Coimbatore- 641402, India.

*E-Mail: mramesh97@gmail.com[2]

**Abstract.** The objective of this work is to analyze the mechanical strengths and water absorption characteristics of sodium hydroxide (NaOH) treated hemp natural fibre (HNF) and eggshell particles (ESP) reinforced polymer composites. The composites were fabricated with and without ESP (0%, 7%, 14% and 21%) by hand-lay process and the tensile, shear, flexural and impact strengths and water absorption characteristics of samples were analyzed. The HNF with 0% ESP composites exhibits the superior tensile strength of 80.46 MPa and holds the maximum shear strength of 1.48 KN, followed by 14% of ESP composites withstands the bending load of 0.19 KN. The samples contains 21% of ESP absorb the high impact energy of 10 joules. The morphological characteristics were analyzed by scanning electron microscopy (SEM) analysis.

***Keyword*:** Hemp fibre; Glass fibre; Eggshell particles; Mechanical properties; SEM studies

## 1. Introduction

Nowadays most of the researchers and industries are focusing their interest to develop the natural fibres based green composites for structural applications in order to reduce the unpleasant environmental conditions [1-4]. The advantages of natural fillers are greater extend to improve the strength of the components and to reduce the cost. The mechanical strength of natural fibre reinforced composites are improved by adding the synthetic fibres and fillers [5-6]. The natural fibre and fillers are bio-degradable, light weight, high density, easily available, low cost and high specific modulus when compared with manmade fibres [6-7]. HNF composites have finding applications in the diverse field ranging from home appliances to aviation industries [1, 10]. ESP can be cost effective natural fillers than others because it should be a commercial and domestic waste. The HNF treatment was carried with 5 wt% of NaoH solution to improve the bonding action between fibre and matrix [4]. The mechanical properties of woven HNF composites are analyzed and results showed the improved tensile and flexural strengths because of silane treatment of fibres. The surface morphological properties were studied by using FTIR spectroscopy and SEM analysis [3].

## 2. Materials and Methods

### *2.1. Materials*

The materials for this experimental work, HNF, E-glass fibres with 300 gsm, Araldite LY556 epoxy resin, Aradur HY 951 were purchased from M/s. Go Green Fibres, Chennai, India. The hardener, thinner, polishing wax and NaOH pellets were purchased from M/s. Sunshine Chemicals, Chennai, India. The eggshells were collected from poultry farms at Namakkal District, Tamil Nadu, India. The properties of long HNF are listed in Table 1.

**Table 1.** Properties of hemp fibres [7, 8, 10, 12, 15-17]

| Properties | Range | Properties | Range |
|---|---|---|---|
| Cellulose (wt%) | 55-90 | Tensile strength (MPa) | 310-1235 |
| Hemicellulose (wt%) | 7-22.4 | Young's modulus (GPa) | 20-70 |
| Lignin (wt%) | 4-13 | Failure strain (%) | 0.9-4.2 |
| Pectin (wt%) | 0.8-1.6 | Specific modulus (GPa-cm$^3$/g) | 0.8 |
| Waxes (wt%) | 0.8 | Micro fibril angle (ɵ) | 6.2 |
| Moisture (wt%) | 9-12 | Length (mm) | 8.3-14 |
| Density (g/cm$^3$) | 68-81 | Diameter (µm) | 17–24 |

### *2.2. Processing*

The fibre constituents like wax, lignin, cellouse and hemicellouse were removed by immersing with 4% of NaOH solution for 12 h in order to increase the bonding between fibre and matrix [1-3, 8-10]. Then the fibres were washed with distilled water to remove the deposited chemicals on the outer surfaces. The HNF and ESP has been dried in order to remove the moistures at 100°C and 200°C in an hot oven. The treated fibres were kept under controlled temperature for 48 h to remove the moisture. Then they have been weaved as bi-directional mat and the weight ratios have been maintained [7, 11]. The ESP was prepared by ball milling process with the particle size of 700 nm to reinforce and increase the strength of the composites. The standard mixing ratio of resin and hardener was maintained as 10:1.

### *2.3. Fabrication*

By hand-lay process, initially the wax coating was applied over a base plate to avoid sticking of laminates then the HNF mat was placed over base plate followed by glass fibre mat [10,12, 15-16]. The ESP has been mixed with resin for getting even distribution of ESP over entire laminates [14]. Then the mixtures were applied evenly over the treated HNF and glass fibre mats and same process were followed for alternate layers. Each laminates consists with 3 layers of HNF mat and 2 layers of glass fibre mat with different weight percentage of ESP. Finally, the fabricated laminates are kept under constant loading for 48 h to get the uniform cross section.

## 3. Result and Discussion

In this study, there are four laminates have been fabricated by reinforcing the treated HNF (0%, 7%, 14% and 21%) of weight of ESP by hand lay process. Three samples from each laminate were prepared as per the ASTM standards. The readings of tensile, shear and flexural strengths were noted by conducting experiments and the energy absorbed by the samples were noted. The observed mechanical strengths and percentage of water absorption are listed in the Table 2.

**Table 2.** Mechanical strengths of treated HNF/glass fibre composites

| Composition | Tensile strength (MPa) | Shear strength (KN) | Flexural strength (KN) | Impact strength (Joules) | Water absorption (%) | | |
|---|---|---|---|---|---|---|---|
| | | | | | Before | After | % |
| HNF+0% ESP | 80.46 | 1.48 | 0.16 | 6 | 2.256 | 2.279 | 1.02 |
| | 53.20 | 1.22 | 0.17 | 6 | 2.217 | 2.288 | 3.20 |
| | 89.32 | 0.80 | 0.18 | 6 | 2.304 | 2.341 | 1.60 |
| HNF+7% ESP | 64.56 | 0.72 | 0.18 | 10 | 2.354 | 2.398 | 1.86 |
| | 66.51 | 0.81 | 0.15 | 8 | 2.241 | 2.290 | 2.18 |
| | 81.92 | 0.83 | 0.19 | 8 | 2.258 | 2.305 | 2.08 |
| HNF+14% ESP | 84.95 | 1.00 | 0.19 | 8 | 2.437 | 2.445 | 0.32 |
| | 52.16 | 0.86 | 0.18 | 8 | 2.498 | 2.538 | 1.60 |
| | 78.67 | 0.96 | 0.19 | 8 | 2.418 | 2.464 | 1.90 |
| HNF+21% ESP | 73.94 | 0.95 | 0.18 | 8 | 2.603 | 2.635 | 0.19 |
| | 74.12 | 0.85 | 0.18 | 10 | 2.630 | 2.668 | 1.44 |
| | 61.91 | 0.79 | 0.15 | 10 | 2.563 | 2.592 | 1.13 |

*3.1 Tensile strength analysis*

Three samples from each laminate were prepared based on ASTM D638 standards for conducting tensile test [2, 3, 6, 7, 10, 13, 15-16]. The tensile strength of all tested samples are listed in Table 2 and compared the results in Figure 1. The results showed that 0% ESP and treated HNF samples holds the tensile strength range of 53.20-89.32 MPa, followed by 14% of ESP and treated HNF samples in the range of 52.16-84.95 MPa. Because of the woven mat the load evenly distributed in all directions which improves the strength during the tensile loading. It was found that the increasing in percentage of ESP, the tensile strength of composite samples was decreased.

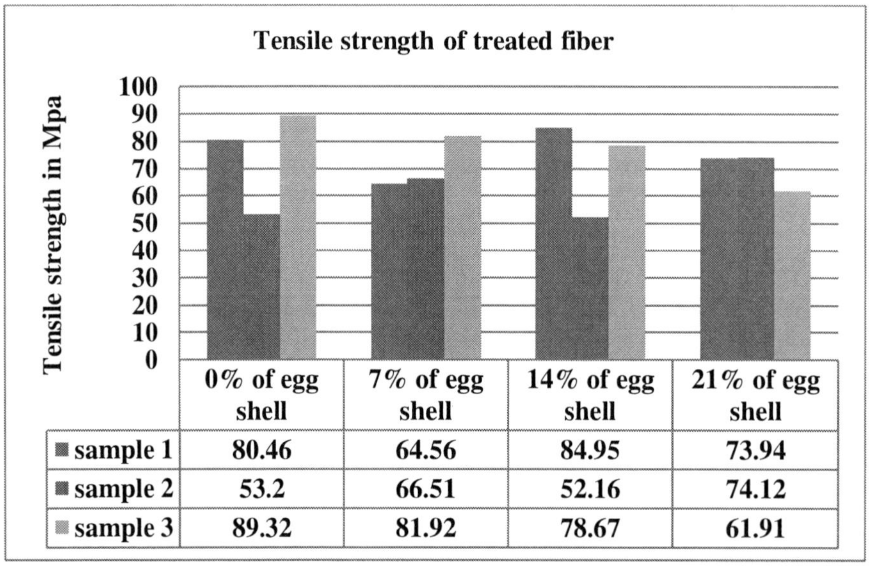


**Figure 1.** Tensile strength comparsion of the composites

*3.2 Shear strength analysis*

The samples were prepared based on ASTM D2344 standards for conducting double shear test [10, 15-17]. The shear strengths of all samples has been noted by applying shear loading. The shear strengths of all tested samples from each laminates were listed in Table 2 and then results are compared in Figure 2. The readings of 0% ESP and treated HNF samples were holds the maximum shear strength in the range of 0.80-1.48 KN, followed by 14% of ESP samples were holds the range of

0.86 -1.0 KN. The distributed ESP between HNF and glass fibres and thus increases the load sharing capability. Similarly it was found that the increasing in ESP in matrix, the shear strength of composite samples has been decreased.

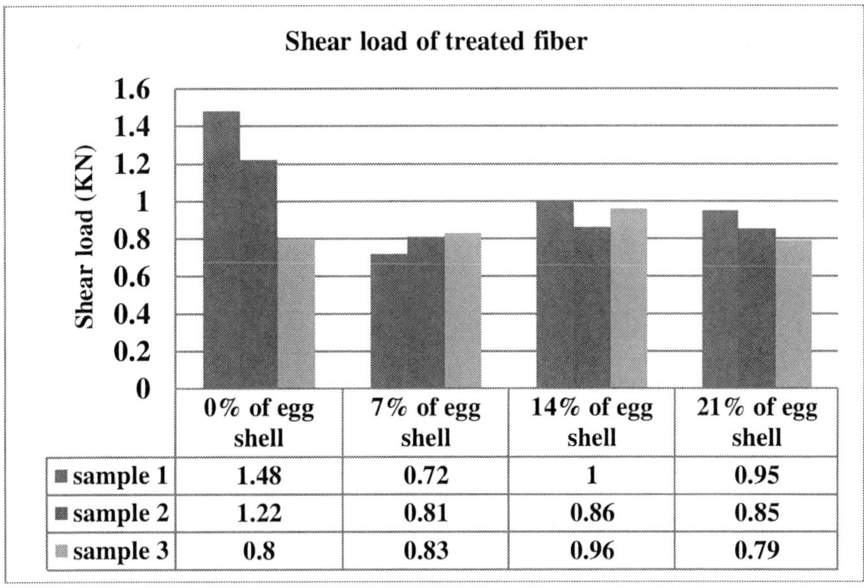


**Figure 2.** Shear strength comparison of the composites

*3.3 Flexural strength analysis*

The samples were prepared based on ASTM D790 standards for conducting flexural test [2, 3, 6, 7, 10, 13, 15-17]. The flexural readings were noted by applying three point bending load. The flexural readings of all samples were listed in Table 2 and the comparison of results is presented in Figure 3. From the results, it is noted that almost all the values are close to each other. The 14% ESP and treated HNF samples were holds the range between 0.18 to 0.19 KN, followed by 7% of ESP and treated HNF samples were holds the range of 0.15-0.19 KN. The flexural readings show the improvement while increasing the percentage of ESP in the composites.

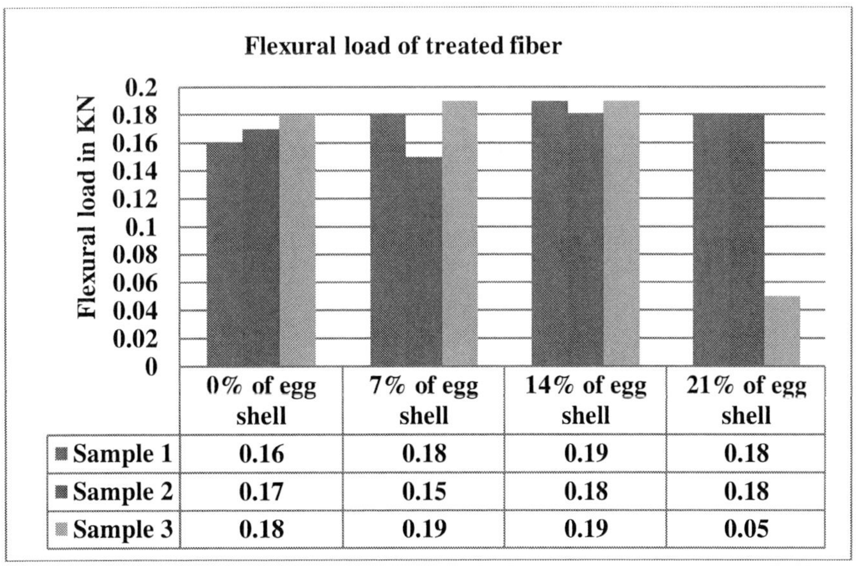


**Figure 3.** Flexural strength comparison of the composites

IOP Conf. Series: Materials Science and Engineering **923** (2020) 012042 doi:10.1088/1757-899X/923/1/012042

*3.4 Impact strength analysis*

The samples were prepared based on ASTM D 256 standards for conducting charpy impact test [6, 10, 15-17]. The V-grooves were cut on the longer side of sample in order to initiate the crack when load was applied. The absorbed impact energies are listed in Table 2 and the results are compared in Figure 4. The results show that the 21% of ESP and treated HNF samples absorbs the maximum energy in the range of 8-10 joules. Similarly, the 7% and 14% of ESP samples holds the average energy of 8 joules. It is found that the increased impact energy of all tested samples while increasing the percentage of ESP in composites.

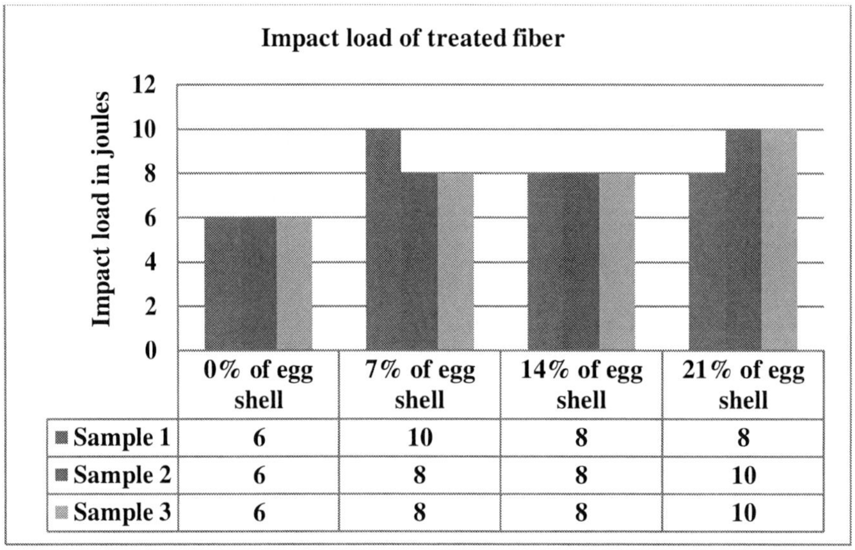


| | 0% of egg shell | 7% of egg shell | 14% of egg shell | 21% of egg shell |
|---|---|---|---|---|
| Sample 1 | 6 | 10 | 8 | 8 |
| Sample 2 | 6 | 8 | 8 | 10 |
| Sample 3 | 6 | 8 | 8 | 10 |

**Figure 4 .** Impact strength comparison of the composites

*3.5 Water absorption analysis*

The samples of size 20 x 20 x 3mm were prepared based on ASTM D5229 standards [2, 6, 7, 10]. The specimens were immersed in water for 24 h in an open atmosphere conditions. Before and after the immersion the specimen weight has been noted. The water intake of all samples was listed in Table 2 and the results are compared in Figure 5. From the results, noted that the 21% of ESP samples absorb less amount of water followed by 14% of ESP samples. The ESP has been increased the bonding between HNF and matrix of the composites. Due to this the energy absorbing characteristics of samples also increased.

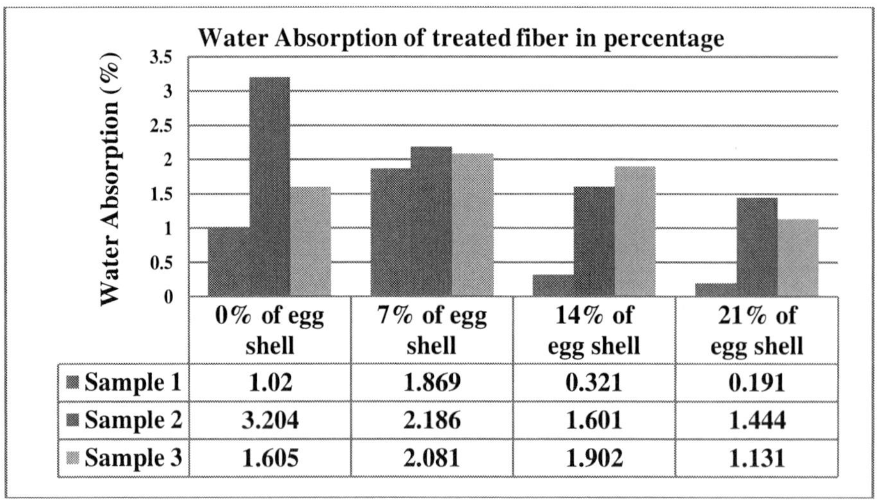


| | 0% of egg shell | 7% of egg shell | 14% of egg shell | 21% of egg shell |
|---|---|---|---|---|
| Sample 1 | 1.02 | 1.869 | 0.321 | 0.191 |
| Sample 2 | 3.204 | 2.186 | 1.601 | 1.444 |
| Sample 3 | 1.605 | 2.081 | 1.902 | 1.131 |

**Figure 5.** Water absorption percentage of composites

*3.6 Morphological analysis*

The interfacial characteristics like matrix failure, fibre fracture, distribution of ESP, pull out of fibre and matrix, voids in the samples were analysed by using SEM analysis [2, 3, 7-10,13, 15-16]. The alternate layers of bi-directional HNF mat are clearly seen and evenly distributed ESP has been clearly visible in the SEM image presented in Figure 6 (a). The fractured epoxy resin matrix and fibres of samples due to loading were visible in the image presented in Figure 6 (b).

**Figure 6.** SEM images of the tested composite samples

## 4. Conclusion

The ESP incorporated HNF composite has been fabricated and the mechanical strengths and water intake characteristics are analyzed. The superior tensile and shear strength of the treated HNF with 0% ESP and 14% ESP has been observed. The increased flexural and impact strengths of samples has been obtained by increasing the ESP in the composites. The water absorption results revealed that decreased water intake while adding ESP in the HNF composites. Thus, the treated HNF composites with 14% of ESP exhibits the superior mechanical strengths and it can be suggested for structural applications in automotive and construction industries. The morphological characteristics were analyzed and noted the distributions of ESP, voids, matrix fractures, pullout of fibre and the orientations by SEM.

## 5. References

[1] Vilaseca F, Rey R D, Serrat R, Alba J, Mutje P and Espinach F X 2018 Macro and micro-mechanics behavior of stifness in alkaline treated hemp core fibres polypropylene-based composites *Composites Part B* **144** pp 118–125

[2] Vaisanen T, Batello P, Lappalainen R and Tomppo L 2018 Modification of hemp fibers (Cannabis Sativa L.) for composite applications *Industrial Crops & Products* **111** pp 422–429

[3] Sepe R, Bollino F, Boccarusso L and Caputo F 2018 Influence of chemical treatments on mechanical properties of hemp fiber reinforced composites *Composites Part B* **133** pp 210 - 217

[4] Sarasinia F, Tirillo J, Sergi C, Seghini M C, Cozzarini L and Graupner N 2018 Effect of basalt fibre hybridisation and sizing removal on mechanical and thermal properties of hemp fibre reinforced HDPE composites *Composite Structures* **188** pp 394–406

[5] Choudhury M R, Srinivas M S and Debnath K 2018 Experimental investigations on drilling of lignocellulosic fiber reinforced composite laminates, *Journal of Manufacturing Processes* **34** pp 51–61

[6] Dayo A Q, Wang A, Kiran S, Wang J, Qureshi K, Xu Y, Zegaoui A, Derradji M, Babar A A and Liu W 2018 Impacts of hemp fiber diameter on mechanical and water uptake properties of polybenzoxazine composites, *Industrial Crops & Products* **111** pp 277–284

[7] Ahmad M A A, Majid M S A, Ridzuan M J M, Mazlee M N and Gibson A G 2018 Dynamic mechanical analysis and effects of moisture on mechanical properties of interwoven hemp/polyethylene terephthalate (PET) hybrid composites, *Construction and Building Materials* **179** 265–276

[8] Ramesh M, Palanikumar K and Hemachandra Reddy K 2017 Plant fibre based bio-composites: Sustainable and renewable green materials *Renewable and Sustainable Energy Reviews* **79** pp 558-584

[9] Ramesh M, Palanikumar K and Hemachandra Reddy K 2016 Evaluation of mechanical and interfacial properties of sisal/jute/glass hybrid fiber reinforced polymer composites *Transactions of the Indian Institute of Metals* **69** pp 1851-59

[10] Ramesh M, Deepa C, Aswin U S, Eashwar H, Mahadevan B and Murugan D 2016 Effect of alkalization on mechanical and moisture absorption properties of Azadirachta indica (Neem tree) fiber reinforced green composites, *Transactions of the Indian Institute of Metals* **70** pp 187-199

[11] Antony S, Cherouat A and Montay G 2018 Experimental, analytical and numerical analysis to investigate the tensile behaviour of hemp fibre yarns *Composite Structures* **202** pp 482-490

[12] Senthilkumar K, Saba N, Rajini N, Chandrasekar M, Jawaid M, Siengchin S and Alotman O Y 2018 Mechanical properties evaluation of sisal fibre reinforced polymer composites: A review *Construction and Building Materials* **174** pp 713–729

[13] Nor A F M, Sultan M T H, Jawaid M, Talib A R A, Azmi A M R, Harmaen A S and Asaari A Z M 2018 The effects of multi-walled CNT in bamboo/glass fibre hybrid composites: tensile and flexural properties *BioResources* **13(2)** pp 4404-15

[14] Dayal A R, Gaikwad A and Pawar A S 2018 Development of a new bio composite material by utilizing walnut shell powder, coir and jute fiber and evaluation of its mechanical properties *International Journal of Mechanical and Production Engineering Research and Development* **8(2)** pp 819-826

[15] Bhoopathi R, Ramesh M, Naveenkumar M, Sanjaybalaji P and Sasikala G 2018 Studies on mechanical strengths of hemp-glass fibre reinforced epoxy composites *IOP Conf. Series: Materials Science and Engineering* **402** 012083, doi:10.1088/1757-899X/402/1/012083.

[16] Bhoopathi R, Ramesh M and Deepa C 2014 Fabrication and property evaluation of banana-hemp-glass fiber reinforced composites *Procedia Engineering* **97** pp 2032-41

[17] Bhoopathi R, Deepa C, Sasikala G and Ramesh M 2015 Experimental investigation on mechanical properties of hemp-banana-glass fiber reinforced composites *Applied Mechanics and Materials* **766-767** pp 167-172

IOP Conf. Series: Materials Science and Engineering **923** (2020) 012043 doi:10.1088/1757-899X/923/1/012043

# Study of graphene coating with metals and metal Nano composites

**A Balaji Krishna bharathi[1,2], P Diyavasanth[1,3] ,S A Srihari[1,4] , M Saravanan[1,5] and P V Sudheer kumar reddy [1,6]**

1. Dept of mechanical engineering R.M.K College of Engineering and Technology
2. balajikrishnabharathimech@rmkcet.ac.in 3. divya16me017@rmkcet.ac.in 4. sriha16me090@rmkcet.ac.in
5. sarav16me081@rmkcet.ac.in 6. pokal16me055@rmkcet.ac.in

**Abstract**—This paper is a study about the graphene coating and their properties so far done. At first, graphene was synthesized a few decades ago with graphene-based devices like supercapacitors transistors and microelectromechanical systems are made. We aim to make a graphene composite that exhibits optimized properties and characters that meet the requirement of the industry field. This paper is a study about properties of graphene when coated with other metals and the application study of graphene with Nanotechnology.

***Keywords***— ***Graphene coating, graphene oxide, corrosion resistance, Nanocomposites, Graphene synthesis.***

## I. INTRODUCTION

Generally, carbon has four allotropes of carbon. They are carbon nanotubes, charcoal and graphite fullerenes. Graphene structure is made of four allotropes of carbon. Graphene has peculiar properties that make it different from other forms of carbon. Though some properties like density are lower in graphene, it's thickness is stronger when compared with steel. Graphene has properties like thermal conductivity. Graphene carbon atoms are packed in the hexagonal structure.

The structure of graphene is studied by electron microscopy transmission by suspending the graphene sheets between the metal bars. The composition of chemical in graphene can generally be modified by oxygen and nitrogen -containing functional groups which are tested using X-ray photoelectron spectroscopy and infrared spectroscopy. Graphene can be synthesized by several methods. They are top -down and bottom -up approaches.[1]

Graphene has several material applications used in several products like solar cells, the light emitting diode(LED), touch panels, smartphones etc...Graphene is often made in powder made of polymer matrix which can be used in advanced composites, batteries, capacitors, thermal management

## II. CORROSION PROTECTION PROPERTIES WITH MILD STEEL

The resistant to corrosion property of Mild Steel can be improved by developing GO nanosheets filled with epoxy coatings the factor which is very essential that influences the performance of coating with graphene oxide in polymer matrix GO to polymer of 0 1% with lower viscosity is added to give required properties like good adhesion properties good quality of dispersion and protection of corrosion under sodium chloride electrolyte[1]

### *A. Properties of Epoxy coatings:*

- Low cost
- Good chemical resistance
- Dimensional stability

### *B. Drawbacks:*

- Brittleness
- Low resistance in crack propagation [2]

### *C. Composition of epoxy coating:*

Epoxy resin is made of hydrophilic hydroxyl groups which make them more permeable to corrosive agents. Nanocomposite coating is a promising method for increasing corrosion resistance in epoxy coating. Graphene increases the corrosion resistance and the barrier properties in the polymer of the matrix are improved. Graphene has strong Vander Waal's force and pi -pi interaction between graphene sheets and high specific surface area.[2]

***D. Materials used:***

Graphite powder (<20 micrometers), KMnO 4 ,Sulphuric acid,Hydrogen peroxide and HCl are used for graphene oxide preparation.Epoxy resin and polyamide hardener are used for epoxy coatings. Mild carbon steel plates (ST -12) are used as a substrate for the epoxy coating. Sandblasting of mild steel and degreasing on acetone is done before the coating of spray-on mild steel [3]

***E. Characterization:***

The characters for the graphene oxide are analyzed by several tests. • The GO crystal structure is characterized by X-ray diffraction • The chemical grafting the polymer of the matrix is found by the Fourier transformation infrared spectroscopy [3] Graphene oxide morphology, quality of dispersion in nanocomposite of GO is found by scanning electron microscope. The adhesion of nanocomposite is found out by the pull-off method.[3]

## III. MECHANICAL ENHANCEMENT OF AN ALUMINUM LAYERBY GRAPHENE COATING

The properties of aluminium coated with the graphene are tested by doing various experiments by numerical tensile methods by dynamics of molecules simulations[4].Based on the results obtained from the results, the aluminium layer coated with the graphene enhances young's modulus, toughness of the material. The loading rate on coating is tested. Though the hybrid coat structure is mostly of the aluminum core[4]. A high loading rate makes structure as amorphous. So, the loading effect and the amorphization temperature are found out by performing various simulations[4].

Among many carbon nanocomposite materials, graphene has many mechanical, chemical and electrical properties. The chemical bonding of graphene is also similar to other carbon nanotubes. The characteristics of Graphene along with several other composites are also studied. Graphene added with several metals like iron, copper -nickel will enhance the strength of those metals [5]. The graphene metal composites have many applications.The metal generally used is aluminum in th e composite and it is made to make with several methods. Difficulties in conducting nanoscale experiments are overcome by conducting numerical experiments with graphene metal composites through the atomistic modeling of these materials. Molecular dynamics simulations are applied for studying the chemical and mechanical properties and its behaviour of nanocomposite materials[6].MD simulations of graphene -nickel and graphene -TiO 2 are studied and their mechanical properties are investigated[6]

***A. Atomistic modeling and simulation details:***

The atomistic model is created by placing the graphene sheet of dimensions 20*20 nm on the free surface of 10 nm thick aluminum block, permitting a gap of 0.224nm which is found to be the bond length between aluminum and carbon atoms. For finding the atomic interactions between the metal atoms, the embedded atom method is used. A formula is also used for finding the atom's total energy. The tensile stress -strain curves of coated as well as uncoated aluminum layer are drawn at a temperature of 300 K[7]. The applications of boundary conditions of the planar direction (X, Y, Z axes) is done. The thermalization is done at a different temperature. The application of Tensile loading in Y direction is done by applying different strain rates [7]. During loading, viral stress is calculated using the formula.

***B. Observation of the coating process:***

It is observed that coating the grapheme on aluminum sheets will enhance several important properties like young's modulus, yield strength. The graphene is coated with the uncoated aluminum surface and so density of the aluminum material increases [8]. Tensile characteristics of aluminium sheet is tested by high loading and graphene plays role in enhancing the properties is also discussed and their results are plotted as graphs[8]. The loading rates have been tested for different temperatures and increasing the loading rate improved the density. The undefined stackings are formed due to the enhancement of loading rate[8].

***C. A suitable method to coat graphene nanoparticles on an aluminum composite:***

Graphene coating is done on the composite. Antibacterial activity, microstructure are investigated[9].The graphene is coated uniformly. Researchers have developed the graphene -coated aluminum composite to avoid all the limitations of organic coatings. [10]

**Properties:**

Excellent bonding strength, superior corrosion resistance, weather resistance [11] The graphene silver

IOP Conf. Series: Materials Science and Engineering **923** (2020) 012043 doi:10.1088/1757-899X/923/1/012043

coating is experimentally prepared by a suitable method and sprayed on the mild steel plate using a cold spray method. X -ray diffraction, scanning electron microscopy, energy dispersive spectroscopy are some of the methods to check the graphene coating on the composite. The antibacterial activities were studied in the bacteriostatic experiment[12]. Through scanning electron microscopy ,the graphene's microstructure which is to be coated has been studied and several curves were drawn through Raman spectroscopy. On the whole, the aluminum matrix composite coated with graphene is chemically synthesized and cold spray. It is also tested using testing methods [12].

## IV. ELECTROPHORETIC COATING DEPOSITION ONCOPPER

Graphene coated composite is synthesized using suitable methods and its application is done on the copper through electrophoretic deposition.By this method, a minimum thickness of 40nm at 10V and at a deposition time of the 30s. By means of electron microscopy the morphological characteristics are studied in copper which describes the distribution of oxides of graphene which is of few nanometers [13].

### *A. Raw materials*

Initially, before the synthesis of the coating, all the necessary raw materials like copper flat sheet, Polymeric isocyanate crosslinked with hydroxy - functional acrylic like adhesive (PIHA) are used.Specific surface charges are measured by particle charge detector (PCD) and the optimum concentration of PIHA is also checked prior to the synthesis of the composite[14].

### *B. Synthesis of GO and suspension of GO, PIHA:*

Then the graphene oxide is prepared by a method called modified hummer. After the synthesis of GO, GO and PIHA are made to suspend each other by using a stirrer.

### *C.GO-PIHA composite coating:*

The EPD setup is set up using two fixed parallel plates. EPD from the GO -PIHA suspension was conducted at constant DC code using source meter. On successful completion of coating, the samples were taken from the setup. Then the GO composite coating is chemically reduced using some chemicals[15].

### *D. Tests to be carried out and characterization :*

Initially, the electrochemical corrosion test is conducted on copper plate coated with the graphene composite coating. Then the graphene composite's microstructure which is to be coated is found out using scanning electron microscopy. Therefore, physical and chemical characteristics of the composite coating will also be understood from the tests being carried out[16].

## V. ENHANCEMENT OFSEAWATER CORROSION RESISTANCE

Acetone Acetone derived graphene coating enhances the efficiency of the copper corrosion in the seawater environment. Copper and alloys of copper are used as components of the seawater system due to enhanced corrosion resistance and biofouling. It not only depends on the inherent cathodi c nobility of metal but also naturally from positive layers [17].

**Layers-Cu2O**

Corrosion rate rely upon parameters like seawater, temperature, the velocity of water flowing to the metal surface and degree and kind of pollution. Many physical barriers are added for preventing corrosion of copper like alloy polymer organic layer but it will change their physical and chemical properties [18].

Graphene has a strong thin metal coating which is protective with considering structural, chemical, thermal stability. Acetone derived from graphene films effectively protect metallic copper growth from corrosion [18]

### *A. Corrosion level checking upper graphene added*

By physical methods like XPS and ATR, F- Diacetone from the graphene layer acts as large cathodic reduction barr iers against dissolved O2and chloride diffusion. It not only helps Cu but also other several metals from corrosion [19]

### *B. Methods of corrosion checking and graphene level checking:*

ATR FT-IR –Used to record reflectance spectra of Cu coil samples which has already undergone annealing using germanium ATR crystal prior and undergone acetone deposition Raman spectroscopy using water project software data analysis is done afterward Field emission (FE -SEM)-energy dispersive X -ray spectrometer mechanically test for acetone work before and after corrosion [19] Acetone drops (Cu foil) on heated at 1000 -degree Celsius gives graphene which is an anti-corrosion barrier as corrosion - resisting coatings in the marine atmosphere.

IOP Conf. Series: Materials Science and Engineering **923** (2020) 012043 doi:10.1088/1757-899X/923/1/012043

Graphene with copper as single layer acts as a large cathodic reduction barrier against dissolved O2 and chloride diffusion into underlying Cu in a seawater. Since partial active oxidation did on the single -layer graphene is powerful so that the multilayered graphene will also effective[19].

Full forms:

RTA-Rapid thermal anneals

XPS-X-ray photoelectron spectroscopy

ATR FT-IR-Attenuated complete reflectance of transforms of Fourier infrared.

## VI. METAL NANOCOMPOSITES: SYNTHESIS, APPLICATIONS AND THEIR CHARACTERIZATION

In recent times nanocomposites found much attention in scientific research due to their improvement in properties when compared with single metal in nanoparticles[20]
The range of nanocomposites will be 1-100nm Nanocomposites are classified based on

- Continuous phase
- Discontinuous reinforcing phase

The basic method of preparation of nanocomposites are

- Metals
- Ceramics
- Polymers.

### *A. Metal Oxide Nanocomposites (MONC)*

The main focus on MONC structure and function is explained by the above three basic methods.

***Synthesis:***
Synthesis is based on optical, magnetic, electrical, biological which helps to determine the properties based on size and dimensions[21].
Another method of approach to synthesis is

- Top-Down approach (physical methods)
- Bottom-up approach (wet methods)

There are different wet methods for synthesis of nanocomposites and metal nanoparticles but primarily hydrothermal co-precipitation and sol- gel methods are less costly when compared with others[21].

***Co-precipitation:***
We use a co precipitation method for the synthesis of nanoparticles mixed with metal ceramics or metal. The precipitate is produced by Nanocomposites that are separated from the solution.

MgO-Al2O3, Mgo-cuo.

*1) Sol-gel method:*
The sol-gel method is used for the synthesis process of nanomaterials by mild reaction and building the molecular variation on properties and material Nano and microstructures can be produced by sol-gel methods[21].

***TiO2–Fe2O3 Ag-TiO2***

*2) Hydrothermal methods:*
This method involves a chemical reaction in an aqueous solvent or no aqueous solvent occurring at more than room temperature at 1atm pressure in a closed system The properties and size can be modified based on requirements[21].

***ZnO-kaolinite Co-MgO***

### *B. Characterization of Metal Oxide Nanocomposites (MONC)*

By different techniques, the Nanoparticles are characterized to analyze the morphology, particle size, phase, composition, thermal stability, optical, magnetic, electrical and thermal properties.

***X-ray Diffraction:***

Powder X-ray Diffraction (XRD) is used for phase determination and unit cell information of the nanocomposites under investigation

$$D=K\lambda\cos\theta$$

D (nm) crystalline domains mean size. K is the shape factor value in which K is near unity and dimensionless.

### *C. Thermal Applications of Metal Oxide Nanocomposites (MONC)*

*1) Application in Environment*

The MONC are frequently used as adsorbents, photocatalysts, and sensors to tackle environmental pollution problems[21].

*2) Application in Agriculture and Food*

The MONC is widely used in the packaging of foodstuff provides strength as a filler material (silicates, clays, TiO2). While in the agriculture sector, MONC used as Nanosensor for pesticide and pathogen detection in plants and source for delivery of genetic material for the improvement of crops[21]

### *D. Application in Health and medicine*

The MONC has many applications in medicine, for delivery of drugs and imaging, diagnosis and screening of diseases, DNA sequencing, in gene therapy and tissue culturing and in cancer treatment. Metal oxide nanocomposites have proven themselves as the materials in the 21st century with the range of applications in every industry sector. The wet chemicals methods mark the easy route for their fabrication that makes them cost- effective as well. The complex matrix which is based on MONC VI is formed by co-precipitation by simple routing[22]

## VII. CONCLUSION

In this paper, the aluminium sheet properties with the graphene is studied by MD simulations for enhancing properties like tensile strength, toughness, yield strength, etc. The loading rates have been tested for different temperatures. The undefined stackings are formed due to the loading rate.

It is observed that coating the grapheme with metals will enhance several important properties like young's modulus, yield strength. Th e graphene is coated with the uncoated aluminum surface and the aluminum material density increases. The aluminium sheet's tensile characteristics are studied using high loading and graphene role in enhancing the properties is also discussed

## VIII. REFERENCES

[1] Exploring corrosion protection properties of solvent based epoxy-graphene oxide nanocomposite coatings on mild steel, Sepideh Pourhashem Mohammad Reza Vaezi Alimorad Rashidi Mohammad Reza Bagherzadeh (November 2016 ) PII: S0010-938X(16)31175-1

[2] Graphene-based materials and their composites: A review on production,applications and product limitations Velram Balaji Mohana, Kin-tak Laub, David Huic , Debes Bhattacharyya (June 2018) PII S1359836817344426

[3] Synthesis, properties, and applications of graphene oxide reduced graphene, Andrew T. Smith, Anna Marie LaChance, Songshan Zeng, Bin Liu, Luyi Sun, Volume 1, Issue 1, March 2019

[4] Mechanical enhancement of an aluminum layer by graphene coating Ahmet Semih Erturk, Mesut Kirca,a) and Levent Kirkayak (September 2018) pp. 2741-2751

[5] M. Rashad, F. Pan, A. Tang, M. Asif, Mechanical properties of nickel-graphene composites synthesized by electrochemical deposition, Nanotechnology 26 (2015) 065706.

[6] K. Gall, M.F. Horstemeyer, M.V. Schilfgaarde, M.I. Baskes, Atomistic simulations on the tensile debonding of an aluminum–silicon interface, J. Mech. Phys. Solids 48 (2000) 2183–2212.

[7] R. Komanduri, N. Chandrasekaran, L.M. Raff, Molecular dynamics (MD) simulation of uniaxial tension of some single-crystal cubic metals at nanolevel, Int. J. Mech. Sci. 43 (2001) 2237–2260.

[8] E. Saether, D.R. Phillips, Molecular-dynamics simulation-based cohesive zone re-presentation of intergranular fracture processes in aluminum, J. Mech. Phys. Solids 54 (2006) 1899–1928.

[9] V. Yamakov, E. Saether, D.R. Phillips, E.H. Glaessgen, Molecular-dynamics simu-lation-based cohesive zone representation of intergranular fracture processes in aluminum, J. Mech. Phys. Solids 54 (2006) 1899–1928.

[10] Z. Yang, Y. Zhou, T. Wang, Q. Liu, Z. Lu, Crack propagation behaviors at Cu/SiC interface by molecular dynamics simulation, Comput. Mater. Sci. 82 (2014) 17– 25

[11] C.R. Dandekar, Molecular dynamics based cohesive zone law for describing Al–SiC interface mechanics, Compos. Part A: Appl. Sci. Manuf. 42 (2011) 355–363.

[12] G. Xu, H. Wang, Molecular dynamics study of interfacial mechanical behavior be-tween asphalt binder and mineral aggregate, Constr. Build. Mater. 121 (2016) 246–254.

[13] AlSi10Mg alloy nanocomposites reinforced with aluminum- coated graphene: Selective laser melting, interfacial microstructure and property analysis Zhanyong Zhao, Peikang Bai, R.D.K. Misra,

Mengyao Dong, Renguo Guan, Yanjun Li, Jiaoxia Zhang, Le Tan, Jianfeng Gao, Tao Ding, Wenbo Du, Zhanhu Guo (April 2019 ) PII: S0925- 8388(19)31271-X

[14] A. Sazgar, M.R. Movahhedy, Development of a molecular dynamic based cohesive zone model for prediction of an equivalent material behaviour for Al/Al2O3 com-posite, Mater. Sci. Eng., A 679 (2017) 116–122.

[15] The production of a corrosion resistant graphene reinforced composite coating on copper by electrophoretic deposition Bimal P. Singh, Sasmita Nayak, Kamala K. Nanda, Bikash Kumar Jena, Sarama Bhattacharjee, Laxmidhar Besra PII: S0008- 6223(13)00372-2

[16] A comprehensive review on graphene-based anti- corrosive coatings Gan Cui∗, Zhenxiao Bi, Ruiyu Zhang, Jianguo Liu, Xin Yu, Zili Li(October 2019) PII S13858947193104

[17] X. Liu, F. Wang, W. Wang, H. Wu, Interfacial strengthening and self-healing effect in graphene-copper nanolayered composites under shear deformation, Carbon 107 (2016) 680–688.

[18] Enhancement of seawater corrosion resistance in copper using acetone-derived graphene coating Jae-Hoon Huh, Seung Hyun Kim, Jae Hwan Chu,a Sung Youb Kim, Ji Hyun Kim and Soon-Yong Kwon (January 2014) DOI: 10.1039/c3nr05997a

[19] Development of oxidation and corrosion resistance hydrophobic graphene oxide-polymer composite coating on copper Bimal P. Singh ∗, Bikash Kumar Jena, Sarama Bhattacharjee, Laxmidhar Besra (October 2013)

[20] Protecting copper from electrochemical degradation by graphene coating Author links open overlay panel, R.K.Singh Ramana, P.Chakraborty Banerjee, Derrek E.Lobob, Hemtej Gullapalli, Madusha Sumandasa (Volume 50, Issue 11, September 2012)

[21] J. Hwang, T. Yoon, S.H. Jin, J. Lee, T.S. Kim, S.H. Hong, S. Jeon, Enhanced me-chanical properties of graphene/copper nanocomposites using a molecular-level mixing process, Adv. Mater. 25 (2013) 6724.

[22] Recent developments in graphene-based polymer composite membranes: Preparation, mass transfer mechanism, and applications Ying-Chen Du,Lin-Jun Huang,Yan-Xin Wang,Kun Yang,Jian-Guo Tang, Yao Wang (January 2019)DOI: 10.1002/app.47761

# Study of compressive behaviour on 3D printed ABS polymer lattice structures infilled with epoxy and polyester resins

**N.Sathishkumar[1], B.Vincent[2], N.Arunkumar[3], K.M. Kumar[4], P.L.Sudharsan[5]**

[1,2]Assistant Professor, Department of Mechanical Engineering, St. Joseph's College of Engineering Chennai 600 119, Tamil Nadu, India.
[3]Professor, Department of Mechanical Engineering, St. Joseph's College of Engineering Chennai 600 119, Tamil Nadu, India.
[4]Associate Professor, Department of Mechanical Engineering, St. Joseph's College of Engineering Chennai 600 119, Tamil Nadu, India.
[5]UG Student, Department of Mechanical Engineering, St. Joseph's College of Engineering, Chennai 600 119, Tamil Nadu, India.

Mail Id: sathishnsk2912@gmail.com , sathishkumarn@stjosephs.ac.in

**Abstract:** The 3D printing process is a design driven manufacturing process. The complex shaped structures can be produced in a short lead time with high precision percentage. The functionally graded polymer oriented cellular structures are possessing great energy absorbing ability. In recent years, the application of cellular structures are highly utilized in 3D printing process for developing functional prototypes without compromising strength to weight ratio. In this study, a novel approach to develop functionally graded uniformly open porous structures of fixed volumetric proportions are created and it is selectively infilled with polyester and epoxy resins. After curing process the samples were tested for analyzing the improvement of compressive behaviour. The results of the compressive behaviour were compared between non-porous structures, resin infilled porous structures and non - resin infilled porous structures. The intensity between bonding of the infilled resins with the boundary of the porous structures are validated with the help of VMS image. The resin infilled lattice structures showed a significant improvement in the compressive strength when compared with the non-porous structures and non - resin infilled lattice structures. The interface between the resin and the base material also exhibited good adherence which promises the usage of this methodology in creating more customized hybrid composite materials for the various applications by using 3D printing technique

**Keywords:** Additive manufacturing, Fused deposition modelling (FDM), Lattice structures, infill of resins, ABS material, Compressive behaviour.

## 1. INTRODUCTION

It is important for industries to reduce the lead time, cost of the industrial products manufactured and to achieve the high strength to weight ratio of the products without compromising the build quality [1]. Additive manufacturing is one of the best solutions to satisfy the above requirements and with the help of additive manufacturing techniques highly energy absorbing, light weight and strong cellular structures can be created [2]. This process involves manufacturing of the product in layer by layer fashion [3]. As per Wohler's report, the application of additive manufacturing widely spreads in many fields like tooling, consumer products, electronics, bio-medical implants, dental, motor vehicles, aerospace, industrial, military and architectural [4]. Cellular structures have been increasingly used in modern engineering and medical applications due to their advantageous mechanical and thermal properties [5]. In addition, cellular structures are widely used in applications, like thermal insulation, shock or vibration damping, acoustic absorption, current collectors in battery electrodes, catalyst supports and biomedical implants [6].The mechanical behaviour of cellular materials e.g. energy absorption capability and plateau load-deformation response mainly depends on the porosity and the material used for its fabrication [7]. For the same porosity, the yield strength could be increased and the build time shortened simultaneously by having a certain cell size control [8]. Additive manufacturing technology enables manufacturing of uniformly regular cellular structures which allows for accurate cell shape control to fulfil some chosen properties for individual applications [9]. The different types of uniformly regular cellular structures generally fabricated through AM process are honey comb, crisscross, triangular, circular and diamond [10]. The tooling produced with AM are suitable for low volume products, rapid changing high volume products and easily adapt design iterations, which are time consuming and expensive in tradition practices [11].The significant advancement in Additive Manufacturing (AM) processes enables the fabrication of innovative parts, characterized by minimum weight and good mechanical Properties [12]. The significant advancement in Additive Manufacturing (AM) processes enables the fabrication of innovative parts, characterized by minimum weight and good mechanical properties [13] This AM process in general is classified into seven types [14]. However this project is mainly emphasized with the FDM type of manufacturing. FDM is highly used polymer based additive manufacturing technique used for creating many complex shaped structures [15].

**The figure 1 shows the fishbone diagram of various 3D printing technologies.**

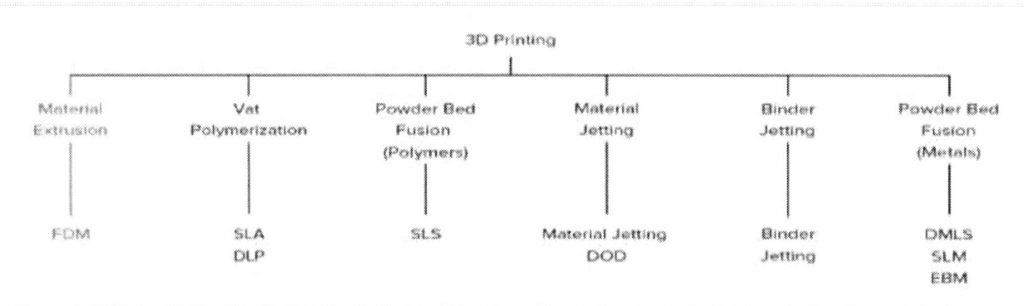


**Figure 1. Various 3D printing technologies**

## 2. METHODOLOGY

In this paper, we discuss a method of improving the mechanical strength of the 3D printed part with the help of resin infilling technique. The design of the part is made such that the volumetric proportions are 60% base material (ABS) and 40% resins filling void. A full solid ABS material is tested whose volume is completely made of ABS and two parts filled with resins of fixed volumetric proportions are tested and a part that is left void for the same proportions is also tested.

## 3. PRINTING AND INFILLING OF THE PART

### 3.1 Design for 3D Printing

The design of the 3D printed part is made with the ASTM standard for compression test of rigid polymers (ASTM D695) [16]. The designing of compressive test specimens were carried out using CATIA V5R20 software. The dimensions of the 3D printed parts are 38.1mm in diameter and 25.4mm in length. Based on the design for the volumetric proportions mentioned earlier, the design is made such that the radius of the void is 4.55mm and has seven through holes in total volume. The figure 2 shows the designed CAD model of the part.

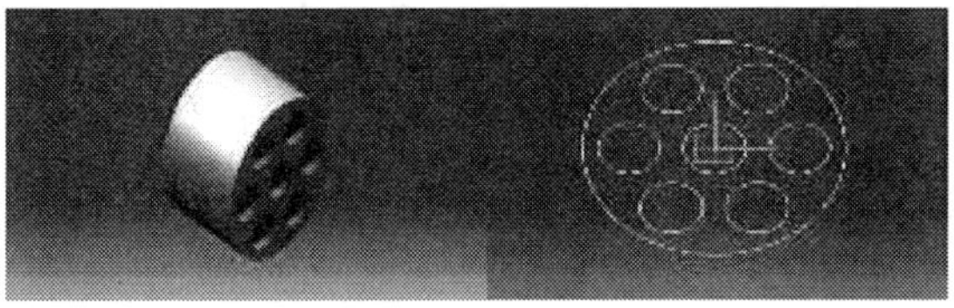

**Figure 2. CAD Model design using CATIA V5.**

### 3.2 Fabrication of Compressive test specimens

The mode of 3D printing is the fused deposition modelling which belongs to the material extrusion family. Here ABS and PLA are the most commonly used printing material. Here for printing the part, infill pattern is chosen to be hexagonal and the infill density is 100%. Other standard parameters like layer height, number of perimeter shell, and the print speed are unchanged and are constant [17, 21]. The extrusion temperature for the filament drawn is 230°c.The properties of the 3D printer used in this study is stated in table 1. The images of the fabricated specimens are shown in figure 3 as follows

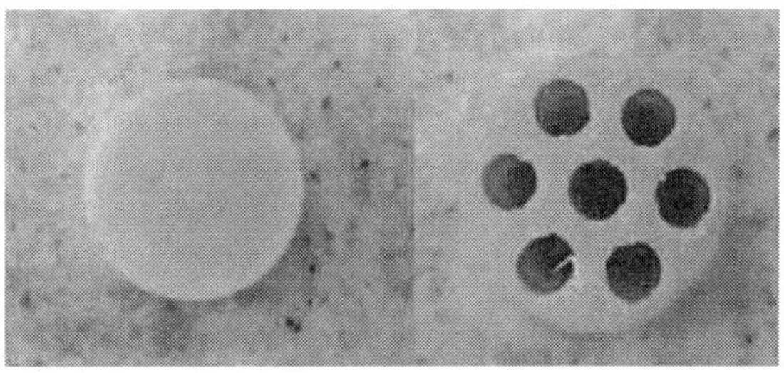

**Figure 3. Fabricated parts.**

### 3.3 Infilling of resins

This study involves two resins namely epoxy and the polyester resin. Although various resins are available for filling the voids, these two are selected based on their meritorious properties, availability and their relatively low cost of nature. These resins are added carefully to the voids only through the manual means [18, 23].

#### 3.3.1 Epoxy resin and Polyester resin preparation

Epoxy resin and polyester resin are mixed with hardener at 2:1 proportion in room temperature. The resin and hardener undergo chemical reaction. This reaction takes until molecular cross linkage occurs. This time is known as curing time. Polyester is an adequate adhesive but not as good as epoxy [19, 22]. As a general rule, the strength of a polyester bond will be around 20 percent weaker than the same bond made with epoxy resins, Polyester resins and MEKP (Methyl Ethyl Ketone Peroxide) catalyst is used as hardener. The mixing proportions used are 30ml of resin and 20drops of MEKP.

**Table 1. Parameters set for 3d printing**

| Printing Parameters | Options Used | Units |
|---|---|---|
| Layer Height | 0.18 | mm |
| Number of Perimeter Shell | 4 | - |
| Fill Density | 100 | % |
| Fill Pattern | Hexagonal | - |
| Print Speed | 40 | mm/sec |
| Travel Speed | 60 | mm/sec |
| Extruder Temperature | 230 | Degree Celsius |

For epoxy and polyester resin the added ratio of resin to hardener is 2:1 and this mixture is allowed to cure for about 24 hours. The figure 4 shows the preparation of resins by mixing with hardener. Though epoxy is general resin without any solvent, the polyester would have very strong smell due to the vaporization of the solvent present in them during the times of curing. To accelerate the phase of curing hardeners are used. The figure 5 shows the cured parts after 24 hours.

**Table 2. Curing time and mixing ratios**

| Resins | Mixing Proportion | | Curing Time (Hours) |
|---|---|---|---|
| | Resin (ml) | Hardener (ml) | |
| Epoxy | 200 | 100 | 24 |
| Polyester | 300 | 150 | 24 |

**Figure 4. Preparation of resins.**

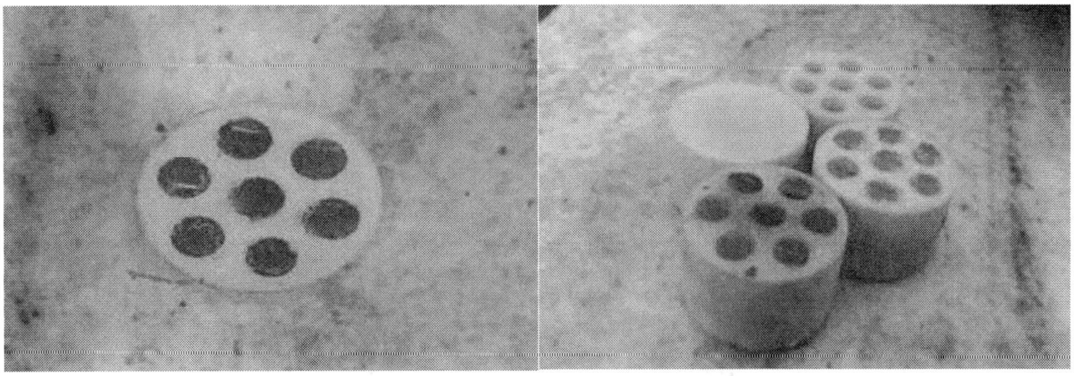

**Figure 5. Specimens after curing**

## 4. TESTING OF SAMPLES

The compression testing is carried out as per ASTM D695 standard at room temperature by using Instron UK compression tester. The ASTM D659 standard is generally used for the rigid plastics [20]. Any plastic under compression will have three phases namely the plastic phase, plateau phase, and the densification phase. The point of conversion of plastic to plateau is called the yield point after which the material disobeys hook's law. At the plateau phase excessive bulging of the material along the length is seen significantly. At the densification stage the material completely disorients and breaks down. The figure 6 shows the specimen after compression test.

### 4.1 Quality of Interface region

The video measuring system was used to ensure the infill quality and it was clearly found that both the resins were perfectly bonded with the base material in the interface region and also the resins infilled the pores without any voids. The figure 7 depicts the enlarged view of quality at interface.

**Figure 6. Specimens after compression test**

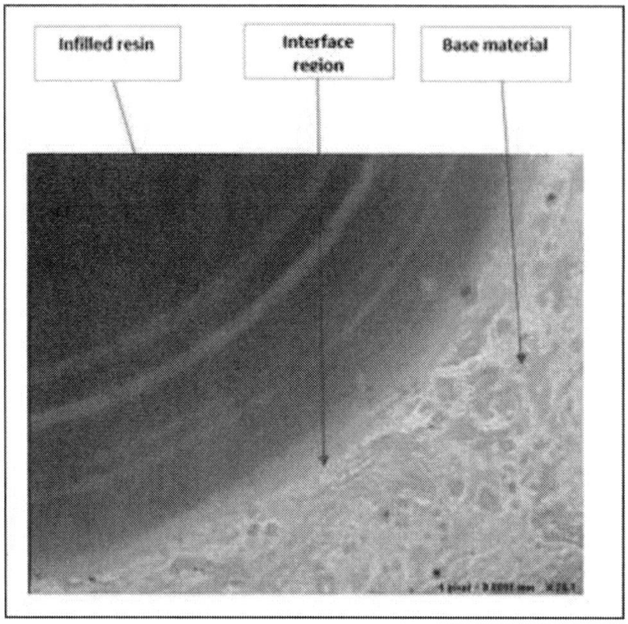


**Figure 7. VMS image showing the quality of interface region**

## 5. RESULTS AND DISCUSSION

The compression test was done for all the four parts and their values are compared. The resulting load vs elongation graph for the various parts are shown in figures 8, 9, 10, 11 respectively**.** The addition of epoxy to ABS parts had improved the strength of the manufactured part for about 25.5%.Epoxy is strong and has great compressive strength. In General thermosetting polymers exhibit more strength than thermoplastic polymer. So this combination provides higher strength than the solid part. The addition of polyester to ABS parts had improved the strength of the manufactured part to about 12.9%.Polyester resins have compressive strength of 94mpa. Polyester resins filled parts have less strength than epoxy and more strength than hollow parts. Addition of polyester resins to ABS have strength improvement than hollow parts but less than epoxy.

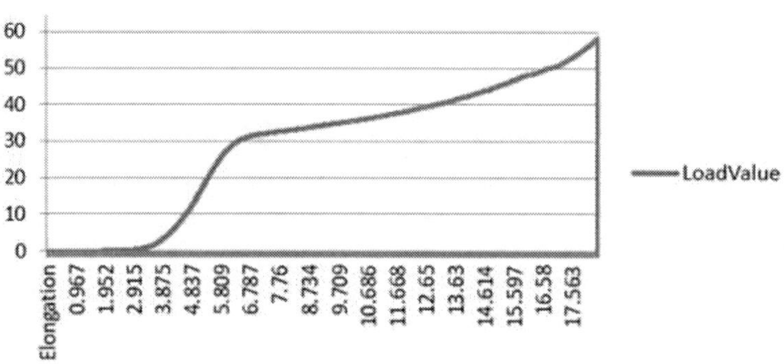


**Figure 8. Load (KN) vs Elongation (%) of solid ABS**

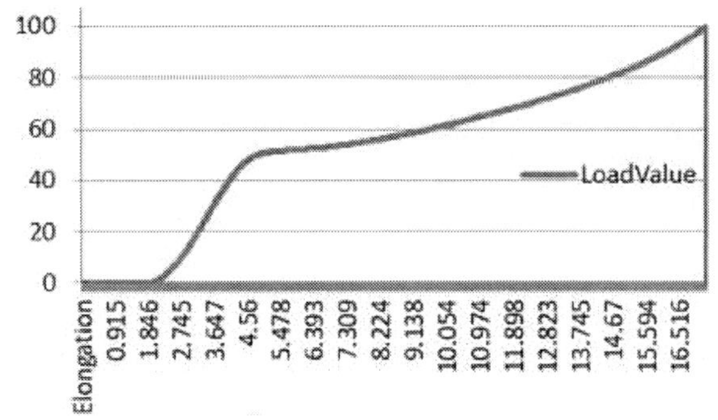


**Figure 9. Load (KN) vs Elongation (%) of epoxy filled ABS**

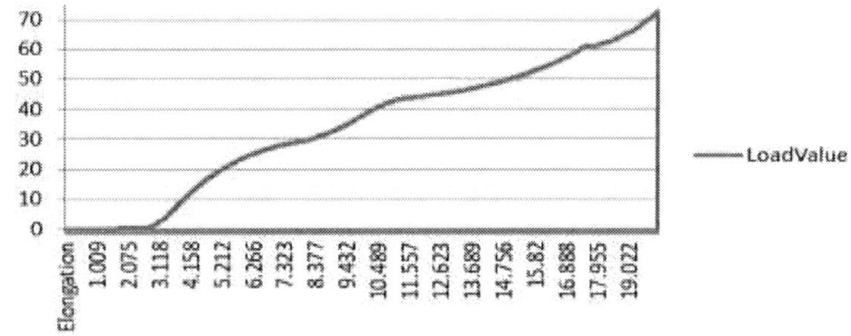


**Figure 10. Load (KN) vs Elongation (%) of polyester filled ABS**

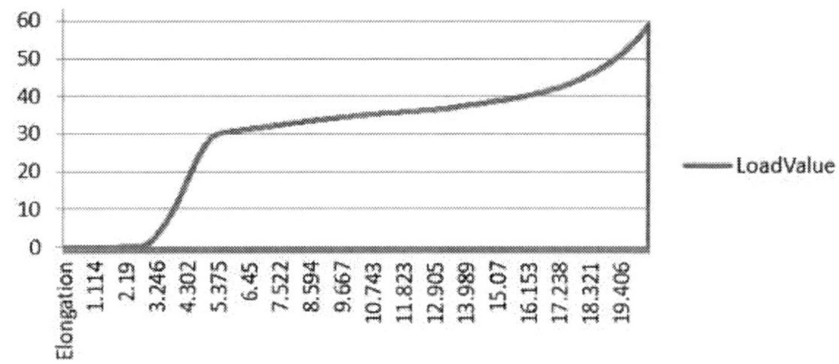


**Figure 11. Load (KN) vs Elongation (%) of void ABS**

From the above graphs the values of stress and strain are calculated. Comparing this with the stress-strain graph the yield and the ultimate stress values are found and are given below in the table 3. The figure 12, 13 and 14 shows the comparison of yield stress, ultimate stress and overall comparison between various specimens used in this study.

**Table 3. Yield stress and ultimate stress of the specimens**

| Category of Specimens | Yield Stress (MPa) | Ultimate Stress (Mpa) |
|---|---|---|
| Solid Sample | 31.19 | 50.96 |
| Epoxy Filled ABS Sample | 41.9 | 85.72 |
| Polyester Filled ABS Sample | 35.82 | 61.85 |
| Void ABS Sample | 23.86 | 51.43 |

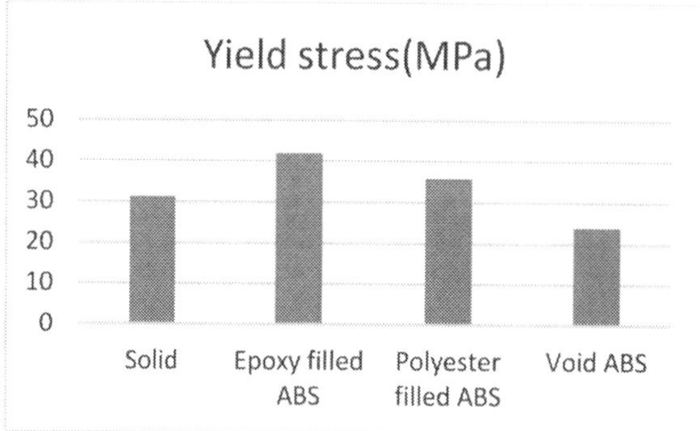


**Figure 12. Comparison of yield stress in Mpa**

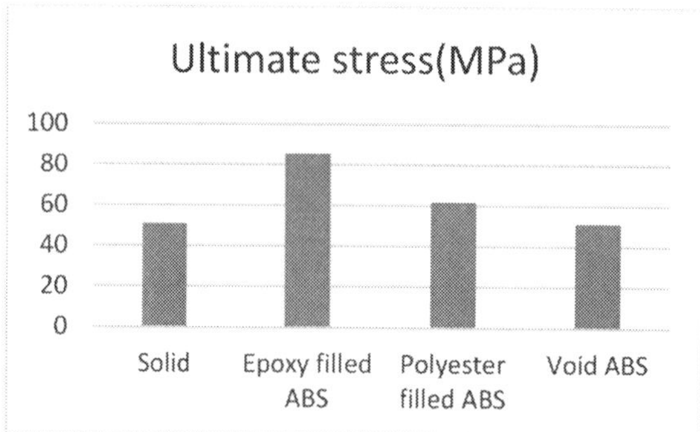


**Figure 13. Comparison of ultimate stress in Mpa**

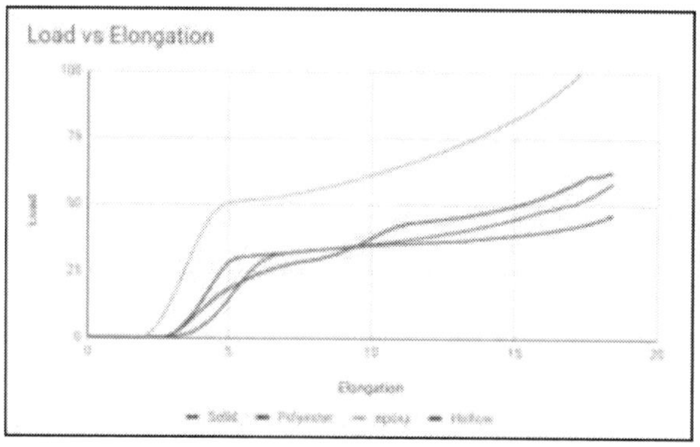


**Figure 14. Overall comparison of Load (KN) vs Elongation (%)**

## 6. CONCLUSION

The open non-stochastic circular cellular structure was designed in cuboid shape as per ASTM standards. The specimens were designed in the volumetric composition of 60% base material density and 40% void density which was filled later with epoxy and polyester resins. This study was carried out as a pre-indicative study to verify the compatibility of creating hybrid functionally graded material which is to be used for direct functional end tooling applications. Four types of specimens which is of different volumetric compositions were tested and the results were inferred. The compression test were carried out to examine the energy absorbing ability of the functionally graded material. The results of this study are concluded as follows

- This novel technique of infilling resins into the additive manufactured cellular structures showed a good bonding interface and also proven to be a better technique in producing functionally graded materials.
- Two types of resins were used in this study (Epoxy – Thermosetting resin) and (Polyester – Thermosetting resin).The base material is ABS (Thermoplastic).The interface layer of ABS-EPOXY and ABS-POLYESTER is good and this promises the possibility of more studies can be carried out by using different materials in future.
- The ABS-EPOXY composition provided higher absorbing capability than ABS-POLYESTER and from other specimens which shows the addition of thermosetting resin reinforced the properties of thermoplastic base material significantly.
- The specimen without any infilled resin (void part) has reduced strength due to less resistance offered by the air.
- The 100% density solid ABS part shows reduced strength than functionally graded structures which promises the utilization of this study in creating customized applications.
- This method of resin addition to strengthen the polymer parts may be further studied under two methods. Firstly, multiple layers of resins may be coated to a certain thickness which would enhance the impact and scratch resistance of the part. Secondly, the resin addition instead of manual method can be done by protected vacuum condition method along with the printing of the base material.

## REFERENCES

1. Olga Ivanova Christopher Williams Thomas Campbell, (2013),"Additive manufacturing (AM) and nanotechnology: promises and challenges", Rapid Prototyping *Journal, Vol. 19 Issue 5 pp. 353 - 364*
2. Mehrdad Mohsenizadeh, Federico Gasbarri, Michael Munther, Ali Beheshti, Keivan Davami, "Additively-manufactured lightweight Metamaterials for energy absorption", *Materials and Design 139 (2018) 521–530.*
3. D.W. Gibson, Rosen, B. Stucker, "Additive manufacturing technologies: rapid prototyping to direct digital manufacturing" *Springer, New York, 2010.*
4. Chunlei Qiu, Sheng Yue, Nicholas J.E. Adkins, Mark Ward, Hany Hassanin, Peter D. Lee, Philip J. Withers and Moataz M. Attallah "Influence of processing conditions on strut structure and compressive properties of cellular lattice structures fabricated by selective laser melting", *Materials Science & Engineering, 2015, Vol. 628, pp. 188–197.*
5. MatejVesenjak, Lovre Krstulovi Opara, Zoran Ren and Zeljko Domazet, "Cell shape effect evaluation of polyamide cellular structures", *Polymer Testing, Vol. 29, 2010, pp. 991–994.*
6. Iyibilgin, O., Yigit, C., and Leu, M.C, "Experimental investigation of different cellular lattice structures manufactured by fused deposition modeling", Proceedings of Solid Freeform Fabrication Symposium, *Austin, TX, 2013, pp. 895-907.*
7. Sathishkumar N, Sugavaneswaran M, Arumaikkannu G, " Investigation of sparse mode build style on material consumption, build time and compressive behaviour of additive manufactured cellular structures", 6th International & 27th All India Manufacturing Technology, Design and Research Conference (AIMTDR-2016), *December 16-18, 2016 at College of Engineering., Pune, Maharashtra, INDIA, ISBN:9789386256270.*
8. Cerardi, Caneri M., Meneghello R., Concheri G., Ricotta M "Mechanical characterization of polyamide cellular structures fabricated using selective laser sintering technologies", *Materials and Design, Vol. 46,2013, pp. 910–915.*
9. Sathishkumar, N., Arunkumar, N., Balamurugan, L., Sabarish, L., & Joseph, A. S. S. (2020). Investigation of Mechanical Behaviour and Surface Roughness Properties on Copper Electroplated FDM High Impact Polystyrene Parts. In Advances in Additive Manufacturing and Joining *(pp. 287-300). Springer, Singapore.*

10. Balamurugan, L., N. Sathishkumar, N. Arunkumar, and G. Aravind. "Investigation of mechanical behaviour and surface roughness properties on electroplated FDM ABS parts." *Int. J. Eng. Res. Mech. Civ. Eng 2 (2017): 60-67.*
11. Sugavaneswaran M., Rajesh N., Sathishkumar N. (2020) Design of Robot Gripper with Topology Optimization and Its Fabrication Using Additive Manufacturing. In: Shunmugam M., Kanthababu M. (eds) Advances in Additive Manufacturing and Joining. Lecture Notes on *Multidisciplinary Industrial Engineering. Springer, Singapore.*
12. Prithivirajan R., Sugavaneswaran M., Sathishkumar N. and Arumaikkannu G. (2019), "Metal bellow hydroforming using additive manufactured die: a case study", Rapid Prototyping Journal, *Vol. 25 No. 4, pp. 765-774.*
13. ang, Li & Harrysson, Ola & Cormier, Denis & West, Harvey & Gong, Haijun & Stucker, Brent. (2015). Additive Manufacturing of Metal Cellular Structures: Design and Fabrication. *JOM. 67. 10.1007/s11837-015-1322-y.*
14. R. Prithvirajan , M. Sugavaneswaran , N. Sathishkumar , G. Arumaikkannu (2017), "Study on feasibility of FDM made ABS die for metal bellow hydroforming", *International Conference on Manufacturing Technology and Simulation, pp. 105-108.*
*15.* Afshara M, Pourkamali Anarakia A, Montazeriana H, Kadkhodapour J, "Additive manufacturing and mechanical characterization of graded porosity scaffolds designed based on triply periodic minimal surface architectures*", Journal of the mechanical behavior of biomedical materials, Vol.62, 2016, pp. 481–494.*
16. Xiaojian Wang, Shanqing Xu, Shiwei Zhou, Wei Xu, Martin Leary, Peter Choong, M. Qian, Milan Brandt, Yi Min Xie, "Topological design and additive manufacturing of porous metals for bone scaffolds and orthopaedic implants: A review", *Biomaterials, Vol. 83, 2016, pp.127-141.*
17. L.J. Gibson, M.F. Ashby, Cellular Solids: Structure and Properties, *Cambridge university press, 1997.*
18. Arun, M., N. Sathishkumar, K. Nithesh Kumar, S. S. Ajai, and S. Aswin. "Development of patient specific bio-polymer incisor teeth by 3D printing process: A case study." *Materials Today: Proceedings* (2020).
19. Ashby, M.F., Evans, A., Fleck, N. A., Gibson, L. J., Hutchinson, J. W., and Wadley, H. N. G., "Metal Foams: A Design Guide". *Woburn, MA: Butterworth-Heinemann, 2000.*
20. Rasoul Mahshid, Hans Nørgaard Hansen, Klaus Loft Højbjerre, "Strength analysis and modeling of cellular lattice structures manufactured using selective laser melting for tooling applications", Materials and Design, *Vol.104, 2016, pp. 276–283.*
21. Sathishkumar, N., Naren Vivekanandan, L. Balamurugan, N. Arunkumar, and Ijaz Ahamed. "Mechanical Properties of Triply Periodic Minimal Surface based lattices made by Polyjet Printing." *Materials Today: Proceedings. Vol.* 22 (2020): 2934-2940.
22. ASTM F2792-12a, "Standard Terminology for Additive Manufacturing Technologies", ASTM *International, West Conshohocken, PA, 2012.*
23. Sathishkumar, N., Udayakumar, ASM., Vincent, B., Ashok Kumar, V., "Study and analysis of 3D printed FDM components by non-destructive testing techniques". *International Journal of Research and Review.* Vol. 7, Issue 5, 2020, pp. 217-222.

IOP Conf. Series: Materials Science and Engineering **923** (2020) 012045 doi:10.1088/1757-899X/923/1/012045

# Utilization of grey Taguchi method to optimize the mechanical properties of hemp and coconut shell powder hybrid composites under liquid nitrogen conditions

**G Velmurugan[1*], K Babu[2], L Immaculyne Flavia[3], C Stories Stephy[4], M Hariharan[5]**

[1, 3,4&5] Department of Mechanical Engineering, Jeppiaar Engineering College, Chennai, Tamilnadu, India-600119.
[2]Department of Mechanical Engineering, SSN College of Engineering, Chennai, Tamilnadu, India-603110.

*Email: velresearch032@gmail.com

**Abstract** Through this work an attempt was made to improve the mechanical properties of composite materials by using a process called Gray-Taguchi. Hand lay-up approach has been adopted for manufacturing. Here, Woven Hemp fiber diameter, Number of hemp layer, coconut shell powder content and cryogenic temperature ranges at different levels were incorporated as result parameter. ASTM specifications were set for the manufactured laminates to perform the mechanical tests. Hemp and coconut shell powder material are exposed to 5 percent NaOH treatment to create the matrix-fiber interface exchange. To optimize composite, the Taguchi method is combined with the Gray relational analysis. The quality targets are set for mechanical properties that include compressive strength, tensile strength, impact strength and flexural strength. This study involved nine experiments on the orthogonal array of the $L_9$ ($3^4$) Taguchi method. A combination of the composite, Gray relational analysis was applied to obtain the optimal parameter. The results of optimum parameters were verified using Conformation test. The maximum prescribed parameters are 0.7 mm woven hemp diameter, two hemp layers, 10 percent CSP content weight, and -130°C cryogenic temperature range provides the maximum energy of impact, tensile strength and flexural strength.

## 1. Introduction

Natural fibers can be widely used for a variety of uses. It evolved over the past few years from an ecological awareness. Present ecological procedures have stressed the engineering sector such as the automobile, wrapping and manufacturing industries to search for new products that can in effect be used as an alternative to traditional composite materials. Such materials consist of plastic matrix and inorganic filler reinforcement material which has an adverse environmental effect. Glass fiber filled thermoplastic composites already use distinct inorganic fillers such as aramid and carbon fibres. There are also limitations on the use of inert fibers as they have issues such as non-biodegradability, health concerns such as skin problems due to inadequate handling and processing. They generate abrasions in the dispensing equipment [1]. The key advantage over traditional artificial fillers is that they are lower in density, less abrasive, cost-effective, reusable and environmentally friendly [2]. Similar to fiber glass, the fiber board obtained from the hemp based composites has greater strength but lighter than other wood composites.

Cellulose is a significant constituent of the fabric, which is the reason for its fabric strength, resilience, and stiffness. The concentration of cellulose in hemp is 70.2-70.4%, which is higher than in other bast fibers [3]. The hybrid composite material is commonly used in low-cost engineering; strength-to - weight ratio, and ease of manufacture [4]. Such hybrid composites provide an alternative for securing property mixtures such as stiffness, ductility and strength, which cannot be consumed by composites reinforced with single fibre. Hybrid composites have increased fatigue life, greater crack resistance and less sensitivity compared to single fiber composites. [5]. Coir fiber is naturally found in plants. Composites made from coconut shell filler particles and epoxy resin, which show the increased flexural and tensile properties of the composites together with the increased content of filler particles [6]. Wear resistance increases during the addition of coir fiber and coconut shell strength to the polyester composite while coir fiber and coconut shell powder is used as reinforcing material [7].

Good compatibility and bonding are very essential for the fiber and matrix in order to obtain optimum properties of the composites. Some modifications on the surface of the fibers were performed on the basis of chemicals used in the treatment. Among the various chemical treatments, NaOH treatment is the successful way to extract oil and wax from the surface of the fibre, and it is the preliminary treatment for expanding the rough surface, as pore structure and WAXS help to evaluate the crystalline shape, density and crystallite thickness. [8-11]. Cryogenic treatments can further enhance the mechanical properties of fibre-reinforced composites. For example, the materials used for aircraft frames can preserve their mechanical properties at extreme temperatures up to -200 °C [12]. Cryogenically treated plastics and composites are stronger, longer lasting and have improved durability and wear resistance [13]. Therefore, cryogenic composite treatment has become an important part of ongoing research and development to enhance the performance of materials in natural compounds.

The main objective of the work is to assess the potential properties of the hemp fibers when hybridized in a thermoplastic composite with coconut shell powder. To achieve this goal, coconut shell powder and hemp fiber based hybrid epoxy compounds were manufactured using simple hand-lay technique. In NaOH solution, natural fibers such as coconut shell powder and hemp fibers were treated to strengthen their adhesion and reduce the absorption of moisture. The composites were subsequently treated with liquid nitrogen for various temperature ranges and their mechanical properties were measured experimentally and compared to optimum processing parameters such as Woven Hemp fiber diameter, number of hemp layer, coconut shell powder content and cryogenic temperature ranges. The Taguchi methods based on Grey have been used for various multi-response routing difficulties and their constraints have been configured effectively [14, 15].

## 2. Experimental Methods

### *2.1 Materials*

In the southern region of Tamilnadu, India, coconut shell powder and hemp required for the present research study was collected from natural fiber industry. All fires were sun-dried for 2 days, and afterwards long fibers were removed by hand sorting. To obtain better properties, the hemp fiber is straightened and compressed by mechanical rolling. The fiber is then graded into different lengths and thicknesses. All specimens are added uniformly with hemp of about 4 grams. Coconut shell powder (CSP) typically contains few moisture content. So, to reduce dampness, CSP is stored at an ambient temperature. Then CSP packs in pockets as percentage and size of weight required. Specific composite materials are produced using the hand-lay technique.

### *2.2 Fiber Treatment*

The raw hemp and coconut shell powder is washed at a temperature of 70°C for about 1 hour with 2 per cent detergent solution, then washed in distilled water and finally dedicated in a vacuum oven at around 70 ° C. The dehydrated fibre was chosen as raw fibre. For surface treatments the organic jute fibers are exposed to Alkali. Initially, the untreated fibers were dewaxed by immersing batches of

IOP Conf. Series: Materials Science and Engineering **923** (2020) 012045 doi:10.1088/1757-899X/923/1/012045

untreated fibers in ethanol and benzene at a ratio of 1:2 for around 72 hours at 50°C. The dewaxed fibers are submerged for 24hrs in 5 percent NaOH at 30°C followed by assiduous washing in distilled water and air drying secures 5 percent of alkali treated fibers.

### *2.3 Fabrication of Bio Composites*

Composite fabrication is done using a stainless steel mold of 300 x 300 x 3 mm. After the molds are cleaned and dehydrated, the epoxy is added. The wax is the releasing agent which is used here. Using hand lay-up technique, the coconut shell powder treated along with the woven hemp fibers were taken for composite manufacture. Coconut shell powder of different contents was first dispersed by intense stirring into the prepared epoxy resin. This combination of matrixes was drizzled onto the layers of fibers in the mouth. When the fiber mats are fully moist by matrix combinations, the mold was bolted at usual hotness and dried. The hybrid composite samples produced were placed in the desiccators to ensure no further moisture absorption. The manufactured samples were then immersed in liquid nitrogen for cryogenic treatment according to the design for different temperature range. The composites were taken out and kept at room temperature after the cryogenic procedure. In accordance with Taguchi design, $L_9$ orthogonal array was selected for 4 parameters each with 3 levels and 9 composite plates were subsequently manufactured for further investigation. Tables 1 and 2 show the parameters and their levels and the $L_9$ orthogonal array (OA).

**Table 1.** Specification and their stages used for experimentation

| Specifications | Symbol | Levels | | |
|---|---|---|---|---|
| Woven hemp Dia (mm) | A | 0.3 | 0.5 | 0.7 |
| Number of hemp layer (no) | B | 1 | 2 | 3 |
| CSP content (%) | C | 10 | 12 | 15 |
| Cryogenic Temp Range (°C) | D | -100 | -130 | -160 |

**Table 2.** L9 Orthogonal Array

| Runs | Stages of A | Stages of B | Stages of C | Stages of D |
|---|---|---|---|---|
| 1 | 1 | 1 | 1 | 1 |
| 2 | 1 | 2 | 2 | 2 |
| 3 | 1 | 3 | 3 | 3 |
| 4 | 2 | 1 | 2 | 3 |
| 5 | 2 | 2 | 3 | 1 |
| 6 | 2 | 3 | 1 | 2 |
| 7 | 3 | 1 | 3 | 2 |
| 8 | 3 | 2 | 1 | 3 |
| 9 | 3 | 3 | 2 | 1 |

## 3. Mechanical Testing

Upon fabrication, the research specimens are applied according to ASTM specifications through variety mechanical checks. The tension test is performed using standardized test method, ASTM D638. Universal testing machine KALPAK performs flexural, tensile and Impact tests. The flexural tests are performed in three-point bending test. The flexural check is conducted in compliance with ASTM standards D790. Flexural force values are calculated using equation (1).

$$\text{Flexural strength} = 3PL/2bd^2 \tag{1}$$

Where, P is applied burden at the purpose of crack, L is the help range length, b is the example width, d is the example thickness. Izod impact testing machine as per ASTM D 256 is used to perform the impact tests. Impact strength (I.S) is calculated using Equation (2).

I.S = Observed Energy / Cross sectional Area (2)

The responses for the experiment sets are tabulated, as shown in Table 3

**Table 3.** Response for the Experimental setup

| Run | Tensile (MPa) | Flexural (MPa) | Impact (KJ/m$^2$) |
|---|---|---|---|
| 1 | 20.19 | 21.54 | 46.36 |
| 2 | 30.19 | 39.47 | 51.31 |
| 3 | 23.43 | 25.49 | 56.66 |
| 4 | 21.12 | 39.87 | 56.04 |
| 5 | 20.17 | 39.95 | 51.92 |
| 6 | 25.19 | 42.98 | 53.31 |
| 7 | 29.73 | 36.04 | 56.24 |
| 8 | 22.72 | 50.28 | 56.29 |
| 9 | 21.92 | 26.72 | 62.83 |

3.1 *Investigation of Experimental outcomes by grey relational method*

Taguchi experiment design method is an effective tool for systematic demonstration, investigation and optimization of various process constraints such as constraint selection, study plan, testing direction, data analysis, optimum arrangement description and confirmation. Using this method, the results of the analysis are converted into the signal-to - noise ratio (S N) for deciding the prominent characteristic. Based on the definitive factor to be improved for the prominent feature, the S / N ratio features were divided into the following phases (i) larger-better, (ii) nominal-better, and (iii) smaller-better.

3.1.1 *Signal to Noise Ratio*

The high S/N ratio represents the enhanced performance characteristic, despite the performance feature criteria. The stage with the greatest S / N ratio is thus an optimal parameter level. In the end, a verification check is conducted to verify the optimal permutation of the restriction sites. Impact, flexural and tensile strength were measured throughout the mechanical testing of natural fiber composites through the larger phase is better. Subsequently, signal to noise fractions of the eminence features (larger is better) are explicit as

$$\text{S/N Ratio} = -10 \log_{10} \frac{1}{k}\sum_{a=i}^{k} \frac{1}{X_a^2} \quad (3)$$

Where $k$ is the repetitions number, $X_{ab}$ is the experiential reaction and a = 1, 2….., k b= 1, 2... k

3.1.2 *Standardized Signal to Noise Ratio*

Standardization is an achievement in altering an individual record response to segment the response equally, and measures it into an adequate series for further investigation. Xab is standardized as $Y_{ij}$

($0 \leq Y_{ij} \leq 1$) by the underlying equation to determine the correct outcome of embracing multiple elements and reduce the inconsistency. For this investigation integrated with Tensile (Larger is good), Flexural (Larger is good), and Impact capabilities (Larger is good), the eminence features are chosen, and these are the enhanced recital of all constraints. Table 4 describes the integration of S/ N ratios and uniform S/N ratios for rising eminence feature.

$$Y_{ij} = \frac{x_{ab} - \min(x_{ab} a = 1,2 \ldots\ldots k)}{\max(x_{ab} a = 1,2 \ldots\ldots k) - \min(x_{ab} a = 1,2 \ldots\ldots k)} \quad (4)$$

**Table 4.** Signal to Ratio and Normalized signal to noise Ratio

| Run | Signal to Noise Ratio | | | Normalized Signal to Noise Ratio | | |
|---|---|---|---|---|---|---|
| | **Tensile** | **Flexural** | **Impact** | **Tensile** | **Flexural** | **Impact** |
| 1 | 26.1 | 26.66 | 32.28 | 0.003 | 0 | 0.135 |
| 2 | 29.6 | 31.92 | 33.14 | 1 | 0.714 | 0.441 |
| 3 | 27.4 | 28.13 | 31.9 | 0.373 | 0.199 | 0 |
| 4 | 26.49 | 32.01 | 34.06 | 0.114 | 0.726 | 0.769 |
| 5 | 26.09 | 32.03 | 33.92 | 0 | 0.729 | 0.719 |
| 6 | 28.02 | 32.66 | 33.54 | 0.55 | 0.814 | 0.584 |
| 7 | 29.46 | 31.13 | 33.89 | 0.96 | 0.607 | 0.708 |
| 8 | 27.13 | 34.03 | 34.71 | 0.296 | 1 | 1 |
| 9 | 26.82 | 28.54 | 32.74 | 0.208 | 0.255 | 0.299 |

### 3.1.3 *Coefficient of Grey Relation*

The Gray Relational Coefficient was calculated using the standard S / N ratio values. The Grey coefficient of relation is represented as follows:

$$\gamma\,(x_a(y), x_b(y)) = \frac{\Delta min + \zeta\, \Delta max}{\Delta_{ab}(y) + \zeta\, \Delta max} \quad (5)$$

Where b = 1, 2. . . *y; y* = 1, 2, . . . , *m; y* is the investigational statistical numbers and *m* is the quantity of retorts; $x_a$ (*y*) is the orientation series ($x_a$ (*y*) = 1, *y*= 1, 2, . . . , *m*); $x_b$ (*y*) is the exact relationship series; ab ={$\Delta x_a$ (y) –$x_b$ (y)} is the entire difference values between $x_a$ (y) and $x_b$ (y); Δmin = min{$x_a$ (y) – $x_b$ (y)} is the least value of $x_b$(y); Δmax = max {$x_a$ (y) – $x_b$(y)} is the maximum value of $x_b$(y); and ζ is the incriminate constant set in between the range of zero to one. In the current research, the process constraints have an equivalent weight and discriminate constant was fixed by 0.5.

### 3.1.4 *Grey Relational Grades*

With the help of equation (6), each Grey relation coefficients feedback is converted to the Grey relationship grade. Any ideal stage of controllable factor is calculated from the grades of the relationship of the gray. The results are outlined in Table 5. In which $\overline{\gamma_j}$ is the $J^{th}$ experiment grades of grey relation and k is performance characteristics numeral. Subsequently, the grades and coefficient of Grey relationship were performed for all $L_9$ ($3^4$) orthogonal experiments; By means of Grey analyzes and Taguchi relationship techniques, the multi-response problem on optimization was changed to a single-response objective function. Grey-based Taguchi approach has been used to achieve the overall grades of Gray Relationship and that it is just the output features. By searching for a parameter setting, you can achieve the highest comprehensive grey relational grade.

$$\bar{\gamma_J} = \frac{1}{k}\sum_{i=1}^{m} \gamma_{ij} \qquad (6)$$

**Table 5**. Grey relational grade values

| Run | Grey Relational Co efficient | | | Grey relational Grades |
|---|---|---|---|---|
| | Tensile | Flexural | Impact | |
| 1 | 0.336 | 0.333 | 0.366 | 0.259 |
| 2 | 1.006 | 0.636 | 0.472 | 0.529 |
| 3 | 0.446 | 0.384 | 0.333 | 0.291 |
| 4 | 0.363 | 0.646 | 0.684 | 0.423 |
| 5 | 0.335 | 0.648 | 0.64 | 0.406 |
| 6 | 0.529 | 0.729 | 0.546 | 0.451 |
| 7 | 0.932 | 0.56 | 0.631 | 0.531 |
| 8 | 0.418 | 1 | 1 | 0.604 |
| 9 | 0.389 | 0.402 | 0.416 | 0.302 |

## 4. Result and Discussions

The Taguchi method and Grey relation investigation examined optimal parameters such as flexural, impact and tensile strength of CSP and hemp based hybrid epoxy composites. The investigation of the Grey Relationship and ANOVA reduces optimal restriction sets and maximum critical feature in natural composite. The Taguchi feedback table methods were utilized to determine the average grade of each factor point. For example, to discover the Grey ratio score of the woven hemp diameter, the ratio value of the grey ratio coefficient of flat (level) 1 in column A would be moderated (i.e., tests 1–3), then the ratio of flat 2 in column A would be moderated (i.e., tests 4–6), then again the ratio of flat 3 in column A would be moderated (i.e., tests 7–9). Similarly, the number of hemp layers, the CSP content and the cryogenic temperature range are planned for the exact flats; this is exposed in Table 6. Fig.1 revealed the response details of the Grade of Gray Relationship at various levels of the composite constraints.

The figures show the optimum level of process constraint produced by the maximum degree of Grey relation. For constraints such as woven hemp diameter, number of hemp layer, CSP content, and cryogenic temperature range respectively, the levels A3, B2, C1 and Grey relational grade value is maximum in D3. The essential values of the hybrid composites were measured from the investigation are woven hemp diameter of 0.7 mm, 2 layers of hemp, CSP content of 10%, and cryogenic temperature of -130°C.

### *4.1 Analysis of Variance*

The ANOVA is formulated using the average value of the Grey relation grade. ANOVA results are used to measure the process prompting parameters of CSP and hemp-based hybrid composites. The square sum deviations from the overall mean grade are used to calculate by isolating the entire variation of the Grey relationship grades. ANOVA can observe the quality objective, which is influenced by process criteria. The impact of process parameter shift on performance characteristics can be assessed through the percentage contribution through the total squared sum deviations [16, 17].

IOP Conf. Series: Materials Science and Engineering **923** (2020) 012045 doi:10.1088/1757-899X/923/1/012045

**Table 6.** Grey Relational Grade

| Levels | A | B | C | D |
|---|---|---|---|---|
| L1 | 0.36 | 0.404 | 0.438 | 0.322 |
| L2 | 0.427 | 0.513 | 0.418 | 0.503 |
| L3 | 0.479 | 0.348 | 0.409 | 0.44 |

The following equations can be used to calculate the percentage contributions of variance

1. Total sum of the square deviations:

$$ss_d = \frac{F_1^2+F_2^2+F_3^2}{3} - \frac{1}{m}[\sum_{i=1}^{m} y_i]^2 \quad (7)$$

2. Sum of square deviations

$$ss_T = \sum_{i=1}^{m} y_i^2 - \frac{1}{m}[\sum_{i=1}^{m} y_i]^2 \quad (8)$$

3. Percentage of contribution

$$\rho = \frac{ss_d}{ss_T} \quad (9)$$

Where F denotes the process factor, and 1, 2, 3 are the corresponding levels. m is the number of levels in the orthogonal array and $y_i$ is the Grey relation grade mean.

**Table 7.** ANOVA outcomes for the Grey Relational Grade

| Factors | Degree of Freedom | Sum of Squares | Mean Sum of Squares | % of Contributions |
|---|---|---|---|---|
| A | 2 | 0.007 | 0.0035 | 18.70 |
| B | 2 | 0.014 | 0.007 | 37.39 |
| C | 2 | 0.00044 | 0.00022 | 1.18 |
| D | 2 | 0.016 | 0.008 | 42.74 |
| Error | 0 | 0 | 0 | 0 |
| Total | 8 | 0.03744 | 0.01872 | 100 |

The effect of each process restriction is defined in the ANOVA table, showing in Table 7 that the diameter of woven hemp has a contribution of 18.70%, the amount of hemp layers has 37.39%, the content of CSP has 1.18% and the temperature range of cryogenics has 42.74% on the various concert physiognomies. This recommends that cryogenic treatments be additional key to natural fiber hybrid-reinforced epoxy composites based on CSP and hemp

*4.2 Confirmation Test*

Confirmation testing was conducted to validate the optimum parameters which yield better composite performance characteristics. The natural fiber composites are produced using the set of parameters that resulted in the best results in the L9 table (woven hemp diameter of 0.3 mm, 2 layers of hemp, 12% of CSP content and -130 ° C of cryogenic temperature range) and the optimum parameters were obtained by means of Grey logical analysis (woven hemp diameter of 0.7 mm, 2 layers of hemp, 10% of CSP content and 130 ° C of Cryogenic temperature) are compared in Table 8. It shows an improvement of 19.72 per cent in tensile strength, 20.61 per cent in flexural strength and 29.55 per cent in impact energy.

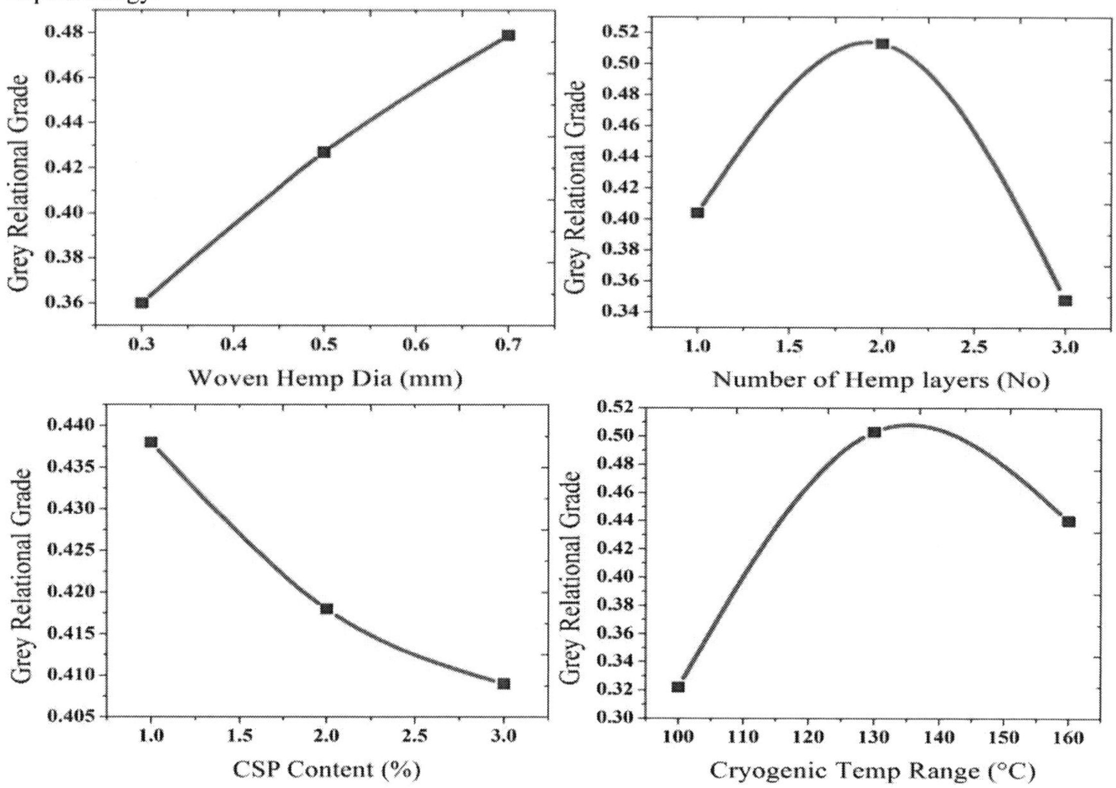


**Figure1**. Shows the Response graph of overall grades of Grey relation for various parameters

**Table 8.** Comparison between initial and optimal results

| Responses | Initial parameters | Final parameters | % of Improvement |
|---|---|---|---|
| **Setting level** | **A2 B2 C3 D1 E1 F2** | **A3 B2 C3 D3 E2 F2** | |
| Tensile strength (MPa) | 28.97 | 31.61 | 8.35 |
| Flexural strength (MPa) | 40.01 | 44.72 | 10.53 |
| Impact energy (KJ/m$^2$) | 2.07 | 2.92 | 29.11 |

## 5. Conclusion

In this article after study is provided the report for the optimization of the composite material having different performance characteristics using Grey relational analysis and Taguchi method. The result reads as follows:

1. The parameter levels prescribed are 0.7 mm woven hemp diameter, two hemp layers, 10 percent CSP content weight, and -130°C cryogenic temperature range obtained from the response graph and table. It gives maximum energy of impact, tensile strength and flexural strength.
2. The Grey- Taguchi method confirms test results showing an increase of 19.72 percent in tensile strength, an increase of 20.61 percent in flexural strength and 29.55 percent in impact energy compared to the introductory parameters.

IOP Conf. Series: Materials Science and Engineering **923** (2020) 012045 doi:10.1088/1757-899X/923/1/012045

3. ANOVA has identified the cryogenic treatment as the most compelling performance enhancement factor, with 42.74 percent. It is wrapped up saying that the composites mainly contribute and have a significant effect on the different performance characteristics.

## References


[1] Espert A, Vilaplana F and Karlsson S 2004 *Composites Part A: Applied Sci and Man* **35** 1267-76.
[2] Bledzki AK, Reihmane S and Gassan J 2006 *Polymer-Plastics Tech and Eng* **37** 451-68.
[3] Kabir MM, Wang H, Lau KT and Cardona F 2012 *Composites Part B: Eng* **43** 2883-92.
[4] Gururaja MN and Hari Rao AN 2012 *Int .J. Soft Comput. Eng* **1** 352–55.
[5] Shahzad A 2011 *J. of Reinf Plast Composites* **30** 1389–98.
[6] Vignesh K, Natarajan U and Anbalagan M 2015 *Int. J. of Applied Eng. Res* 10, issue **13** 11066-79.
[7] Vignesh K, Natarajan U and Vijayasekar A 2014 *international conference on recent trends in engineering and management IOSR.J. Mecha. Civil Engi 53-57.*
[8] Le Troedec M, Sedan D, Peyratout C 2008 *Composites Part A: App Scie and Mang* **39** 514-22.
[9] Sawpan MA, Pickering KL and Fernyhough A 2011 *Composites Part A: App Sci and Manu* **42** 888-95.
[10] Mwaikambo LY and Ansell MP 2002 *J. of. App. Pol. Sci* **84** 2222-34.
[11] Hermans PH and Weidinger A 1949 *J .of Poly Sci Part A: Polymer Chemistry* **4** 135-44.
[12] Liu X, Cheng L, Zhang L 2011*Mate Sci and Eng: A* **528** 7524-28.
[13] Kalia S 2010 *J.of Low Temp Physics* **158** 934-45.
[14] Haq AN, Marimuthu P and Jeyapaul R 2008 *The Int. J. of Adv Manu Tech* **37** 250-55.
[15] Kuo C-FJ and Wu Y-S 2006 *The Int.J. of Adva Manu Tech* **27** 525-30.
[16] Navaneethakrishnan S and Athijayamani A 2015 *Int J .of Plas .Tech* **19** 227-40.
[17] Sri Siva R, Mohan Lal D and Jaswin MA 2012 *Tribology Transactions* **55** 854-62.

# CONVENTIONAL CONCRETE OVER POLYMER IMPREGNATED CONCRETE USING SILICA FUMES

**GutthiAswini*[1] and Mrs. Kalpana[2]**

* [1]UG Scholar ,Saveetha School of Engineering , Saveetha Institute of Medical and technical sciences, Chennai, India

[2]Assistant Professor, Saveetha School of Engineering ,Saveetha Institute of Medical and Technical Sciences, Chennai , India

*E-mail: [1]gutti.aswiniashu@gmail.com

E-mail: [2]kalpanamohancivil@gmail.com

**Abstract**: Present research targets finding compacted quality as well as scraped spot opposition capacity for polymer tar changed cement with silica rage. While exploratory program, solid blends are planned with allotted estimations of water-to-binder proportion at 0.7, volume of total at 67%, sand level at 45% volume of total, as well as polypropylene measure fiber at 0.6 % volume of cement. Trial factors incorporated the silica smolder sum shifted at 0 %, 5% and 10% of powder volume, and the measure of unadulterated acrylic polymer pitch differed at 0%, 5% and 10% load of powder. The exploratory aftereffects of study shows when strength of cement diminished by 22.4 %~34.2 % with the expansion of the polymer content. The weight decrease of the polymer altered cement incited by scraped area assault diminished 6.78%~21% and 14.68%~25.82% in dry and wet conditions, individually.

Keyword: Silica, polymer, cement

## 1. Introduction:

Concrete was named ananisotropic substances unsatisfactorily selected forspecific states of work pressure, for example ductile or bowing. Preservationist plan of solid design typically considers compactive energy of concrete as urgent features for planning procedure due to differentiation for tractable or twisting energy compressive quality of solid is prevalent. Selecting for solid format for holding on for elastic or twisting burden with providing good quality,preprocessing for coupling framework to upgrade opposition of solid to pliable pressure is significantly noticed. Between current specialized strategies, polymers usage for expanding flexibility of solid format had gotten normal methods. In any case, the polymers impact expansion on building as well as especially scraped area obstruction of cement ought to be additionally found because of the constrained result in those areas of research .

With polymer micro mixture as electrical protection, dielectric spectroscopy context madein 2003 [5] as well asslightly release research categorized in 2005 as well as in 2007[6]. That is discovered that permissive of epoxy diminished because of expansion of titania micro fillers till specific sum. Besides, micro mixtures displayed an intense increment in opposition by micro fillers expansion. Those discoveries has pulled in a lot of consideration [7-8]. From that point, 10 years had gone, as well as numerous important publishings are distributed till a moment. Current study manages epoxy as well as alumina micro mixtures with various types of qualities from point view of similar characteristics as well as obstruction [8-9] to research their likelihood that are utilized for twisting protection.

A voltaic tree is fractional breakdown organised shaped in matured polymer insulator. That is significant reasons for protection rejection, as well as broad researches has completed [1-5]. By as well as large, procedure of dielectrical treeing incorporates a tree hatching (or commencement) duration as well as a tree developing duration [6]. From brooding time frame, that was release that is holding channels to other with directing dividers in polymeric voltaic, for example, epoxy tars as well as XLPE [10, 11]. Lately, micro meter shaped objects are broad utilized as fillers to generatedielectrical attributes of regular polymers [12-17]. Voltaic trees in micro mixtures are additionally affirmed to stifled where trees breadthis impressively large as well as surpasses a few number as well as few mm [18-21]. To additionally study tree instruments for micro mixtures, that are critical to explore attributes of beginning minor trees in beginning period as well as that provides another knowledge to collaboration of trees till micro fillers objects. Subsequently, it is important to additionally lessen breadth of a tree. Current explorationcenters around beginning trees of breadth in epoxy silica mixtures so as to explain the attributes, for example qualities to create introductory trees, fillers scattering in mixtures as well as communications of starting trees with micro objects.

## 2. Literature Review:

Polymer materials are broadly utilized by silicone rubbers, when differentiated with the various utilized polymer especially top voltage protection implementation to prevalent electrifying, lukewarm, as well as biochemical effects. To acquire prevalent effects of mechanics, elastoma of silica are vigorously loaded up with ragedparticles of silica. To the superior comprehension of collaboration among $SiO_2$ small particles as well as elastic form of silicones similar characteristics of dielectric of seethed silica and estimations of spectroscopy are completed over a huge recurrence goes from close to DC with a limit of 9MHz. Elastic composite tests of Silicones loaded up with exterior handled seethed particles of silica are set up with warm squeezing as well as impacts of centring of fillers and exterior managed, as well as blending technique on charactersof dielectric are researched. Impact of polymer filled collaboration was broke down by contrasting spectroscopic dielectric estimation information and filtering electron microscopy (SEM) pictures.

Ultra-leastload warmth opposing rage nanospheres of silica are effectively arranged through dissolution polymers. Filtering microscopic molecules showed the silicone seethe hybrid nanospheresfurthermore, seethe substancesof silica are equally scattered in nanospheres as well as framed loads of microporous substances. Moreover, thickness procedural outlined that thickness of compound is about 2.07–1.38 g/cm3 with great reasonableness with conveying of water.Warm strength of silica seethe compound substances are essentially superior to

IOP Conf. Series: Materials Science and Engineering **923** (2020) 012046 doi:10.1088/1757-899X/923/1/012046

unadulterated with silica. Here rage composite macrospheres of silica have an altogether achievable implementation in utilization of oil as ultra-least load proppens.

Raged as well as hastened silica as well as micro mixtures of alumina are assessed for attributes of dielectric as well as incomplete release (PD) obstruction. Attributes of dielectricpermissiveness as well as scattering factor are estimated in recurrence measure from 45 Hz to 15 MHz. Obstruction of PD is surveyed utilizing a pole to-plane terminal along with air hole 0.6 mm under 8 kVrms. Another revelation is prepared for dampness and compound buildup impacts for attributes of dielectric. Normal dampness wontinfluence a lot of perpetual attributes of dielectric. That is explained that fillers of macro expansion will upgrade obstruction, true to form. Moreover, that has discovered that macro mixtures of silica is preferred in obstruction over macro mixtures of alumina.

Numerous researches are done on breakdown properties to achieve 4 to 6 mm before. Rather, tree intimation those attributes are seriously explored in the current study that has been communication of trees along with micro-fillers in the initial development time. Beginning trees are characterized as modest trees of 40 to 70 µm width which are distinguished by an as of late created discovery technique. That is demonstrated the tree inception qualities shows similar patterns by voltage as well as handled as well as unhandled micro fillers like attributes however those are less in period. Due to attributes accompanying three outcomes are gotten. (1) Tree inception period is delayed that epoxy is loaded up by 7% micro silica of various kind all in all. (2) That is larger for 70 nm silica than 90 nm silica. (3) Raged silica tends to larger time for tree commencement than melded silica. By SEM perception that has shown that fillers accumulation is bigger in epoxy that melded silica in epoxy by unique silica, as well as that was bigger in epoxy by unhandled smoldered silica then epoxy with untreated combined silica. Attachment of micro fillers with epoxy ought to becompelling to build tree inception period compared to filler accumulation. Those are significant discoveries is placed connection among beginning trees as well as micro fillers objects. SEM pictures holds vast majority of beginning trees are immaculate with micro fillers objects, when a less introductory trees are contacted for them. Those report is made on how beginning trees that connects with micro fillers inway of the development.

## 3. Conclusion:

Impact of rages of silica as well as polymer sap augmentations on compactive energy as well as scraping execution of solidified cement with triplebinding mixture of standard Portland concrete, less $ca_2$ debris is basically directed. As per test outcome, silica seethe expansion expanded compactive energy of solidified cement however didn't improve more drawn out compactive energy of the cement. At any instant, silica seethe expansion brought about the cement particles with expanded fragility related with extreme harm of ground cement experienced the scraping assault. Compactive energy of cement diminished by 28.6~37.9% with expanded substance of the polymer sap. Weight reduction of polymer adjusted cement prompted by scraping assault decreased 8.93~26% and 18.42~32.65% in particular conditions, individually.

## References:

[1] Hoang-Anh Nguyen, Jia-LunWun,Ta-Peng Chang 2017 Compressive Strength of Silica Fume-Polymer Resin Modified Concrete *International Conference on Information, Communication and Engineering.*

[2] M. Roy, J.K. Nelson, R.K. MacCrone, L.S. Schadler, C.W. Reed, R. Keefe , W. Zenger 2009 Polymer Nanocomposite Dielectrics – The Role of the Interface *Institute Of Electrical And Electronics Engineers.*

[3] J. Seiler and J. Kindersberger 2014 Insight into the Interphase in Polymer Nanocomposites *Institute Of Electrical And Electronics Engineers.*

[4] M. Ezzat, N. A. Sabiha and M. Izzularab 2014 Accurate model for computing dielectric constant of dielectric nanocomposites **4**.

[5] 2017 Standard Specification for Portland Cement *American Standard for Testing and Materials.*

[6] T. Tanaka A. Greenwood Effects of Charge Injection and Extraction on tree Initiation in Polyethylene *Institute of Electrical and Electronics Engineers.*

[7] M. Todd and F. Shi 2004 Dielectric Characteristics of Complex Composite Systems Containing Interphase Regions.

[8] L.A. Dissado, P.J.J. Sweeney and J.C. Fothergill 1992 The field dependence of electrical tree growth *Dielectric Materials, Measurements and Applications,* 13-16.

[9] Huang Xingyi, Liu Fei, Jiang Pingkai 2010 Effect of Nanoparticle Surface Treatment on Morphology, Electrical and Water Treeing Behaviors of LLDPE Composites *Institute Of Electrical And Electronics Engineers.*

[10] NaohiroHozumi, Tatsuki Okamoto and Hiromasa Fukagawa 1988 TEM Observation of Electrical Tree Paths and Microstructures in Polyethylene **27** 1230-1233.

[11] T. J. Lewis 2004 Interfaces are the dominant feature of dielectrics at the nanometric level *Institute Of Electrical And Electronics Engineers.*

[12] P. Barber, S. Balasubramanian, Y. Anguchamy, S. Gong, A. Wibowo, H. Gao, H. J. Ploehn, and H. C. Loye 2009 Polymer Composite and Nanocomposite Dielectric Materials for Pulse Power Energy Storage.

[13] Y. Yamano, M. Iizuka 2009 Suppression of electrical tree initiation in LDPE by additives of polycyclic compound *Institute Of Electrical And Electronics Engineers.*

[14] P. Barber, S. Balasubramanian, Y. Anguchamy S. Gong, A. Wibowo, H. Gao, H. J. Ploehn, and H. C. Loye 2009 Polymer Composite and Nanocomposite Dielectric Materials for Pulse Power Energy **2** 1697-1733.

[15]M.G. Todd and F.G. Shi 2005 Complex permittivity of composite systems: A comprehensive interphase approach *Institute Of Electrical And Electronics Engineers* **12** 601-611.

[16] D. Tan, Y. Cao, E. Tuncer, and P. Irwin 2013 Nanofiller Dispersion in Polymer Dielectrics *Materials Sciences and Applications* **4** 615.

[17] T. Tanaka, M. Kozako, N. Fuse and Y. Ohki 2005 Proposal of a Multicore model for Polymer Nanocomposite Dielectrics *Institute Of Electrical And Electronics Engineers.*

[18] Y. Yamano, M. Iizuka 2009 Suppression of electrical tree initiation in LDPE by additives of polycyclic compound *Institute Of Electrical And Electronics Engineers* 16 189-198.

[19] T.Iizuka, K.Uchida, T.Tanaka, 2009 Voltage endurance characteristics of epoxy/silica nanocomposites IEEJ **129** 123-130.

[20] S.Raetzke,Y.Ohki, T.Imai, T.Tanaka and J.Kindersberger 2009 Tree initiation characteristics of epoxy resin and epoxy/clay nanocomoposite *Institute Of Electrical And Electronics Engineers.*

IOP Conf. Series: Materials Science and Engineering **923** (2020) 012047 doi:10.1088/1757-899X/923/1/012047

# A REVIEW ON POLYMER IMPREGNATED CONCRETE WITH STEEL FIBRE

**S. Indra Teja *[1] and G.B. Ramesh kumar [2]**
* [1]UG Scholar , Saveetha School of Engineering , Saveetha Institute of Medical and technical sciences, Chennai, India

[2]Assistant Professor, Saveetha School of Engineering ,Saveetha Institute of Medical and Technical Sciences, Chennai , India

*E-mail: indratejaindra123@gmail.com
E-mail: rameshkumargb@gmail.com

**Abstract:** Work of steel strands has been taken out from the waste tires in fibre generation fortified fresh cement is another test in diminishing the effect of unused components on earth. These research plans to evaluate impact of RSF on compacting strength as well as reduction conduct of SCC by remembering various strands length for expansion to utilize different contexts. Six blends of reused steel filaments fresh cementwas created. Control blend, single RSF length content blends as well as mixed substance blends (substitution of small filaments by large strands at substance of 20%, 40% as well as 90%) has been done. Complete RSF context is placed steady at 45 kg/m3. Outcomes tells that expansion ofRSF influences reduction of SCC blendsespecially following 23 days. Expansion of RSF defers breaking time not with Standing decrease width of splits.

Keyword: Cement, RSF, SCC

## 1.Introduction:

Fresh Concrete had recognised as kind of solid that had created to accomplish some pivotal prerequisites [1]. The solid kind gives extra advantages past those of common cement [2]. The advantages that are simpleas well as stream wherever that too in the zone clogged with strengthened steel bars, stream under the original weight (not needed for the sudden moment)that makes to nonappearance of clamor of vibrations device.Therefore,monetary related advantages for example diminishing development period as well as price.Cement displays decrease because of reduction of dampness after drying producing inward ductile burdens where might be bigger than elasticity of cement. Inferable from the large capacity of concrete glue which is important to give smoothness, SCC displays top reduction than typical cement [3]. It can build occurrence of creating reduction actuated splitting in SCC behind limited situations. These conducts diminishes breaking opposition of cement [4]. Various methodologies that has utilized to relieve impact of reduction actuated breaking. Moreover, utilization of discrete, small as well as optinally appropriated steel strands that are taken as perhaps better methodologies. Strands can give a critical commitment in deferring break proliferation as well as restricting split width. ISF with different shaped attributes has utilized for a long time to decrease breaking of cement because of limited reduction. Moreover, the use of reused steel filaments (RSF) has taken from waste tires to improve shrinkage splitting obstruction of cement hasn't generally found[5]. Work of steel filaments has taken from a waste tires in generation strands strengthened SCC is another test in upgrading manageability of cement. Utilization of unwanted things in SCC that are preserve regular assets as well as diminish effect of unwanted objects on earth. Numerous looks into have demonstrated that utilization of RSF separated from scratch tires that may upgrade some attributes of mechanical of cement [8-12]. Specialists explored impact of RSF on numerous characteristics of SCC, however impact of such strands on reduction conduct is uncommon [12]. Along these lines, the research plans to inspect impact of RSF on reductionconduct of SCC under limited as well as over top situations. Research additionally explores impact of utilizing RSF on compactive quality of SCC.

From various periods, humans has attempted to discover methods for just holding promptly accessible objects e.g., straws, rocks, and so on joined to accomplish some auxiliary trustworthiness. Bond possible with procedure of hydration is probably the soonest arrangement. At present, along with advanced technological field, that has boron filaments in framework of metal, glass strands or glass-fortified crystallines in epoxy, or sands reinforced with a epoxy. Current project use of advanced innovation to old quality of using water powered bond for preparing mixtures. Explicitly we will examine a portion of improvements relating to usage of exceptionally prepared cement for open air protection.

For quite a while it wascreating solid mixtures as well as procedures for different mechanical as well as electrical devices all through Westinghouse. Previous periodic time it has introduced to EPRI the PIC ought to give a minimal effort material framework as well as assembling strategy for covers. Conspicuous serious issue zones to be stood up to relate to that are what happened to liquid substance , what impact pores that has reduced.

## 2.Literature Review:

Depends on the examined output of 16 steel fiber fortified solid corbels, numbered using by not linear limited component technique is embraced to examine their practices as well as to explore impact of steel strands capacity proportion, lower length to profundity proportion as well as quality of cement with shear practices, split as well as distortion of solid corbels. Outcomes show thatlower practices increments with expansion of capacity proportion of steel strands solid quality. Lower practices diminishes with expansion of lower length proportion. Steel strands improves splitting obstruction limit of the steel strands strengthened solid corbels, diminishes width of break as well as confines parallel twisting of cement. Moreover, the consequences of numbered reenactment as well as output of pertinent researches joins well, as well as it demonstrates the discernment of units as well as pressure of the material stress relationships as well asloss of criteria utilized in unbounded component research shapes.

Cementassumes a significant job in constructionarea. Moreover , cement is powerless in strain, so as to beat this, regularly fortifications are utilized in development. Because of , these fortifications willn't have the option to quit starting elastic breaks as well as different imperfections in cement. Point of the research is to locate ideal capacity level of steel strands are essential to expand elasticity of cement with research facility tests. So, six gatherings of evaluation 25 cement are delivered with expansion of steel filaments at various volumetric rates: 0.72%, 0.98%, 0.94%, 1.08%, 1.03%, 1.58% by supreme load of cement. Steel filaments with angle proportion, l/d = 40 with 0.98 mm width as well as snared at the two closures, are utilized right now. Impacts of including Steel strands in cement are estimated for the 3D square compactive quality (fcu) as well asparting elasticity (fct). Output information plainly shows higher rate increment in 9 and 32 days compactive quality as well as rigidity for 0.94% steel filaments included cement.

This is certainly not a total arrangement of calculation as well as logical strategy on steel strands cement basically Wall-pillar, steel strands cement just upheld Wall-shaft execution as well as working situations is just few examine output through investigation got. Depends on examine hardware, test techniques as well as different steps legitimately influence unwavering quality of test outcomes, in this manner, these are various researches steel strands concrete basically Wall-shaft, yet just subjective examination as well as quantitative investigation isn't sufficient, it is important to consider. The figuring model of steel strands cement just Wall-bar is established by utilizing rule essentials of mechanics. Considering technique for steel strands cement just upheld Wall-pillar examined with not linear limited component. It is demonstrated by model that not linear limited component broke down attributes of mechanic of steel strands concreteWall-shaft.

Most essential attributes of reused steel strands total are tried, for example, surface highlights, water assimilation, clear ensity, smashing file, block compactive quality, etc. Reused steel strands total are utilized to supplant regular total by 5%, 28%, 45%, 86% extent to make cement. Subsequently, compactive quality, parting elasticity as well as flexural quality are researched at ages 6, 9and 17 days. Outcomes demonstrated thistle look of reused steel strands total makes pressing thickness decline, pounding worth as well as retention increment. Compactive quality, parting rigidity as well as flexible quality of reused cement rely upon the blend extents. For the most part, quality of reused cement that are 18-45% less than cement that are made.

Current project investigates an estimation of area impact testing information for steel strands strengthened solid utilizing fast information securing framework. Right now field impact test is led by the BRU of University to explore conduct of steel strands fortified solid board exposed to air impact stacking. Steel strands strengthened solid boards are exposed to air impact stacking utilizing plastic touchy (PE4) weighing 4kg each. Variables estimated are air impact force, clear area impact force as well as furthermore quickening of chunk utilizing fast information obtaining framework as well as outcome is displayed in the proposed work.

## 3.Conclusion:

Joining of RSF does not remarkable effect compacting quality in SSC. Mixture contains large RSF described a border decrease in compactive quality wherever Compactive quality in SSC freely enhance where RSF joined along with mixture context.

Inclusion of RSF has been decreased clear reduction stress thereafter 30 days by utilizing individual strands length makes to increment of clear decrease stress. Thereafter this is reverse affect of mixture content on clear decrease.

Joining of RSF increments splitting obstruction. Small RSF described the best obstruction to splitting even than strands. Externally, that decreases length of splitting. Mixture which contains mixed strands that doesn't displays splitting when the trail schedule goes on.

## References:

1. Alireza Khaloo, et al 2014 Mechanical performance of self-compacting concrete reinforced with steel fibers *Construction and Building Materials* **51**.
2. Khaleel, H.Y., et al., 2017 Feasibility of using recycled steel fibres to enhance the behaviour of recycled aggregate concrete **310** 113-123.
3. Ganesan, N., J.B. Raj, and A. Shashikala 2013 Flexural fatigue behavior of self compacting rubberized concrete *Construction and Building Materials* **44** 7-14.
4. Long, W.J., et al. 2014 Mechanical Properties of Fiber Reinforced Self-Compacting Concrete *Applied Mechanics and Materials*.
5. 2006 Concrete subjected to restrained shrinkage: experimental study and modeling. *Journal of Materials in Civil Engineering* **18** 46-54.
6. Younis, K.H., 2014 Restrained Shrinkage Behaviour of Concrete with Recycled Materials.
7. Pilakoutas, K. and R. Strube 2001 Re-use of TyresFibres in Concrete. in Proceedings of the International Symposium on Recycling and Reuse of Used Tyres.
8. Younis, K.H. and K. Pilakoutas, 2013 Strength prediction model and methods for improving recycled aggregate concrete *Construction and Building Materials* **49** 688-701.
9. Younis, K.H., 2016 Mechanical Performance of Concrete Reinforced with Steel Fibres Extracted from Post-Consumer Tyres in *International Engineering Conference on Developments in Civil and Computer Applications*.
10. Rossli, S.A. and I.S. Ibrahim 2012 Mechanical Properties of Recycled Steel Tire Fibres in Concrete.
11. Wang, Y., H. Wu, and V.C. Li 2009 Concrete reinforcement with recycled fibers. Journal of materials in civil engineering **4** 314-319.
12. Kareem, A.S 2016 Rheological Properties, Mechanical Strength And Structural Behavior Of Reused Steel Fiber Self Compacting Concrete.
13. Moaveni, S. 2008 Finite element analysis theory and application with ANSYS *Publishing house of electronics industry*.
14. Gao D.Y.,Zhao J., Zhu H.T 2002 Design and application of steel fiber reinforced concrete *China building industry press*.
15. Banthia N 2012 FRC: Milestone in international research and development.
16. Ibrahim I.S. and Che Bakar M.B 2011 Effects on mechanical properties of industricalised steel fibres addition to normal weight concrete *Conference on structural engineering and construction*.
17. Rana A. 2013 Some studies on steel fiber reincorced concrete *International journal of emerging technology and advanced engineering* **3** 120-127.
18. Xu B.W. and Shi H.S 2009 Correlations among mechanical properties of steel fibre reinforced concrete *Construction and building materials* **23** 3468-3474.
19. Romualdi J.P. and Mandel J.A 1964 Tensile Strength of Concrete Affected by Uniformly Distributed Closely Spaced Short Lengths of Wire Reinforcement **61**, 657-671.

IOP Conf. Series: Materials Science and Engineering **923** (2020) 012047 doi:10.1088/1757-899X/923/1/012047

20. Johnston C.D 1982 Definitions and Measurement of Flexural Toughness Parameters for Fiber Reinforced Concrete **4** 53-60.
21. J.O.Berger, P.J.Heffernan, R.G. Wight 2008 Blast testing of CFRP and SRP strengthened RC Columns Structures under Shock and Impact X **96**.
22. T.Ngo, P. Mendis, T. Krauthammer 2007 Behavior of Ultrahigh – Strength Prestressed Concrete Panels Subjected to Blast Loading *Journal of Structural Engineering* 1582 - 1590.
23. Alias, Norazman, Ariffin, Fauzi, Risby, Ng Choy Peng, 2009 Field blast test on Concrete Structures Subjected to Blast Loading *Asia pacific conference on defence and technology*.
24. S.Lan, Tat senglok, H.Leonard,Composites 2005 structural panels subjected to explosive loading, *Construction and building materials* **19** 387397.

IOP Conf. Series: Materials Science and Engineering **923** (2020) 012048 doi:10.1088/1757-899X/923/1/012048

# Comparative Study On Mechanical Behaviour Of Synthetic, Natural And Hybrid Composite Laminates.

**N Rajamurugu[1],P Karthikeyan[2], A Aakash[3], A Abhilash Meshak[4],K.Ajithkumar[5]**
[1,2] Asst Prof , Department of Aeronautical Engineering, Apollo Engineering College ,Chennai, India.
[3,4,5] U.G Student, Department of Aeronautical Engineering, Apollo Engineering College, Chennai, India.
{n.rajamurugu,propkarthik.12,Aakash.anandan3330,meshakabi,ajithaeroak}@gmail.com.

**Abstract.** Composites play a significant role as an engineering material and their use has been increasing day by day due to their specific properties such as high strength to weight ratios, high modulus to weight ratio, corrosion resistance and wear resistance. In composite material matrix and fiber play an important role; as the load is taken by matrix and transmitted to the reinforcement. Reinforcement can be either synthetic or natural fiber. Manufacturing of synthetic fibers is harmful from the environmental point of view; hence recent research work is focused on the use of natural fiber as reinforcement. In this current work jute as natural fiber taken into account for reinforcement along with synthetic glass fiber along with LY556 Epoxy resin and HY951 hardener. Three types of composites materials were prepared such as Glass-Epoxy unidirectional, Jute-Epoxy, Glass & Jute Epoxy hybrid composite laminates had been prepared using hand lay-up process and the testing was done by using ASTM Standards. The major properties that govern the strength of composites materials are tensile strength, compressive strength, flexural strength, and vibration characteristics. The content of natural and synthetic reinforcement was found to be 60% and 40% respectively. The experimental results show that the Glass & jute Epoxy composites give the major withstanding capacity for tension and flexural test. Vibration test exhibits higher natural frequency, but in compression test Glass epoxy reflects its higher mechanical strength, whereas in compression test delamination occurs. The additional use of jute with glass reduces the cost and thus it is eco-friendly.

## 1.Introduction

Material science is an interdisciplinary field involving the properties of matter and its applications to various areas of science and engineering. It includes elements of applied physics as well as mechanical property, electrical engineering and civil. In material science rather than haphazardly looking for and discovering materials and exploiting their properties fundamentally so that new materials with the desired properties can be created. It involves relating the desired properties and relative performance of a material in a certain application to the structure of the atoms and phases in that material through

IOP Conf. Series: Materials Science and Engineering **923** (2020) 012048 doi:10.1088/1757-899X/923/1/012048

characterization. A composite material is a material made from two or more constituent materials with significantly different physical or chemical properties that, when combined, produce a material with characteristics different from the individual components. The individual components remain separate and distinct within the finished structure, differentiating composites from mixtures and solid solutions. The new material may be preferred for many reasons. Common examples include materials that are stronger, lighter or less expensive when compared to traditional materials. Typical engineered composites include reinforced concrete and masonry, composite wood such as plywood, reinforced plastics, such as fiber-reinforced polymers or fiberglass, ceramic and matrix composites, Metal Matrix composites and other advanced composites materials. Hybrid materials are composites consisting of two constituents at the nanometer or molecular level. Commonly one of these compounds is inorganic and the other one organic in nature. Thus, they differ from traditional composites where the constituents are at the macroscopic (micrometer to the millimeter ) level. Mixing at the microscopic scale leads to a more homogeneous material that either show characteristics in between the two original phases or even new properties. Many natural materials consist of inorganic and organic building blocks distributed on the nanoscale. In most cases, the inorganic part provides mechanical strength and an overall structure to the natural objects while the organic part delivers bonding between the inorganic building blocks and/or the soft tissue. Typical examples of such materials are bone or nacre. The literature survey has to be conducted and as per our survey, we took a survey from the following paper. (1) Ramesh.et.al. investigated on the hybrid composites and the effect of various parameters on the performance of the hybrid composites are subjected to mechanical testing, flexural and impact tests. The results indicated that the jute composite material shows maximum tensile strength and the jute composite material shows the incorporation of sisal – jute fiber with GFRP can improve the properties and as an alternate material for glass fiber reinforced polymer composites. (2) Boopalan.et.al made a comparative study on the mechanical properties of jute and sisal fiber-reinforced polymer composites and concludes that the mechanical properties of the jute fiber-reinforced composites are higher than the sisal fiber-reinforced composites. (3)M. Muthuvel.et.al investigated that the jute and glass fiber hybrid composite leads to the successful fabrication of Glass, Jute fiber and chopped fiber reinforced polyester composites with different fiber lengths is possible by simple hand lay-up technique. The mechanical properties of the composites like tensile, flexural and impact strength of the composites are also greatly influenced by the fiber lengths. (4 M Alamgir Kabir.et.al studied that with the increase of fiber loading the values of flexural strength and Charpy impact strength of composite material have been decreased. But the values of other mechanical properties remain almost the same. (5) Sanjay M R .et.al This experimental investigation on Jute - Glass Fiber Reinforced Polyester composites leads to the high tensile strength for the 50%-50% of glass & jute for flexural specimen 60%-40% of glass & jute shows higher flexural strength and for impact strength, 50%-50% of glass & jute shows higher impact energy. (6) Zhanh.et.al The tensile properties of the flax/glass fiber reinforced hybrid composites were improved with the increasing of glass fiber content. The interlaminar shear strength and the interlaminar fracture toughness of flax/glass fiber reinforced hybrid composites were higher than those of GFRP.(7) D. Dash.et.al Mechanical properties are decreased with decreasing in the total percentage of the reinforcement of the glass. (8) T. Munikenche Gowda.et.al concluded that although the mechanical properties of jute/polyester composites do not possess strengths and modules as high as those of other conventional composites, they can process better strengths than wood composites and some plastics. (9)Satish Pujari.et.al explores the potentiality of jute &banana fiber composites, emphasizes both mechanical and physical properties and their chemical composition. The utilization and application of the cheaper goods in the high-performance appliances are possible with the help of this composite technology. (10) Md Rafiquzzaman .et.al. glass-jute fiber polymer composites were studied, results showed that by incorporating the optimum amount of natural fibers, the overall strength of glass fiber reinforced hybrid composites can be increased and cost saving of more than 30% can be achieved. The utilization and application of cheaper goods in higher performance appliances are possible with the help of this composite technology. Combining the higher properties of two different materials, lower manufacturing

IOP Conf. Series: Materials Science and Engineering **923** (2020) 012048 doi:10.1088/1757-899X/923/1/012048

cost, versatility, etc., Hence with this background, it is concluded that, the composites stand the most wanted technology in the fast-growing trend.

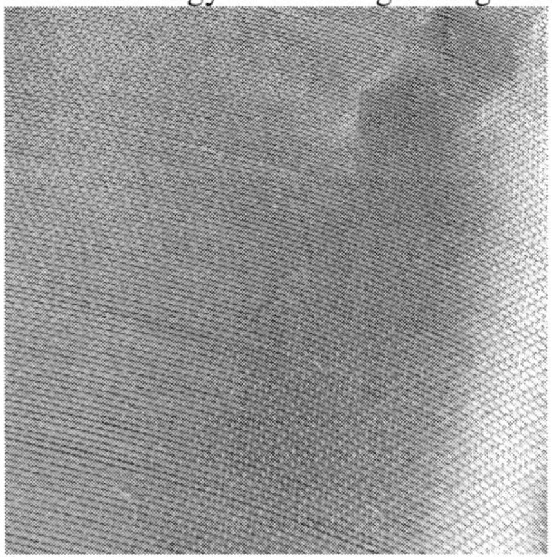

**Figure 1.1.** *Glass fiber*

**Figure 1.2.** *Jute fiber*

**2 Experimental Study**

The material used in the project is selected based on their properties The materials are classified as synthetic and natural fibers based on their nature. Here we chose glass fiber as synthetic, jute as natural and Glass-Jute fiber composition as hybrid composites. Jute material shows high corrosive resistance and comparatively low cost. Glass material is also well known for its heat resistance and corrosive resistant properties it shows high strength to less weight ratio thus combining both the materials shows combined mechanical properties for the materials obtained. To manufacture the composites there are so many methods that are used currently some of the manufacturing procedures are mainly classified into two types open mold and closed mold in open mold hand layup, spray up and filament winding processes are used. In closed mold vacuum bag molding, vacuum infusion processing, Resin transfer molding, compression molding, pultrusion molding, reinforced reaction injection molding, centrifugal casting, continuous lamination process are used. We selected the hand lay-up process Since it is an easier process and a more effective one. The fabrication process includes unidirectional laminate it includes the basic principle of hand lay-up process with the binding of epoxy resin the process is done in three separate processes as Glass epoxy composites, Jute epoxy composites, and Glass-Jute epoxy composites. The fabrication of glass epoxy includes 12 plies of fiber mixed with 10:1 ratio of resin and binder, the geometry of jute epoxy includes 6 ply of jute fiber and the hybrid laminate is fabricated in the of 60 & 40 % as 6 plies of jute fiber and 4 plies of glass fiber, thus the fabricated composite is set to dry for 4 days and it is bought for testing. The final fabricated plates were cut into required dimensions for testing.

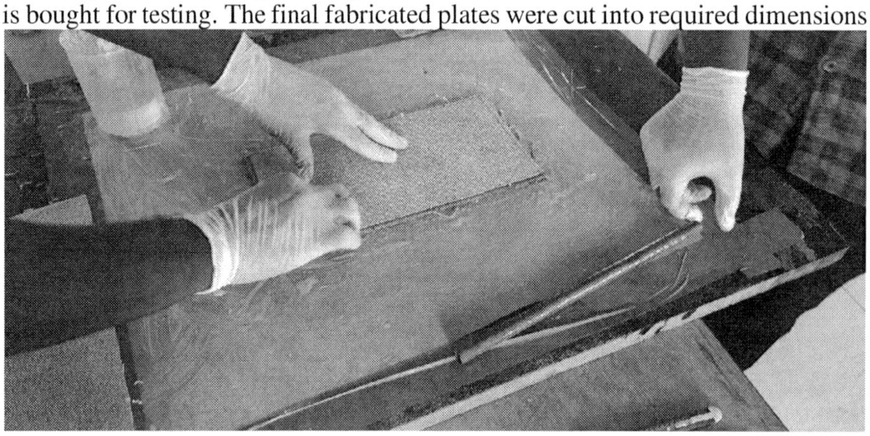

**Figure 2.1.** *Hand lay-up Technique*

IOP Conf. Series: Materials Science and Engineering **923** (2020) 012048 doi:10.1088/1757-899X/923/1/012048

| Specimen | Dimension (in mm) | | |
|---|---|---|---|
| | Glass-epoxy | Jute-epoxy | Glass& Jute epoxy |
| Tension test | 100x20x3.2 | 100x20x5.4 | 100x20x6.3 |
| Compression test | 30x30x3.2 | 30x30x5.4 | 30x30x6.3 |
| Flexural test | 100x15x3.2 | 100x15x5.4 | 100x15x6.3 |
| Vibration test | 250x25x3.2 | 250x25x5.4 | 250x25x6.3 |

**Table 2.1** *Specimen Dimensions*

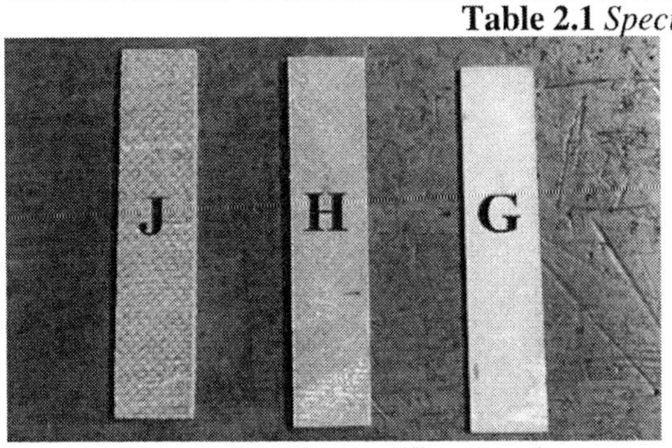


**Figure 2.2.** *Tension Test Specimen*

**Figure 2.3.** *Compression Test Specimen*

**Figure 2.4.** *Flexural Testing Specimen*

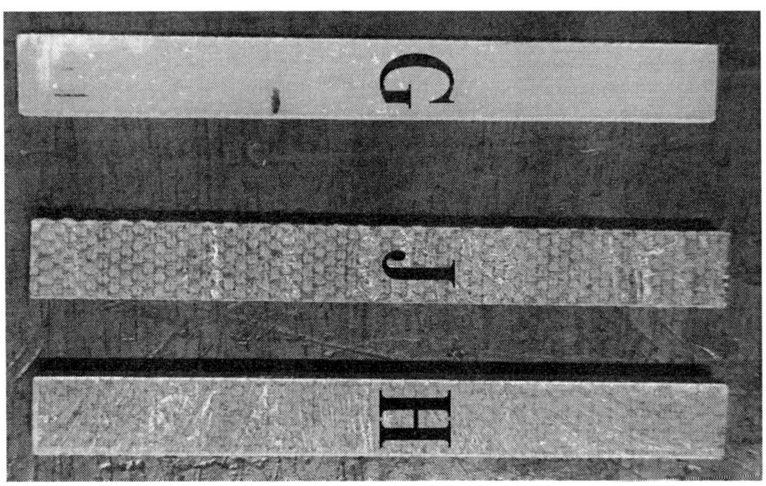

**Figure 2.5.** *Vibration Testing Specimen*

## 3 Testing Procedure

The various mechanical properties of the fabricated composites are carried out by different instruments for testing. The specimen was sized as per the ASTM standards for composite materials testing methods.

IOP Conf. Series: Materials Science and Engineering **923** (2020) 012048 doi:10.1088/1757-899X/923/1/012048

The shimadzu AD-Xplus universal testing machine with stroke rate of 0.0005 to 1000 mm/min and top load cell of capacity up to 50KN.

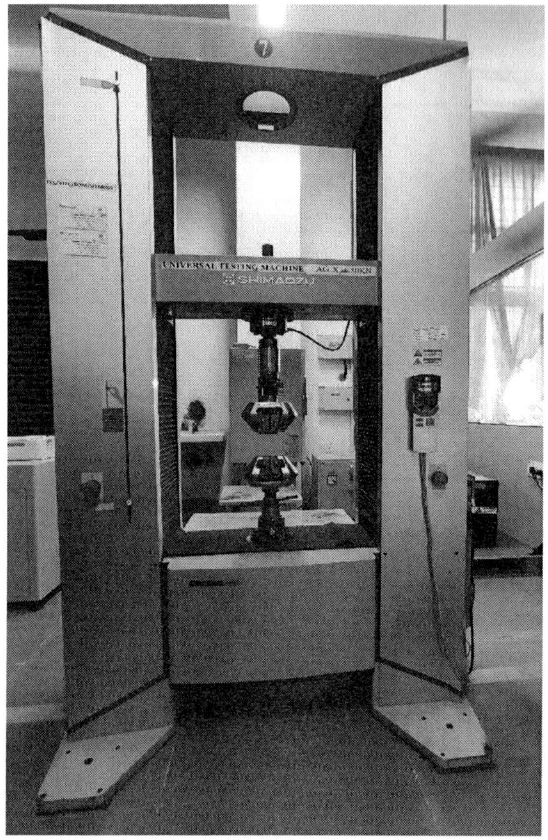


**Figure 3.1.** *Shimadzu AG-Xplus Universal Testing Machine*

*3.1 Tension Test*

The tensile test was carried out by sizing the specimen as per the ASTM D3039 standard. tensile testing is used to measure the force required to break a polymer composite specimen and the extent to which the specimen stretches or elongates to that breaking point. The specimen is loaded in the universal testing machine and load is applied until the specimen breaks. The maximum tensile load, tensile stress, displacements are noted and the stress-strain curve is generated.

**Figure 3.1.1.** *Tensile Test Setup*

3.2 *Compression Test*

The compression test was carried out by sizing the specimen as per the ASTM D3410 standard. The specimen is loaded in the universal testing machinc and the compressive load is applied until the specimen breaks. The maximum compressive load, compressive stress, displacements are noted and the stress-strain curve is generated.

**Figure 3.2.1**. *Compression Test Setup*

3.3 Flexural Test

The flexural or three-point bending test was carried out by sizing the specimen as per the ASTM D7264 standard. testing of flexural properties of polymer matrix composites using a bar of rectangular cross section supported on a beam and deflected at a specified rate. The specimen is lied on the support span and the load is applied to the center by the loading nose producing three-point bending at a specified rate.

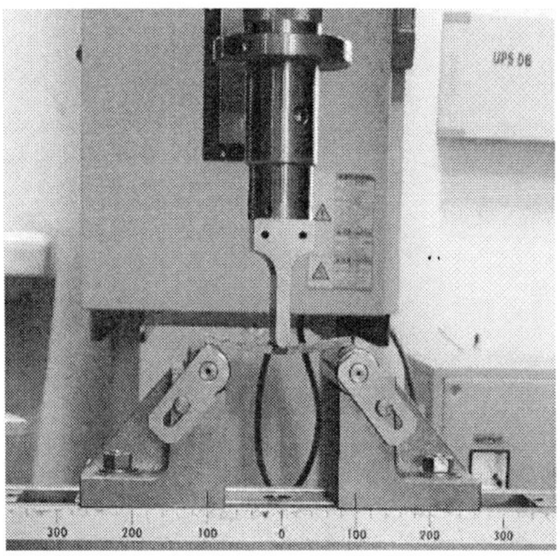


**Figure 3.3.1**. *Flexural Test Setup*

3.4 Vibrational Test

The vibrational test was carried out as per the ASTM standard. The accelerometer are placed on X and Y axis of the specimen and the test is carried out at each node. The natural frequency and the damping ratio is noted at each nodes.

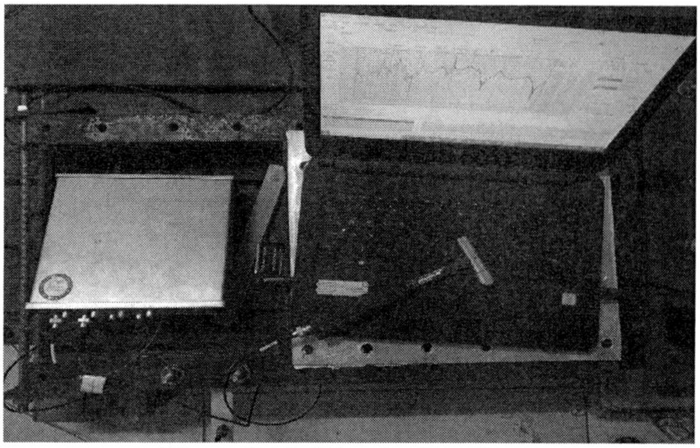

**Figure 3.4.1.** *Vibration Test Setup*

## 4 Results And Discussion

4.1 Tension Test

The tensile test results shows that the glass-jute epoxy laminate has higher tensile load carrying capacity and maximum tensile stress.

| Composite material | Maximum force(N) |
|---|---|
| Glass epoxy | 2805.77 |
| Jute epoxy | 4977.84 |
| Glass-jute epoxy | 6953.42 |

**Table 4.1.1.** *Experimental Values*

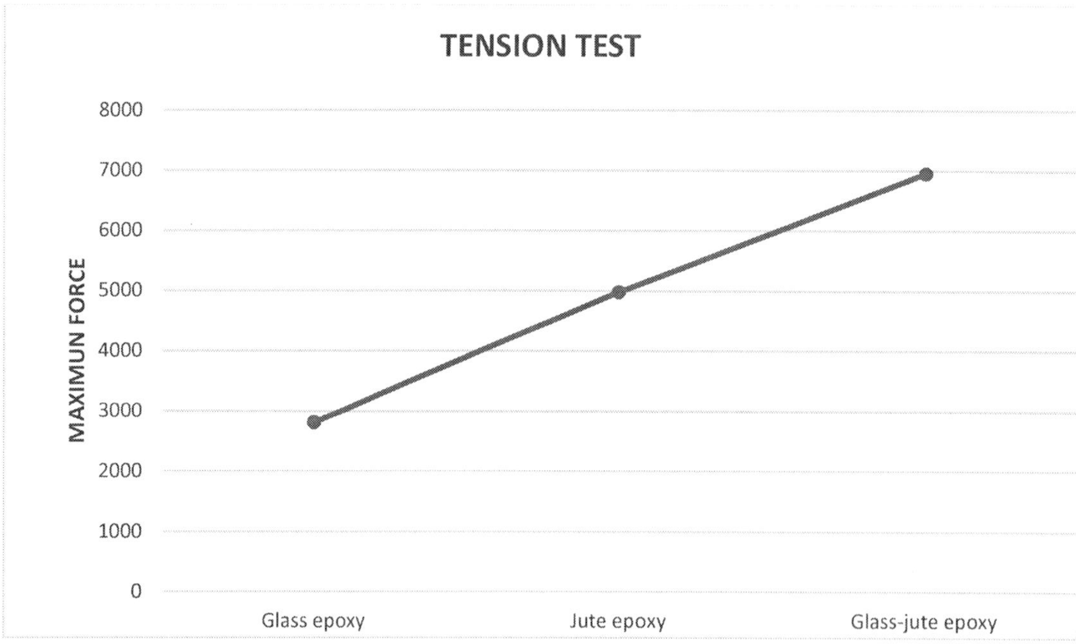


**Graph 4.1.1.** *Comparative Stress-Strain Curves For Tensile Test*

IOP Conf. Series: Materials Science and Engineering **923** (2020) 012048 doi:10.1088/1757-899X/923/1/012048

4.2 Compression Test

The compression test results shows that the glass fiber laminate has the higher compressive load carrying capacity and stress.

| Composite material | Maximum force(N) |
|---|---|
| Glass epoxy | 11221.1 |
| Jute epoxy | 6657.24 |
| Glass-jute epoxy | 10762.9 |

**Table 4.2.1.** *Experimental Values*

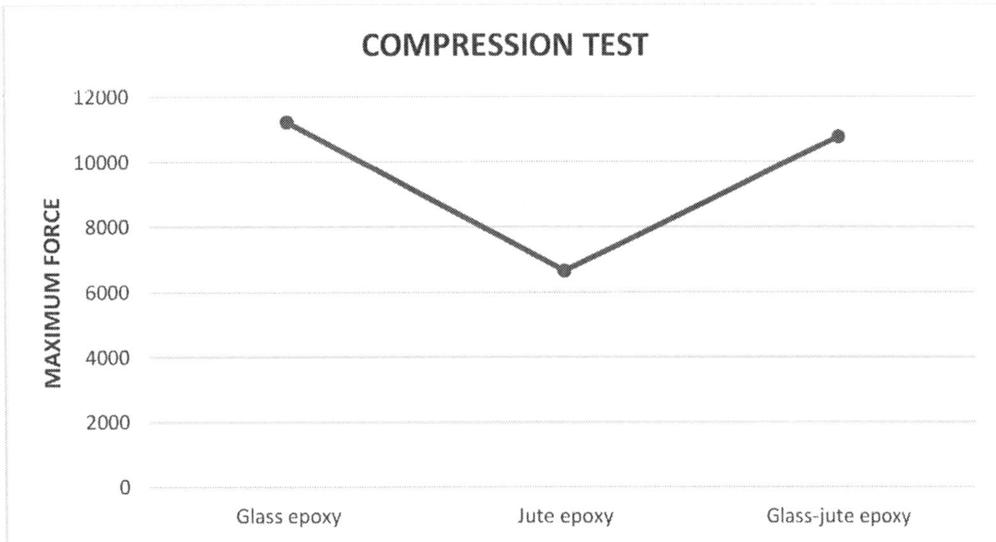


**Graph 4.2.1.** *Comparative Stress-Strain Curves For Compressive Test*

4.3 Flexural Test

The flexural test results shows that the glass-jute epoxy laminate has higher load carrying capacity and the glass epoxy has higher flexural modulus.

| Composite material | Maximum force(N) |
|---|---|
| Glass epoxy | 98.1808 |
| Jute epoxy | 204.285 |
| Glass-jute epoxy | 317.955 |

**Table 4.3.1.** *Experimental Values*

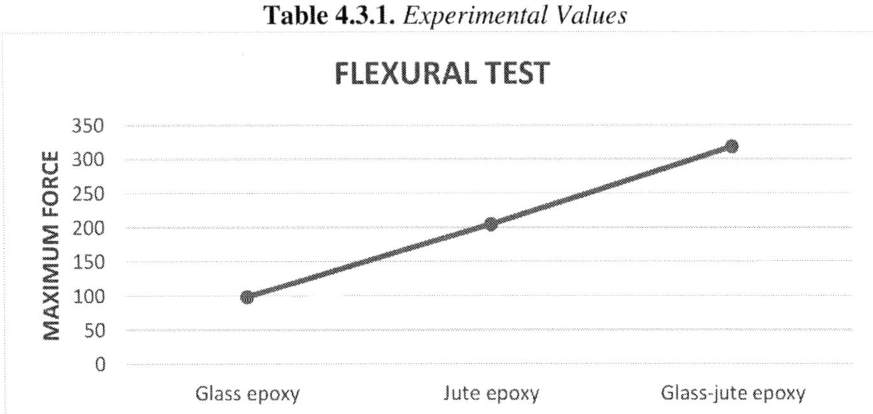


**Graph 4.3.1.** *Comparative Stress-Strain Curves For Flexural Test*

4.4 Vibration Test

The vibration test results shows that the glass-jute epoxy laminate has higher natural frequency and its damping ratio has 0.0045.

| **Composite material** | **Natural frequency** | **Damping ratio** |
|---|---|---|
| Glass epoxy | 655 | 0.1445 |
| Jute epoxy | 799 | 0.1446 |
| Glass-jute epoxy | 843 | 0.0045 |

**Table 4.4.1.** *Experimental Values*

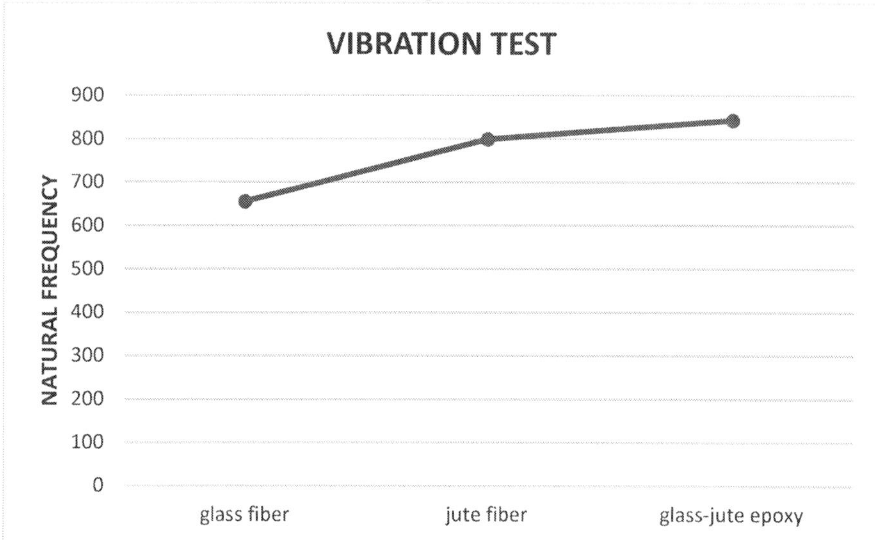


**Graph 4.4.1.** *Comparative Graph On Natural Frequency*

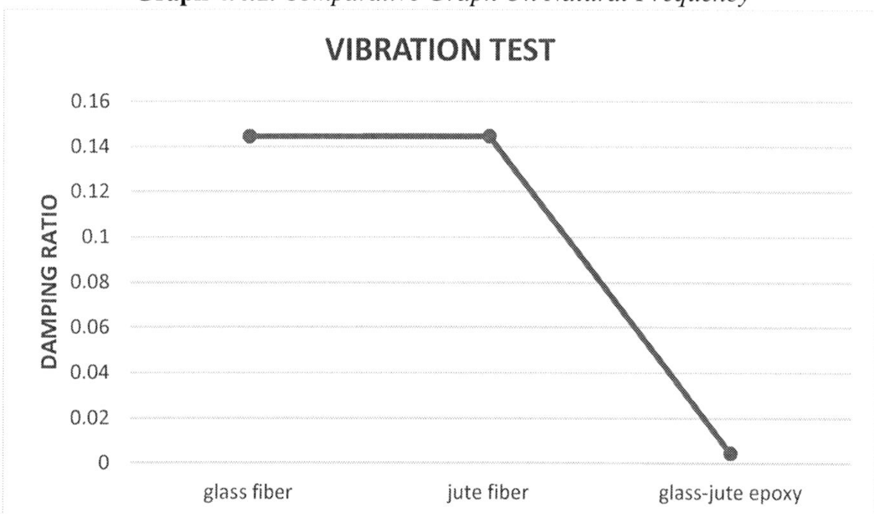


**Graph 4.4.2.** *Comparative Graph On Damping Ratio*

**5 Conclusion**

The experimental investigation on glass, jute, glass-jute fiber reinforced polymer composites with different fiber composition is possible by simple hand lay-up technique, it has been noticed that the mechanical properties of the glass-jute fiber reinforced polymer composites exhibit the higher tensile, flexural load carrying capacity and also exhibits higher natural frequency on vibration test. While compression glass fiber reinforced composites exhibits higher compressive force. During the

IOP Conf. Series: Materials Science and Engineering **923** (2020) 012048 doi:10.1088/1757-899X/923/1/012048

compression test, the glass-jute fiber doesn't withstand compressive force due to the delamination of plies. The addition of jute fiber to the glass fiber enhances its tensile load carrying capacity by 60% and flexural load carrying capacity by 70%. In the vibration test, glass-jute fiber reinforced polymer composites exhibits the 22% natural frequency. The results showed that by incorporating the optimum amount of natural fibers, the overall strength of glass fiber reinforced hybrid composites can be increased and a considerable amount of mechanical properties could be achieved.

**References**

(1) M.Ramesh, Palanikumar K, Hemachandra Reddy K, Mechanical Property evaluation of sisal-jute-glass Fiber Reinforced Polyester Composites, Composites Part B,vol48 pp.1-9, 2012.
(2) M. Boopalan, M. Niranjana, M.J. Umapathy, Study on the mechanical properties and thermal properties of jute and banana fiber reinforced epoxy hybrid composites, Composites: Part B 51 (2013) 54–57.
(3) M. Muthuvel, G. Ranganath , K. Janarthananand K. Srinivasan, Characterization Study of Jute and Glass Fiber Reinforced Hybrid Composite Material, International Journal of Engineering Research & Technology, ISSN: 2278-0181 Vol. 2 Issue 4, April – 2013.
(4) M Alamgir Kabir, M. Monimul Huque, M. Rabiul Islam, Andrzej k., Bledzki, Treated jute-PP composite, Bio Resources 5(3), 1618-1625, 2010
(5) Sanjay, M.R., Arpitha, G.R. and Yogesha, B. (2015) Study on Mechanical Properties of Natural—Glass Fiber ReinforcedPolymer Hybrid Composites:Review.*MaterialsToday:Proceedings*,**2**2967.http://dx.doi.org/10.1016/j.matpr.2015.07.264
(6) Zhang, Y.L., Li, Y., Ma, H. and Yu, T. (2013) Tensile and Interfacial Properties of Unidirectional Flax/Glass Fiber Reinforced Hybrid Composites. *Composites Science and Technology*, **88**, 172-177.
http://dx.doi.org/10.1016/j.compscitech.2013.08.037
(7) D.Dash, S. Samanta, S.S. Gautam, M. Murlidhar. Mechanical Characterizations of Natural Fiber Reinforced Composite Materials, Advanced Materials Manufacturing & characterization Vol 3 Issue 1,2013
(8) T. Munikenche Gowda, A.C.B. Naidu, Rajput Chhaya, Some mechanical properties of untreated jute fabric-reinforced polyester composites, Received 10 December 1997; accepted 1 July 1998.)
(9) Satish Pujari, A. Ramakrishna, M. Suresh Kumar, Comparison of Jute and Banana Fiber Composites: A Review, Accepted 10 January 2014, Available online 01 February 2014, Special Issue-2, 2014.
(10) Md. Rafiquzzaman "Mechanical property evaluation of glass–jute fiber reinforced polymer composites" polymer advanced technologies 2016.

**Acknowledgments**

We would like to express our sincere and deepest gratitude to the management of Apollo Engineering College and to The Department Of Aeronautical Engineering for the valuable support and suggestion, we wish to place our sincere thanks to Mr. N. Rajamurugu , head of the department and our project guide Mr. P. Karthikeyan for his excellent guidance in completing our research work well on time. We place our sincere thanks to our friends, family members and staffs for the completion of this research work.

# Improved hardness characteristics of Mg-3Zn-1Cu-0.7Mn alloy composites reinforced with $Al_2O_3$ synthesized by powder metallurgy

**S Silan Tharakan[1] and M Sivapragash[2]**

[1] Research Scholar, Dept. of Mechanical Engineering, Noorul Islam Centre of Higher Education, Kumaracoil, Kanyakumari Dist., Tamil Nadu, India
[2] Professor, Dept. of Mechanical Engineering, Universal College of Engineering and Technology, Vallioor, Tirunelveli Dist., Tamil Nadu, India

E-mail: silantharakans@gmail.com

**Abstract.** A novel composite Mg-3Zn-1Cu-0.7Mn reinforced with $Al_2O_3$ particulates was synthesized using the powder metallurgy route and extruded at $400^oC$. The investigation mainly focuses on the influence of $Al_2O_3$ on the hardness characteristics of the composites. The results indicate the microhardness of the composite is enormously enhanced by incorporating the $Al_2O_3$ particulates to Mg alloy. The homogeneous distribution of the hard ceramic reinforcement particulates and the presence of intermetallic phases such as $Cu_5Zn_8$, MgZnCu, and $Al_{0.93}Cu_{1.07}Mg$ attributed the improved hardness. It is recognized that the stress-relieving process further improves the hardness of the composites. The extrusion parameters also influence the hardness characteristics of the composites.

## 1. Introduction

Magnesium (Mg) and Mg -based alloys are extensively utilized in structural applications including aerospace, military, automotive, electrical, and biomedical sectors due to its lightweight, high strength to weight ratio, better machinability, and excellent damping capacity. But its applications are limited at present in the corresponding fields due to the insufficient elastic modulus, strength, and hardness [1, 2]. The mechanical characteristics of the Mg metal can be improved with alloying and heat treatment. The recent studies show that the inclusion of ceramic particles is an efficient method for enhancing the mechanical characteristics of magnesium alloys. In situ composite consists of multiple phases, including the primary and secondary phases. The secondary reinforcing phase is formed inside the metal matrix by specific chemical reactions which is in the form of fine particles and thermally stable in the metal matrix. It is also emerging in thermodynamic agreement at the matrix/reinforcement interface [3, 4]. Thus the Mg alloy based composites fabricated through in-situ methods display excellent mechanical properties.

In the field of lightweight materials, Al matrix composites fabricated through various in-site processes has been drawing attention from researchers for use in industrial utilisation due to its excellent mechanical properties. At the same time, Mg-matrix in situ composites couldn't find much attraction from researchers. Research studies reveal that Mg -matrix in situ composites can be a potential candidate for both structural applications and biodegradable biomaterials [5, 6]. The literature survey reveals that Silicon Carbide ($SiC_p$) are the most preferred reinforcement among the researchers. The high hardness, wear resistance, elastic modulus, wear resistance, performance at elevated temperatures and low price of the $SiC_p$ cause for the further studies of the composites reinforced with SiC particles [7, 8]. But recently, $Al_2O_3$, CNTs, $Y_2O_3$, etc. are also emerged as the reinforcements to strengthen the Mg matrix. The pattern of research on the composites is based on the improvement in characteristics by varying the size, weight fraction, morphology and distribution of the reinforcement particles and by altering the fabrication techniques [9–13]. In the last couple of years, several alloys were developed based on Mg-Zn, Mg–Si , Mg–Ni, Mg–Ca, Mg–Sc, and Mg–Cu–Mn [14–17]. Nishiyama et al. fabricated Mg-Cu-Mn alloys through Powder Metallurgy (PM) technique followed by sintering. The properties of the alloys are found to be superior compared to the pure Mg metal. At the same time, Mg–Cu–Mn alloys prepared through simple conventional melting process exhibited better damping properties, even though the tensile strength was reduced [18].

Hot extrusion is a secondary process for improving the mechanical characteristics of the alloys. The microstructure of the alloys is altered during the hot extrusion process. Several new grains are formed, and secondary intermetallic phases are precipitated after the hot extrusion process. The dynamic recrystallization and dynamic precipitation are processes resulting in the abovementioned activities. Gupta et al. have done a lot of work in improving the mechanical properties of the Mg alloys by reinforcing with different reinforcements, particularly $Al_2O_3$ [19–22]. The composites based AZ81 reinforced with $Al_2O_3$ exhibited superior tensile and compressive characteristics compare to the AZ81 base alloy. The composites strengthened by nano SiC particles by Ferker and Mordike reported improvement in tensile and creep properties [23]. In the present work, a new Mg-3Zn-Cu-0.7Mn based alloys and the alloy-based composites reinforced with $Al_2O_3$ particulates are synthesized. The microstructure, metallurgical, and hardness characteristics of the composites are analyzed. The influence of stress-relieving on the hardness characteristics is also investigated here.

## 2. Experimental procedure

Elemental metal powders of Mg (99.99 *wt.%* purity, 200 mesh), Zn (99.99 *wt.%* purity, 200 mesh), Cu (99.99 *wt.%* purity, 200 mesh) and Mn (99.99 *wt.%* purity, 200 mesh) are employed for the fabrication of the alloy billet using the PM technique. The ceramic metal powder $Al_2O_3$ (99.99 *wt.%* purity, 200 mesh) is added to the alloy to synthesize the composite. The blend-press-sinter PM technique is used for the fabrication of the alloys and composites. A ball milling machine is exercised for the homogeneous mixing and blending of the metal powders. The ball mill is operated for 24 hours at 100 rpm with a ball feed ration of 3:1. The mixture is compacted utilizing hydraulic power press to form the alloys and composites billet, applying a load of 1500 $kg$. The sintering of the billets is executed in the Argon atmosphere at a temperature of $400^oC$ for 1 hour. Secondary strengthening of the billets is done with hot extrusion. The billets are hot extruded at $400^oC$ to form rods of small diameter with an extrusion ratio 5.4:1 using an extrusion die with a hydraulic power press at a speed of 0.2 to 0.5 $mm/s$. The rods thus formed are stress-relieved at $260^oC$ for 15 minutes in the Argon atmosphere in an electronic muffle furnace. The properties of the composite samples are analyzed with two categories, one without the stress-relieving, i.e., as-extruded (represented as E1 and F1) and the other with stress-relieving (designated as E2 and F2). An etchant prepared using 1.5 $g$ picric acid, 25 $ml$ ethanol, 10 $ml$ acetic acid, and 10 $ml$ distilled water is applied for capturing the optical microstructure of the specimens. X-Ray Diffraction (XRD) is used for the detailed phase analysis of the alloys and composites.

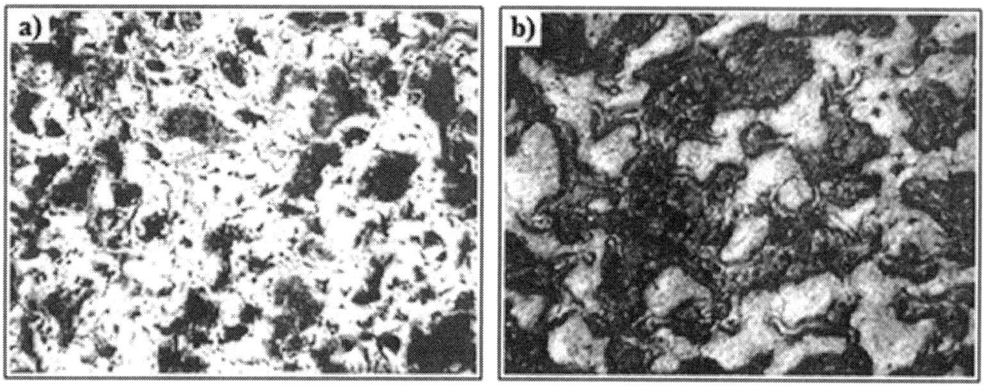


**Figure 1.** Optical micrographs of the extruded Mg-3Zn-1Cu-0.7Mn / $Al_2O_3$ composite sintered at $400^oC$ (a) as extruded alloy composite (F1) (b) stress relieved alloy composite at $260^oC$ (F2)

## 3. Results and Discussion

### *3.1. Microstructural analysis*

The microstructural analysis of the samples reveals the morphological changes that happened during the manufacturing process. Figure 1 represents the optical microstructure of the polished, etched composite specimens F1 and F2. According to microstructural analysis, the $Al_2O_3$ particles are distributed uniformly across the entire microstructure. Minimum pores are observed in the microstructure. It indicates the strong bonding between the alloys matrix and reinforcement phase. The blending parameters used and the higher extrusion ratio employed resulted in uniform distribution of the secondary phases with minimal pores. The limited number of microvoids and the interfacial reaction happened between the primary and secondary phases in the composites caused the extraordinary bonding between the particles [19-21]. The micrographs show equiaxed and well-refined grains. It may be due to the DRX happened following the hot extrusion process. The internal energy stored in the material is dissipated during the hot extrusion at elevated temperature, which causes the formation of DRXed grains. The higher thermal and strain energy provided during the hot extrusion process lead to dynamic precipitation also. Therefore the grains become well refined followed the hot extrusion process through DRX and DP. The number of microvoids and micropores are found to be reduced after the stress-relieving process. The grain refinement took place may be fundamentally due to the ability of $Al_2O_3$ particles to nucleate Mg grains during the recrystallization process. The pinning effect of $Al_2O_3$ particulates on Mg grains that caused for restricted grain growth during the recrystallization process. It may also be a reason for the grain refinement.

The research exposes the following facts, (i) homogeneous distribution of reinforcement across the entire microstructure (ii) the bonding between the alloy matrix and the reinforcement phase (iii) minimal porosity of the composites and (iv) the interfacial characteristics of matrix and reinforcement.

### *3.2. Density and Porosity Results*

Table 1 gives the theoretical densities, actual densities, and porosities of the composites fabricated. The experimental density of the stress-relieved composites (F2) is observed to be less than that of the as-extruded composites (F1). It may be due to the limited number of micropores and microvoids after the stress-relieving process. The entrapped gases during the blending process result in the porosity in composites following the PM technique. Since the actual densities and theoretical densities are very close, the porosity values are found to be tiny figures. The limited porosity is compatible with the results of the microstructural analysis. The hydrogen in composites also causes the porosity in the composites. The porosity values less than 0.5% indicates the synthesis of the composite was successful, and it is also proved that PM is a proper method for the fabrication of Mg-based composites.

**Table 1.** Density and porosity of the composite samples at room temperature

| Sl. No. | Material | Theoretical density ($g/cm^3$) | Experimental density ($g/cm^3$) | Porosity (%) |
|---|---|---|---|---|
| 1. | Mg-3Zn-1Cu-0.7Mn /$Al_2O_3$ composite (F1) | 1.834 | $1.825 \pm 0.003$ | $0.49 \pm 0.006$ |
| 2. | Mg-3Zn-1Cu-0.7Mn /$Al_2O_3$ composite (F2) | 1.834 | $1.829 \pm 0.003$ | $0.27 \pm 0.005$ |

### *3.3. X-ray diffraction studies*

The composite specimens in powder form were applied with Cu-K$\alpha$ radiation of wavelength 1.5406$\mathring{A}$ to recognize the different phases associated with the microstructure of the composite fabricated. The scanning speed was set $2^o$/min with a beam current $30mA$ and a voltage of $40kV$. The lattice spacing ($d$) and the Bragg angles of the phases are compared with related phases of Mg, $Al_2O_3$, Zinc, Manganese, and Copper. The X-ray diffractograms obtained is shown in Figure 2. From the XRD analysis, it is revealed that the following phases of Mg and $Al_2O_3$, $Cu_5Zn_8$, MgZnCu, and $Al_{0.93}Cu_{1.07}Mg$ are present as dispersed in the microstructure. The intermetallic phases are found to be in the dispersed form near the grain boundaries of the alloy matrix. The dynamic precipitation occurred due to the hot extrusion process contributed to this much number of phases in the composites, as found in the microstructural analysis.

### *3.4. Hardness studies*

The microhardness of pure Mg, AZ31B, fabricated alloys, and composites are given in Table 2. The maximum microhardness is observed for the stress-relieved composites (F2). The comparison between the alloys and composites are indicated in Figure 3. The microhardness of both the alloys and composites are much improved than pure Mg metal. Multialloying is proved to be a suitable technique to improve the microhardness of the monolithic Mg metal. It is observed that the microhardness of the Mg alloys is further enhanced by reinforcing with $Al_2O_3$ particulates. Hence it is unveiled that incorporating $Al_2O_3$ particles to the Mg alloy matrix through powder metallurgy improves the microhardness. Thus PM technique can be used to synthesis Mg composites with superior microhardness.

The influence of $Al_2O_3$ in improving the microhardness of the Mg composites is remarked here. The primary reason for the improvement is due to the hard ceramic nature of incorporated $Al_2O_3$ particles. The refined grains attained by the dynamic recrystallization and dynamic precipitation through the hot extrusion process also contributed the improved hardness. Hence

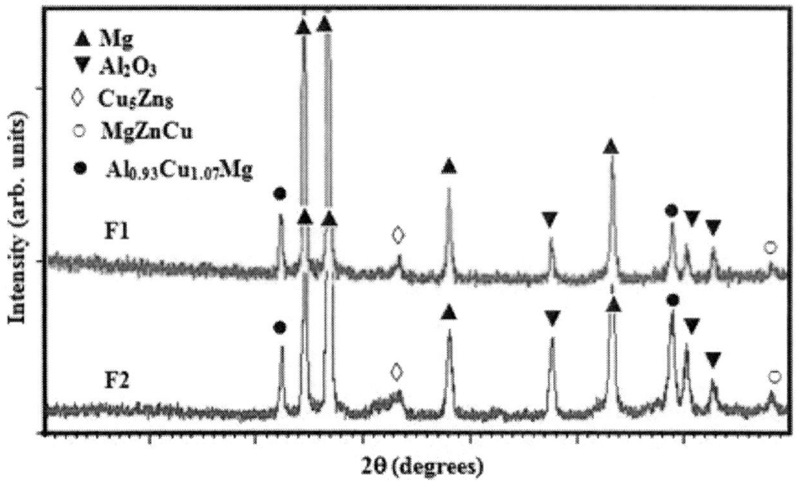


**Figure 2.** XRD Patterns of as-extruded Mg-3Zn-1Cu-0.7Mn / $Al_2O_3$ composite (F1) and stress-relieved alloy composite at $260^oC$ (F2) sintered at $400^oC$

**Table 2.** Microhardness observed for the alloys and composites

| Sl. No. | Materials | Micro hardness (HV) |
|---|---|---|
| 1. | Pure Mg | $48 \pm 1$ |
| 2. | AZ31B | $63 \pm 1$ |
| 3. | Mg-3Zn-1Cu-0.7Mn alloys – as-extruded (E1) | $74 \pm 3$ |
| 4. | Mg-3Zn-1Cu-0.7Mn alloys – Stress-relieved (E2) | $78 \pm 2$ |
| 5. | Mg-3Zn-1Cu-0.7Mn /$Al_2O_3$ composites – as-extruded (F1) | $80 \pm 3$ |
| 6. | Mg-3Zn-1Cu-0.7Mn/$Al_2O_3$ composites – Stress-relieved (F2) | $87 \pm 2$ |

the influence of the hot extrusion process on the microhardness of the composites is also revealed here. The hard intermetallic phases such as $Cu_5Zn_8$, MgZnCu, and $Al_{0.93}Cu_{1.07}Mg$ found in the XRD analysis also contributed the superior microhardness. These intermetallic phases restrict the deformation of the alloy matrix upon indentation. The limited porosity of the composites, as per the results of porosity analysis, also provided added hardness to the composites compared to pure Mg and the fabricated alloys.

From Figure 3, the potential influence of the stress-relieving process on the hardness of alloys and composites can be recognized. The hardness of the as-extruded composite is reported to be 80.45 HV, which is improved to 87.19 HV after stress-relieving. More than 8% of hardness is achieved through the stress-relieving process. The micropores in the composites are formed due to the entrapment of gases during the blending and mixing process. These entrapped gases create micropores upon the compaction process. Due to the liberation of the trapped gases in the

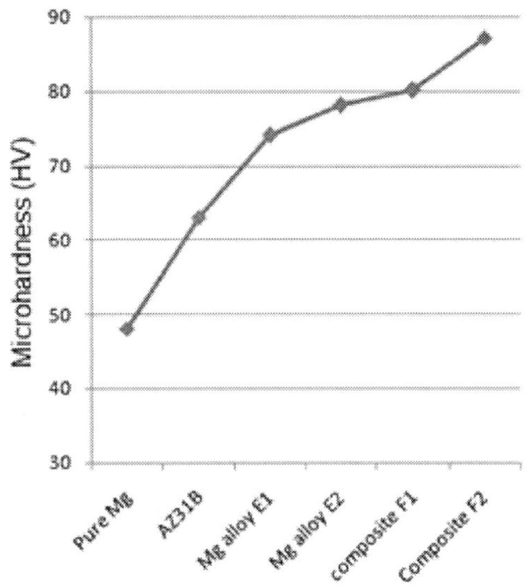


**Figure 3.** The comparison of microhardness of the alloys and composites fabricated

composites, the size and number of micropores are reduced on stress-relieving. The improvement in microhardness has resulted by the reduced number of micropores and microvoids following the stress-relieving process. The analysis of the optical microstructure validates the minimal amount of micropores in the stress-relieved composites.

## 4. Conclusion

The fabrication of the composite Mg-3Zn-1Cu-0.7Mn /$Al_2O_3$ was successful, employing the blend-press-sinter powder metallurgy route accompanied by a hot extrusion process for superior mechanical characteristics. The investigation is mainly focussed on the role of $Al_2O_3$ particles in improving the microhardness of the Mg alloy matrix composites and the mechanism affiliated with it. The study also examined the influence of the stress-relieving process on the hardness characteristics of the composites. The following parameters are concluded from the research.

(i) Powder metallurgy is proved to be a proper technique for the manufacture of the Mg alloy based composites.

(ii) The grain refinement, followed by dynamic recrystallization achieved through the hot extrusion process enormously, influences the hardness characteristics of the Mg composites.

(iii) The $Al_2O_3$ particulates incorporated to the Mg alloy enhances the microhardness of the Mg composites due to the homogeneous distribution of the reinforcement of particles over the entire microstructure.

(iv) The presence of hard intermetallic phases such as $Cu_5Zn_8$, MgZnCu, and $Al_{0.93}Cu_{1.07}Mg$ adds improved hardness to the composites.

(v) The stress-relieving process is found to influence the hardness of the composites strongly. The reduced number of micropores and microvoids after the stress-relieving process result in the improved hardness of the composites.

## References

[1] Cui Z, Zhang Y, Cheng Y, Gong D and Wang W 2019 *Materials Science and Engineering: C* **99** 1035–1047

[2] Lim P, Lam R, Zheng Y and Thian E 2016 *Materials Letters* **172** 193–197

[3] Cui C, Shen Y, Meng F and Kang S 2000 *Cailiao Kexue Yu Jishu(Journal of Materials Science & Technology)(China)(USA)* **16** 619–626

[4] Aikin R 1997 *JOM* **49** 35

[5] Zreiqat H, Howlett C, Zannettino A, Evans P, Schulze-Tanzil G, Knabe C and Shakibaei M 2002 *Journal of Biomedical Materials Research: An Official Journal of The Society for Biomaterials, The Japanese Society for Biomaterials, and The Australian Society for Biomaterials and the Korean Society for Biomaterials* **62** 175–184

[6] Li L, Gao J and Wang Y 2004 *Surface and Coatings Technology* **185** 92–98

[7] Mallick A, Tun K S and Gupta M 2012 *Materials Science and Engineering: A* **551** 222–230

[8] Ye H Z and Liu X Y 2004 *Journal of materials science* **39** 6153–6171

[9] Zhao W, Huang S J, Wu Y J and Kang C W 2017 *Metals* **7** 293

[10] Zulkoffli Z, Syarif J and Sajuri Z 2009 *International Journal of Mechanical and Materials Engineering* **4** 156–159

[11] Zhang Z Y, Peng L M, Zeng X Q and Ding W J 2008 *Transactions of Nonferrous Metals Society of China* **18** s55–s58

[12] Wu Y, Wu K, Deng K, Nie K, Wang X, Hu X and Zheng M 2010 *Journal of alloys and compounds* **506** 688–692

[13] Srikanth N, Ugandhar S, Gupta M and Srivatsan T S 2005 *Advanced Engineering Materials* **7** 735–743

[14] Hu X, Wu K and Zheng M 2006 *Scripta materialia* **54** 1639–1643

[15] Hu X, Zhang Y, Zheng M and Wu K 2005 *Scripta Materialia* **52** 1141–1145

[16] Wan D, Wang J, Lin L, Feng Z and Yang G 2008 *Physica B: Condensed Matter* **403** 2438–2442

[17] Trojanova Z, Riehemann W, Von Buch F, Lukáč P and Mordike B 1999 *physica status solidi (a)* **175** 555–560

[18] Nishiyama K R M Y I S N and Sakaguchi T 2003 *Journal of Alloys and Compounds 355* 22–25

[19] Hassan S, Tan M and Gupta M 2008 *Materials Science and Engineering: A* **486** 56–62

[20] Habibnejad-Korayem M, Mahmudi R and Poole W J 2009 *Materials Science and Engineering: A* **519** 198–203

[21] Paramsothy M, Tan X, Chan J, Kwok R and Gupta M 2012 *Journal of alloys and compounds* **545** 12–18

[22] Goh C, Gupta M, Wei J and Lee L 2007 *Journal of composite materials* **41** 2325–2335

[23] Ferkel H and Mordike B 2001 *Materials Science and Engineering: A* **298** 193–199

IOP Conf. Series: Materials Science and Engineering **923** (2020) 012050 doi:10.1088/1757-899X/923/1/012050

# EXPERIMENTAL INVESTIGATION OF PROSOPIS JULIFLORA AND MANGO TREE WITH EPOXY RESIN

**Elayaraja.R[1],Sasikumar.R[2],Gokul.M[3],Gowrisankar.T[4]**

[1]Assistant Professor,Department of Mechanical Engineering, Mahendra Engineering College, Mallasamudram, Namakkal,Tamilnadu, India

[2] Associate professor, Department of Mechanical Engineering, Mahendra Engineering College, Mallasamudram, Namakkal,Tamilnadu, India

[3,4] UG Students, Department of Mechanical Engineering, Mahendra Engineering College, Mallasamudram, Namakkal,Tamilnadu, India

E-mail : elayarajamech509@gmail.com

**Abstract:** Research on the development of composites has been going on long. But growing concerns about global warming, waste generation & management, increasing environmental awareness, increasing pressure on fossil fuel resources have led to the development of green composites. Even, there are some natural fibers where the properties can be good compared to the synthetic fibers which have contributed to this development. The work aims to develop a polymer composite with Prosopis juliflora and mango tree as reinforcements of a natural composite epoxy resin matrix. Composite plates were produced using a compression mold method with a composition ratio of 60:40,65:35 and 70:30. The resin and hardener proportions were 10:1 respectively. The manufactured composites were tested following ASTM standards to assess mechanical characteristics such as tensile strength, compressive strength, flexural strength, impact strength, hardness value, and water absorption test.

***Keywords:*** *Prosopis juliflora, mango tree, natural fibers, fossil fuel, mechanical characteristics.*

## 1. Introduction

A composite is an aggregate of materials, considered one of that is called the reinforcement stage, that's known as the strands, sheets, or debris, and is inserted into one of a kind materials known as the community level. Strengthening fabric and grid fabric may be metallic, clay, or polymer. Composites have a fibrous or molecular phase that is stiffer and more grounded than a coherent lattice section and fills because of the primary burden of transporting human beings. The network is going about as a heap pass medium between the filaments, and in less ideal situations in which the heaps are puzzling, the framework may additionally even want to endure masses transverse to the fiber pivot. The network is greater bendy than the strands and is therefore a source of composite energy. Also, the grid serves to defend the filaments from environmental harm previous to, in the course of and after composite processing. At the right time whilst scheduled, the sparkling attached cloth indicates the desired cost of normal materials. Composites are used for their auxiliary traits, however also for electrical, warm, tribological, and ecological packages. Composites are multifunctional cloth frames that offer characteristics that are not possible from any discreet material. They are long-lasting systems produced by the physical integration of at least remarkable components, unique in layout and satisfactory, and occasionally in composition. Composites should not be regarded as essential as a combination of two materials. In an extra large centrality, the blend has its own unique houses. As some distance as the unity of safety from warmth or a few different appealing first-class is involved, it's far superior to both of the components by myself or no longer to the identical volume as both. Composites are compounds that differ from combinations in that the male or female additives nevertheless retain their traits and are so mixed into composites that they make the best use of their traits and not their inadequacies, he said, as a way of achieving advanced substances. Composite materials as heterogeneous substances, which include at the least powerful levels, that are in close touch with every other on a tiny scale. They also can be considered as homogeneous materials on a small scale, as any part of them may have the same bodily properties. Current work is directed in the direction of the manufacture of Prosopis juliflora and mango tree hybrid composites the use of an epoxy matrix. The versions in mechanical homes (Compression, Flexural, hardness, impact, and water absorption) are studied[6].

The study of mechanical properties of Kenaf and Fiberglass Hybrid Composite laminates by Hand layup method was investigated by

Z. Salleh, et al. [16]. Short, Long fiber and powder fiber are used. Variation of the tensile strength of distinct kinds of fibers has been researched. Test findings show that the highest tensile strength and tensile modulus are found for long hybrid fiber composites. From the SEM assessment, the surface failure is due to the fracture of the matrix, the removal of the fiber, the debonding of the matrix, and the fracture of the fiber.

MohdSuhairilMeon, et al. [17]used high-density Polyethylene (PE) and polypropylene (PP) separately to prepare two sets of short kenaf composite. The kenaf fibers were treated with 3%, 6% & 9% of NaOH for a day and then dried for 24 hours at 80o C. The tensile properties of treated kenaf were improved when compared with the untreated kenaf fibers. Further coupling agents like MAPP, MAPE have amplified the tensile properties of both treated and untreated cases.

S Sivasaravanan, et al. [18] made an investigation on glass fiber/epoxy/nano clay composite by varying Nano clay from 1 to 5 wt% by hand layup process. E- Glass fiber is of bi-directional: 45o orientation is used. The average value of 5% wt of nano clay showed good impact results when compared to the other combinations used.

A. Atiqah, et al. [19] developed kenaf –glass reinforced unsaturated polyester hybrid composite. Kenaf and glass fibers are used in mat form. The kenaf fiber was treated with 6% NaOH using the Mercerization method for 3 hours. The test results showed that high tensile, flexural and impact strength were obtained using treated kenaf with 15/15 v/v kenaf/glass fibers. The adhesion between the matrix and surface of fiber was enhanced, which plays a key role in improving the mechanical properties of KG- UPE hybrid composites. The development of hybrid composite using jute and E- glass mats reinforced with epoxy (LY-556) resin and HY951 hardener is carried out by M. R. Sanjay, et al. [20]. Laminates were prepared by varying both mat layers. The test results show that hybrid composite has better mechanical properties and leads to an increase in the utilization of natural fibers in various applications.

## 2. Experimental

### *2.1 Material*

The raw materials involved in the production were epoxy tar LY 556, Prosopis juliflora, and mango tree, Hardener HY 951. The Prosopis juliflora and the mango tree are arranged haphazardly in polymer lattice. Tests are to be organized with the soluble base of treated strands. Prosopis juliflora and mango tree were salt-treated in 1% NaOH response for 30 min expel any oily material and hemicellulose and then dried in daylight. The second fiber was used to make Prosopis juliflora and mango tree epoxy composites.

### *2.2 Material Preparation and Fabrication process*

The shape is produced of wood in the measurements set out in the ASTM Standards for Pre-parathion Specimens. The OHP sheet is mounted and the discharging specialist is linked over it and equipped with the inside of the form, allowing it to dry. A glass container and a glass pole or stirrer shall be removed and well washed with running water and then with hot water. From that point on, the determined amount of epoxy tar and the deliberate amount of hardener (proportion 10:1) are added to the breaker and the mixture is mixed for almost 25 min. The reason for this mixing is to create a homogeneous mix. At that point, the determined amount of filaments (proportion 1:1) is included after the blending has been completed and the mixing procedure has proceeded for the following 45 min. At that stage, the mixture filled in the shape and softly pushed into a uniform settlement.

**Fig.1.Fabricated plate**

## 3. Characterizations of Composite Materials

### *3.1 Determination of Water Absorption Behavior of Composite*

Water ingestion of Prosopis juliflora and mango tree fortified polymer composite was investigated by inundation in refined water at room temperature for 12, 24, 48, 72 hours. The sample instance size (5 mm X 5 mm) is set according to ASTMD-570 for the water retention test. The edges of the instance were attached to the polyester rubber. Tests were dried at 50oC for 24 hours. After 24 hours of testing, it was exactly weighed. Fixed examples were then immersed in refined water at room temperature for 12, 24, 48, 72 hours. Tests were expelled from the water after a lawful time and washed with tissue paper to remove surface water. Ingestion of clamminess in the composite was evaluated by the incidental inspection of the fabric between moments[ 13-14]. The volume of increase in the mass of the manual to the fundamental mass is indicated by the rate of clamminess ingestion.

The weight gain of the sample was measured as an increase in weight % using the formula.

Increase in weight %= ((wet weight- sample weight) /sample weigh)*100

### *3.2 Tensile and compressive test*

The ductile experiments were conducted by the ASTM D 3039 standard of the Mechanized Universal Testing Machine. The stacking course of action for the instance and the picture of the device used were shown in the figure. Examples with length measurements of 300 mm and a width of 25 mm have been used. The experiment was carried out at a crosshead velocity of 5 mm/min using a load cell of 10 kg. For each situation, 3 examples were used and the normal quality was taken into account. The compressive quality is usually obtained through the use of UTM compressive test techniques.

### *3.3 Flexural and Impact Test*

The flexural quality of the composites was determined by the three-point twisting technique. It tends to be finished in the adapted UTM machine as per ASTM D790. The stacking course of action for the instance and the picture of the device used were shown in the figure. All the composite examples were of a rectangular shape with a length of 80 mm x 15 mm x 5 mm. Trials were conducted at a crosshead velocity of 0.5mm / min. At that stage, the flexural quality was determined to use a straight forward minute outline of the fundamentally bolstered bar at the primary problem load. Effect quality was determined by the Izod Effect Test using examples of measurements.

## 4.Result and Discussion

### *4.1 Water Absorption behavior of Composite*

Water absorption is one of the main issues in the use of natural fiber composites in many applications. Water absorption in hybrid composites rises at a pace of 1 %-5.5 % in 24 hours, maximum and minimum water uptake after 10 hours.

### *4.2 Impact strength*

The variety of effect quality with the fiber content, if there should be an occurrence of Prosopis juliflora composite is displayed in the figure. For this situation, the Prosopis juliflora and mango tree composites displayed better sway properties. The effect quality increments with the expanding volume part of filaments, arriving at a most extreme incentive at 30%. Past 30%, the effect quality demonstrates a diminishing pattern. The most extreme effect quality of the composites fluctuates between 0.8 Joules to 2.3 Joules. Antacid treated Prosopis juliflora and mango tree indicated improved effect quality.

### *4.3 Compressive strength*

The variety of compressive quality with the fiber content on the antacid treated composites has appeared in the figure. The Prosopis juliflora and mango tree composite material were tried and the compressive quality was determined. Three examples were tried, the normal compressive quality was accounted for. The compressive quality was expanding relentlessly up to 30% and past that, the change was negligible. The compressive quality of the Prosopis juliflora and mango tree from 19.77 MPa to 35.63 MPa.

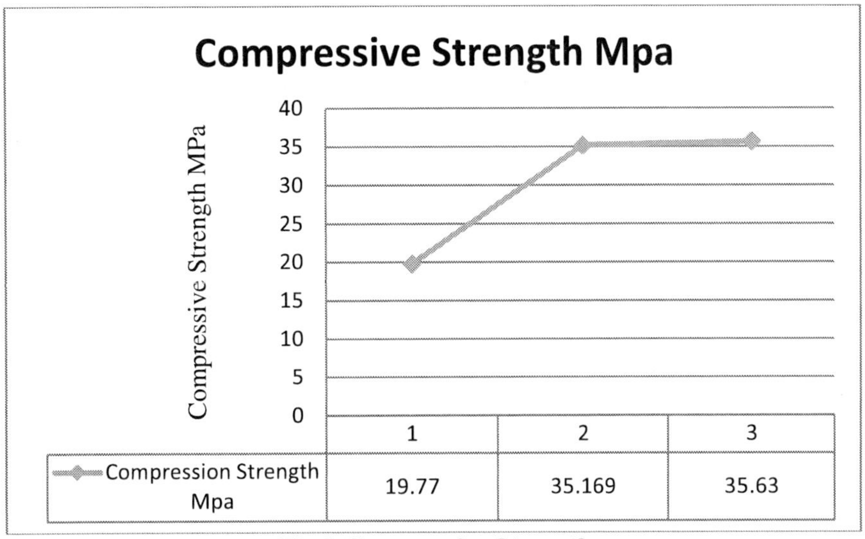


**Fig.2 Compressive Strength**

### *4.4 Tensile and Flexural strength*

Fiber substance and fiber quality are affecting parameters for the quality-related properties of the composite. Subsequently, the quality variety of fiber stacking demonstrated in an unexpected way. This variety in the ductile and flexural quality of the composites with 55%,65%, and 75 of fiber substance has appeared in figures 9 and 10 separately. These figures obviously show rigidity and flexural quality for half and 60% fiber content. Anyway, there is a decline in both the pliable and flexural quality of the composite with 70% fiber content.

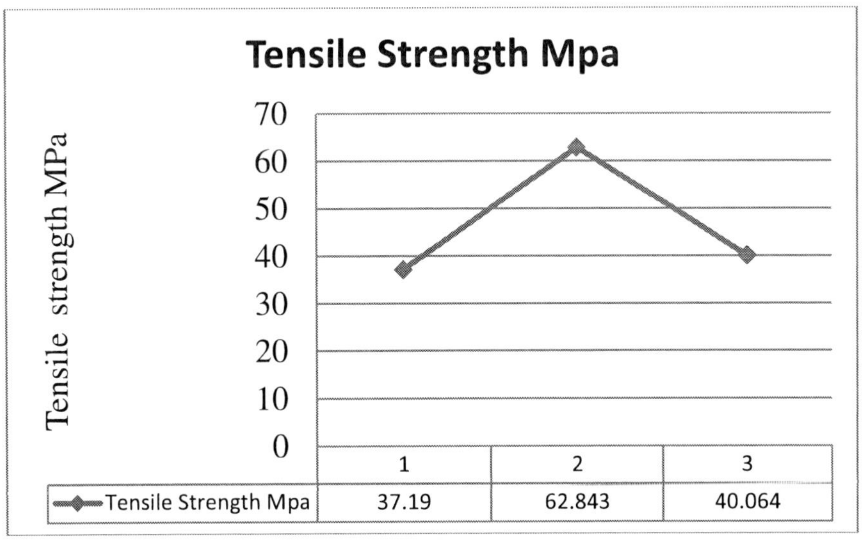


**Fig.3 Tensile strength**

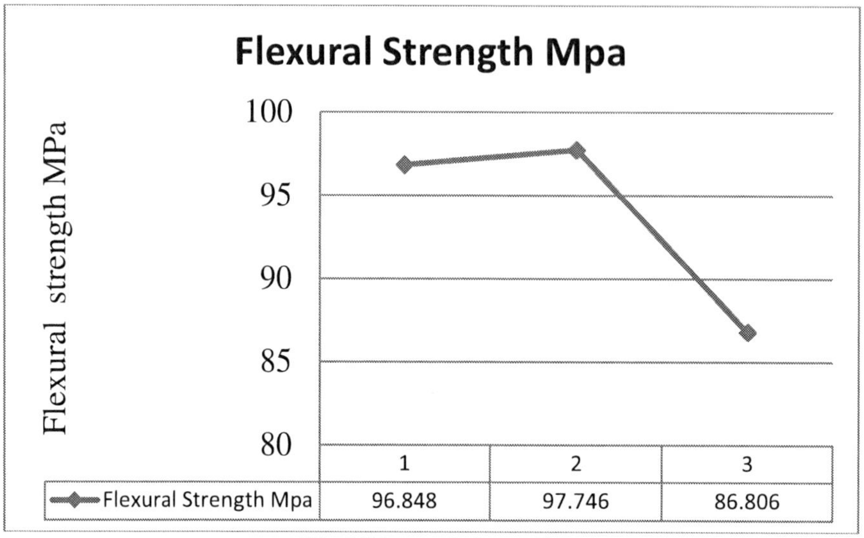


**Fig.4 Flexural strength**

## 5. Conclusion

This whole project explains the mechanical property of the composite material (tensile test, Compression, flexural test, impact test) while it can be both Prosopis Juliflora and mango trees. From the results of mechanical characteristics, 65:35 of composition shows a better value in the tensile test, Compressive, flexural test and impact test. Parallel the test values are shows nearly to the synthetic fiber composite values from based the literature review. From the result it clearly indicates, synthetic fiber can be replaced by natural fiber.

## Reference

[1] Furqan Ahmad, Heung Soap Choi, Myung Kyun Park, A review: natural fiber composites selection in view of mechanical, light weight, and economic properties, Macromol. Mater. Eng. 300 (1) (2015) 10–24.

[2] C.S. Boland, R. De Kleine, G.A. Keoleian, E.C. Lee, H.C. Kim, T.J. Wallington, Lifecycle impacts of natural fiber composites for automotive applications: effects of renewable energy content and light weighting, J. Ind. Ecol. 20 (1) (2016) 179–189.

[3] N. Saba, M. Jawaid, O.Y. Alothman, M.T. Paridah, A. Hassan, Recent advances in epoxy resin, natural fiber-reinforced epoxy composites and their applications, J. Reinf. Plast. Compos. 35 (6) (2016) 447–470.

[4] D. Mohana Krishnudu, D. Sreeramulu, P.V. Reddy, Alkali treatment effect: Mechanical, thermal, morphological, and spectroscopy studies on abutilon indicum fiber-reinforced composites, J. Natural Fibers (2019) 1–10.

[5] D.M. Krishnudu, D. Sreeramulu, P.V. Reddy, H.R. Rao, Effect of alkali treatment on mechanical properties of prosopis juliflora hybrid composites, Int. J. Appl. Eng. Res. 13 (5) (2018) 2933–2935.

[6] Elayaraja.R, Saravanan.P, Experimental investigation of E-Glass and Kenaf fibre with epoxy resin, IRJET, Vol 6(11) (2019) 303–306.

[7] D.M. Krishnudu, D. Sreeramulu, P.V. Reddy, Optimization the mechanical properties of coir-luffa cylindrica filled hybrid composites by using Taguchi method, AIP Conference Proceedings vol. 1952 (2018).

[8] P.M. Borba, A. Tedesco, D.M. Lenz, Effect of reinforcement nano particles addition on mechanical properties of SBS/Curauá fiber composites, Mater. Res.17 (2) (2014) 412–419.

[9] N. Saba, M.T. Paridah, K. Abdan, N.A. Ibrahim, Effect of oil palm nano filler on mechanical and morphological properties of kenaf reinforced epoxy composites, Constr. Build. Mater. 123 (2016) 15–26.

[10] A. Shalwan, B.F. Yousif, Influence of date palm fibre and graphite filler on mechanical and wear characteristics of epoxy composites, Mater. Des. 59 (2014) 264–273.

[11] S. Biswas, A. Satapathy, A. Patnaik, Effect of ceramic fillers on mechanical properties of bamboo fiber reinforced epoxy composites: a comparative study, in: Advanced Materials Research, Trans Tech Publications, 2010, pp. 1031–1034.

[12] S. Dalbehera, S.K. Acharya, Effect of cenosphere addition on the mechanical properties of jute-glass fiber hybrid epoxy composites, J. Ind. Text. 46 (1) (2016) 177–188.

[13] B.C. Samata , T.Maity, S.Dalai, A.K.Banthia, Mechanical properties of modified epoxy: effect of chain length, Journal of Pigment and Resin technology, 35 (2006) 16-23.

[14] K.Sewda, S.N.maiti, Mechanical properties of teak wood flour reinforced HDPE composites. Journal of Applied Polymer Science, 18 (2009) 26 - 34.

[15] Elayaraja.R, Experimental investigation of natural composite with epoxy resin,International journal of Research and Analytical reviews,02 (2019) 458-464.

[16] Z. Salleh, M.N. Berhan, Koay Mei Hyie and D.H.Isaac, "Cold-pressed Kenaf and Fiber glass Hybrid Composites Laminates: Effect of Fibre Types", International journal of chemical, Molecular, Nuclear and Metallurgical Engineering, Vol:6, No:11,2012.

[17] MohdSuhairilMeon, Muhamad Fauzi Othman, Hazaran Husain, Muhammad FairuzRemeli, MohdsyaharMohdSyawal, "Improving tensile properties of kenaf fibers treated with sodium hydroxide", International Symposium on Robotics and Intelligent Sensors,Elsevier,2012,pp. 1587-1592.

[18] S.Sivasaravanan, V.K.Bupesh Raja, "Impact properties of epoxy/glass fiber/nano clay composite materials",IOSR Journal of Mechanical and Civil Engineering, 2014,pp39-41.

[19] A. Atiqah, M.A. Maleque,M. Jawaid, M. Iqbal, "Development of kenaf-glass reinforced unsaturated polyester hybrid composite for structural applications", Elsevier,2013,pp. 68-73.

[20] M. R. Sanjay, B. Yogesha, "Studies on Mechanical Properties of Jute/E-Glass Fiber Reinforced Epoxy Hybrid Composites", Journal of Minerals and Materials Characterization and Engineering, 2016, Vol:4, pp. 15-25.

IOP Conf. Series: Materials Science and Engineering **923** (2020) 012051 doi:10.1088/1757-899X/923/1/012051

# Review on Fiber reinforced /modified Bismaleimide resin composites for Aircraft Structure Application

N.Prasanaa Iyer[1,a], Dr.N.Arunkumar [2,,b]

[1]Assistant professor St Joseph's College of Engineering, OMR,Chennai,India

[2]Professor and Head of Students Affairs St Joseph's College of Engineering, OMR,Chennai,India

Email:[a]prasanaaiyer@gmail.com, [b]narunkumar72@gmail.com

**Abstract:** Bismaleimide (BMI) is a high performance thermoset material which possess similar characteristics as that of Polyimides which include high strength and rigidity at high temperature. It can be used for high temperature application upto 260° C.It stands between epoxy and polyimide in its thermal stability and moisture absorption properties. The resin is brittle in nature which necessitated modification with various modifiers viz epoxy, phenol Novolac, Polydimethylsiloxane (PDMS),Allyl phenol, Amines, Carbon nanotubes(CNT), ceramic materials etc. at chemical level to reduce its brittleness. The bismaleimide is used as a matrix in composite materials is reinforced with fibers viz carbon, Kevlar, Basalt, glass fibers. Various authors studied the bismaleimide composite material for its application in Aircraft and missile structures which requires thermal stability and high strength at high temperature. is the resin. In this article the fiber reinforced modified bisamleimide resin composite materials are reviewed with an intend to explore the wide application of the resin system.

## 1.Introduction:

The demanding needs of the Engineering materials for high strength and high temperature application made an insight into the advanced materials. The advanced materials are Phenolics, Novalac epoxy,Bismaleimide,Polyimides.Bismaleimide are thermoset materials which inherently possess superior properties than epoxy resin and finds application in aircraft structure, missile structures atc.Bismaleimide resin is known for its excellent mechanical and thermal properties. The BMI has a brittle network hence necessitated the resin modification with comonomers viz aromatic amines, Phenolic resin etc to impart improved impact strength and toughness into it without affecting the thermal properties of it. This has created a large scope for investigating the materials and manufacturing composites based on the resin with modification , and with fiber reinforcement. The resin has been often copolymerized with Allylphenol which makes reduces the brittleness in it. Futher to the development of modified resin it been incorporated by fibers to futher enhance the thermo mechanical properties in it. In this journal the bismaleimide resin , its modifier , and fiber reinforced resin for composite materials is reviewed. Also the advancement in the resin composite is also reviewed which includes the Carbon nanotubes grafting into the fibers thereof .

IOP Conf. Series: Materials Science and Engineering **923** (2020) 012051 doi:10.1088/1757-899X/923/1/012051

## 1.1 Bismaleimide Resin

Bismaleimide is an important compound with maleimide groups connected to nitrogen atoms on either side by a linker which can be used as reagents for crosslinking thermoset polymers. It is synthesized by condensation of Pthalic anhydride with an aromatic diamine at a molar ratio of 2:1.The most important bismaleimide monomer is 4-4' bis(maleimido) diphenylmethane (BMI) has a melting point of 155° C, often copolymerized with 2-2' diallylbisphenol A(DBA).

DABA

BMI-P

BMI-M

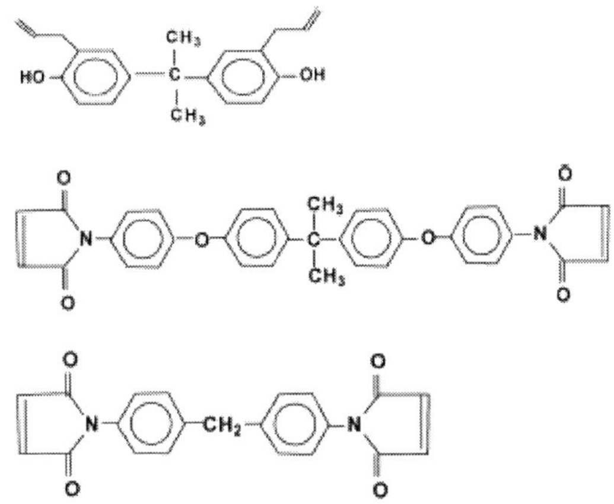


Fig 1 Bismaleimidophenoxyphenylpropane(BMI-P), Diallylbisphenol A (DABA), Bismaleimidodiphenyl Methane(BMI-M),

These resin are usually insoluble in ordinary organic solvent and soluble in high boiling polar solvents. The most curatives are aromatic diamines.

## 1.2 Toughening of Bismaleimide:

The presence of highly ordered stacking of polar carbonyl group in BMI makes it more brittle with less fracture toughness which can be modified with suitable modifiers for enhancing the properties. Hence BMI has been toughened by various chemicals and by various methods to improve the properties of the resin system for its enhanced application. The common toughening agents/mechanisms are allylphenol[47] copolymers, Micheal's addition copolymers, thermoplastic ,thermoset,elastomer,hyperbranched polymer additions ,particle strengthening.

With allyl phenyl and allyl phenol compounds, the rigid and brittle aromatic units in BMI is reduced by its reaction by Diels-Alder type reaction in which adducts of bi and tris are obtained from excess BMI which is commonly called as Alder-ene reaction. Various work on it have been done by various researchers for improving the toughness of the resin system. With the

Michael reaction the nucleophilic addition of a carbon ion or a nucleophile to an unsaturated carbonyl compound forming mild C-C bond. The reaction between diamines and BMI is by Michael's addition reaction where the diamines are nucleophiles.

### 1.3 Bismalemide with Thermoplastic/ thermoset/ elastomer/Hyperbranched polymer Addition.

BMI are modified with thermoplastic modifiers like polyetharimides,poly ether ketone, poly ether sulfone [49] for reducing its brittleness and improving the toughness of the system thereof.Thermoset like epoxy [48]/Novolac epoxy[52] can increase the processability of bismaleimide with suitable diamine addition to it and hence enhance the thermo mechanical properties of the system.Elastomers [50] modifiers include the flexibility, lap shear in the system but with reduced strength. The properties are also improved by the addition epoxy [58] and siloxane [51]to BMI resin system.Also BMI has been modified with high performance polymers and nano particles [54] for retaining its properties at high temperature. Continuous Fiber is reinforced with modified BMI composites for its use in advanced application opened a diverse study of the composites which is reviewed in the journal.

## 2.Fiber reinforced BMI composites

BMI resins are used to manufacture composites through Autoclave and Out of Autoclave techniques like Resin Transfer moulding, Hot melt impregnation, through solvents. Different types of fibers are reinforced in BMI system exploring the wide possibility of its structures for different application. Though the resin infusion is enabled by Resin transfer moulding process, the resin can also be dissolved in solvents and prepegs are made which is then compressed to form composite materials. In this various work on fiber reinforced BMI composites are reviewed for its properties and application.

The fiber that are reinforced with BMI resin system are carbon fibers, Kevlar fibers, Glass fibers, Basalt fiber, quartz fiber, Carbon nana fibers and hybrid fibers, each contributing to diverse properties and application. The fiber reinforced are used as received or it treated with acids, silane, plasma or by grafting carbon nanotubes etc in it.

### 2.1 Carbon Fiber Reinforced modified BMI composites:

**Bismaleimide has been commonly modified with** diallylbisphenol A[56] and the resin system is incorporated into fiber reinforcement for manufacturing composites.Various investigation were done on carbon fiber reinforced BMI composite material of which **John Montesano et al,(2012)** [3]reported that the off-axis tensile fatigue behaviour of a woven carbon fiber/bismaleimide laminate is investigated at various temperatures. Fatigue tests at the same test temperatures all showed that the fatigue response developed in three distinct stages. During the initial stage of cycling at the off-axis side of the malinate there is a sudden increase in permanent strain and mean strain with slight increase in the axial stiffness. The author showed that the mean

IOP Conf. Series: Materials Science and Engineering **923** (2020) 012051 doi:10.1088/1757-899X/923/1/012051

strain developed is not a function of temperature change which is a significant finding and unique to this study.

**Li Yuan et al,** [4]reported that ,The three-component resin system consisting of 4,4'-bismaleimidodiphenylmethane (BMI), O,O -diallylbisphenol A(BA) and organic rectorite (OREC) was prepared to develop a novel fiber reinforced BMI/BA/OREC composite.The morphologies of composites were characterized using a scanning electron microscope (SEM) which indicated that the appropriate content of OREC can significantly improve the mechanical properties such as the flexural strength, the interlaminar shear strength and the modulus of flexural elasticity of fiber reinforced BMI/BA composite. The proper addition of OREC could enhance the water resistance and the hot-air aging property of the fiber reinforced BMI/BA composites due to the barrier and high thermal stability of OREC.

Author **X.J Xian and C.L. Choy [6]** studied the fracture toughness and fatigue fracture behaviour of carbon-fiber-reinforced with modified bismaleimide (BMI) composites . These composites were found to have higher fracture toughness, better damage tolerance and longer fatigue life than carbon-fiber composites with epoxy matrices. The improved performance is due to the presence of thermoplastic particles in the modified BMI matrix leading to better adhesive bond between them with pronounced deformation by plasticity. The change of fiber orientation in the composites reduced the crack propagation and hence increasing its fatigue life.

Author **P T. CURTIS**, [8]conducted studies on twelve different carbon fibers (each of them with different fiber failure strain percentage) reinforced with epoxy, bismaleimide and thermoplastic matrices its mechanical properties. The fiber with intermediate modulus leads to the improvement of Tensile strength in both plain and notched specimen with an increase in stiffness. Also he inferred that the larger diameter fiber has lead to the improvement of compressive strength and improving its hot/wet performance which is desirable in aerospace applications.

The effects of environmental changes on the thermal and mechanical properties of the composite were studied by various authors under different conditions. The change in moisture and hygrothermal effects in mechanical properties as well as its Glass transition temperature has been studied by Author **Michelle Leali Costa et al [9] who** conducted a comparative studies the two different laminate consists of Epoxy/Carbon fiber and Bismaleimide/Carbon fiber. He observed that in both the laminate system with an increase in moisture content at a specified temperature there is reduction in interfacial bonding and hence reduced Interlaminar shear strength ILSS.the glass transition temperature has also been reduced with increasing moisture content due to greater mobility of molecules leading to the transition in both the resin system by analyzing its Dynamic mechanical behaviour.
It is also being studied by Author **M. Akay et al**,[15]the effect of ageing in the carbon fiber reinforced bismaleimide (Cycom 5250-4) composites on the thermo mechanical behaviour. The thermal ageing process involves placement of cut specimens in air circulating oven set at 210°, 230° and 250 °C for various periods of time up to 2000 h.He observed the activation energy in the range of 151–270 kJ/mol lead to the twofold deterioration of flexural and interlaminar shear strengths as a result of matrix loss and microcrack but with an increase of Tg.(325° C) for both

laminate types(54% and 58%fibre volume content laminates) from 245° and 275° C respectively, beyond 500 h ageing.
Futher Author **L F M da Silva[7]** et al, investigated the effect of thermal variation from -55° to 200° C twill weave carbon fiber/ BMI composite. The strength and modulus haven't varied in the temperature range making it suitable for temperature applications upto 200° C.The transverse strength when compared to longitudinal strength is low which is undiserable when it is subjected to loading in through thickness direction.

A comparison between 2D and 3D carbon fiber fabric with BMI resin composites is made by Author **Shen Chou et al [10]** reported that due to the Z directional fibers in 3D fabric network the crack propagation is restricted which lead to the improvement of shear strength and also the Impact strength has been increased to a level of 66% to 68% more than that of the 2D composites .The experiments were conducted at varying temperatures ranging from 25° C to 275° C.

Author **Divya K. Chakravarthi** et al,[12] reported the potential importance of lightweight application of carbon fiber composites for next-generation structures. In their work they optimized the addition of Ni coated Single walled carbon nanotubes (Ni-SWNTs) with bismaleimide (BMI)–carbon fiber composite based on their dispersion, surface area interaction and its resistance. Atomic force microscopy analysis improved dispersion of Ni-SWCNT in carbon/BMI composites improving the electrical conductivity with an optimal value of less than 4% of addition.Also showing good lightning strike protection with minimal fiber pullouts under mechanical testing.
Author **Vernon T. Bechel et al,** [13]reported the microcracks developed in crass ply and quasiistropic laminates of carbon/BMI composites where the specimen is thermally treated at cryogenic temperature with liquid nitrogen LN.The cycle of immersion is varied and the micro crack formation is studied.The micro cracks were formed on the surface and edged which sparsely penetrated the centre after 400 cycles of its immersion in LN and returning back to room temperature.The tensile strength showed a decline in the layup and adjacent one having same orientation.
**Author Li-RongBao et al**,[14] 2002, studied the effect of temperature on Moisture absorption and hygrothermal aging in a woven and woven/uni-weave composites. The samples are experimented at different at different temperature. The diffusion is of two stage process. In the first stage there is diffusion followed by relaxation of the chemical in the second stage. It is found that the structural relaxation occurs below 70° C and degrades above 90° C signifying the change in temperature effecting the moisture absorption.

The authors **Yuan-Fei Fu** et al,[16] used Silane surface modification method for the surface treatment of carbon fiber to improve the interfacial adhesion of the carbon fiber reinforced bismaleimide(BMI) composite. Experimental results revealed that silane treatment largely reduced the friction and wear of CF/BMI composites. The silane treatment method binds organic functional groups on carbon fiber surfaces, which increase the interlock between the fiber and BMI matrix, accordingly the TPB properties of CF/BMI composite are improved. Under the same test conditions, silane treated CF/BMI composite registers low friction coefficient and wear.

The authors **M. I. Faraz** et al,[17] fabricated Novel carbon nanofiber (CNF) -filled bis-malemide composites by a thermokinetic mixing method. The thermal and mechanical properties of composites containing 1 wt % and 2 wt % CNFs were investigated. A new BMI prepolymer was reinforced successfully with CNF by using an innovative melt compounding technique. The reinforcing of CNF caused restriction of polymer chain mobility which was confirmed by the increase of the storage modulus and the glass transition temperature as measured by DMA.

Author **Satoshi Kobayashi** et al,[18] studied the Microscopic damage behavior of high temperature CFRP, carbon/BMI (bismaleimide) under tensile loading was investigated experimentally.10 kinds of laminate configurations were tested at both 25ºC and 180º C.Microscopic damage in laminates at room temperature were observed by optical microscopy. Microscopic damages in both [0/90°]s and [0/90°$_2$]s laminates grow in the same manner. The number of matrix cracks increased as the laminate strain increased.

Author **Pei Sun**,et.al (2011) [19] studied the hygrothermal aging on carbon fiber /Bismaleimide composites. The Interlaminar shear test was conducted on the samples at different temperature and it is found that the ILSS decreased with increasing temperature with hygrothermal aging attributing to the plasticization, voids, micro cracks and interfacial debonding.

**Toshiyuki Shimokawa** et al, [20] as part of their investigation of the potential of high temperature composite materials for the next-generation SST(Supersonic transport), the authors' group assessed open-hole (OH) tensile and compressive static strengths at RT(Room temperature) and 120°C . The Authors further investigated the fatigue strength or *S-N* relationships of CF/BMI composite materials with a quasi-isotropic stacking sequence at RT(Room temperature) and 150°C. He concluded that the Furthermore, the OHT fatigue strength is less dominated by the number of load cycles, while the OHC and OHTC fatigue strengths moderately depends on the number of load cycles. An open hole greatly reduced static strength attributing to the presence of hole either for tension load or compression load which was smaller at 150°C than at RT.

**C. L. Heisey** et al, [22]studied the adhesive behaviour of plasma treated polyacrylonitrile based carbon fiber with phosphorous containing Bismaleimide resin composites using micrbond pull-out test. It's been concluded that the BMI adhered well with the surface treated fiber than with the untreated fiber due to increase in surface free energy, wettability with more active site for chemical interaction with the matrix.

**Bor Z. Jang**, [23] improved the interfacial adhesion of the carbon fiber reinforced with Bismaleimide resin by using different gases for cold plasma treatment. He proposed that Ammonia and Ammonia/Argon plasmas provides better strengthening of fiber than the other plasma. The matrix system has Bismaleimide ,epoxy with model polypropylene.The plasma polymerized polypropylene improved the adhesive strength of the matrix.

**Sergey Evsyukov** et.al.,2019 [24] developed a ternary blend of Bismaleimide (BMI) monomers, viz m-xylyene bismaleimide,4,4'-bismaleimidodiphenylmethane and 2,2-4 bisbenzophenone and o-o'-diallylbisphenol A as co-monomer is processed by Resin infusion/transfer is in the temperature range of 90-110°C. The modified resin is used in the fabrication of carbon fiber laminate resulted in Flexural strength of 3650 Kg/cm² ,(157 MPa for BMI:DABA at 100:80 weight ratio [2])with Fracture toughness of 0.63 MPa/m² mechanical properties ,Tg of 285 ° C .

**Yuan-Fei Fu et al**, [16]used Silane surface modification method for the surface treatment of carbon fiber to improve the interfacial adhesion of the carbon fiber reinforced bismaleimide(BMI) composite. Experimental results revealed that silane treatment largely reduced the friction and wear of CF/BMI composites. The silane treatment method binds organic functional groups on carbon fiber surfaces, which increase the interlock between the fiber and BMI matrix, accordingly the Tribological behaviour of CF/BMI composite are improved. Under the same test conditions, silane treated CF/BMI composite registers low friction coefficient and wear.

**SiyangLiu** et al (2018) [3]investigated Bismaleimide containing Pthalide cardo and Cyano group PCBMI is mixed with o-o' diallyl Bisphenol A (DABA) for its thermal properties by varying the molar ratio of PCBMI/DABPA. The resultant composites exhibited excellent thermal stability, heat resistance and mechanical properties and also had an excellent interfacial bonding and aging resistance.

**M. Simpson et al** [27] reported that results of thermogravimetric analysis simultaneously done with mass spectrometry which shows that the curing volatiles are entrapped in PMR -15 laminates which were post-cured with and without bagging films. These volatiles have an impact on thermomechanical response of the laminate, which increases the thermal strain with formation of thermal microcracking. The Post-cured laminates in the bagging films have more volatiles and in the case of laminate post cured in the absence of these films, the rate of desorption of cyclopentadiene lead to a variability in cross-linking density and hence thermomechanical response of the matrix through the thickness of the laminate.

**S.P.Wilkinson**, et al, (1993), [28] improved the toughness of Bismaleimide by thermoplastic modifier of 15 % to 20 % polyether sulphone. The modified resin was impregnated with unidirectional carbon prepreg. Mode I and mode II fracture toughness test was performed using end notched specimen. The values of $G_{IC}$ improved from 359±17 J/m2 which is for unmodified resin to 489±25 J/m2.The interlaminar strengthening is attributed to the resin rich zone within the interlayer thus toughening the composites.

Author **Jing Zhou** et al.(2016), [29] stated that Mode-I interlaminar fracture toughness could be significantly improved in carbon fiber/bismaleimide composites using a microwave curing which brings about remarkable 133.5% and61.2% increase in fracture toughness and fracture Resistance.This is due to enhanced interfacial adhesion which have big potential applications in low-cost manufacturing of high performance aerospace structures.

IOP Conf. Series: Materials Science and Engineering **923** (2020) 012051 doi:10.1088/1757-899X/923/1/012051

**R.Biji** et al, [30]studied the shape recovery of the carbon fiber reinforced 2,2-bis 4[(4-maleimidophenoxy)phenyl]propane (BMIP) and Diallylbisphenol A with poly(tetramethyleneoxide)(PTMO). The ternary blend is chemical characterized with FTIR. The polymer had good shape recovery due to the presence of PMTO which increased the factor by its increasing content and showed a recovery of 88% to 98% in a minute.

**Qi Yu** et al,[40] studied the impact of vacuum thermal cycling on the properties of the carbon/BMI composite where they conducted tests on finding the dynamical mechanical behaviour, Flexural strength, Inlerlaminar Shear strength, Tg, Surface roughness and Tensile strength under different thermal cycles (48,95,198,283).The DMA showed that there is increase in Tg from 48 cycles to 283 cylces with increase in cross linking density with a reduction in toughnes and tensile strength of the composite. The Flexural strength and ILSS showed a similar trend of increase in strength to a certain cycle and decreasing after 198 cycles. The surface roughness also increased initially and decreased to futher increase in thermal cycles.

**Lirui Yao** et al, [34] compared the interfacial adhesion of carbon fiber with epoxy and bisamleimide resin. With each of the resin system the author included three different carbon fiber which are sized, desized and partially sized. The partially sized carbon fiber showed better adhesion with improved toughness than the other two variation in the fiber with epoxy resin system whereas all the three variation of carbon fiber in BMI has shown no effects in improvement of adhesion as well the properties.

**Joshua W.Ehrlich** et al [55] produced carbon fiber/BMI composite with resin infusion moulding by optimizing the moulding process parameters. He studied thecomposits with different layups of carbon fiber and analysed the tensile performance parameters.The layer with $[0°]_4$ outperformed than the other composite layup of$[90°]_4$ and $[+45°,90°,90°,-45°]_4$in Tensile test with increased tensile stress and strain.

**Mohammad H.Haque** et al [57]studied the behaviour of unidirectional carbon fiber bismaleimide composite's weight loss through TGA nad its flexural strength at 260 C in air for 3000h.The thermal ageing and the flexural strength of $[0°]_{16}$ outweighed the $[90°]_{16}$

The first specimen showed good consistency in its properties till 100h at 260 C and lost its flexural strength by 95% ,whereas the same amount of loss in properties is observed in the $[90°]_{16}$ after 1200h of thermal ageing.

## 2.2 Glass Fiber Reinforced modified bismaleimide composites:

IOP Conf. Series: Materials Science and Engineering **923** (2020) 012051 doi:10.1088/1757-899X/923/1/012051

Author **K. Ambika Devi et al, (2008)[1]** [46]reported that the novolac epoxy (EPN)–2-2-diallyl bisphenol-A (DABA) system was modified by reactive blending with bisphenol-A bismaleimide (BMI) to improve its high temperature properties. The incorporation of BMI into the epoxy–phenol system(EPB) into glass fibers yielded a matrix system with improved glass transition temperature and comparable mechanical properties. The toughening of the system using polyether sulphone improved its mechanical properties with a marginal penalty on its glass transition temperature. The reactive blending of bismaleimide with an epoxy–allylphenol system provided a co-cured matrix system with improved high temperature resistance in comparison to the epoxy–phenol system.

**E. Drukker** et al, [21]studied the effect of thermal gradient in glass fiber reinforced BMI composite on the mechanical , physical and chemical properties. He conducted Interlaminar shear strength test by varying the depth to thickness ratio. The thermal gradient of 150 is introduced between the top and bottom plates and is analyzed for its interlaminar strength, glass transition temperature. Its been concluded by the authors that the at higher curing temperature the glass transition temperature increases with decrease in mechanical properties due to the micracks, brittleness due to the decrease in ether band leading to reduction in flexible link.

Author **Anna Razgon** et al,[36]reported that Bismaleimide polymer composites were coated with adherent, uniform, crack-free, titania films. The deposition was made on the glass fiber MBI prepreg by adopting liquid phase deposition technique (LPD).The use of Thermogravimetric (TGA) analysis revealed of the effect of such coatings and the ageing phenomenon on BMI/Glass fiber stability and the scope of finding the difference between thermo-oxidative from purely thermal degradation processes in the coated composite provided a good model for future work.

**Yehai Yan** et al,[37] modified BMI resin with allyl novolac which reduced the viscosity of the resin system for its use in Resin transfer moulding and impregnated with woven glass fiber. From the DSC study the curing cycle of the resin system is derived. The author compared the Flexural strength of the neat resin with fiber reinforced resin at varying temperature (Room temperature, 200 C, 300 C where the latter one proved with improved result.The modified resin provided higher glass transition temperature proving its retention at high temperature.

**Yian Zhao** et al, [38] studied the effect of E-glass/BMI composites immersed in sea water maintained at room tempearature, 50° C and 80° C. The sample is aged in the sea water and the samples are tested for Mode I and II and mixed mode of interlaminar strength. The fracture toughness,R curve and its glass transition temperature are noted.Also J integral method is employed to determine the delamination length to avoid the discrepancy through naked eye inference.Its been concluded by the author that there is loss in properties at 80° C due to plasticisation of the matrix with initial resistance to delamination due to the same. The viscoelastic response of the E-glass/BMI [39] is studied by the author and showed that below 50°

C , the moisture absorbed dried and cured specimen rebound its Tg and its flexural properties as that of the dried sample.

**Ping Liu** et al [45]modified the BMI/DABA resin with polysiloxane in suitable amount and reinforced with glass fiber. The composite is tested for is wettability of resin in fiber, dielectric properties and interfacial adhesion.The resultant Glass fiber/modifies BMI (G/mBMI) showed an increase of 20-40% increased interfacial adhesion. The modified resin has Si-OH groups in polysiloxane increasedbthe wettability of resin to fibers.

## 2.3 Kevlar Fiber Reinforced modified bismaleimide composites:

**T.K.Lin** et al [42] studied the effect of Kevlar fiber treated with chlorosulfonation/allyamine with three different bismaleimides of which BMIPP has provided proven results with IFSS at 0.2% for 150s with chlosulfonic acid treatment.At higher level of acid with greater tiem of exposure the composites yielded lesser results.The T peel strength has improved due to ethylenediamine introduction into the fiber, increases the reaction sites with BMI due to the presence of amino group in it.

**Min Su** et al [43] studied the effect of Kevlar fiber treated with oxygen plasma for its better bonding and adhesion with BMI matrix by controlling the plasma treatment power and time and optimized it. The plasma treatment Kevlar fiber/BMI has proven results in ILSS, dielectric properties and its moisture resistance.the author has given 700 W for 5 min as the optimal value for Kevlar plasma treatment.

**Wei Chen** et al, [44] investigated a model BMI composite with carbon nanotubes grafted Kevlar fiber for its mechanical performance. The Tensile strength and ILSS of the modified fiber has increased by 12% and 30% respectively than the original fibers.

## 2.4 Hybrid Fibers/ Novel fibers Reinforced modified bismaleimide composites:

Author **Dong Liu et al [5]** reported that, the radio frequency oxygen plasma was used to strengthen the interfacial adhesion of PBO fiber/ bismaleimide composite. The interlaminar shear strength (ILSS) was improved to 57 MPa or by 29% after treatment at 50 W. Oxygen atoms and polar groups were increasingly introduced onto the fiber surface, which contributed much to better wettability. Physical interactions between the fiber and the matrix might play leading roles in strengthening the adhesion. The modification effects did not degrade within the storage time of 1-50 days, indicating that the oxygen plasma created stable structures on the fiber surface. The composites had low water absorptions and high retentions of more than 90% in the ILSS, therefore the composites had good humid resistance before and after treatment.The author also studied the argon plasma treated .

**Hao Jiang** et al,[35] studied the effect of curing initiators Dicumyl peroxide (DCP) or hexamethylenetetramine (HTMA) on BMI /Allyl phenol (BAN) with and without PVA fibers oin reducing the curing time as well on its mechanical properties. The Results revealed that the BAN/ PAN showed less curing time with the addition of DCP along with the increased Flexural strength when compared with the HTMA initiator. Also the BAN/PVA fiber with DCP has reduced viscosity and hence improved the processability of the resin system with no holes/bubbles in the cured matrix.

**L.R. Grace** et al [41] studied moisture adsorption behaviour of Quartz fiber/BMI composite with three different ply of six, twelve, forty where they deviated from fickian diffusion model but well agreed with 3D HDM model irrespective of the laminate thickness. The diffusion occurs 13 times faster along the edges of the fiber rather than through the thickness. The recovery of moisture absorption parameters after 16.5 months before reaching the saturation is also done.

**Joung-Man** et al [31] studied the Interfacial shear strength (IFSS) of the carbon, SiC, Glass fiber/BMI with two stage and three stage curing by conducting fragmentation test on tensile and compressive test specimens with acoustic emission(AE) monitor and obtained the stress-strain curve. The three stage curing of the fibers with BMI showed an increased IFSS with a wider vertical micro cracks. The tensile mode of failure is a sharp diamond in three stage curing when compared to the dull failure observed in 2 stage curing due to its incomplete curing whereas in the compression, there is no matrix failure but the slippage of carbon fiber occurred. In the case of Glass and sic fiber there is no fiber fracture due to its high strain to failure ratio. With AE the amplitude distribution of separated and closer one in tension and compression is found respectively.

**Nannan Ni et al,(2015)[13]** reported that a new type of hybrid composite with aramid nonwoven fabrics (ANF) modified with semi-crystalline polyvinylidene fluoride (PVDF) between carbon-fibre reinforced bismaleimide (BMI) laminate which improved the damping properties with higher flexural modulus and loss factor simultaneously improving mechanical properties. The interlaminar shear strength of the co-cured composite with PVDF film was 26.2% more than that of the control sample.The flexural Strength and flexural modulus is 4.1% and 10.8% more than that of control samples.

**Y.Z.Wan** et.al [32] studied the Mechanical performance of hybrid bismaleimide composites reinforced with three dimensional, hybridized carbon(T300) and Kevlar 49 fabrics.The mechanical properties like load – displacement behaviour, flexural strength and modulus, shear strength and impact properties of the composite system were studied in room temperature and dry condition. The experimental study revealed that the flexural property increased with loading of carbon in carbon-Kevlar in the ratio of 3:2 and then dropped. The maximum strength of 665±33.3MPa and highest flexural modulus of 42.5±1.4GPa at carbon-Kevlar ratio of 3:2 was observed.

## 3 Conclusion:

The potential modification of the resin and its incorporation in different fiber with an extent of different manufacturing processes opened a greater scope of future work for researchers in developing Novel material as well as Novel composites for advanced applications.

## References:

1. K. Ambika Devi, C. P. Reghunadhan Nair & K. N. Ninan.*Studies on bismaleimide co-cured Novolac epoxy–diallyl bisphenol-A system*, Composite interfaces Vol. 15, pp. 807–827 (2008).

2.Nannan Ni , Yuefang Wen, Delong He , Xiaosu Yi , Chengcheng Wang , Yahong Xu.*Synergistic reinforcement effect of aramid nonwoven fabrics and PVDF on mechanical and damping properties of bismaleimide matrix composites*, Composites: Part A 79 (2015) 176–182.

3.John Montesano , Marina Selezneva, Zouheir Fawaz, Cheung Poon, Kamran Behdinan.*Elevated temperature off-axis fatigue behavior of an eight-harness satin woven carbon-fiber/bismaleimide laminate*, Composites: Part A 43 (2012) 1454–1466.

4.Li Yuan, Aijuan Gu, Guozheng Liang, Xiaoyan Ma and HongxiaYan ,*Properties of Fiber Reinforced Bismaleimide(BMI)/O,O'-diallyl bisphenol A(BA)/Organic Rectorite(OREC) Composites* , 2009.

5.Dong Liu, Ping Chen, Mingxin Chen, Zhe Liu , *Improved interfacial adhesion in PBO fiber/bismaleimide composite with oxygen plasma plus aging and humid resistance properties* , 2012,532,76-83

6.X.J Xian and C.L. Choy. *Fatigue Fracture behaviour of carbon-fiber reinforced modified bismaleimide composites*, 1994.

7.L F M da Silva and R D Adams ,*Effect of temperature on the mechanical and bonding properties of a carbon-reinforced bismaleimide*, J. Materials: Design and Applications2007, , Vol. 222 216–226

8.P T. CURTIS, *An Investigation of the Mechanical Properties of Improved Carbon Fibre Composite Materials*, Journal of COMPOSITE MATERIALS, Vol 21. 1987.

9.Michelle Leali Costa, Sérgio Frascino Muller de Almeidab , Mirabel Cerqueira Rezendea ,*Hygrothermal Effects on Dynamic Mechanical Analysis and Fracture Behavior of Polymeric Composites*,Materials Research, April 6, 2005,Vol. 8, 335-340 .

10.Shen Chou, Hong-Chu Chen &Chorng-Cherng Wu, *BMI resin composites reinforced with 3D carbon-fibre fabrics* , Composites Science and Technology 43 (1992) 117-128 .

11.R. Chandra & L. Rajabi, *Recent Advances in Bismaleimides and Epoxy- Lmide/ Bismaleimide Formulations and Composites*, Journal of Macromolecular Science, Part C,2006, 37:1, 61-96.

12.Divya K. Chakravarthi , Valery N. Khabashesku , RanjiVaidyanathan , Jeanette Blaine, ShridharYarlagadda , David Roseman , QiangZeng and Enrique V. Barrera ,*Carbon Fiber–Bismaleimide Composites Filled with Nickel-Coated Single-Walled Carbon Nanotubes for Lightning-Strike Protection*, Adv. Funct. Mater. 2011, 21, 2527–2533.

13.Vernon T. Bechel, Mark B. Fredin, Steven L. Donaldson, Ran Y. Kim, John D. Camping,*Effect of stacking sequence on micro-cracking in a cryogenically cycled carbon/bismaleimidecomposite*,Applied science and manufacturing : Part A 34 (2003) 663–672.

14.Li-RongBao , Albert F. Yee , *Moisture diffusion and hygrothermal aging in bismaleimide matrix carbon fiber composites: part II—woven and hybrid composites*,
Composites Science and Technology,62 ,(2002), 2111–2119.

15.M. Akay , G.R. Spratt ,*Evaluation of thermal ageing of a carbon fibre reinforced bismalemide*,Composites Science and Technology 68 (2008) 3081–3086.

16.Yuan-Fei Fu , Kan Xu , Zhao-Yao Sun , Fu-Qiang Zhang, De-Min Chen & Jing Li, The *Effect of Silane Surface Treatment of Carbon Fiber on the Tribological Properties of Bismaleimide(BMI) Composite*, Polymer-Plastics Technology and Engineering, July 2012.

17.M. I. Faraz, S. Bhowmik,C. De Ruijter, F. Laoutid, R. Benedictus, Ph. Dubois,J. V. S. Page, S. Jeson, *Thermal, Morphological, and Mechanical Characterization of Novel Carbon Nanofiber-Filled Bismaleimide Composites,* Journal of Applied Polymer Science, Vol.117,2159-2167(2010).

18.Satoshi Kobayashi, Nobuo Takeda,*Experimental characterization of microscopic damage behavior in carbon/bismaleimide composite—effects of temperature and laminate configuration*,Composites: Part A 33 (2002) 1529–1538,August 2002

19.Pei Sun, Yan Zhao, Yunfeng Luo, Lili Sun, *Effect of temperature and cyclic aging on the interlaminar shear strength of Carbon fiber / Bismaleimide (BMI) composites*.Materials and design 32 (2011) 4341-4347.

20. Toshiyuki Shimokawa, Yoshiaki Kakuta and Takenori Aiyama, *G40-800/5260 Carbon Fiber/Bismaleimide Composite Material: High Temperature Characteristics of Static and Fatigue Strengths*,Advances in Composite materials-Ecodesign and Analysis,

21. E. Drukker, A.K. Green, G.Marom,*Mechanical and chemical consequences of through thickness thermal gradients in polyimide matrix composite materials*, Composites: Part A 34 (2003) 125–133, December 2002.

22. C. L. Heisey, P. A. Wood, J. E. McGrath & J. P. Wightmany, *Measurement of Adhesion Between Carbon Fibers and Bismaleimide Resins*,The Journal of Adhesion, 1995, vol.53,pp 117-147.

23. Bor Z. Jang, *Control of interfacial adhesion in continuous carbon and Kevlar fiber reinforced polymer composites*, Composites Science and Technology 44 (1992) 333-349, August 1991

24. Sergey Evsyukov, Ronald Klomp-de Boer, HD Stenzenberger, Tim Pohlmann and Matthijs ter Wiel., *"A new m-xylyene bismaleimide- based high performance resin for Vacuum –assisted infusion and resin transfer molding"*, Journal of Composite Materials 2019 ,0(0),1-10,

25. Yuan-Fei Fu , Kan Xu , Zhao-Yao Sun , Fu-Qiang Zhang, De-Min Chen & Jing Li, *The Effect of Silane Surface Treatment of Carbon Fiber on the Tribological Properties of Bismaleimide(BMI) Composite*, Polymer-Plastics Technology and Engineering, July 2012.

26. SiyangLiu,Yizi Chen, Ping Chen, Dongwei Xu, Jing Wang, *"Properties of Novel Bismaleimide Resins and Thermal ageing effects on the ILSS performance of their carbon fiber-Bismaleimide composites"* ,Polymer composites -2018, DOI 10.1002/pc

27. M. Simpson, P.M. Jacobs and F.R. Jones. *Generation of thermal strains in carbon fibre-reinforced bismaleimide (PMR-15) composites*, volume 22, number 2,1991.

28. S.P.Wilkinson,T.C.Ward and J.E.Mcgrath (1992),*Effect of thermoplastic modifier variables on toughening a bismaleimide matrix resin for high performance composite materials*,Polymer 1993,Volume 34, Number 4

29. Jing Zhou, Yingguang Li, Nanya Li and Xiaozhong Hao. *Enhanced interlaminar fracture toughnessof carbon fiber/bismaleimide compositesvia microwave curing*,Journal of Composite Materials,2016,0(0),1-11

30.R.Biju and C.P.Reghunadhan Nair, *High Transition temperature shape memory polymer composites based bismaleimide resin*, High performance polymers I-II 2013.

31.Joung-Man Park,Jin-Won Kim, Dong-Jin Yoon, *Interfacial evaluation and microfailure mechanisms of single carbon fiber/bismaleimide (BMI) composites by tensile and compressive fragmentation tests and acoustic emission*, composites sciencc and Technology 62 (2002) 743-756.

32. Y.Z.Wan, Y.L.WANG,F.He, Y.Huang,H.Jiang, *Mechanical performance of hybrid bismaleimide composites reinforced with three dimensional braided carbon and Kevlar fabrics,* Composites: Part A 38 (2007) 495-504

34.Lirui Yao, Min Li, Qing Wu,Zhishuang Dai, Yizhuo Gu, Zuoguang Zhang, *Comparison of sizing effect of T700 grade carbon fiber on interfacial properties of fiber/BMI and fiber/epoxy*, Applied surface science 263 (2012) 326-333.

35.Hao Jiang, Rumin Wang, Shameel Farhan, Shuirong Zheng, *Properties of Dicumyl peroxide Initiated Bismaleimide modified Allyl Novolac and its Molded Composites*,Polymer composites 2016, DOI 10.1002/pc

36.Anna Razgon and Chaim N. Sukenik. *Ceramic coatings for fiber matrix composites: Titania thin films on bismaleimide-glass fiber composites*, J. Mater. Res., Vol. 20, Sep 2005.

37.Yehai Yan, Xianming Shi, Jinge Liu, Tong Zhao, Yunzhao Yu, *Thermosetting Resin system nased on Novolak and bismaleimide for Resin-Transfer Molding*, Journal of Applied Polymer Science, Vol.83, 1651-1657 (2002).

38.Yian Zhao, Wei Liu, Leong Keey Seah, Gin Boay Chai, *Delamination growth behavior of a woven E-glass/bismaleimide composite in sea water*, Composites Part B 106 (2016) 322-343.

39.Zhao Yian, Wang Zhiying, Seah Leong Keey, Chai Gin Boay, *Long-Term Viscoelastic Response of E-glass/Bismaleimide composite in seawater Environment*,Applied Composite material, DOI 10.1007/s 10443-014-9431-2.

40.Qi Yu, Ping Chen, Yu Gao, Jujie Mu, Yongwu Chen, Chun Lu, Dong Liu, *Effects of Vacuum thermal cycling on mechanical and physical properties of high performance carbon/bismaleimide composite*, Materials Chemistry and Physics 130 (2011) 1046-1053.

41. L.R. Grace, M.C. Altan, *Three-Dimensional Anisotropic Moisture Absorption in Quartz-Reinforced Bismaleimide Laminates*, Polymer Engineering and Science- 2014.

42.T.K. Lin, S.J. Wu, J.G. Lai, S.S. Shyu, *The Effect of chemical treatment on reinforcement/matrix interaction in Kevlar-fiber/bismaleimide composites*,Composites Science and Technology 60 (2000) 1873-1878.

43.Min Su, Aijuan Gu, Guozheng Liang, Li Yuan, *The effect of oxygen-plasma treatment on Kevlar fibers and the properties of Kevlar/bismaleimide composites*, Applied Surface Science 257 (2011) 3158-3167.

44.Wei Chen, Xin-Ming Qian, Xue-Qiu He, Zhen-Yi Liu, Ji-Ping Liu, *Surface Modification of Kevlar by Grafting Carbon Nanotubes*, Journal of Applied Polymer Science , Vol.123, 1983-1990(2012).

45.Ping Liu, Qingbao, Aijuan Gu, Guozheng Liang, Li Yuan, Jianfei Chang, *Interface and its effects on interlaminate shear strength of novel glass fiber/hyperbranched polysiloxane modified maleimide-triazine resin composites*, Applied Surface Science 258 (2011) 572-579.

46. K. Ambika Devi, C. P. Reghunadhan Nair & K. N. Ninan.*Bismaleimide modified epoxy–diallyl bisphenol system- Effect of Bismaleimide Nature on Properties*,Polymer and Polymer CompositesVol. 17, No. 3 (2009).

47. Gu Aijuan &Liang Guozheng (1997), *Effects on dielectric properties of modified Bismaleimide Resins*, International Journal of polymeric materials and polymeric Biomaterials, 35:1-4,29-37.

48. K.Dinakaran, R,Suresh Kumar, and M.Alagar, *Bismaleimides (N,N' – Bismaleimide- 4-4'Diphenylmethane and N,N' – Bismaleimide- 4-4'Disulphone) modified Bisphenoldicyanate epoxy matrices for Engineering Applications*,Materials and Manufactuirng Processes, 20:299-315, 2005

49. Rajangam Rajasekaran, Muthukaruppan Alagar, *Bismaleimides modified polyethersulfone/ epoxy matrices for engineering applications,* Journal of composite Materials, Vol 42, No.10/2008

50. R.Suresh Kumar, M.Alagar (2005),*Studies on Mechanical,Thermal and Morphology of Diglycidylether- Terminated Polydimethylsiloxane modified Epoxy-Bismaleimide Matrices*, Journal of Applied polymer Science, Vol 668-674(2006)

51. A.Ashok kumar, M.Alagar, R.M.V.G.K.Rao, *Preparation and Characterization of Siliconized Epoxy/Bismaleimide (N-N'- Bismaleimido- 4-4' diphenyl methane) Intercrosslinked Matrices for Engineering Applications*,Journal of Applied Polymer Science, Vol. 81,38-46(2001)

52.Whan Gun Kim,Tae Young Nam, *Curing characteristics of o-cresol Novolac Epoxy Resin modified by Bismaleimide*.Jornal of polymer Science: Part A: Polymer Chemistry Vol 34,957-962 (1996).

53.A.Vanaja, R.M.V.G.K.Rao, *Synthesis and Characterisation of epoxy-novolac/bismaleimide networks*, European Polymer Journal 38 (2002) 187-193.

54.Yan Hongxia, Li Pengbo, Zhang Jungping, Ning Rongchang,*Mechanical Properties of Novel Bismaleimide Nanocomposites with Si3N4 Nanoparticles*. Journal of Reinforced plastics and composites, vol.29,no.10/2010.

55.Joshua W.Ehrlich, LaNetra C.Tate, Sarah B. Cox, Brian J.Taylor,M.Clara Wright, Anne J.Caraccio, Jeffery W.Sampson, *Process Optimization of Bismaleimide(BMI) Resin Fiber Composite*, NASA, Kennedy Space Center , FL 32899.

56.Xiangyu Liu, Ling Li, Zibing Chen, Xianfa Duan, Yongijan Yu and Lizhuan Sun, *Curing behavior, thermal and mechanical properties of N-N'(4-4'-diphenylmethane)bismaleimide/2-2'-diallyldisphenol A/3-allyl-5,5-dimethylhydantoin resin system*, High performance polymers1-14,2019.

57.Mohammad H.Haque, Priyank Upadhyaya, Samit Roy, Taylor Ware, Wlater Voit, Hongbing Lu, *The changes in flexural properties and microstructure of carbon fiber bismaleimide composite after exposure to a high temperature*, Composite Structures 108 (2014) 57-64.

58.Xuhai Xiaong, Ping Chen, Jinxiang Zhang, Qi Yu, Baichen Wang, *Preparation and properties of High Performance Pthalide containing Bismaleimide modified Epoxy Matrices*, Journal of Applied Polymer Science , DOI 10.1002/app.33588.

IOP Conf. Series: Materials Science and Engineering **923** (2020) 012052 doi:10.1088/1757-899X/923/1/012052

# Development of basalt based nanoparticles reinforced composites

**S Saravanan[1]*, N Parkunam[1], G Navaneethakrishnan[1], K Karthikeyan[2], S T Jaya Suthahar[3].**

[1]Department of Mechanical Engineering, K.Ramakrishnan College of Technology, Tichy-621112, India.
[2]Department of Mechanical Engineering, CK College of Engineering and Technology, Cuddalore-607003, India.
[3]Department of Mechanical Engineering, University College of Engineering, Anna University, Thirukkuvalai-610204, India.
*Corresponding Author: nanosaran007@gmail.com.

**Abstract.** Epoxy resin is used in a various fields, from transport to automotive industrial components due to their outstanding mechanical and chemical indisposition properties, the formulation is very easy and their cost also very low. Basalt fiber is very popular because of its ease in manufacturing and having good mechanical properties. To ensure the proper dispersion, the epoxy filled filler mixed with the hardener using a mechanical stirrer. The composites properties such as density, tensile, compressive, impact strength, hardness and wear rate required dimensions as per the ASTM standards. The wear rates are decreased with increased in TiC weight percentage in basalt fiber, because of high hardness. The values of wear rates are started from 2.179 to 1.337 m3/Nm.
Keywords: Basalt fiber, Epoxy, TiC, Hardness, Wear rate.

## 1. Introduction

Polymer composites are frequently used in various engineering fields as well as structural components where these are subjected to wear. Wear resistance involved in polymer composites is an important area in various applications such as automobile, marine and mining industries. This is due to the eco friendly and reusability of these composites. Among the various natural fibers, basalt fiber is very popular because of its ease in manufacturing and having good mechanical properties. For getting best filler to matrix ratio, filler shape and size and the well suited interfacial bonding between the filler and matrix, the composite properties can be improved number of fold as compared to monolith polymer materials [1]. Thus, the designing great materials, the selection of fillers and loadings are the crucial process for this consideration by the fabricator. The investigations carried out in frictional and wear properties of polymer composites with fibers combined with organic and inorganic fillers [2, 3]. Incorporating of nano fillers in the polymers increased for the past few years has led to the exigency of these materials in tribological application. Nano fillers not only change the material properties but also tend to the significant improvement in thermal, mechanical and abrasive wear resistance of the filler filled polymer composites. Various researchers [1-6] investigated the tribological properties of nano fillers in polymers and analyzed the mechanical and tribological performance of filled composites. Mursel Ekrem et al [5] investigated the effects of epoxy resin with boron nitrate Nano particles on friction and wear behaviour. The tribological behaviour of epoxy resin was investigated by determined the addition of optimum percentage of boron nitride. Incorporation of boron nitride nano platelets resulted in improvement in wear performance of epoxy resin. Upadhyay et al [4] improved the tribological behaviour of epoxy matrix by mixing fullerene C70 and multiwalled carbon nano tubes. They concluded that the wear rate is in the range of $10^{-3}$-$10^{-2}$ $mm^3$/N-m. The incorporation of $Ti_3SiC_2$ and graphite to PTFE composites reveal better tribological and mechanical properties than unfilled matrix. In the $Ti_3SiC_2$/PTFE composites the addition of only one weight percent of $Ti_3SiC_2$ improved the wear resistance by two orders of magnitude [7]. Due to the relative motion between the surface and the substances, the increment in the loss of materials ends with wear damage in solid surface [11]. The tribological performance of the short fiber-reinforced polymers filled with nano particles was investigated. It was found that the load-carrying capacity of the nano-sized $TiO_2$ particles was improved due to their friction reducing abilities of the nano particles [8]. The inclusion of nano silica particles into epoxy polymer is currently one of the most common ways to make polymer nano composites. Aviles et al. [12] investigated wear behaviour of polymer matrix composites reinforced with non-oxide ceramic called TiC which having the better wear, high strength and hardness and corrosion resistance. The TiC has the better filler material used in polymer composites because of this properties make the composite useful in the various applications like cutting tools, metal forming dies and wear resistant parts. The work on mechanical and wears behaviours of nano TiC filled basalt epoxy composite are not done previously and it is clearly shown from the literature review [9-15]. In view of literatures mentioned above, the study of mechanical and wear behaviour of TiC Nano particles incorporated with epoxy and basalt fiber is investigated to know the clarification in wear performance and other mechanical properties

## 2. Experimental work

Basalt woven fabric (diameter 20 μm) was obtained from Austria and used as reinforcement as shown in Figure 1a. Multifunctional epoxy and hardener (HY951) are used as a matrix for produce the composite. TiC nanoparticles are added to the epoxy for improving the mechanical properties of composites. The hardener was mixed into the nanoparticles filled epoxy resin using a mechanical stirrer to ensure the proper dispersion. The 10:1 weight percentage was followed for the epoxy and hardener mixing. To enhanced the consolidation and also the voids and spots reduction the epoxy was smeared manually on to the basalt fabric as described elsewhere [13]. The curing and post curing were done at the room temperature for 24 hours under the pressure of 14 psi and 3 hours at 100 $^0$C. The details of the various percentages of TiC nanoparticles reinforced composites such as Sample A (0 wt. % TiC), Sample B (1 wt. % TiC), Sample C (3 wt. % TiC), Sample D (5 wt. % TiC) and Sample E (7 wt. % TiC). The laminate of dimensions 300×300×3 mm was fabricated (Figure 1b) and the specimens for the properties such as density, tensile, compressive, impact strength, hardness and wear test required dimensions as per the ASTM standards. As per the ASTM-G65 standard, the wear test specimens are prepared for conducting the three-body abrasive wear test at a rotational speed 300 rpm, load 10 N and distance 100 m. The digital electronic balance weighing machine was used to record the weight before it was mounted on the specimen holder. The initial and final weight of the specimen gives the wear loss by its difference. Repeatability of the test data ensured by conducting minimum of three trials.

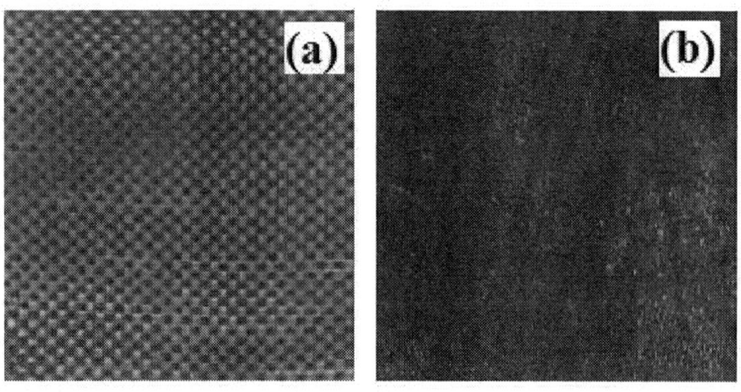


**Figure 1**(a) Basalt fiber (b) Basalt nanocomposites.

## 3. Results and discussion

IOP Conf. Series: Materials Science and Engineering **923** (2020) 012052 doi:10.1088/1757-899X/923/1/012052

Figure 2(a) shows densities of the TiC reinforced basalt nanocomposites. The knowledge of void content is confirming the quality of the composites and it is clear that the fewer voids composites are the good one for get the best mechanical properties. However, done the fabrication in hand lay-up, the voids are unavoidable and density values are decreased but in this work the density values are slightly increased with increase in weight percentage of TiC, because of fewer voids. The tensile strength of the TiC based nanocomposites is more effective than the other nanocomposites. The values are in the range of 330 to 354 MPa, due to uniform distribution of TiC nanoparticles in epoxy and basalt fiber. The Figure 2(b) shows the perfect result of improvement on tensile strength in the 5 wt. % of TiC and then decreased. Figure 2(c) shows the compressive strength of the basalt reinforced nanocomposites. The compressive strength of the composites are clearly shows the improvement up to 5 wt. % of TiC. This is because of the less cracks and flaws present in the composites. The impact strength was totally co-relate with its overall toughness. The presence of hard reinforced particles in basalt fiber improves the impact strength of the nanocomposites in Figure 2(d). So, the values in the range of 4.44 to 7.02 J are clearly shows the best result in the field of development of impact strength in nanocomposites. The optimized value of impact is got at the TiC weight percentage of 5. After increase the weight percentage of TiC from 5 to 7, the impact strength was slightly down to 5.44 J. This will indicates the poor impact strength occurred for added more than 5 wt. % of TiC in nanocomposites. The high hardness of TiC nanoparticles improves the hardness value of nanocomposites as shown in Figure 2 (e). The hardness values are in the range of 65 to 75 Hv. This will shows the improvement of hardness value up to 5 wt. % of TiC. The wear rates are decreased with increased in TiC weight percentage in basalt fiber, because of high hardness. The values of wear rates are started from 2.179 to 1.337$m^3$/Nm. Due to the surface process of TiC is harder than matrix, the higher wear resistance in nanocomposites. The matrix and TiC particles are well bonded and contribute to increase the wear resistance. The wear surface encloses of shear deformed matrix covers broken TiC particles of the metallic surface and act as a transfer layer.

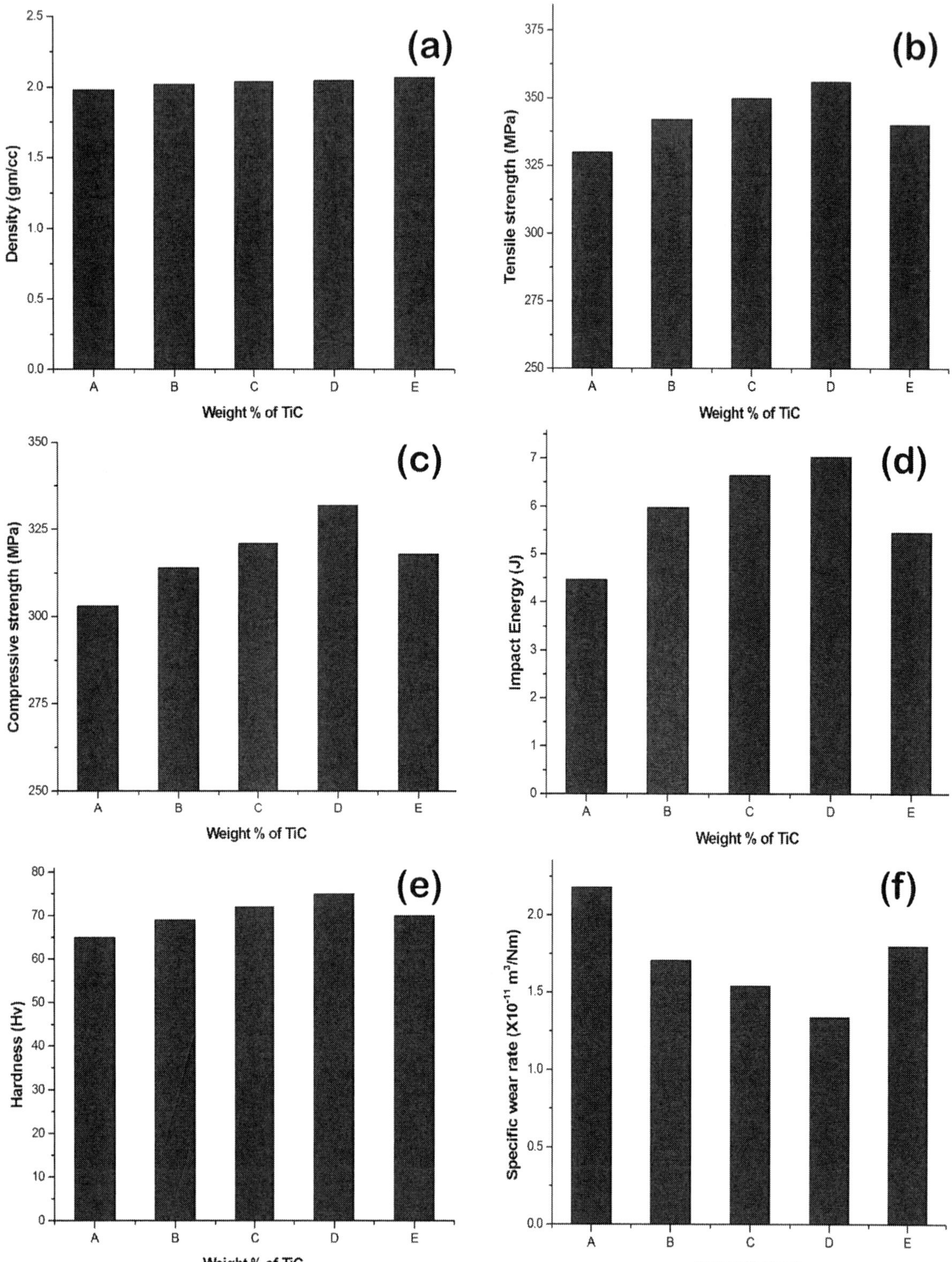


**Figure 2** (a) Density, (b) Tensile strength, (c) Compressive strength, (d) Impact strength, (e) Hardness, (f) Specific wear rate vs weight % of TiC.

## 4. Conclusion

From development of basalt based nanoparticles reinforced composites studies, the following conclusions are drawn

- The mechanical properties of the nanocomposites has linearly increased up to % and then decreased.
- The homogenous distribution of TiC particles in the matrix has improved the properties.
- The wear rates are decreased with increased in TiC weight percentage in basalt fiber, because of high hardness.
- The abrasive dominant wear mechanisms of these nanocomposites were observed.

## References

[1] Fiore V, Scalici T, Di Bella G, Valenza A. 2015 A review on basalt fibre and its composites, *Compos Part B Eng* **74**:74–94.

[2] Lingaraju, D., Ramji, K., Devi, M.P., Lakshmi, U.R. 2011 Mechanical and Tribological Studies of Polymer Hybrid Nanocomposites with Nano Reinforcements. *Bu. Mat. Sci.,* **34(4)**, 705–712.

[3] Dhand V, Mittal G, Rhee K.Y, Park S-J, Huic D. 2015 A short review on basalt fiber reinforced polymer composites, *Compos Part B Eng* **73**:166–180.

[4] Upadhyay R K, Kumar A, 2018 A novel approach to minimize dry sliding friction and wear behavior of epoxy by infusing fullerene C70 and multiwalled carbon nanotubes, *Tribol. Int.* **120**: 455–464.

[5] Mursel Ekrema, Hayrettin Duzcukoglub, Muhammet A. Senyurtc, Omer S Sahinc, Ahmet Avcıc, 2017 Friction and Wear Performance of Epoxy Resin Reinforced with Boron Nitride Nano platelets *J. Tribol* **140(2)**: 022001.

[6] Banibayat P, Patnaik A, 2014 Variability of mechanical properties of basalt fiber reinforced polymer bars manufactured by wet-layup method, *Mater Des.,* **56**: 898–906.

[7] Dehkordi M T, Nosraty H, Shokrieh M M, Minak G, Ghelli D, 2013 The influence of hybridization on impact damage behavior and residual compression strength of intraply basalt/nylon hybrid composites, *Mater Des.,* **43**: 83-90.

[8] Hemanth R, Suresha B, Sekar M, 2014 Physico-mechanical behaviour of thermoplastic co-polyester elastomer/polytetrafluroethylene composite with short fibers and micro fillers, *J Comp Mat.,* **49**: 2217-2229.

[9] Kumar S, Chauhan S R, 2012 Mechanical and Dry Sliding Wear Behavior of Particulate Fillers $CaCO_3$ and $CaSO_4$ Filled Vinyl ester Composites. *Int J Comp Mat,* **2(5)**: 101-114.

[10] Chang, Q., Wang, K., and Zhaoc, H. 2013 The Friction and Wear Properties of Short Carbon/CNT/PA6 Hybrid Composites under Dry Sliding Conditions, *Int J Poly Mat Bio.,* **62**: 540-543.

[11] Saurín N, Avilés M D, Espinosa T, Sanes J, Carrión F J, Bermúdez M D, Iglesias P, 2017 Carbon nanophases in ordered nanofluid lubricants, *Wear* **376-377**:747–755.

[12] Avilés M D, Carrión F J, Sanes J, Bermúdez M D, 2018 Effects of protic ionic liquid crystal additives on the water-lubricated sliding wear and friction of sapphire against stainless steel. *Wear* **408–409**: 56–64.

[13] Wei B, Cao H, Song S, 2010, Environmental resistance and mechanical performance of basalt and glass fibres, *Mater Sci Eng A-Str.,* **527**: 4708-15.

[14] Jiang Y F, Liu H W, Jia T F, 2014 Improving mechanical properties of wood composites using basalt glass powder and basalt fiber, *Adv Mater Res,* **1048**: 400-403.

[15] Huaiyuan W, Li C, Xiaoshuang Y, Lixiang Y, Lin Y, Yanji Z, Andrew T H, Andrew I M, Patrick T, Klaus F, 2014 Anisotropy in tribological performances of long aligned carbon nanotubes/polymer composites, *Carbon,* **67** :38–47.

# Tribological behaviour of aluminum alloy (AA7075) based hybrid composites

**S Saravanan[1]*, A Sreenithi[2], M Ajitha[2], R Ilakkiyal[2], S R Narmatha[2], S Rajkumar[2], K Karthikeyan[3].**

[1]Department of Mechanical Engineering, K.Ramakrishnan College of Technology, Tichy-621112, India.
[2]Department of Aeronautical Engineering, Srinivasan Engineering College, Perambalur-621212, India.
[3]Department of Mechanical Engineering, CK College of Engineering and Technology, Cuddalore-607003, India.

*Corresponding Author: nanosaran007@gmail.com.

**Abstract.** All over the world, numerous applications in the field of aerospace and structural constructions are carried out using metal matrix composites as a suitable alternate for traditional materials. The preparation of AA7075/$TiB_2$/$MoS_2$ hybrid composites was carried by stir casting technique. The mechanical properties of the fabricated hybrid composites have been improved by the addition of reinforcement. The aerospace applications involve high load bearing capacity and corrosive resistance in their components. Wear performance of the hybrid composite was examined by using pin-on disc machine for identifying influential process parameters and optimum results of lower wear rate were observed for the combination of higher reinforcement. The SEM image of wear surface analysis of hybrid composites describes the abrasive wear mechanism.
Keywords: AA7075, $TiB_2$, $MoS_2$, stir casting, wear rate.

## 1. Introduction

Nowadays composites are progressively more used for structural applications since of their outstanding strength to weight ratio and attractive properties. When two or more materials with hugely varying physical and chemical properties are mixed to fabricate another material with the significant character change from base materials, then the fabricated material is known as composite materials [1-5]. Aluminum is the most bounteous metal and forms around 8 % of the worlds outside layer. It happens normally as Bauxite which contains 30-50 % hydrated alumina. AA7075 (Al–Zn–Mg–Cu) which combines in definite proportions to form $Mg_2Si$, making them heat treatable, it posses good strength, formability, weldability and corrosion resistance apart from low cost of processing [6-9]. It is a weightless metal with a density of 2700 kg/m$^3$. The superior extrudable property, corrosion resistance and prolonged annealing time make AA7075, to be used in structural, marine and automotive applications [10-12]. Surappa [13] studied aluminium composites has more strength to weight ratio in the materials. Gopalakrishnan et al [14], has explained the titanium carbide reinforcement AA6061composite by in-situ reaction techniques. In the addition of TiC particles, the hardness of the AA6061 composites has improved. The Al-$B_4$C composites were examined the mechanical properties including UTS, YS, % of elongation and hardness [10]. Titanium diboride ($TiB_2$) is a white solid inorganic substance is preferred due to its good wettability, high hardness, melting point (3230 °C), density (4.52 g/cm$^3$), poisson's ratio (0.27), and excellent wear resistance, mechanical and chemical stability [11]. Molybdenum disulfide ($MoS_2$) is similar to graphite and it widely used as a dry lubricant because of its low friction and robustness. Rai et al [17] fabricated in situ aluminium titanium carbide composites with $K_2TiF_6$ and graphite powder at 1200 $^0$C. In stir casting process the occurrence of titanium carbide were isolate to the matrix grain edge. The synthesised Al-5Ti-0.8C and Al-5Ti-1.2C alloys with titanium carbide particles for the grain enhancement of Al-7Si alloy [16]. The Al-$TiB_2$ composites to adding of Al-Ti-B powder to the liquefy aluminum at 900$^0$C and deliberated the method of progress of $TiB_2$ [19]. The commonly used reinforcement is silicon carbide particulates in cast alloy matrix (Aluminium alloys 7075) and zirconia particulates in wrought alloy matrix [20]. Albiter et al., fabricated and heat treated the composites. By doing so, enslavement on the ceramic content is well-built [12]. Later than heat treatment, the 55 % titanium carbide composites illustrated a raise of hardness from 28.5 to 38.5 HRC; for now ultimate tensile strength enhanced from 379 to 480 MPa. Ranjit Bauri has explained aluminum titanium carbide composites are developed at various temperatures from 700 °C to 1200 °C and with various TiC percentages to get the outcome of these two factors on the development of titanium carbide particles [15]. The temperature has raised the blocky nature of $Al_3Ti$ which varies to needle like representing let go of Ti to a better level and as an outcome more and more titanium carbide particles form. The composites are produced through self-propagating high-temperature synthesis method through a casting method supported by self-propagating high-temperature synthesis; fluxes were used to make active in effect in the aluminum melt and support superior Ti and C yield in the composites [18]. From the above literature process the limited work has progressed in AA7075 hybrid composites. In this research, the effect of molybdenum disulfide and titanium diboride on heat treatable aluminum alloy 7075 composites were synthesised by stir casting techniques.

## 2. Experimental work

Figure 1 shows the stir casting process, AA7075 rod was made to melt in a closed graphite crucible of an electrical resistance furnace for to prevent oxidation during the casting process and the molten metal was sustained at a temperature of 750°C. The required quantity of $TiB_2$ particles were accurately weighed and mixed in the ratio of 0 (Sample A), 1(Sample B), 3(Sample C) and 5 (Sample D) wt. % with 1 wt. % $MoS_2$ on AA7075. The $TiB_2$ particles are preheated at 1000 °C to remove wetness and introduced into the molten metal at 750 °C to get unvarying and fine $MoS_2$ particles. The stirring of the mixture was performed for 3-5 minutes with a motor controlled stirrer. Then the molten metal was poured into mould and the casted samples are machined as per ASTM standards for conducting mechanical and wear test. Based on the available literature, the 9.81 N load, 1000 m sliding distance and 1 m/s sliding velocity were identified and used and wear test performed in pin on disc setup as shown in Figure 2 .

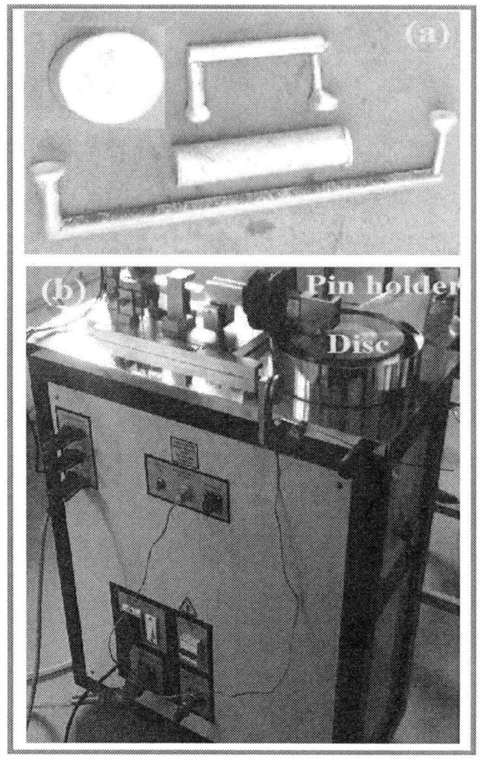


**Figure 1** Stir casting process setup **Figure 2** (a) casted samples (b) Pin on disc machine

## 3. Results and discussion

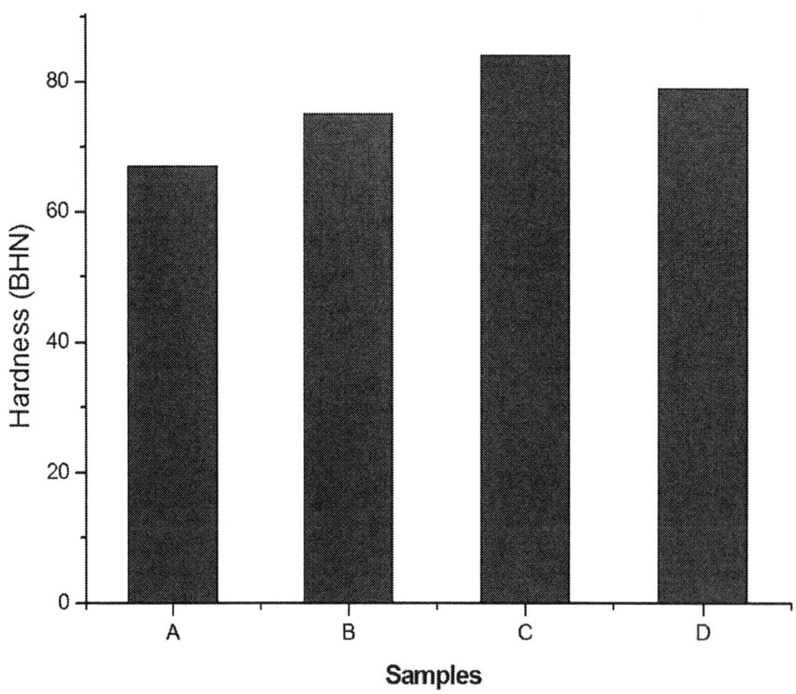


**Figure 3** Hardness of the AA7075 hybrid composites

In Figure 3, it is evident that the hybrid composite with 3 wt. % has highest hardness because of the presence of multiple reinforcements and weight proportion of the $TiB_2$. Increase in the weight percentage of the $TiB_2$ and $MoS_2$ leads to rise in hardness of the hybrid composites. At a point $TiB_2$ and $MoS_2$ acts as a non homogenous nucleation in the metal pool of AA7075. Without any about, it is clear that the enhancement of hardness can be attributed to the intrinsic ceramic properties of $TiB_2$ and $MoS_2$ reinforcements present in the AA7075 matrix results in increase the load bearing capacity of the hybrid composites. Because of the wettability and homogeneity of the hybrid particulates in AA7075, causes strong intermetallic bonding which is responsible for the strengthening mechanism.

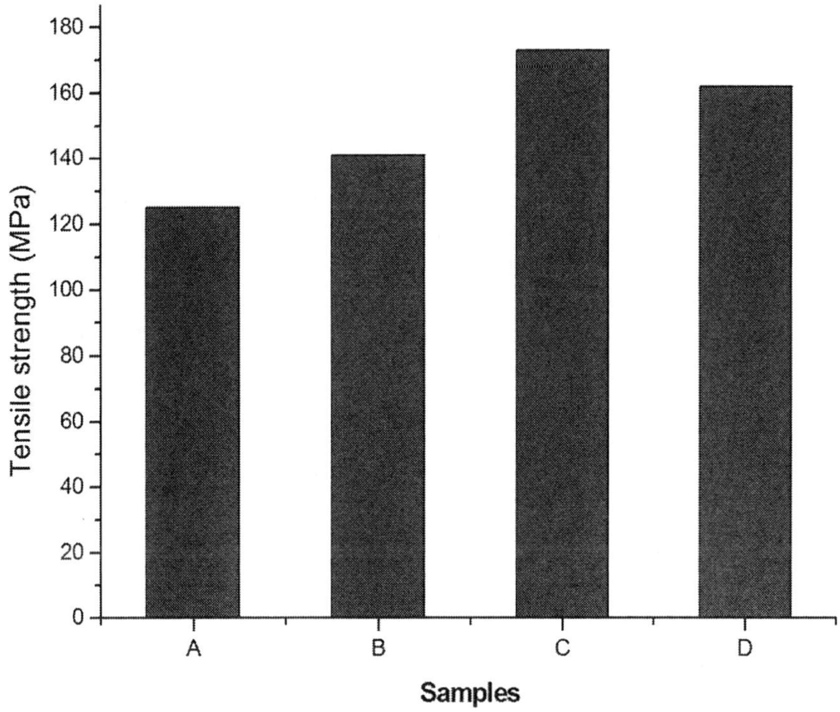


**Figure 4** Tensile strength of the AA7075 hybrid composites

The significant improvement in the ultimate tensile strength can be understood by the results. From the results it is clear that the increase in hybrid particulates has adverse effect in the increase of hardness of the hybrid composite. Figure 4 shows the variation of ultimate tensile strength of hybrid composites with respect to the various weight proportions of $TiB_2$ particles. Highest tensile strength is obtained for the sample C of 3 wt. % which is 173MPa. The dislocation due to increased stress concentration near the AA7075 matrix reinforcements can be restricted by introducing the hybrid reinforcements, $TiB_2$ and $MoS_2$ into the AA7075 matrix. The individual interfacial properties of matrixes, these properties also depend on the intermetallic bonding of hybrid composites. Effect on impact strength by adding $TiB_2$ and $MoS_2$ are shown in Figure 5. From the Figure 5 it can be understood that the impact strength increases with wt% $TiB_2$ increase and then decrease of hybrid composite also increases brittle nature of it. Based on the energy consumed by the material when a sudden load is applied, the result of the impact test depends. Further addition of more than 3 wt. % of $TiB_2$ particles on the hybrid composites the impact strength decreases due to the agglomeration of hybrid reinforcements. Here the poor wettability between the matrix and the particulate constituents the hybrid composites strength has declined.

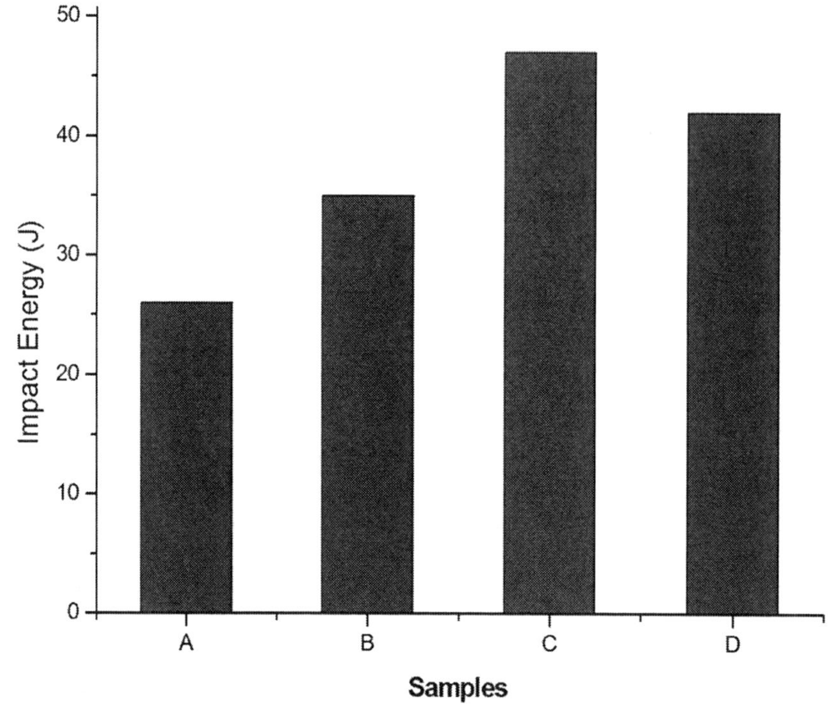


**Figure 5** Impact strength of the AA7075 hybrid composites

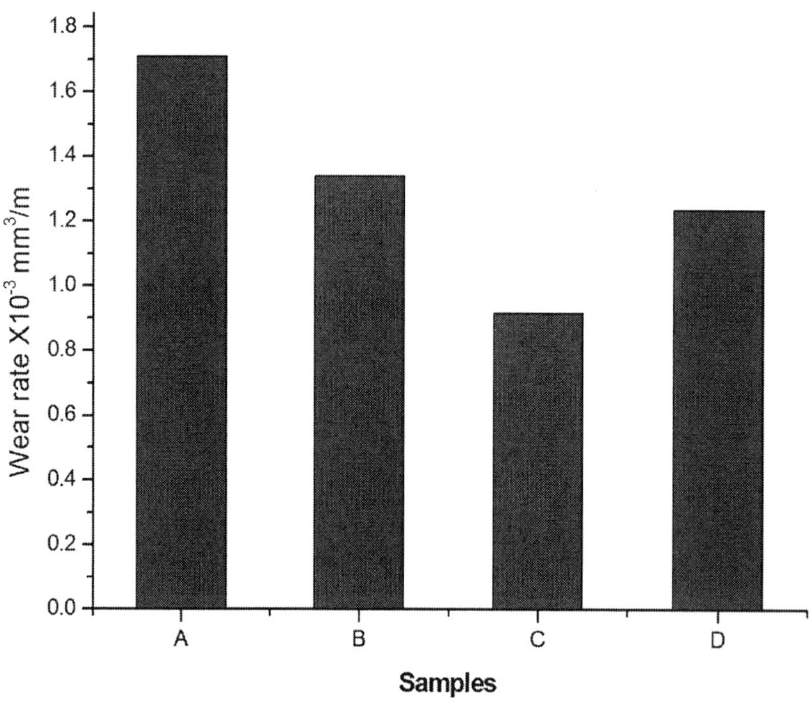


**Figure 6** Wear rates of the AA7075 hybrid composites

IOP Conf. Series: Materials Science and Engineering **923** (2020) 012053 doi:10.1088/1757-899X/923/1/012053

Figure 6 shows the wear rate of combination of AA7075 hybrid composites with respect to different weight percentage of $TiB_2$ particles. Wear rate is observed to be decreased in 3 wt. % of $TiB_2$ reinforcements and then starts to increased. It is observed that the increase in load increased the wear rate at lower levels of sliding velocity as because of the increase in pressure amid the contact surface. The soft AA7075 on the pin obtain harsh plastic deformation and directs to develop the patches to be making fine sliding of the buddy surface there by dropping the wear rate as observed. The SEM image of wear surface analysis of hybrid composites describes the abrasive wear mechanism as shown in Figure 7.

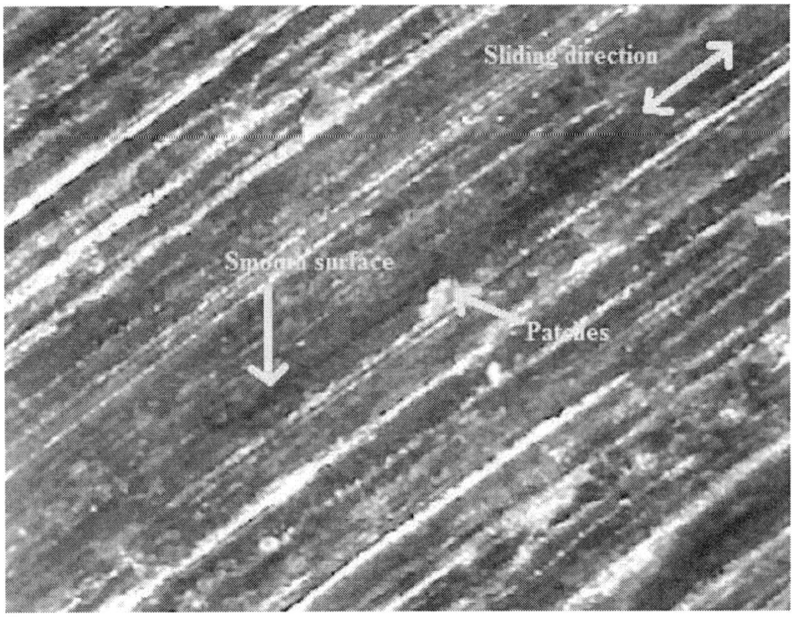


**Figure 7** Worn surface of the AA7075 hybrid composite

## 4. Conclusion

AA7075 based hybrid composites were fabricated via stir casting process successfully. The tribological behaviour of hybrid composites was analysed using pin on disc machine. The fabricated hybrid composites exhibits enhanced properties like hardness, tensile and impact strength comparing to base metal, which have improved by the addition of $TiB_2$ reinforcement. The maximum tensile strength is obtained for the sample C of 3 wt. % which is 173 MPa. The addition of weight percentage of $TiB_2$ and $MoS_2$ reduces the wear rate of the hybrid composites. The SEM image of worn surfaces of hybrid composite samples analysis of describes the abrasive wear mechanism.

## References

[1] Kok M, 2005 Production and Mechanical Properties of $Al_2O_3$ particle Reinforced 2024 Aluminium Composites, J Mat Proc Tech., **16**: 381-387.

[2] Geeta Rani M, Parameswara Rao C V, Rama Kotaiah K, 2017 Studies on characterization of Al 6061/$MoS_2$ metal matrix composite, *Int J Mech Eng Tech.,* **8**: 998–1003.

[3] Kalaiselvan K, Murugan N, Siva Parameswaran, 2011 Production and characterization of AA6061–$B_4C$ stir cast composite, *Mat Des.,* **32**: 4004-4009.

[4] Sivaprasad K, Narayanasamy R, Kumaresh Babu S P, 2008 Study of abrasive and erosive wear behavior of Al6063-$TiB_2$ in situ composites, *Mat Sci Engg A,* **498**: 495-500.

[5] Suresh S, Shenbaga Vinayaga Moorthi N, Vettivel S C, Selvakumar N, 2014 Mechanical behavior and wear prediction of stir cast Al–$TiB_2$ composites using response surface methodology, *Mat Des.,* **59**: 383–396.
[6] Venkataraman B, Sundararajan G, 2000 Correlation between the characteristics of the mechanically mixed layer and wear behaviour of aluminium, Al-7075 alloy and Al-MMCs, *Wear* **245**: 22–38.
[7] Subba Rao E, Ramanaiah N, 2017 Influence of heat treatment on mechanical and corrosion properties of Aluminium metal matrix composites (AA6061 reinforced with $MoS_2$). *Mat Today Proc.,* **4**: 11270–11278.
[8] Suresh S, Shenbaga Vinayaga Moorthi N, Vettivel S C, Selvakumar N, Jinu G R, 2014 Effect of graphite addition on mechanical behavior of Al7075/$TiB_2$ Hybrid composite using acoustic emission, *Mat Sci Engg A,* **61**: 16–27.
[9] Tong X C, 1998 Fabrication of in situ TiC reinforced aluminum matrix composites: part I microstructural characterization, *J Mat Sci.,* **33**: 5365-5374.
[10] Kerti I, Toptan F, 2008 Microstructural variations in cast $B_4C$-reinforced aluminium matrix composites (AMCs), *Mat Let,* **62**(8–9): 1215–1218.
[11] Yang B, Wang Y Q, Zhou B L, 1998 The mechanism of formation of $TiB_2$ particulates prepared by in situ reaction in molten aluminum, *Met.Mater.Trans.B,* **29**: 635-640.
[12] Albiter A, Leon C A, Drew R A L, Bedolla E, 2000 Microstructure and heat-treatment response of Al-2024: TiC composites, *Mat Sci Engg A,* **289**: 109-115.
[13] Surappa MK, 2003 Aluminium matrix composites: challenges and opportunities, *Sadhana*, **28** (1-2), 319-334.
[14] Gopalakrishnan S, Murugan N, 2012 Production and wear characterization of AA6061 matrix titanium carbide particulate reinforced composite by enhanced stir casting method, *Comp: Part B*, **43**(2): 302-308.
[15] Ranjit Bauri, 2009 Synthesis of Al -TiC in-situ composites: Effect of processing temperature and Ti:C ratio, *Trans. IIM*, **62**(4-5): 391-39.
[16] Vinod Kumar G S, Murty B S, Chakraborty M, 2005 Development of Al-Ti-C grain refiners and study of their grain refining efficiency on Al and Al-7Si Alloy, *J. Alloys Compd.,* **396** : 143-150.
[17] Rai R N, Rao A K P, Dutta G L, Chakraborty M, 2013 Forming behaviour of Al-TiC in-situ composites, *Mater. Sci. Forum,* **765**: 418-422.
[18] Peijie Li, E.G.Kandalova, V.I.Nikitin, A.G.Makarenko, A.R.Luts, and Zhang Yanfei, 2003 Preparation of Al-TiC composites by self-propagating high-temperature synthesis, *Scripta Materialia* **49**: 699-703.
[19] Christy V, Murugan N, Kumar S, 2010, A comparative study on the microstructures and mechanical properties of Al 6061 alloy and the MMC Al 6061/$TiB_2$/$12_P$, *J Mine Mat Charact Engg.,* **9**: 57-65.
[20] Ranjan A, Shanmugasundaram S, 2019 Experimental investigation of mechanical and Tribological properties of Al 7075-$MoS_2$/$ZrO_2$/Ni hybrid composite. *Adv Mater Metal:* 299–309.

IOP Conf. Series: Materials Science and Engineering **923** (2020) 012054 doi:10.1088/1757-899X/923/1/012054

# EXPERIMENTAL INVESTIGATION OF FLEXURAL STRENGTH OF BANANA FIBER REINFORCED COMPOSITES

**Harish Kumar J[1] & Alagar Pandi K[2]**

[1] AMET University, East Coast Road, Chennai – 603112, India.
[2] Solamalai College of Engineering, Veerapanjan, Madurai - 625 020, India.

E-Mail: harishkumarj20@gmail.com.

**Abstract:** The present generation widely makes use of en number of composites in the field of materials technology, with a very high performance and costs appropriate for the applications such as structural and automotives etc. Fiber reinforced composites are those composites that are called as the oldest and frequently used composite materials, in which Banana fibers are the most common ones. Nowadays consciousness and environmental awareness about using natural fibers as a serious alternative has been keep on increasing globally. And also in these present days natural fiber reinforced polymer composites are being widely used in various engineering applications. Natural fibers such as jute, sisal, palm, coir and banana are the most commonly used reinforcements. In vinyl ester resin matrix banana fibers have been used as reinforcement. This work deals with the study of effect of sodium hydroxide (NaOH) on fibers in addition with alumina ($Al_2O_3$) particulate. It is seen that the flexural strength value of UT, UT+ $Al_2O_3$, T+ $Al_2O_3$ has increased with increase in the fiber content.

**Keywords:** *Composite Fabrication, Vinyl ester, Mechanical properties, Scanning electron microscopy, UT - Untreated, T - Treated*

## 1. Introduction

In this Paleolithic age all humans are well aware and well known about composites. Clay mixed with finely chopped straw was used in constructing the 300 ft high ziggurat or temple tower in the city center of Babylon [1, 2]. In recent years, many applications such as automotives, sporting goods, marine, electrical, industrial, construction, household appliances, etc makes use of polymeric based composite materials. These types of composites have very high stiffness and strength, light in weight, and high corrosion resistance. An extensive research work has been carried out in the past decades on the natural fiber reinforced composite materials. It was found that natural fibers which are available in abundant can be used to reinforce the polymers in order to obtain light and strong materials from them. The researchers have given results as natural fibers from plants are in the beginning stage to find their way into commercial applications such as automotive industries, household applications, etc.

Lot of investigations has been made on several types of natural fibers to study the effect of these fibers on the mechanical properties of composite materials [4–7]. Mansur and Aziz [6] has studied about bamboo-mesh reinforced cement composites and found that this reinforcing material could enhance the ductility and toughness of the cement matrix, and increase its tensile, flexural, and impact strengths significantly. On the other hand, in order to evaluate the mechanical properties and compare it with wood composites, the jute fabric-reinforced polyester composites were tested [7], and the test came up with good results that the jute fiber composite has better strengths than wood composites. Also there was an investigation made with pulp fiber reinforced thermoplastic composite and it was found to have an increase in stiffness by a factor of 5.2 and increase in strength by a factor of 2.3 relative to the virgin polymer [8]. Information on the usage of banana fibers in reinforcing polymers is limited in the literature.

In dynamic mechanical analysis, Laly et al. [9] has made an investigation on banana fiber reinforced polyester composites and found that the optimum content of banana fiber is 40%. Mechanical properties of banana–fiber–cement composites were investigated physically and mechanically by Corbiere-Nicollier et al. [10]. It was reported that Kraft pulped banana fiber composite has good flexural strength. In addition, short banana fiber reinforced polyester composite was studied by Pothan et al. [11]; the study concentrated on the effect of fiber length and fiber content. The maximum tensile strength was observed at 30 mm fiber length while maximum impact strength was observed at 40 mm fiber length. Incorporation of 40% untreated fibers Provides a 20% increase in the tensile strength and a 34% increase in impact strength tested with banana fiber and glass fiber with varying fiber length and fiber content as well.

## 2. Experimental Details

This research work aims in developing the flexural strength of Banana Fiber/Particulate Reinforced Epoxy Hybrid Composites. It was found that only few researches were made using this material and hence to explore the mechanical properties such as flexural strength of the material this work is proposed which may be a platform for the future development.

### *2.1. Specimen Preparation*

The mould should be well cleaned and dried using releasing agent (wax) before laying the vinyl ester resins on the mould. The first layer of the mould is uniformly laid with vinyl ester mixture using a special brush.. The first layer of non-woven fiber mat is placed into the mould. Another layer of Vinyl ester will be applied uniformly on the first layer of non-woven fiber mat. The second layer of non-woven fiber mat is then placed. In order to get the required laminates all these procedures are continuously followed. The mould is then closed and the composite laminates are pressed uniformly for 24 hours to obtain required thickness and curing. Once the composite is completely dried out then it is separated from the mould.
The composite laminates are cut in to specimen size as per ASTM standard for testing.

## 3. Testing of Flexural Property

A measure of the force that is required to bend a beam under three point loading conditions is known as flexural strength. These datas are often used to select materials for parts that will support loads without flexing. An indication of a material's stiffness when it is flexed is called as flexural modulus. Since the physical properties of many materials (especially thermoplastics) can vary depending on ambient temperature, it is sometimes appropriate to test materials at temperatures that simulate the intended end user environment.

### *3.1 Flexural Test Procedure:*

- Most commonly the specimen will be lying on a support span.
- The load is applied at the center by the loading nose, producing three points which in turn starts bending at a specified rate.
- We have to consider certain parameters such as support span, speed of loading, and the maximum deflection.
- These parameters are based on the test specimen thickness and these are well defined differently by ASTM and ISO. For ASTM D 790, the test is completed when the specimen reaches 5% deflection or the specimen breaks before 5%.
- For ISO 178, the test is completed when the specimen breaks.

IOP Conf. Series: Materials Science and Engineering **923** (2020) 012054 doi:10.1088/1757-899X/923/1/012054

- If the specimen does not break, then the test has to be continued as far as it is possible.

**Table 1.** Various composite samples and its Flexural strength.

| Samples | Flexural Strength in Mpa |
|---|---|
| **Sample 1** | |
| UT | 4 |
| UT+$Al_2O_3$ | 3 |
| T | 3 |
| T+$Al_2O_3$ | 5 |
| **Sample 2** | |
| UT | 4 |
| UT+$Al_2O_3$ | 5 |
| T | 4 |
| T+$Al_2O_3$ | 4 |
| **Sample 3** | |
| UT | 4 |
| UT+$Al_2O_3$ | 4 |
| T | 3 |
| T+$Al_2O_3$ | 4 |

**Figure 1.** Shows the loading of the specimen in UTM for Flexural Test.

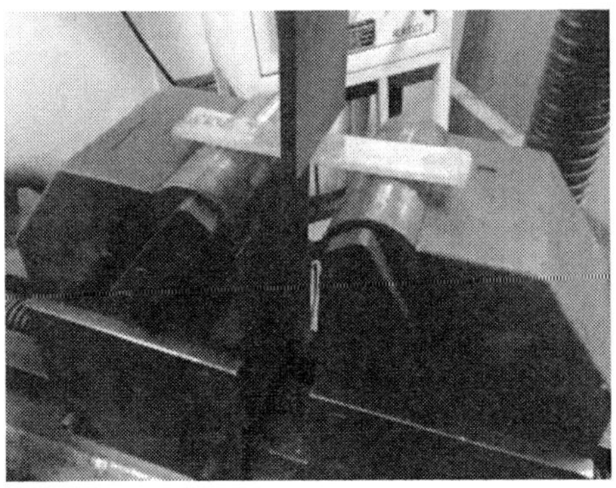

**Figure 2.** Specimen before Flexural Test.

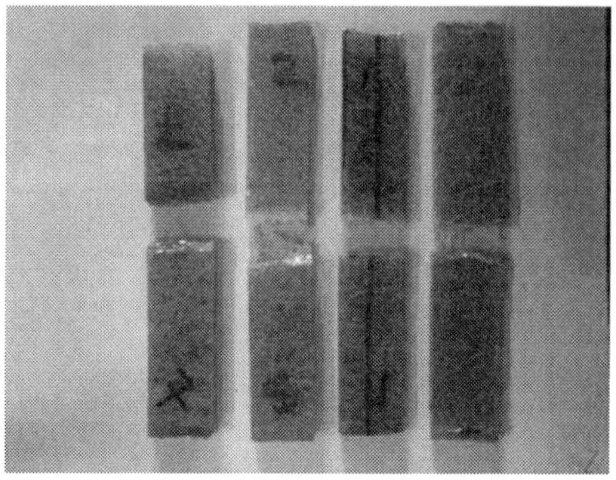

**Figure 3.** Specimen after Flexural Test.

## 4. Result and Discussion

### *4.1 Flexural strength*

The bending properties of composites were tested by applying 3-point bending method. Sample 1- The flexural strength of T+$Al_2O_3$ has been increased by 25% when compared with UT. Sample 2 - The flexural strength of composite faring UT+ $Al_2O_3$ of 5 Mpa, has been increased by 25% when compared with T+ $Al_2O_3$. Sample 3 – The flexural strength of UT, UT+ $Al_2O_3$ and T+ $Al_2O_3$ are same as show in figures 4 (a) & (b), 5 (a) & (b) and 6 (a) & (b).

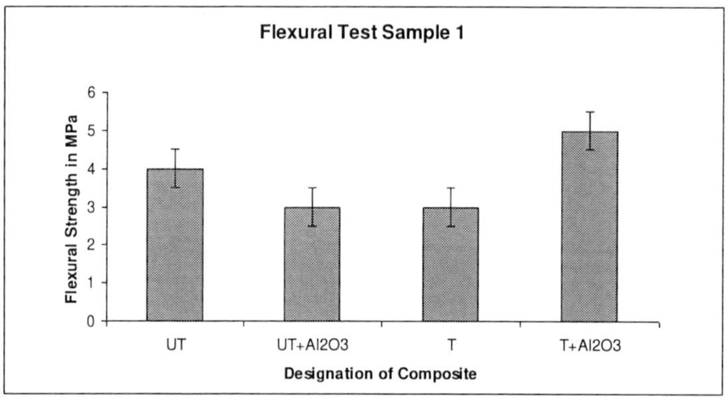


**Figure 4 (a).** Flexural Test Result of Sample 1

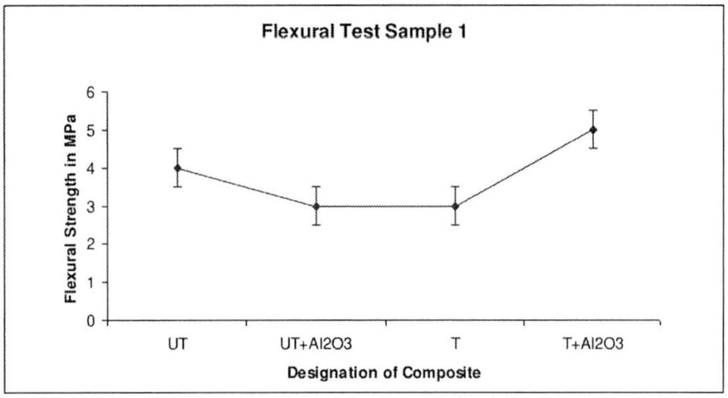


**Figure 4 (b).** Flexural Test Result of Sample 1

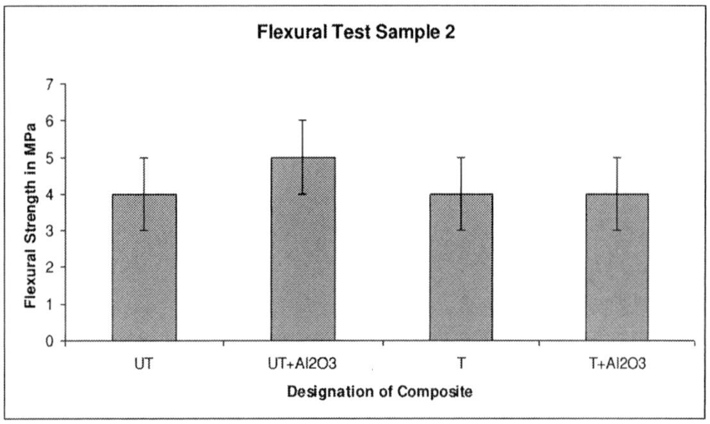


**Figure 5 (a).** Flexural Test Result of Sample 2

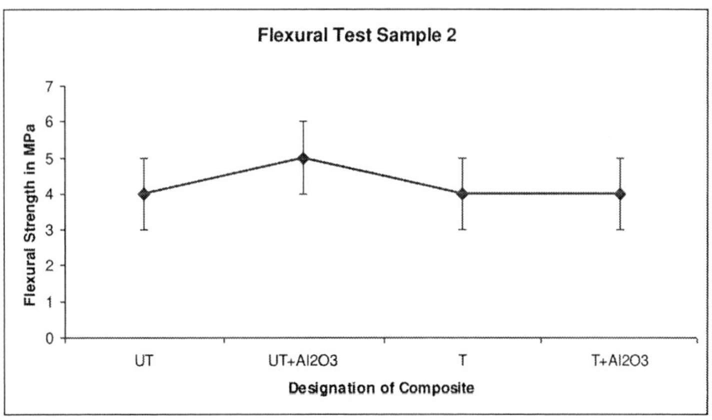


**Figure 5 (b).** Flexural Test Result of Sample 2

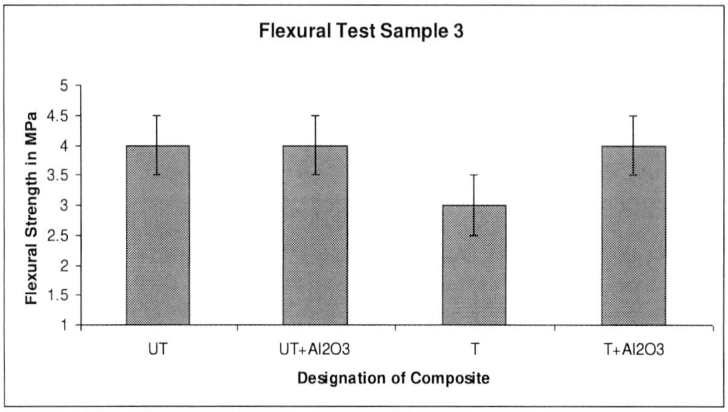


**Figure 6 (a).** Flexural Test Result of Sample 3

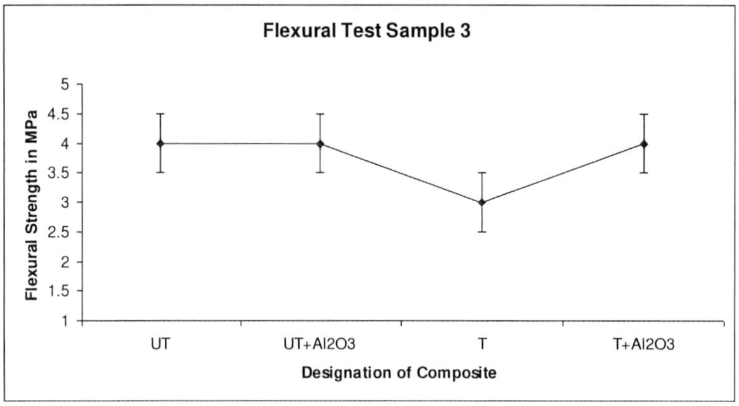


**Figure 6 (b).** Flexural Test Result of Sample 3

The variations in the flexural strength of samples 1,2 & 3 in single layer composite laminate, and Samples 1,2 & 3 in double layer composite laminate are shown in figures 4 (a) & (b), 5 (a) & (b) and 6 (a) & (b) respectively.

## 5. Conclusion

In this research work, by adding $Al_2O_3$ and modifying the surface some of the mechanical properties are experimentally evaluated and the following conclusions have been drawn:

- ✓ Flexural strength has been increased invariably.
- ✓ In addition of $Al_2O_3$ when it is treated with NaOH it has provided better results in the increase/improvement of flexural strength and other mechanical properties.
- ✓ The flexural strength can be even increased more when these hybrid composites undergoes with surface treatment.
- ✓ Composite materials that are light in weight are made using banana fiber with high specific strength which in turn can be used in making automotive interiors.
- ✓ Hybridized synthetic and natural fiber polymer composites, can be utilized in various fields like structure, marine, consumer articles and in industries also.

Thus it has been concluded that when a systematic and persistent research is carried out it will bring lot of scope and a better future is developed for synthetic and natural fibre – polymer hybrid composites in the upcoming years.

**References**:

[1] A.Laly Pothan, Zachariah Oommenb and Sabu Thomas, *Dynamic Mechanical Analysis of Banana Fiber Reinforced Polyester Composites, Composites Science and Technology,* **63(2)(2003)**, pp.283–293.

[2] P.A.Sreekumar, Pradeesh Albert, G.Unnikrishnan, Kuruvilla Joseph and Sabu Thomas, *Mechanical and water sorption studies of ecofriendly banana fiber reinforced polyester composites fabricated by RTM.* J. App. Poly. Sci, **109, (2008)** pp.1547-1555.

[3] L.A.Pothan, T.Sabu, and Neelakantan, *Short Banana Fiber Reinforced Polyester Composites: Mechanical, Failure and Aging Characteristics, J. Reinforced Plastics and Composites,* **16(8)(1997)** pp.744–765.

[4] Sherely Annie Paul, et.al., *Effect of fiber loading and chemical treatments on thermo physical properties of banana fiber/polypropylene commingled composite materials , Composites Part A Applied Science and Manufacturing,* **Volume 39, Issue 9, September 2008**, Pages 1582-1588.

[5] D. Feng, D.F.Caulfield, and A.R. Sanadi. *Effect of compatibilizer on composites, Polymer Composites,* **22:506** (2001).

[6] L. Y. Mwaikambo and M.P.Ansell., *Chemical modification of hemp, sisal, jute, fibers by alkalisation,* Journal of Applied Polymer Science., **84(12):2222–34(2002).**

[7] M. A. Mansur and M. A. Aziz., *Study of Bamboo – Mesh Reinforced Cement Composite,* International Journal of Cement Composites and Lightweight Concrete, **Vol 5. No.3. August 1983**, Pages 165-171.

[8] Dhakal, S., Keerthi Gowda, B.S., 2017. *An experimental study on mechanical properties of banana polyester composite.* Materials Today: Proceedings **4 (8),** 7592–7598.

[9] Jordan, W., Chester, P., 2017. *Improving the properties of banana fiber reinforced polymeric composites by treating the fibers.* Proceedia Engineering **200 (2017),** 283–289.

[10] Ramesh, M., et al., 2014. *Processing and mechanical property evaluation of banana fiber reinforced polymer composites.* Proceedia Engineering **97 (2014),** 563–572.

[11] H. Ku, H. Wang, N. Pattarachaiyakoop, M. Trada, *A review on the tensile properties of natural fiber reinforced polymer composites,* Composites **Part B 42 (4) (2011)** Page No. 856–873.

[12] Amir, N., Abidin, K.A.Z., Shiri, F.B.M., 2017. *Effects of fibre configuration on mechanical properties of banana fibre/PP/MAPP natural fibre reinforced polymer composite.* Procedia Engineering **184 (2017**), 573–580.

[13] Balaji, D. S., S. Prabhakaran, and K. Umanath. *"Multi-Objective Optimization in Abrasive Water Jet Drilling of AA6061 alloy."***2017.**

[14] C. Rajesh, N. Venkata Niranjan Kumar and G. Gowthami. *"Evaluation of Wear Behavior OFPLA and AVS Parts Fabricated by Operate FDM Technique with Distinct Orientations"* International Journal of Recent Technology and Engineering, **Vol. 7. Issue 5S3 February 2019**.

[15] S. Muthubaskaran *"Development and Testing of Jute Fibre Epoxy Composites"* Test Engineering and Management, **Vol.82. Issue: January – February 2020**, Pg. No. 11583 – 11587.

# Tribological Behaviour of Aluminum Metal Matrix Composites – A Review

**Ashok Kumar R [a], Karthick K[b], Jayasuriya R[c], Aswin Kumar J[d], Karthik Balaji V[e]**

[a], Assistant Professor, RMK College of Engineering and Technology, Tiruvallur-601206, Tamil Nadu, India.
[b, c, d, e & f], UG Students, RMK College of Engineering and Technology, Tiruvallur-601206, Tamil Nadu, India.

Email: ashok.kumar19872003@gmail.com

**Abstract**. An examination of Aluminum Metal Matrix Composites (AMMCs) dry sliding wear activities reinforced with reinforcements along with carbides, oxides, and nitrides. However, from the analysis, the method adopted for the composite processing plays a crucial role in achieving a uniform homogeneous composition. Preparation of base metal and reinforcement powders plays a vital role in the process of powder metallurgy, improving wettability between reinforcement particles and molten base metal in liquid metallurgy is a difficult task. AMMCs wear activity is based on different test parameters. As mechanical and tribochemical are the process involved in substance removal during dry sliding wear is subdivided into groups. The loss of reinforcement from the AMMCs through sliding wear decreases the wear rate due to abrasion by three parts. The use of solid lubricant decreases wear severity through the formation of a tribolayer between the rubbing surface. Optimizing the wear intensity is through the use of suitable replacements with optimal lubricants

**Keywords:** Aluminium Metal Matrix Composites (AMMCs); Dry sliding wear; solid lubricant particles; scanning electron microscopy

## 1. Introduction:

Metal matrix composites (MMCs) are interfusion of two or more materials, which expose properties that are hard to obtain from a single material with high strength and stiffness with less density[1]. Hybrid matrix composite is the preferred form of metal matrix composites used for lightweight high strength application in aerospace and automobile field[2]. The present engineering applications need materials that are less expensive, stronger and lighter so the aluminium is the perfect example in the development of materials. In aluminium, matrix composites the attributes like high stiffness, low density and strength and superior wear resistance [3]. It has been proclaimed that the tribological properties of Al alloys can be enriched by adding ceramic reinforcements. AMMC reinforced with silicon carbide ($3.18gm/cm^3$) particles are used to increases the hardness of the composite. Since we are producing a hybrid metal matrix composite there must be another reinforcement material added to the composites, so in this study we introduce quarry dust as the secondary reinforcement material for the enhanced properties. It had been found that the use of AMMCs in engine application can reduce the overall weight, fuel consumption and pollution in automobile field [4,5]. Aluminium alloys have an effective application in various mechanical components because of the extraordinary combination of properties like good thermal conductivity, high corrosion resistance and the strength to weight ratio also being high. Followed by the composite and reinforcement material selection, the casting techniques should be selected wisely[6]. From different casting practices, we prefer stir casting technique for its clarity, resilience and applicability to large quantity production with cost leverage. The stir casting process has opted for the production of AMMCs which will reduce the cost of the composites and it is economical so this technique is used massively. Finally, the casted Al composites are directed to the Scanning Electron Microscope (SEM) analysis for the study of the grain and microstructure. On the other hand cast aluminium is subjected to wear testing. Here the pin-on-disc is preferred for the wear testing when investigating the crack of a coating at the boundary of a wear track. The result of this test gives the tribological behaviour of the hybrid AMC's[7].

## 2. Material Selection

Al alloy is extensively used in aerospace and marine industries for their high corrosion resistance and for their superior welding features. In this process, we use Al alloy as a base metal in the metal matrix composites.

### *2.1. Matrix Material*

Aluminium is widely used as matrix material in MMCs.

#### *2.1.1. Al 6061*

Al 6061 is precipitation-hardened aluminium alloy, containing magnesium and silicon as its major alloying element widely called as "Alloy 61s". The composition of Al6061 is given in Table 1. It is developed in 1935. It has good mechanical properties. It is the most common alloy of aluminium and general-purpose use. In this study, we prefer Al7075 than Al 6061. The primary difference between 6061 and 7075 is exhibits immense strength to weight ratio. Al 7075 provides the greatest strength available in aluminium bar. Al 7075 contains more zinc than Al 6061 alloy[8].

**Table.1.** Composition of Al6061

| S.No | Composition | Alloying element |
|---|---|---|
| 1. | 0.65% | Silicon |
| 2. | 1.0 – 1.5 % | Magnesium |
| 3. | 0.19 – 1.2 % | Copper |
| 4. | 0.64% | Iron |
| 5. | 0.5% | Manganese, chromium, lead |

#### *2.1.2. Al 7075*

The first AA7075 were introduced by Japanese in the year 1935 by the Japanese company "Sumitomo Metal". And it was consistently used in aerospace engineering from 1947. In 7075 Aluminium alloy (AA7075), zinc is the primary alloying element which deserves the properties like good ductility, high strength, toughness, and good corrosion resistance than other 2000 alloys. The composition of Al7075 is given in Table 2. It is the most generally used alloy for high stressed structural applications and it is primarily used in aircraft structural parts[9,10].

**Table.2.** Composition of Al7075

| S.No | Composition | Alloying element |
|---|---|---|
| 1. | 5.1 – 6.1% | Zinc |
| 2. | 2.1 – 2.9% | Magnesium |
| 3. | 1.2 – 2% | Copper |
| 4. | 0.5% | Silicon, Iron, Manganese |

### *2.2. Reinforcement*

Reinforcements are the hard particles added in MMCs to improve the properties of base matrix material

#### *2.2.1. Boron Carbide ($B_4C$)*

Boron Carbide is blackish powder-like material which is odourless in nature. Also known as "Tetrabor", it is one of the hardest materials similar to Diamond. It exhibits better toughness compared to other materials and also stable under radioactive conditions. In this powder, reinforcement chemical reaction ensues at 800°C to 1400[11]. Al alloys, when reinforced with boron carbide, results in higher strength with lower ductility materials when compared with other reinforcements. There are various techniques of reinforcement with Al alloy composites such as Hot iso-static pressing(HIP) followed by High-strain rate forging(HSRF), HIP followed by dual-step quasi-isostatic forging (QIF) and quad-step QIF resulting similar mechanical properties. Among these methods, the first two are used widely due to their improved characteristics[12].

#### *2.2.2. Aluminium Oxide ($Al_2O_3$)*

Aluminium Oxide, also commonly known as "Alumina", is a white odourless crystalline powder. It is amphoteric in nature, which occurs commonly in corundum form. It is used as an abrasive material and as refractory material due to its good hardness and high melting point respectively. Increase in a volume fraction of aluminium oxide as reinforced phase results in lowered fracture toughness of the metal matrix composites[11]. The durability, hardness, wear and tensile fortitude of the aluminium composites are found to be increased with smaller grain size when the typical stir casting process. By vortex method, the wettability and the binding between $Al_2O_3$ particles are improved by applying pressure .It also increases the wear resistance of the MMC[13].

#### *2.2.3. Silicon Carbide (SiC)*

Silicon Carbide (SiC) is a prominent abrasive and contemporary engraver due to endurance and low cost. It is acknowledged as carborundum incorporated with silicon and carbon. Silicon carbide particle grains can be constrained together by sintering to contour very hard ceramics that are universally used in applications like car brakes. Large sole crystals of silicon carbide could be developed by Lely Method and they can be downsized into gems recognized as synthetic moissanite. The reinforcement of Silicon Carbide into AMCs had proved that increase in tensile strength and the hardness. [11,14]. The microstructure of SiC is shown in Figure 1. The major conclusion derived from the former entirety carried out tells us

- Silicon carbide reinforced with Al-MMCs possess higher wear protection compared to ($Al_2O_3$)
- SiC reinforced Al-MMCs are convenient material for brake drum as they have more advanced wear hindrance properties but cannot be used in the brake linings.
- The wear resistance of SiC reinforced Al MMCs is superior to boron carbide reinforced MMCs.

IOP Conf. Series: Materials Science and Engineering **923** (2020) 012055 doi:10.1088/1757-899X/923/1/012055

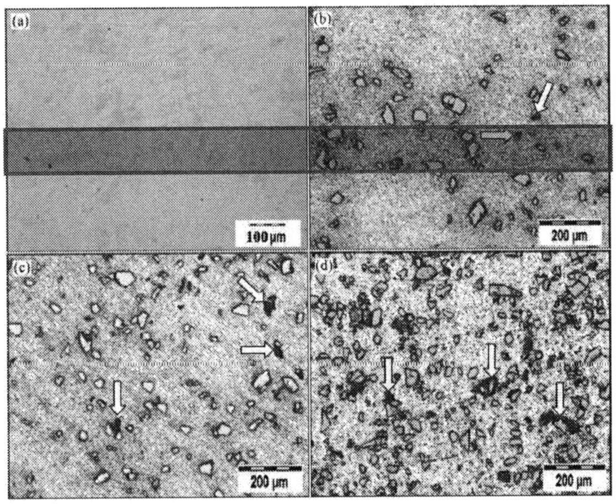


**Figure 1.** Microstructures of pure Al (a) and Al/5SiC (b), Al/10SiC (c) and Al/20SiC (d) composites (Some porosities are marked with arrows) [13]

*2.2.4. Quarry Dust (Waste material used as secondary reinforcement)*

Quarry Dust which is an outcome from pulverised rock/stone as a waste by-product. The EDAX result is shown in Figure 2. It is proposed in research as an alternate for SiC and with the objective, to ample, the gaps spotted in wastes that are already in use as reinforcement in AMMCs. The studies had proved that the addition of quarry dust in AMCs will increase macro hardness and tensile properties of the composite. When dry sliding wear performance of AMC is investigated with the help of a pin-on-disc wear testing setup, it is found that inclusion of quarry dust particles as reinforcements elevates the wear resistance of the composites[15]

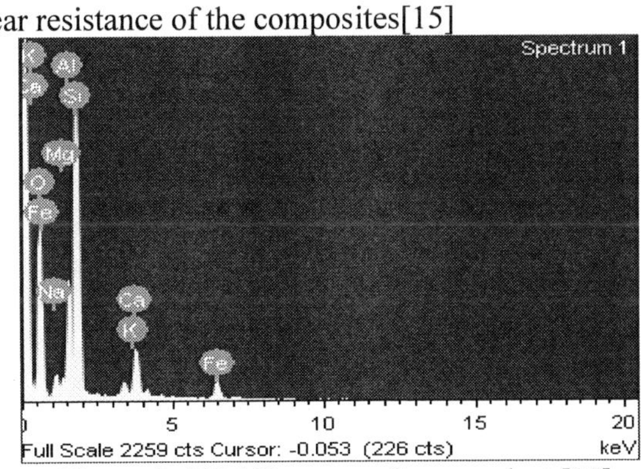


**Figure 2.** EDAX report of quarry dust [15]

## 3. Fabrication Technique

The most important thing with Al composite is the uniform distribution of reinforcement particles within the matrix alloy hence ball milling is the pre-process carried out. Two generic techniques of fabrication can be used for the composite production (i.e.) powder metallurgy and molten metal method. These both processes have their own dominance and deprivation.

*3.1. c*

A ball mill is one kind of grinding used to grind and mixing of the materials in ceramics and the mineral dressing processes. It works on the principle of attrition, as grain size decrement is done by the impact as the balls drop from near the top of the cylindrical shell. If we have to reduce the size of the particles in a (SiC) we can do ball milling and thus we can reduce the micro size of reinforcement materials like SiC,$Al_2O_3$ etc [16]. A sample of ball milled microstructure is shown in Figure 3.

**Figure 3.** Influence of the mechanical milling time on the morphology of M10Sn nanocomposite powders: (a) 1 h, (b) 3 h (c) 5 h (d) 15 h (e) 20 h and (f) 25 h [17].

*3.2. Powder Metallurgy*

Powder Metallurgy (PM) greatly reduce the need to use metal removal processes. It is used to make unique materials, impossible to get from melting in other ways. In this metallurgy, Al alloy powder is blended with reinforcement particles to obtain a homogenous mixture. The appropriate size ratio of alloy and reinforcement particles will help to obtain uniform distribution in the composites[18]. The procedure of developing an aluminium composite in Figure 4.

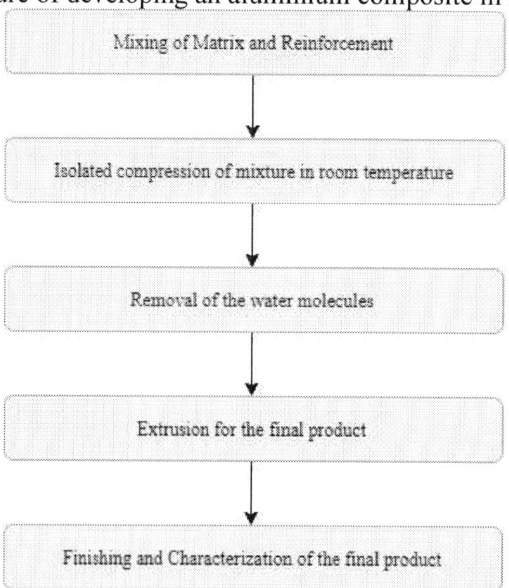


**Figure 4.** Developing of AMMCs by PM Technique

Powder Metallurgy has a number of benefits, one of that is any type of alloy and reinforcement can be done for the fabrication of the composite, but it also has many defects, that is long mixing time is required for the uniform distribution of the particles[19].

### *3.3. Liquid Metallurgy*

This method is also known as molten metal vortex method. It is a highly attractive method because it is very cheap and highly productive. The procedure of developing an aluminium composite in Figure 5.

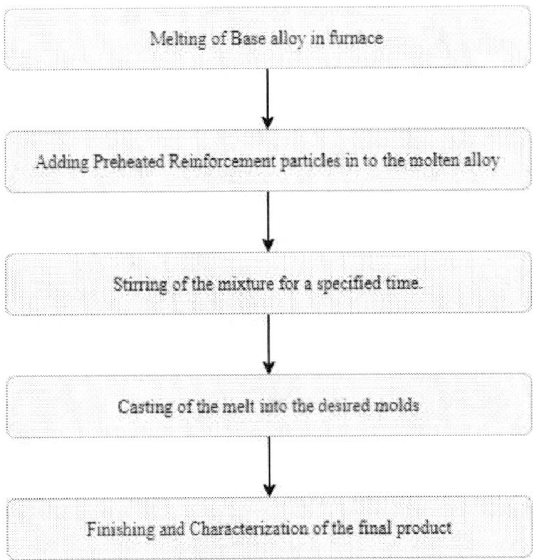


**Figure 5.** Developing of AMMCs by Liquid Metallurgy Technique

#### *3.3.1. Stir Casting Technique*

Stir casting is a liquid state technique for the production of composite materials, in this technique dispersed phase of reinforcement is blended with a molten matrix metal by mechanical stirring. It is a simple and cost-effective method for composite material fabrication. Aluminium metal matrix composites are fabricated by stir casting method parameters, which will affect final microstructure and properties of the composites. AMMCs was developed by using two different techniques, conventional stir casting and ultrasonic stir casting, conventional stir casting is preferred for low cost production [7,19].

In the conventional stir casting, the reinforcement SiC powder should be carefully encapsulated in an aluminium foil packer for insertion in molten aluminium in ordered to fabricate composites with SiC as 3% weight. The SiC powder should be pre-heated at 350 °C for 4 hours before the casting process for the removal of moisture and impurities. The aluminium should be heated up to the temperature of 850 °C within a furnace and stirring periods should be above 6 minutes, the factors of matrix interface, ceramic incorporation were evaluated by Scanning Electron Microscope (SEM) and it had been concluded that shorter stirring is required for the ceramic incorporation to achieve the maximum metal bonding at the interface [20]. The higher stirring temperature can give improved ceramic embodiment. A microstructure sample is shown in Figure 6.

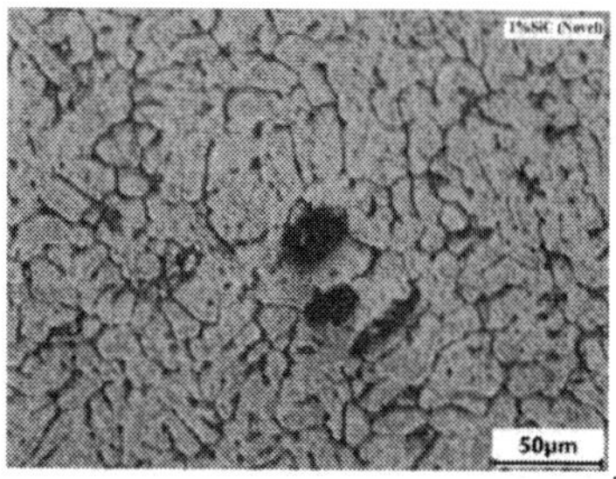


**Figure 6.** SEM image of the specimen after stir casting [21]

## 4. Tribological Properties

Tribology is often defined as a study handling application and subjects of friction, wear and lubricating applications, and all together of the most aspects of tribological work on wear. Wear are often defined because the phase-out of materials on solid surfaces. The form, size and style of the reinforcement are key factors for the Al-based MMCs ' wear characteristics [22].

### *4.1. A12219 / Gr / $B_4C$*

It has been demonstrated that the abrasion behavior of hybrid A12219/Gr/B4C AMMC had been fortified by utilizing reinforcement, Al 2219 with 8% of B4C and cross breed composite (8% B4C+ Al 2219 + 3%Gr). because of the presence of earthenware support particles, the hybrid composite had indicated more noteworthy wear spot opposition which is appeared in Figure 7. during this hybrid composite, the reinforcement particles had been haphazardly and finely scattered in the matrix alloy. During this examination, it had been discovered that tensile strength and yield pressure were diminished by the expansion of auxiliary support, yet it had been demonstrated that in absence of Boron carbide reinforcement (B4C), wear pace of the Al combination is more, by expansion of this support diminishes wear, Figure 8 shows a down surge in the wear rate by this composite. [12].

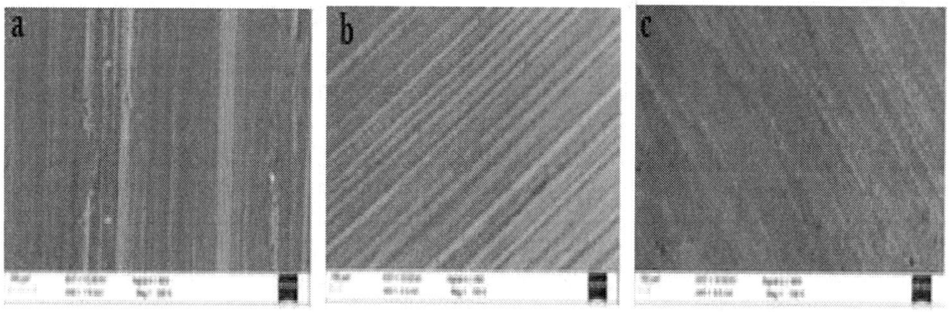


**Figure 7.** SEM Analysis for different composition (a) Micrograph at the wear surface of alloy[12][12][11] (b) Al + 8% $B_4C$ (c) Al + 8% $B_4C$ +3%Gr [12]

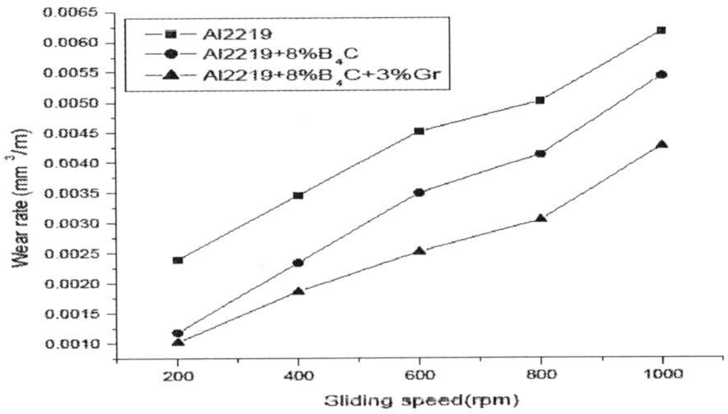


**Figure 8.** Wear Rate Vs Sliding speed [12]

### *4.2. Aluminium reinforced with the Silicon Carbide particles (SiC)*

The wear conduct of the aluminum fortified with the Silicon Carbide has been examined, It has been indicated that the reinforcement weight percentage was the most significant factor in the composite wear rate, sliding velocity and burden were second and third factors. Right now, wear conduct and friction attributes of aluminum combinations and Al MMCs containing (SiC) aluminum composites have been concentrated in their work., it has been demonstrated that wear pace of the aluminum lattice in Figure 9 [23].

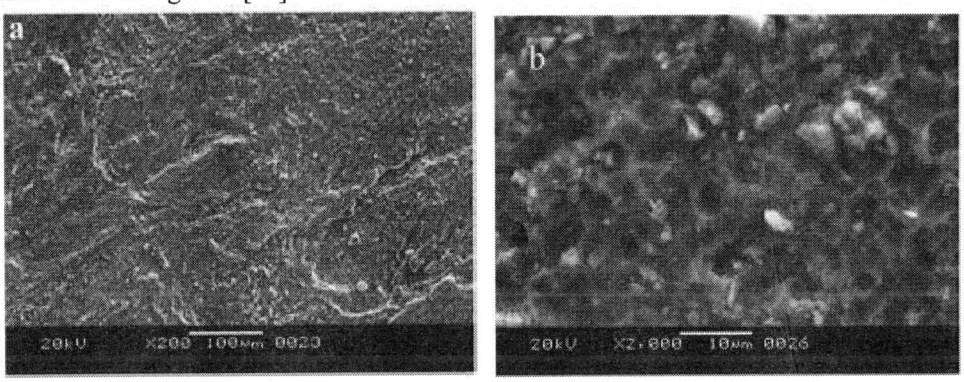


**Figure 9.**Images of Al-SiC worn out sample under an alkaline environment at mid-levels of normal load and sliding speed: (a) SEM (200X), (b) SEM (2000X)[23]

### *4.3. Aluminium- Silicon alloys with graphite particulates*

As per the review, the engine chambers produced using Al alloy/graphite composite will show a drop in the pace of wear, higher corrosion obstruction and lower coefficient of rubbing and in correlation with aluminium matrix composites with earthen ceramic particles and the looks into has indicated that graphite will advance tribological properties by delivering a graphite rich lubrication between the territory of sliding. From the study, it had been discovered that co-efficient of friction will drop with increment of graphite particles content, because of that a thick graphite lubrication film on the sliding surface. Tribological qualities had been going down when the substance of graphite in the composite is high because of the holes in bunched graphite particulates is appeared in Figure 10 &11 [24].

**Figure 10.** Optical micrographs of fabricated composites (a) Al2024/5%SiC/5%Gr and (b) Al2024/5%SiC/10%Gr composites.[24]

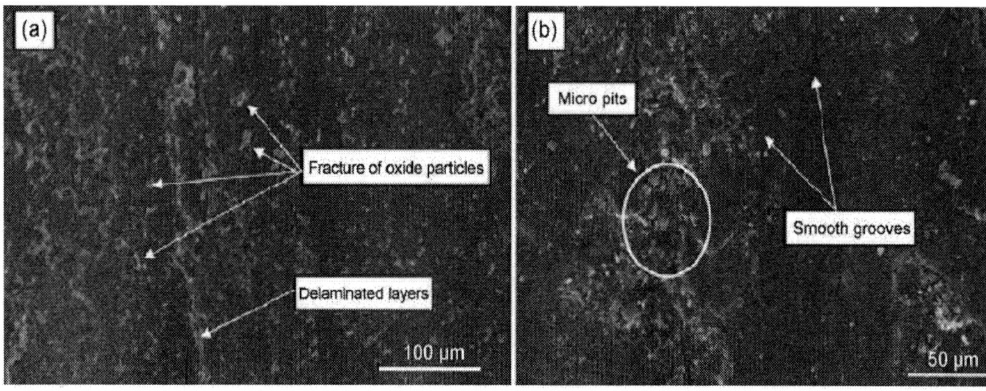


**Figure11.** SEM morphologies of the worn surface of Al/5%SiC/10%Gr composite at an applied load of 20 N at (a) low magnification and (b) high magnification.[24]

*4.4. Aluminium- Silicon alloys with graphite particulates*

Quarry dust is an extremity from the devastating of rock/stone. The XRD of Quarry dust is shown in Figure 11. It is having a density (1.7 – 1.9 g/cm3) which is not as much as that of aluminum (2.7 g/cm3) and SiC from the examination [3,4], it had been demonstrated that specific wear rate is seen to be raised somewhat with an expansion in the appropriation of quarry dust in the MMC in Figure 12 [6,25]. The incorporation of quarry dust downturn the wear opposition rate underneath 7% which show us that the as fortification in Al base compound won't speedy the wear properties of the composite. The overview has demonstrated that the fuse of the quarry dust in Al-Mg-SiC combination will bring about a minor decrease on the thickness of composite, the diminishing in the wear and hardness of 5% were seen with an upsurge in quarry dust in Figure 13 and 14 [25].

IOP Conf. Series: Materials Science and Engineering **923** (2020) 012055 doi:10.1088/1757-899X/923/1/012055

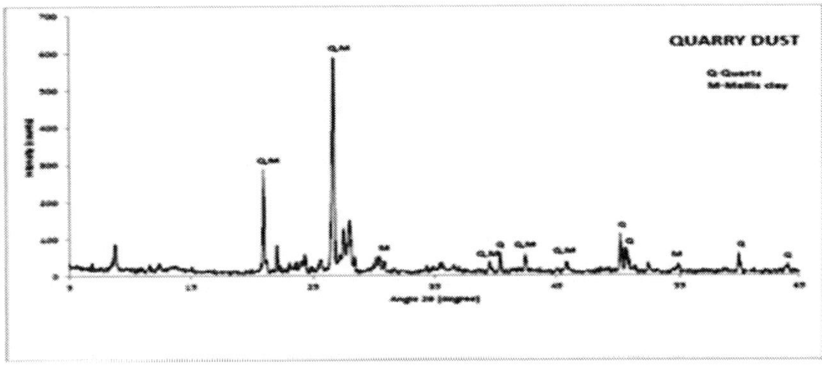


**Figure12.** X-ray Diffraction pattern of quarry dust[25].

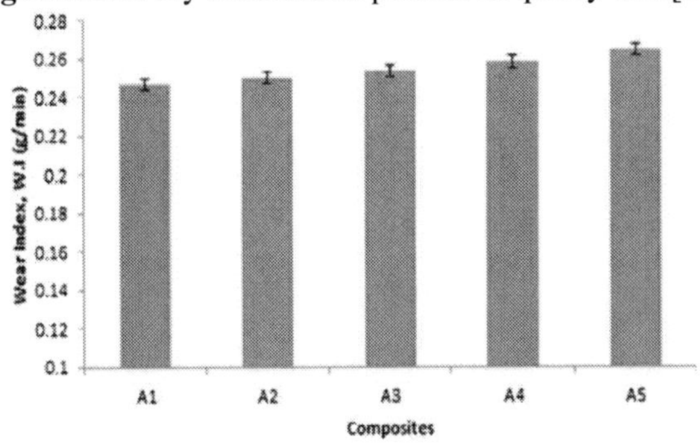


**Figure13.** Wear index for the Silicon Carbide and Quarry Dust reinforced Al-Mg-Si Composites [25]

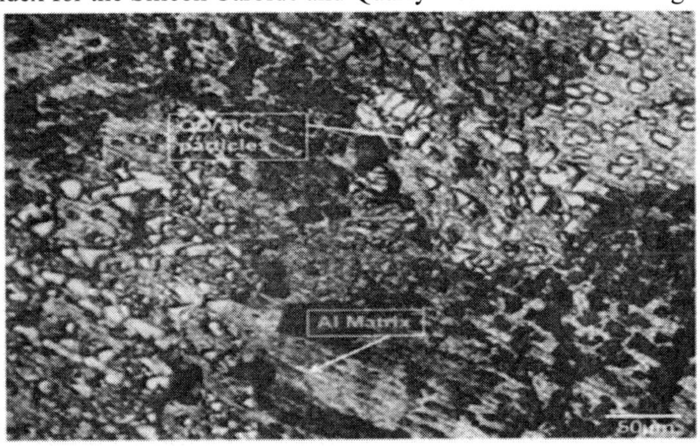


**Figure14.** Optical micrographs of Hybrid Al/SiC/Quarry Dust (1:1) reinforced[25]

## 5. Conclusion

This analysis indicates the theories, the experimental results obtained and the observations of numerous investigators over the years in the study of particle reinforced Al-MMCs. This paper also sets out the manufacturing techniques for the composites and tribological properties of SiC reinforced Al-MMCs. SiC reinforced Al-MMNCs exhibit stronger tribological properties over unreinforced base alloys. The widely used technique for the production of Al-MMCs is stir casting, as it is the cheapest process. Ultrasonic casting and spark plasma sintering are more complex methods that can be used for high delicacy units. From the analysis of the abrasion tolerance and wear index composites. It is assumed that the abrasion resistance and wear index of the composites are also increased if the reinforcement fillings are indicated in the matrix material. The analysis of the tribological properties was carried out with regard to the wear ratio and the outcome of the wear test process

**References**

[1] Bauri R and Yadav D 2018 Introduction to Metal Matrix Composites *Met. Matrix Compos. by Frict. Stir Process.* 1–16
[2] Kumar R A, Devaraju A and Arunkumar S 2018 *Experimental Investigation on Mechanical Behaviour and Wear Parameters of TiC and Graphite Reinforced Aluminium Hybrid Composites* vol 5 (Elsevier Ltd)
[3] Lalmuan S K, Das S, Chandrasekaran M and Tamang S K 2017 Machining Investigation on Hybrid Metal Matrix Composites- A Review *Mater. Today Proc.* **4** 8167–75
[4] Singh K K, Singh S and Shrivastava A K 2017 Comparison of Wear and Friction Behavior of Aluminum Matrix Alloy (Al 7075) and Silicon Carbide based Aluminum Metal Matrix Composite under Dry Condition at Different Sliding Distance *Mater. Today Proc.* **4** 8960–70
[5] Ravi B, Balu Naik B B and Udaya Prakash J 2015 Characterization of Aluminium Matrix Composites (AA6061/B4C) Fabricated by Stir Casting Technique *Mater. Today Proc.* **2** 2984–90
[6] Sunil Ratna Kumar K, Ratnam C and Nagababu B 2019 Fabrication and mechanical behavior of Al 2024-B4C MMCs and Al 2024- B4C -GR hybrid MMCS through powder metallurgy technique *Mater. Today Proc.* **18** 219–29
[7] Kumar A, Singh R C and Chaudhary R 2019 Recent progress in production of metal matrix composites by stir casting process: An overview *Mater. Today Proc.*
[8] Fahimpour V, Sadrnezhaad S K and Karimzadeh F 2012 Corrosion behavior of aluminum 6061 alloy joined by friction stir welding and gas tungsten arc welding methods *Mater. Des.* **39** 329–33
[9] Babu K A, Venkataramaiah P and Yerrathota S 2018 Material Selection for Preparation of Aluminium Hybrid Mmcs *Mater. Today Proc.* **5** 12209–22
[10] Imran M and Khan A R A 2019 Characterization of Al-7075 metal matrix composites: A review *J. Mater. Res. Technol.* **8** 3347–56
[11] Ravikumar M, Reddappa H N and Suresh R 2018 Aluminium Composites Fabrication Technique and Effect of Improvement in Their Mechanical Properties - A Review *Mater. Today Proc.* **5** 23796–805
[12] Ravindranath V M, Shiva Shankar G S, Basavarajappa S and Siddesh Kumar N G 2017 Dry sliding Wear Behavior of Hybrid aluminum Metal Matrix composite reinforced with Boron carbide and graphite particles *Mater. Today Proc.* **4** 11163–7
[13] Bharath V, Nagaral M, Auradi V and Kori S A 2014 Preparation of 6061Al-Al 2 O 3 MMC's by Stir Casting and Evaluation of Mechanical and Wear Properties *Procedia Mater. Sci.* **6** 1658–67
[14] Sijo M T and Jayadevan K R 2016 Analysis of Stir Cast Aluminium Silicon Carbide Metal Matrix Composite: A Comprehensive Review *Procedia Technol.* **24** 379–85
[15] Soorya Prakash K, Sathiya Moorthy R, Gopal P M and Kavimani V 2016 Effect of reinforcement, compact pressure and hard ceramic coating on aluminium rock dust composite performance *Int. J. Refract. Met. Hard Mater.* **54** 223–9
[16] Tang J, Qiao J, Liu Z, Zhou X, Yu G and Zhao J 2018 Mechanism characteristic analysis and soft measuring method review for ball mill load based on mechanical vibration and acoustic signals in the grinding process *Miner. Eng.* **128** 294–311
[17] Kamrani S, Penther D, Ghasemi A, Riedel R and Fleck C 2018 Microstructural characterization of Mg-SiC nanocomposite synthesized by high energy ball milling *Adv. Powder Technol.* **29** 1742–8
[18] Patil O M, Khedkar N N, Sachit T S and Singh T P 2018 A review on effect of powder metallurgy process on mechanical and tribological properties of Hybrid nano composites *Mater. Today Proc.* **5** 5802–8
[19] Ramanathan A, Krishnan P K and Muraliraja R 2019 A review on the production of metal

IOP Conf. Series: Materials Science and Engineering **923** (2020) 012055 doi:10.1088/1757-899X/923/1/012055

matrix composites through stir casting – Furnace design, properties, challenges, and research opportunities *J. Manuf. Process.* **42** 213–45

[20] Nirala A, Soren S, Kumar N and Kaushal D R 2019 A comprehensive review on mechanical properties of Al-B4C stir casting fabricated composite *Mater. Today Proc.* 10–3

[21] Madhukar P, Selvaraj N, Gujjala R and Rao C S P 2019 Production of high performance AA7150-1% SiC nanocomposite by novel fabrication process of ultrasonication assisted stir casting *Ultrason. Sonochem.* **58** 104665

[22] Watson M, Christoforou P, Herrera P, Preece D, Carrell J, Harmon M, Krier P, Lewis S, Maiti R, Skipper W, Taylor E, Walsh J, Zalzalah M, Alhadeff L, Kempka R, Lanigan J, Lee Z S, White B, Ishizaka K, Lewis R, Slatter T, Dwyer-Joyce R and Marshall M 2019 An analysis of the quality of experimental design and reliability of results in tribology research *Wear* **426–427** 1712–8

[23] Pradhan S, Barman T K, Sahoo P and Sahoo S 2018 Tribological behavior of Al-5% SiC metal matrix composite in sodium hydroxide environment *Mater. Today Proc.* **5** 8057–64

[24] Singh J 2016 Fabrication characteristics and tribological behavior of Al/SiC/Gr hybrid aluminum matrix composites: A review *Friction* **4** 191–207

[25] Alaneme K K and Bamike B J 2018 Characterization of mechanical and wear properties of aluminium based composites reinforced with quarry dust and silicon carbide *Ain Shams Eng. J.* **9** 2815–21

IOP Conf. Series: Materials Science and Engineering **923** (2020) 012056 doi:10.1088/1757-899X/923/1/012056

# Fabrication and Corrosion Behaviour of Aluminium Metal Matrix Composites – A Review

**Ashok Kumar R [a], Akash S J[b], Arunkumar S[c], Balaji V[d], Balamurugan M [e], Jeevan Kumar A[f]**

[a & c], Assistant Professor, RMK College of Engineering and Technology, Tiruvallur-601206, Tamil Nadu, India.
[b, d, e & f], UG Students, RMK College of Engineering and Technology, Tiruvallur-601206, Tamil Nadu, India.

Email: ashok.kumar19872003@gmail.com

**Abstract.**
Metal matrix composites (MMCs) corrosion nature is discussed with significance in regard to MMCs corrosion behavior with MMCs electrochemical and chemical properties. Galvanic corr osion between base metal and reinforcement influences the corrosion process in MMCs. The corrosion behaviour of Aluminium Metal Matrix Composites (AMMCs) reinforced with carbide and other solid waste is reviewed. AMMCs are capable of admitting the assault of corrosion over the base metal. AMMC's corrosion type of behaviour is localized with galvanic corrosion, the intermetallic phase between base alloy and reinforcement, micro cracks, voids, are the corrosion-influencing factors. The corrosive action of solid carbide-reinforced AMMCs (Aluminum Metal Matrix Composites) and other solid waste is analysed. The AMMCs have strong mechanical and tribological characteristics but no achievement in the effects of corrosion. Coatings methods are used to enhance the behaviour of AMMCs in corrosion.

**Keywords:** AMMCs, Stir Casting, Corrosion, Pitting, Potentiodynamic Polarization.

## 1. Introduction:

Metal Matrix Composite [MMC] consists of two or more constituent materials in which one acts as a matrix and other being reinforcement. During the fabrication of MMC, the stress applied over the matrix materials are distributed to the reinforcement materials. Subsequently, the required mechanical strength is provided to the composite in a preferential direction by the reinforcement materials. A good MMC should possess brittleness and ductility in the form of reinforcement and matrix respectively. MMCs have numerous advantages when compared to monolithic materials. They have lightweight, superior high-temperature properties, low coefficient of thermal expansion, better abrasion, improved strength and stiffness when compared with monolithic materials [1,2]. Nowadays, MMCs replace various conventional alloys in applications such as automobiles, marine, recreation industries and defense [3].

If a Metal Matrix Composite is made up of more than two constituent materials, it is termed as hybrid composites. Generally, metals such as Aluminum, Steel and Magnesium are used as matrix materials for composites. In MMCs, reinforcement materials are mostly used in the form of wires, whiskers or particulates, continuous/discontinuous fibres in various volume fraction percentages. However particulate metal matrix composites were found to be having superior properties than their monolithic alloys. Callister found out that ceramic material such as carbides, nitrides and oxides when used as reinforcement's exhibit improved stiffness and specific strength at both ambient and elevated temperatures.

MMCs can be fabricated through various production methods such as powder metallurgy, vacuum hot pressing, co-spray deposition and stir casting. The method of production of MMCs also plays an important role in the behaviour of the MMC as it influences the level of porosity, distribution of particles and matrix/reinforcement interfacial properties. It was reported that Stir Casting improves the strength and mechanical characteristics of fabricated composites [4]. Aluminium based metal matrix composites attract much attention in today's world due to their potential to be used as top choice material in various engineering applications. They have good mechanical properties, good corrosion and wear resistance [5–7]. The production of Aluminium metal matrix composites is also cheaper when compared to other metal matrix composites. Nowadays, industrial wastes (fly ash, red mud) and ago based materials (rice husk ash, bagasse) are used as reinforcement materials in Aluminium metal matrix composites [8].

AMMCs suffer from corrosion in various environments due to the presence of the reinforcement phase. Generally, AMMCs exhibits poor corrosion resistance than the base metal. This is due to the reason that aluminium alloys have neutral strong oxide layers on its surface which when subjected to breakdown create a passive film which protects the aluminium alloys from corrosion initiation and crevice. However, in aluminium based metal matrix composites, the passive film gets breakdown due to the reinforcement phase which will make them more susceptible to corrosion. This article reviews various literature works done on the fabrication and corrosion behaviour of aluminium metal matrix composites.

## 2. Material Selection

### *2.1. Reinforcement Selection*

The selection of reinforcements plays an important role in the fabrication of metal matrix composites. Among all the reinforcements, ceramic reinforcements provide MMCs with superior properties. The various factors which are considered for selecting ceramic reinforcements for MMC are

**(i)** Elastic modulus
**(ii)** Tensile strength
**(iii)** Density
**(iv)** Melting temperature
**(v)** Thermal stability
**(vi)** Coefficient of thermal expansion
**(vii)** Compatibility with a matrix material
**(viii)** Size, shape and cost

The thermal mismatch strain ($\varepsilon$) between matrix and reinforcement material are also one of the criteria considered mainly for MMCs that are subjected to thermal cycling.

$$\varepsilon=\Delta\alpha\Delta T$$

Where,

$\Delta\alpha$ = coefficient of thermal expansion,

$\Delta T$=change in Temperature.

In order to minimise the thermal mismatch strain ($\varepsilon$), the coefficient of thermal expansion should be minimal. It was also evident that the relaxation of these strains will alter the thermomechanical processing response of MMC significantly when compared to unreinforced alloys. The investigation done on the size ratio of SiC reinforcement particle to aluminium matrix powder in

IOP Conf. Series: Materials Science and Engineering **923** (2020) 012056 doi:10.1088/1757-899X/923/1/012056

the fabrication of MMCs during Powder Metallurgy (PM) has revealed that there is a relationship between the toughness of fabricated composite and the SiCp-Al powder size ratio.

### *2.2. Matrix Selection*

There is a wide range of metallic alloys available which can be used as matrices in the fabrication of MMCs. However low density, high thermal conductivity and other requirements promote the use of aluminium and magnesium as a predominant matrix element in fabrication of MMCs. For using the MMCs at elevated temperatures, the presence of thermodynamically stable dispersions is required. This can be achieved by using an alloy dispersoids system which minimises the coarsening and interfacial reactions. several investigations have found out that low matrix alloying additions provide MMCs with superior properties.

### *2.3. Matrix Ceramic Interface*

The interface between the matrix and the ceramic reinforcements plays an important role in the load transfer and deformation characteristics of MMCs during deformation. The studies which were done in the early 1960s revealed that the interfacial bond strength between the matrix and the reinforcements can be increased by promoting wetting and by minimizing the formation of oxides [9]. It was also found that the wetting occurs between the matrix and the reinforcements when the interfacial bonding strength between the matrix and the reinforcements exceeds the liquid's surface tension. However, in molten metal-ceramic reinforcement MMC, it is difficult to achieve wetting. This is because the surface tension of molten MMCs is higher. This theory is justified by an investigation which revealed that the aluminium and its alloys exhibit a poor wetting effect on SiC, $Al_2O_3$, Carbon and $B_4C$ below 950°C.

The Wetting behaviour of molten metal systems can be improved in the following ways.

(a) By alloying the matrix material with other reactive materials.
(b) By subjecting the ceramic particulates to heat treatment.
(c) By coating the ceramic particulates with metallic coatings [10].

### *2.4. Other Effects of Reinforcement in the Matrix*

The ageing response of MMCs is different from the equivalent unreinforced alloys. This is due to the presence of Ceramic particulate reinforcements [11–13]. The age-hardening sequence of 6061 aluminium alloy reinforced with SiC and $B_4C$ particulates was similar to the age-hardening sequence of unreinforced Al6061 alloy. Researches also revealed that the formation of intermediate phase and precipitation, the GP zone formation and dissolution were increased in the presence of reinforcement materials. It was also found that the quench insensitive alloys such as Al6061 become quench sensitive alloys due to the effect of SiC particulates reinforcements. However, the quench sensitivity of more quench sensitive Al7475 alloy remains unaltered with reinforcement addition [14]. These findings reveal that the effect of reinforcement depends upon the nature of matrix material.

The presence of high diffusivity paths due to the thermal expansion that exists between the matrix and the reinforcement particulates was reported to be the main reasons for the acceleration of ageing and quench sensitivity in reinforced alloys. Quench sensitivity was also reported in PM monolithic materials but to a lesser degree [15]. Research done on segregation revealed that there is severe segregation of Magnesium and formation of MgO precipitates occurs at Al-SiC interfaces in Al2124-SiC MMCs [12].

## 3. Fabrication

The methods which are most commonly used for the fabrication of MMCs are Solid-phase processes and liquid-phase processes. The selection of the fabrication process for MMCs depends on

1) The temperature of the matrix during processing
2) The desired degree of microstructural integrity
3) Reinforcement loading

The solid-state process which is used for MMC fabrication are isostatic pressing (Powder Metallurgy (PM) Processing), diffusion bonding and spray deposition techniques. Melt stir casting, Spray Casting, melt infiltration and in situ processing are some of the liquid state processes used in MMC Fabrication.

### *3.1. Fabrication methods for Aluminium Composites*

The uniform distribution of reinforcement particle within the matrix alloy is a major issue which is encountered in the fabrication of aluminium composites. In order to overcome this issue, proper selection of processing techniques and operating parameters are required. Powder Metallurgy (PM) and molten metal method are the two methods which are commonly used for the fabrication of aluminium composites [1,14]. Based on the operating parameters, both fabrication techniques have certain limitations.

### *3.2. Powder Metallurgy*

The first step in the fabrication of aluminium composites through Powder Metallurgy (PM) is to blend the aluminium alloy powder with reinforcement particles in order to prepare a homogeneous mixture. The size of the aluminium powder and reinforcement particle should be selected properly. Otherwise, it will cause agglomeration at the end of the blending process [14]. It was also inferred from the studies that the selection of proper size ratio of alloy and reinforcement particle is essential to get uniform distribution in the final product [1]. For the Al/SiC composite, the appropriate size ratio was found to be of the order of 7:10. The hard reinforcement particle suffers from particle fracture during blending. Flaw density and aspect ratio are the two important factors which are responsible for particle fracture during blending. The appropriate aspect ratio for Al alloy and reinforcement powder should be less than 5:1 with a size range of Al alloy and reinforcement powder be 20-40 and 30-20 μm respectively [14]. In Powder metallurgy technique, the initial process involves a blending of matrix and reinforcement powders in order to produce a homogeneous mixture [1]. Then the homogeneous mixture which is obtained is subjected to compression. The compression is done at an isotactic pressure in room temperature. The water molecules present in the mixture may cause porosity during consolidation. So, these water molecules must be removed before the final consolidation process. Vacuum hot-pressing method can be used during the final consolidation process either below alloy's solidus point or in a liquid-solid state to produce densified composites. Extrusion is the final process which is performed at the end of the PM technique to obtain the final product. The extrusion ratio should be greater than 20:1[14]. A high extrusion ratio is essential to obtain a uniform distribution of reinforcement particle in a matrix alloy and superior interfacial bonding between matrix and reinforcement particles.

Powder metallurgy has many advantages [1,14,15]. It allows the usage of any type of matrix alloy and reinforcement for composite fabrication. Through PM, we can also produce composites which contain higher reinforcement content. However, the requirement of long mixing time for obtaining a uniform distribution of particles is a major drawback of Powder Metallurgy (PM) technique [16,17].

### *3.3. Stir Casting*

Stir casting is a liquid metallurgy technique which is most commonly used for the fabrication of particulate reinforced metal matrix composites. This process consists of melting the matrix alloy in the furnace of the stir casting machine. Then the reinforcement materials are added to the melt of the matrix alloy and the mixture is stirred by using a mechanical stirrer. The obtained molten metal mixture is transferred to a mould of the desired shape. After transferring the molten metal, the mould is allowed to cool. After solidification, the required composite is ejected from the mould and subjected

IOP Conf. Series: Materials Science and Engineering **923** (2020) 012056 doi:10.1088/1757-899X/923/1/012056

to various heat treatments. The major challenge in stir casting is the wettability of reinforcement material in the molten metal. It was found that the use of magnesium ribbons provides sufficient wettability while using ceramic materials as reinforcement.

### *3.4. Stir Casting of Aluminium Metal Matrix Composites*

Aluminium metal matrix composites were fabricated using SiC as reinforcement material and magnesium as wettability agent. The findings revealed that the mechanical properties of the composites namely yield strength, hardness and tensile strength improved when the addition of reinforcement particle to the base alloy was less than 20 percentage [18]. A study was done on Aluminium metal matrix composites reinforced with SiC having 60μm particle size. Mechanical stirrer consisting of mild steel impeller was used to mix the alloy and reinforcement mixture in this study. The stirring was done with different stirrer speed (500 rpm,600 rpm,700 rpm) and different stirring time (5 min,10 min,15 min). It revealed that higher stirring speed and stirring time leads to the production of composites with homogeneous particle distribution and greater hardness [19]. Fabrication of Al-$Al_2O_3$ composites through Stir casting revealed that the ductility of the composites decreased with the addition of reinforcements to the base alloy. However other mechanical properties such as yield strength, tensile strength and hardness were considerably increased with reinforcement addition [20]. Preheated Borax was preheated to 250°C for 20 minutes and used as a wetting agent during the fabrication of Al6063-SiC Composite. It was found that composite containing SiC reinforcement between 9 to 12 percentage by volume exhibits improved strength. However, the ductility of MMC remains unchanged [21]. the microhardness and wear behaviour of the aluminium metal matrix composites improved with the addition of $TiB_2$ reinforcements to the Al6061 base alloy [22]. A study done on the effect of stirrer parameters on the fabrication of metal matrix composites revealed that the formation of clusters can be prevented by modifying the stirrer design thereby the uniform distribution of base alloy and reinforcement can also be achieved. It was also found that changing the feeding mechanism in MMC productions provides a uniform spray of particulates [23]. The development of Al-SiC-$Al_2O_3$ hybrid metal matrix composite using an automatic agitator revealed that the reinforcements were distributed homogeneously in the Al6061 based alloy [24].

The strengthening derivable from load transfer in MMC from the matrix to the reinforcements can be decreased by using reinforcements such as rice husk which is softer when compared to SiC [25]. The use of argon gas as a protective covering for a furnace in which AZ91 was charged prevented the effects of agglomeration in the fabricated AZ91-SiC Composites [26]. The blending of reinforcement particle with a considerable amount of flux during stir casting removed slag from the molten mixture. As a result, wettability was increased [27].

## 4. Corrosion Behaviour

The major limitations in the applications of MMC are their corrosion behaviour. The heterogeneities in MMC tends to affect its corrosion characteristics. Galvanic Coupling, micro crevices, porosity, voids and high reactivity interfacial phases are some of the corrosion-related problems in MMC. The reported works on corrosion behaviour of MMCs is also less when compared to other behaviours of composites. Various reporters had used various tests namely Stress Corrosion Crack (SCC), immersion test and Potentiodynamic test to study the corrosion behaviour of MMCs.

### *4.1. Corrosion behaviour of Aluminium-based particulate metal matrix composites*

The initiation of pitting on the PMMCs interface is considered to be the main reason for the corrosion of aluminium-based PMCs. Researchers reinforced Al2024, Al5466 and Al6061 with SiC through Powder metallurgy and studied their pitting potential and polarization behaviours in solutions containing 0.1 N and 0.8 N NaCl. The polarization behaviour of the fabricated composites was found to be in similar to their matrix alloys. Similarly, the pitting potentials of the fabricated MMCs were similar in accordance with the pitting potentials of respective alloys except in cases when Al2024 was used as a base alloy. The pitting potential value of Al2024-SiC composite was found to be more negative than the base alloy. However, this doesn't affect the matrix structure of the Al2024-SiC composites as well as the base alloy [28]. Studies also revealed that the SiC reinforced Al6061-T6 exhibits better corrosion resistance than Al6061-SiC composites [29]. The studies done on the corrosion behaviour of Al6061 reinforced with carbon fibres and alumina fibres or SiC whiskers revealed that the Al6061 based MMC and Al6061 exhibits similar pitting Potentials. It was also found that SiC whiskers reinforced Al6061 MMC Showed greater pit initiation than the base alloy. It was also found that the cathodic current of Al6061-SiC whiskers was higher than base alloys due to the presence of interfacial layer [30]. The study done on the polarization behaviour of Al1050 reinforced with $SiC_p$ of particle size 3μm and 30 μm composite revealed that the pitting attacks on the composites are higher when compared to the base alloys. It was due to the presence of voids at the interface between matrix and reinforcement. However, polarization behaviour remains unaffected by the residual stress that exists between matrix and reinforcements. The author also concluded that the 20μm $SiC_p$ were found to be more prone to localized corrosion than the 60 μm $SiC_p$ [31]. The research done on the stress corrosion behaviour of Al2014-T6, Al6061-T6, Al6061-T6 reinforced with $SiC_p$ & $SiC_w$ and Al2014-T6 reinforced with $Al_2O_3$ revealed that the corrosion rate of Al6061 based MMCs in 3.5% NaCl was considerably lesser than the Al2014 alloy and its composites [32].

A study done on localized corrosion and its effect on the Al2024-$SiC_p$ MMC Containing $SiC_p$ of particle size 14 μm in varying volume fractions revealed that Al2024-$SiC_p$ MMCs exhibits less corrosion resistant when compared to the unreinforced base alloy. It also revealed that the increase in the addition of up to the base alloy decreases the pitting potential. At the same time, it increases the number of active sites on the composite's surface [33]. Yue fabricated Al6013-$SiC_p$ composite through PM technique. He used Nd: YAG laser to change the surface microstructure of fabricated composite thereby improving the corrosion resistance of the composites. His findings revealed that the pitting resistance of the fabricated composites was improved by using Nd: YAG laser during PM technique. Researchers studied the corrosion characteristics of heat-treated Al6063 containing $SiC_p$ reinforcements in various weight percentage. They used a two-step stir casting process to fabricate the composites. They conducted the corrosion tests in 5% NaCl Solution. Their findings revealed that the addition of $SiC_p$ increases the corrosion rate of the composites [34]. A Study done on the corrosion behaviour of Pressure infiltrated Al-Mg alloy/$SiC_p$ reported that the corrosion attacks on pure Al matrix were three folds more when compared to Mg containing aluminium matrices. The corrosion resistance of the composites is shown in Figure1 and Figure 2 was found to be increased with the presence of Mg content in the Al matrix [35]. The corrosion resistance of Al-SiC composites fabricated through PM technique revealed that the corrosion resistance increases with the increase in a volume fraction of $SiC_p$ reinforcement particles. However, the reduction in particle size of $SiC_p$ considerably improves the corrosion resistance of the fabricated composites [35].

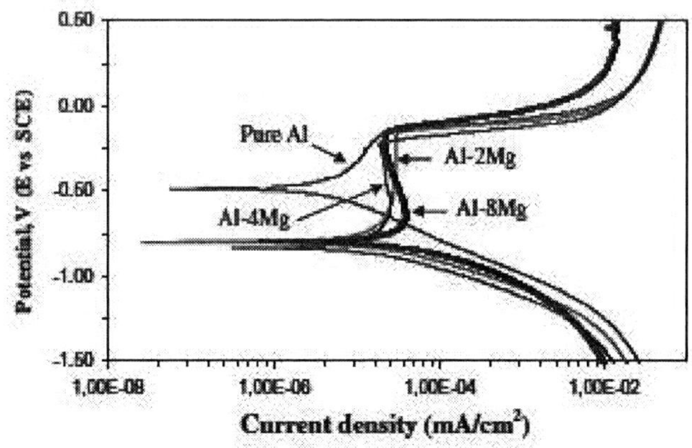


**Figure 1.** Polarization curves of pure Al, Al-2Mg, Al-4 Mg, Al-8 Mg in 3.5 wt.% NaCl for 1 hr [36]

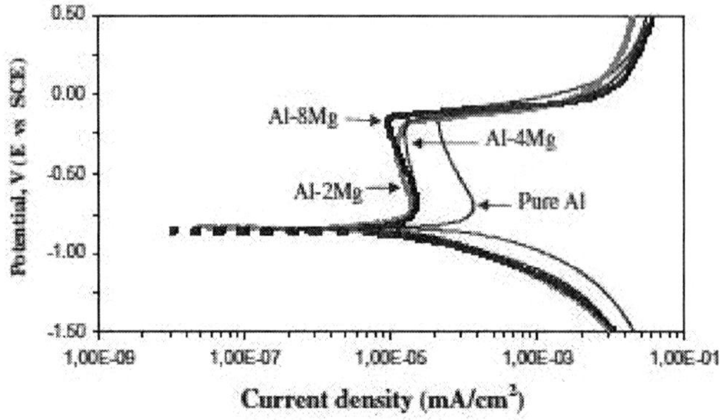


**Figure 2.** Polarization curves of pure Al, Al-2Mg, Al-4 Mg, Al-8 Mg in 3.5 wt.% NaCl for 7 days [36]

The fabrication of LM13-albite MMC by reinforcing LM13 with 2 to 6 wt.% of albites through liquid metallurgy revealed that the corrosion rates of both matrices and fabricated MMCs decreased with the increase in exposure time and increased linearly with increase in temperature [37]. A corrosion study was done on the Al2024-TiC MMC in 3.5% NaCl solution revealed that the size and number of pits in the composites decreases with increase in anodic current density [38]. The corrosion crack behaviour of Al7075-Zrp MMC fabricated through Stir casting was studied. The particle size of Zr ranges from 40-50 µm. Natural seawater was used as a corroding medium. Findings revealed that the corrosion characteristics of Al7075-Zr P and base alloys increase with the increase in temperature of seawater. This is due to an increase in the diffusion rate of hydrogen. A detailed report is done on the stress corrosion cracking (SCC) behaviour of pure Al/Si$C_w$ and Al2024/Si$C_w$ composites, treated annealing and fabrication by squeeze casting [39]. The sample was taken as a double cantilever specimen, which was immersed in simulated seawater i.e., 3.5% NaCl solution for evaluating SCC properties. Annealing temperature majorly affects the steady propagation rate of SCC for both composites. The dislocation density and stable propagation rates of SCC were found to be decreased, on an increase in annealing temperature, but the threshold stress intensity factor increases. Reports show that annealing at high temperature resists the SCC behaviour. Al 2009/Si$C_p$ composite was investigated based on its corrosion behaviour in de-aerated neutral 3.5% NaCl solution at 25°C by Potentiodynamic polarization [40]. This polarization test showed that carbide turned sample had less resistance to corrosion when compared to WEDM and diamond-turned samples. A detailed report was done on the corrosion properties of Al-Si-Mg-Si$C_p$ composite, with addition of scandium in minimal amount, fabricated by powder process technique [41]. The Potentiodynamic polarization test was carried out at room temperature in electrolyte consisting of 0.1% and 3% NaCl with pH variation of 2, 7, 12. With addition of 1.5% $Na_2SO_3$ to 3% NaCl, the corrosion study was made. The study revealed that pitting behaviour was more observed in presence of 3% NaCl solution than 0.1% NaCl solution. Investigations done on the pitting phenomenon of aluminium MMC's (A 360/10% Si$C_p$, A360/20% Si$C_p$, A380/10% Si$C_p$, A380/20% Si$C_p$) fabricated by pressurized die casting technique in 1-3.5 wt% NaCl solution at room temperature using Potentiodynamic polarization revealed that the corrosive attack in Al/Si$C_p$ composite materials was observed to be started at the interfacial layers of matrix/Si$C_p$ and matrix/intermetallic phases[42].

### *4.2. Corrosion Inhibitors*

It is necessary to protect the aluminium based metal matrix composites from corrosion to prolong their usage period. Various studies done by researchers revealed that corrosion inhibitors such as polymer coating and anodizing help to protect the composites from corrosion attacks in a better way. Researchers used rare-earth metal ions such as Ce3+, Y3+, La3+, Pr3+, Nd3+ for preventing pitting corrosion in composites [43].

Investigation performed on the action of anodizing coating on Al2024 alloy and $SiC_p$ / Al2024 PMMC as an inhibiting agent in 3.5% NaCl solution revealed that the anodizing coating Al2024 alloys possess good protection from corrosion when compared to Al2024/$SiC_p$ PMMC. It was due to the presence of SiC particulates in Al2024/$SiC_p$ [44]. A Study was performed using $K_2Cr_2O_7$, Sodium molybdate and Cerium chloride as coatings for preventing corrosion in Al6013/$SiC_p$ Composites in 3.5% NaCl solution. The study concluded that cerium chloride treatment provides better protection to the composites than the other coatings. The anti-corrosion film prepared by using micro-arc oxidation (MAO) on Al2024/$SiC_p$ composite revealed that the MAO technique inhibits corrosion attacks on the samples in 3.5% NaCl Solution [45].

## 5. Conclusion

The above cited review for Aluminium metal matrix Composites gives the following Conclusion

- Aluminium alloys are more susceptible to localized pitting corrosion in chloride environment.
- Aluminium based MMCs reinforced with SiC, B4C and fly ash have more tendency to corrode when compared to TiC.
- The volume fraction and Size of the particulates influence the corrosion behaviour of the composites.
- The MMCs which are subjected to heat treatments after its fabrication exhibits improved corrosion resistance.

## References

[1] Surappa M.K. 2003 Aluminium Matrix Composites: Challenges and Opportunities // Sadhana. Department of Metallurgy, Indian Institute of Science, Bangalore 560 012, India **28** 319–334

[2] Waku Y and Naeasawa T 1994 Future Trends and Recent Developments of Fabrication Technology for Advanced Metal Matrix Composites *Mater. Manuf. Process.* **9** 937–63

[3] Kumar R A, Devaraju A and Arunkumar S 2018 *Experimental Investigation on Mechanical Behaviour and Wear Parameters of TiC and Graphite Reinforced Aluminium Hybrid Composites* vol 5 (Elsevier Ltd)

[4] Inegbenebor A O, Bolu C A, Babalola P O, Inegbenebor A I and Fayomi O S I 2018 Aluminum Silicon Carbide Particulate Metal Matrix Composite Development Via Stir Casting Processing *Silicon* **10** 343–7

[5] Hao S and Xie J 2013 Tensile properties and strengthening mechanisms of SiCp- reinforced aluminum matrix composites as a function of relative particle size ratio *J. Mater. Res.* **28** 2047–55

[6] Kaveendran B, Wang G S, Huang L J, Geng L and Peng H X 2013 In situ (Al3Zr + Al2O3np)/2024Al metal matrix composite with novel reinforcement distributions fabricated by reaction hot pressing *J. Alloys Compd.* **581** 16–22

[7] Sajjadi S A, Ezatpour H R and Beygi H 2011 Microstructure and mechanical properties of Al-Al2O3 micro and nano composites fabricated by stir casting *Mater. Sci. Eng. A* **528** 8765–71

[8] Alaneme K K, Akintunde I B, Olubambi P A and Adewale T M 2013 Fabrication characteristics and mechanical behaviour of rice husk ash - Alumina reinforced Al-Mg-Si alloy matrix hybrid composites *J. Mater. Res. Technol.* **2** 60–7

[9] Lindroos V K and Talvitie M J 1995 Recent advances in metal matrix composites *J. Mater. Process. Tech.* **53** 273–84

[10] Rajan T P D, Pillai R M and Pai B C 1998 Reinforcement coatings and interfaces in aluminium metal matrix composites *J. Mater. Sci.* **33** 3491–503

[11] Papazian J M 1988 Effects of SiC whiskers and particles on precipitation in aluminum matrix

composites *Metall. Trans. A, Phys. Metall. Mater. Sci.* **19 A** 2945–53

[12] Nutt S R and Carpenter R W 1985 Non-equilibrium phase distribution in an Al-SiC composite *Mater. Sci. Eng.* **75** 169–77

[13] Lea C and Molinari C 1984 Magnesium diffusion, surface segregation and oxidation in Al-Mg alloys *J. Mater. Sci.* **19** 2336–52

[14] Lloyd D J 1994 Particle reinforced aluminium and magnesium matrix composites *Int. Mater. Rev.* **39** 1–23

[15] Kumar S and Balasubramanian V 2008 Developing a mathematical model to evaluate wear rate of AA7075/SiCp powder metallurgy composites *Wear* **264** 1026–34

[16] Ted Guo M L and Tsao C Y A 2002 Tribological behavior of aluminum/SiC/nickel-coated graphite hybrid composites *Mater. Sci. Eng. A* **333** 134–45

[17] Smith L N, Midha P S and Graham A D 1998 Simulation of metal powder compaction, for the development of a knowledge based powder metallurgy process advisor *J. Mater. Process. Technol.* **79** 94–100

[18] Ghazi J H 2013 Production and Properties of Silicon Carbide Particles Reinforced Aluminium Alloy Composites *Mater. Sci. Eng.* **1** 191–4

[19] Prabu S B, Karunamoorthy L, Kathiresan S and Mohan B 2006 Influence of stirring speed and stirring time on distribution of particles in cast metal matrix composite *J. Mater. Process. Technol.* **171** 268–73

[20] Bharath V, Nagaral M, Auradi V and Kori S A 2014 Preparation of 6061Al-Al 2 O 3 MMC's by Stir Casting and Evaluation of Mechanical and Wear Properties *Procedia Mater. Sci.* **6** 1658–67

[21] Alaneme K K and Aluko A O 2012 Fracture toughness (K 1C) and tensile properties of as-cast and age-hardened aluminium (6063)silicon carbide particulate composites *Sci. Iran.* **19** 992–6

[22] Suresh S and Moorthi N S V 2013 Process development in stir casting and investigation on microstructures and wear behavior of TiB2 on A16061 MMC *Procedia Eng.* **64** 1183–90

[23] Thomas A T, Parameshwaran R, Muthukrishnan A and Kumaran M A 2014 Development of Feeding & Stirring Mechanisms for Stir Casting of Aluminium Matrix Composites *Procedia Mater. Sci.* **5** 1182–91

[24] Umanath K, Selvamani S T, Palanikumar K K and Niranjanavarma D 2014 Metal to metal worn surface of AA6061 hybrid composites casted by stir casting method *Procedia Eng.* **97** 703–12

[25] Alaneme K K and Ajayi O J 2017 Microstructure and mechanical behavior of stir-cast Zn–27Al based composites reinforced with rice husk ash, silicon carbide, and graphite *J. King Saud Univ. - Eng. Sci.* **29** 172–7

[26] Viswanath A, Dieringa H, Ajith Kumar K K, Pillai U T S and Pai B C 2015 Investigation on mechanical properties and creep behavior of stir cast AZ91-SiCp composites *J. Magnes. Alloy.* **3** 16–22

[27] Raviraj M S, Sharanprabhu C M and Mohankumar G C 2014 Experimental Analysis on Processing and Properties of Al-TiC Metal Matrix Composites *Procedia Mater. Sci.* **5** 2032–8

[28] Trzaskoma P P 1983 Corrosion Behavior of SiC/Al Metal Matrix Composites *J. Electrochem. Soc.* **130** 1804

[29] Aylor D M and Moran P J 1984 Effect of Reinforcement on the Pitting Behavior of Aluminum-Base Metal Matrix Composites. *Proc. - Electrochem. Soc.* **84–9** 584–95

[30] Shimizu Y, Nishimura T and Matsushima I 1995 Corrosion resistance of Al-based metal matrix composites *Mater. Sci. Eng. A* **198** 113–8

[31] Trowsdale A J, Noble B, Harris S J, Gibbins I S R, Thompson G E and Wood G C 1996 The influence of silicon carbide reinforcement on the pitting behaviour of aluminium *Corros. Sci.* **38** 177–91

[32] Monticelli C, Zucchi F, Brunoro G and Trabanelli G 1997 Stress corrosion cracking behaviour of some aluminium-based metal matrix composites *Corros. Sci.* **39** 1949–63

[33] Feng Z, Lin C, Lin J and Luo J 1998 Pitting behavior of SiCp/2024 Al metal matrix composites *J. Mater. Sci.* **33** 5637–42
[34] Alaneme K 2011 Corrosion Behaviour of Heat - Treated Al-6063 / SiC p Composites Immersed in 5 wt % NaCl Solution *Leornardo J. Sci.* 55–64
[35] Zakaria H M 2014 Microstructural and corrosion behavior of Al/SiC metal matrix composites *Ain Shams Eng. J.* **5** 831–8
[36] Candan S 2009 An investigation on corrosion behaviour of pressure infiltrated Al-Mg alloy/SiCp composites *Corros. Sci.* **51** 1392–8
[37] Sharma S C 2001 A study on stress corrosion behavior of A16061/albite composite in higher temperature acidic medium using autoclave *Corros. Sci.* **43** 1877–89
[38] Albiter A, Contreras A, Salazar M and Gonzalez-Rodriguez J G 2006 Corrosion behaviour of aluminium metal matrix composites reinforced with TiC processed by pressureless melt infiltration *J. Appl. Electrochem.* **36** 303–8
[39] Hu J, Ren W and Yao Z 2002 Effect of aging treatment on stress corrosion cracking behavior of alumina borate whisker reinforced aluminum composite *Corros. Sci. Prot. Technol.* **14** 136–8
[40] Yu H, Dong Q, Chen Y and Chen C 2018 Influence of silicon on growth mechanism of micro-arc oxidation coating on cast Al–Si alloy *R. Soc. Open Sci.* **5** 1069–72
[41] Datta J, Datta S, Banerjee M K and Bandyopadhyay S 2004 Beneficial effect of scandium addition on the corrosion behavior of Al-Si-Mg-SiCp metal matrix composites *Compos. Part A Appl. Sci. Manuf.* **35** 1003–8
[42] Pardo A, Merino M C, Merino S, Viejo F, Carboneras M and Arrabal R 2005 Influence of reinforcement proportion and matrix composition on pitting corrosion behaviour of cast aluminium matrix composites (A3xx.x/SiCp) *Corros. Sci.* **47** 1750–64
[43] Mishra A K and Balasubramaniam R 2007 Corrosion inhibition of aluminum alloy AA 2014 by rare earth chlorides *Corros. Sci.* **49** 1027–44
[44] He C, Lou D, Wang J and Cai Q 2011 Corrosion protection and formation mechanism of anodic coating on SiC p/Al metal matrix composite *Thin Solid Films* **519** 4759–64
[45] Xue W, Wu X, Li X and Tian H 2006 Anti-corrosion film on 2024/SiC aluminum matrix composite fabricated by microarc oxidation in silicate electrolyte *J. Alloys Compd.* **425** 302–6

IOP Conf. Series: Materials Science and Engineering **923** (2020) 012057 doi:10.1088/1757-899X/923/1/012057

# Experimental investigation of the effect of Nano-silica on the mechanical properties of concrete

**S.Thenmozhi, V.Gowri, A.Vignesh and S.Suganthan**

Department of Civil Engineering, St. Joseph's Institute of Technology, OMR, Chennai 600119.

E-mail: thenmozhicivil@yahoo.com

**Abstract**. Construction Industry is one of the leader in deterioration of environment. The production of cement consumes large quantities of raw materials, energy, heat and releases greenhouse gases. This paper focus to minimize the use of cement in the production of concrete. Concrete cubes and cylinders were prepared with 0,2, 4 and 6% nano-silica and cured. Mechanical properties such as compressive strength, Split tensile strength, Flexural Strength and Sulphate attack of concrete were obtained and compared with conventional concrete for 7 days, 14 days and 28days curing. The tests results shows 4% of nano-silica in concrete increases the Compressive and Split tensile strength of concrete.

Keywords *Keywords:* Nano-silica, Compressive strength, Split tensile strength, Concrete, Flexural Strength, Sulphate attack

## 1. Introduction

In construction industry the use of high strength concrete has increased to a large extend. The strength of the concrete can be increased by adding additives and admixtures exhibiting similar characteristics as that of cement. Nano sized particles contribute to the technological advancements. Nano particles are added to the concrete whose properties influence the strength characteristics of concrete. Nano silica is an additive which is used in concrete. The unique properties of nano silica in the nano scale impart higher mechanical and durability properties than conventional concrete. The very fine nano silica fills the pores of the concrete matrix increasing the binding strength of concrete. Nano silica also enhances the production of calcium silicate hydrate during hydration process which contributes to the improvement in strength and durability. The nano fillers reduce the porosity of concrete and increase the early pozzolonic strength to the concrete. This paper attempts to study the optimum percentage of nano silica that provides optimum mechanical properties of concrete incorporating Nano-Silica.

The addition of nano silica in the concrete and its strength behavior has been reported by many researchers. Hongjian Du et al., (2014) investigated the durability properties of concrete containing nano silica and reported that at early stages of hydration nano silica incorporated concrete exhibited higher pozzolonic reaction. This showed the early increase in compressive strength. SEM analysis depicted that pore sizes were homogeneous resulting in lesser rate of water and chloride ion penetration. Based on the study investigated by Anjusha and Bindhu (2014) compressive strength, flexural strength, split tensile strength and modulus of elasticity improved with 2% addition of nano silica. The impact strength also showed an improvement. The durability studies with the addition of nano silica showed improvement towards the sulphate resistance and water absorbtion [2]. The nano silica has a pore filling effect and pozzolonic activity of nano silica with cement towards improvement of mechanical property and durability characteristics, this is stated by Maheswaran et al (2013). Jonbi et al., (2012) on the investigation of incorporating nano silica along with silica fume in concrete reported that the percentage on nano silica to be incorporated in concrete should not be more than 10%. Based on the inference from SEM analysis it is observed that nano silica readily react with calcium hydroxide to produce new forms of calcium silicate hydrate which leads to improvement in strength and durability [4].

The effect of nano silica for the chemical composition, mechanical characterization and chloride permeability was studied by Gopinath et al., (2012). It was concluded after investigation that nano silica on hydration reduced the development of calcium hydroxide indicative of the formation of calcium silicate hydrate gel. Compressive strength was observed to be increased with addition of nano silica. The addition of 2% by weight of nanosilica with 50% GGBS reported an increase of compressive strength by 18% and the setting time also showed variation when compared to slag concrete [6].

## 2. Materials

Concrete was cast with Ordinary Portland cement 53 grade in accordance to IS 12269-1987., natural river sand with fraction passing through 4.75mm sieve and retained on 60micron sieve was used. 20mm coarse aggregate was used for the study. Nano silica of particle size 50nm is used. The specific gravity of nano silica is determined as 1.03. Super plasticiers confined to IS9103-1999 was used. M30 grade of concrete was designed as per IS 10262-2009, with mix ration of 1:1.61:2.68 and the water cement ratio of 0.45 was adopted for this study.

## 3. Experimental Procedure

Compression tests were carried out for conventional concrete and for concrete with varying percentages of nano silica (2%, 4%, 6%) as partial replacement of cement. The tests were done in accordance with IS 516-1999 standards conducted on concrete specimen size 150mm x 150mm x 150mm. The specimens were tested respective to their days in compression testing machine.

### *3.1. Split Tensile Strength*

Cylinder specimens of 150 mm diameter and 300 mm length were prepared with designed mix concrete for varying percentages of nano silica and split tensile this tests were carried out in accordance with IS 516-1999.

### *3.2. Flexural Strength Test*

Flexural tests were done with specimen size of 500x100x100 in accordance to IS 519-1999. The specimen testing should be done with two rollers of 38mm diameter and specimen placed along lines spaced 13.3 cm apart. The line of fracture measured on the center line of the specimen

### *3.3. Sulphate Resistance Test*

An experimental study was undertaken to evaluate the relative chemical resistance of concrete made with 4% of nano silica to the following solution; 1 % $H_2SO_4$, 1% HCl and 5% ammonium sulphate. Ammonium sulphate was used for sulphate resistance determination. Acid resistance study was conducted using sulphuric acid and hydrochloric acid. Time taken for 25% weight loss of the specimen fully submerged is taken as a criterion for failure. Concrete cube specimen of 150mm were used for testing and specimens were subjected to alternate wetting and drying process for a 15 days interval. The weight loss of the specimens is determined in terms of percentage.

## 4. Results and Discussion

### *4.1. Compressive Strength Test*

The cube specimens were tested for their compressive strength. Nano silica incorporated concrete with 2%, 4% and 6% by weight of cement has been casted for compression tests in compression testing

IOP Conf. Series: Materials Science and Engineering **923** (2020) 012057 doi:10.1088/1757-899X/923/1/012057

machine. It is inferred that the compression strength increases till 4% replacement of nano silica and then drops down. The pozzolonic action of nano silica within the pores of matrix, accelerates the hydration and increases the strength of concrete. Calcium silicates formation during hydration process reduces the generation of cracks and exhibits the increase in strength. The increase in addition of the nano silica does not participate in the strength increase after 4% because excess formation of calcium silicates does not contribute to the improved strength. Table 1 shows the compressive strength of concrete specimens for 7, 14 and 28 days.

**Table 1.** Compressive strength of 7, 14 & 28 days having different proportion.

| **% of Nano Silica** | **7 DAYS** (N/mm²) | **14 DAYS** (N/mm²) | **28 DAYS** (N/mm²) |
|---|---|---|---|
| Conventional Concrete (0 %) | 21.62 | 26.72 | 32.35 |
| 2 % | 22.08 | 27.08 | 34.23 |
| 4 % | 24.64 | 29.23 | 36.81 |
| 6 % | 22.13 | 27.98 | 35.12 |

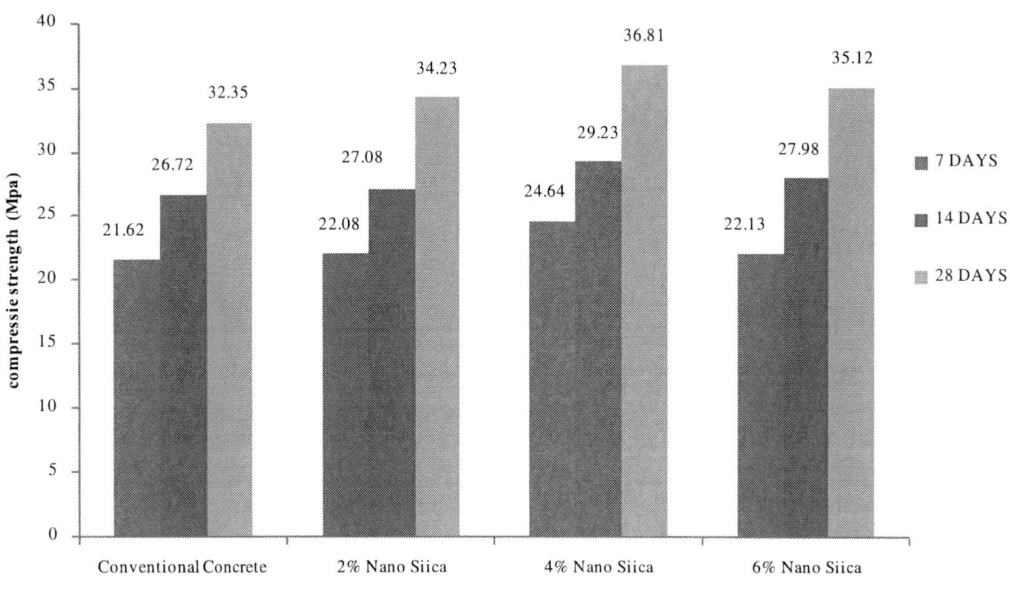


**Figure 1.**.Compressive Strength of concrete

*4.2. Split Tensile Strength*

To evaluate the tensile strength of concrete the cylindrical samples were tested for their split tensile strength as per IS 5816:1999. Table 2 reports the 7, 14 and 28 days Split tensile strength with respect to varying percentages of nano Silica in the concrete specimens. It is inferred that tensile strength of nano silica incorporated specimens increase and it is higher than conventional concrete.

**Table 2.** Split Tensile strength of 7, 14 & 28 days having different proportion.

| % OF NANO SILICA | 7 DAYS ($N/mm^2$) | 14 DAYS ($N/mm^2$) | 28 DAYS ($N/mm^2$) |
|---|---|---|---|
| Conventional Concrete(0 %) | 2.35 | 2.64 | 2.80 |
| 2 % | 2.47 | 2.70 | 2.94 |
| 4 % | 2.67 | 2.79 | 3.15 |
| 6 % | 2.40 | 2.61 | 2.98 |

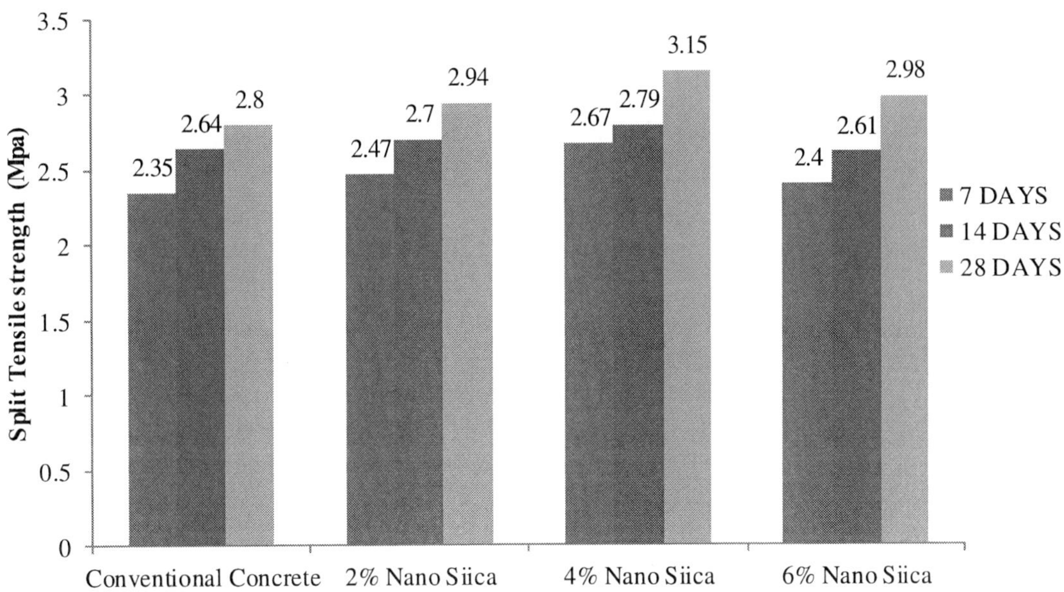


**Figure 2.**. Split Tensile Strength of concrete

*4.3. Flexural Strength of Concrete*

The Prism samples were tested for theirflexural strength. Table 3 reports the flexural specimen at 4% shows maximum flexure strength.

**Table 3.** Flexural Strength of 7, 14 & 28 days having different proportion.

| % OF NANO SILICA | 7 DAYS ($N/mm^2$) | 14 DAYS ($N/mm^2$) | 28 DAYS ($N/mm^2$) |
|---|---|---|---|
| Conventional Concrete (0 %) | 2.46 | 3.13 | 3.48 |
| 2 % | 2.50 | 3.21 | 3.52 |
| 4 % | 2.55 | 3.34 | 3.55 |
| 6 % | 2.52 | 3.29 | 3.50 |

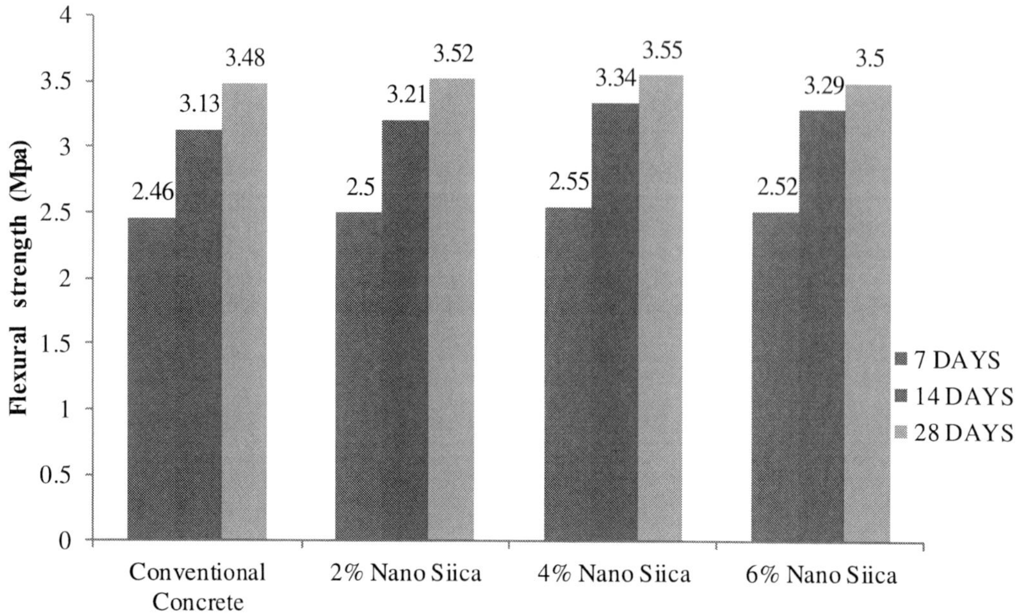


**Figure 3.** Flexural Strength of concrete

### *4.4. Sulphate Attack Test*

The specimens of size 100mmx100mmx100mm were tested and the dried weight of the specimen before and after sulphate attack is taken. It can be observed from the test that the average weight loss for nano silica incorporated specimen is less compared to conventional concrete.

**Table 4.** Sulphate attack test for 40% nano silica concrete.

| Mix ID | Initial weight (Weight after 28 days) | Final weight (After 3 days immersion in $MgSo_4$) | % of Weight loss |
|---|---|---|---|
| | 8.678 | 8.240 | 5.04 |
| Conventional concrete | 8.646 | 8.242 | 4.67 |
| | 8.652 | 8.247 | 4.68 |
| Concrete made with 4% of Nano silica | 8.435 | 8.128 | 3.63 |
| | 8.457 | 8.154 | 3.58 |
| | 8.402 | 8.198 | 3.43 |

## 5. Conclusion

Based on experimental research for Concrete made with 4% of Nano silica is used as partially replacement of cement and the harden properties of compressive strength, flexural strength and split tensile strength of concrete was tested at7days,14day and 28 days.

The test results shows that the use of 4% of Nano silica in concrete,its increase the compressive strength, split tensile strength and also increase in the flexural strength of concrete as compared with conventional concrete.

The durability test are to be conducted on concrete made with 4% of Nano silica, its shows greater resistance against chemical attach as compared with conventional concrete. The following benefits are drawn when the Nano silica used in concrete,

- Concrete with high initial and final compressive and tensile strengths.
- Concrete with good workability.

- Cessation of super plasticizing utilization
- Cessation of silicosis risk
- Accelerates the hydration.
- Better bond between aggregates and cement paste.
- Improves the toughness, shear, tensile strength and flexural strength of concrete

**References**

[1] Hongjian Du, Suhuan Du, Xuemei Liu, Durability performance of concrete with Nano silica, Construction and building material 73 (2014) 705-712.
[2] Anjusha R, Bindu S, Influence of Nano silica on properties of ternary blended concrete, International journal of scientific and engineering research, Volume 5, Issue 7, July 2014.
[3] Maheswaran S, Bhuvaneshwari B, Palani G S, Nagesh R Iyer and Kalaiselvan S, An overview on the influence of Nano silica in concrete and a research initiative, Research journal of recent sciences, Vol 2 (ISC 2012), 147-24 (2013).
[4] Jonbi, Ivindra Pane, BinsarHarindja, Iswandi Imran, The use of Nano silica for improving of concrete Compressive strength and Durability, Applied mechanics and materials, Vol.204-208 (2012) pp 4059-4062.
[5] S.Gopinath, P.C.H. Mouli, A.R.Murthy, N.R.Iyer, S.Maheswaran, Effect of Nano silica on mechanical properties and durability of normal strength of concrete, Archives of Civil Engineering, LVIII,4, 2012.
[6] Min-Hong Zhang, Jahidul Islam, 2012, Use of nano silica to reduce setting time and increase early strength of concretes with high volumes of fly ash or slag, Constr. Build Mater., 29, 573–580.
[7] Bureau of Indian Standards IS 10262: 2009, Concrete Mix Proportioning-Guidelines.
[8] IS: 516-1959, Indian standard code of practice- Methods of test for strength of concrete, Bureau of Indian Standards, New Delhi, India, 1979.
[9] Bureau of Indian Standards, IS: 5816-1999, Indian standard code of practice- Method of Test for Splitting Tensile Strength of Concrete New Delhi, 1999.

# INVESTIGATION ON MECHANICAL PROPERTIES OF ALUMINUM METAL MATRIX COMPOSITES – A REVIEW

**Karthick K[a], Bharathidhasan D[b], Ashok Kumar R[c], Mohamed Jaffarsha F[D], Sreeraam M[e], Surya Prakash K[f]**

[a & c], Assistant Professor, RMK College of Engineering and Technology, Tiruvallur-601206, Tamil Nadu, India.
[b, d, e & f], UG Students, RMK College of Engineering and Technology, Tiruvallur-601206, Tamil Nadu, India.

Email: ashok.kumar19872003@gmail.com

**Abstract**. Light weight Aluminium alloys are extensively used in aircraft, supercars, and marine applications owing to its corrosion resistance and density. However, the mechanical properties like stiffness, tensile properties and tribological characteristics are far inferior than steels and titanium alloys. The tribomechanical properties of existing Aluminium alloys can be improved by adding suitable reinforcement agent and heat treatment techniques. In this review, effect of different reinforcement on mechanical properties and tribological characteristics of Aluminium Metal Matrix Composites (AMMC) were discussed in detail. Also, various fabrication processing techniques like solid state processing, liquid state processing and in-situ processing on tribological behavior of AMMC is discussed.

**Keywords:** AMMCs, Tribological Behaviour, Mechanical Properties, Processing Techniques

## 1. Introduction:

At present in worldwide there is a surge of need and requirement for the advanced materials in order to obtain desired properties. Metal Matrix Composites (MMCs) reinforced with reinforcement like Sic, Quarry dust are very auspicious materials for aerospace, industry and defense applications. Metal matrix composites blend the metallic alloys and reinforcement's properties which results in wards superior characteristics. Vast amount of research is done on aluminium based metal matrix composites, due to its extraordinary properties like good strength to weight ratio, improved corrosion resistance, good mechanical properties and wide range of applications. MMCs has reduced the overall weight, pollution and fuel consumption within the aircraft and automobiles. It has found that processing techniques for MMCs such as powder metallurgy, vacuum hot pressing, co-spray deposition process, squeeze casting, and stir casting methods are generally selected for the preparing MMCs. But of all these methods, stir casting method is selected to other methods because it is simple, easy and parameters can be readily varied and monitored [1].

Stir Casting method is the inexpensive and economical method available for the fabrication of Aluminium Metal Matrix Composites (AMMCs). The advantages of stir casting method are flexible, simple and mass production. The micro-sized SiC particles are used as reinforcement

IOP Conf. Series: Materials Science and Engineering **923** (2020) 012058 doi:10.1088/1757-899X/923/1/012058

particles for the fabrication of AMMCs. The addition of Mg improves the wettability between molten metal and reinforcement [2]. Stir-squeeze bottom pouring is a new method used for fabrication of SiC nanoparticle with aluminium matrix composites. Step one is melting the Aluminium in furnace. Next, stirring the molten metal and adding the reinforcement particles. Then reheating the matrix and mixing the already heated reinforcements. Then again stirred for homogeneous distribution of the reinforcements into matrix and then, squeeze casting process was done for aluminium composites [3]. The Hybrid metal matrix composites have specific special properties. The ceramic reinforcements like aluminium oxide, SiC, zirconia, Boron carbide are added to aluminium matrix to boost mechanical properties. The intention of the research is to find out the mechanical behaviour of Al7075 alloy strengthened with three types of ceramic reinforcements, viz. SiC, $Al_2O_3$, and 1 wt. % of Mg. The results of the mechanical properties like compressive strength, tensile strength, hardness and impact strength are compared. It has been found that the results are better than standard Al7075. The tensile strength and impact strength of the metal matrix nano-composites are tested under the universal testing machine. Hardness test is done on a Vickers hardness machine [4].

Friction stir processing (FSP) is the process of developing surface properties that can vary grain size, strength and surface roughness, etc. In this process, a hard-rotating tool pass through the workpiece and advances forward. The surface composites can be formed by friction stir processing. It is a highly efficient green process. To produce this composite, the reinforcing ceramic powder is inserted in the aluminium metal matrix. In this process, the reinforcing particles go through the surface of metal at a certain depth is shown in Figure 1. The benefits of friction stir processing are the refinement of the grain matrix and hardness of the composites. The other benefit is quality of workpiece is increased [5].

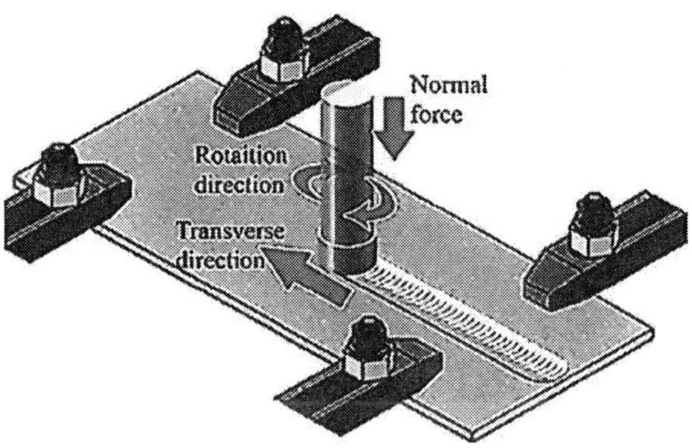


**Figure 1** Schematic of Friction Stir Process

Mica and SiC reinforcement particles were fused into Aluminium 356 alloy by stir casting process. SiC reinforcement is combined with Al356 and mica to produce tough hybrid aluminium composites. The mechanical properties of Al356 alloy composites are calculated [6]. There was a 10% rise in tensile strength and no modification in fracture toughness by addition of 1.5% of Mg. Density, impact strength and stress properties of 5, 10 and 15 wt.% of SiC reinforced Al MMC, the effect of SiC reinforcement proportion on machining, and the goods of machining specification such as cutting speed, rate of feed and depth of cut on wear of tools and surface roughness has been reviewed [7]. The another method of fabricating MMC is vacuum stir casting process. The mechanical and micro structural results of the composites produced by this technique are investigated. This process is isotropic and homogeneous in nature. It has a strong chemical bond between a matrix and reinforcement. The schematic view of vacuum stir casting sctup is shown in Figurc 2. Thc homogeneous distribution of the reinforcement in composite is major factor in preparing AMMCs[8].

**Figure 2** Vacuum Stir Casting setup

A mixture of $ZrO_2$, $MoS_2$ and Ni with differing weight percentages are used as the reinforcement. In terms of self-lubricating properties, $MoS_2$ has the ability to withstand more loads than graphite. The superior hardness and wear resistance are achieved by fabrication of hybrid aluminium composites. The composites having more than one reinforcement are called Hybrid Composites. The cost of the MMC product is minimized [10]. The impact of the machining parameters like cutting speed and forces, surface finish is studied. The surface condition of the workpiece and the wear of the tool and the cutting forces are inspected and compared. Disordered particulate reinforced aluminium metal matrix composites bars of 80 mm in diameter is used for experimentation [11]. The dislocation formation in the composites undoubtedly increase the strength of the aluminium composites. The necessary dislocation produced during initial matrix plastic deformation also strengthens the composite. The void initiation is avoided by the generation of dislocations and reserved in the matrix to allow for deformation. The particle size effects and distribution of particle in the matrix and the corresponding mechanical properties are studied [12].

Age Hardening is the process of heat treatment which increases the strength of metals. It is also called as Precipitation Hardening. It increases the tensile and yield strength. But the precipitates forbid the movement dislocations in the matrix. The precipitate hardening influence of the Al7075 MMCs on mechanical properties is studied [14]. The fly ash reinforced aluminium 6061 alloy exposed that tensile strength and ductility of composite are decreased when the fly ash weight percentage is increased [18]. The nano composite of Al7075 and SiC is also processed by cold and hot dynamic compaction. The investigations are carried out under quasi-static loading conditions [19].

## 2. Fabrication Techniques

The processing techniques of aluminium composites are grouped into liquid state and solid-state process. The various types of liquid State process are listed below

- Stir Casting [20]
- Melt Infiltration [21]
- Squeeze Casting [22]
- Melt Oxidation [23]
- Compo Casting [24]

Some of the solid-state process are given below

- Powder Metallurgy [25]
- Spray Deposition [26]
- Chemical Vapour Deposition [27]

- Electroplating [28]
- Diffusion Bonding [29]

The Commonly used manufacturing process are Stir Casting [20], Powder Metallurgy [25], Spark Plasma Sintering [30], Ultrasonic assisted casting [31], Ball Milling [32]. The detailed discussion of the above process are done.

### 2.1. *Powder Metallurgy Method*

Powder metallurgy is a method in which particulate materials are developed into finished and semi- finished products. The basic steps of the technique are manufacture of powders, powder blend, compressing of powder in a die and sintering. Powder metallurgy method is applicable for materials in the form of powder. It reduces scrap loss and also machining time. This technique is suitable for large number of components. Lauri et al. [33] made nano composites of Al and (1vol. % of size 30nm) SiC through high energy planetary milling and then by powder metallurgy method and sintering. They are tested the formation of nano composites by changing the parameters of milling. The arrangement of the composites is elaborated in Figure 3. The distribution of the ball size of 10mm used for 20 hours milling better than the ball size of 20mm for same time at the speed of 360 rpm. Gallardo et al. [34] formed nano composites of Al2024 and (0 to 5) wt % nano SiC through the ball milling followed by the method of powder metallurgy. The process of milling took place about 2h. It is found that there is a uniform distribution of the reinforcements of SiC in the Al2024 alloy matrix and there is no formation of crack. These results are found from the SEM and TEM micrographs is shown in Figure 4.

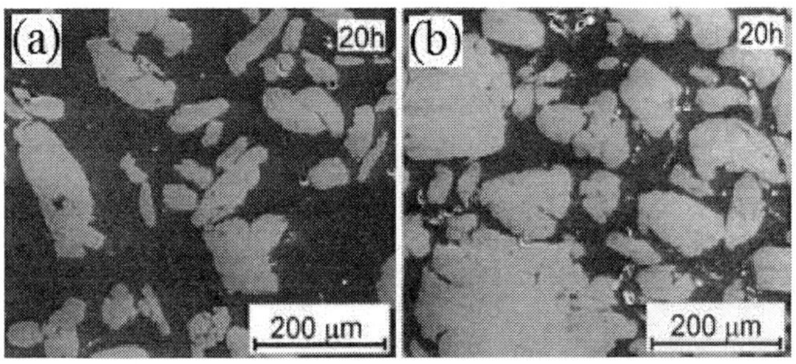


**Figure 3.** Composites 20 h milled and sintered (a) 360 rpm, 10mm balls (b) 360 rpm, 20mm balls [33]

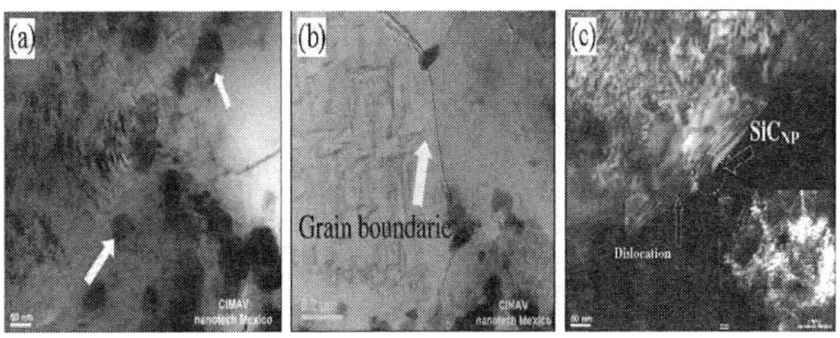


**Figure 4** TEM images of composites (a) SiC distribution in the matrix (b) grain boundaries of SiC (c) interaction of dislocation lines with SiC [34]

### 2.2. *Ball Milling*

Ball milling is a process in which grinding balls squash and break down the material into small sizes. The important factors of the ball milling are the rotation speed, sample quantity, size, time of grinding and number of grinding balls. The sample homogeneity can be upgraded by increasing the number of grinding balls. Hot isostatic pressing method enhances the uniform distribution of ball milled powder. Feng et al. [35] made nano composites of Al5083 and (6.5 vol. % of 25nm size) SiC in

IOP Conf. Series: Materials Science and Engineering **923** (2020) 012058 doi:10.1088/1757-899X/923/1/012058

three steps. The first step is cryomilling, next step is HIP consolidation, and the last step is hot rolling process. The SEM and TEM images in Figure 5 shows that the reinforcement was distributed uniformly in the matrix, the interface bonding between SiC and Al5083 matrix was good. Knowles et al. [38] formed nano composites of Al 6061 and (10, 15) wt % of 500 nm size SiC through three steps. The initial step is through high energy ball milling followed by hot isostatic pressing and the last step is extrusion process. The SEM images in Figure 6 exposed that particles in the Al6061 matrix were distributed homogeneously with less number of clustering of the nano particles.

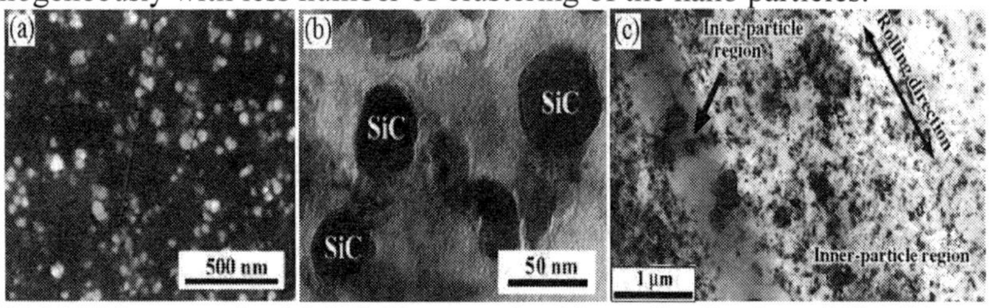


**Figure 5** Al5083/SiC composite (a) SEM image of uniform distribution of SiC (b) TEM image (c) TEM image of inter particle region in rolling direction [35]

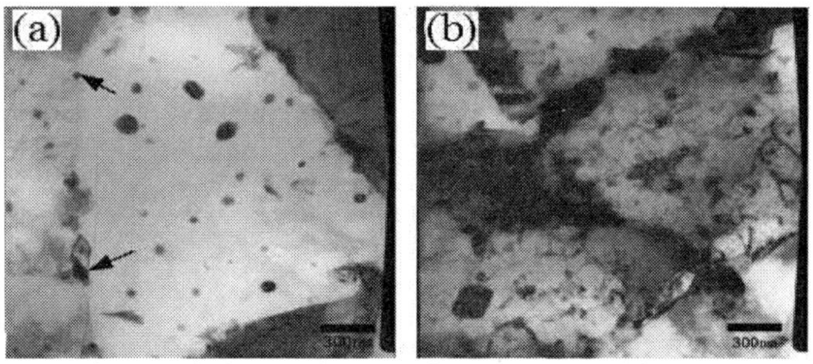


**Figure 6** TEM image of extruded (a) Al 6061 (b) Al 6061 with 15 wt. % SiC composite [36]

### *2.3. Ultrasonic Assisted Casting*

Ultrasonic assisted casting is a process of melting the materials by ultrasonic waves during or after the adding the reinforcement and then followed by casting. The principle of this method is creation of the high intensity ultrasonic waves which creates transitory cavitation and acoustic streaming in liquids. This method is more fortunate in preparing the composites with uniform dispersion of reinforcements in small scale at laboratories and also suitable for large scale production. Bulk production requires high source of power. By this technique Dehnavi et al. [37] fabricated metal matrix composites of A413 and (0.5, 1, 1.5, 2) wt. % SiC and 0.5 wt. % Mg at nano level. The A413 was melted in the crucible of graphite with the assistance of resistance furnace under argon atmosphere. The SiC reinforcements are preheated in a muffle furnace for one hour at 1000°C. Then the preheated reinforcements were added to the melt and stirred mechanically by using stirrer which is made of graphite for 30 min at a 650-rpm speed. The slurry was treated ultrasonically at a frequency of 20 kHz and 350W power for 10 to 20 min and then drained into a preheated grey cast iron die. Yong et al. [38] synthesized metal matrix composites of A356 with (2 wt. % of size 30nm) SiC by this method. The SEM images of composites as casted showed that there is no uniform distribution in the matrix of SiC particles and formation of clusters of SiC particles and the reinforcement particles were distributed uniformly along the A356 matrix alloy in a composite fabricated from ultrasonic assisted casting method is shown in Figure 7.

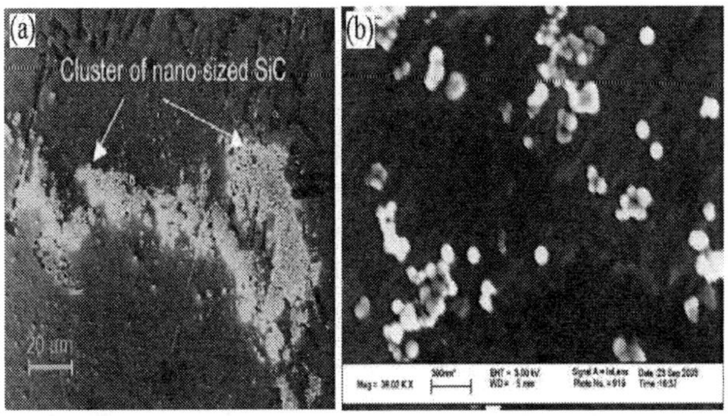


**Figure 7** SEM images of A356 and SiC composites (a) as casted (b) as ultrasonic cavitation [38]

### *2.4. Spark Plasma Sintering*

Spark plasma sintering is a high-speed powder strengthening sintering process. This process is capable of treating nonconductive as well as conductive materials. The basic idea of spark plasma sintering process is based on the phenomenon of electrical spark discharge. It passes low voltage pulse current temporarily which generates spark plasma at high energy in local areas between the particles. The temperatures in this process were lower than the traditional sintering temperatures. The processing time of the material is 5 to 25 min. It establishes strong control over microstructure and grain growth. Sivaiah et al. [39] fabricated composites of Al 5083 with 10 wt.% SiC through high energy ball milling and then by spark plasma sintering process. They studied the samples of composites with SEM images in Figure 8 and XRD plots for the ball milled Al 5083 matrix which has crystalline grain size, it is found to be 25nm and after spark plasma sintering it was coarsened up to 30nm.

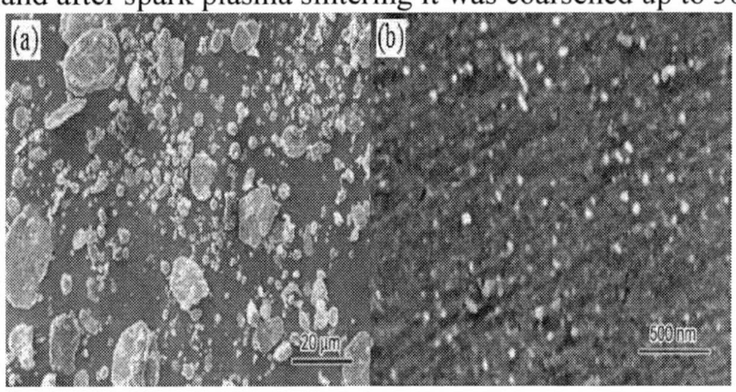


**Figure 8** SEM images of Al 5083 with SiC (a) powders after ball milling for 15h (b) after spark plasma sintering process [39]

Sherif et al. [40] synthesized Al composites with (2, 5, 7, 10) vol. % of size 5nm SiC by this spark plasma sintering process. The SEM images revealed the uniform distribution of the reinforcement particles in the matrix. The XRD analysis in Figure 9 after the ball milling time 0ks and 86ks showed no formation of Al4C3 and Si.

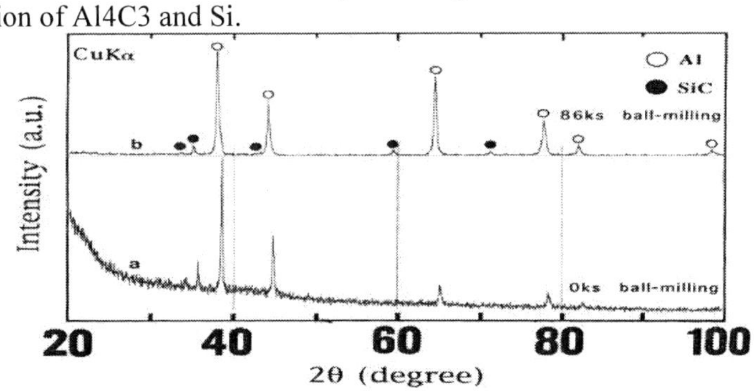


**Figure 9** XRD patterns of powder after (a) 0 ks and (b) 86 ks of the ball milling time [40]

IOP Conf. Series: Materials Science and Engineering **923** (2020) 012058 doi:10.1088/1757-899X/923/1/012058

### *2.5. Stir Casting Method*

Stir casting is a liquid state process for fabrication of composite materials. In this process the mechanical stirrer is used to mix the reinforcement particles with metal matrix which is in molten state. Then the composite material which is in liquid state is casted by traditional casting methods and conventional metal forming techniques. Ali et al. [41] fabricated nano composites of A356 and (0 to 4.5 vol. %) SiC and 1 wt.% Mg by stir casting method. The aluminium matrix is melted to a temperature of 750°C and at a stirring speed of 600rpm for 15 min by stirring with graphite impeller. From the SEM image in figure 10, it is found that the reinforcement particles in the A356 alloy matrix are distributed uniformly.

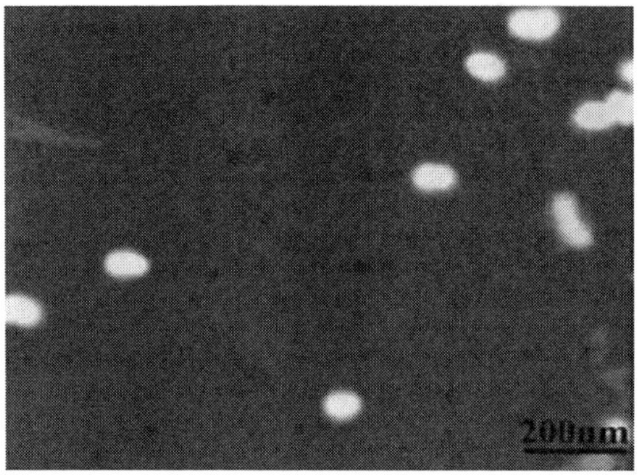


**Figure 10** SEM image of 2.5 vol. % SiC [41]

## 3. Mechanical Properties

### *3.1. Hardness*

Hardness is defined as the ability of the material to resist plastic deformation. The resistance to indentation or scratch is termed as hardness. Vickers, Brinell and Rockwell hardness testers are the commonly used hardness measuring instruments. Hc = VrHr + VmHm is the rule used to find hardness of the composite theoretically where c stands for composites, r stands for reinforcement, m stands for matrix and V stands for volume, H stands for hardness. It helps in calculating approximate hardness values [42]. Several researchers have contributed regarding the reinforcement effects on hardness of composites which are summarized as follows;

The microhardness increases (linearly) considerably in the content of SiC and Al2O3 particle reinforcement relative to Al 7075 alloy. It can be found that the microhardness of composites was greater than that of its base Al7075 alloy is shown in Figure 11. This is due to the presence of aluminium oxide ($Al_2O_3$) and silicon carbide (SiC) particles in the Al7075 matrix [4].

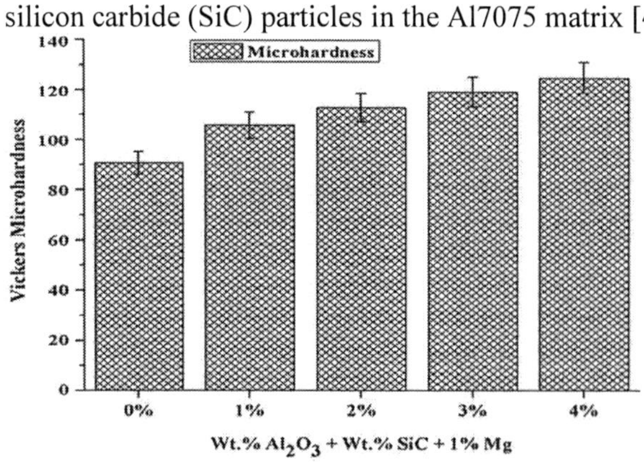


**Figure 11** Wt.% of reinforcement vs microhardness values [4]

The hardness value of the composites reinforced with varying mass fractions of mica. It is found that the hardness of Al/10SiC–6mica composite is lower when compared with Al/10SiC–3mica composite. Thus, the hardness of the MMCs rises more or less with the fraction of volume of reinforcement particulates in the matrix alloy [6]. The reinforcements such as SiC, aluminide and Al2O3 [43-45] are mostly preferred to enhance higher hardness in the composite. TiC when diffused in Al matrix, improves the hardness to weight ratio. Also, it transmits thermodynamic stability to the composites [46-48]. Abdulhaqq et al. [49] and Lloyd et al. [50] found the importance of hard ceramic particles in improving the hardness of Al-MMCs. Deuis et al. analyzed that the increase in the hardness of the MMCs having ceramic reinforcement not only depends on the reinforcement size but also on the composite structure and excellent interface bonding [51].

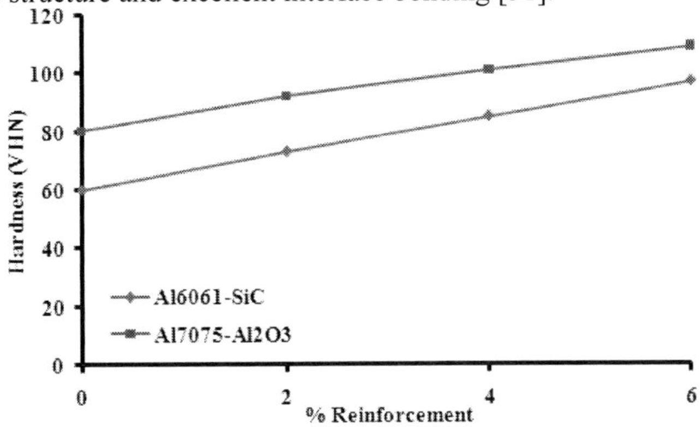


**Figure 12** Variation of Vicker's Hardness Al6061-SiC and Al7075-$Al_2O_3$ Composites

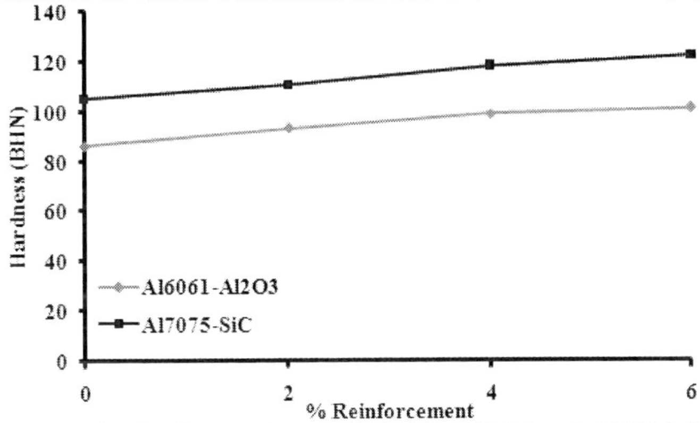


**Figure 13** Variation of Brinell's Hardness of Al6061-Al2O3 and Al7075-SiC Composites

The composites produced with respect to the above explanation are subjected to hardness test which are done in the Vicker's and Brinell's hardness testing machines and the results are calculated. From the Figures 12 and13, it can be found that the hardness of composites was higher than that of its base alloy. Also, the hardness of the composite is found to boost with increased filler content.

### *3.2. Tensile Strength*

The tensile properties of the Aluminium 7075 alloy and its composites is inspected on a computerized universal testing machine which has the capacity of 50KN. As per ASTM E8 standards, the specimens are prepared for the testing of the product. Figure 14 explains the difference of tensile strength of the composites with the different weight portions of $Al_2O_3$ (20–30 nm) and SiC (50 nm) reinforcement particulates. It is observed that the tensile strength increased with a rise in the weight fraction of $Al_2O_3$ and SiC [4].

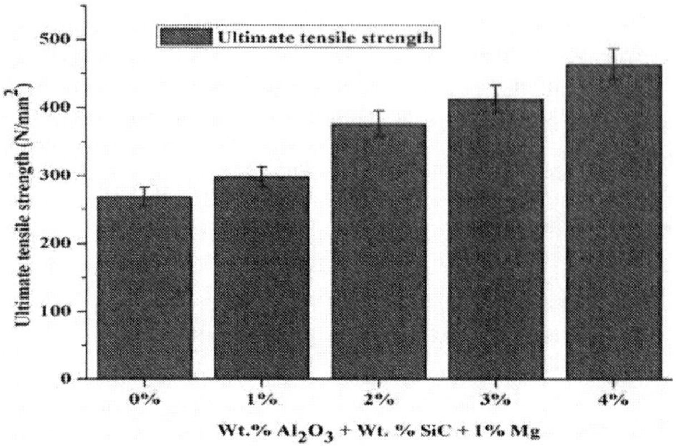


**Figure 14** Wt.% of reinforcement versus UTS [4]

The results from the tensile tests of composites which are mica reinforced shows that the aluminium composites reinforced with 10% SiC and 3% mica particle have the greater tensile strength of 150 MPa and the tensile strength of 148 MPa is found in the aluminium composite with 10% SiC and 6% mica particle. The outcomes reveal that when the mass fraction of mica increases, the tensile strength of the composites rises up to certain value and then decreases [6]. The mechanical properties of the composites depend on the structure and properties of the reinforcement materials. The strong interface between matrix and reinforcement is achieved due to increase in value of elastic modulus and strength of the composites [52]. The strength of $Al_2O_3$, SiC, TiC [56], and TiB2 particulate reinforced Al-MMCs is increased at the loss of reduced ductility, by hiking the volume percentage of ceramic phase and by reducing the reinforcement size in the composite [52-55]. The heat treatment of the composites under aged condition before and after fabrication improves mechanical properties [57]. The theoretical expression for tensile yield stress of matrix and composites are given by Pm = cαy where Pm is hardness or yield pressure, c is constant value and αy is tensile yield stress [57]. The tensile strength of $AlSiMg_2$ composites depending on reinforcement ratio is described in Fig. 15. There had been hike in the tensile strength of produced composites when a 180MPa of tensile strength for metal mould casting is considered. Specifically, in 10 wt.% of SiC reinforced composite there had been 100% and 56% rise in testing with the matrix material of sand casting and matrix material of metal casting, respectively. However, in 15 wt.% of SiC reinforced composite material the tensile strength of composite is decreased because of broken particles in the structure

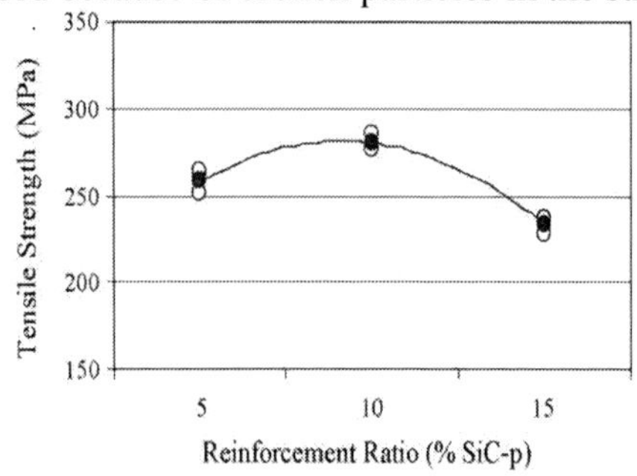


**Figure 15** Tensile strength values of composite material [7]

*3.3. Compressive Strength*

The investigations revealed that the compressive strength progressively increased by the increase in wt.% of the reinforcement i.e., $Al_2O_3$ and SiC which contributed the metal matrix. The ideal compressive strength was noticed at 4% aluminium oxide and silicon carbide. The graph of increase in compressive strength is shown in Figure 16. The inclusion of the reinforcements in the matrix particle creates cores which lead to the bulk development of grains. It enhances better strength of the composites [4].

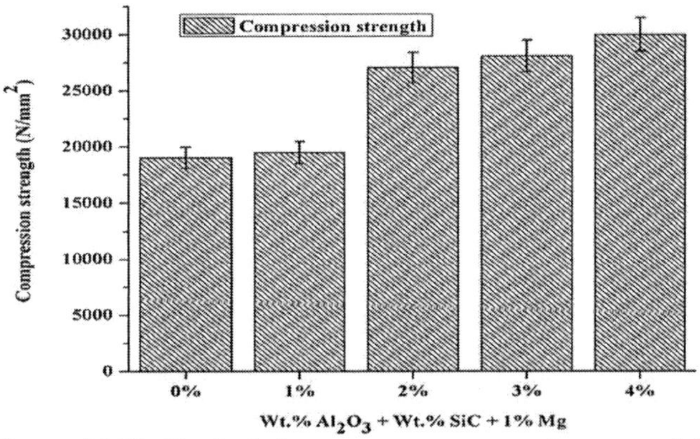


**Figure 16** Wt.% of reinforcement vs compression strength [4]

The aluminium composites which has 10 wt. % SiC reinforced shows better compressive strength than aluminium composites having 5wt. % SiC and pure base aluminium alloy. The aluminium metal matrix having 70 nm size of SiC particles have outstanding mechanical properties. In the Figure 17 the samples of different sizes containing 10 wt. % SiC are shown. From that 70 nm size of SiC particles are having maximum compressive strength of 601MPa [59].

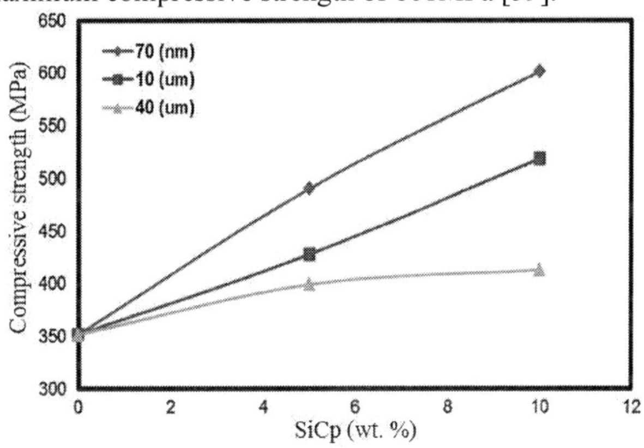


Figure 17 variation of the compressive strength of composites having different sizes of SiC particles [59]

## 4. Conclusion

This review suggests the ideas, experimental results gathered, and conclusions made over the years by various investigators in the work of particle reinforced Al-MMCs. This paper also exposes the fabricating techniques of the composites and mechanical properties of SiC particle reinforced Al-MMNCs. The SiC reinforced Al-MMNCs exhibits better mechanical properties over the unreinforced base alloys. The commonly used fabrication technique in the making of Al-MMNCs is stir casting as it is the cheapest method. The ultrasonic assisted casting and spark plasma sintering are more expensive processes which can be used for high sensitivity parts. The hardness, compressive strength and tensile strength of the composites were reviewed. It is concluded that when the reinforcement contents are increased in the matrix material, the hardness and tensile strength of the composites are also increased. The review on mechanical properties were done with respect to strength.

## References

[1] A. O. Inegbenebor, C. A. Bolu, P. O. Babalola, A. I. Inegbenebor, O. S. I. Fayomi1, "Aluminum Silicon Carbide Particulate Metal Matrix Composite Development Via Stir Casting

Processing", Springer.
[2] R.A. Kumar, A. Devaraju, S. Arunkumar, Experimental Investigation on Mechanical Behaviour and Wear Parameters of TiC and Graphite Reinforced Aluminium Hybrid Composites, Elsevier Ltd, 2018.
[3] Pooja Verma, Prabha Kumari, Joyjeet Ghose, Vijay Pandey, "Investigation of Mechanical Properties and Microstructure of Pure Al-SiC-Nanocomposite Casted by Stir-Squeeze Casting Process", Springer.
[4] S. Suresh, G, Harinath Gowd, M. L. S. Deva Kumar, "Mechanical Properties of AA 7075/Al2O3/SiC Nano-metal Matrix Composites by Stir-Casting Method", Springer.
[5] Navid Molla Ramezani, Behnam Davoodi, Mohammad Aberoumand, Mojtaba Rezaee Hajideh, "Assessment of tool wear and mechanical properties of Al 7075 nanocomposite in friction stir processing (FSP)", Springer.
[6] T. Rajmohan, K. Palanikumar, S. Ranganathan, "Evaluation of mechanical and wear properties of hybrid aluminium matrix composites", Elsevier.
[7] Tamer Ozben, Erol Kilickap, Orhan, "Investigation of mechanical and machinability properties of SiC particle reinforced Al-MMC", Elsevier.
[8] Jagbir Singh, C. S. Jawalkar, R. M. Belokar, "Analysis of Mechanical Properties of AMC Fabricated by Vacuum Stir Casting Process", Springer.
[9] Mohammed Imran, A.R. Anwar Khan, "Characterization of Al-7075 metal matrix composites: a review", Elsevier.
[10] Anish Ranjan, A. Shanmugasundaram, "Experimental Investigation of Mechanical and Tribological Properties of Al 7075—MoS2/ZrO2/Ni Hybrid Composite", Springer.
[11] Alakesh Manna, B. Bhattacharayya, "Influence of machining parameters on the machinability of particulate reinforced Al/SiC–MMC", Springer.
[12] Chao Sun, Min Song, Zhangwei Wang, Yuehui He, "Effect of Particle Size on the Microstructures and Mechanical Properties of SiC-Reinforced Pure Aluminum Composites", Journal of Materials Engineering and Performance.
[13] Omid Yaghobizadeh, Hamid Reza Baharvandi, Ali Reza Ahmadi, Ermia Aghaei, "Development of the Properties of Al/SiC Nano-Composite Fabricated by Stir Cast Method by Means of Coating SiC Particles with Al", Springer.
[14] Din Bandhu, Ashish Thakur, Rajesh Purohit, Rajesh Kumar Verma, Kumar Abhishek, "Characterization & evaluation of Al7075 MMCs reinforced with ceramic particulates and influence of age hardening on their tensile behaviour", Springer.
[15] G. S. Pradeep Kumar, R. Keshavamurthy, Vijay Tambrallimath, Riya Biswas, Yogesh Sahebrao Andhale, "Mechanical Properties of Al7075-SiC-TiO2 Hybrid Metal Matrix Composite", Springer.
[16] Parmesh Kumar, Vijay Kumar B P, RaghavendraJoshi, T H Manjunatha, R Suresh kumar, "Evaluation of Al7075 Reinforced with SiC for its Mechanical Properties & Surface Roughness by Drilling", Elsevier.
[17] S.Ajay Kumar, A. Prabhu Kumar, B. Balu Naik, B. Ravi, "Production and investigation on Mechanical Properties of TiC reinforced Al7075 MMC", Elsevier.
[18] Divya Chandra, Nathi Ram Chauhan, S. Rajesha, "Hardness and toughness evaluation of developed Al metal matrix composite using stir casting method", Elsevier.
[19] A. Atrian, G.H. Majzoobi, M.H. Enayati, H. Bakhtiari, "Mechanical and microstructural characterization of Al7075/SiC nanocomposites fabricated by dynamic compaction", Springer.
[20] L. ulmer, F. Pitard, D. Poncet, O. Demolliens, "Formation of Al3Ti during physical vapour deposition of titanium on aluminium", Microelectronic Engineering.
[21] Weiping Chen, Yixiong Liu, Chao Yang, Dezhi Zhu, Yuanyuan Li, "Al7075/( SiC+Ti) hybrid composites with high strength and large plasticity fabricated by squeeze casting" Materials

Science and Engineering.

[22] B. N. Sarada, P. L. Srinivasa Muthy, G. Ugrasen, "Hardness and wear characteristics of Hybrid Aluminium Metal Matrix Composites produced by stir casting technique", Materials Today: Proceedings.

[23] Ilija Bobic, Jovana Ruzic, Biljana Bobic, Miroslav Babic, Aleksandar Vencl, Slobodan Mitrovic, "Microstructural characterization and artificial aging of compo-casted hybrid A356/SiC/Grp composites with graphite macroparticles" Materials Science and Engineering.

[24] E. S. Eardley, H. M. Flower, "Infiltration and solidification of commercial purity aluminium matrix composites", Materials and Engineering.

[25] Basak Dogru Mert, "Corrosion protection of aluminum by electrochemically synthesized composite organic coating" Corrosion Science.

[26] S. C. Tjong, Z. Y. Ma, "High-temperature creep behaviour of powder metallurgy aluminium composites reinforced with SiC particles of various sizes", Composites Science and Technology.

[27] Jia-lei ZHAO, Jin-chuan JIE, Fei CHEN, Hang CHEN, Ting-ju LI, Zhi-qiang CAO, "Effect of immersion Ni plating on interface microstructure and mechanical properties of Al/Cu bimetal", Trans. Nonferrous Met. Soc. China.

[28] X. P. Zhang, L. Ye, Y. W Mai, G. F. Quan, W. Wei, "Investigation on diffusion bonding characteristics of SiC particulate reinforced aluminium metal matrix composites (Al/SiC-MMC)", Composites Part A: applied science and manufacturing.

[29] J. S. Sureshbabu, P. K. Nair, C. G. Kang, "Fabrication and characterization of aluminium based nano-micro hybrid metal matrix composites", 16th International conference on composite materials.

[30] Xie Bin, Wan Xiaogang, "Thermo-physical properties and reaction process of SiC/Al-7Si-5Mg aluminum matrix composites fabricated by pressureless infiltration" Rare Metal Materials and Engineering.

[31] G. A. Sweet, M. Brochu, R. L. Hexemer Jr, I. W. Donaldson, D. P. Bishop, "Consolidation of aluminum-based metal matrix composites via spark plasma sintering", Materials Science and Engineering.

[32] A. Urena, E. E. Martinez, P. Rodrigo, L. Gil, "Oxidation treatments for SiC particles used as reinforcement in aluminium matrix composites", Composites Science and Technology.

[33] Lauri Kollo, Marc Leparoux, Christopher R Bradbury, Christian Jaggi, Efrain Carreno Morelli, Mikel Rodriguez Arbaizar, "Investigation of Planetary Milling for Nano Silicon Carbide Reinforced Aluminium Metal Matrix Composites", Journal of Alloys and Compounds.

[34] C. Carreno Gallardo, I. Estrada Guel, C. Lopez Melendez, R. Martinez Sanchez, "Dispersion of Silicon Carbide Nano Particles in AA2024 Aluminium Alloy by a High Energy Ball Mill", Journal of Alloys and Compounds.

[35] Feng Tang, Masuo Hagiwara, Julie M Schoenung, "Microstructure and Tensile Properties of Bulk Nano Structured Al5083/SiC Composites Prepared by Cryomilling", Materials Science and Engineering.

[36] A. J. Knowles, X. Jiang, M. Galano, F. Audebert, "Microstructure and Mechanical Properties of Al 6061 Alloy Based Composites with SiC Nano Particles", Journal of Alloys and Compounds.

[37] M. R. Dehnavi, B. Niroumand, F. Ashrafizadeh, P. K. Rohatgi, "Effects of Continuous and Discontinuous Ultrasonic Treatments on Mechanical Properties and Micro-Structural Characteristics of Cast Al413/SiCnp Nano Composite", Materials Sciencc & Engineering.

[38] Yong Yang, Jie Lan, Xiaochun Li, "Study on Bulk Aluminium Matrix Nano Composite Fabricated by Ultrasonic Dispersion of Nano sized SiC Particles in Molten Aluminium Alloy", Materials Science and Engineering.

[39] Sivaiah Bathula, Saravanan M, Ajay Dhar, "Nano Indentation and Wear Characteristics of Al 5083/SiC Nano Composites Synthesized by High Energy Ball Milling and Spark Plasma Sintering", Journal of Material Science and Technology.

[40] M. Sherif EI. Eskandarany, "Mechanical Solid-State mixing for Synthesizing of Al/SiCp Nano Composites", Journal of Alloys and Compounds.
[41] Ali Mazahery, Mohsen Ostad Shabani, "Plasticity and Microstructure of A356 Matrix Nano Composites", Journal of King Saud University Engineering Sciences.
[42] S.C. Sharma, "The sliding wear behavior of Al6061–garnet particulate composites", Wear 249 (2001).
[43] I.M. Hutching, "Wear by particulates", Chemical Engineering Science.
[44] F. M. Husking, F. Folgar Portillo, R. Wunderlin, R. Mehrabian, "Composites of aluminium alloys: fabrication and wear behaviour", Journal of Material Science.
[45] Debdas Roy, Bikramjit Basu, Amitava Basu Mallick, "Tribological properties of Tialuminide reinforced Al-based in situ metal matrix composite", Intermetallics.
[46] Rajnesh Tyagi, "Synthesis and tribological characterization of in situ cast Al–TiC composites", Wear 259 (2005).
[47] P.H. Shipway, A.R. Kennedy, A.J. Wilkes, "Sliding wear behaviour of aluminium-based metal matrix composites produced by a novel liquid route", Wear 216 (1998).
[48] S.K. Chaudhury, A.K. Singh, C.S. Sivaramakrishnan, S.C. Panigrahi, "Wear and friction behavior of spray formed and stir cast Al–2Mg–11TiO2 composites", Wear 258 (2005).
[49] Abdulhaqq A. Hamid, P.K. Ghosh, S.C. Jain, Subrata Ray, "The influence of porosity and particles content on dry sliding wear of cast in situ Al(Ti)–Al2O3(TiO2) composite", Wear 265(2006).
[50] Kim CK, Park SY, "A study on the fabrication and mechanical properties of SiC fiber aluminum alloy composites", J Korean Institute of Metallurgy and Materials.
[51] R.L. Deuis, C. Subramaniun, J.M. Yellup, "Abrasive wear of aluminium composites—a review", Wear 201 (1996).
[52] Rang Chen, Akira lwabuchi, Tomoharu Shimizu, Hyung Seop Shin, Hidenobu Mifune, "The sliding wear resistance behavior of NiAI and SiC particles reinforced aluminum alloy matrix composites", Wear 213 (1997).
[53] Ma ZY, Tjong SC, "In situ ceramic particle-reinforced aluminum matrix composites fabricated by reaction pressing in the TiO2 (Ti)–Al–B (B2O3) systems", Metallurgy and Materials Transitions.
[54] Yoshiro Iwai, Hidetomo Yoneda, Tomomi Honda, "Sliding wear behavior of SiC whisker-reinforced aluminum composite", Wear 181-183 (1995).
[55] Heguo Zhu., Hengzhi Wang, Liangqi Ge, "Wear properties of the composites fabricated by exothermic dispersion reaction synthesis in an Al–TiO2–B2O3 system", Wear 264(2008).
[56] Kassim S. Al-Rubaie, Humberto N. Yoshimura, Jose Daniel Biasoli de Mello, "Two body abrasive wear of Al–SiC composites", Wear 233–235 (1999).
[57] J. LLorca, "Failure micro-mechanisms in particulate-reinforced metal matrix composites", Journal of Physics.
[58] H.C. How, T.N. Baker, "Characterization of sliding friction-induced subsurface deformation of Saffil-reinforced AA6061 composites", Wear 232 (1999).
[59] Omyma EI Kady, A. Fathy, "Effect of SiC Particle Size on the Physical and Mechanical Properties of Extruded Al Matrix Nano Composites", Materials and Design.

# Optimization of drilling parameters on hybrid natural fiber reinforced polymer matrix composites

**R.Selvam[1],N.Sathishkumar[2],K.M. Kumar[3],N.Arunkumar[4],Khaleelur Rahman.T[5]**

[1]Professor, Department of Mechanical Engineering, St. Joseph's College of Engineering Chennai, Tamil Nadu, India.
[2]Assistant Professor, Department of Mechanical Engineering, St. Joseph's College of Engineering Chennai, Tamil Nadu , India.
[3]Associate Professor, Department of Mechanical Engineering, St. Joseph's College of Engineering Chennai, Tamil Nadu , India.
[4]Professor, Department of Mechanical Engineering, St. Joseph's College of Engineering Chennai, Tamil Nadu , India.
[5*]PG Student, Department of Mechanical Engineering, St. Joseph's College of Engineering Chennai,Tamil Nadu India.

Mail Id: selvamr@stjosephs.ac.in

**Abstract:** The natural fiber reinforced composites are highly utilized in the applications of building and construction industry, storage containers, transportation industry and in other industrial applications. The natural fibers are preferred for its distinct properties than synthetic fibers. Natural fibers are economically cheaper, recyclable and ecologically biodegradable. Many studies were proposed by the researchers to improve the machinability and mechanical behaviour of natural fiber reinforced composites. The addition of filler material is a kind of approach to improve the property of natural fiber reinforced composites. In this study the composite laminates are created by reinforcing bi-directional bast type jute and unidirectional leaf made sisal fibers into an epoxy LY556 resin matrix material. The silicon carbide (SiC) is used as filler material. Two types of composite laminates are manufactured. The first sample is considered in the ratio of 70% epoxy resin to 30% Jute + Sisal fiber composition without any filler material. The second composition includes the ratio of 70% epoxy resin to 30% Jute + Sisal fiber along with 10% SiC as filler material. The hardner HY951 of 10% is used in both compositions. The machining operation was performed by using HSS twist drill for three different machining parameters (Drill bit type, Spindle speed and Feed rate). The circularity and surface roughness are considered as output parameters. The machining is performed in the range of 20 different combinations as directed by the RSM software. The conformation tests are carried out to validate the RSM results. The optimization results shows that the sample with SiC filler exhibits better machinability behaviour than sample without filler material.

***Keywords:*** *Hybrid polymer composite; Natural fibers; Epoxy resin; Silicon carbide filler; Drilling parameters; Response Surface Methodology.*

## 1.Introduction

The natural fiber reinforced composites are highly researched to find out its capability for replacing the use of conventional fibre reinforced petroleum based composites [1]. Numerous natural fibre compositions are extracted from different bio materials. These biocomposites are highly utilized in many critical applications [2]. In most of the cases instead of natural resins, the synthetic resins provided better bonding strength and strength to weight ratio when it is mixed with natural fibers like sisal, jute and coir etc..[3]. Natural fibre reinforced composites exhibits low density lightweight property and other specific properties which makes them ideal for many industrial applications.[4]. The cost of the natural fibre extracts are very low comparing to synthetic fibres which is mostly extracted from depleting crude oil based refining process as a by product.[5]. The natural fibres are classified into two broad categories as non wood natural fibers and wood based natural fibers. The examples of non wood based natural fibers are straw fibers, bast fibers like flax, kenaf, jute, hemp, fruit based, seed based and grass based fibers. The examples of wood based fibers are like soft wood fibers and hard wood fibers etc...[6]. S.Harish et.al [7] developed an epoxy based composite reinforced with coir material. The mechanical properties and the machining properties of the epoxy-coir based composite was found good than equivalent synthetic fiber composition.The qualitative evaluation of the fractured surface was carried out by using scanning electron microscope and the interfacial properties are better than conventional synthetic composite. Wang wei et.al [8] prepared a natural fibre reinforced composite with a coir fiber stack. The characteristics of the coir fiber is initaially determined. The length of the fibers used in the study are at the range of 8 to 337 mm long fibers. The rubber is used as matrix material. The fabrication of composite is done by using heat press machine. The tensile properties of coir-epoxy composites are very high when comparing to other compositions.Drilling of banana galss fibre hybrid composites of two different proportions was carried out by using a standard twist drill. The feed rate and speed was considered as parameters. The hole quality was studied by using ANNOVA method and it was found that the feed rate has major influence than drilling speed on the hole quality [9] {V.Santhanam et.al}. The Coir–polyester composites were prepared by using handlayup method. The prepared composite laminates were subjected to study their mechanical and machinability characteristics. The expriments were carried out by using a twist drill of 6mm diameter. The minimum value of thrust force, torque and tool wear are found at spindle speed of 600 rpm and feed rate of 0.3 mm/rev [10]. The drilling behavior of unidirectional sisal-epoxy and sisal-polypropylene composite laminates has been experimentally investigated. Chip formation characteristics in the context of both thermoset and thermoplastic natural fiber composite laminates have been discussed. Further, the analysis of drilling force signals has been reported and the mapping of drilling forces has been proposed. Influence of drill geometry, feed, and spindle speed on drilling forces and a comparative analysis of damage characteristics of drilled hole have also been reported.[11]. The traditional drilling process carried out by using a twist drill is the highly used machining process for creating a hole or riveting or fastening process of structural component assemblies in the aerospace and automotive industries. [12]. The important objective of developing composite materials are environment friendliness and light weight, with high specific properties. For satisfying these criteria the natural fibers are incorporated into manmade fibers, and partially ecofriendly hybrid composites have been developed by using glass, sisal and jute fibers as reinforcing material in the polymer resin matrix. The drilling of composite materials is difficult when compared to metals, because the tool has to pass alternately through the matrix and reinforcement, which have different properties [13]. An effort to utilize the advantages offered by renewable resources for the development of biocomposite materials based on biopolymers and natural fibers has been made through fabrication of Natural fiber powdered material (Sisal (Agave sisalana), Banana (Musa sepientum), and Roselle (Hibiscus sabdariffa)) reinforced polymer composite material by using bio epoxy resin. The present work focuses on the prediction of thrust force and torque of the natural fiber reinforced polymer composite materials, and the values, compared with the Artificial Neural Network [14]. Recently, Jute fiber is being used as a reinforcement material in the development of reinforced plastics for various engineering applications. Its biodegradability, low cost, and moderate mechanical properties make it a preferable reinforcement material in the development of polymer matrix composites. [15] . The results reveal that, the mechanical properties of polypropylene based composites are substantially improved on account of the addition of the Jute fiber reinforcement. It has also been observed that the significance of the enhancement of the mechanical properties increased as the weight percentage of the Jute fiber reinforcement increased up to 40% [16]. The results show that the jute reinforced epoxy composite exhibited better mechanical properties than Jute-polyester composite. [17]. Composite prepared with 0.02% $K_2Cr_2O_7$ treated jute fabrics showed the highest values of the mechanical properties. [18] . Natural fibers have the advantage that they are renewable resources and have marketing appeal [19]. The objective of this work is to perform drilling analysis of all the composites are studied by using Response surface experimental design Fabrication of natural fiber (sisal & jute) reinforced epoxy based composite. Valuation of mechanical properties (tensile strength, flexural, hardness, impact strength etc.). Besides the above all the objective is to develop new class of composites by incorporating sisal and jute fiber reinforcing phases into a polymeric resin Also this work is expected to introduce a new class of polymer composite that might find many engineering applications.

## 2.Experimentation

### *Materials Used*

The materials used in this work are

1. Sisal Fiber
2. Jute Fiber
3. Particulate Filler Material (SiC 600mesh)

4. Epoxy Resin (LY556)
5. Hardener (HY951)
6. Mould for sample preparation
7. Releasing Agent

***Matrix Material***

Matrix materials are of different types like metals, ceramics and polymers. Polymer matrices are most commonly used because of cost efficiency, ease of fabricating complex parts with less tooling cost and they also have excellent room temperature properties when compared to metal and ceramic matrices. Polymer matrices can be either thermoplastic or thermoset. Thermoset matrices are formed due to an irreversible chemical transformation of the resin into an amorphous cross-linked polymer matrix. Due to huge molecular structures, thermoset resins provide good electrical and thermal insulation. They have low viscosity, which allow proper fiber wet out, excellent thermal stability and better creep resistance. Normally, these resins can be formulated to give a wide range of properties upon the requirement. The most commonly used thermoset resins are epoxy, polyester, vinyl ester and phenolics. Among them, the epoxy resins are being widely used for many advanced composites due to their excellent adhesion to wide variety of fibers, superior mechanical and electrical properties and good performance at elevated temperatures. In addition to that they have low shrinkage upon curing and good chemical resistance. Due to several advantages over other thermoset polymers as mentioned above, epoxy (LY 556) is chosen as the matrix material for the present work.

***Reinforcements***

The present investigation employs sisal and jute as the natural fiber in the epoxy matrix to fabricate a series of hybrid composites.The pictorial views of bi-directional jute fiber mats and unidirectional sisal fiber are used for composite fabrication for this study are given in figure 1 as follows. Conventional filler chosen for this work is SiC, which has a great potential to be used in various polymeric matrices. It is the only chemical compound of carbon and silicon. It was originally produced by a high temperature electro-chemical reaction of sand and carbon. Today the material has been developed into a high quality technical grade ceramic with very good mechanical properties. It is used in abrasives, refractories, and ceramics and in numerous high-performance structural and wear applications. SiC is composed of tetrahedral of carbon and silicon atoms with strong bonds in the crystal lattice. Due to its bonding nature SiC produces a very hard and strong material. SiC is not attacked by any acids, alkalis or molten salts up to 800°C. It has low density of about 3.1 gm/cc, low thermal expansion, high elastic modulus, high strength, high thermal conductivity, high hardness, excellent thermal shock resistance and superior chemical inertness. This can also be made an electrical conductor and has applications in resistance heating, flame igniters and electronic components.

**Figure 1. Bi-directional Jute Fiber and Unidirectional Sisal Fiber**

***Fabrication of Composites***

The fabrication of the various composite materials is carried out through the hand lay-up technique. The schematic diagram of hand layup technique is shown in figure 2 as follows. Sisal and Jute fibers are reinforced with Epoxy LY 556 resin, chemically belonging to the 'epoxide' family is used as the matrix material. Its common name is Bisphenol A Diglycidyl Ether. The low temperature curing epoxy resin (Araldite LY 556) and corresponding hardener (HY951) are mixed in a ratio of 10:1 by weight as recommended. Two different types of composites has been fabricated with same orientations with and without the addition of filler material. Each composite consisting of 30% of fiber and 70% of epoxy resin. Epoxy resin and hardener are weighed and mixed vigorously. It is fabricated with 300gms of fiber (2 layers of sisal 60gms & 3 layers of jute 90 gms), resin 700gms and hardener 10%. The orientation of both hybrids are shown below in figure 3. Suitable mould of dimension 300x300mm is taken. First of all releasing agent is applied for easy removal of finished composite. The mixture of resin and hardener is applied over the layer of releasing agent. Jute bidirectional mat fiber is placed over the layer of resin mix, once again resin is applied over the jute fiber and rolled using a roller to remove air bubbles. Unidirectional sisal fiber is placed above and once again resin is applied following the same procedure to remove air bubbles. Likewise alternate layer of bidirectional layer of jute fiber and sisal fibers are stacked one above the other. After the completion of fabrication, the composite is cured under a load of about 50 kg for 24 hours before it removed from the mould. Then this cast is post cured in the air for another 24 hours after removing out of the mould. Specimens of suitable dimension are cut using a diamond cutter for mechanical testing. Utmost care has been taken to maintain uniformity and homogeneity of the composite. The same procedure is followed for fabricating the second sample, the only difference is that, in this fabrication SiC filler material of about 10% is added to the resin mix to improve the mechanical properties. It is fabricated with 300gms of fiber (2 layers of sisal 60gms & 3 layers of jute 90 gms), resin 700gms and hardener 10% and SiC filler material 10%. The figure 4 shows the various processing steps involved in the preparation of samples. The figure 5 and 6 shows the prepared sample after hand layup technique.

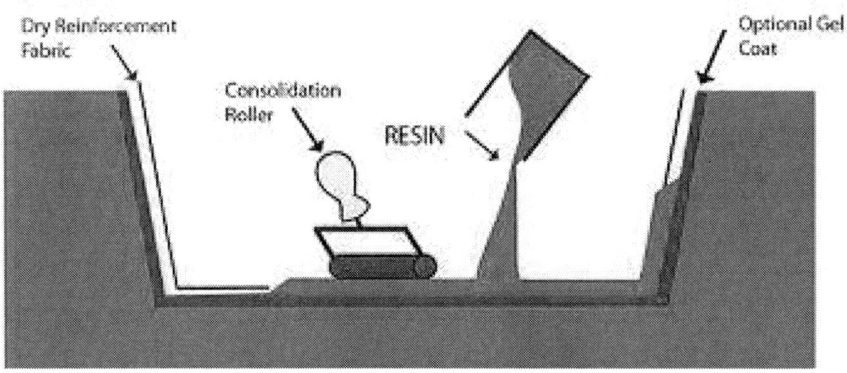


**Figure 2. Handlayup technique**

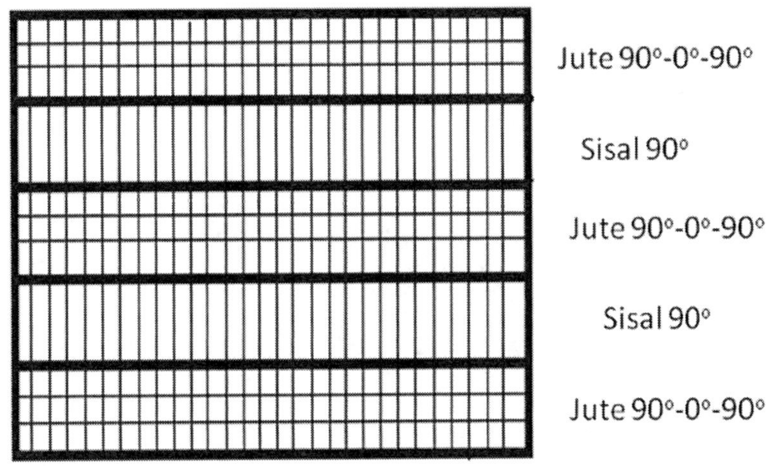


**Figure 3. Ply Orientation of Samples**

Application of releasing agent

Place the jute fiber on the mould and apply resin

Place the sisal fiber over the jute and apply resin

**Figure 4. Preparation of Samples.**

**Figure 5. Sample1 without filler material**

**Figure 6. Sample 2 with filler material**

IOP Conf. Series: Materials Science and Engineering **923** (2020) 012059 doi:10.1088/1757-899X/923/1/012059

**Drilling of composites**

Conventional drilling machine was used for drilling the composites specimens. Three types of drill tools such as spur, brad and straight flute of 10mm diameter is used for drilling the holes. The speed was varied as 500 rpm, 1000rpm and 1500 rpm. Since 8mm diameter tool was used in the drilling operation the cutting speed (peripheral) was kept in between 20m/min up to 60m/min. Most of the conventional machines operate in this range of cutting speed. Also if higher cutting speed is used, the matrix material which is soft may get softened due to the heat generated. The feed rate was also varied as 0.2mm/rev, 0.4mm/rev and 0.6mm/rev in this experiment. The tool material used in this experiment was HSS twist drill.The table 1 shows the drilling parameters used for machining and the corresponding responses after machining.The figure 7 shows the drilling setup used for this study. The drilled samples were shown in figure 8 and 9 respectively. Circularity of both the hybrids were measured by using video measuring machine as shown in figure 10. The surface roughness was measured by using surface roughness tester.

**Table 1 Drilling parameters and responses**

| Exp No | Tool Type | Spindle Speed (rpm) | Feed Rate (mm/rev) | Circularity | | Surface Roughness | |
|---|---|---|---|---|---|---|---|
| | | | | With SiC | Without SiC | With SiC | Without SiC |
| 1 | Spur | 500 | 0.2 | 4.94/9.88 | 4.96/9.92 | 0.72 | 1.44 |
| 2 | Spur | 1500 | 0.6 | 4.93/9.86 | 4.98/9.97 | 0.69 | 2.02 |
| 3 | Spur | 500 | 0.6 | 4.99/9.98 | 5.07/10.15 | 0.92 | 1.69 |
| 4 | Spur | 1000 | 0.4 | 4.95/9.90 | 4.95/9.90 | 0.6 | 1.33 |
| 5 | Spur | 1500 | 0.2 | 4.96/9.92 | 4.91/9.82 | 0.98 | 0.97 |
| 6 | St.Flute | 1500 | 0.4 | 5.20/10.41 | 5.01/10.02 | 1.21 | 0.98 |
| 7 | St.Flute | 1000 | 0.4 | 5.17/10.34 | 5.09/10.18 | 1.41 | 0.90 |
| 8 | St.Flute | 1000 | 0.4 | 5.13/10.27 | 5.10/10.21 | 0.71 | 0.68 |
| 9 | St.Flute | 1000 | 0.4 | 5.07/10.14 | 5.10/10.21 | 1.63 | 1.21 |
| 10 | St.Flute | 1000 | 0.4 | 5.13/10.26 | 5.12/10.24 | 1.98 | 0.62 |
| 11 | St.Flute | 500 | 0.4 | 5.14/10.28 | 5.02/10.05 | 0.98 | 0.66 |
| 12 | St.Flute | 1000 | 0.4 | 5.10/10.21 | 5.07/10.14 | 0.62 | 2.06 |
| 13 | St.Flute | 1000 | 0.2 | 5.10/10.20 | 5.04/10.08 | 1.36 | 1.68 |
| 14 | St.Flute | 1000 | 0.6 | 4.86/9.73 | 5.10/10.20 | 1.02 | 1.22 |
| 15 | St.Flute | 1000 | 0.4 | 5.14/10.29 | 5.00/10.00 | 0.98 | 0.64 |
| 16 | Brad | 1500 | 0.2 | 4.99/9.98 | 4.92/9.85 | 2.57 | 0.82 |
| 17 | Brad | 500 | 0.6 | 4.98/9.96 | 4.96/9.92 | 0.57 | 0.24 |
| 18 | Brad | 1500 | 0.6 | 5.0/10.01 | 5.01/10.03 | 0.65 | 1.13 |
| 19 | Brad | 1000 | 0.4 | 4.98/9.97 | 4.95/9.91 | 0.72 | 0.52 |
| 20 | Brad | 500 | 0.2 | 4.93/9.86 | 4.94/9.89 | 0.63 | 0.58 |

**Figure 7. Drilling Machine setup**

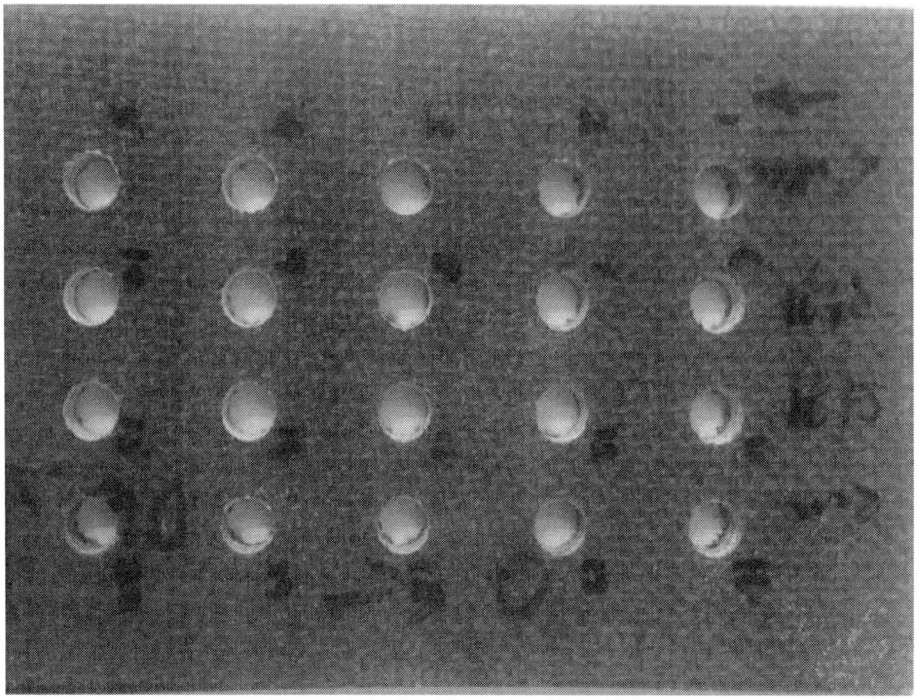

**Figure. 8 Drilled Sample (Hybrid 1)**

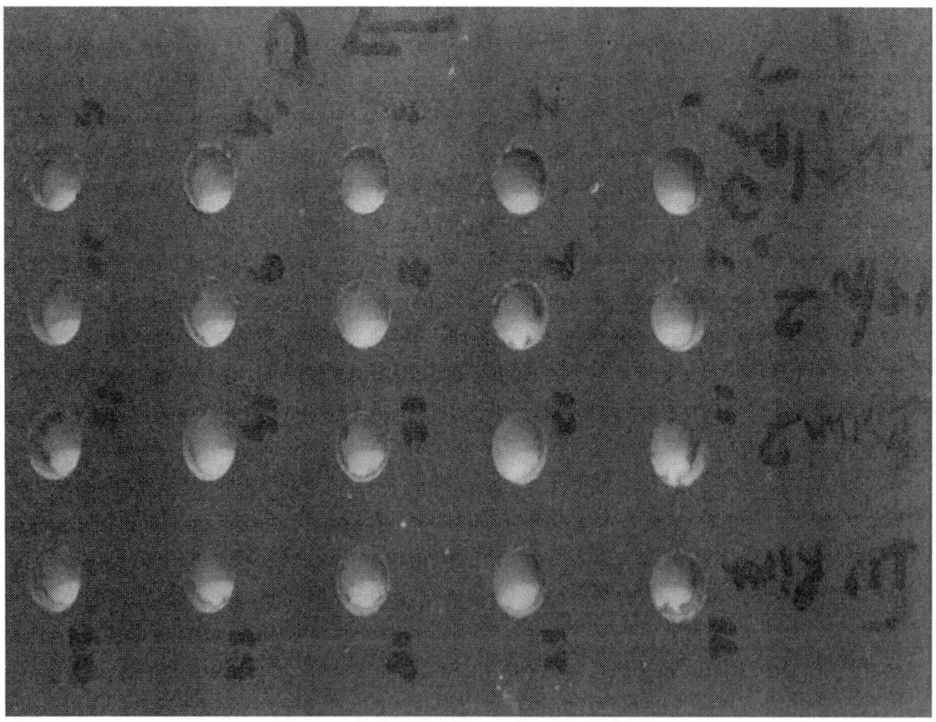

**Figure. 9 Drilled Sample (Hybrid 2)**

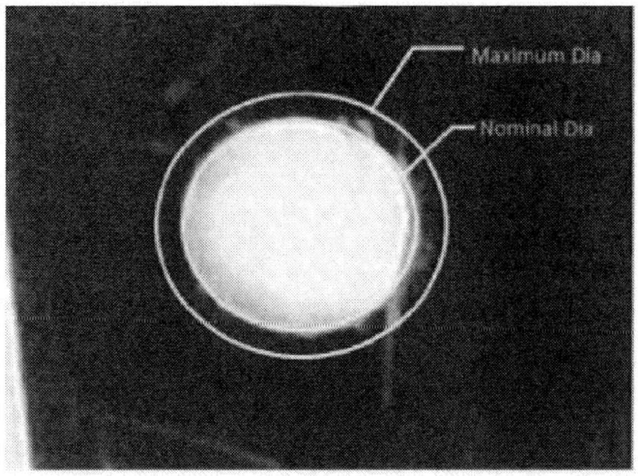


**Figure 10. VMS Digital Image of Drilled Hole**

**3.Results and Discussion**

The parameters are optimized using Response Surface Methodology. The following steps are to be followed to generate optimum input parameters for machining of Hybrid laminated sheet using drilling. of all preliminary tests are done and journals are studied in order to find out the machining parameters which are most influential in machining Hybrid laminated sheet. The levels of the input parameters are derived. After finding levels of parameters, design matrix as shown in table was prepared with the help of Design Expert V 7.0 software. Experiments were done as per the design matrix. Circularity is measured after making the drills with respective input parameters. Then the Surface Roughness was measured after each trial and all the data was recorded. The control parameters used in this study are shown in table 2. Table 3 shows the factors with level values. Table 4 shows the ANNOVA for circularity with SiC. The figure 11 to 14 shows the graphs showing predicted vs actual values.

**Table 2. Control Parameters**

| | | |
|---|---|---|
| 1 | **Factor A** | Tool Type (mm) |
| 2 | **Factor B** | Speed (rpm) |
| 3 | **Factor C** | Feed (mm/rev) |

**Table 3. Factors with level values**

| No. | Symbols | Machining parameters | Levels | | | Units |
|---|---|---|---|---|---|---|
| 1 | A | Tool Type | Spur | St.Flute | Brad | mm |
| 2 | B | Speed | 500 | 1000 | 1500 | rpm |
| 3 | C | Feed | 0.2 | 0.4 | 0.6 | mm/rev |

The Model F-value of 18.92 implies the model is significant. There is only a 0.01% chance that a "Model F-Value" this large could occur due to noise.Values of "Prob > F" less than 0.0500 indicate model terms are significant.In this case $A^2$, $B^2$, $C^2$ are significant model terms.Values greater than 0.1000 indicate the model terms are not significant.If there are many insignificant model terms (not counting those required to support hierarchy), model reduction may improve your model.The "Lack of Fit F-value" of 0.51 implies the Lack of Fit is not significant relative to the pure error. There is a 76.27% chance that a "Lack of Fit F-value" this large could occur due to noise. Non-significant lack of fit is good -- we want the model to fit.

**Table 4. ANOVA for circularity (With SiC)**

| Source | Sum of Squares | df | Mean Square | F Value | p-value Prob > F | |
|---|---|---|---|---|---|---|
| Model | 0.613 | 9 | 0.06811 | 18.9168 | < 0.0001 | significant |
| A-Tool Type | 0.00784 | 1 | 0.00784 | 2.17746 | 0.1708 | |
| B-Speed | 0.01444 | 1 | 0.01444 | 4.01053 | 0.0731 | |
| C-Feed | 0.00225 | 1 | 0.00225 | 0.62491 | 0.4476 | |
| AB | 0.00781 | 1 | 0.00781 | 2.16982 | 0.1715 | |
| AC | 1.25E-05 | 1 | 1.25E-05 | 0.00347 | 0.9542 | |
| BC | 0.00451 | 1 | 0.00451 | 1.25329 | 0.2891 | |
| $A^2$ | 0.19445 | 1 | 0.19445 | 54.0049 | < 0.0001 | |
| $B^2$ | 0.0571 | 1 | 0.0571 | 15.8577 | 0.0026 | |
| $C^2$ | 0.111 | 1 | 0.111 | 30.8295 | 0.0002 | |
| Residual | 0.03601 | 10 | 0.0036 | | | |
| Lack of Fit | 0.01212 | 5 | 0.00242 | 0.50755 | 0.7627 | not significant |
| Pure Error | 0.02388 | 5 | 0.00478 | | | |
| Cor Total | 0.649 | 19 | | | | |

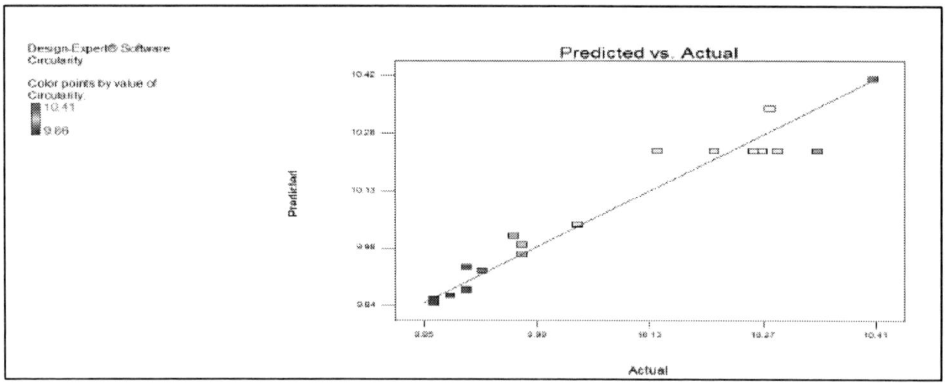


**Figure 11. Graph representing predicted Vs actual values of circularity**

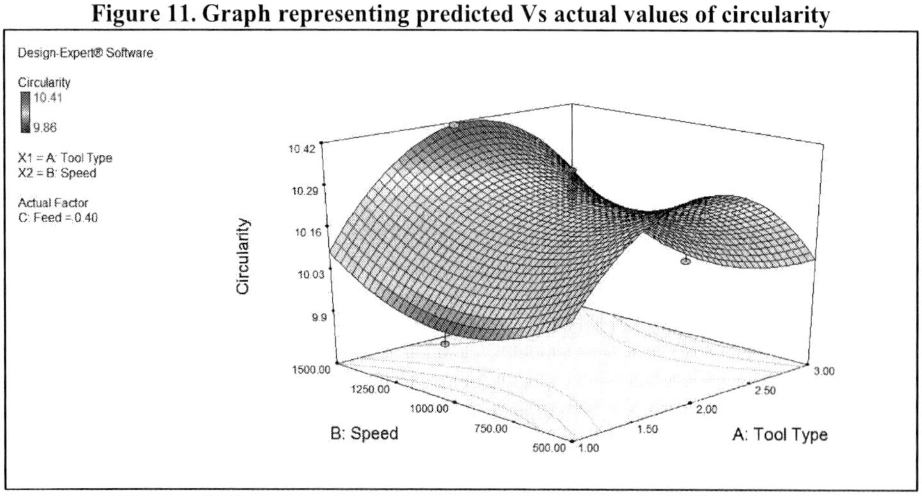


**Figure 12.Graph representing circularity vs Speed and Tool Type**

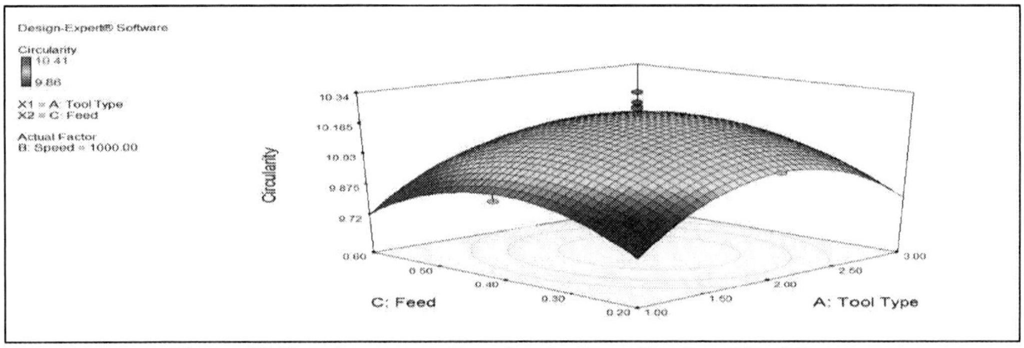


**Figure 13.Graph representing circularity vs Feed and Tool Type**

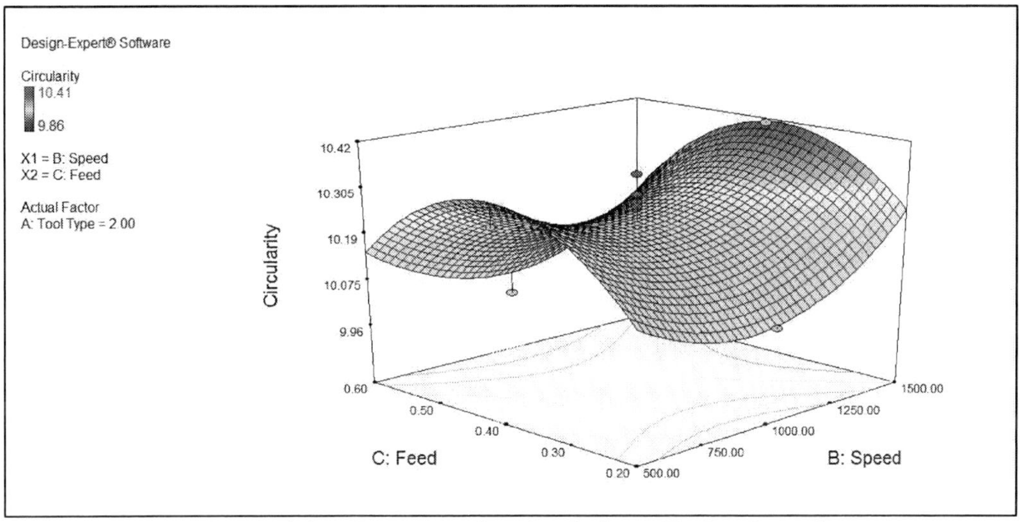


**Figure 14.Graph representing circularity vs Feed and speed**

The Model F-value of 3.06 implies the model is significant. There is only a 4.29% chance that a "Model F-Value" this large could occur due to noise. Values of "Prob > F" less than 0.0500 indicate model terms are significant. In this case BC are significant model terms. Values greater than 0.1000 indicate the model terms are not significant. If there are many insignificant model terms (not counting those required to support hierarchy), model reduction may improve your model.

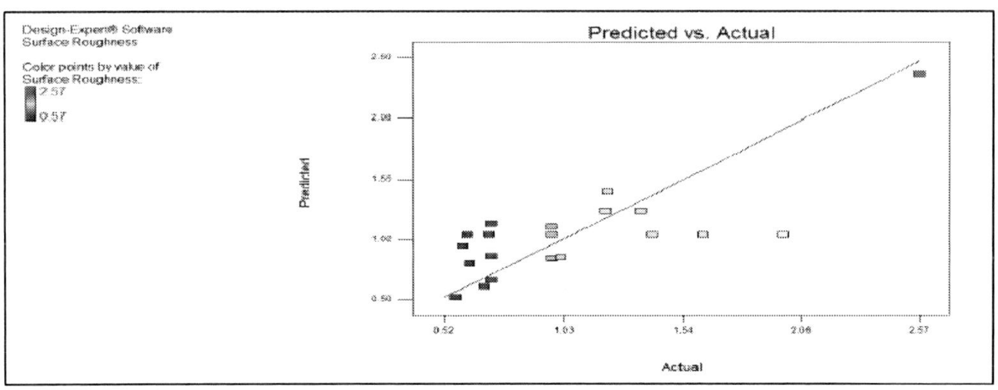


**Figure 15. Graph representing predicted Vs actual values of circularity.**

The "Lack of Fit F-value" of 0.29 implies the Lack of Fit is not significant relative to the pure error. There is a 94.01% chance that a "Lack of Fit F-value" this large could occur due to noise. Non-significant lack of fit is good -- we want the model to fit.

**Table 5. ANOVA for surface roughness (With SiC)**

| Source | Sum of Squares | df | Mean Square | F Value | p-value Prob > F | |
|---|---|---|---|---|---|---|
| Model | 3.02646 | 6 | 0.50441 | 3.061283 | 0.0429 | significant |
| A-Tool Type | 0.1 | 1 | 0.1 | 0.606904 | 0.4499 | |
| B-Speed | 0.42025 | 1 | 0.42025 | 2.550513 | 0.1343 | |
| C-Feed | 0.41616 | 1 | 0.41616 | 2.525691 | 0.1360 | |
| AB | 0.6962 | 1 | 0.6962 | 4.225264 | 0.0605 | |
| AC | 0.5618 | 1 | 0.5618 | 3.409585 | 0.0877 | |
| BC | 0.83205 | 1 | 0.83205 | 5.049743 | 0.0426 | |
| Residual | 2.14202 | 13 | 0.164771 | | | |
| Lack of Fit | 0.682537 | 8 | 0.085317 | 0.292285 | 0.9401 | not significant |
| Pure Error | 1.459483 | 5 | 0.291897 | | | |
| Cor Total | 5.16848 | 19 | | | | |

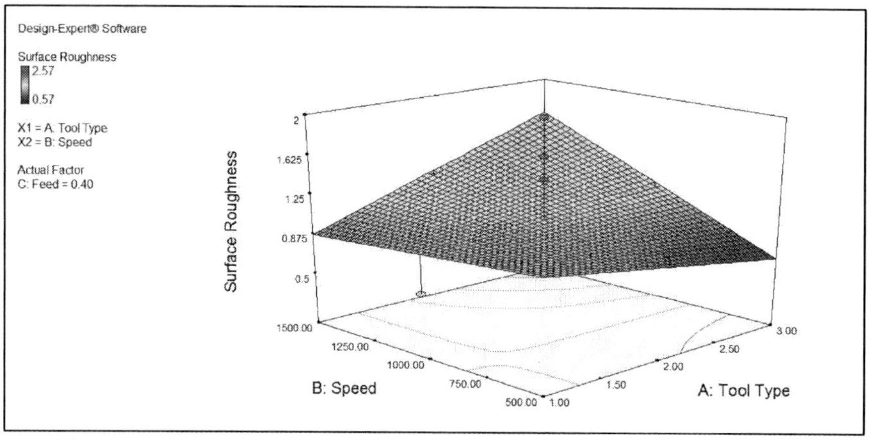


**Figure 16. Graph representing Surface Roughness vs Tool Type and Speed**

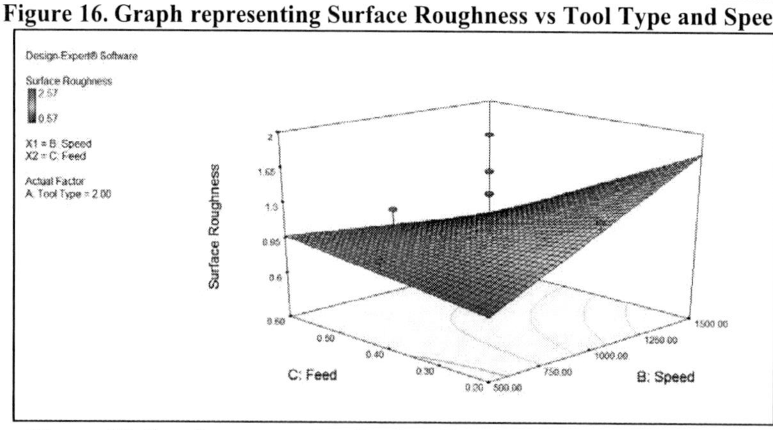


**Figure 17. Graph representing Surface Roughness vs Feed and Speed**

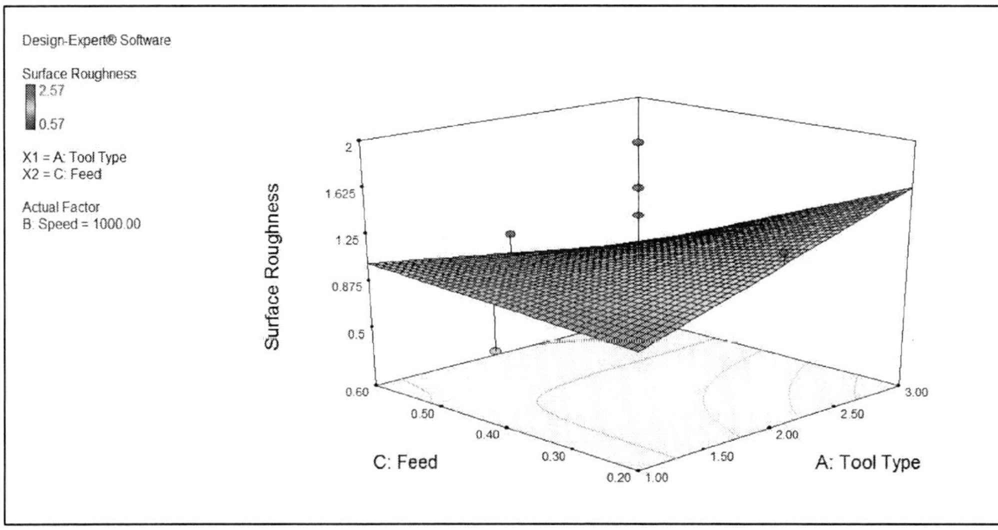


**Figure 18.Graph representing Surface Roughness vs Tool Type and Feed**

**Table 6. Solution for Hybrid with SiC**

| Number | Tool Type | Speed | Feed | Circularity | Surface Roughness | Desirability |
|---|---|---|---|---|---|---|
| 1 | 2.99 | 500 | 0.53 | 10 | 0.577164 | 0.998 |

**Table 7. ANOVA for Circularity (Without Sic)**

| Source | Sum of Squares | df | Mean Square | F Value | p-value Prob > F | |
|---|---|---|---|---|---|---|
| Model | 0.303754 | 9 | 0.03375 | 5.973464 | 0.0050 | significant |
| A-Tool Type | 0.00784 | 1 | 0.00784 | 1.387594 | 0.2661 | |
| B-Speed | 0.01296 | 1 | 0.01296 | 2.293778 | 0.1608 | |
| C-Feed | 0.03481 | 1 | 0.03481 | 6.160988 | 0.0324 | |
| AB | 0.006613 | 1 | 0.006613 | 1.17034 | 0.3047 | |
| AC | 0.010513 | 1 | 0.010513 | 1.860597 | 0.2025 | |
| BC | 0.000312 | 1 | 0.000312 | 0.055309 | 0.8188 | |
| A^2 | 0.104082 | 1 | 0.104082 | 18.42134 | 0.0016 | |
| B^2 | 0.011457 | 1 | 0.011457 | 2.027731 | 0.1849 | |
| C^2 | 0.004501 | 1 | 0.004501 | 0.796551 | 0.3931 | |
| Residual | 0.056501 | 10 | 0.00565 | | | |
| Lack of Fit | 0.018767 | 5 | 0.003753 | 0.497368 | 0.7692 | not significant |
| Pure Error | 0.037733 | 5 | 0.007547 | | | |
| | | | | | | |
| Cor Total | 0.360255 | 19 | | | | |

The Model F-value of 5.97 implies the model is significant. There is only a 0.50% chance that a "Model F-Value" this large could occur due to noise. Values of "Prob > F" less than 0.0500 indicate model terms are significant. In this case C, A2 are significant model terms. Values greater than 0.1000 indicate the model terms are not significant.If there are many insignificant model terms (not counting those required to support hierarchy),model reduction may improve your model.The "Lack of Fit F-value" of 0.50 implies the Lack of Fit is not significant relative to the pure error. There is a 76.92% chance that a

"Lack of Fit F-value" this large couldoccurdue to noise. Non-significant lack of fit is good -- we want the model to fit.

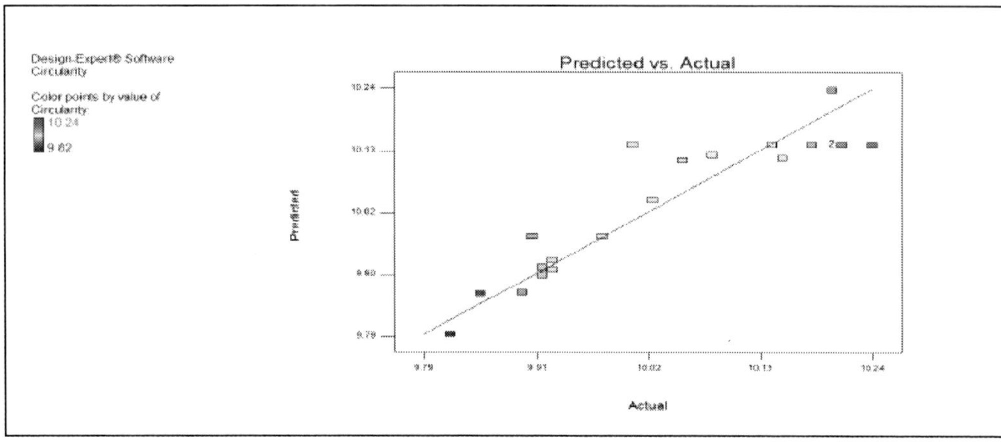


**Figure 19. Graph representing predicted Vs actual values of Circularity**

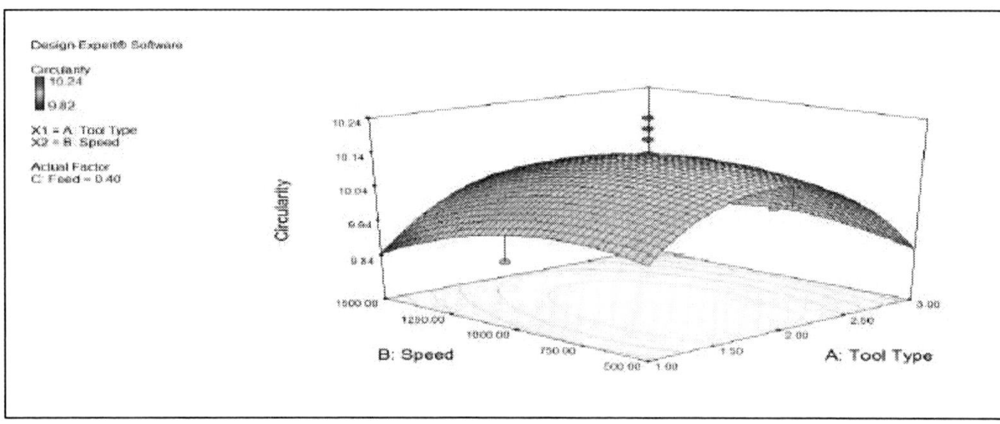


**Figure 20. Graph representing Circularity vs Tool Type and Speed**

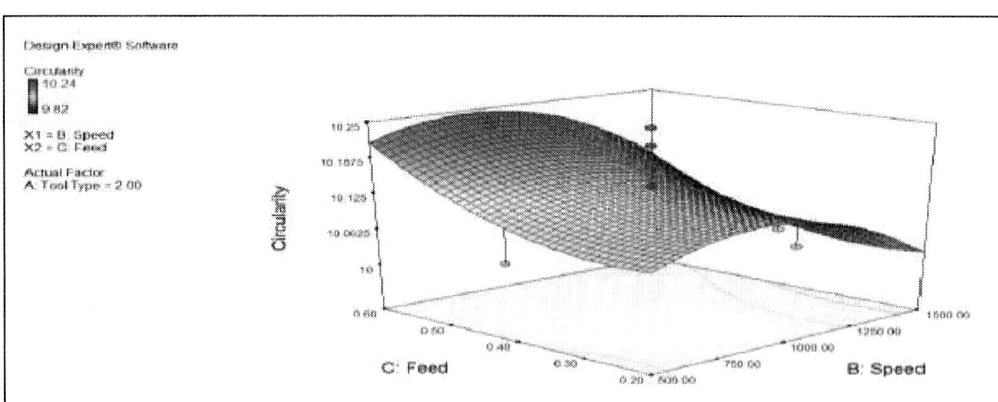


**Figure 21. Graph representing Circularity vs Feed and Speed**

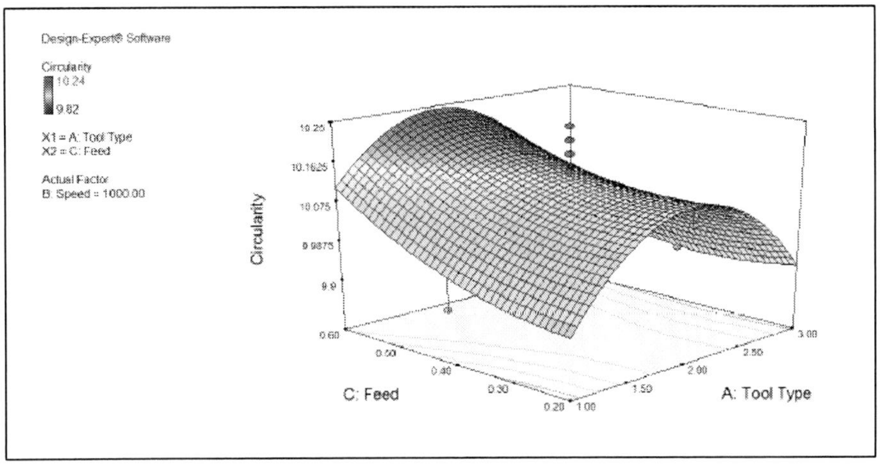


**Figure 22. Graph representing Circularity vs Tool Type and Feed**

**Table 8. ANOVA for Surface Roughness (Without Sic)**

| Source | Sum of Squares | df | Mean Square | F Value | p-value Prob > F | |
|---|---|---|---|---|---|---|
| Model | 2.32529 | 3 | 0.775097 | 4.322316 | 0.0206 | significant |
| A-Tool Type | 2.27529 | 1 | 2.27529 | 12.68812 | 0.0026 | |
| B-Speed | 0.049 | 1 | 0.049 | 0.273248 | 0.6083 | |
| C-Feed | 0.001 | 1 | 0.001 | 0.005576 | 0.9414 | |
| Residual | 2.86919 | 16 | 0.179324 | | | |
| Lack of Fit | 1.317107 | 11 | 0.119737 | 0.38573 | 0.9127 | not significant |
| Pure Error | 1.552083 | 5 | 0.310417 | | | |
| Cor Total | 5.19448 | 19 | | | | |

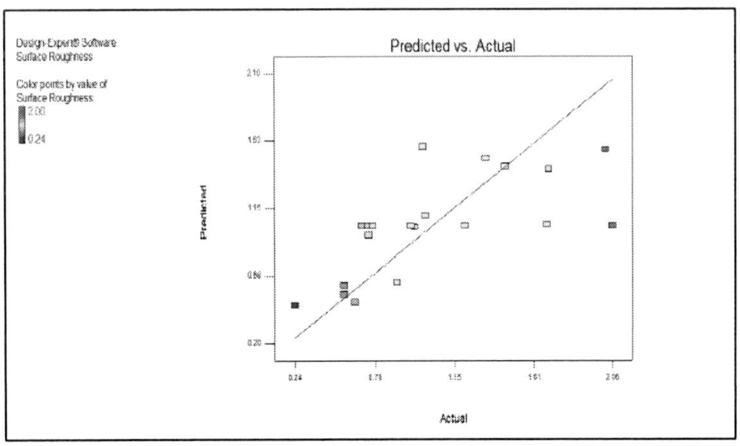


**Figure 23. Graph representing predicted Vs actual values of Surface Roughness**

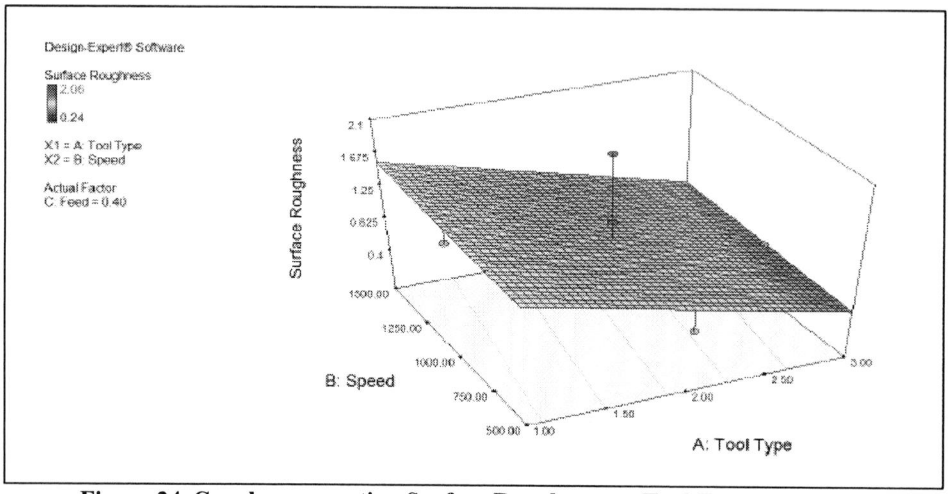


**Figure 24. Graph representing Surface Roughness vs Tool Type and Speed**

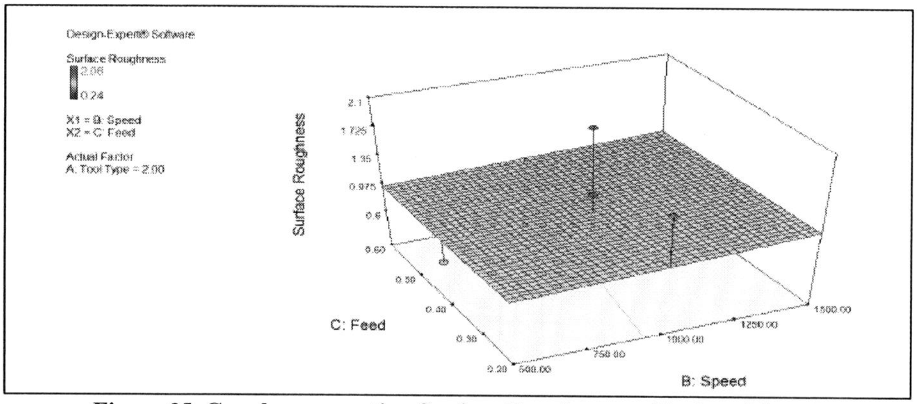


**Figure 25. Graph representing Surface Roughness vs Feed and Speed**

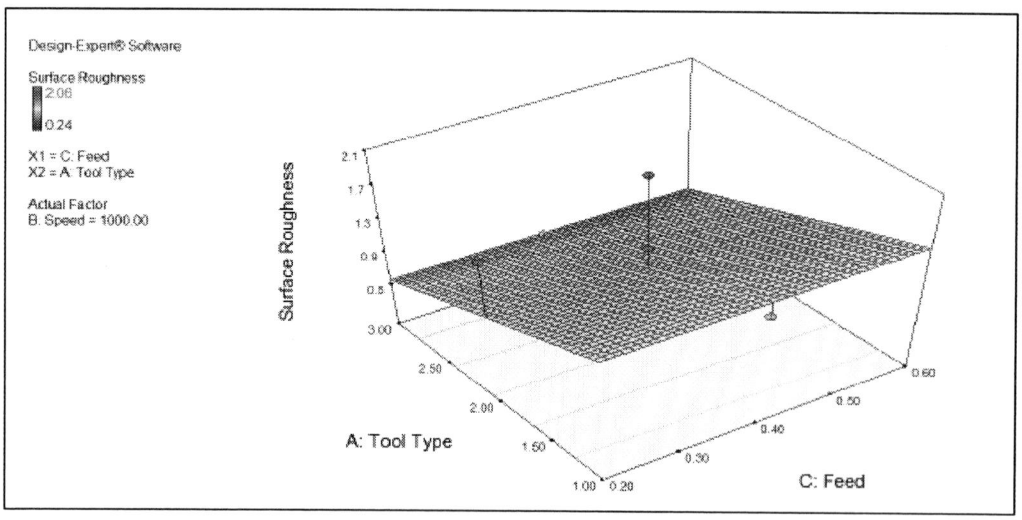


**Figure 26. Graph representing Surface Roughness vs Tool Type and Feed**

The Model F-value of 4.32 implies the model is significant. There is only a 2.06% chance that a "Model F-Value" this large could occur due to noise. Values of "Prob > F" less than 0.0500 indicate model terms are significant. In this case A are significant model terms. Values greater than 0.1000 indicate the model terms are not significant. If there are many insignificant model terms (not counting those required to support hierarchy), model reduction may improve your model.The "Lack of Fit F-value" of 0.39 implies the Lack of Fit is not significant relative to the pure error. There is a 91.27% chance that a "Lack of Fit F-value" this large could occur due to noise. Non-significant lack of fit is good -- we want the model to fit.

**Table 9. Solution (Without SiC)**

| Number | Tool Type | Speed | Feed | Circularity | Surface Roughness | Desirability |
|---|---|---|---|---|---|---|
| 1 | 2.92 | 710.91 | 0.6 | 9.999983 | 0.533522 | 0.915819 |

**Table 10. Confirmation Test Results**

| Parameters | Units | H1 | Conformation Test | H2 | Conformation Test |
|---|---|---|---|---|---|
| Tool [Type | mm | 2.92 | 3 | 2.99 | 3 |
| Speed | Rpm | 710.91 | 700 | 500 | 500 |
| Feed | mm/rev | 0.6 | 0.6 | 0.53 | 0.6 |
| Circularity | Mm | 9.99 | | 10 | |
| Surface Roughness | µm | 0.533 | | 0.577 | |

**Conclusion**

In this study the composite laminates are created by reinforcing bi-directional bast type jute and unidirectional leaf made sisal fibers into an epoxy LY556 resin matrix material. The silicon carbide (SiC) is used as filler material. Two types of composite laminates are manufactured. The first sample is considered in the ratio of 70% epoxy resin to 30% Jute + Sisal fiber composition without any filler material. The second composition includes the ratio of 70% epoxy resin to 30% Jute + Sisal fiber along with 10% SiC as filler material. The hardner HY951 of 10% is used in both compositions. The machining operation was performed by using HSS twist drill for three different machining parameters (Drill bit type, Spindle speed and Feed rate). The circularity and surface roughness are considered as output parameters. The machining is performed in the range of 20 different combinations as directed by the RSM software. The conformation tests are carried out to validate the RSM results. The optimization results shows that the sample with SiC filler exhibits better machinability behaviour than sample without filler material.

**References**

[1] T. Ramalingam,S.Bhaskar,K.Seshumadhav,K.Viswanath Allamraju 2018 *Volume 5, Issue 9, Part 3 Pages 18933-18940*
[2] Vamsi Krishna Balla,Kunal H. Kate, Jagannadh Satyavolu, Paramjot Singh, Jogi Ganesh Dattatreya Tadimeti 2019 *Engineering Volume 1741 October Article 106956*
[3] Wang wei, Huang Gu, 2009 *journal Volume 30, Issue 7August Pages 2741-2744*
[4] Anas Ahmed Ansari,S. K. Dhakad,Pankaj Agarwal 2020 *Volume 26, Part 2 Pages 2400-2404*
[5] Saket Saurabh,Tanmay Tiwari,Akash Nag,AR Dixit,Ashish Kumar Srivastava 2018 *Volume 5, Issue 9, Part 3 Pages 18061-18069*
[6] K. Siva Prasad,G. Chaitanya *Available online 1 July 2020*
[7] J. Schwartzentruber,M. Papini, J. K. Spelt 2018 *Volume 112September Pages 299-314*
[8] S.Harish, D. Peter Michael, A. Bensely, D. Mohan Lal, A.Rajadurai 2009 *journal Volume 60, Issue 1January Pages 44-49"*

[9] Puneet Kumar,Ravi Kant 2020 *Volume 28, Part 2Pages 1164-1169*
[10] M. A. Azmir,A. K. Ahsan 2009 *Volume 209, Issue 2 November Pages 6168-6173*
[11] Paul Wambua, Jan Ivens and Ignaas Verpoest 2003 *Volume 63, Issue 9 Pp 1259-1264*
[12] .C.C. Tsao and H. Hocheng 2008 *Volume 203, Issues 1–318 Pages 342-348*
[13] Marta Fernandes and Chris Cook 2006 *Volume 46, Issue 1January Pages 76-79*
[14] S.C. Lin and IK. Chen 1996 *Volume 194, Issues 1–2Pages 156-162*
[15] K. Murugan,S. Venkatesh,R. Thirumalai,S. Nandhakuma *Available online 2020*

# Investigation of Tensile and wear properties of aluminium metal matrix composite

M.Arumugasamy[1], M. Suba pradha[2] and V.Vaitheeswari[3]

[1] Assistant Professor, M.A.M School of Engineering, Siruganur, Tamilnadu, India
[2] Assistant Professor, M.A.M School of Engineering, Siruganur, Tamilnadu, India
[3]U.G Student, M.A.M School of Engineering, Siruganur, Tamilnadu, India

E-mail: cool.pradha@gmail.com

**Abstract**. The recent development of piston materials in internal combustion engines has contributed to the overall improvement in terms of durability and operation reliability. R&D and scientific centers both at home and abroad are conducting studies aimed at increasing the net power, torque and reducing the fuel consumption. The main function of piston is to convert thermal energy into mechanical work. Therefore selection of best material for piston plays a vital role in terms of extending its life, engine efficiency and low fuel consumption. Recent research has come up with a concept of composite materials, which shows that it has good mechanical properties compared to existing materials.
**Keywords**: automobiles, heavy transport, light transport, medium transport.

## 1. INTRODUCTION:

Materials that are most commonly used for manufacturing pistons include cast iron, alloy steel and Aluminium alloys, Aluminium-silicon (Al- Si) alloys and Aluminium-copper Aluminium alloys are distinguished by good formability during casting and good machinability (machine cutting). (1) The major drawbacks of these alloys include: a large thermal expansion coefficient, low hardness and low strength indices at elevated temperatures. (2) Cast-iron pistons are less and less often used. They can be found in low-speed self-ignition engines. They are characterized by the slide properties, retaining good mechanical properties at the elevated temperatures, and a small thermal expansion coefficient. The main disadvantage is that limit the use of cast-iron piston in contemporary high-speed engines a low thermal conductivity coefficient and high density .It results in a large piston mass and considerable inertia forces. (3) Due to high strength and low thermal expansion, pistons are increased oftenly and made of alloy steels. In spite of high density of steel, to take advantage of good mechanical properties of the material, designers give pistons appropriate shapes with small overall dimensions, which reduces the piston mass and make it compare to the mass of Aluminium alloy pistons.(4). The main objective of the study is to understand the suitability of Aluminium alloy (8090). The properties of Al alloy (8090) shows that it has better resistive properties to withstand the thermal stress in combustion chamber.

## 2. PISTON FAILURES

The most frequently occurring engine breakdowns are the failure of the engine pistons. The Engine piston occur at various mileage and due to different cause. This failure is caused by the material defect and engineering operational error.

The common causes of piston failures include: their insufficient cooling, insufficient lubrication of the piston guiding part, thermal fatigue of the piston head surface, failures due to an incorrect combustion process, and mechanical damage. During the engine operation, piston attain higher temperature than the cylinder, which will result in different thermal expansion of piston and cylinder. The thermal expansion of piston is much larger than the cylinder. In addition, the thermal expansion coefficient of Aluminium alloys is two times greater than the grey cast iron.

IOP Conf. Series: Materials Science and Engineering **923** (2020) 012060 doi:10.1088/1757-899X/923/1/012060

## 2.1 TYPES OF FAILURES

**Cracked piston**: Detonation is most likely caused of a cracked or broken piston. The hammer- like blow of detonation is literally beat a piston to death. It causes similar to those that can burn a piston: the lean fuel mixture, the over- advanced spark timing, bad knock sensor, a low octane fuel that cause the engine to run hotter.

**Scuffed piston**: Scuffed pistons is caused by too much heat in the combustion chamber, the engine overheat or inadequate lubrication. The piston-to- cylinder clearance is late-model engines which is much less and they are used to reduce piston rock and noise. If the piston or cylinder get too hot, the clearance will go away and get metal-to-metal contact.

**Destroyed piston**: If the piston is shattered and self-destructed, it will hit a valve. On interference engines, there is not enough clearance between piston and valves to avoid direct contact, the timing belt breaks. In most instance, the piston will survive a timing belt failure but bend it valves. But, some instance, the impact of a valve against a piston is more than casting can handle the piston to shatter like hand grenade. The debris goes into crankcase and it cause additional damage to bearings or piston.

## 2.2 WEAR PROPERTY OF A MATERIAL

Wear is progressive loss of material from operating surface of solid occurring as a result of relative motion between two surfaces. It appears as it occurs due to the relative motion between two bodies which are solid. There can be different mode or form of wear such as abrasion, adhesion, erosion and fatigue. When the two surface appears smooth macroscopically are brought together then the contact occurs at the isolated asperities on the surface. Ingeneral, wear can be classified into following types and it is explained in the following sections.

Abrasive wear, Fretting wear, Corrosive wear, Adhesive wear, Fatigue wear

## 3. PROPOSED METHOD

The in-depth study is carried out to understand the performance of Al alloy (8090) reinforced with the fly ash and zirconia under various test (tensile and wear). The comparative analysis of study between Al alloys (8090) reinforced with different compositions of zirconium and fly ash has been carried in this project work.

## 3.1 MATERIAL SELECTIONS

A composite material is the material made from two or more constituent material with significantly different physical or chemical properties, when combined, produce the material with characteristics different from individual components. The individual components remains separate and distinct with the finished structure. The new material is preferred for many reasons: common examples include materials which are stronger, lighter, or less expensive when compared to traditional material. Composites is made up of individual material referred to as constituent materials. There are two main categories of the constituent materials: matrix and reinforcement. At least one portion of type is required. The matrix material surrounds and support the reinforcement materials by maintaining the relative position. The reinforcements impart their special mechanical and physical property to enhance the matrix properties. A synergism produces the material properties unavailable from the individual constituent materials, the wide variety of matrix and strengthening materials allow the designer of product or structure to choose an optimum combination.

IOP Conf. Series: Materials Science and Engineering **923** (2020) 012060 doi:10.1088/1757-899X/923/1/012060

## 3.2 METAL MATRIX COMPOSITES (MMCs)

Metal matrix composites (MMCs) are consist of a metal or an alloy as continuous matrix and a reinforcement that can be a particle, short fiber or whisker, or continuous fiber. There are three kind of MMCs such as particle reinforced MMCs, short fiber or whisker reinforced MMCs, continuous fiber or sheet reinforced MMCs. Important metallic matrices are aluminium alloys, titanium alloys, magnesium alloys, copper etc. Many process for fabricating the metal matrix composites are available. For most of the part, these processes involve processing in the liquid and solid state.

### 3.2.1 MATRIX

The matrix is the material in which the reinforcement is embedded, and is completely continuous. This means that it is a path through the matrix to any point in the material, unlike two materials sandwiched together. In structural application, the matrix is usually lighter metal such as aluminium magnesium and it provides a compliant support for reinforcement. In high- temperature application, cobalt and cobalt–nickel alloy matrices both are common.

### 3.2.2 REINFORCEMENT

The reinforcement material is embedded with a matrix. It does not always serve purely structural task (reinforcing the compound), but is also used to change the physical properties such as coefficient, or. The reinforcement. It can be either continuous, or discontinuous. Discontinuous MMCs can be worked with the standard metalworking techniques, such as extrusion, forging, or rolling. In addition, it may be machined using conventional techniques, but commonly need the use of polycrystalline diamond tooling (PCD).

### 3.2.3 ALUMINIUM ALLOY 8090 (MATRIX)

Aluminium 8090 alloy is a lithium-based wrought alloy. Addition of lithium to aluminum will helps to reduce the density and increase stiffness. When it is properly alloyed, aluminum- lithium alloys can have an excellent combination of strength and toughness. The alloy has an excellent mechanical properties and is designed to promote the less density with increased stiffness. It is used for an advanced material in the design of high technology parts for the aerospace sector. It is used in the defense and weaponry systems.

| | |
|---|---|
| Density | 2540 $kg/m^3$ |
| Melting Point | 600-655 °C |
| Elastic Modulus | 77 GPa |
| Tensile Strength | 450 MPa |
| Yield Strength | 370 MPa |
| Percent Elongation | 7% |
| Thermal Expansion Coefficient | $21.4 \times 10^{-6}$/K |

**Table 3.1 Properties of Aluminium Alloy 8090**

| CONSTITUENT | PERCENTAGE |
|---|---|
| Aluminum, Al | 93 - 96.2 |
| Lithium, Li | 2.2 - 2.7 |
| Copper, Cu | 1- 1.6 |
| Magnesium, Mg | 0.60 - 1.3 |
| Iron, Fe | $\leq$ 0.30 |
| Zinc, Zn | $\leq$ 0.25 |
| Silicon, Si | $\leq$ 0.20 |
| Titanium, Ti | $\leq$ 0.10 |
| Chromium, Cr | $\leq$ 0.10 |
| Manganese, Mn | $\leq$ 0.10 |
| Zirconium, Zr | 0.040 - 0.16 |

**Table 3.2 Compositions of Aluminium Alloy 8090**

### 3.2.4 ZIRCONIA (REINFORCEMENT)

Zirconia has an excellent resistance to the chemicals and corrosion without typical brittleness common in technical ceramics. When compared to the other advanced ceramic materials, zirconia has exceptional strength at room temperature. The principal properties of the material include high fracture toughness, high density, high hardness and wear resistance, good frictional behavior, high temperature capability up to 2400°C, non-magnetic, low thermal conductivity, electrical insulation, coefficient of the thermal expansion is similar to iron, and modulus of elasticity which is similar to steel.

### 3.2.5 FLY ASH (REINFORCEMENT)

The preference is to use fly ash as a filler or reinforcement in metal and polymer matrices is Fly ash which is a byproduct of coal combustion, available in very large quantities (80 million tons per year) at very low cost since much of it is currently land filled. Currently, the use of the manufactured glass microspheres has limited applications due to their high cost of production. Therefore, the material costs of composites can be reduced significantly by the incorporating fly ash into matrices of polymers and metallic alloys. However, very little information is available on aid in the design of the composite materials, even though the attempt is made to incorporate the fly ash in both polymer and metal matrices. Cenosphere fly ash has a lower density than the talc and calcium carbonate, but it slightly higher than hollow glass. The cost of the cenosphere is likely to be much lower than the hollow glass.

Cenosphere may be turn out to be one of the lowest cost fillers in terms of cost per volume. The high electrical resistivity, the low thermal conductivity and low density of fly- ash may be helpful for making the light weight insulating composites. Fly ash is a filler in Al casting which reduces cost, decreases density and increase hardness, stiffness, wear and abrasion resistance. It also improve the Mach inability, damping capacity, coefficient of friction etc. which are needed in the various industries like automotive etc. As the production of Al is reduced by utilization of fly ash. It reduces the generation of greenhouse gases as they are produced during the bauxite processing and the alumina reduction.

| COMPOSITION | BITUMINOUS | SUB BITUMINUOS | LIGNITE |
|---|---|---|---|
| SiO2% | 20-60 | 40-60 | 15-45 |
| Al2O3% | 5-35 | 20-30 | 20-25 |
| Fe2O3% | 10-40 | 4-10 | 4-15 |
| CaO% | 1-12 | 5-30 | 15-40 |
| LOI% | 0-15 | 0-3 | 0-5 |

**Table 3.3 Chemical Composition of Fly Ash**

## 4. METHODOLOGY

### 4.1 COMPOSITE MATERIAL PREPARATION

Recent trends in the fabricating metal matrix composite is stir casting. Stir casting is a liquid state method of the composite material fabrication, in which a dispersed phase (ceramic particles, short fibers) mixed with the molten matrix metal by means of the mechanical stirring. Stir casting is simple and the most cost effective method of the liquid state fabrication. The liquid composite material is cast by the conventional casting methods and it may also be processed by the conventional Metal forming technologies.

#### 4.1.1 STIR CASTING PROCESS

- Pre- heating time of die- 500°C for 90 minutes.
- Pre- heating time of reinforcement- 600°C for 90 minutes.
- Temperature of the furnace- 800°C.
- Stirring speed- 700 rpm to 1200 rpm for 10 minutes.
- Weight of the matrix material- 650 grams per sample.

### 4.1.2 DIFFERENT COMPOSITIONS OF SPECIMENS

- Specimen 1: Al+3% fly ash

  650ams of Al +19.5 grams fly ash

- Specimen 2: Al+5% fly ash

  650 grams Al+32.5 grams fly ash

- Specimen 3: Al+ 5% zirconia

  650 grams Al+32.5 grams zirconia

- Specimen 4: Al+3% fly ash +3% Zirconia

  650 grams Al+19.5 grams fly ash+19.5 grams zirconia

- Specimen 5: Al+ 5% fly ash+5% zirconia

  650 grams Al+ 32.5 grams fly ash+ 32.5 grams zirconia

**Experimental setup**

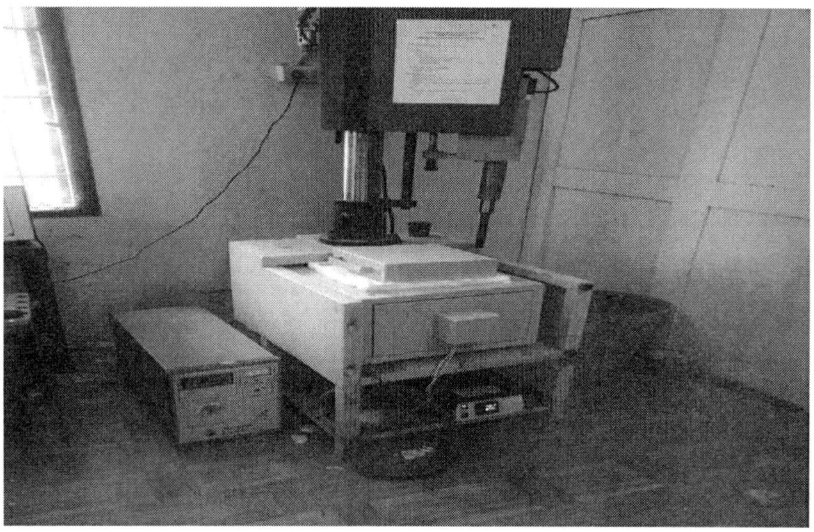

**Figure.4.1 Pre- Heating Furnace**

Muffle furnace is used to heat the material to the desired temperatures by conduction, convection, Or Black body radiation from the electrical resistance of heating elements. A muffle furnace is a Furnace in which the subject material is isolated from fuel and all the products of combustion is including gases and flying ash. This furnace has a capacity to achieve the maximum temperature of 1000°C, which is shown in the Figure.4.1

### 4.2 STIRRING PROCES

**Figure.4.2. Stirring Process**

The stirrer is used in the process is made of stainless steel. Because of the high temperature the crucible, the stirrer should be capable of withstanding the temperature. Hence the stainless steel is used for making the stirrer in Figure.4.2 A crucible container that can be withstand at very high temperatures and is used for the metal, glass, and pigment production as well as number of modern laboratory processes is used. The crucible is used here and is made up of graphite.

### 4.3 SPECIMEN PREPARATION

Pure Al 8090 of 650 grams placed in the graphite of crucible and is kept in an electric Resistance furnace. The furnace is set to the maximum temperature of 800°C. The melting point of Aluminum starts at 750°C after an elapsed time of 90 minutes. The furnace is kept at untouched for more than one hour at 800°C.3% Fly Ash is taken and preheated to temperature of 600°C in order to ensure the moisture of free surface. The die is also placed in the separate furnace at a temperature Of 500°C for pouring the molten metal. Then Fly Ash is added to the molten metal and the mixture is stirred for 10 to 15 minutes at 700 rpm using stirrer. Then the slurry is poured to the die and allowed to solidify. After solidification, the die is opened and the casted piece is also taken out. The above said procedure is repeated for the casting and for remaining specimens. The casted piece is 180mm in length and 20 mm in diameter. The fabricated specimens are shown in the Figure 4.3 below

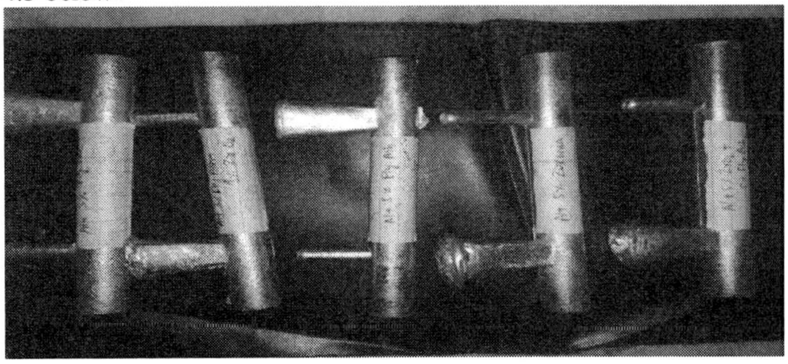

**Figure.4.3. Casted Specimens**

IOP Conf. Series: Materials Science and Engineering **923** (2020) 012060 doi:10.1088/1757-899X/923/1/012060

## 4.4 MATERIAL TESTINGPROCEDURE

**Tensile Testing**

The specimen for tensile testing is prepared from casted specimen. This specimen has a length of 180 mm and diameter 20mm. The Tensile testing specimen should have gauge length of 5*D of the casted specimen. So the length of 100 mm is taken as gauge length where the elongation takes place. Knurling is done on remaining portion of the specimen on both sides to hold the specimen in Universal Testing Machine (UTM), which is shown in the Figure.4.4.

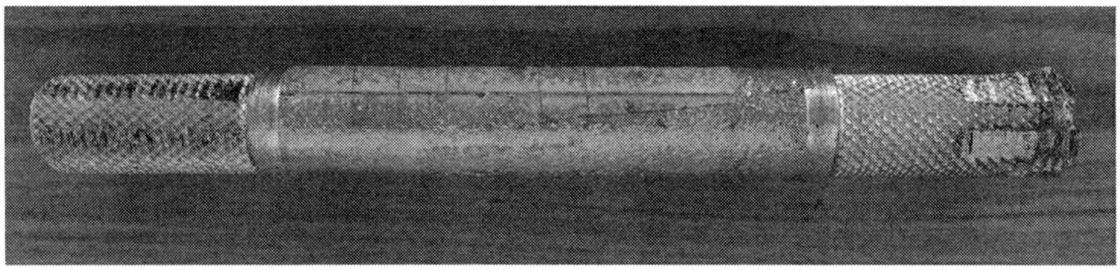

**Figure.4.4 Specimen for Tensile Test**

Four specimens are casted in stir casting is subjected to tensile test one by one. The specimen should be mounted on machine carefully by rotating the wheels placed here. After the confirmation of the correct placement of specimen, the machine is switched ON. The UTM setup is shown in the Figure.4.5

**Figure.4.5 Specimen Mounted in Universal Testing Machine**

### 4.4.1.UTM SETUP

The initial load in the machine is kept as the tare load. From the tare load it starts to increase as the load is applied in the specimen gets increased. At a certain point the load slightly decreases and increases which will represent the breaking point of the specimen. The specimen gets destructed at the point and the damaged pieces are collected which is shown below Figure.4.6.The change in the length after the tensile testing is measured for percentage elongation of specimen.

**Figure.4.6 Specimens after tensile test**

### 4.5.WEAR TESTING

For testing the wear in the specimen, a pin from the specimen and a disc of suitable material is required. In this work, the composite material is proposed for the piston application. So it is important to select the disc material for application. The piston is in the cylinder reciprocates from TDC to BDC during the Combustion process. Due to the movement, wear occurs between the piston and cylinder walls. Most of the cylinders in the Internal Combustion engines are made of the cast iron. Hence in this work the composite material is used for making Pin while the cast iron is used to make the disc to perform wear test. The pin is made with dimensions of length 50mm and diameter 7mm. The disc is made with dimensions of diameter 30 mm which is shown above in Figure.4.7.

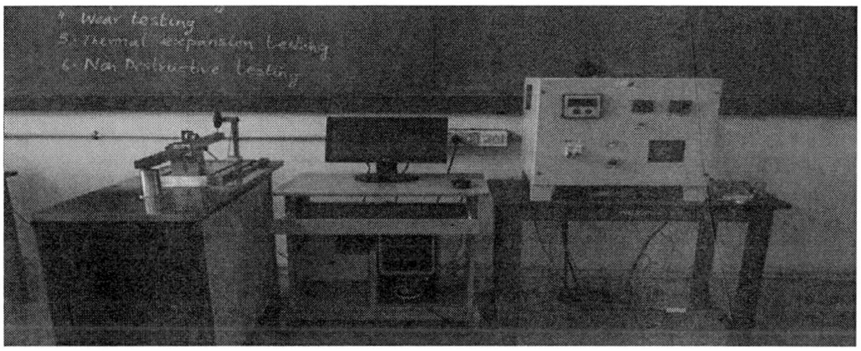

**Figure.4.7 Pin on Disc Apparatus**

**Figure.4.8 Disc and Pin for Wear Test**

For the piston application, the high temperature wear test was carried in the project work using pin on the disc apparatus shown in the Figure.4.8. Three specimens are used here for performing wear test. With the help of the diameter of disc and distance to be covered by pin, the time for the test is calculated. The distance is to be covered by the pin is estimated as 1000 meters. The load is applied for each specimen is 25 kN. With the help of the function, the time in the apparatus is set to 22 minutes for covering the distance.

By brief survey of literature on the wear testing, it is noted that the friction rate starts to increase at the temperature of 300°C. With the help of this, the temperature for the wear test is set to 300°C. It is difficult to predict the temperature inside combustion chamber of an I.C engine. From the literature survey, many of the authors had conducted wear and analysis the starting temperature of 250°C. The speed of the disc is set to 300 rpm for all the three specimens. Hence this temperature is kept at constant for all three specimens which are subjected to wear test. The friction rate will be different for each specimen. After comparing the reduction of weight in the specimens, graphs for the different parameters is to be plotted.

## 5 .RESULTS AND DISCUSSION

### 5.1 BREAKING LOAD OF SPECIMENS

The tensile strength is determine the load bearing capacity of the material. In this work, the fabricated composite material is allowed to tensile stress until there is destruction in the specimen. The above test is performed at room temperature. Two graphs are drawn using the parameters such as load vs displacement and stress vs strain. The displacement of the particular specimen is given load shown below in Figure.5.1

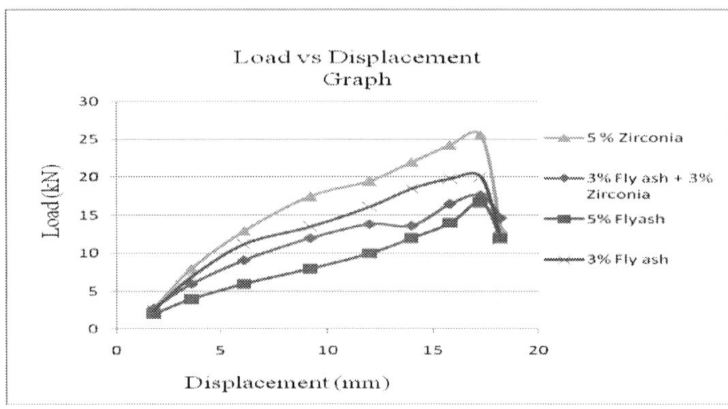


**Figure.5.1 Load vs Displacement Graph**

The reinforcement addition is the metal matrix contributes to the various results in the tensile stress of material. From the above graph it is noted that to increase in the reinforcement decreases the chance of material to be ductile. Only small increment in the deformation is seen when the material is subjected to the loading. 5% fly ash in the matrix shows least breaking load on other specimens. This is because of agglomeration in the composite material due to addition of fly ash. While 5% zirconia addition tends to the withstand some loads more than other specimens. The maximum displacement for specimen reinforced with 5% zirconia is at 25.5 mm, the percentage of the elongation is about 3% with 25.580 KN breaking load.

For 3% fly ash + 3% zirconia in the matrix, the maximum displacement is to 21.1 mm. The percentage of the elongation is about 3%. The breaking load is at 17.660 kN. The addition of the fly ash with zirconia shows disadvantages over tensile strength of specimen. When zirconia and fly ash added separately it show good responses in properties individually. The maximum displacement among all the specimens is exhibited by3% fly ash. 27.9 mm is the displacement of the specimen when breaking with 9% elongation. 5% fly ash addition holds average breaking load of 17.26 KN with a maximum displacement of 18.6mm. Fly ash is individually does not hold the property of stiffness and hardness seen from the results above.

## 5.2 TENSILE STRENGTH OF SPECIMENS

The stress vs strain curve describes the maximum tensile strength of specimen that can hold before breaking. The 5% zirconia shows major tensile strength of about 0.081 $kN/mm^2$ with the total strain of 24 mm at breaking. 5% addition of Zirconia holds the maximum stress during the testing the other specimens are failed to reach maximum value. The stress vs strain curve is shown in Figure.5.2

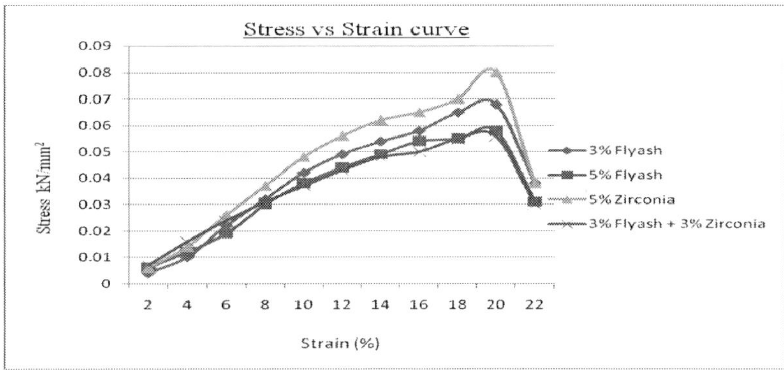


**Figure 5.2 Stress vs Strain Curve**

## 5.3 WEAR RATE OF THE SPECIMENS

The Friction force and co-efficient of friction are two important parameters which will influence the wear rate in specimen. When the friction between pin and the disc is less, the resistance to wear is also enhanced. The co- efficient friction is also contribute in the wear behavior of the material. The pin weight of 5% zirconia reinforcement is reduced to the 5.34 gmfrom 5.39 gm. The reduction in weight after testing is 0.05 gm. For 3% fly ash+ 3% zirconia addition, the weight will reduced to 5.23 gm from 5.30 gm. The difference in the weight before and after testing is 0.07 gm. This specimen shows less contribution to wear resistance. Even though fly ash is added for wear resistance property, the reaction between the fly ash and zirconia when added together shows adverse effect on material. 5% fly ash + 5% zirconia shows greater wear resistance among all specimens. The weight of the specimen is reduced to 5.22 gm from 5.25 gm. This is because the addition of more amount of the fly ash and the previous specimen. 2% fly ash + 2% zirconia addition shows positive effect on wear resistance. Due to the hardness of zirconia, the first specimen holds average wear resistance than second specimen. The friction force vs time graph is shown in the Figure.5.3.

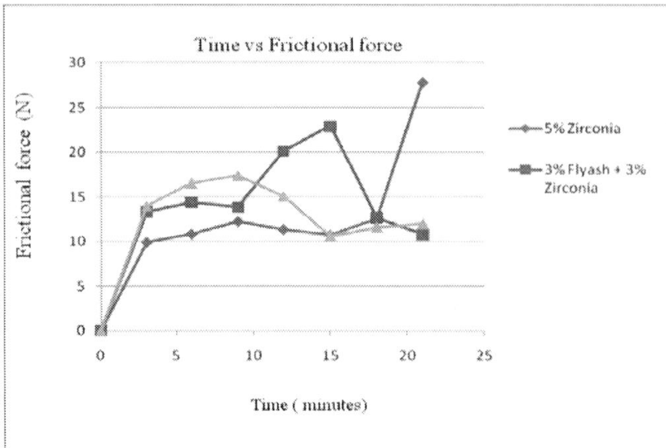


**Figure.5.3 Time vs Frictional force**

The red curve in above graph represent the least frictional force which depends on the wear property of fly ash. 5% fly ash specimen holds the properties- efficient of friction. There is difference on the co-efficient for 5% zirconia specimen. After 12 minutes, the friction co- efficient start in the increasing rapidly than other specimens. There is a delay in frictional force increase the same specimen in the above graph. This is because of sliding velocity of pin in the disc which increases the coefficient at certain points. Even though this velocity influence the friction rate, the 5% fly ash specimen shows the steady wear track. For 3% fly ash addition, there is a sudden increase in friction coefficient at the time of interval of 9 to 18 minutes ending to a normal wear track near 5% fly ash + 5% zirconia specimen. Addition of fly ash in matrix that clearly exhibits wear resistance and wear track which is constant throughout the testing.

## 6.CONCLUSION

Addition of the fly ash in matrix material shows poor response in the case of tensile strength. Where as in case of wear, fly ash contributes the great influence in the resistance to wear rate. As fly ash is easily available it can be used for the dominant property. While other properties does not suit for modern application. Zirconia addition contribute its actual property of hardness in term of tensile strength. Wear property is also a good when it is added together with fly ash. Zirconia holds a good reinforcement in the metal matrix. Sometimes the casting defects like cracks, blow holes occur during composite material preparation. These defects might show negative effects on the property of material. The casting is to be done carefully to avoid these effects. Improper reinforcement in the matrix material causes adverse effect in individual property of composite material. This is eliminated by the proper dispersion of reinforcement in the matrix material during casting. The mechanical behavior of ZrO2 reinforced composites showed the improved results when compared with double reinforcement with fly ash. Finally it is noted that separate addition of reinforcement will be improve the mechanical behavior in case of tensile properties. While wear the resistance of fly ash with ZrO2 is due to ability to restrict the deformation and preventing the hard asperities from causing abrasive wear.

**REFERENCES**

1. John J. Lewandowski, [2002], „Effects of processing conditions and test temperature on fatigue crack growth and fracture toughness of Be - Al metal matrix composites", Materials Science and Engineering, Vol, A344, pp. 215-228.

2. Jufu Jiang, [2015], „Compression Mechanical Behavior of 7075 Aluminum Matrix Composite Reinforced with Nano-Sized SiC Particles in Semisolid State", Journal of Materials Science & Technology, Vol, 16, pp. 302-324.

3. Li Longbiao, [2015], „cyclic fatigue behavior of carbon fiber reinforced ceramic – matrix composites at room and elevated temperatures with different fiber perform", Materials Science and Engineering, Vol, A654, pp. 368-378.

4. Meandri .C [1994], „High temperature friction and wear testing of silicon nitride ceramics", Tribology International Vol, 28, pp.403-413.

5. Pramanik. A [2015], „The Effects of reinforcement on wear resistance of aluminum matrix composites", Trans. Nonferrous Met. Soc. China, Vol, 26, pp. 348- 358.

6. Radhika. N [2015], „Development of functionally graded aluminium composites using centrifugal casting and influence of reinforcements on mechanical and wear properties", Trans. Nonferrous Met. Soc. China, Vol, 26, pp. 305-316.

7. Ravi.B [2015], „Characterization of Aluminium Matrix Composites (AA6061/B4C) fabricated by Stir Casting Technique", Materials Processing and Characterization, Vol, 2, pp. 298

IOP Conf. Series: Materials Science and Engineering **923** (2020) 012061 doi:10.1088/1757-899X/923/1/012061

# Analysis of Green Fuel for Diesel Engine

**N S Senthur*, H Imamulhasan , H RamGanesh, and S Shafquath ibn sulthan**
Department of Mechanical, Einstein college of Engineering, Tirunelveli-627012 India.

*E-mail: senthur.ns@gmail.com

Abstract. We are One of the inconsumable oils that could be used to produce green fuel is Tamanu oil. It is derived from Tamanu green plant seeds. This research describes about the features, analysis that have been performed by an empirical study on using Tamanu oil as a raw material to generate green fuel in diesel engine. The green fuel Tamanu Methyl Ester Tamanu100 and its blends Tamanu30 and Tamanu60 are used in engine to test the efficiency of the fuel. The characteristics of the engine that are impacted such as brake thermal efficiency, brake specific energy consumption, unburned hydrocarbon, carbon monoxide, and release of $NO_X$ have been measured using Tamanu100 green fuel and the output of the tests have been contrasted with conventional fuel. It is noted that the green fuel resulted in trivial reduction in brake thermal efficiency. Release of hydrocarbon and carbon monoxide has lessened on using green fuel with substantial release of oxides of nitrogen. Also, the ignition characteristics such as cylinder gas pressure, ignition delay period have been detailed. The output from the experiments has been contrasted with conventional diesel fuel by applying various load conditions.

*Keywords:* Tamanu Methyl Ester, Emulsion, Brake Thermal Efficiency, Specific Energy Consumption.

## 1. Introduction

The raw material for engine that is fuel is getting exhausted as the consumption has increased.There is a widening gap between the supply and demand of fuel across the globe. Using petrol and diesel causes hazardous effect on the environment that causes respiratory disorders. Hence a look out for another fuel has risen [1].Green fuels are accepted across the globe due to depletion of petroleum and diesel. Green fuels are decomposable, imperishable and very much habitat to the environment as they do not cause any harm in contrast to petroleum-based fuels. Green fuel is a combination of fatty acid methyl ester that are abstracted from any type of vegetable oil and animal fats [2]. Due to higher dependence on vegetable resources in India using that for green fuel has reduced hence an alternative fuel in nature such as in consumable oil resources from seeds have been analyzed. Inconsumable resource is obtaining overall consideration because they are found various countries across the globe specifically in wastelands that are inappropriate for cultivation food crops which removes disputes for food crops and are higher in efficiency and environment friendly they are also cost effective in contrast to consumable oils [3].A significant drawback of Tamanuis it has higher level of free fatty acids that has to be cleaned while trans-esterification procedure is applied. The resultant of the transesterification procedure is very sticky oil.

that reduces the green fuel quality. The next stage in production of green fuel resulted in the green fuel that met the ASTM standards that had characteristics identical to petroleum diesel [4].Low volumetrically energy capacity, high kinematic viscosity, poor cold-flow properties, inferior oxidation stability and high release of $NO_X$ are some demerits of utilizing green fuel that reduces its worldwide use. Nonetheless numerous research works have been done to evaluate the engine effectiveness, ignition and release of toxic gas features while using green fuel in CI engine. A Tamanu seed is an easily cultivable and widely accessible resource for green fuel trans-esterification in developing countries like India. There are no literatures available that informs about setting up Tamanu with diesel blends with 30%, 60% and 100% in a CI engine. Therefore, the current research focuses on the engine effectiveness, release of toxic gas and ignition features of direct injection diesel engine using various blends of Tamanu Like 30% Tamanu, 60% Tamanu and 100% Tamanu. The output of the investigation has been contrasted with conventional diesel propellant. Certain important features like brake specific energy consumption, brake thermal efficiency, heat release rate, in-cylinder pressure, unburned hydrocarbons, carbon monoxide, carbon dioxide, release of $NO_X$ and smoke density have been assessed.

## 2. Material and methods

### *2.1 Green fuel preparation*

Predominantly trans-esterification process is utilized to lessen the stickiness of vegetable oil to generate green fuel [5]. Triglyceride is made of three esters of fatty acid chain connected to glycerol backbone. Glycerol part of Triglyceride adds to maximum stickiness in vegetable oil. Free Fatty Acid portion has 10 times smaller amount of stickiness in contrast to the stickiness of vegetable oil. It is pivotal that the component in the raw material utilized for alkali-catalyzed trans-esterification must have free fatty acids. Fatty acids in the raw material might substantially have an impact on the ester yield and glyceride conversion in alkali-catalyzed trans-esterification process [6]. The fatty acids are changed by using methanol and acid catalyst to generate ester. Molar ratio of alcohol to oil is one of the main parameters that impacts the conversion efficiency, green fuel yield and production. In addition, it also raises the capability of blending and enhances the contact among alcohol and tri-glyceride. Besides, the surplus quantity of alcohol improves the green fuel purity. The maximum conditions for the reactants were found to be 16:1methanol to oil ratio, 10 ml of $H_2SO_4$ for 100 ml of oil, reaction temperature of 60 LC, and 45 min of reaction time along with a constant stirring rate on a magnetic stirrer. A trans-esterification method is applied using alkali catalyst. The maximum conditions for the reactants have been 16:1 methanol to oil ratio, 0.5 g concentration of Sodium hydroxide for 100 ml of oil, reaction temperature of 60 LC, 30 min of reaction time along with a constant stirring rate. Oil is refined by eliminating methanol when heating green fuel at 75 LC for about 30 min to eliminate trapped methanol and the output of green fuel had been around 85%. The varied physical and chemical properties of Tamanu have been assessed and a comparison has been done with diesel as shown in Table 1

**Table 1.** Properties of fuels

| Properties | Diesel | Tamanu |
|---|---|---|
| Acid Number (mg of KOH/g) | - | 0.38 |
| Calorific value (MJ/kg) | 42.5 | 37.8 |
| Cetane Index | 38.5 | 58.7 |
| Density at 15m$^0$ (kg/m$^3$) | 827 | 874 |
| Kinematic viscosity at 40$^0$ (mm$^2$/s) | 2.28 | 5.3 |
| Flash point $^0$C | 54 | 168 |

### 2.2 *Experimental setup*

To test the green fuel Kirloskar TAF-1 direct injection four-stroke single cylinder diesel engine that has certain specification as given in Table 2.The engine is joined with a Swing field electric dynamometer to apply load that is used to vary with a controller in the dynamometer as illustrated in Figure 1. The air flow rate is calculated by measuring the pressure difference of the U-tube manometer connected to the path of airflow pipe in the engine and the time taken for 10 cc of fuel is also noted using stop watch and weighted fuel tank for calculating fuel flow rate. AVL digas 444 exhaust gas analyzer is used for measuring engine exhaust emission such as HC, CO, $CO_2$, and $NO_X$. Nevertheless, the release of toxic gas is computed in g/kW h according to the emission standard method used in non-road application engines. The AVL DIGAS 444 exhaust gas analyzer is used to measure the release of CO and CO2 in percentage and release of HC and $NO_X$ as ppm. The measured CO, HC and $NO_X$ release are changed to ppm and further changed to g/kW h according to the standard release of toxic gas in the testing procedure.

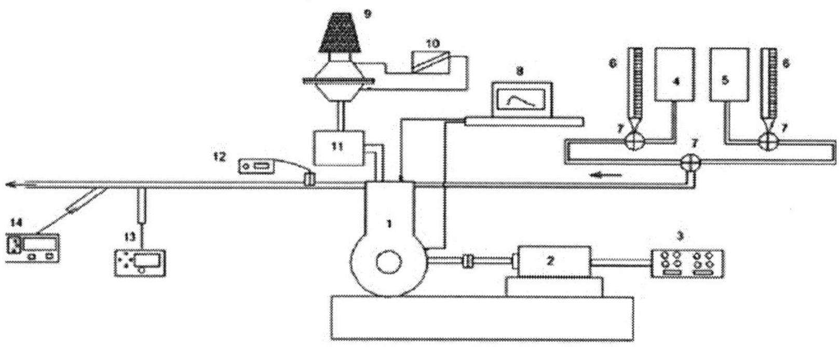


**Figure 1.** Experimental Setup Layout

The Exhaust Gas Temperature is interpreted from the ChromelAlumel (K-Type) thermocouples which has a range of 200 LC to +1260 LC. AVL 437C Smoke meter that has a measurement range of 0–10 FSN is used to calculate the amount of smoke in the exhaust in FSN which is measured by passing a specific amount of exhaust gas through a white filter paper. A microprocessor is used to assess the filter paper blackening that is sensed by a photoelectric measuring head. An electro optical sensor is utilized to sense the TDC position by supplying voltage signals that are sent to A/D converter and then to data acquisition system to record the data. The pressure signals are also recorded by the data acquisition system. A Kistler transducer that has a sensitivity of about 79.5 pC/bar is used to measure the in cylinder pressure and is mounted flush on the cylinder head for avoiding passaging effects. The in-cylinder pressure of the engine was measured using an AVL Pressure transducer. A high-speed computer based Digital Data Acquisition System is utilized for measuring the signals of in-cylinder pressure and TDC position. A 12-bit analog to digital converter is used to convert analog signals to digital form. The A-D converter had external and internal triggering facility with sixteen single ended channels. 100 continuous cycle data is recorded and the optimum increase in the rate of pressure, peak pressure has been ascertained by processing the signals. The uncertainty of engine parameters is ascertained by using square root method and the output is shown in Table 3

**Table 2.** The Engine Specifications

| | |
|---|---|
| Stroke length (mm) | 110 |
| Bone diameter (mm) | 87.5 |
| Connecting rod (mm) | 255 |
| Compression ratio | 17.5 : 1 |
| Rated power (kW) | 4.41 |
| Speed (rpm) | 1500 |
| Displacement volume (cc) | 600 |
| Stroke type | 4 stroke engine |

**Table 3.** The percentage of uncertainty parameter

| S. No. | Parameter | % of Uncertainty |
|---|---|---|
| 1 | Brake power (kW) | 0.6 |
| 2 | Brake Thermal efficiency (%) | 0.7 |
| 3 | Air flow rate (kg/s) | 0.5 |
| 4 | Brake specific fuel Consumption (kg / kWh) | 1.2 |
| 5 | Exhaust gas temperature ($^0$C) | 0.5 |
| 6 | Hydro carbon emission (ppm) | 0.3 |
| 7 | Carbon monoxide (%) | 0.2 |

## 3. Results and discussion

Experiments for the study have been done by allowing the engine to warm up for a certain period of time. The investigations are done at injection timings of 23L CA b TDC and injection pressure of 200 bar. The engine effectiveness, ignition and release of toxic gas have been acquired at varied loads of 0%, 25%, 50%, 75% and 100%.

### *3.1 The brake thermal efficiency*

A comparison of the varied green fuel blends has been done to access the engine features such as engine effectiveness, brake thermal efficiency, brake specific energy consumption and brake specific fuel consumptions. An engine's capability to change the fuel's heat energy into mechanical energy denotes the Brake thermal efficiency of the engine. Figure 2 illustrates the brake thermal efficiency of Tamanu green fuel by using different blends of green fuel. The brake thermal efficiency of CI green fuel is lesser than that of diesel since it is highly dense, sticky and has lesser calorific value than that of green fuel. The brake thermal efficiency of green fuel decreases when there is increase in the percentage of green fuel in the green fuel-diesel blends as there is lesser calorific value with increasing concentration of green fuel in the blends. This happens because of the uneven ignition due to less atomization and vaporization of blends since the stickiness and density is high [7].

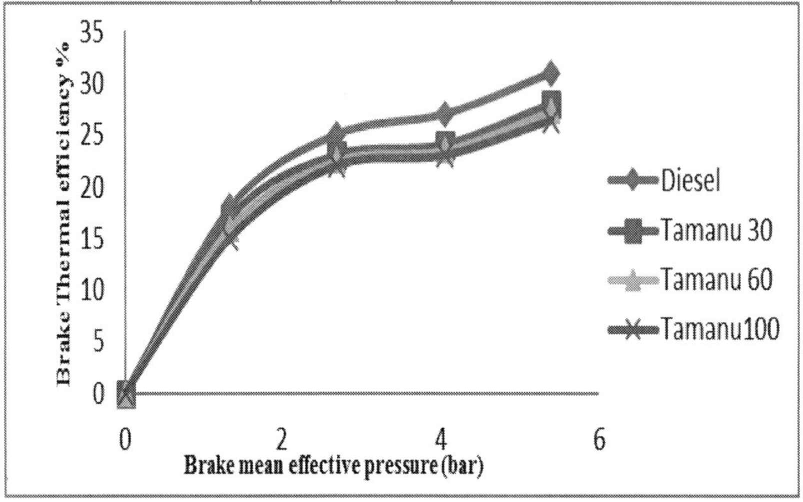


**Figure 2.** Variation of BTE Brake versus BMEP

*3.2. The Brake Specific Fuel Consumption*

A higher level of BSFC has been noted than that of diesel because of the low calorific value of Tamanu green fuel and the modification in ignition features. Higher level of propellant is has to be supplied to the engine such that the same input energy is sustained. The level of Tamanu green fuel sprayed by the injector is higher than that of diesel by suing the same volume due to its high density [8]. Figure 3 illustrates the deviation of BSFC for various percentages of green fuel-diesel blends. BSFC of green fuel blends surges up on increasing the ratio of green fuel in the green fuel-diesel blend since lesser calorific value of green fuel on increasing the proportion of green fuel. The lesser calorific value ensues increased supply of fuel to the engine to generate same amount of brake power consequently higher level of fuel consumption.

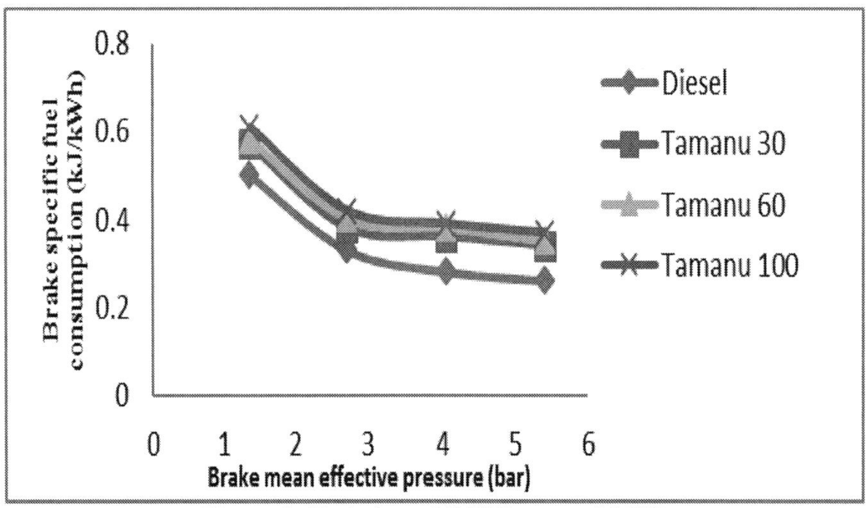


**Figure 3.** Variation of BSFC versus BMEP

*3.3. Brake Specific Energy Consumption*

BSEC indicates the engine effectiveness to acquire energy from the propellant to generate unit power. The specific energy consumption is an exact measure in contrast to specific fuel consumption to compare fuels with varied calorific values [9]. An illustration of the deviation in brake specific energy consumption under varied injection timings is shown in Figure 4. The BSEC of green fuel is high in contrast to that of diesel fuel. Possible reasons for higher brake specific energy consumption are the low calorific value, high viscosity and boiling point. The brake specific energy consumption is an exact parameter to compare fuels with varied calorific values. The upsurge in the proportion of blends in the green fuel results in increase of BSEC since calorific value of the fuel decreases on increasing the proportion of blend.

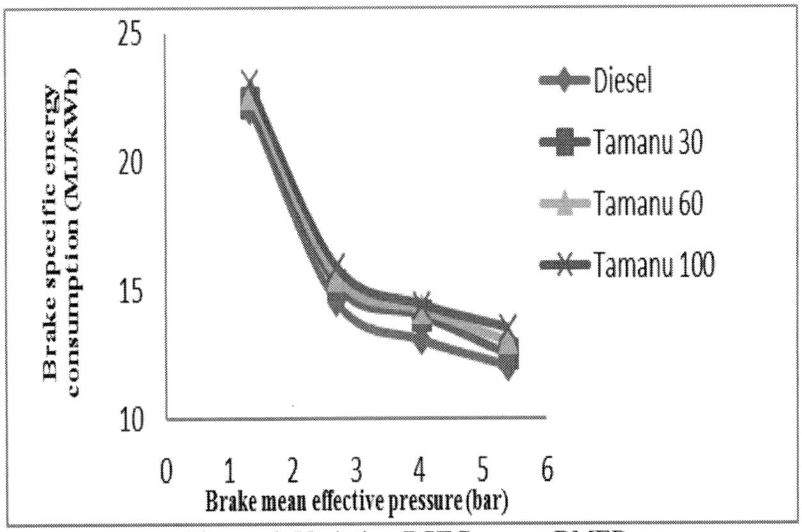


**Figure 4.** Variation BSEC versus BMEP

### *3.4. Release of Unburned Hydrocarbon Emission*

Values of HC, CO, $NO_X$ and smoke are acquired on using diesel and Tamanu blends of fuel. The cause of HC release is due to the incomplete ignition of fuel. The release of UBHC is lesser for green fuel due to improved ignition. There is an improvement in ignition due to high cetane index and higher level of gas temperature of the green fuel that causes shorter ignition delay [10]. The UBHC release from green fuel reduces on increasing the level of Tamanu in diesel blends as depicted in Figure 5. Higher level of molecular oxygen and cetane content in green fuel causes complete ignition of the fuel.

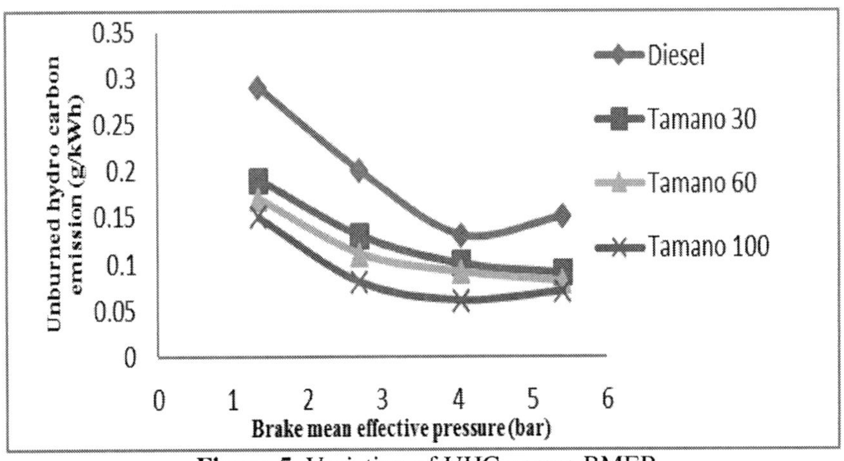


**Figure 5.** Variation of UHC versus BMEP

### *3.5. Release of Carbon Monoxide Emission*

Partial oxidation of elements in carbon causes formation of carbon monoxide that is very hazardous in nature. CO contributes to a major percentage in forming ground level ozone. Figure 6 illustrates the deviation in release of CO. The release of CO from Tamanu green fuel is lesser in contrast to that of diesel. This is because availability of high level of oxygen concentration in the green fuel that facilitate complete oxidation of CO [11].There is a very minute upsurge in release of CO that is caused due to shorter ignition delay; ignition is delayed because of the high temperature and cylinder pressure and the inferior premixed combustion phase with respect to retarded injection timing. The release of CO is due to lower temperature along with insufficient air supply during ignition. The variation in release of CO has the same trend as that of release of HC. On increasing the green fuel proportion in the blends there is lesser level of CO release. On increasing the proportion of green fuel leads to an increase in the oxygen concentration of the green fuel that aids complete ignition.

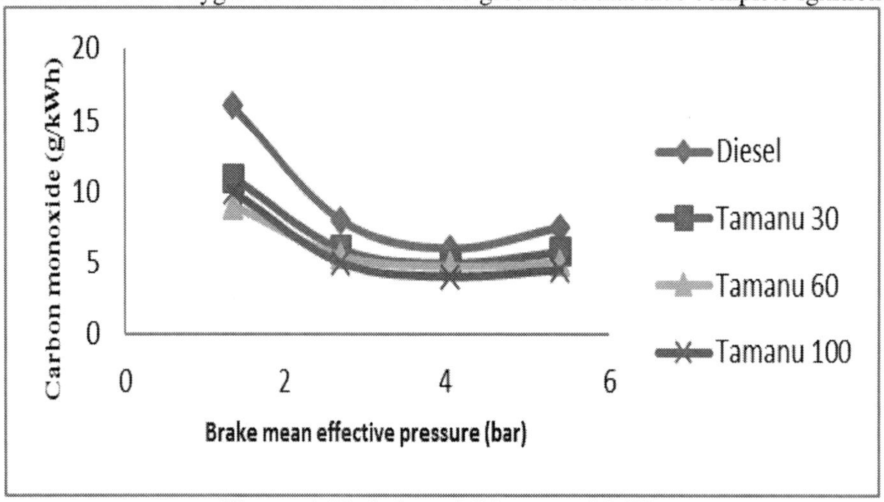


**Figure 6.** Variation of CO versus BMEP

### *3.6. Release of Oxides of Nitrogen Emission*

Figure 7 illustrates the deviation in release of $NO_X$. The $NO_X$ is formed due to the existence of higher level of oxygen content and high temperature. The proportion in release of $NO_X$ upsurges on increasing the proportion of green fuel in the blends.

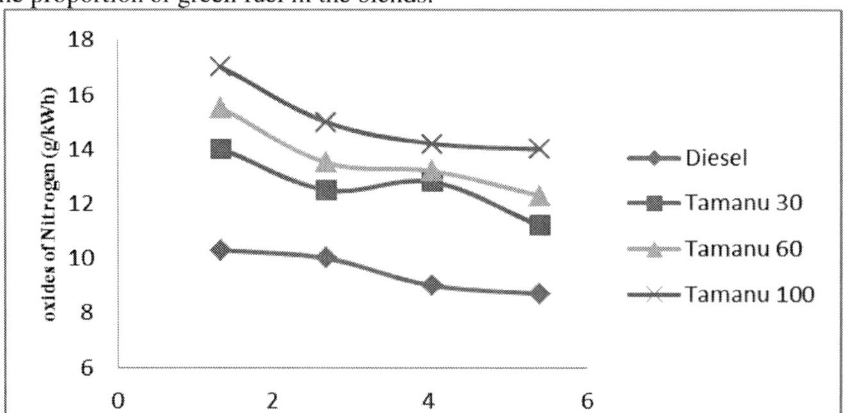


**Figure 7.** Variation of $NO_X$ versus BMEP

IOP Conf. Series: Materials Science and Engineering **923** (2020) 012061 doi:10.1088/1757-899X/923/1/012061

The elevated density of green fuel too plays a role in the formation of $NO_X$ release [12]. In contrast to diesel the release of $NO_X$ on using Tamanu green fuel is high. The reason for higher emission is because of higher level of ignition temperature that causes improved ignition as high molecular oxygen exists in the green fuel. Due to reduced mixing time and curtailed delay in ignition slow burning rate and slow rise in pressure and temperature subsist.

*3.7. Release of Smoke Opacity Emission*

The release of smoke on using green fuel is less as compared with diesel on applying higher load. At higher load due to the existence of high level of oxygen content in green fuel and higher BTE that completely converts large sized carbon atoms into CO because of complete ignition causes lesser smoke [13]. Figure 8 depicts the deviation in smoke on using varied concentrations of green fuel blends. The level of smoke released lessens on increasing the proportion of green fuel in the blend of green fuel and diesel but for Tamanu100. The release of smoke by using diesel is high in contrast to green fuel and its blends with diesel. The lesser smoke opacity of green fuel is because of existence of aromatic compounds in the green fuel and a lesser carbon to hydrogen ratio in contrast to conventional diesel fuel.

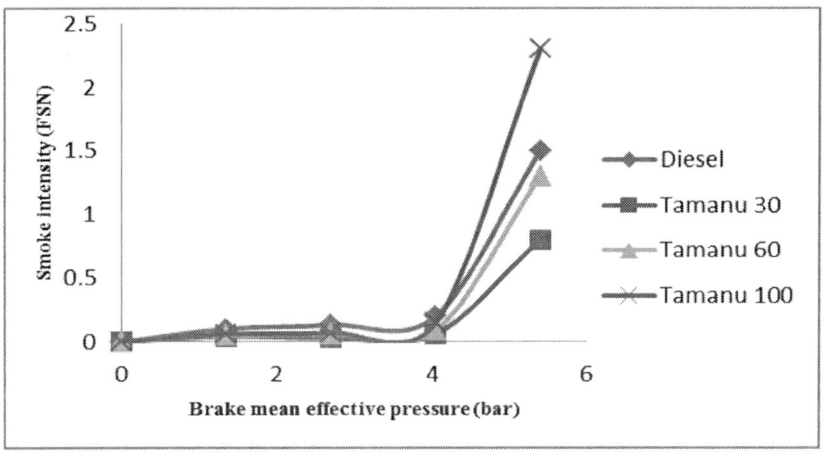


**Figure 8.** Smoke Opacity Verses Brake power

*3.8. In Cylinder Pressure*

Ignition parameters such as in-cylinder pressure, ignition delay and heat release rate are discussed in the following section. The increase in the rate of pressure establishes the smooth movement of gas pressure that pushes to the crank-shaft and smooth functioning of the engines. The proportion of cylinder pressure by suing diesel is high in contrast to using green fuel that is depicted in Figure 9 and there is lesser pressure on the cylinder on increasing the proportion of green fuel in the green fuel-diesel blend. The calorific value of the fuel determines the in-cylinder gas pressure [14]. In general, the in-cylinder gas pressure while using green fuel is lesser in

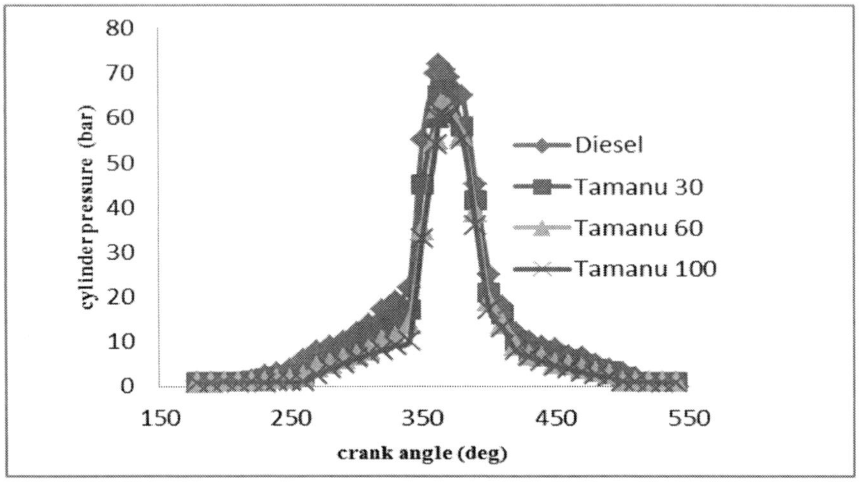


**Figure 9.** Variation of Cylinder Pressure Versus Crank angle

*3.9 Combustion delay*

The disparity in the start of the injection and start of combustion causes the ignition delay. The ignition delay on using Tamanu green fuel is lesser as compared with diesel due the high cetane number depicted in Figure 10. The ignition delay reduces on increasing the load. The reason for increasing the load is the reduced exhaust gas dilution and higher cylinder pressure and temperature [15]. The delay in ignition lessens on increasing the proportion of green fuel in the blends since there is decrease in cetane number.

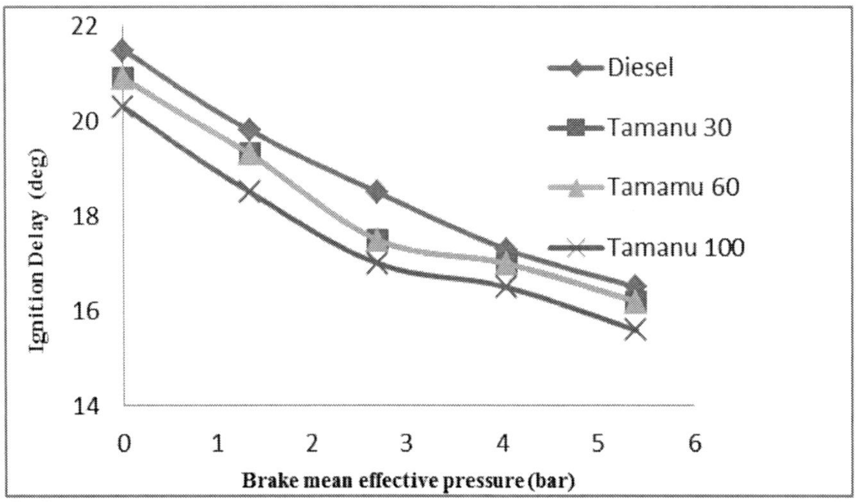


**Figure 10.** Variation of Ignition Delay versus BMEP

## 4. Conclusion

Tamanum ethyl ester has been acquired from the plants and the procured green fuel is blended with diesel in the proportion of Tamanu30, TAMANU60 and TAMANU100 to test engines. The engine effectiveness, releases of toxic gas and ignition characteristics of different blends of Tamanu green fuel have been contrasted with those of the fossil fuel diesel. The Blends of Tamanu can be used in engines without modifying the parameters in the diesel engine because properties of Tamanu and diesel fuel are same. The BTE is high on using diesel in contrast to Tamanu green fuel whilst it reduces on increasing the proportion of green fuel in the blends. BSFC and BSEC raises on increasing the ratio of Tamanu green fuel in the blends. On increasing the ratio of Tamanu green fuel in the blends causes lesser release of HC, CO and smoke emissions whilst the release of $NO_X$ is high. Combustion parameters of the engine like in-cylinder pressure, ignition delay and heat release rate are acquired by using Tamanu fuel and the outcome of the usage are according to the norms in contrast to diesel. It has been concluded that Tamanu is an appropriate alternative fuel for conventional diesel. It is widely acceptable as it does not cause any harm to the environment. Moreover, research o using varied green fuel blends with diesel at varied ratios could decide the best situation to use green fuel.

## 5. Acknowledgement
The authors sincerely thank our Einstein College of Engineering, Tirunelveli. For facilitating the laboratory to carry out the experimental investigations.

## 6. References

[1] Shrirao P.N, and Pawar A.N 2011 Experimental investigation on the performance of single cylinder diesel engine with Mullite as thermal Barrier coating *Journal of Engineering and Technology* **3(1)** 45-54
[2] Shrivastava J.K and Vermont M 2008 Methyl ester of Karanja oil as an alternative renewable source of energy *Fuel* **87(7)** 1673-1677
[3] Singh S.P and Dipti Singh 2010 Biodiesel production through the use of different sources and characterization of oils and their esters as the substitute of diesel: A review. *Renewable and Sustainable Energy Reviews* **14** 200-216
[4] Sivakumar A, Maheswar D and Vijayakumar R.K 2011 Analysis of LHR extended expansion engine with variable speed operation for different compression ratios *International Journal of Advanced Research in Science, Engineering and Technology* **11(1)** 121-128
[5] Sivaramakrishnan K 2018 Investigation on performance and emission characteristics of a variable compression multi-fuel engine fuelled with karanja biodiesel-diesel blend *Egyptian Journal of Petroleum* **27** 177-186
[6] Soo Y.N 2011 Non-edible vegetable oils and their derivatives for alternative diesel fuels in CI engines: A review. *Renewable and Sustainable Energy Reviews* ***15*(1)** 131-149
[7] Hirner F.S, Hwang J, Bae C, Patel C, Gupta T and Agarwal A.K 2019 Performance and emission evaluation of asmall-bore biodiesel compression-ignition engine *Energy* **183** 971–982
[8] Labeckas G, Slavinskas S and Kanapkien E 2019 Study of the effects of Bio fuel-Oxygen of various origins on a CRDI diesel engine combustion and emissions *Energies* **12** 1241-1245
[9] Soudagar S and ManzooreElahi M 2018 The effect of nano-additives in diesel-biodiesel fuel blends: A comprehensive review on stability, engine performance and emission characteristics *Energy Conversion and Management* **178** 146-177
[10] Nigam P.S and Singh A 2011 Production of liquid biofuels from renewable resources *Progress in Energy and Combustion science* **37** 52-68
[11] Caspeta L, Buijs, N.A.A and Nielsen J 2013 The role of biofuels in the future energy supply *Energy and Environment Science* **6** 1077-1082
[12] Singh S.P and Dipti S 2010 Biodiesel Production through the Use of Different Sources and Characterization of Oils and Their Esters as the Substitute of Diesel: A Review. *Renewable and Sustainable Energy Review* **14** 200-216

[13] Pratas M.J, Freitas S.V.D, Oliveira M.B, Monteiro S.C, Lima A.S and Coutinho A.P 2011 Biodiesel density: Experimental measurements and prediction models *Energy Fuels* **25** 2333–2340
[14] AyatallahGharehghani, Mostafa Mirsalim and Reza Hosseini 2017 Effects of waste fish oil biodiesel on diesel engine combustion characteristics and emission *Renewable Energy* **101** 930-936
[15] Varatharaju Perumal and Ilangkumaran M 2017 Experimental analysis of engine performance, combustion and emission using pongamia biodiesel as fuel in CI engine *Energy* **129** 228- 236

# PERFORMANCE ANALYSIS OF HEAT TRANSFER PARAMETERS IN SHELL AND TUBE HEAT EXCHANGER WITH CIRCUMFERENTIAL TURBULATOR

Raghulnath D[1*], Robin Britto Antony A[2], Veeramanikandan S[3], Vignesh C[4], Vijaya kumar G[5]
[1]Assistant professor, [2,3,4,5] UG Students, K. Ramakrishnan College of Technology, Trichy, Tamilnadu, India.

## Abstract

Performance of heat exchanger is evaluated by the value of heat transfer coefficients, Reynolds number, Nusselt number, temperature distribution along the length, residence time and pressure drop. Improving these parameters for better performance is the main focus in designing the heat exchanger. In that view, helical coil gives more turbulence to the fluid flowing through it that leads to improved tube side heat transfer coefficient. For increasing the turbulence in cold flow, turbulators can be used in the circumferential area between helical coil and shell. In this work, heat transfer rate of shell and helical coil heat exchanger is improved by using circumferential turbulators. For the same mass flow rate of cold fluid different hot fluid mass flow rates (0.03 kg/s, 0.06 kg/s, 0.09 kg/s and 0.12 kg/s) were fixed for taking readings and the improvement is obtained by calculation of performance parameters. The temperature distribution along the length of the heat exchanger is plotted for different flow rates. Results show that, increasing the mass flow rate of cold fluid above the mass flow rate of hot fluid decreases the performance.

**KEYWORDS:** Shell and helical coil heat exchanger; Turbulator; overall heat transfer coefficient; heat transfer enhancement.

## 1. Introduction

Transferring heat from one part to another part is major concern in many industries. For transferring heat between liquids, heat exchangers can be used. Based on the area of application any one type of heat exchanger can be chosen. Selecting a particular type of heat exchanger is more important to make the heat transfer process more effectively. Helical coils have more advantages in heat transfer applications because of its circular pathway. It creates more turbulence comparing with straight tubes which is desirable in the heat exchangers. It is important that the heat transfer coefficient in the shell side should be high to absorb heat from the hot fluid. In that view, turbulator is given in the shell side between helical coil and shell. That helps to increase the turbulence and reduce the residence time of cold water in the shell. Corrugated tube can be used to improve heat transfer in tube side. Deep corrugations making disturbance improving the heat transfer of the shell side [1]. Realizable K model was adopted for a shell and tube heat exchanger to run numerical simulations were investigated [2]. Thermal and exergy behaviour of air -water non boiling two phase flow at a horizontal flow start helically corrugated tube. The corrugated tube was put under constant heat flux. The improvement of gas and liquid superficial Reynolds number cause increment in pressure drop, heat transfer coefficient and exergy destruction [3]. To improve thermal performance in a typical set heat exchanger for exhaust heat recover, effects of the numbers and heat transfer and flow characteristics of MTT is inserted in SRT heat exchanger which is numerically investigated and compared with those of typical SIT and SRT [4]. Demonstration has been done for hydrothermal behaviour to depict the impact of variable magnetic field on Nanofluid transportation with square perforated twisted tape [5]. Numerical simulation on the Drift flux model is formed to investigate the process parameters of Shell and-tube waste heat boiler that can be used in industry exhaust recovery. [6]. One fluid flows inside cylindrical shell of shell and tube heat exchangers, transferring between fluids without mixing. The optimal value of heat transfer and pressure drop in a STHX is a major need in the industrial applications. CFD analysis has been performed to investigate the effect of baffle cut and inclination angle. [7]. The investigations were conducted for a triangular wire inserts in a circular tube for Reynolds numbers 2731 to 27,732 resulted that entropy generation increases with increasing of Reynolds number [8]. The influence of PR; k and Rea for water to air heat exchanger on hydrothermal behaviour are studied. [9]. Techniques for developing the heat transfer in systems which need a high thermal efficiency in heat transfer applications. Perforated baffles were used with pores axis ratios of 0.190, 0.425 or 0.660. The obtained results by varying Reynolds number result that the vortex flows made by baffles help to stabilize the stagnant flow in the region of downstream baffles leading to enhance the heat transfer in the channel [10]. The overall heat transfer coefficient and heat transfer rate for different mass flow rates results in increased performance with minimum amount for mass flow rate 0.86kg/s whereas it is increases with maximum amount if the mass flow rate is 1.71kg/s. [11]. Performance of shell and tube heat exchanger and helical coil heat exchanger with counter flow arrangement, helical coil heat exchanger was more effective than shell and tube heat exchanger. The Overall heat transfer coefficient increases with increase in mass flow rate of hot water. As the curvature radius is increased, the centrifugal forces lowers which results in reduction of heat transfer coefficient and pressure drop. [12]

The convective heat transfer coefficients of a helical coil heat exchanger, in his study three helical coils with different curvature ratio and pitch are investigated in a shell and are tested for counter flow arrangement. His results shows that maximum heat transfer characteristics was exhibited by the helical coil with 18 turns and curvature radius of 90 mm and pitch ratio of 0.13 with Characteristic Shell length of 0.5m and Shell Diameter of 15.24cm [13]. The forced laminar fluid flow in rectangular coil pipes with circular cross section and characterized by pipe straightness, curvature and torsion. Calculations were done on the performance parameters for forced laminar fluid flow in rectangular coil pipe for four rectangular coiled pipes at different Reynolds's number. The deviation obtained in maximum velocity area occurs simultaneously from outer wall through the straight tubes [14]. The performance and pressure drop of the helical coil heat exchanger with and without helical crimped fins. The experimental are done at the cold and hot water mass flow rates ranging between 0.10 and 0.22kg/sand between 0.02 and 0.12kg/s, respectively. The result is due to the heat transferred from the hot water increases with increasing its mass flow rate [15].

## 2. Experimental Setup

Shell and helical coil heat exchanger with circumferential turbulator is used to analyze the heat transfer performance in this work. Figure 2.1 shows the shell of the heat exchanger where the cold fluid is passed. Inside the shell, helical coil and turbulator are fixed. Specifications of shell dimensions are tabulated in Table 1. Helical coil made of copper tube is used to pass hot water at different flow rates. Helical coil is shown in Figure 2.2. For increasing the turbulence in the cold flow, circumferential turbulator is placed between the shell and helical coil, shown in Figure 2.3. Final experimental setup is arranged for counter flow configuration with pressure gauges.

Figure 2.1 Shell

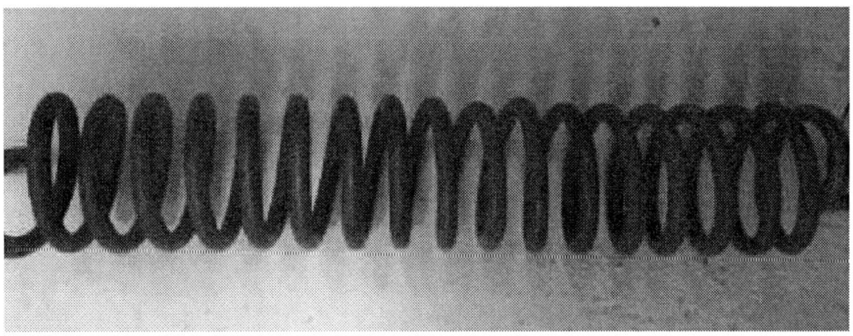

Figure 2.2 Helical coil

Figure 2.3 Helical coil

## 3. Experimental procedure

Experiment on shell and helical coil heat exchanger fitted with turbulator is experimented for effectiveness and overall heat transfer coefficient. Starting from hot fluid flow rate of 0.03 kg/s to 0.12 kg/s is fixed for the four different cold fluid flow rates. Totally sixteen type of flow rate combination were used and temperature distribution of hot fluid, inlet and outlet temperature of cold fluid, pressure at inlet and outlet of hot fluid were taken.

## 4. Result and discussion

Comparison between the shell and tube heat exchanger with helical coil and shell and tube heat exchanger with helical coil and turbulator are graphically represented. Effectiveness, Nusselt number and overall heat transfer coefficient in all the flow rate combinations were compared.

### 4.1. Effectiveness

Effectiveness for 0.03 kg/s cold fluid flow rate and hot fluid flow rates ranging from 0.03 kg/s to 0.12 kg/s is shown in figure 4.1. Effectiveness is closely travelling in this mass flow rate combination except in the first combination (0.03 kg/s). Addition of turbulator is not showing much deviation in the effectiveness.

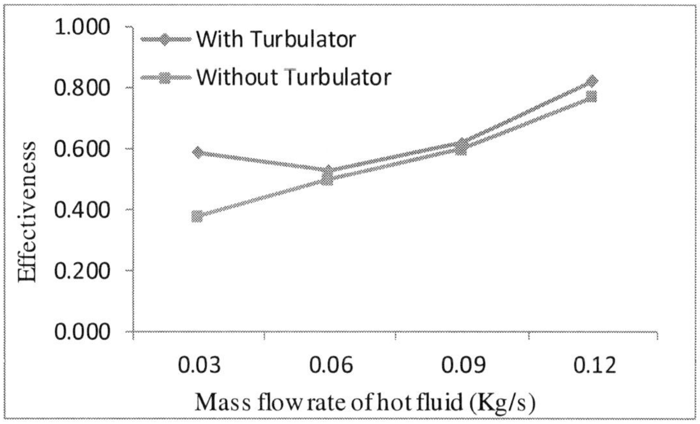


Figure 4.1 Effectiveness ($m_c$= 0.03 kg/s)

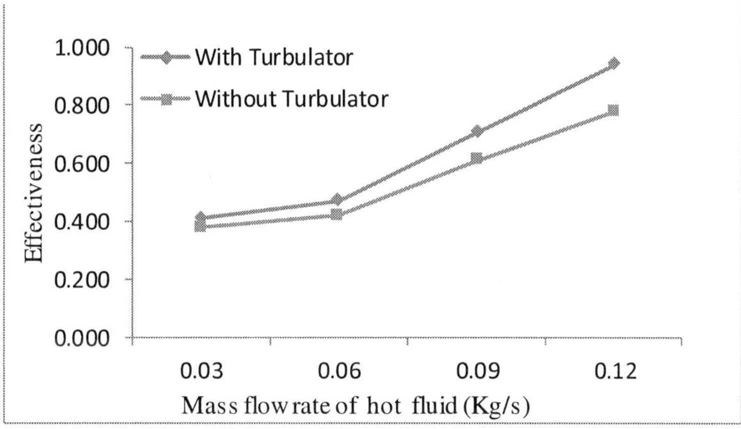


Figure 4.2 Effectiveness ($m_c$ = 0.06 kg/s)

Figure 4.2 shows the effectiveness for the cold mass flow rate of 0.06 kg/s and for hot flow rate of 0.03 to 0.12 kg/s. the effectiveness is increasing as the mass flow rate hot fluid is increasing in both the cases i.e. with and without turbulator.

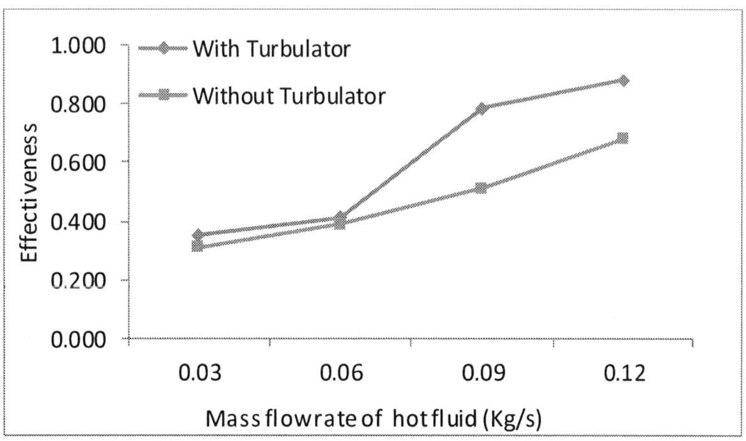


Figure 4.3 Effectiveness ($m_c$ = 0.09 kg/s)

Figure 4.3 shows the effectiveness for the cold mass flow rate of 0.09 kg/s and for hot flow rate of 0.03 to 0.12 kg/s. the effectiveness is increasing as the mass flow rate hot fluid is increasing in both the cases i.e. with and without turbulator. For the combination of 0.09 kg/s in both hot and cold fluid with turbulator is giving higher effectiveness and it is higher than the without turbulator. This difference is more comparing with other flow rate combinations.

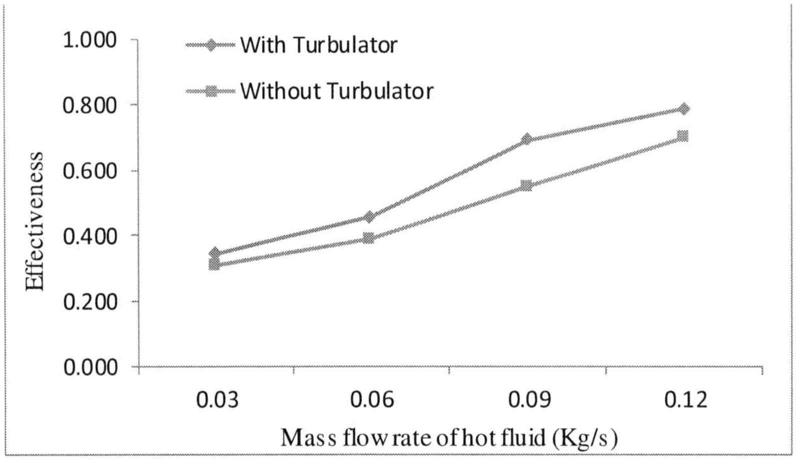


Figure 4.4 Effectiveness ($m_c = 0.12$ kg/s)

Figure 4.4 shows the effectiveness value of the cold flow rate of 0.12 kg/s and four different values of hot fluid flow rates. In this combination, effectiveness of heat exchanger in both with and without turbulator cases showing the same manner of results. In all the above twelve flow rate combinations, the 0.09 kg/s in both hot and cold flow rates is giving the best effectiveness value.

### 4.2. Nusselt number

Nusselt number is the major focusing parameter in the heat exchanger analysis which decides the heat transfer performance. The Nusselt number variation for the different hot and cold flow rate combinations are discussed in this section.

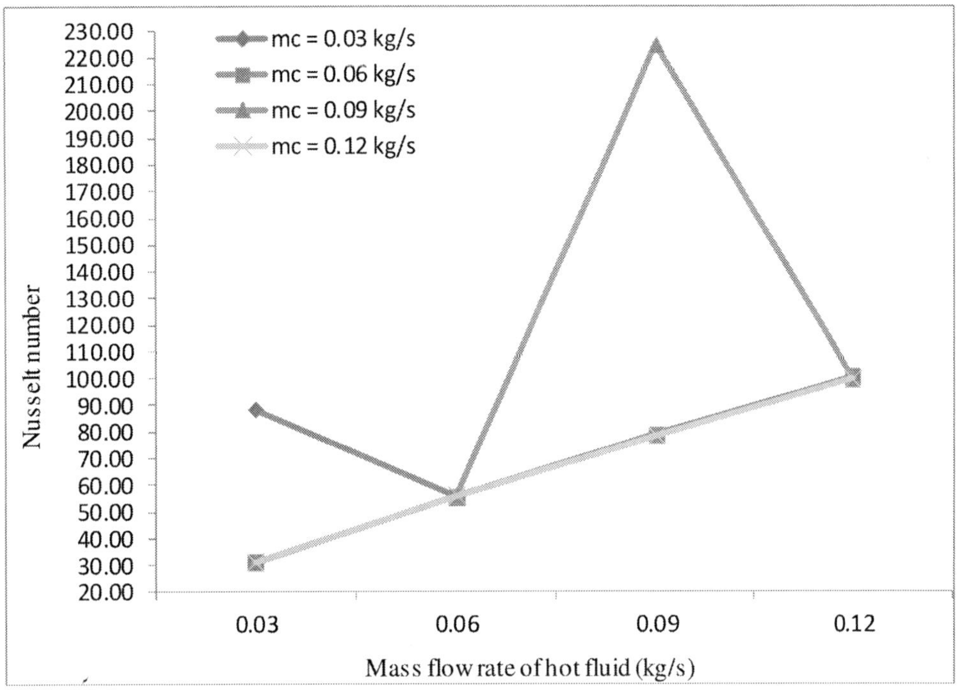


Figure 4.5 Nusselt number

Figure 4.5 represents the Nusselt number variations for the sixteen combinations of hot and cold fluid flow rates with turbulator. Except in the 0.03 and 0.09 flow rates of hot fluid, all the other flow rate combinations are showing the same nature of values. 0.03 flow rates in hot and cold flow rate results in some higher Nusselt number and 0.09 flow rate in hot and cold flow rate gives the highest value of Nusselt number.

## 4.3. Overall heat transfer coefficient

Overall heat transfer coefficient value for the different flow rate combinations of hot and cold fluid flow rates are plotted in the figure 4.6. From the figure it is clear that the cold flow rate of 0.06 kg/s for different flow rates of hot fluid is giving better overall heat transfer performance.

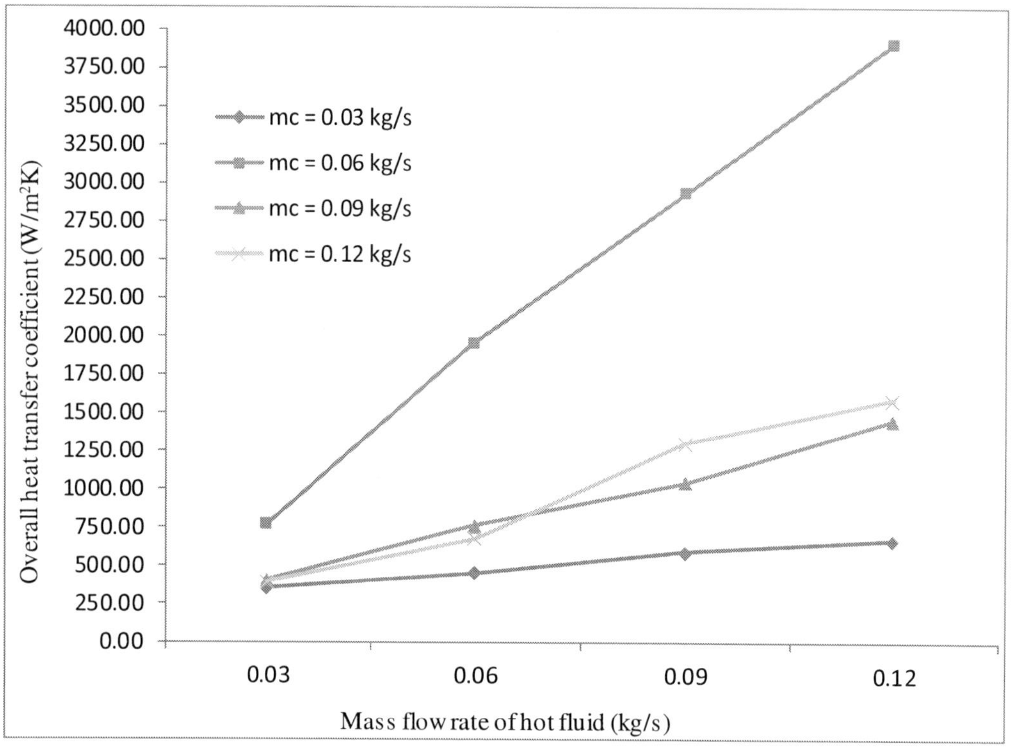


Figure 4.6 Overall heat transfer coefficient

## 5. Conclusion

Heat transfer characteristics of the shell and helical coil heat exchanger with turbulator arrangement have been analyzed experimentally. Analysis has been carried out under counter flow configuration. The attachment of turbulator resulted that fluid particles undergoes turbulent flow in the shell and tube this causes fluctuation in the heat transfer rates. By observing the performance of the heat exchanger the following results were concluded. The maximum effectiveness is achieved at the flow rate combination of $m_h$=0.12 kg/s and $m_c$=0.09 kg/s. Overall heat transfer coefficient is maximum in the mass flow rate combination of $m_h$=0.12 kg/s and $m_c$=0.12 kg/s. Nusselt number is maximum in the flow rate combination of $m_h$=0.09 kg/s and $m_c$=0.0.09 kg/s. From the above results, the best combination of flow rates for better heat transfer performance is 0.09 kg/s in both the fluids.

### *Reference*

[1] Amol Andhare and Kriplani V M, *Thermal analysis of a helical coil heat exchanger*, international journal of research in advanced engineering, **volume(1)** issue 12, December 2014.

[2] Sahiti N, *Heat transfer enhancement by pin elements*, International Journal of Heat and Mass Transfer **48** (2005) 4738–4747

[3] Paisarn Naphon, *Thermal performance and pressure drop of the helical-coil heat exchangers with and without helically crimped fins,* **19** December 2006, International Communications in Heat and Mass Transfer

[4] Amitkumar Puttewar S and Andhare A M, *Design and thermal evaluation of shell and helical coil heat exchangers, international journal of research in engineering and technology*. Volume: **04** Issue: 01, Jan-2015, eISSN: 2319-1163 | pISSN: 2321-7308.

[5] Nasser Ghorbani, *An experimental study of thermal performance of shell-and-coil heat exchangers*, international communication in heat and mass transfer,**37**(2010) 775-781.

[6] Sivakumar K, *Performance analysis of heat transfer and effectiveness on laminar flow with effect of various flow rates*, International Journal of Chem Tech Research CODEN (USA): IJCRGG ISSN: 0974-4290**Vol.7**, No.6, pp 2580-2587, 2014-2015

[7] Mehdi Bahiraei, *A novel application for energy efficiency improvement using nanofluid in shell and tube heat exchanger equipped with helical baffles*, ELSEVIER, Accepted 24 October 2015, Energy **93** (2015).

[8] Ya-Ping Chen, *Experimental investigation on performances of trisection helical baffled heat exchangers for oil-water heat transfer*. Accepted 16 May 2015, Energy Conversion and Management **101** (2015).

[9] Ming Pan, *Improving heat recovery in retrofitting heat exchanger networks with heat transfer intensification, pressure drop constraint and fouling mitigation*, 20 September 2015, Applied Energy (2015)

[10] Zhang-Jing Zheng, *Optimization of porous insert configurations for heat transfer enhancement in tubes based on genetic algorithm and CFD* , 5 April 2015 ,International Journal of Heat and Mass Transfer **87** (2015) 376–379

[11] Jie Yi, *Heat transfer characteristics of the evaporator section using small helical coiled pipes in a looped heat pipe*, 5 July 2002, Applied Thermal Engineering **23** (2003) 89–99,

[12] ShiveDayal Pandey, *Experimental analysis of heat transfer and friction factor of nanofluid as a coolant in a corrugated plate heat exchanger*, 2 January 2012, Experimental Thermal and Fluid Science **38** (2012) 248–256.

[13] Jian-Zhong Lin, *Flow and heat transfer characteristics of nanofluids containing rod-like particles in a turbulent pipe flow*, 16 September 2015,International Journal of Heat and Mass Transfer **93** (2016) 57–66

[14] Mohammed Reza Salimpour, *Heat transfer coefficient of shell and coiled tube heat exchangers*, Experimental Thermal and Fluid Science **33** (2009) 203–207. Doi: 10.1016.

# Biodiesel Production from Mixed *Elengi* and *Pongamia* Oil using Calcined Waste Animal Bone as a Novel Heterogeneous Catalyst

**M Anand[1], B Deepanraj[2], J Ranjitha[3] and M M Noor[4]**

[1]Department of Science & Humanities, Kingston Engineering College, Vellore, India.
[2]Department of Mechanical Engineering, Jyothi Engineering College, Thrissur, India.
[3]$CO_2$ Research and Green Technologies Centre, Vellore Institute of Technology, Vellore, India.
[4]Faculty of Mechanical & Automotive Engineering Technology, Universiti Malaysia Pahang, Pahang, Malaysia.

Email: deepanrajb@jecc.ac.in

**Abstract:** Recently, researchers have shown more interest towards biodiesel production from non-edible vegetable oils. The main advantages of biodiesel as a fuel includes biodegradability, non-toxicity, renewability and low emission profiles. In this study, crude mixed oil was used as feedstock for biodiesel production using Heterogeneous Catalyst synthesized from waste animal bone. Initially, mechanical extraction process was used to extract the crude mixed oil from the seeds of Mimusops elengi, and Pongamia pinnata. The crude oil collected from different plant species was characterized using GC-MS spectral data to identify their fatty acid composition. Consequently, the mixed crude oil was converted into biodiesel in the presence of calcinated heterogeneous catalyst obtained from waste animal bone and the catalyst was characterized by SEM, XRD and FTIR spectral data. The effect of variables including methanol to oil molar ratio, catalyst concentration, reaction temperature, reaction time and rate of mixing on the biodiesel yield was evaluated and optimized. The characteristics biodiesel obtained from mixed oil were close to commercial diesel fuel and used as an alternative to diesel in near future.

**Keywords:** *Mimusops elengi*, *Pongamia pinnata*, Biodiesel, CaO Heterogeneous Catalyst.

## 1. Introduction

Most of the researchers have gain interest and entered the field of alternative fuel is due to the depletion of fossil fuel resources and the increasing cost of petroleum product. The main reason is greenhouse gas effects and the environmental concerns [1-2]. Some of the hazardous gasses emitted from vehicles are always threats to the human beings. Qiu *et. al.* [3] studied the biodiesel production from the mixed oil of rapeseed and soybean through transesterification reaction in the presence of base and acid catalyst and obtained the optimized conditions as mole to methanol ratio is 5.1:1, reaction time is 2 hours and temperature is 55°C with the maximum biodiesel yield of 94%. In the transesterification of soybean oil - Mg–Al hydrotalcites used as a catalyst along with methanol, resulting in the negative effect of Mg/Al ratio on the basicity and catalytic activity. In addition, the catalyst reuse and the effect of calcinations temperature were addressed [3-5]. The calcium oxide supported on mesoporous silica to enhance the catalytic activity, as determined by the percentage yield and rate of the reaction involved in the transesterification of soybean oil [6]. The advantage of base (alkali) catalyst is requiring a small quantity in catalysis and to increase the activity of KI support catalyst by using mesoporous silica as a supporter [7, 8].

Various catalysts were synthesized in order to study the conditions which give maximum activity for biodiesel production [9-14]. Leevijit *et. al.* [15] investigated the production and utilization of degummed/esterified mixed crude palm oil-diesel blends in turbocharged automotive engine and ound that DgMCPO and EMCPO can be easily produced with cost comparable to diesel. At a speed of 2400 rpm under load range of 5–32.5 kW, the blends (DgMCPO20 and EMCPO30) were able to run the test engine in the same manner as diesel. Their oxygen content promoted easier combustion and resulted in lower emissions of black smoke (−36% and −61%) and carbon monoxide (−25% and −16%) than diesel. Wong *et. al.* [16] evaluated the biodiesel production through transesterification of palm oil using CaO–CeO2 mixed oxide catalysts. The catalysts were synthesized in order to improve the physico-chemical properties, catalytic activity and reusability of the bulk CaO in transesterification of palm oil. CaO–$CeO_2$ mixed oxide catalysts with diferent CaO compositions were prepared via wet impregnation method. Impacts of different CaO loadings, the relationship between BET surface area and basicity with biodiesel yield were studied. In the present study, *mimusops elengi and Pongamia pinnata* seeds were collected in and around of Vellore City, Tamilnadu and then oil is extracted by the mechanical extraction process. Further, the bio-oil is converted into biodiesel using novel heterogeneous catalyst calcined waste animal bone. The characterization studies were done by SEM, XRD FTIR and GC-MS spectral data.

## 2. Materials and Methods

### *2.1. Chemicals*
All the chemicals and reagents were purchased from SRL and Sigma Aldrich Chemicals Private Limited., India.

### *2.2. Collection of the seed materials*
The non-edible elengi seed is obtained from the plant nursery located inside the VIT University. The seeds was segregated from unwanted waste and then cleaned thoroughly with water. After washing, the seeds were dried and the kernel present inside the seeds was collected for further processing. The seed kernels are kept in air dryer to remove the moisture content completely. The air-drying process is carried out in the air dryer for about 5 hours. In the similar manner, pongamia seeds were collected from the plant nursery located inside in and around Vellore district. The seeds were removed from the non-edible fruit, after drying in an air dryer. The seed kernel was removed from the seeds by cracking the outer shell. These kernels were washed thoroughly to remove the unwanted foreign bodies (or) dust particles. The rinsed seed was dried in an air dryer for about 5 hours to remove the moisture content.

### *2.3. Extraction of elengi and pongamia oil*
The dried *m. elengi* seeds were then subjected to mechanical oil mill process for the extraction of oil. The 'x' grams of seeds were taken and then grinded using the mechanical oil mill processor. The mill has a central shaft driven by a high-power electric motor; that rotates vertically and crushes the seeds, because of this application of high pressure and constant bruising, the oil starts coming out of the seeds. The extracted oil was then collected in the basin at the bottom of the mill. Finally, the residues in the oil was made to settle down, and then filtered to remove any impurities. In the similar manner, bio-oil is collected from Pongamia seeds.

### *2.4. Pre-treatment of elengi and pongamia oil*
The mixed oil was used for pre-treatment analysis using the following protocol. Initially, mixed oil was treated with potassium hydroxide to maintain alkaline nature, so that, free fatty acid (FFA) value of the mixed oil was neutralized and then purified with water. The resulting mixture was heated up to 100°C to remove all water content present in the oil. The FFA content was analysed by a standard titration method [17].

*2.5. Determination of acid value and saponification value*

The acid and saponification value of bio-oil was determined according to the standard AOCS methods [18].

*2.6. Preparation of Catalyst*

The novel heterogeneous catalyst CaO (calcium oxide) used in this study was synthesized from goat bone. The waste animal bones were collected from local meat supplier in Vellore district. The bones were washed with dilute solution of $H_2SO_4$ to remove the dirty material. The bone was then powdered and treated for hydrothermal process [19]. The powdered material was thermally decomposed in a furnace at 800°C for about 2 hours and then refluxed in distilled water for 24 about hours. The sample was then filtered and dried at 120°C in hot air-oven. Further, the sample was recalcined at 900°C for 2 hours using muffle furnace to produce fine calcium oxide. The synthesized CaO was kept in a vacuum desiccator for further use.

*2.7. Catalyst characterization*

The XRD analysis was performed on a Siemens D-500 diffractometer equipped – consists of X-ray tube (CuK$\alpha_1$ radiation ($\lambda$ = 1.54060 Å), power conditions (40 kV / 30 mA)). The source light is germanium monochromatic X-rays collected from linear PSD (opening angle: $2\theta = 6°$). XRD patterns were measured in the range of 20° to 90°, $2\theta$ with the step size of 0.02° and 30 s counting per step at room temperature (25°C). Similarly, SEM analysis was carried out for the synthesized heterogeneous catalyst - JEOL; JSM-6330 LA operated at 20.0 kV and 1.0000 nA.

*2.8. Synthesis of Biodiesel*

In a 500mL round bottom flask connected with reflux condenser, magnetic stirrer and thermometer added 200 ml of bio-oil and then, 0.5 % w/w CaO, 60 ml $CH_3OH$ was added under stirring condition. After 4 hours, the mixture was cautiously transferred into a separating funnel and allowed to stand 24 hours. The lower layer contains glycerol, methanol and catalysts was decanted. The upper layer contains methyl esters, little amount of methanol, n-hexane and trace amount of the catalyst was washed thoroughly with de-ionized water in order to remove the impurities. The fatty acid methyl ester was washed twice with dilute HCl 0.5 mol/L solution until to get a clear phase. The percentage conversion of biodiesel was calculated by the standard formula; biodiesel conversion yield (%) = (experimental yield)/ (theoretical yield) × 100%. The produced biodiesel was further characterised using GC-MS analytical technique [3, 20].

*2.9. Characterization studies for mixed oil and biodiesel*

FTIR spectroscopy was carried out using a Nicolet 510 - FTIR instrument equipped with a DTGS detector. Spectra recorded in the 4000-400 $cm^{-1}$ range with a resolution of 2 $cm^{-1}$. The measurements were made using dried KBr pellets, which were prepared by mixing and pressing 1 mg/ 1ml of the sample with 100 mg KBr.

The qualitative and quantitative analysis of fatty acids was analyzed using GC-MS analytical technique. The detailed protocol used for GC-MS sample analysis is mentioned below; *GC operating condition:* An Agilent 6890 gas chromatograph - straight deactivated 2 mm direct injector liner and a 15m. All tech EC-5 column (250µ I.D., 0.25µ film thickness). The helium carrier gas was set to 2 ml/minute flow rate. *MS operating condition:* JEOL GC mate II bench top double-focusing magnetic sector mass spectrometer operating in electron ionization (EI) mode with TSS-20001 software. The different fatty acid compounds were integrated with the recorded spectra in the data bank mass spectra of NIST library V 11 provided by the GC-MS instrument.

## 3. Results and Discussion

The acid and saponification value of elengi oil were found to be 0.9248 mg KOH/g and 189.21 mg KOH/g. Similarly, the pongamia oil acid and saponification values were analysed in the range of 1.925 mg KOH/g and 184.45 mg KOH/g, respectively. If the acid value is less than 2.0 mg KOH/g, in the trans-esterification reaction, saponification interface will not significantly affect the mixture oil for biodiesel production and quality standard. Correspondingly, the total free fatty acid composition for the mixed oil was found to be in the level of 13.6%. Based on the above results, the trans-esterification reaction for mixed oil was carried out by using a novel heterogeneous catalyst. Initially, oil/ Methanol molar ratio (1:9) and reaction were conducted at 70°C under constant stirring at 700 rpm for 4 hours. After the reaction, it was transferred into separating funnel allow to settle down for 2 hours to form two layers. The organic layer is collected and concentrated in a simple distillation process to get the product as biodiesel.

FTIR spectra of mixed oil clearly confirm different functional groups present in the bio-oil sample. Based on the IR spectral data, the following functional groups such as –CH stretching bond 3005.17, 2852.72, C=O carbonyl gruop1741.72cm$^{-1}$. These functional groups represent the presence of aromatic and aliphatic compounds in the bio-oil sample as shown in the Figure 1.

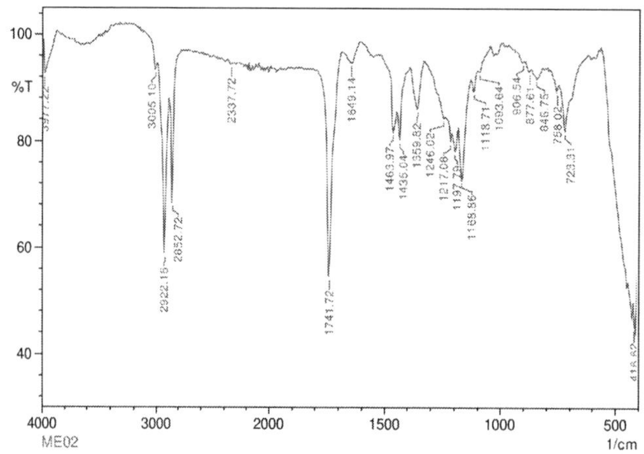


**Figure 1.** FTIR spectrum of bio-oil

The mixed oil was analysed using Perkin Elmer GC-MS in total scan mode to identify the fatty acid composition. The individual peaks identification was done in GC-MS, based on the NIST inbuilt library search as shown in the Figure 2. The relative percentage of fatty acid esters were calculated from total molecular ion peaks obtained by computerized integrator. Totally nine fatty acid compounds were identified in the mixed oil. These fatty acids are only responsible for the production of bio-diesel using trans-esterification reaction. The fatty acids are caprylic, lauric, arachidic, stearic, palmitic, margaric, nonadecylic, pentadecylic, behenic acids were identified based on the GC-MS spectral data.

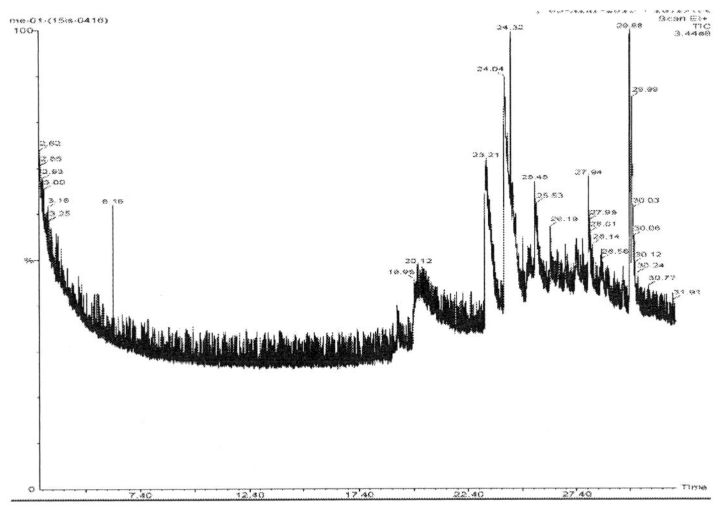


**Figure 2**. GC-MS Spectrum for mixed oil

The XRD spectrum explains clearly about the initiation of the stable thermal decomposition of $CaCO_3$ into CaO. It can be observed that the spectrum produces a sharp peak beyond 800°C below which the chance of peak formation responsible for CaO was very slim. The peak intensity was at its threshold limit of 1100°C, making it the optimum temperature for thermal decomposition of animal bones as shown in the Figure 3.

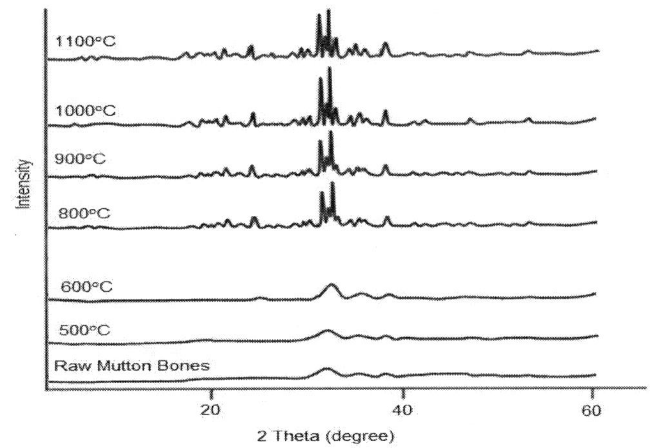


**Figure 3.** XRD spectrum of CaO Heterogeneous catalyst obtained from animal bone

The microscopic image of the Calcium Oxide was obtained by examining them under the scanning electron microscope (SEM) which displays their porosity behavior. The image of the calcium oxide was recorded at a magnification level of 100 μm which clearly displayed the surface of the calcium oxide. Figure 4 show the SEM image of the calcium oxide respectively. Bio-diesel obtained from mixed oil was analyzed using Perkin Elmer GC-MS in total scan mode to identify the fatty acid methyl esters composition. The individual peaks identification was done in GC-MS, based on the NIST inbuilt library search as shown in the Figure 5. The relative percentage of fatty acid esters were calculated from total molecular ion peaks obtained by computerized integrator and results obtained was compared with standard fatty methyl esters retention time.The biodiesel consists of methyl esters

IOP Conf. Series: Materials Science and Engineering **923** (2020) 012063 doi:10.1088/1757-899X/923/1/012063

of caprylic, lauric, arachidic, stearic, palmitic, margaric, nonadecylic, pentadecylic, behenic acids were identified based on the GC-MS spectral data. The GC-MS scan also indicated that the concentration of fatty acid ethyl ester in the biodiesel is 99.98%. Conversion of fatty acids into fatty acid methyl esters were achieved using a novel CaO heterogeneous catalyzed transesterification processs. The best percentage yield given by solid catalyst – calcium oxide 82.3%, compared with acid catalyst sulphuric acid. The comparison of physio-chemical properties of biodiesel with ASTM D0975, ASTM D6751 and EN 14214 standards are given in Table 1.

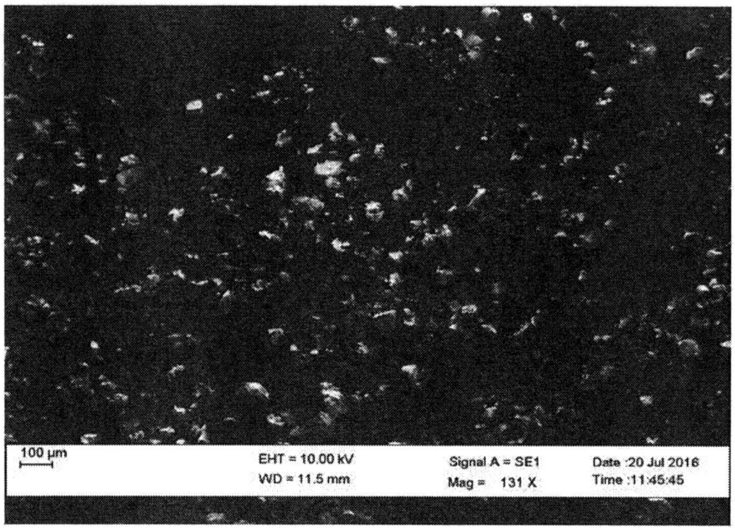


**Figure 4.** SEM image of calcium oxide synthesized from goat bone

**Table 1.** Physcio-chemical properties of biodiesel

| Properties | Biodiesel | ASTM D0975 | ASTM D6751 | EN 14214 |
|---|---|---|---|---|
| Density at 30°C (g/cm$^3$) | 0.88 | 0.876 | 0.875-0.90 | 0.86 to 0.90 |
| Viscosity at 40°C (mm$^2$/sec) | 3.532 | 1.9 to 4.1 | 1.6 to 6 | 3.5 to 5 |
| Specific gravity | 0.815 | 0.850 | 0.88 | --- |
| Flash point (°C) | 114 | 60-80 | 100 to 170 | >120 |
| Pour point (°C) | -11 | -35 to -15 | -15 to 16 | --- |
| Cloud point (°C) | 8 | -15 to 5 | -3 to 12 | --- |
| Cetane number | 60 | 40 to 55 | 47 to 65 | --- |
| Acid value (mg KOH/g) | 0.30 | 0.35 | <0.8 | <0.5 |
| Iodine value ($I_2$/g) | 86.3 | --- | --- | <120 |
| Ash content (%) | Nil | 0.01 | <0.02 | <0.02 |
| Water content (%) | Nil | 0.02 | 0.03 | <0.05 |

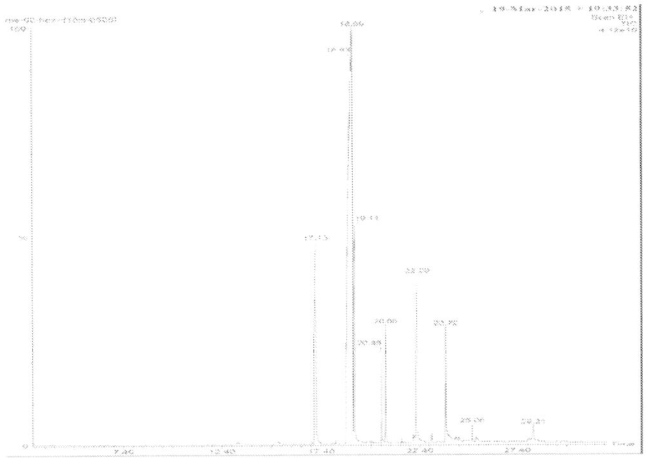

**Figure 5.** GC-MS Spectrum for biodiesel

## 4. Conclusion

A higher yield 82.3% was achieved for the production of biodiesel using a novel soil catalyst calcium oxide synthesized from goat bone and the results were compared with acid catalyst $H_2SO_4$ by monitoring the various parameters like molar ratio, quantity of catalyst, reaction time & methodology and temperature conditions. The synthesis of bio diesel was initiated by carrying out esterification process between oil and methanol in presence of Conc. Sulphuric acid as catalyst in proportional amount under regulated condition of temperature 60°C. The obtained product was analysed by GC-MS chromatography and found to have a reasonable amount of fatty acid methyl ester, which indicates the formation of bio diesel. The physiochemical properties and performance analysis will be carried to check its characteristics and more improvisation will be enabled in future works.

## Acknowledgement

The authors would like to express their deep gratitude to Vellore Institute of Technology, Jyothi Engineering College, Kingston Engineering College for providing the laboratory facilities and Universiti Malaysia Pahang for providing the financial support through research grant RDU190386.

## References

[1] Kannan M, Karthikeyan R, Deepanraj B and Baskaran R (2014) Feasibility and performance study of turpentine fueled DI diesel engine operated under HCCI combustion mode, *Journal of Mechanical Science and Technology*, **28** (2), pp.729-737.
[2] Deepanraj B, Sankaranarayanan G, Senthilkumar N and Pugazhvadivu M (2017) Influence of dimethoxymethane addition on performance, emission and combustion characteristics of the diesel engine, *International Journal of Ambient Energy*, **38**, pp. 622-626.
[3] Qiu F, Li Y, Yang D, Li X and Sun P (2011) Biodiesel production from mixed soybean oil and rapeseed oil, *Applied Energy*, **88** (6), pp. 2050–55.
[4] Silva Carla Cristina C M, Aranda Donato A G, Ribeiro Nielson F P and Souza Mariana M V M, (2010) Biodiesel production from soybean oil and methanol using hydrotalcites as catalyst, *Fuel Processing Technology*, **91** (2), pp. 205 10.
[5] Deepanraj B, Srinivas M, Arun N, Sankaranarayanan G and Abdul Salam P (2017) Comparison of jatropha and karanja biofuels on their combustion characteristics, *International Journal of Green Energy*, **14** (15), pp. 1231-1237.

[6] Samart C, Chaiya C and Reubroycharoen P (2010) Biodiesel production by methanolysis of soybean oil using calcium supported on mesoporous silica catalyst, *Energy Conversion and Management*, **51** (7), pp. 1428–31.

[7] Samart C, Sreetongkittikul P and Sookman C (2009) Heterogeneous catalysis of transesteri Fi cation of soybean oil using KI / mesoporous silica, *Fuel Processing Technology*, **90** (8), pp. 922–25.

[8] Abdullah A Z, Razali N and Lee K T (2009) Optimization of mesoporous K / SBA-15 catalyzed transesteri Fi cation of palm oil using response surface Methodology, *Fuel Processing Technology*, **90** (8), pp. 958–64.

[9] Zabeti M, Ashri Wan Daud W M and Aroua M K (2009) General optimization of the activity of $CaO/Al_2O_3$ catalyst for biodiesel production using response surface methodology, *Applied Catalysis A*, **366**, pp. 154–59.

[10] Obadiah A, Swaroopa G A, Vasanth Kumar S, Jeganathan K R and Ramasubbu A (2012) Biodiesel production from palm oil using calcined waste animal bone as catalyst, *Bioresource Technology*, **116**, pp. 512–516.

[11] Corro G, Sanchez N, Pal U and Banuelos F (2016) Biodiesel production from waste frying oil using waste animal bone and solar heat, *Waste Manag*ement, **47**, pp. 105–113.

[12] Nisar J, Razaq R, Farooq M, Iqbal M, Ali Khan R, Sayed M, Shah A and Rahman I (2017) Enhanced biodiesel production from Jatropha oil using calcined waste animal bones as catalyst, *Renewable Energy*, **101**, pp. 111-119.

[13] Deepanraj B, Lawrence P, Sivashankar R and Sivasubramanian V (2016) Analysis of pre-heated crude palm oil, palm oil methyl ester and its blends as fuel in a diesel engine, *International Journal of Ambient Energy*, **37** (5), pp. 495-500.

[14] Deepanraj B, Dhanesh C, Senthil R, Kannan M and Santhoshkumar A (2011) Use of palm oil biodiesel blends as a fuel for compression ignition engine, *American Journal of Applied Sciences*, **8** (11), pp. 1154-1158

[15] Leevijit T, Prateepchaikul G, Maliwan K, Mompiboon P, Okaew S and Eiadtrong S (2016) Production, properties and utilization of degummed / esterified mixed crude palm oil-diesel blends in an automotive engine without preheating, *Fuel*, **182**, pp. 509–16.

[16] Wong Y C, Tan Y P, Taufiq-Yap Y H, Ramli I and Tee H S (2015) Biodiesel production via transesterification of palm oil by using $CaO – CeO_2$ mixed oxide catalysts, *Fuel*, **162**, pp. 288–93.

[17] Corro G, Pal U, Tellez N (2013) Biodiesel production from *Jatropha curcas* crude oil using $ZnO/SiO_2$ photocatalyst for free fatty acids esterification, *Applied Catalysis B Environmental*, **129**, pp. 39–47.

[18] Official Methods and Recommended Practices of the American Oil Chemists' Society, fifth ed., AOCS, Champaign, 1998.

[19] Yoosuk B, Udomsap P, Puttasawat B and Krasae P (2010) Improving transesterification acitvity of CaO with hydration technique, *Bioresource Technology*, **101** (10), pp. 3784–3786.

[20] Deepanraj B, Senthilkumar N, Sivaramakrishnan V, Santhoshkumar A, Lawrence P and Valarmathi R (2012) Transesterified palm oil as an alternate fuel for compression ignition engine, *IEEE International Conference on Advances in Engineering, Science and Management (ICAESM 2012)*, India, pp.389-392.

# Productivity and Bioenergy Potential Assessment of Microalgae from Waste Stabilization Ponds

**Padmapriya P[1], K. M. Kumar[2]**
[1]Assistant Professor, Department of Civil Engineering, St. Joseph's College of Engineering.
[2]Associate Professor, Department of Mechanical Engineering, St. Joseph's College of Engineering.
**Email**: padmapriya174@gmail.com

**Abstract:** This study focused on the characterization of microalgae for bioenergy production and waste water treatment nutrient removal (from waste stabilization ponds). The dry biomass, lipid content and fatty acids in the microalgae were evaluated using centrifuge, Solvent extraction method and Gas Chromatography analysis method respectively. The *Diatom numulloides sp* with high lipid content and high productivity makes it suitable for future development by growing it in photo bioreactors to treat both municipal and industrial waste water and to produce biofuel making more benefits to human beings.
***Keywords:*** Microalgae from waste stabilization ponds, *Diatom numelloids*, Bioenergy from Microalgae, Photo Bioreator

## 1. Introduction

Microalgae are group of unicellular or simple multicellular photosynthetic microorganisms. It's divided in four groups (red, green, brown and diatoms microalgae). Microalgae are one of most important plant in kingdom plants, it has high photosynthetic process. So algae can be grown very fast and it has high reproduction.

Microalgae can be used for pollution control and also as an energy source as well as food, fertilizers, cosmetics, pharmaceuticals and aquacultures. So microalgae cultivation is one of a good technology for wastewater treatment. Microalgae cultivation in wastewater is for wastewater treatment, pollution control and production of energy from biomass. Microalgae cultivation is also useful in solving the environmental problems such as global warming as it consumes carbon dioxide in Photosynthesis process to produce oxygen and glucose.

Algae can be grown in open ponds system, closed ponds system, photo bioreactors, marine environment and wastewater. However, closed pond system can be control by all the conditions such as CO2, gases, water, sun light supply and fertilizers (for example: photo bioreactors). Additionally, the microalgae can be grown (green and brown) in both seawater and wastewater in closed pond system when all the conditions are under control. Studies on microalgae have been carried out since the 1980s [1], but in recent years their importance has grown fast, the reason is that microalgae appear to be the only source of biodiesel that has the potential to completely displace fossil diesel. Moreover, compared to other biofuels sources, such as traditional crops and wood, microalgae have several advantages: they grow extremely fast, reaching high areal and volumetric productivities, they do not require arable land, being able to use waste water, they can directly capture $CO_2$ released by industries and they overtake the food vs. fuel debate and their ability to grow in a wide range of conditions, resisting to severe temperatures, pH, and salinity makes them even more attractive. Recently, algae have received a lot of attention as a new biomass source for the production of biofuels.

## 2 Microalgae sample collection

The microalgae species were collected from different waste stabilization ponds are listed in the Table

**Table 2.1 List of waste stabilization ponds for Algal Sampling**

| Sl.No | Location of the pond | Capacity of the pond ($m^3$) | Purpose of the pond |
|---|---|---|---|
| 1 | Oxidation pond 1 IITM, Chennai | 120000 | Sewage Treatment |
| 2 | Oxidation pond 2 IITM, Chennai | 120000 | Sewage Treatment |
| 3 | VIT – Pond Vandalur, Chennai | 11113 | Waste Water Treatment |
| 4 | Mangalam Pond Pulikundram | 183143 | Sewage Treatment |
| 5 | Kankyo Cleantech Aminjikarai, Chennai | 132000 | Waste Water Treatment |
| 6 | Sewage Treatment Plant, Chithathur | 150000 | Sewage Treatment |
| 7 | Parry Nutraceuticals Bose road, Chennai | 3581 | Waste Water Treatment |

### 2.1 Identification of Microalgae Species

Identification of microalgae was carried out by means of microscopic analysis, using an optical microscope. The classification of species were identified by placing a drop of water to fill the well and place coverslip on top of slide and it were identified by recording the observations utilizing the pond identification sheet for quick reference.

### 2.2 Isolation of Microalgae Using Suitable Algal Media

All the algae strains for the present study were isolated and kept in Phychospectrum Environmental Research Centre (PERC) Laboratory. These algal strains were grown in 200 mL glass tubes (2.5 cm in diameter) with different nutrient medium at room temperature, and light intensity was set at 150 μmol/($m^2$·s) from the 4th day. The $CO_2$ concentration was set as 0.5% (by volume) in the bubbling air, and initial cell concentration was set at 0.5 ($OD_{680}$). Microalgae were isolated using algal media listed in Table 2 depending on the type of microalgae species identified. Stock solutions were made up by accurately weighing the prescribed amount of nutrient and dissolving in a specified volume of distilled water in a volumetric flask. Some nutrients readily dissolved and others need heat and stirring to fully dissolve.

IOP Conf. Series: Materials Science and Engineering **923** (2020) 012064 doi:10.1088/1757-899X/923/1/012064

**Table 2.2 BBM, BG-11 & PAAP Nutrient Medium Ingredients**

| Ingredients | BBM Media (mg/L) | BG- 11 Media (mg/L) | PAAP Media (mg/L) |
|---|---|---|---|
| NaNO3 | 025.00 | 150.00 | 002.00 |
| CaCl2.2H2O | 002.50 | 036.00 | 00.540 |
| MgSO4.7H2O | 007.50 | 075.00 | 00.500 |
| K2HPO4 | 007.50 | 040.00 | |
| KH2PO4 | 017.50 | - | 00.780 |
| NaCl | 002.50 | - | - |
| Citric acid | - | 006.00 | - |
| Ferric ammonium citrate | - | 006.00 | - |
| EDTA (disodium salt) | - | 001.00 | - |
| Na2CO3 | - | 020.00 | - |
| **Trace metals solution 1ml/1L Distilled** | **BBM Media(g/L)** | **BG- 11 Media(mg/L)** | **PAAP Media (mg/L)** |
| ZnSO4.7H2O | 008.82 | 022.20 | 0.008 |
| MnCl2.4H2O | 001.44 | 181.00 | 0.060 |
| MoO3 | 000.71 | - | - |
| CuSO4.5H2O | 001.57 | 079.00 | 0.002 |
| Co(NO3)2.6H2O | 000.49 | 049.40 | - |
| H3BO3 | - | 286.00 | - |
| NaMoO4·2H2O | - | 039.00 | - |
| Na/SiO3 | - | - | 0.700 |
| fl-glycerophosphoric acid | - | - | 0.100 |
| FeC13 - 6H20 | - | - | 0.020 |
| CoCI2. 6H2O | - | - | 0.004 |
| Na2MoO4.2H2O | - | - | 0.138 |
| Cyclo acid (buffer) | - | - | 50.00 |

The composition of f/2 Nutrient Media are represented in Table 3 with the concentration of Trace Metal Solutions

**Table 2.3 f/2 Nutrient Medium Ingredients**

| Ingredients | f/2 Media (mg/L) |
|---|---|
| $NaNO_3$ (75.0 g/L dH2O) | 01.00 |
| $NaH_2PO_4 \cdot H_2O$ (5.0 g/L dH2O) | 01.00 |
| $Na_2SiO_3 \cdot 9H_2O$ (30.0 g/L dH2O) | 01.00 |
| f/2 Trace Metal Solution | 01.00 |
| f/2 Vitamin Solution | 00.50 |
| **Vitamin Solution 0.5ml/1L Distilled water** | **f/2 Media (mg/L)** |
| Vitamin B12 (1.0 g/L dH2O) | 001.00 |
| Biotin (0.1 g/L dH2O) | 010.00 |
| Thiamine HCl | 200.00 |
| **Trace Metal mix 1ml/1L Distilled water** | **f/2 Media (mg/L)** |
| $FeCl_3 \cdot 6H_2O$ | 03.15 |
| $Na_2EDTA \cdot 2H_2O$ | 04.36 |
| $CuSO_4 \cdot 5H_2O$ (9.8 g/L dH2O) | 01.00 |
| $Na_2MoO_4 \cdot 2H_2O$ (6.3 g/L dH2O) | 01.00 |
| $ZnSO_4 \cdot 7H_2O$ (22.0 g/L dH2O) | 01.00 |
| $CoCl_2 \cdot 6H_2O$ (10.0 g/L dH2O) | 01.00 |
| $MnCl_2 \cdot 4H_2O$ (180.0 g/L dH2O) | 01.00 |

## 2.3 Growth of Microalgae

In this study, the microalgae were grown in conical flasks where unicellular algae grow continuously by a process known as cell division. Each cell enlarges and divides into two daughter cells that subsequently grow and divide yielding a culture that increases exponentially. Growth slows as the algal population becomes more
crowded. Most oil-producing microalgae have an optimal growth temperature of 25 to 30$^0$C. Selection of algal strains grown at high temperatures, such as 40$^0$C, may be needed, for example, in South China, Thailand, etc. The lipid content of microalgae is strongly influenced by temperature and depends on the strains. An increase in temperature from 20 to 25$^0$C will double the lipid content of Nannochloropsis oculata (from 7.9 to 14.9%), while an increase from 25 to 35$^0$C caused a decrease in the lipid content of Chlorella vulgaris, from 14.71 to 5.90% [8].

Nutrients are expected to be depleted, metabolites build, and light penetration decreases because of self-shading. The cultures that have reached their stationary phase for the current conditions will not increase in density. The time for algae to divide and the maximum density attained depend upon several factors such as pH, Water temperature, Salinity, Nutrients and

$CO_2$ Aeration. Growth of unicellular algae was measured in terms of cell numbers in a given volume of water.

## 2.4 Characterization of Microalgae Species

The algae was isolated and transferred to culture room and a day later the growth of algae was monitored by counting the cells through a hemocytometer. Microalgae were characterized by quantifying the Lipid content, Protein and carbohydrates quantity. It produces many different kinds of lipids, triglycerides and diglycerides, phospholipids and glycolipids, hydrocarbons and others [19].
The dry biomass weight was determined by obtaining pellet of the species grown in the nutrient media in cake formation after 20 days and 40 days from isolation procedure from which the dry biomass weight is obtained after evaporation of moisture content with the help petri plates and that dry biomass is weight in mg/L.

## 2.5 Solvent Extraction of Lipids

The biomass pellet was transferred to an 8mL vial glass using methanol acetyl chloride. After adding 1mL hexane, the vials were capped tightly and the lipids were extracted at 100$^0$C for 60 minutes. The content was transferred to 15mL centrifuge tube and centrifuged at 1800g for 5 min at room temperature. The upper layer is separated and weighed to 0.1mg precision to determine the amount of lipids extracted.

## 2.6 FAME Analysis

To determine the fatty acids, analysis was carried out using Gas Chromatography equipped with a flame ionization detector with Mass spectrometer. 30mg biomass was extracted with 2mL of methanol acetyl chloride as internal standard (IS) to the extraction mix and the samples were heated at 100$^0$C for 60min. the upper layer was collected and filtered using syringe filter prior to analysis by Gas Chromatography. Separation was achieved on a capillary column and the temperature was programmed to hold at 50$^0$C for 1 min and then rise to 175$^0$C.

## 3. Results and Discussions

The type of microalgae species obtained from different waste stabilization ponds are listed in Table 3.1. The microscopic image of each species is shown in Figure 3.1

**Table 3.1 Microalgae Species obtained from different waste stabilization ponds.**

| Name of the Pond | Name of the species obtained |
|---|---|
| Oxidation pond, IIT Madras | *Chlorococcum sp* |
| VIT Sewage treatment pond, Vandalur | *Diatom numulloides sp* |
| Sewage Treatment Plant, Perungudi | *Chlorococcum sp*<br>*Diatom Muelleri sp*<br>*Diatom Scutellum sp*<br>*Diatom Platycephala sp*<br>*Diatom Lyra sp*<br>*Anthrospira sp* |

IOP Conf. Series: Materials Science and Engineering **923** (2020) 012064 doi:10.1088/1757-899X/923/1/012064

| | |
|---|---|
| Sewage Treatment Plant, Chithathur | *Oscillatoria sp* |
| Sewage Treatment Plant, Pulikundram | *Chlorella Vulgarisis sp* |
| Parry Nutraceuticals, Chennai | *Chlorella Vulgarisis sp*<br>*Spirulina sp* |

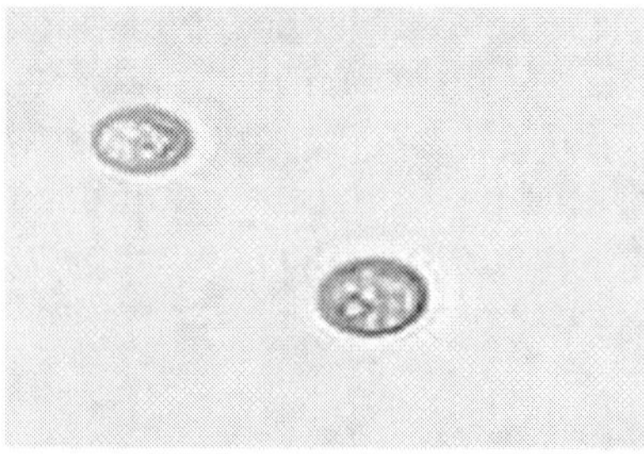

**(a) *Chlorococcum sp***

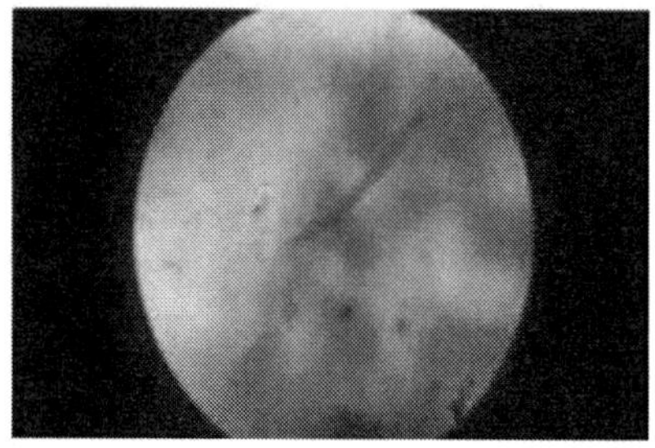

**(b) *Spirulina &Diatom sp***

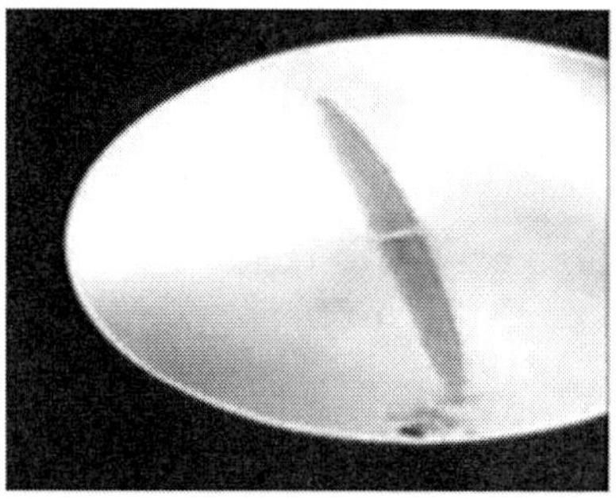

**(c) *Chlorococcum sp***

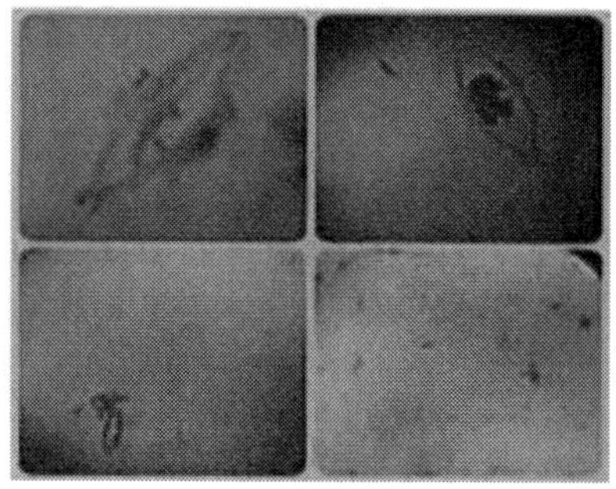

**(d) *Diatom sp***

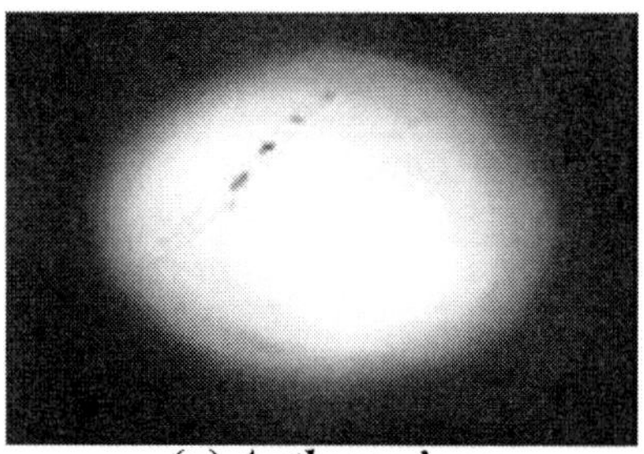

**(e) *Anthrospira sp***

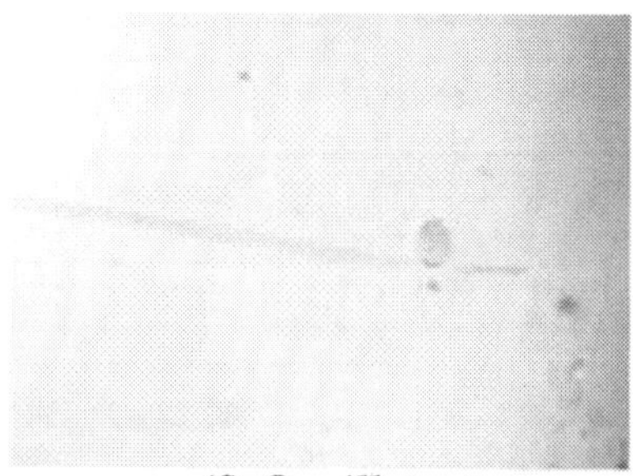

**(f) Oscillotoria sp**

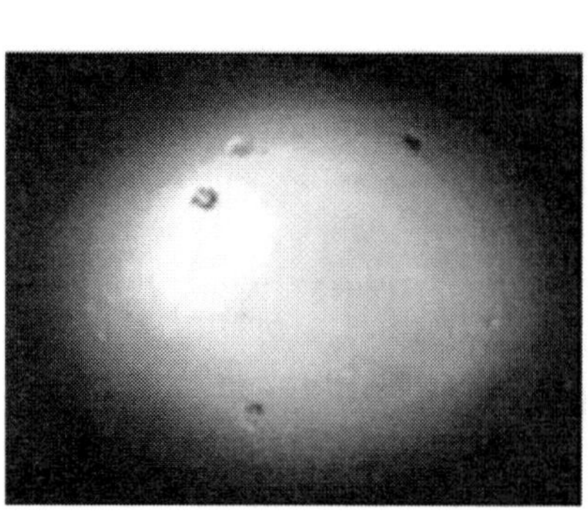

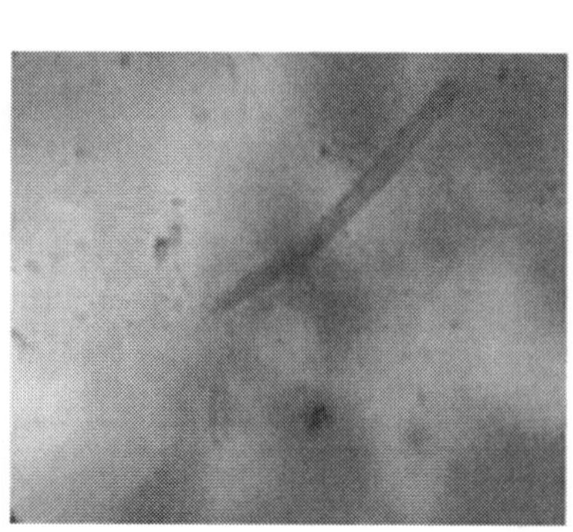

**(g) *Chlorella Vulgarisis sp*** **(h) *Spirulina sp***

**Figure 3.1 Microalgae Species from different waste stabilization ponds**

The microalgae species obtained from different waste stabilization ponds varies with respect to its growth factor such as Nitrogen and Phosphorus concentration. It is observed that *Diatom* species and *Chlorococcum* species are commonly found in waste stabilization ponds which are rich in nutrients as shown in Table 4. The *Diatom* species was obtained diversely in the Sewage Treatment Pond in Perungudi which has Total Nitrogen content and Phosphorus of 92.80mg/L and 26.30mg/L respectively which is the highest among all the other ponds in this study. Similarly the Total Nitrogen content and Phosphorus in IIT Madras Oxidation Pond is 49.30mg/L and 19.20mg/L which stands next to Perungudi sewage treatment pond in Nutrient Concentration with the same Chlorococcum species.

## 3.1 Cell Growth Count of Microalgae Species

The growth of algae was monitored by counting the cells through a hemocytometer where the count was observed and represented in Table 3.2 from average of 4 quadrants

**Table 3.2 Cell Growth count of Microalgae species**

| Name of the Species | (Initial stage) cells/m L | (20 days Growth) cells/mL | (40 days Growth) cells/m L |
|---|---|---|---|
| *Chlorococcum sp* | 64.00 x $10^4$ | 92.40 x $10^4$ | 153.00 x $10^4$ |
| *Diatom numulloides sp* | 25.50 x $10^4$ | 66.20 x $10^4$ | 106.00 x $10^4$ |
| *Chlorococcum sp* | 62.00 x $10^4$ | 86.00 x $10^4$ | 92.00 x $10^4$ |
| *Diatom mulleri sp* | 36.00 x $10^4$ | 41.00 x $10^4$ | 149.00 x $10^4$ |
| *Diatom surtellum sp* | 39.00 x $10^4$ | 52.00 x $10^4$ | 162.00 x $10^4$ |
| *Diatom Plateycephala* | 43.00 x $10^4$ | 62.00 x $10^4$ | 96.00 x $10^4$ |
| *Diatom Lyra sp* | 52.00 x $10^4$ | 82.00 x $10^4$ | 106.0 x $10^4$ |
| *Anthrospirasp* | 19.00 x $10^4$ | 26.00 x $10^4$ | 86.00 x $10^4$ |
| *Oscillotoriasp* | 10.00 x $10^4$ | 32.00 x $10^4$ | 97.00 x $10^4$ |
| *Chlorella sp* | 72.00 x $10^4$ | 106.00 x $10^4$ | 162.00 x $10^4$ |
| *Spirulina sp* | 62.00 x $10^4$ | 86.00 x $10^4$ | 102.00 x $10^4$ |

## 3.2 Dry Biomass Productivity of Microalgae Species

The dry biomass weight was determined after 20 days and 40 days from isolation procedure and weighed in mg/L as tabulated in Table 3.3

**Table 3.3 Dry Biomass Weight of Microalgae species**

| Name of the Species | (20 days Growth) mg/L | (40 days Growth) mg/L |
|---|---|---|
| *Chlorococcum sp* | 00.01 | 132.00 |
| *Diatom numulloides sp* | 04.30 | 109.50 |
| *Chlorococcum sp* | 18.50 | 175.50 |
| *Diatom mulleri sp* | 41.55 | 408.82 |
| *Diatom surtellum sp* | 20.45 | 374.40 |
| *Diatom Plateycephala* | 14.96 | 198.70 |
| *Diatom Lyra sp* | 21.91 | 302.45 |
| *Anthrospirasp* | 10.25 | 182.75 |
| *Oscillotoriasp* | 05.55 | 217.34 |
| *Chlorella sp* | 27.50 | 475.23 |
| *Spirulina sp* | 15.99 | 132.41 |

The dry biomass weight of *Chlorococcum* species after 40 days growth is 132.00mg/L which was obtained from Oxidation pond in IIT, Madras. The dry biomass weight of *Chlorococcum* species is 280.00mg/L [1]. The dry biomass weight completely depends upon the cell growth count and the concentration of Nutrients in the media. The cell growth count of *chlorococcum* species in BBM Media is 153.00 cells/mL x $10^4$ as shown in Table 3.3

### 3.3 Lipid Content of Microalgae Species

The lipid content in each microalgae species were weighed to 0.1mg precision and determined in mg/L as shown in Table 3.4

**Table 3.4 Lipid content of Microalgae Species**

| Name of the Species | Lipid Content (mg $g^{-1}$ dwt) (Trial 1) | Lipid Content (mg $g^{-1}$ dwt) (Trial 2) | Average Lipid Content (mg $g^{-1}$ dwt) |
|---|---|---|---|
| *Chlorococcum sp* | 055.00 | 056.00 | 054.00 |
| *Diatom numulloides sp* | 130.00 | 122.00 | 126.00 |

| | | | |
|---|---|---|---|
| *Chlorococcum sp* | 043.21 | 063.49 | 053.34 |
| *Diatom mulleri sp* | 052.85 | 056.18 | 054.51 |
| *Diatom surtellum sp* | 093.98 | 083.31 | 088.64 |
| *Diatom Plateycephala* | 112.09 | 115.87 | 113.98 |
| *Diatom Lyra sp* | 103.03 | 091.34 | 097.19 |
| *Anthrospirasp* | 049.55 | 026.17 | 037.85 |
| *Oscillotoriasp* | 114.73 | 133.68 | 124.20 |
| *Chlorella sp* | 111.06 | 068.74 | 089.91 |
| *Spirulina sp* | 036.97 | 027.95 | 030.92 |

The lipid content of *Diatom numulloides* sp was found to be the microalgae species with maximum lipid content among other microalgae species in this study with 126.00 mg $g^{-1}$ dwt in 40 days growth with dry biomass productivity 109.50mg/L. The lipid content of same species is 92.30 mg $g^{-1}$ dwt with dry biomass productivity 112.00 mg/L [1]. The study shows that the lipid content of the species is proportional to the dry biomass weight obtained

### 3.4 Fatty Acid Content of Microalgae Species

The Fatty acid content in each microalgae species were weighed to 0.1mg precision and determined in mg/L as shown in Table 3.5

The Fatty acid content of *Diatom* sp was found to be the Microalgae species with maximum Fatty Acid content among other microalgae species in this study with 85.23 mg $g^{-1}$ dwt in 40 days growth with dry biomass productivity 109.50 mg/L. The Fatty acid content of same species is 32.60 mg $g^{-1}$ dwt with dry biomass productivity 170mg/L [1]. The study shows that the Fatty Acid content of the species is proportional to the Dry biomass weight obtained.

The Lipid content and Fatty Acid analysis of obtained microalgae species varies with respect to the dry biomass weight gained after each day. The dry biomass weight depends on the growth of the same microalgae species and it is clear that higher the Nutrient concentration in media, higher the productivity.

**Table 3.5 Fatty Acid content of Microalgae Species**

| Name of the Species | Fatty Acid Content (mg $g^{-1}$ dwt) (Trial 1) | Fatty Acid Content (mg $g^{-1}$ dwt) (Trial 2) | Average Fatty Acid Content (mg $g^{-1}$ dwt) |
|---|---|---|---|
| *Chlorococcum sp* | 055.00 | 056.00 | 054.00 |
| *Diatom numulloides sp* | 130.00 | 122.00 | 126.00 |
| *Chlorococcum sp* | 043.21 | 063.49 | 053.34 |
| *Diatom mulleri sp* | 052.85 | 056.18 | 054.51 |
| *Diatom surtellum sp* | 093.98 | 083.31 | 088.64 |
| *Diatom Plateycephala* | 112.09 | 115.87 | 113.98 |
| *Diatom Lyra sp* | 103.03 | 091.34 | 097.19 |
| *Anthrospirasp* | 049.55 | 026.17 | 037.85 |
| *Oscillotoriasp* | 114.73 | 133.68 | 124.20 |
| *Chlorella sp* | 111.06 | 068.74 | 089.91 |
| *Spirulina sp* | 036.97 | 027.95 | 030.92 |

## 3.5 Characteristics of The Microalgae Species From Waste Stabilization Ponds

The bioenergy production potential is measured by the parameters such as lipid content and fatty acids for all the microalgae species. The characteristics of microalgae from different waste stabilization ponds are tabulated in Table 3.6 and represented as a chart in Figure3.2 to compare its bioenergy production potential and thus the species with maximum Lipid content as well as Fatty acid content is considered as highly potential microalgae for Bioenergy production among the obtained species from the waste stabilization ponds.

**Table 3.6 Characteristics of the Microalgae species from waste stabilization ponds**

| Name of the Species | Media | Average Lipid Content (mg $g^{-1}$ dwt) | Average Fatty Acid Content (mg $g^{-1}$ dwt) |
|---|---|---|---|
| *Chlorococcum sp* | BG11 | 54.00 | 39.39 |
| *Diatom numulloides sp* | f/2 | 126.00 | 85.23 |

| | | | |
|---|---|---|---|
| *Chlorococcum sp* | BG11 | 53.34 | 51.00 |
| *Diatom mulleri sp* | f/2 | 54.51 | 18.00 |
| *Diatom surtellum sp* | f/2 | 88.64 | 33.35 |
| *Diatom Plateycephala* | f/2 | 113.98 | 69.00 |
| *Diatom Lyra sp* | f/2 | 97.19 | 62.00 |
| *Anthrospirasp* | BBM | 37.85 | 21.00 |
| *Oscillotoriasp* | f/2 | 124.20 | 77.00 |
| *Chlorella sp* | BBM | 89.91 | 45.00 |
| *Spirulina sp* | BBM | 30.92 | 21.00 |

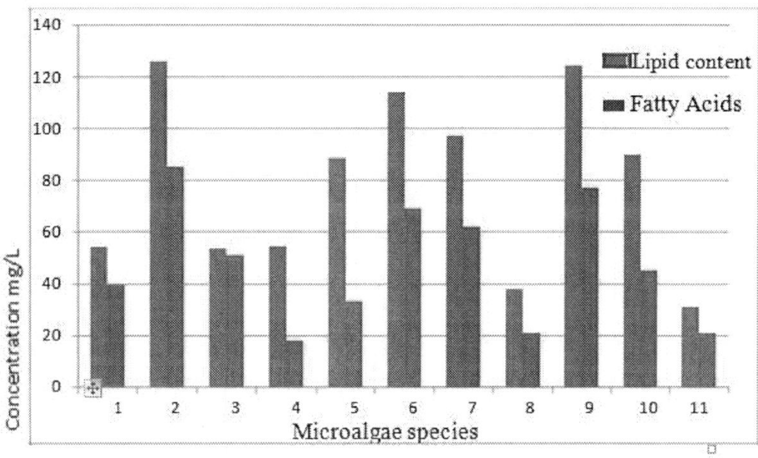


**Figure 3.2 Characteristics of the Microalgae species from waste stabilization ponds**

### 3.6 Nutrient Removal in Waste Water By Microalgae Treatment

The Nutrient concentration such as Total Kjeldahl Nitrogen and Phosphorous are tabulated in Table 3.7 for the respective Sewage Treatment Plant. The sewage treatment pond in Perungudi was found with highest nutrient of about 92.80 mg/L Nitrogen and 26.30 m g/L Phosphorus content before algal treatment among all the ponds in this study. The species obtained from this pond was also found with maximum cell growth when compared with other species in the study with 162.00 x $10^4$cells/mL.

**Table 3.7 Nutrient removal in waste water by Microalgae treatment**

| Name of the/pond | Before Algal Treatment (Inlet) | | During Algal Treatment (Pond) | | After Algal Treatment (Outlet) | |
|---|---|---|---|---|---|---|
| | TKN (mg/L) | P (mg/L) | TKN (mg/L) | P (mg/L) | TKN (mg/L) | P (mg/L) |
| Oxidation pond 1 IITM, Chennai | 49.30 | 19.20 | 33.60 | 17.60 | 20.19 | 10.62 |
| VIT – Pond Vandalur, Chennai | 21.80 | 10.20 | 16.60 | 9.17 | 9.23 | 6.28 |
| STP, Perungudi | 92.80 | 26.30 | 66.48 | 16.20 | 19.19 | 6.12 |
| Sewage Treatment Plant, Chithathur | 46.21 | 19.56 | 29.60 | 10.61 | 17.20 | 9.60 |
| Mangalam Pond Pulikundram | 12.20 | 5.72 | 9.27 | 4.20 | 6.60 | 4.61 |
| Parry Nutraceuticals, Chennai | 23.12 | 12.63 | 17.62 | 8.05 | 7.91 | 2.92 |

Next to Perungudi pond, the Oxidation Pond in IIT, Madras was found with highest nutrient of about 49.30 mg/L Nitrogen and 19.20 mg/L Phosphorus content before algal treatment. Similar to the species obtained in Perungudi pond, the species from IIT pond has maximum cell growth of about 106.00 x $10^4$ cells/mL.

## 4. Conclusions

From the experiments conducted for the microalgae stains it is seen that the biomass growth was always directly related to the removal in nutrients and algal lipid content. From the results obtained it was identified that *Diatom numulloides sp* has the highest lipid content and Fatty acid content among the other species and thus it is considered to be the microalgae species with greater bioenergy productive potential microalgae species. It was also observed that the Pond in which the *Diatom numulloides sp* was obtained has the highest nutrient concentration and that concentration was removed effectively by the respective algal growth.

The *Diatom numulloides sp* exhibited the highest growth rate associated with nutrient removal. Finally the algal lipid content and Fatty acid varied with time and higher values were observed on day 40 of cultivation compared to day 20. The dry biomass weight of *Diatom numulloides sp* after 40 days growth is 109.50 mg/L which was obtained from Sewage Treatment Pond, Perungudi. The Dry biomass weight completely depends upon the cell growth count and the concentration of Nutrients in the media. The cell growth count of *Diatom numulloides sp* in f/2 Media is 106.00 x $10^4$ cells/mL at $25^0$C.

*Diatom numulloides* sp was found to be the microalgae species with maximum lipid content among other microalgae species in the study with 126mg $g^{-1}$ dwt in 40 days growth with dry biomass productivity 109.50mg/L. The Fatty acid content of *Diatom* sp was found to be the microalgae species with maximum Fatty Acid content among other microalgae species in this study with 85.23 mg $g^{-1}$ dwt in 40 days growth. The study shows that the lipid content of the species is proportional to the dry biomass weight obtained.

References

[1] Ahmed Al Darmaki, L Govindrajan (2016), Cultivation and Characterization of Microalgae for Wastewater Treatment *World Congress on Engineering 2012* Vol. I, pp. 1-94.

[2] Andriana F.Aravantinou, Marios A. Theodorakopoulos, laonnis D. Manariotis (2016), Selection of microalgae for waste water treatment and potential lipid production, *The journal, Frontiers in Energy Research-Front. Energy*, pp. 8-52.

[3] Anuja J and Chandra S (2015), *Studies on fresh water algae in relation to chemical constituents of Thiruneermalai temple tank near Chennai, India* ,Sci 2012, 4:21-29.

[4] Anyue, Wei Chen, Ling zheng (2016), 'Identification of high-lipid producers for biodiesel production from forty-three green algal isolates in China', *Progress in natural science*, pp. 269-271.

[5] Cynthia Victoria, Maria del Carmen, Fransico Gabriel (2016), 'Protein measurements of microalgal and cyanobacterial biomass', *Bioresource Technology* 101 (2010), pp. 7587–7591.

[6] Deusa A. M. P. da Ponte, Marcelo M. Werneck, Donato A. Aranda(2016), 'Advances for Opaque PBR Internally Illuminated for Fiber Optic for Microalgae Production Natural Science', *Natural science*- 8, pp. 341-358.

[7] Ikrat M Hassan, Ibtsam F. Aljbory, Essam F. Al Jumaily (2015), 'Stimulation of biodiesel production from two algae: Chlorella vulgaris and Nitzschiapale and study their some growth parameters under different light intensity', *IOSR Journal Of Environmental Science, Toxicology And Food Technology*, Vol. 6, pp 31-42

[8] Madhavi latha, Rajkumar and Saikrishna (2014), 'Choice of culture media for isolation of algae from soils of some rice-fields', *Phykos* 44 (2): pp. 44-53

[9] Mata T, Martins A, Caetano N (2013), 'Microalgae for biodiesel production and other applications', *Sustain. Energy Rev*. 14, pp. 217–232

[10] Matteo Marsullo, Alberto Mian, Adriano Viana Ensinas, Giovanni Manente, Andrea Lazzaretto and Francois Marecha(2015), 'Dynamic modeling of the microalgae cultivation phase for energy production in open raceway ponds and flat panel photobioreactors', *The journal, Frontiers in Energy Research-Front. Energy Res*, pp. 3-41.

[11] Miguel Ballen-Segura1, Luisa Hernandez Rodríguez, David Parra Ospina, Asly Vega Bolanos, Karen Perez(2016), 'Using Scenedesmus sp. for the Phycoremediation of Tannery Wastewater', *Tecciencia*, volume 21. Pp.11-22

[12] Paran Gani, Norshuhaila Mohamed Sunar, Hazel Matias-Peralta, Ab Aziz Abdul Latiff, Umi Kalthsom Parjo, Abdul Rafiq Ab. Razak(2015), 'Phycoremediation of Wastewaters and Potential Hydrocarbon from Microalgae: A Review', *Advances in Environmental Biology*, 9(20) , pp. 1-8.

[13] Rakesh R. Narala, Sourabh Garg, Kalpesh K. Sharma, Skye R. Thomas-Hall, Miklos Deme, Yan Li and Peer M. Schenk(2016), 'Comparison of microalgae cultivation in PBR, Open raceway pond and a two stage hybrid system', *Front. Energy Res.* Pp. 4-29.

[14] Sanders, W.R (1991), 'Mixotrophic protists in marine and freshwater ecosystems' J. Protozool. 38, pp. 76–81.

[15] Shahrbanoo Hamedi, Mahmood A. Mahdavi, Reza Gheshlaghi,(2016), 'Improved lipid and biomass productivities in Chlorella vulgaris by differing the inoculation medium from the production medium', *Biofuel Research Journal* 10, pp. 410-416.

[16] Shakeel A Khan, Rashmi, Hussain (2015), 'Prospects of biodiesel production from microalgae in India' *Renewable and Sustainable Energy Reviews*, Volume II, pp. 2361–2372.

[17] Tsuyashi takeshita, Kohei Takeda (2015), ' A simple method for measuring the starch and lipid content in the cell of microalgae', Cytologia 80(4), pp. 475–481.

[18] XinZhang,JunfengRong,HuiChen,ChenliuHeand QiangWang(2014),'Current status and outlook in the application of microalgaein biodiesel production and environmental protection',*Frontiers in Energy Research* ,volume 2, article 32.

[19] Yano Surya Pradanaa, Hanifrahmawan Sudibyoa, Eko Agus Suyonob, Indartoc, Arief Budimana,(2017), 'Oil Algae Extraction of Selected Microalgae Species Grown in Monoculture and Mixed Cultures for Biodiesel Production', *Energy Procedia*, pp. 105-277.

[20] Yuchi Han, Qinxue Wen, Zhiqiang Chen, Pengfei Li (2014), 'Review of Methods Used for Microalgal Lipid-Content Analysis' *Energy Procedia*, pp. 12- 944 – 950

IOP Conf. Series: Materials Science and Engineering **923** (2020) 012065 doi:10.1088/1757-899X/923/1/012065

# A Numerical Investigation of Variably Heated Enclosure to Enhance its Heat Transfer Rate using CFD

**P. Ragupathi[1], S. Aravind[1], D. Sathish Kumar[1], G. Vignesh[1]**

[1]Department of Mechanical Engineering, Karpagam Academy of Higher Education, Coimbatore-641021, Tamil Nadu, India

ponragupathi@gmail.com

**Abstract**. The flow simulation with the help of computational tools helps in solving many engineering problems related to fluid dynamics. This in turn reduces the time and cost incurred in doing real time experimental investigations. In this present research work, computational investigation was carried out in a variably heated enclosure to enhance the life span of the electronic components inside the enclosure. The investigation was performed over a twilight enclosure system, the purpose of this system is switch on or off the street lights with support of sensors automatically. This particular enclosure is subjected to heating as it is continuously exposed to sunlight through the day. This enclosure consists of delicate electronic components which are sensitive to heat. The heat transfer can be enhanced in this system by the prevailing natural flowing air in the surrounding by free convection process. To enhance and optimize this heat transfer process, numerical simulation using computational tools was carried out by changing the physical layout of the enclosure. To perform this numerical simulation, flow simulation module available in the SolidWorks was used. Also, to model this system, the same modeling software was used. From the outcome of this numerical flow simulation, it showed a drop of 36.3% in the heat transfer rate when compared to the existing model to the proposed model used for the simulation. By using the modified and proposed model, the life span of the entire system will be improved.

Key words: Computational tool, Enclosure, Free convection, Heat Transfer, Simulation.

## 1. Introduction

In the modern world, automotive revolution is going on in a fast pace. The automotive sectors are trying in recent days, the use of sensor based electronic components enclosed in an enclosure is widely used in many places and also, they are exposed to the surrounding conditions. The surrounding conditions here it indicates the heat transfer effect. All electronic components work properly upto the indicated temperature range. If heat accumulates owing to radiative heat transfer, then the temperature goes beyond the operating temperature limit suggested by the component manufacturer and leads to malfunction of such electronic components. One such example selected for this study is the twilight switch box which is generally used in street lights on or off automatically based on sensor at an average height of 6 to 6.5 m from ground level. This switch box is subjected to radiative heat transfer as it is exposed to sunlight for long duration. Such radiative heat transfer is accompanied with the natural convection, which tries to transfer the heat from the switch box to the surrounding thereby enhancing the life of switch box. The rate of convection depends on many parameters one such parameter can be considered to be the design of the switch box. Markatos et al. [1] in their study performed numerical analysis on a square cavity subjected to natural convection under laminar and turbulent cases. They also compared the Rayleigh numbers up to 106 with the standard bench mark results. Ostrach et al. [2] in his study explained the importance of the scaling analysis and experiments that are conducted to get the flow details. Numerical approaches with their deficiency in providing the exact solution for various ranges of problems is discussed in his work. Dehghan et al. [3] performed a numerical analysis on a discretely heated open cavity for combined free convection, conduction and radiative heat transfer. They compared the local heat transfer coefficient of the conjugate analysis and found that the radiative heat transfer does not show much difference in the performance of heat transfer of the heat source. But when compared with the experimental results the radiative hat transfer had considerable effect on the same. Manz et al. [4] conducted an investigation over the rate of heat transfer by free convection in the air enclosed inside cavities of vertical rectangular building façade elements for different aspect ratios using CFD code. They compared the calculated Nusselt number as a function of Rayleigh number with five different correlations and found in line for 4 correlations. Their study concluded that the CFD code so created can be starting point for the future applications on complex façade elements for which no experimental data are found. Chan et al. [5] performed a numerical study for a laminar steady state free convection over 2-D rectangular open cavity for Rayleigh number up to 106 and noticed that the heat transfer is accompanied by the strong characteristics of heated cavities. The result goes in par with the experimental result for Rayleigh number equaling 106. Polat et al. [6] in their work conducted a numerical study for laminar steady state free convection in shallow cavities which are inclined and with side facing the opening subjected to constant heat flux. The numerical result obtained revealed that the heat transfer is governed by the Rayleigh number, angle of inclination and the aspect ratio. Bligen et al. [7] carried out a mathematical study on an inclined square cavity which is partially opened with laminar steady state free convection for various range of Rayleigh number, dimensionless aperture size and the inclination angle. It resulted with the flow rate of volume and the Nusselt number as a growing function of Rayleigh number and the remaining two parameters. Also, the heat transfer rate can be increased or decreased by altering those parameters.

Huelz et al. [8] studied numerically the transfer of heat in a square cavity with two walls opposite to each other in a differential manner and the other two walls completely insulated with flow in the laminar region as there is a change in Rayleigh number and the inclination angle. It showed that a basic power law influence in the Nusselt number related to the Rayleigh number with the coefficient and exponent as functions of the angle of inclination. It is found that for a constant Rayleigh number, hysteresis as the inclination angle grows or declines. The limit of inclination angles for hysteresis be subject to the Rayleigh number. Sahu et al. [9] analyzed the cylinders having equilateral triangles of dissimilar dimensions inside a square enclosure for its heat transfer and flow owing to free convection in air. This is computationally modelled by Ansys Fluent 13 commercial CFD package. The analysis is carried out for Rayleigh number range of 104 to 106 with varying size of enclosure, cylinders of varying dimensions with respect to aspect ratios from 0.2 to 0.5 with a step of 0.1. And the Results are obtained as contours of isotherm and stream function. Pinto et al. [10] performed a numerical investigation on heat transfer by free convection of a fluid flowing inside a square hole with two inner bodies with help of FORTRAN code using finite element method for solving the concerned governing equations. In this work the study is extended for Grashof number and thermal diffusivity ranging from 2 x 105 to 108 and 0.1 to 100.

Several such numerical studies with help of different software's are conducted to study the effect of heat transfer by free convection over different bodies. All such studies resulted in par result with the experimental studies. In this present work one such numerical study is conducted to compare the heat transfer enhancement by design modification in a twilight switch box which is differentially heated up by suns radiation. The computation model is created using SolidWorks and the same is tested for heat transfer by natural convection in SolidWorks simulation module. The test results revealed that the modified switch box has effective heat transfer.

## 2. Problem Description

The twilight switch box selected for the study is made of ABS plastic and glass material which is placed in a lamp post for sensing the light intensity and automate the ON/OFF of the street light. Exposing of such switch box with sensing element in sunlight for the entire day results in high radiation phenomenon, as the major part is made of glass. This results in the faulty working of the sensor due to prolonged exposure to such radiation heat transfer. In this present work, it is tried to reduce such radiation heat transfer by two types of modified switch boxes. The model is created using SolidWorks and analysed using SolidWorks flow simulation. And the results are compared between the prevailing model (figure.1) and the proposed model (figure.2).

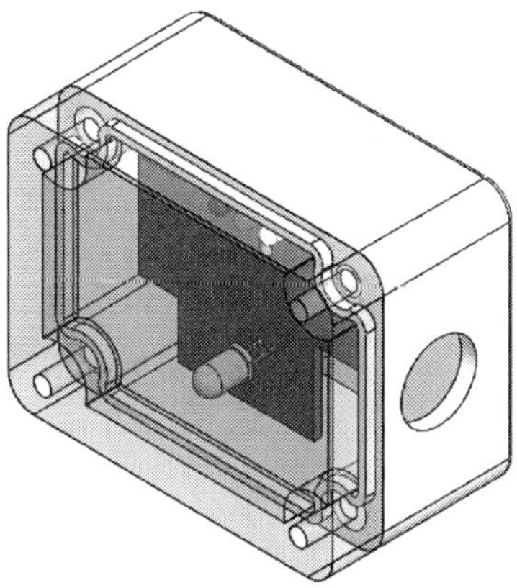

**Figure. 1** Existing model of switch box

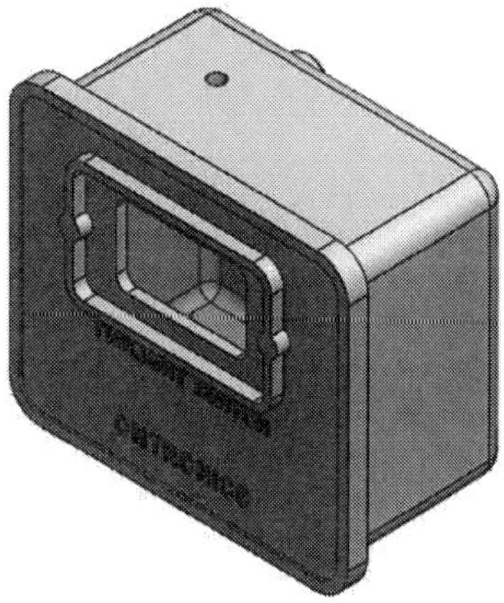

**Figure. 2** Proposed model of switch box

## 3. Methodology

The created model is exported to the SolidWorks flow simulation module. Any CFD packages have these three stages and they are the pre-processing, Solution and Post processing. The pre-processing stage includes the selection of geometry, type of analysis to be carried out, materials selection, boundary conditions and mesh settings [11, 17]. The existing model and the proposed models are corrected in order to avoid the meshing error and to have a reduced time while it undergoes analysis. Then the type of analysis is selected to be external flow heat transfer along with radiation. The materials are set to be ABS plastic for the switch box cover and the front panel as glass material, the fluid medium is selected as air.

The boundary condition for air is set to be with flow velocity of 8, 5 and 3 m/s along x, y and z directions. The air is set to standard atmospheric pressure of 1.01325 bar and temperature around 27°C. The boundary condition for radiation is set with solar intensity of 1000 W/m$^2$ along the y and z directions based on the sensor box position. And the corresponding surfaces subjected to radiative heat transfer is selected. The initial mesh level is set to be at third level with narrow channel refinement. The mesh produced is shown in figure. 3. With these initial steps involved in pre-processing stage, the models are subjected to analysis. That is the solver is made to run to solve the governing equations related to the above set of conditions.

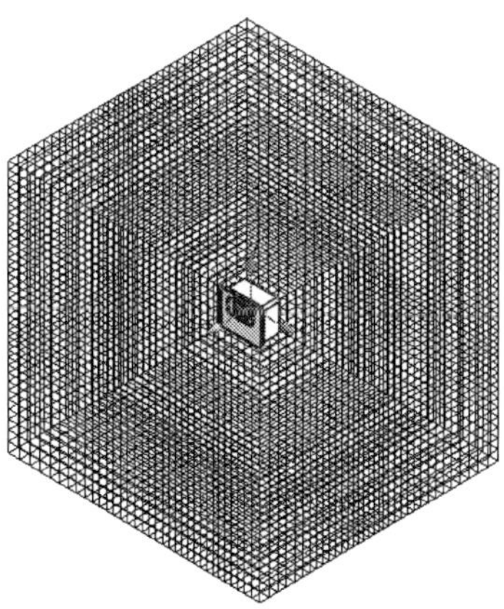

**Figure. 3** Mesh Generated

IOP Conf. Series: Materials Science and Engineering **923** (2020) 012065 doi:10.1088/1757-899X/923/1/012065

## 4. Results & Discussion

The SolidWorks flow simulation solver takes some time to analyse the computational domain and ends up with the computational results in the post processing stage. The following figures 4 & 5 shows the air flow vector along x, y & z directions for the existing and proposed model.

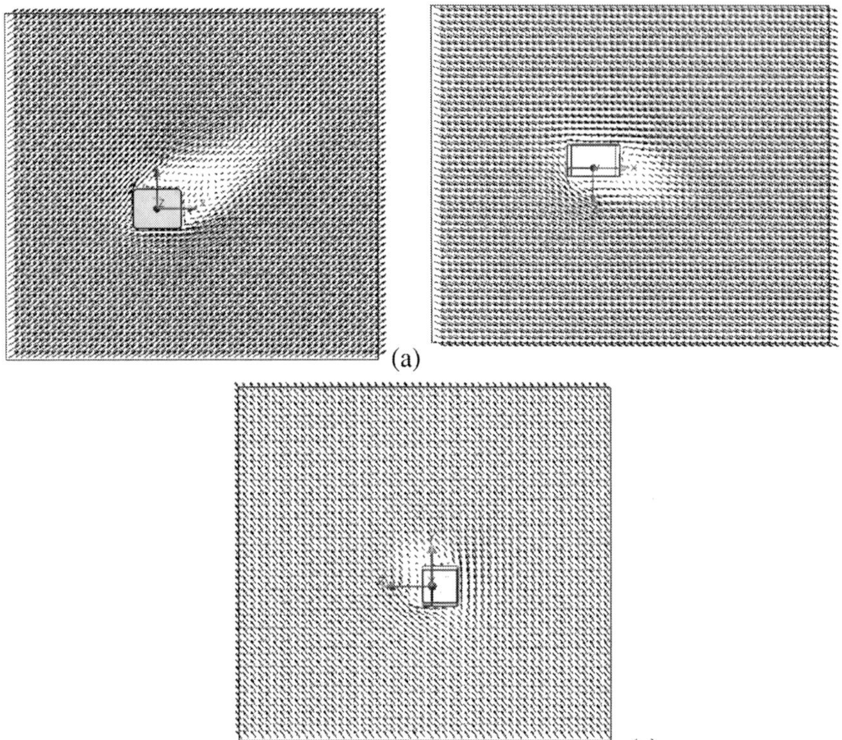

(a) (b) (c)

**Figure 4.** Air flow vector in (a) X axis (b) Y axis (c) Z axis around the existing assembly

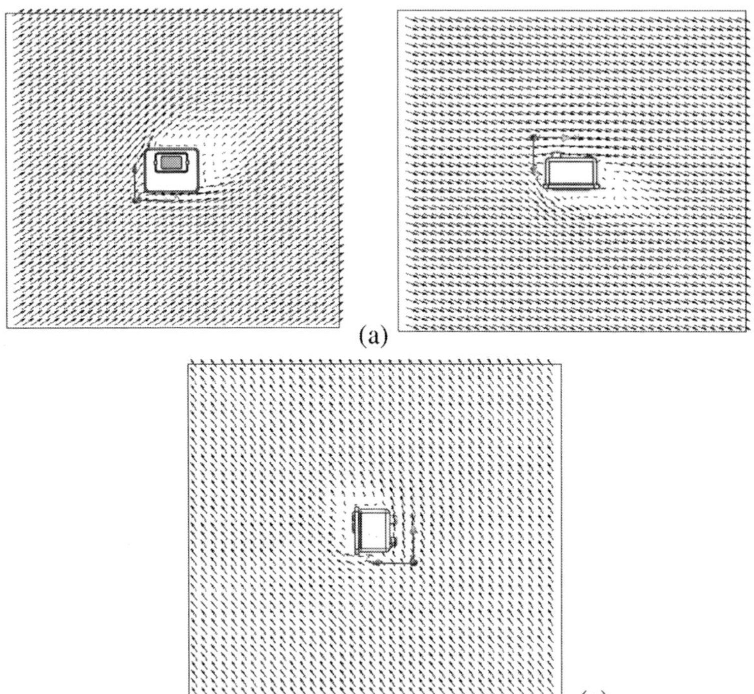

(a) (b) (c)

**Figure. 5** Air flow vectors in (a) X axis (b) Y axis (c) Z axis around the proposed assembly

IOP Conf. Series: Materials Science and Engineering **923** (2020) 012065 doi:10.1088/1757-899X/923/1/012065

The following contours in the figures 6 & 7 shows the surface temperature of the switch box for the existing model and the proposed model. The surface temperature contours show that the surface temperature is more for the prevailing model when related to the proposed model. Around 25.4 % more than the proposed model when related to the prevailing model [15].

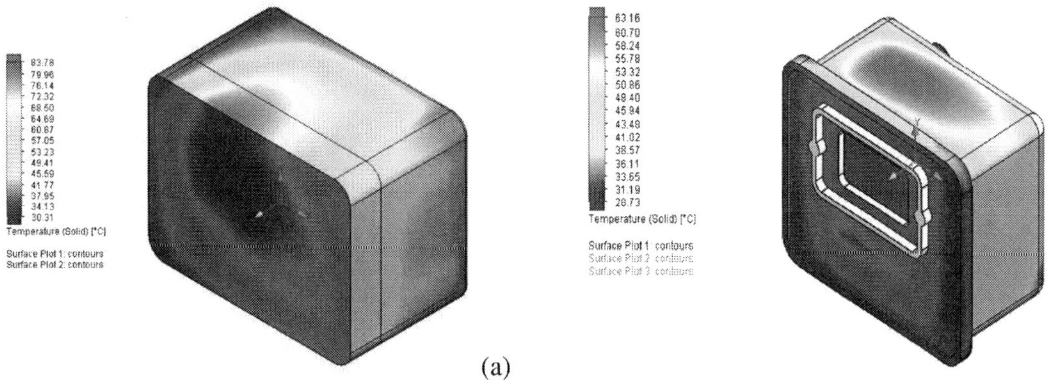


(a) (b)

**Figure. 6** Surface temperature of (a) Existing model (b) the proposed model

The maximum total heat flux for the existing model is about 1060 $W/m^2$, where as it is around 1036 $W/m^2$ for the proposed model. But the minimum heat flux is 42.2% higher for proposed model when compared to the existing model. The total heat flux contours for the existing & proposed model is shown in the figure 6.

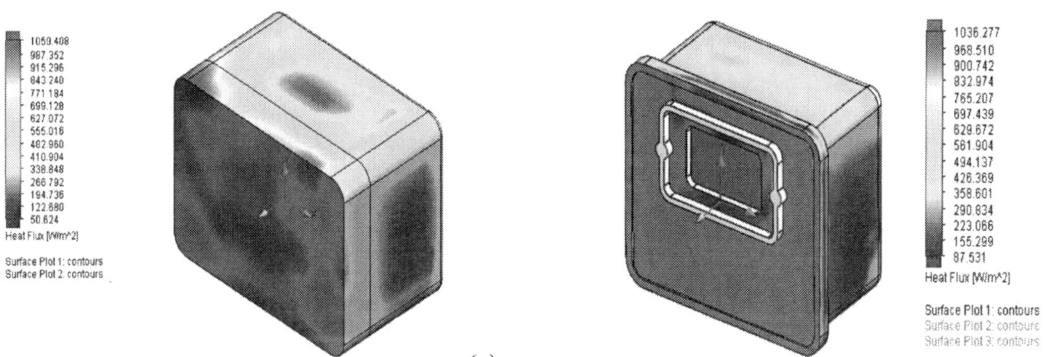


(a) (b)

**Figure. 7** Total Heat flux of (a) Existing model (b) the proposed model

The figure 8 depicts the resulted surface heat flux for the described conditions which shows that the surface heat flux has reduced from a maximum of 1611.5 $W/m^2$ for existing model to 1247 $W/m^2$ for the proposed model. Similarly, the minimum surface heat flux for the proposed model has reduced to about 27.8 % when compared to the existing model.

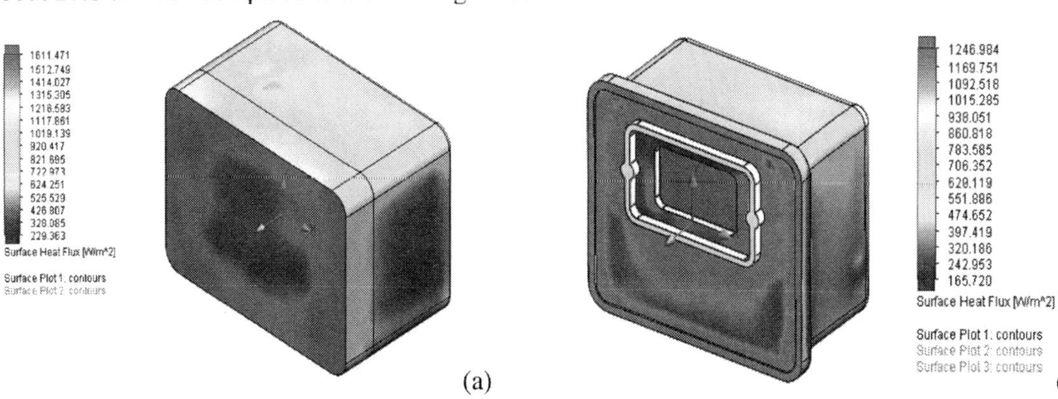


(a) (b)

**Figure. 8** Surface Heat flux of (a) Existing model (b) the proposed model

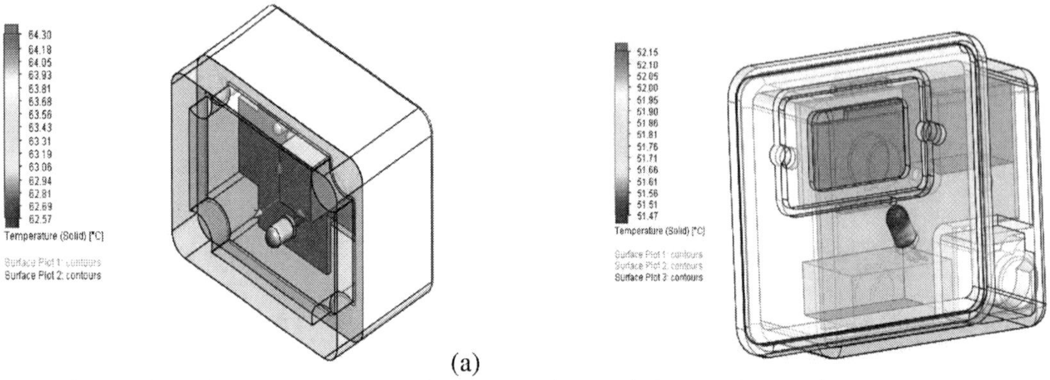

(a) (b)

**Figure. 9** Surface temperature on the sensor of (a) Existing model (b) the proposed model

The surface temperature of the sensor inside the enclosure has reduced from a maximum value of 64.3°C for existing model to 52.2°C for the proposed model, this in turn definitely improves the efficiency of the sensor in rendering its need. This also increases the life time of the sensor [14-17]. The figure 9 shows the surface temperature contour of the sensor in existing & proposed model.

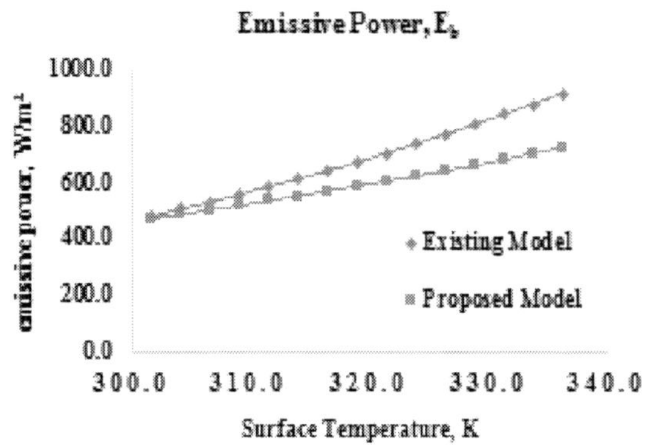


**Figure. 10** Emissive power of the existing & proposed model

From the graph shown in the figure. 10 shows the substantial decrease in the emissive power for the proposed model from the existing model.

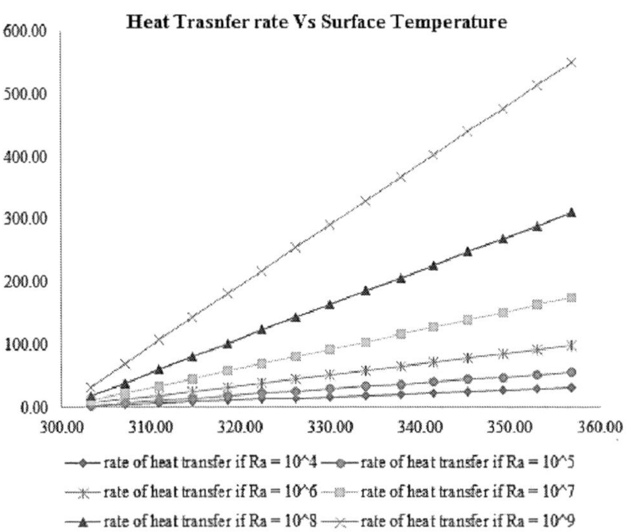


**Figure. 11** Heat transfer rate for increasing Rayleigh number from $10^4$ to $10^9$ for existing model

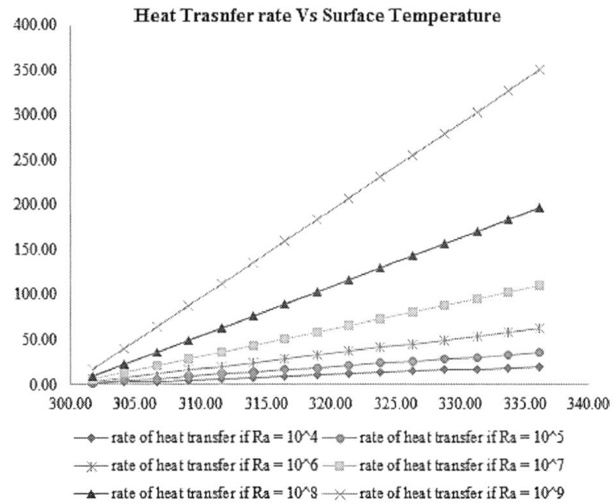


**Figure. 12** Heat transfer rates for increasing Rayleigh number from $10^4$ to $10^9$ for proposed model

The graphs shown in the figure 11 & 12 shows the heat transfer rate for increasing Rayleigh number in the range of $10^4$ to $10^9$. The average Nusselt number varied from range of 6 to 105 for different Rayleigh number. The average heat transfer coefficient varied from 247.8 to 4406.0 W/m$^2$K for varying Nusselt number. These properties where calculated based on the co-relations available in the standard heat transfer data tables.

From the graphs, it is found that the heat transfer rate to the switch box has a drop of 36.3% from the existing switch box model to the proposed switch box model [15-17]. The heat transfer rate calculated analytically based on the existing co-relations are in par with the heat flux values that are found numerically using the SolidWorks flow simulation module.

## 5. Conclusion

The present study performed on the twilight switch box with the objective of reducing the rate of radiative heat transfer to avoid malfunction of the sensor for automated ON/OFF of the street lights has resulted in a considerable alteration of changing the existing switch box model to the proposed model. As the proposed design has better heat transfer characteristics when related to the existing design. Various heat transfer parameter like the surface temperature, total heat flux, surface heat flux, emissive power and the heat transfer rate for the proposed design is better than the prevailing design. Hence it can be concluded that by replacing the existing model with the proposed model will certainly result in increased sensitivity as well as increased life of the sensor.

The proposed model has been manufactured and the numerical study performed has helped to increase the quality of the twilight switch box.

## References

[1] Markatos, N.C. and Pericleous, K.A., 1984. Laminar and turbulent natural convection in an enclosed cavity. International Journal of Heat and Mass Transfer, 27(5), pp.755-772.

[2] Ostrach, S., 1988. Natural convection in enclosures, 1175-1190.

[3] Dehghan, A.A. and Behnia, M., 1996. Combined natural convection–conduction and radiation heat transfer in a discretely heated open cavity, 56-64.

[4] Manz, H., 2003. Numerical simulation of heat transfer by natural convection in cavities of facade elements. Energy and Buildings, 35(3), pp.305-311.

[5] Chan, Y.L. and Tien, C.L., 1985. A numerical study of two-dimensional laminar natural convection in shallow open cavities. International Journal of Heat and Mass Transfer, 28(3), pp.603-612.

[6] Polat, O. and Bilgen, E., 2002. Laminar natural convection in inclined open shallow cavities. International Journal of Thermal Sciences, 41(4), pp.360-368.

[7] Bilgen, E. and Oztop, H., 2005. Natural convection heat transfer in partially open inclined square cavities. International Journal of Heat and Mass Transfer, 48(8), pp.1470-1479.
[8] Huelsz, G. and Rechtman, R., 2013. Heat transfer due to natural convection in an inclined square cavity using the lattice Boltzmann equation method. International journal of thermal sciences, 65, pp.111-119.
[9] Sahu, K.B. and Singh, R.K., 2014. Analysis of heat transfer and flow due to natural convection in air around heated triangular cylinders of different sizes inside a square enclosure. Procedia Engineering, 90, pp.550-556.
[10] Pinto, R.J., Guimarães, P.M. and Menon, G.J., 2016. Numerical Study of Natural Convection in Square Cavity with Inner Bodies Using Finite Element Method. Open Journal of Fluid Dynamics, 6(02), p.75.
[11] Ji, Y., 2014. CFD modelling of natural convection in air cavities. CFD letters: An International Journal, 6(1), pp.15-31.
[12] Mondal, B. and Mishra, S.C., 2008. Simulation of natural convection in the presence of volumetric radiation using the lattice Boltzmann method. Numerical Heat Transfer, Part A: Applications, 55(1), pp.18-41.
[13] Yu, E. and Joshi, Y.K., 2000. Natural convection air cooling of electronic components in partially open compact horizontal enclosures. IEEE Transactions on Components and Packaging Technologies, 23(1), pp.14-22.
[14] Arik, M., Becker, C.A., Weaver, S.E. and Petroski, J., 2004, January. Thermal management of LEDs: package to system. In Third international conference on solid state lighting (Vol. 5187, pp. 64-75). International Society for Optics and Photonics.
[15] Anton, R., Jonsson, H. and Moshfegh, B., 2009. Detailed CFD modelling of EMC screens for radio base stations: a parametric study. IEEE Transactions on Components and Packaging Technologies, 32(1), pp.145-155.
[16] Galloway, J.S., 1998, May. Thermal analysis of portable electronic products. In ITherm'98. Sixth Intersociety Conference on Thermal and Thermomechanical Phenomena in Electronic Systems (Cat. No. 98CH36208) (pp. 1-7). IEEE.
[17] Lee, T.Y., Chambers, B. and Mahalingam, M., 1995. Application of CFD technology to electronic thermal management. IEEE Transactions on Components, Packaging, and Manufacturing Technology: Part B, 18(3), pp.511-520.

IOP Conf. Series: Materials Science and Engineering **923** (2020) 012066 doi:10.1088/1757-899X/923/1/012066

# Experimental and Numerical Studies of Thermal Energy Storage using Paraffin Wax Phase Change Materials

**R.R. Thirumaniraj[1*], K. Muninathan[2], V. Ashok Kumar[2], B. Jerickson Paul[1], Rahul R Rajendran[3]**

[1*, 1] Student, St. Joseph's College of Engineering, Anna University, Chennai, India
[2] Assistant Professor, St. Joseph's College of Engineering, Anna University, Chennai, India
[3] Research Scholar, Lehigh University, Bethlehem, USA

*E-mail: thirumaniraj97@gmail.com

**Abstract:** The main idea of this work is to design and analyze efficient storage of thermal energy using phase change material. Solar energy is a readily available and renewable source of energy. It is also a clean energy as it does not emit carbon dioxide. However maximum utilization of solar energy is not possible without the use of thermal energy storage (TES). This thermal storage system can form an integral part of solar heating system. In this work a TES tank is designed and fabricated. Paraffin wax is the phase change material used. It is encased in stainless steel balls. The high specific heat capacity of PCM is used to store latent heat which can be used any time. Temperature measurements are taken using T-type thermocouple along with indicator. Numerical analysis is done using the CFD software Ansys Fluent.
This work mainly focuses to find the solution for the mismatch of abundant solar energy available during daytime and non-availability of energy during night-time. This project work also forms a part of a wider study to explore the possibility of PCMs to store the energy during off-peak load and releases the same during peak load. In several countries, during the daytime the load will be at peak level which leads to high cost for production. Therefore, cost saving can be achieved if heat pumps can be operated during the off-peak periods to store heat for use during the peak periods

***Key words:*** TES, Phase Change Materials, paraffin wax, thermal storage

*List of Symbols and Nomenclature*

| **Symbol** | **Description** |
|---|---|
| Q1 | Heat given by hot water |
| ṁh | Mass flow rate of hot water |
| Thi | Specific heat of hot water |
| Tho | Inlet temperature of hot water |
| Thi | Average outlet temperature of hot water |
| Q2 | Heat gained by cold water |
| ṁc | Mass flow rate of cold water |
| cpc | Specific heat of cold water |
| Tco | Average outlet temperature of cold water |
| Tci | Inlet temperature of cold water |
| ηstorage | Efficiency of Storage system |

## 1. Introduction

Energy storage is a major issue to be faced to allow intermittent energy supply, typically renewable sources, to match energy supply with demand. Most of Technologist are facing a problem in storing the thermal energy. It cannot be economical used if we store in batteries and so it should be stored in the appropriate form with suitable procedure. Xiang Lu et al. conducted an experiment on Nano sheets for thermal energy storage using polyurethane phase change materials in which he proposed energy storage is a major issue to be faced to allow intermittent energy supply, typically renewable sources, to match energy supply with demand is Thermal energy [1]. H. Wei et al. done a study on shell composite phase change materials for high temperature thermal energy storage, he said that it was reported that the 80% of utilization of energy is in the form thermal energy and 45% of energy is wasted. Most of Technologist is facing a problem in storing the thermal energy. It cannot be economical used if we store in batteries and so it should be stored in the appropriate form with suitable procedure. In many regions of the world, direct solar radiation is considered to be one of the most prospective sources of energy. The major disadvantages of solar energy are its intermittent supply. To overcome this drawback, it is necessary to develop energy storage devices [2]. Chuanfei Shen et al proposed that most of the experiments carried out in this issue results in 11.18%-36.17%. They used molten salts and phase change materials generally. The molten salts like Sodium sulphate dehydrate, sodium chloride, chlorides, silicates and other inorganic salts [4]. Vivek Tiwari et al. has done a SWOT analyses of high –temperature phase change materials for thermal energy storage, he says that the thermal energy storage is classified into two types. They are

- Latent heat thermal energy storage (LHTES)
- Sensible heat thermal energy storage (SHTES)

N.Nallusamy et al. has done an Experimental investigation on a combined sensible and latent heat storage system integrated with constant/varying (solar) heat sources which resulted that the Latent thermal storage materials are very effective. They store 5 to 14 times more heat per unit volume than sensible heat storage materials such as water, masonry, or rock. Phase change materials (PCMs) absorb and emit heat energy whenever supply energy is surplus and not able to meet the demand. This may be major reason to suggest PCM as an alternative to conventional storage mediums. Thermal gradients during charging and discharging are small and hence simultaneous charging and discharging is possible with appropriate selection of heat exchanger [6]. Samira Golestani Ranjbar et al. has done a study on Fabrication and Characterization of phase change material –SiO2 for Thermal energy storage in buildings they proposed that there are some basic qualities to be seen while selecting the phase change material as storage medium which includes the specific heat capacity, thermal and chemical stability, non-toxic, non-flammable and phase transition temperature should be practically and economically attained. The main aim is to store and use the thermal energy efficiently [7]. There is variety of phase change materials can be used but the PCM which is economically beneficial is paraffin. The paraffin has a phase transition temperature near to the normal temperature 20° to 30°C. Other PCM's are having higher range of about 350°C-1500°C which is not economically beneficial to store the thermal energy.

## 2. Material Selection

*2.1 TES Tank*

For both TES tank and spherical balls Stainless steel is selected because it has more corrosion resistance and high tensile strength. It also has higher degree of reusability.

*2.2Paraffin Wax*

Paraffin wax has high heat trapping capacity and it is insoluble in water also it is not affected by any chemical reagents. Its Heat of combustion is 42MJ/Kg. Hence it is used a Phase Change Material in this process.

*2.3 Geometry of Balls*

Paraffin wax is stored in a spherical shaped hollow ball. Spherical shape is chosen because it has more surfaces to volume ratio which permits uniform heat transfer between the bonds of paraffin wax.

*2.4 Grade of Stainless Steel*

High chromium ferritic grades (446) of stainless steel are used as TES tank material in this process. Austenitic grades of stainless steel have high strength at elevated temperatures.

*2.5 Melting Point of Wax*

The paraffin wax having melting point in the range of 40-degree Celsius to 43 degree Celsius is selected for this process.

*2.6 Thermocouples*

Six T -type thermocouple are used to obtain the temperature values. A channel for measuring inlet water temperature, a channel for measuring outlet water temperature and four channels for measuring the layer by layer temperature of paraffin wax are connected to get digital output.

## 3. Layout of Project

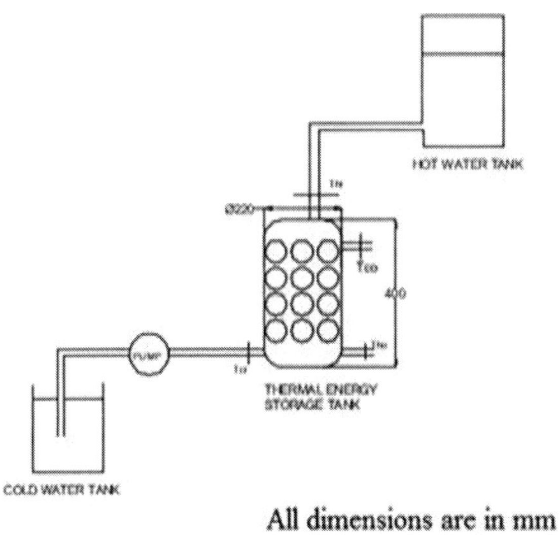


**Fig. 1.** Experimental Setup of Thermal Energy Storage

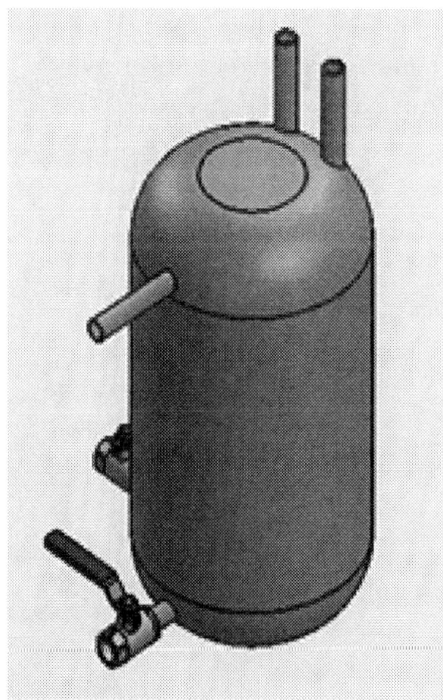

**Fig. 2.** Tes Tank Isometric View

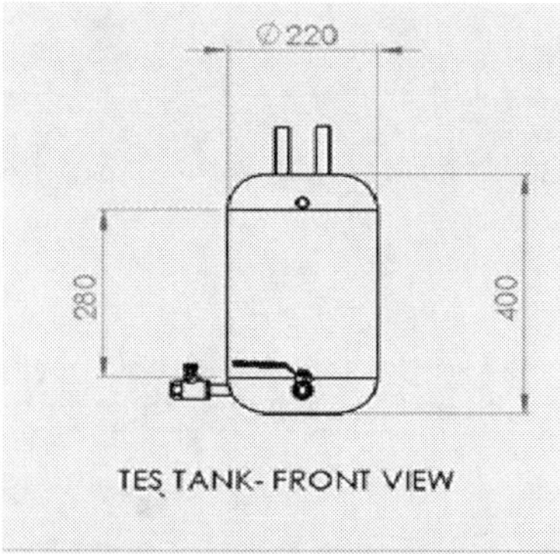


**Fig. 3.** Tes Tank – Front View with Dimensions

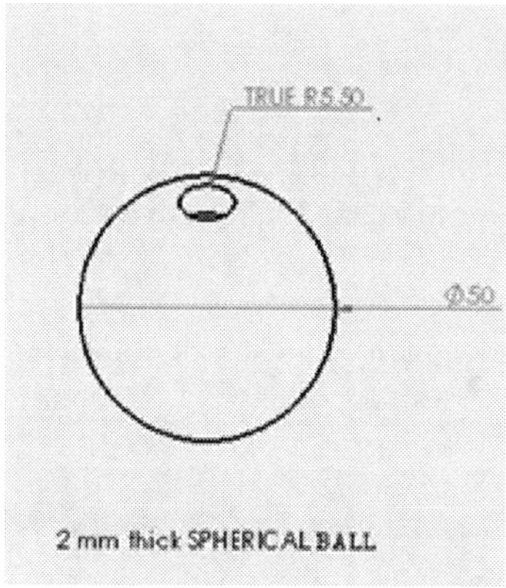


**Fig. 4.** Spherical Ball

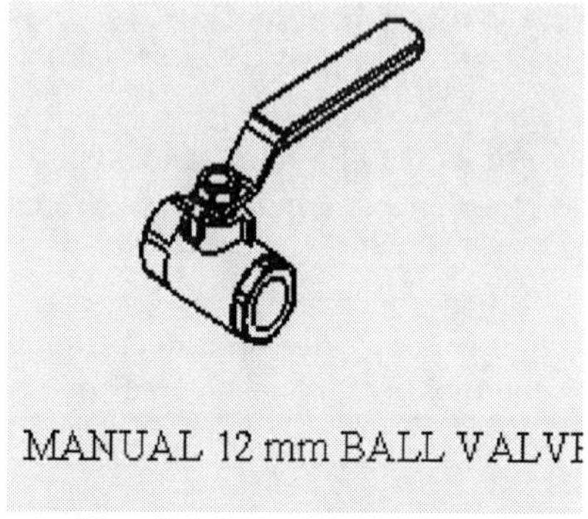


**Fig. 5.** Manual Ball Valve

## 4. Experimental Work

The experimental work consists of the following major components:
1. Thermal energy storage tank
2. Hot water supply tank for charging
3. Cold water supply tank for discharging
4. Pump
5. Control valves
6. Flexible hoses
7. Thermocouple for temperature measurements

*4.1 Thermal Storage Tank*

The thermal storage tank made of stainless steel of 400 mm height and 220 mm diameter. The PCM are filled in spherical stainless balls and arranged in six rows with the help of wire mesh. Each row contains 13 spherical balls. Hence 72 numbers of spherical balls are used to store thermal energy. The tank has also provision to admit hot water at top and the hot water leaves at bottom. The cold water is admitted at bottom and leaves at the top. Valves are provided at both entrance and exit to control the flow rate of waters.

**Fig. 6.** TES tank before and after insulation

Two thermocouple wires are embedded in the PCM materials to sense the temperature of the PCM during charging and discharging.

*4.2 Hot Water Supply Tank for Charging*
The hot water supply tank is located at top of thermal storage device in order to avoid pumping. An electric heater is installed in the hot water tank to heat the water and maintain the water temperature. This tank serves as a source of heat which is like solar water heating system. The main reason for using the electric heater is to simulate the process of solar heater in order to maintain constant temperature of hot water. It also incorporated with control valves and thermocouple to regulate and measure the hot water temperature.

*4.3 Cold Water Supply Tank for Discharging*
The cold-water supply tank is located at the bottom of thermal storage device to enable the entry of cold water at bottom of thermal storage tank. The cold water takes the heat energy stored by the PCM and leaves at top. A pump is used to circulate the cold water through the PCM storage tank.

*4.4 Pump*
A centrifugal pump of small discharge is used to pump the cold water from cold water tank to thermal storage tank.

*4.5 Control Valves*
Spherical ball type control valves of 12 mm size are used to regulate the hot and cold-water entry and exit.

*4.6 Flexible Hoses*
Flexible hoses are used to carry the water from hot and cold-water tank to thermal storage tank.

*4.7 Thermocouple*
There are 6 number of T type thermocouple used to obtain temperature values. These thermocouples are connected to a six-channel temperature indicator to get digital output.

## 5. Experimental Procedure

The experimental procedure consists of the following four steps

1. Charging
2. Draining the hot water
3. Discharging
4. Temperature data acquisition

*5.1 Charging*
The electric heater provided in the hot water tank is switched on and the temperature of hot water is observed. When the temperature reaches to a predetermined level, the control valves are opened, and the water is allowed to enter into the thermal storage tank. This hot water increases the temperature of PCM and then melts it. The charging process continues until the PCM attains a steady state.

*5.2 Draining the Hot Water*
The hot water available in the thermal storage tank is drained so that the storage tank is ready for discharging process.

*5.3 Discharging*
The cold water available in the cold-water tank is pumped into thermal storage tank. The heat energy stored in PCM is released and this heat is absorbed by the cold water and hence the cold-water temperature rises. The discharging process continues until the PCM reaches to its original temperature.

IOP Conf. Series: Materials Science and Engineering **923** (2020) 012066 doi:10.1088/1757-899X/923/1/012066

## 6. Temperature Data Acquisition

The temperature readings were recorded with the thermocouples connected to a temperature indicator. There are six thermocouple wires used to acquire data at the following locations during charging and discharging:

1. Hot water inlet
2. Hot water outlet
3. Cold water inlet
4. Cold water outlet
5. Top layer of PCM
6. Bottom layer of PCM

## 7. Efficiency Calculation of Thermal Storage System

### *7.1 Heat given by Hot Water*

The heat given by the hot water during charging is calculated as follows:

$Q_1 = \dot{m}_h \ c_{ph} ( T_{hi} - T_{ho} )$ (7.1)

$Q_1 = 0.307 * 4186 * (80 - 35)$

$Q_1 = 57829.59$ J/s

$Q_1 = 57829.59$ W

### *7.2 Heat gained by Cold Water*

The heat gained by the cold-water during discharging is calculated as follows:

$Q_2 = \dot{m}_c \ c_{pc} ( T_{co} - T_{ci} )$ (7.2)

$Q_2 = 0.307 * 4186 * (60 - 32)$

$Q_2 = 35982.85$ W

### *7.3 Efficiency of Storage System*

$\eta_{storage}$ = (Heat gained by cold water/Heat given by hot water )* 100

$\eta_{storage} = (Q_2/Q_1)* 100$ (7.3)

$\eta_{storage} = (Q_2/Q_1)* 100$

$= (35982.85 / 57829.59) * 100$

$\eta_{storage} = 62.22$ %

## 8. Experimentation

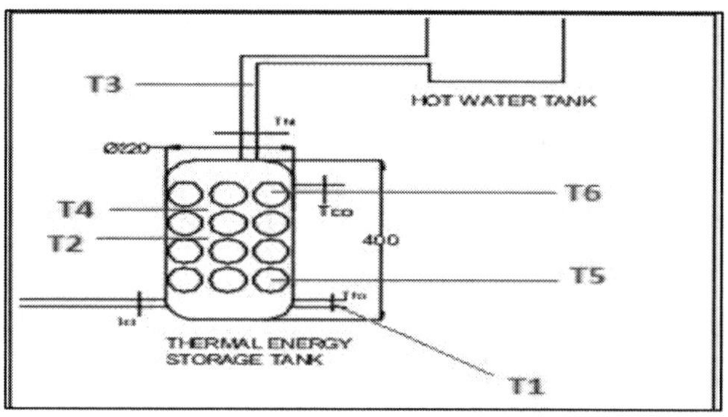


**Fig. 7.** Thermocouple connections

*8.1 Experimental Readings*

The Thermocouples are connected as shown in the Figure. The experiment is started by switching on the water heater and the pump. During the charging the temperature readings were noted down from the six channel temperatue indicator at a regular interval of 5 minutes. The charging process is continued till the temperature of the paraffin wax reaches to about 90 $^0$C. The readings are tabulated as shown in the table 1.

Discharging is done by allowing water at room temperature into thermal energy storage tank. Here the temperature readings are recorded for a regular interval of 3 minutes. The discharging process is continued till the temperature of the Paraffin wax reaches to room temperature. The readings are tabulated in the table 2.

**Table 1.** Temperature values observed during charging

| S. No | Time | $T_1$ | $T_2$ | $T_3$ | $T_4$ | $T_5$ | $T_6$ |
|---|---|---|---|---|---|---|---|
| | Min | °C | °C | °C | °C | °C | °C |
| 1 | 5 | 28 | 28 | 28 | 27 | 27 | 28 |
| 2 | 10 | 29 | 29 | 29 | 28 | 28 | 29 |
| 3 | 15 | 29 | 29 | 30 | 28 | 28 | 29 |
| 4 | 20 | 29 | 30 | 31 | 29 | 29 | 30 |
| 5 | 25 | 30 | 31 | 35 | 30 | 29 | 30 |
| 6 | 30 | 30 | 33 | 42 | 33 | 29 | 32 |
| 7 | 35 | 32 | 40 | 47 | 41 | 31 | 40 |
| 8 | 40 | 46 | 50 | 56 | 51 | 37 | 44 |
| 9 | 45 | 49 | 51 | 59 | 53 | 39 | 45 |
| 10 | 50 | 52 | 57 | 66 | 54 | 43 | 50 |
| 11 | 55 | 56 | 60 | 70 | 63 | 45 | 52 |
| 12 | 60 | 59 | 64 | 71 | 66 | 47 | 55 |
| 13 | 65 | 62 | 67 | 74 | 68 | 49 | 57 |
| 14 | 70 | 65 | 69 | 76 | 71 | 52 | 61 |
| 15 | 75 | 67 | 72 | 80 | 74 | 55 | 62 |
| 16 | 80 | 71 | 76 | 82 | 78 | 60 | 75 |
| 17 | 85 | 72 | 77 | 82 | 79 | 63 | 76 |
| 18 | 90 | 76 | 80 | 87 | 82 | 68 | 81 |
| 19 | 95 | 78 | 82 | 90 | 84 | 74 | 83 |
| 20 | 100 | 81 | 85 | 93 | 87 | 78 | 86 |
| 21 | 105 | 84 | 88 | 95 | 89 | 82 | 89 |
| 22 | 110 | 88 | 91 | 98 | 92 | 88 | 92 |
| 23 | 115 | 85 | 91 | 93 | 92 | 87 | 93 |
| 24 | 120 | 83 | 91 | 88 | 91 | 87 | 93 |
| 25 | 125 | 78 | 90 | 80 | 90 | 86 | 93 |
| 26 | 130 | 77 | 90 | 77 | 89 | 85 | 92 |

**Table 2.** Temperature values observed during Discharging

| S. No | Time | $T_1$ | $T_2$ | $T_3$ | $T_4$ | $T_5$ | $T_6$ |
|---|---|---|---|---|---|---|---|
| | Min | °C | °C | °C | °C | °C | °C |
| 1 | 0 | 77 | 90 | 77 | 89 | 85 | 92 |
| 2 | 3 | 36 | 81 | 90 | 89 | 69 | 91 |
| 3 | 6 | 30 | 37 | 88 | 88 | 62 | 84 |
| 4 | 9 | 30 | 33 | 77 | 78 | 51 | 77 |
| 5 | 12 | 30 | 31 | 68 | 60 | 50 | 62 |
| 6 | 15 | 30 | 31 | 40 | 39 | 47 | 59 |
| 7 | 18 | 29 | 30 | 39 | 36 | 46 | 58 |
| 8 | 21 | 29 | 30 | 37 | 33 | 43 | 57 |
| 9 | 24 | 29 | 29 | 35 | 32 | 41 | 56 |
| 10 | 27 | 29 | 29 | 33 | 31 | 40 | 56 |
| 11 | 30 | 29 | 29 | 32 | 31 | 39 | 55 |
| 12 | 33 | 29 | 29 | 31 | 31 | 38 | 54 |
| 13 | 36 | 29 | 29 | 30 | 30 | 36 | 54 |
| 14 | 39 | 29 | 29 | 30 | 30 | 36 | 53 |
| 15 | 42 | 29 | 29 | 30 | 30 | 34 | 52 |
| 16 | 45 | 29 | 29 | 30 | 30 | 34 | 51 |
| 17 | 48 | 29 | 29 | 30 | 30 | 32 | 48 |

*8.2 Graphs Obtained from Charging*

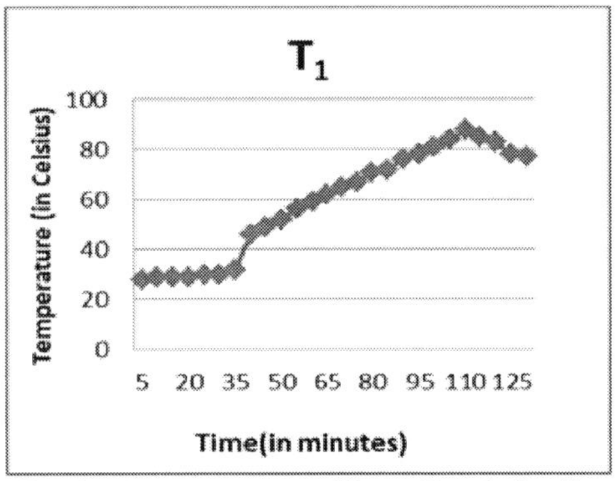


**Fig. 8.** Temperature vs Time Graph for $T_1$

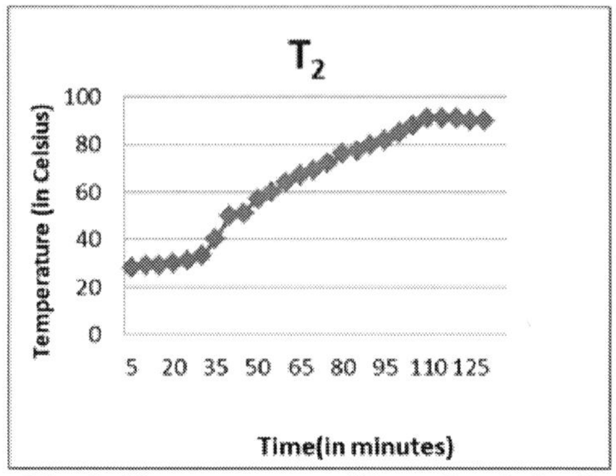


**Fig. 9.** Temperature vs Time Graph for $T_2$

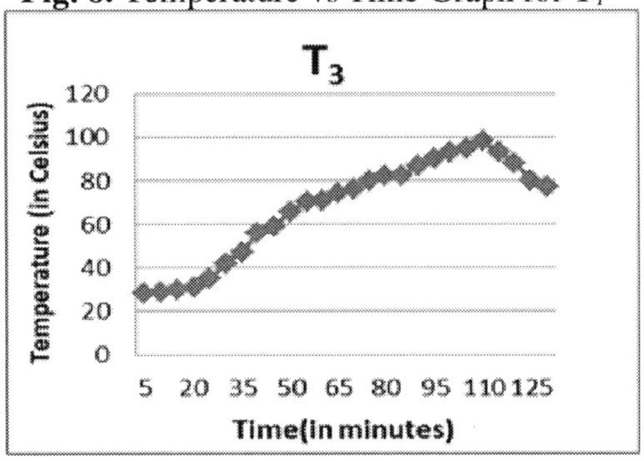


**Fig. 10.** Temperature vs Time Graph for $T_3$

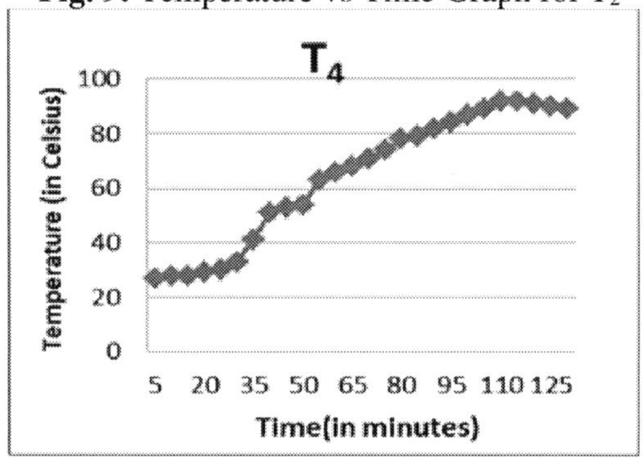


**Fig. 11.** Temperature vs Time Graph for $T_4$

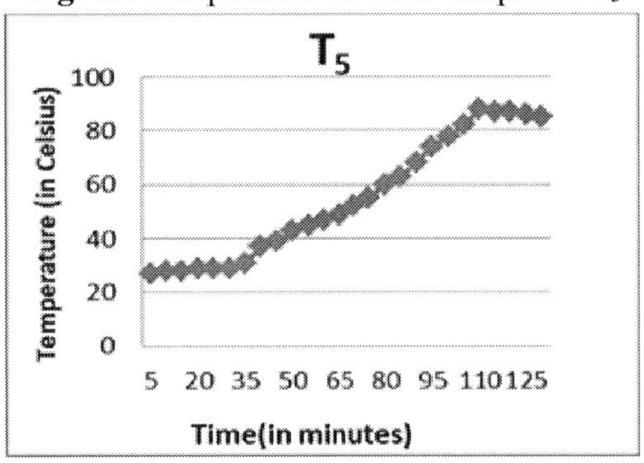


**Fig. 12.** Temperature vs Time Graph for $T_5$

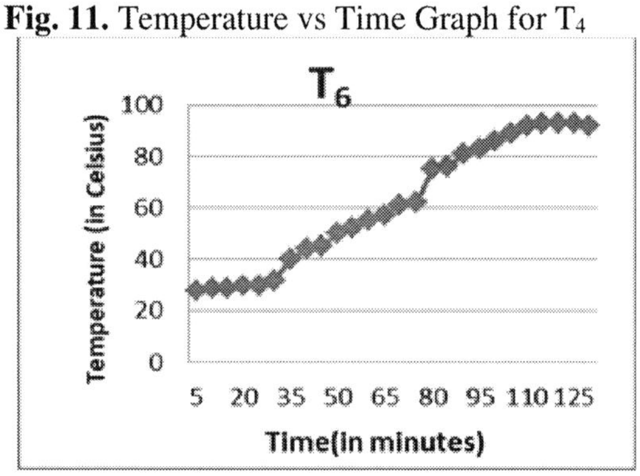


**Fig. 13.** Temperature vs Time Graph for $T_6$

*8.3 Graphs Obtained from Discharging*

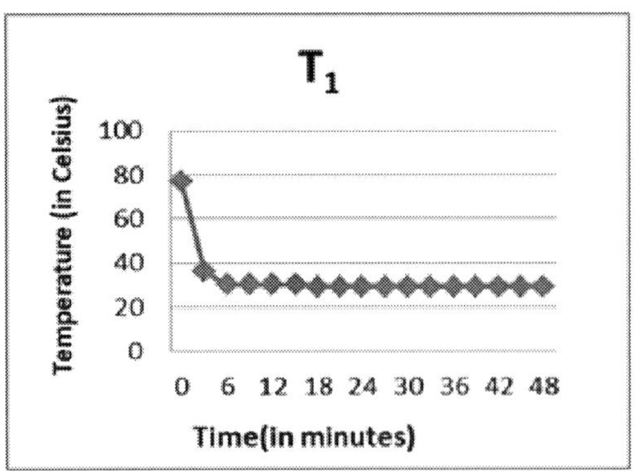


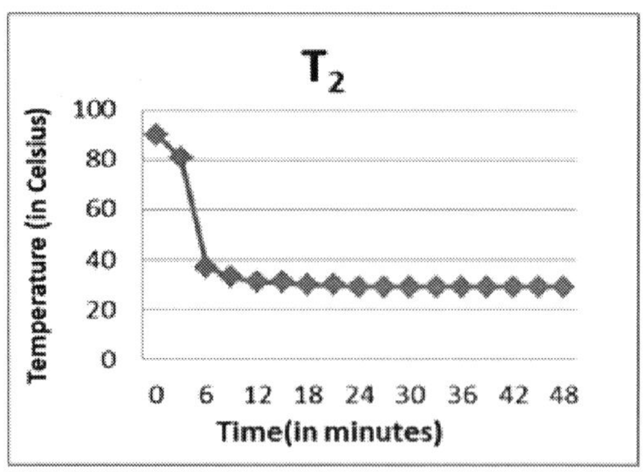

**Fig. 14.** Temperature vs Time Graph for $T_1$

**Fig. 15.** Temperature vs Time Graph for $T_2$

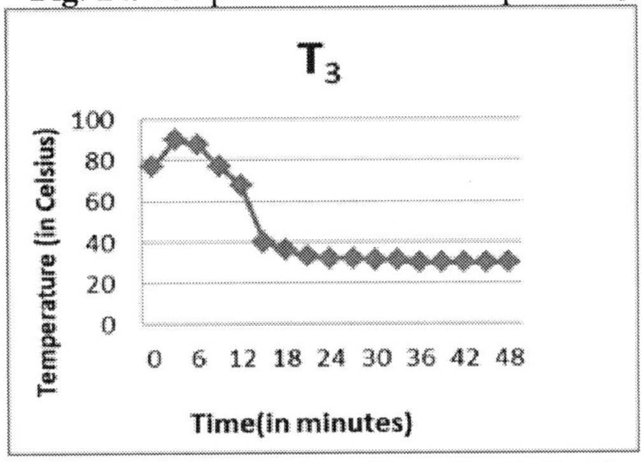


**Fig. 16.** Temperature vs Time Graph for $T_3$

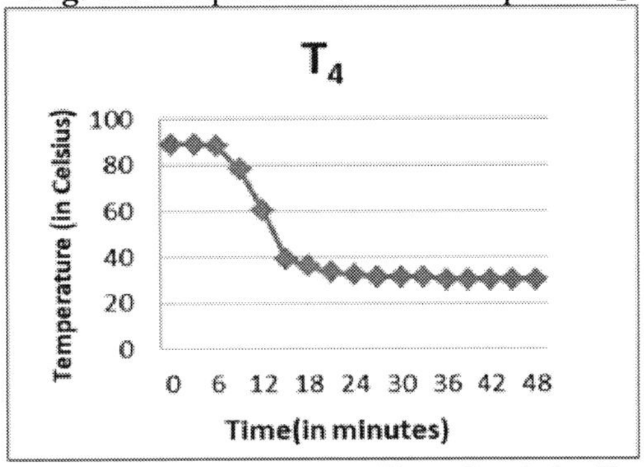


**Fig. 17.** Temperature vs Time Graph for $T_4$

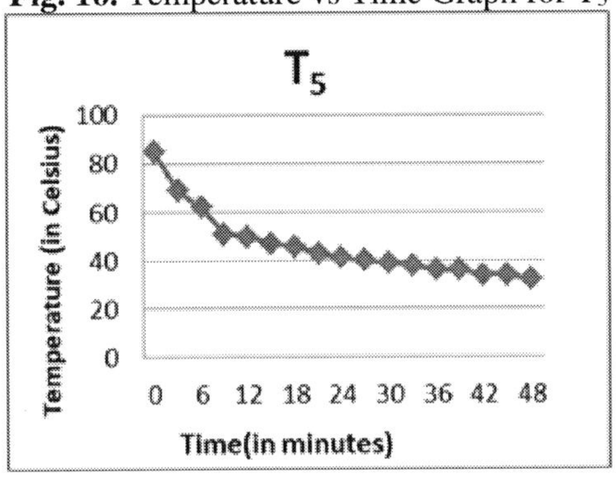


**Fig. 18.** Temperature vs Time Graph for $T_5$

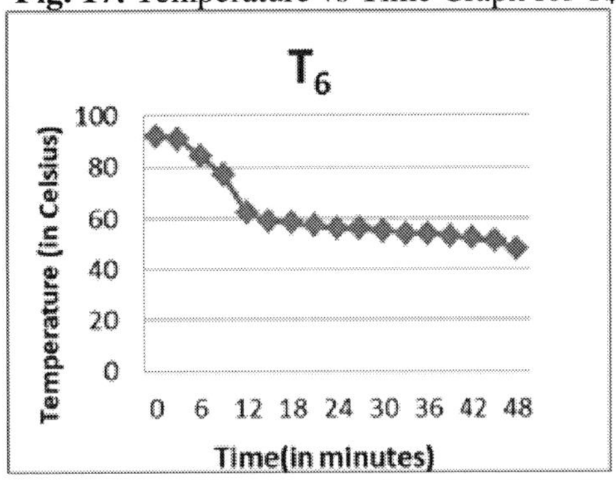


**Fig. 19.** Temperature vs Time Graph for $T_6$

## 9. Computational Details of The Present Study

*9.1 Numerical Model and Mesh Formation*

The first task in the process of simulation is the creation of a computational mesh to represent the flow domain of the geometry. The model of the diffuser representing the fluid domain has been modelled using the Ansys Fluent Geometry Modeller. In this chapter, the details of the fluid domain and the mesh generated for the diffuser models will be discussed.

*9.2 Model Geometry*

The Tank Model is created in the Fluent Geometry Modeller. The six layers of PCM are also given in the tank. Boolean operations are given for the required parts. The different parts of geometry are joined to form a new Part. Inlet and outlet pipes were also given in the tank.

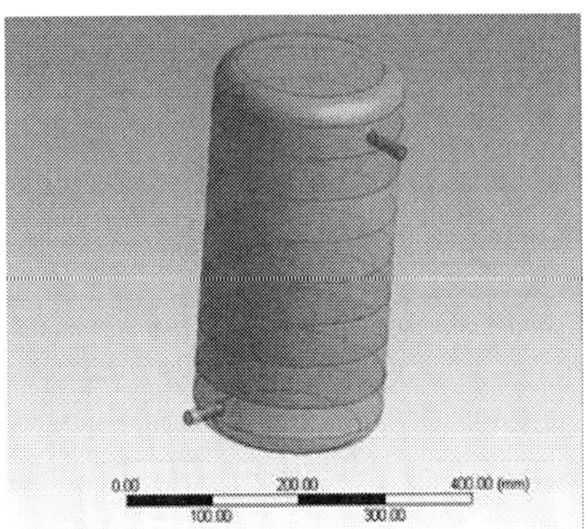

**Fig. 20.** Input Model of the Tank

**Fig. 21.** Meshed model

## 10. Results of Numerical Analysis

### *10.1 Charging*

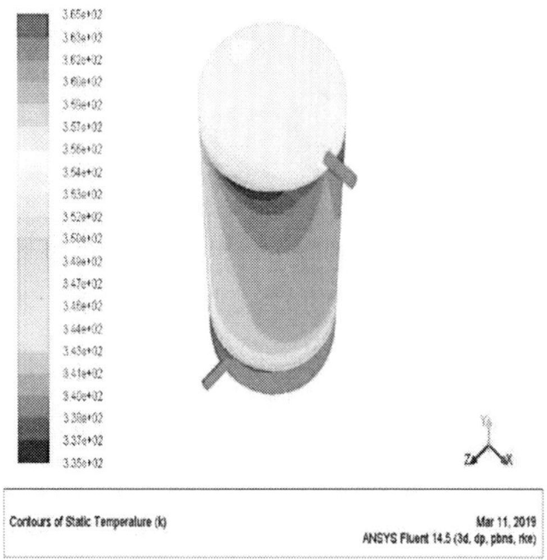


**Fig. 22.** Static temperature contours during charging

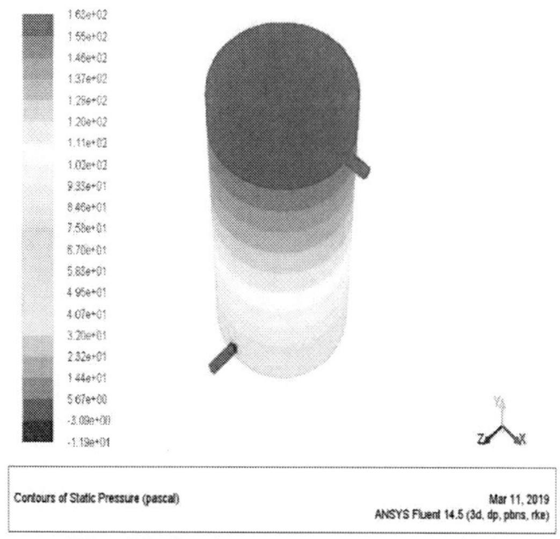


**Fig. 23.** Static pressure contours during discharging

### *10.2 Discharging*

From the Static temperature contours during discharging (Figure 9.2.1) it can be seen that cold water entering at the bottom of the tank gains heat from the charged PCM layers and leaves the TES tank as hot water. The water is discharged in reverse manner only for the purpose of experimentation.

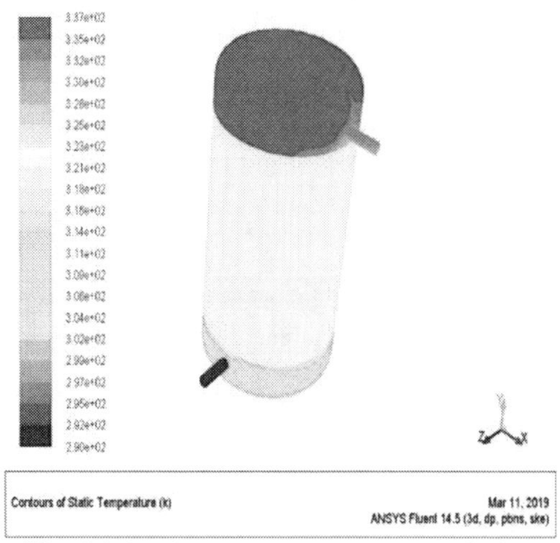


**Fig. 24.** Static temperature contours during discharging

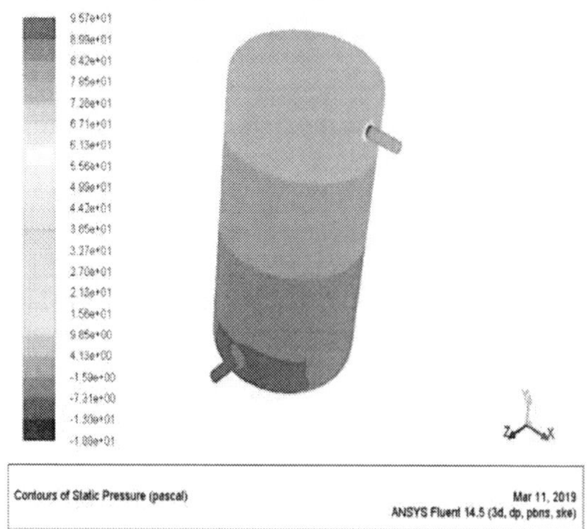


**Fig. 25.** Static pressure contours during Discharging

## Conclusion

In this project work experimentation is done on Thermal storage system using Paraffin wax as Phase Change Material and water as heat transfer fluid. This storage system helps for the simulation of continuous supply of heat energy when the source is intermittent such as solar water heating system. In this work a method to calculate the efficiency of storage system is arrived. By this project the overall efficiency of the experimental setup amounts to 62%. The efficiency can be increased by using different PCM with higher latent heat and better insulation materials such as glass wool. CFD analysis is also done by giving the properties of paraffin wax and temperature and pressure contours are obtained during both charging and discharging.

## REFERENCES

[1] N.Nallusamy, S.Sampath, R. Velraj,"Experimental investigation on a combined sensible heat and latent heat storage system integrated with constant/ varying solar heat sources. Renewable energy, April 2006.
[2] Atul Sharma, V.V. Tyagi, C.R.Chen, D.Buddhi. " Review of thermal energy storage with phase change materials and applications". Renewable and Sustainable energy Reviews, 2009
[3] G. Senthil Kumar, D Natarajan, L.A. Chidambaram, V. Kumeresan, Y.Ding, R.Velraj. "Role of PCM addition on stratification behavior in a thermal storage tank - An experimental study". Energy 2016. Vol-115, Pgs-1168-1178.
[4] A. Felix Regin, S.C.Solanki, J.S.Saini "An analysis of a packed bed latent heat thermal energy storage system using PCM capsules: Numerical Investigation". Renewable Energy, December 2009.
[5] Velraj R., et al.(1997), heat transfer enhancement in a latent heat storage system. Solar energy Vol.65, No.3, pp.171-180.
[6] K. Muninathan, R R Thirumaniraj, and M A Suryanarayanan, "Heat Transfer Modeling and Analysis Of Annular Diffuser with and without Double Tapered Struts", 2018 IEEE International Conference on System, Computation, Automation and Networking (ICSCA), Pondicherry, 2018, pp.1-5. doi: 10.1109/ICSCAN.2018. 8541227
[7] Arasu, A.V., Sasmito, A.P., Mujumdar, A.S. (2013). Numerical performance study of paraffin wax dispersed with alumina in a concentric pipe latent heat storage system. Thermal Science 17 (2) : 419-430. ScholarBank@NUS Repository.https://doi.org/10.2298/TSCI110417004A.
[8] K. Muninathan, R. Lakshminarayanan and Binesh S. Kumar. Design, Heat Transfer Modeling and Analysis Of Environmentally Benign Charcoal Kiln, IJEP 39 (2) : 173-177 (2019).

IOP Conf. Series: Materials Science and Engineering **923** (2020) 012067 doi:10.1088/1757-899X/923/1/012067

# Experimental Investigation of Eco-Friendly and Energy Efficient Auto and Allo Thermal Charcoal Kiln

**K. Muninathan [1*], M. Venkataramanan [2], G. Ashwin Prabhu[1], K. Vikneshpriya[3] , C. Daarthi[3],**

[1*],[1] Assistant Professor, St. Joseph's College of Engineering, OMR, Chennai -119
[2] Professor, Institute for Energy Studies, Anna University, Chennai – 600025
[3] Student, St. Joseph's College of Engineering, OMR, Chennai -119

*E-mail: muninathankota@gmail.com

**Abstract:** Like many countries in India, Charcoal production is the primary employment for rural masses in most of South Tamilnadu. It is a foremost source of income and environmental degradation in rural regions. The yield and transport of charcoal are a precarious process with energy and economic cycle. This process emits greenhouse gases to the atmosphere. It frequently made in traditional charcoal method that are harmful to the environment. It requires efficient man power for completing the process. The quality and yield for the conventional method is very low because of prolonged time. This method is mostly practised in rural areas. People over there will be more concern about quantity not the quality. So this reduces the yield of the charcoal. In this eco-friendly concept time consumption is reduced and hence yield will be more than the conventional method. In this paper, the indirect heating kiln has been tested with *prosopis juliflora* wood and the results were the lowest ash content of less than 2% and fixed carbon content of greater than 75%. The temperature was in range of 400-600° C inside the kiln which helps to understand the process of pyrolysis reactions.

***Keywords:*** Charcoal, Emissions, Proximate analysis, Ultimate analysis, Thermogravimetric analysis, Biomass

**Nomenclature:**

| | | |
|---|---|---|
| $T_f$ | – | Average flue gas temperature (°C) |
| $T_a$ | – | Ambient temperature (°C) |
| M | – | kg of moisture in 100kg of fuel |
| $H_2$ | – | kg of hydrogen present in fuel on 100 kg basis |
| $C_p$ | – | Specific heat in kCal/kg°C |
| V.M | – | kg of volatile matter present in fuel on 100 kg basis |
| $V_m$ | – | Wind velocity in m/s |
| $T_s$ | – | Surface temperature (K) |
| $T_a$ | – | Ambient temperature (K) |

## 1. Introduction

Due to increasing concerns about global warming, the development of accessible, clean and sustainable resources has arrived much attention. Effectiveness of forest management can be enhanced via energy utilization from biomass when obtaining bioelectricity and biofuels throughout direct combustion, thermochemical or bio-chemical processes, respectively. One of the appropriate biofuels is charcoal. Wood feedstocks which possess energy value of around 3EJ are transformed to charcoal which holds an energy value of 0.7-1 EJ (even in the cases of minimal efficiency). Its energy density is such that it can be transported long distances and, with many reduced emissions in cookstoves, it is a fuel that is better suited to urban use in developing countries than fuel-wood [10]. One of the oldest technology existing from the medieval times is the production of charcoal by using earth-mound kiln and its production involves the burning of raw material without the presence of air (oxygen). Using this ancient technology, charcoal production was probably with a pit kiln, positioned in the forest, close to the point of wood collection. This involves stacking the wood to be used inside a shallow level pit and its tightly sealed by surrounding the pit by straw, vegetation and earth. This wood was lit and the burning allowed to progress from one end of the pit to the other end, a process taking around 10-15 days [1]. Unpredictable fires and heat loss through radiation occurs during the above-mentioned time period required for charcoal production using an earth-mound kiln. As a consequence, the efficiency rating is lowered. An experimental method is illustrated in this paper for producing charcoal of high quality from long stick woody biomass. Within 3 to 5 hours of time taken for reaction, with respect to the moisture content of the woody biomass, the outcome is a yield of 38% experimentally.

For this study, to address the above-mentioned points, we have designed and constructed a kiln to determine the production and yield of charcoal and to obtain the temperature profiles inside the kiln [10]. During the carbonisation process, smoke is partly burned off, resulting in reduced air pollution of 75% [7]. The CO released due to burning of wood will be reused by circulating them back inside the walls of the kiln and it will be made to burn by supplying a fire inlet (Fire + Oxygen). By this method of re-circulating, the CO the amount of wood to be burned at the inlet can be reduced. Quality of charcoal depends on the factors such as the size of the material, type of kiln, and the carbonization process [6]. The amount of charcoal is only 8%-25% at a temperature of 500°C. The usual cycle will be like 7 days of heating and 5-7 days of cooling, ash content will be 4%-6% and the other tarry residues will constitute 30% of weight [5]. This method of charcoal production using kiln is expected to give an overall yield of 35%-40% and decreased ash content.

## 2. Allo Thermal Retort

The Allo Thermal Retort was designed and fabricated which as shown. The Allo Thermal Retort is positioned in the horizontal and hot gases is used for the carbonisation process. The four experiments were conducted and results are shown below. All heat required for pyrolysis is transferred through the walls of the reactor. The transfer of heat is relatively slow inside the biomass bed but large reactors cannot depend only on this heating method. For large reactors, heating through the external walls should be accompanied by the use of internal direct heating.

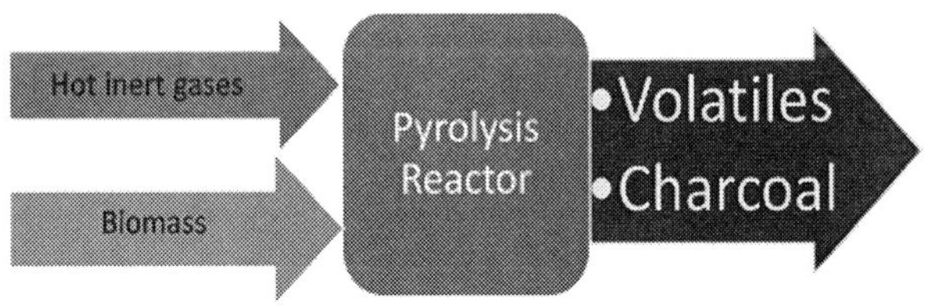


**Fig. 1.** Heating Through Reactor Wall

## 3. Modelling of Kiln

- In this the hot gas which is obtained by burning wastes in the fire chamber is allowed to pass inside the furnace through the inlet valve V1.
- Then the hot gas flows through the wood logs and goes out through the outlet chimney through valve V2.
- During this first phase of the process the valve V3 will be closed.
- In the second process the valve V1 and V2 are closed, now due to high pressure and temperature inside the furnace all the CO gas will be passed through the hollow space of the kiln.

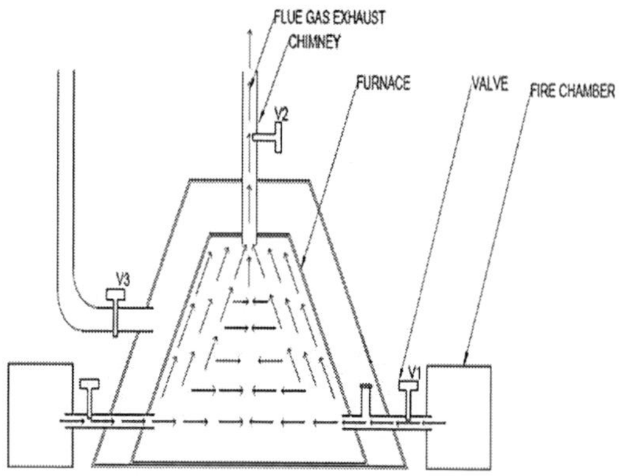


**Fig. 2.** Heating Wood with hot gase

- The CO gas will be re-circulated through the hollow space and a fire inlet will be given from the bottom side.
- So the outer surface of the furnace will be heated through burning of CO gas around it.

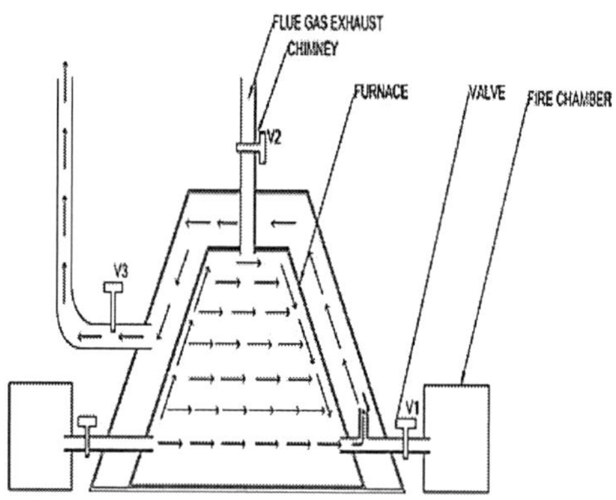


**Fig. 3.** Heating chamber with wood gase

## 4. Energy Balance

Quality of charcoal depends largely on its end-use. There are various opinions on how much-fixed carbon content charcoal must possess and the minimum amount of volatile matter that should be present. These opinions have to be taken into account before being regarded as good quality charcoal. Good commercial charcoal must have a volatile matter content(moisture-free)of less than 30%, the higher the volatile content, the lower will be the calorific value of charcoal and the optimum value of the volatile content depends on the end use. For example, charcoal for domestic use can have a minimum volatile matter of 20%-30% while charcoal for metallurgy must have a fixed volatile content of 85%-90%. Ash content of good quality charcoal must lie in between 0.5%-5% and it should have a heat value of about 28-33MJ/kg.

### *4.1 Heat loss due to evaporation of moisture present in fuel*

$$Q_m = M * \{ 584 + C_p * ( T_f - T_a ) \}$$
$$= 11.6 * \{ 584 + 1 * ( 100-30) \}$$
$$= 7\,645 \text{ kcal / 100kg of wood}$$

Latent heat corresponding to the partial pressure of water vapour is 584

IOP Conf. Series: Materials Science and Engineering **923** (2020) 012067 doi:10.1088/1757-899X/923/1/012067

### *4.2 Heat loss due to formation of water from $H_2$ in fuel*

$Q_H = 9 * H_2 * \{584 + Cp ( T_f - T_a ) \}$

$= 9 * 0.0529 * \{584 + 1 (100\text{-}30)\}$

= 314 kcal / 100 kg of wood

### *4.3 Heat loss due to volatile matter in fuel*

$Q_{VM} = V.M * C_p * ( T_f - T_a ) \}$

= 55.4 * 0.45 * ( 350 - 30 )

= 7977.6 kcal / 100kg of wood

### *4.4 Heat loss due to radiation and convection*

$Q_{RC} = 0.548 \text{ x } [ (Ts / 55.55)^4 - (Ta / 55.55)^4] + 1.957 \text{ x } (Ts\text{- } Ta)^{1.25}$ xsq.rt of [(196.85 Vm + 68.9) /68.9]

$= 0.548 \text{ x } [ (50 / 55.55)^4 - (30 / 55.55)^4] + 1.957 \text{ x } (50\text{- }30)^{1.25}$ x sq.rt of [(196.85*5 + 68.9) / 68.9]

= 399.9 kcal / $m^2$

= 1199.7 kcal / 100kg of wood

### *4.5 Heat required to heat wood*

$Q_R = Ms * Cp * ( T_f - T_a )$

= 33 *0.45 * (400-30)

= 5494.5 kcal / 100 kg of wood

### *4.6 Heat loss in flue gas*

$Q_f = Ms * Cp * ( T_f - T_a )$

= 3542 kcal / 100kg of wood

### *4.7 Total heat loss*

Total Heat Loss

$Q_T = Q_m + Q_h + Q_{vm} + Q_{rc} + Q_r + Q_f$

= 26 173 kcal/ 100kg of wood

CV = 3000 kcal / kg of wood

Efficiency of HAG = 60%

Wood required = 26 173 / (3000 X 0.6)

= 14.55 kg / 100 kg of wood

### *4.8 Auto thermal method ( Direct heating )*

Moisture = 11.6 kg

Volatile matter =55.4 kg

Wood required for heating =14.55 kg

Yield of Charcoal = 100 - 11.6 - 55.4 - 14.55 = 18.5 %

### *4.9 Allo thermal process (Indirect heating)*

Moisture = 11.6 kg
Volatile matter = 55.4 kg
Wood required for heating = Nil (Burning waste only )
Yield of Charcoal = 100 - 11.6 - 55.4
= 33 %

## 5. Results and Discussion

### *5.1. Charcoal Yield*

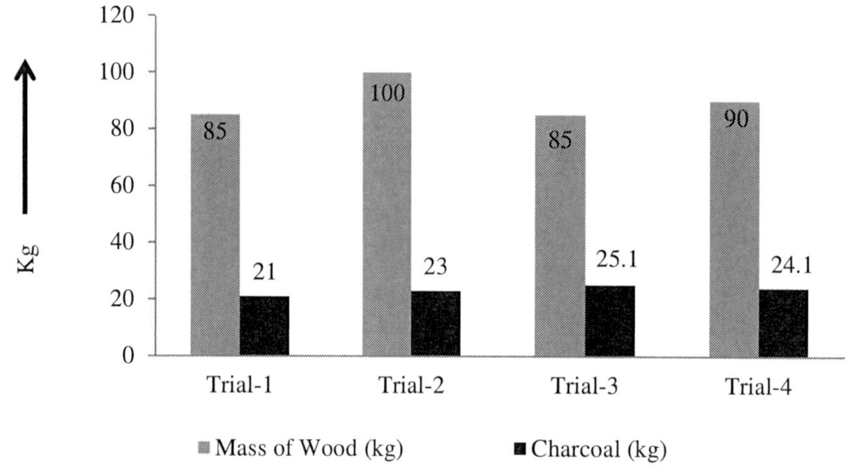


**Fig. 4.** The Allo Thermal Retort Charcoal Yield

As we observed in the trial 1, the temperature of the kiln was exceeds above 500 °C and the maximum volatile were released which results as low yield charcoal and the un-burnt quantity also minimum. In the trail 2 , the maximum wood was loaded which results much un-burnt and lower yield.

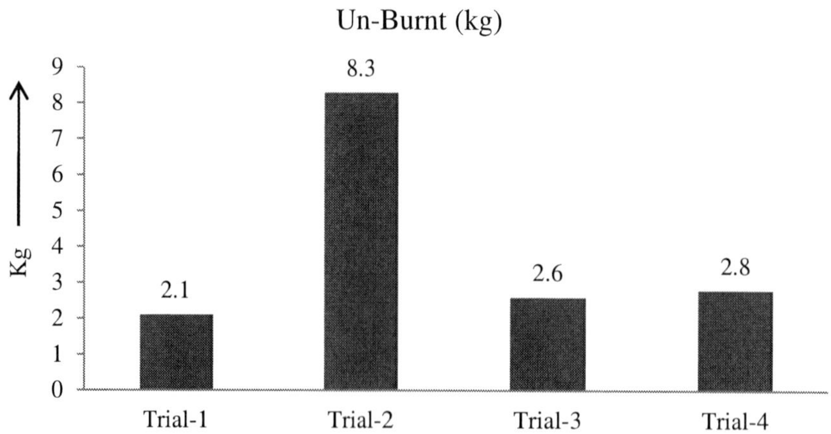


**Fig. 5.** Un-Burnt of Allo Thermal Retort Trials

Even the temperature of the trial 2 was attain 520 °C but in the trial 2 & 3 was well controlled inside kiln temperature by the heating rate.

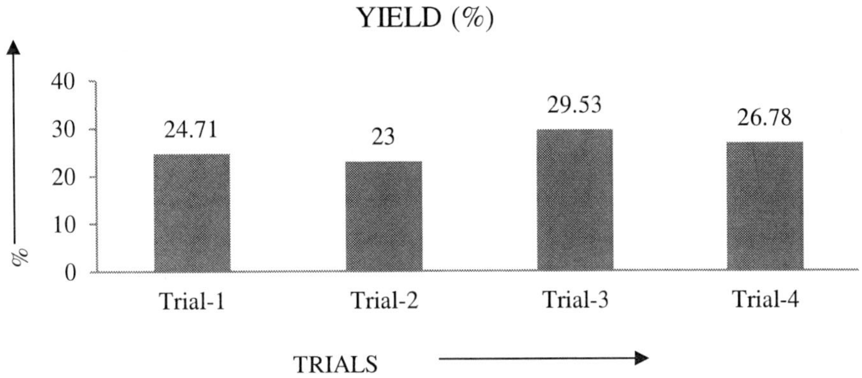


**Fig. 6.** Percentage Yield of Allo Thermal Retort

The temperature of the kiln monitored every 15mins as well as both wood chamber and chimney flue gas temperature. Even though trial 2 & 3 was used high moisture content, its only given high yield when compared to other trials.

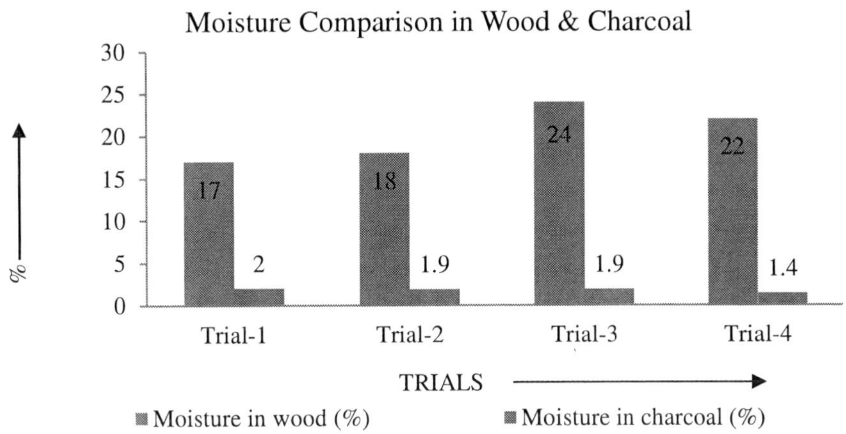


**Fig. 7.** Moisture Content of Wood Used in the Allo Thermal Retort

*5.2 Temperature distribution in the wood chamber*

The peak temperature is reached during the carbonization process. This temperature largely has an impact on the quality (i.e., the volatile matter content) and other properties of the charcoal product.

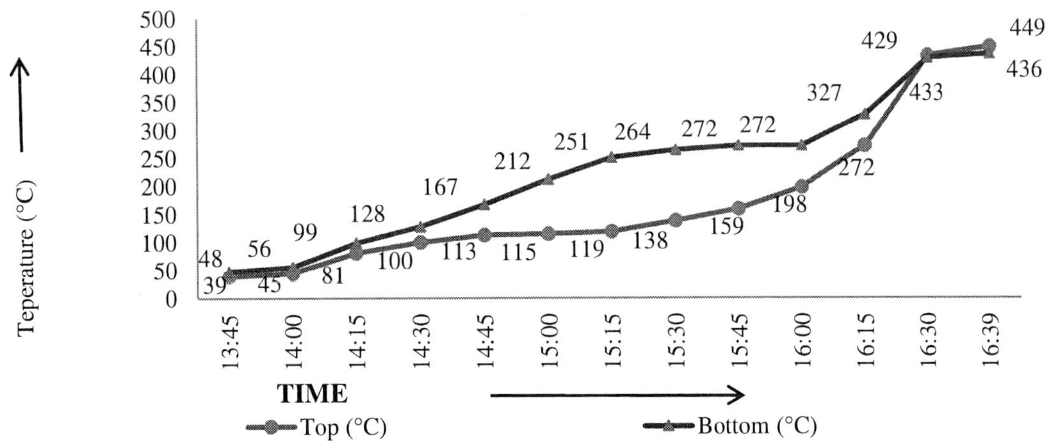


**Fig. 8.** Temperature Distribution in the Wood Chamber

Loosely bound water is evaporated from the wood within the temperature range of 100 to 180° Celsius, Pyrolysis oil releases gases in the temperature range 180-270°C containing CO, CO2 and condensable vapours. An Exothermic reaction starts above 270° and 288° C and it can be detected by the increasing temperature of CO and CO2 gases along with a spontaneous generation of heat. The mass of condensable vapours will intensify. Cellulose loses weight below 220° C via formation of water. But at a slightly elevated temperature than that of the former (i.e) at 250°C, CO2 and CO are also produced.

*5.3 Temperature distribution in the chimney flue gas*

The carbonization process upon entering the exothermic phase requires no more external heating and this condition is sufficient to reach a temperature of 500° C. By the above continuous process, the gas-vapour mixture which is consistent, is let out. The charring unit is kept in the high temperature mentioned in the former for a fairly

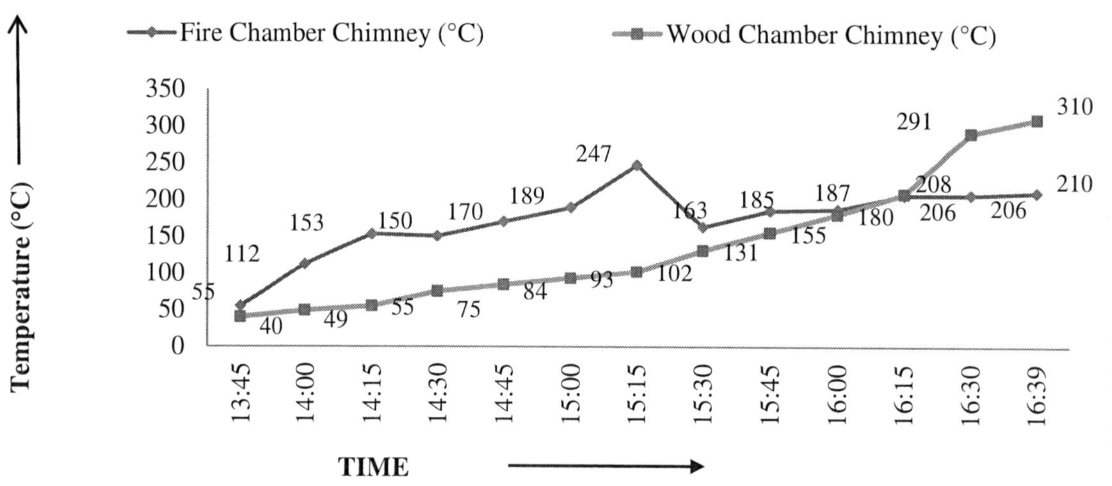


**Fig. 9.** Temperature Distribution in the Chimney flue gas

short time and the ceaseless operation rely on the residence time of the charring unit. A Time Vs Temperature graph is drawn for the wood charring process. The maximum pyrolysis temperature in the bottom, middle and top zone are found to be 450 ° C,420 °C,430 °C respectively. Thus the pyrolysis temperature is not uniform in all zones due to unsteady state of burning of wood in these zones in the kiln. This method resulted in the highest yield of charcoal(30%) when compared to the methods used in other trials

*5.4 Influence of carbonization temperature on yield and composition of charcoal*

**Table 1.** Effect of Carbonisation Temperature on Yield and Composition of Charcoal

| Trials | Temperature | Chemical analysis of Charcoal | | yield |
|---|---|---|---|---|
| | °C | % of Fixed Carbon | % Volatile | Yield (%) |
| 2 | 520 | 72 | 26 | 23 |
| 1 | 500 | 70 | 27 | 24.7 |
| 4 | 450 | 66 | 29.5 | 26.8 |
| 3 | 400 | 64 | 32 | 29.5 |

IOP Conf. Series: Materials Science and Engineering **923** (2020) 012067 doi:10.1088/1757-899X/923/1/012067

Higher yield of charcoal is observed in low carbonization temperatures, but low-grade charcoal is obtained. Acidic tars are present in low-grade charcoal and its corrosive which leads to burning with smoke production. A relatively high temperature of 500°C is the carbonizing temperature of good commercial charcoal which possess a standard carbon content of about 75%.

### *5.5 Proximate analysis of charcoal sample*

**Table 2.** Proximate Analysis of Charcoal Samples in ATR

| Composition | Trial 1 | Trial 2 | Trial 3 | Trial 4 |
|---|---|---|---|---|
| **Moisture (%)** | 0.99 | 0.5 | 1.5 | 1.5 |
| **Volatile Matter (%)** | 27 | 26 | 32 | 29.5 |
| **Ash (%)** | 2.01 | 1.5 | 2.5 | 3 |
| **Fixed Carbon (%)** | 70 | 72 | 64 | 66 |
| **CV Kcal/kg)** | 8160 | 8354 | 7856 | 7873 |

### *5.6 Ultimate analysis of charcoal sample*

**Table 3.** Ultimate Analysis of Charcoal Samples in ATR

| Composition | Trial 1 | Trial 2 | Trial 3 | Trial 4 |
|---|---|---|---|---|
| **C (%)** | 86.1 | 87.6 | 83.4 | 83.6 |
| **H (%)** | 4.8 | 4.8 | 5.0 | 4.9 |
| **N2 (%)** | 1.6 | 1.6 | 1.5 | 1.5 |
| **Mineral Matter(%)** | 2.21 | 1.65 | 2.75 | 3.3 |
| **Moisture(%)** | 0.99 | 0.5 | 1.5 | 1.5 |
| **Oxygen (%)** | 4.3 | 3.8 | 5.8 | 5.2 |

## 6. Comparison of charcoal sample analysis with conventional kilns

### *6.1. Comparision proximate analysis of charcoal samples*

**Table 4.** Comparison Proximate Analysis of Charcoal

| Composition | Conventional | Igloo | Allothermal Retort |
|---|---|---|---|
| Moisture (%) | 5.07 | 2.8 | 0.5 |
| Volatile Matter (%) | 30.56 | 22.8 | 26 |
| Ash (%) | 4.9 | 3.2 | 1.5 |
| Fixed Carbon (%) | 59.47 | 71.2 | 72 |
| CV(Kcal/kg) | 6810 | 7725 | 8354 |

### *6.2. Comparision ultimate analysis of charcoal samples*

**Table 5.** Comparison of Ultimate Analysis of Charcoal Samples

| Composition | Conventional | Igloo | Allothermal Retort |
|---|---|---|---|
| C (%) | 75.9 | 83.2 | 87.6 |
| H (%) | 4.6 | 4.5 | 4.8 |
| N2 (%) | 1.5 | 1.6 | 1.6 |
| Mineral Matter (%) | 5.39 | 3.52 | 1.65 |
| Moisture (%) | 5.07 | 2.80 | 0.5 |
| Oxygen (%) | 7.46 | 4.37 | 3.8 |

## 7. Advantages of allo thermal retort

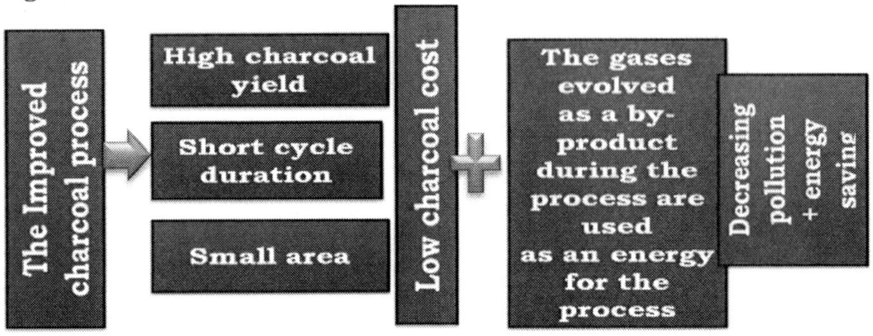


**Fig. 10.** Advantages of Allow Thermal Retort

## CONCLUSION

The traditional earth mound kiln is used to make charcoal and this technique persists in Southern Province of Tamilnadu mainly because it is flexible to use and cheap.

However, this traditional method results in a meagre yield (typically 15-20%). Inconsistency in the quality of charcoal is observed because it is unfeasible to maintain uniform carbonization. Environment pollution is also on the rise due to the emission of tar and poisonous gases.

- The processing time is quick. Duration of the cycle is about 5 hours along with an additional duration of 5-6 hours for cooling. The overall time consumption is less when compared to the other kilns.
- Allo thermal Retort imparts flexible design due to the absence of mechanical or electrical components. Therefore, customised kilns can be locally manufactured in mass production which is not possible with the other kilns.
- Allo Thermal Retort requires minimal supervisions of its working conditions since flame avoids contact with the chamber and it avoids the burning of wood ( Heat rate 14000 - 15000 kcal/hours)
- Allo Thermal Retort has a high charcoal yield efficiency (30-35%) depending upon the moisture of wood whereas the efficiency of other kilns is (20-25%)

## REFERENCES:

[1] Anders Rooma Nielson, Rasmus Wochnik Aniol, Morten Boberg Larsen, Peter Glarborg, Kim Dam-Johnsen. Mixing large and small particles in a pilot scale rotary kiln. Powder Technology 210 (2011) 273-280.
[2] Hanne K. Sjolie. Reducing greenhouse gas emission from households and industry by the use of charcoal from sawmill residues in Tanzania. Journal of cleaner production 27 (2012) 109-117.
[3] Stefan Kern, Michael Halwachs, Gerhard Kampichler, Christoph Pfeifer, Tobias Proll, Hermann Hofbauer. Rotary kiln pyrolysis of straw and fermentation residues in a 3 MW pilot plant – Influence of Pyrolysis product performance. Journal of Analytical and Applied Pyrolysis 97 (2012) 1-10.
[4] Sheau-Horng Lin, Ling-Ying Hsu, Chuen-Shii Chou, Jhih-wei Jhang, Ping Wu. Carbonization Process of Moso bamboo (Phyllostachys pubescene) charcoal and its governing thermodynamics. Journal of Analytical and Applied Pyrolysis xxx (2014) xxx-xxx.
[5] Jeng-Chyan Muti Lin. Development of a high yield and low cycle time biomass char production system. Fuel Processin Technology 87 (2006) 487-495.
[6] K. Muninathan, R R Thirumaniraj, and M A Suryanarayanan, "Heat Transfer Modeling and Analysis Of Annular Diffuser with and without Double Tapered Struts", 2018 IEEE International

Conference on System, Computation, Automation and Networking (ICSCA), Pondicherry, 2018, pp.1-5. doi: 10.1109/ICSCAN.2018. 8541227.

[7] Veronica Bustamante-García, Artemio Carrillo-Parra, Humberto González-Rodríguez, Roque Gonzalo Ramírez-Lozano, José Javier Corral-Rivas, Fortunato Garza-Ocanas. Evaluation of a charcoal production process from forest residues of Quercus sideroxyla Humb., & Bonpl. in a Brazilian beehive kiln. Industrial crops and Products 42 (2013) 169-174.

[8] Thomas Griessacher, Jurgen Antrekowitsch, Stefan Steinlechner. Charcoal from agricultural residues as alternative reducing agent in metal recycling. BIOMASS AND BIOENERGY 39 (2012) 139-146.

[9] K. Muninathan, R. Lakshminarayanan and Binesh S. Kumar. Design, Heat Transfer Modeling and Analysis Of Environmentally Benign Charcoal Kiln, IJEP 39 (2) : 173-177 (2019).

[10] Y.Schenkel, P.Bertaux, S.Vanwijnbserghe and J.Carre. An Evaluation of the mould kiln carbonization technique. PII: S0961-9534(97)10033-2.

[11] A. Saravanakumar, T.M. Haridasan, and R. Kasturi Bai. Technical and feasibility study of conversion of long-stick wood to charcoal in a partial combustion metal kiln.

IOP Conf. Series: Materials Science and Engineering **923** (2020) 012068 doi:10.1088/1757-899X/923/1/012068

# Performance and Emission Evaluation of Diesel Engine Fueled with Karanja Methyl Ester

**K Rajesh, P K Devan, R Saran, S Vasanthakumar, K Vignesh**

Department of Mechanical Engineering, RMK College of Engineering and Technology, Thiruvallur - 601206, Tamil Nadu, India.

*E-mail: rajeshmechh@gmail.com

**Abstract**. Petroleum based fuels are the main source of energy for the transportation vehicles. It is consumed in a very faster rate due to continuous use. Major drawback of fossil fuel usage is exhaust gas emission and it is non-renewable. The search for alternate energy source has been increasing; one of the better alternatives is biodiesel. The biodiesel can be effectively used as an alternate fuel instead of diesel as its properties is very close to diesel. The biodiesel extracted from Karanja oil is tested in diesel engine as B100 and B20 blend and its results were compared with diesel. It was found that the brake thermal efficiency is low for biodiesel and its blend compared to diesel. The specific fuel consumption increases for B100 and B20 than diesel due to its reduced heating value. Carbon monoxide and hydrocarbon emission is reduced considerably for B100 compared to B20 and neat diesel. At the same time the NOx emission is also observed to be decreasing for B100 and the NOx of B20 is closer to diesel. From the results obtained the Karanja oil biodiesel can be directly used in direct injection diesel engines and can be a better alternate fuel for the conventional one.

***Keywords: Karanja oil, methyl ester, biodiesel, transesterification, performance, emission***

## 1. Introduction

Due to increasing population and transport requirement there is huge demand for petroleum-based fuels. Many works have been carried out to find a better alternative to the existing problem. There are many alternative energy sources like LPG, alcohol, natural gas, hydrogen, biodiesel, etc. Among these biodiesel stands as a best solution for replacing diesel fuel. The properties of the biodiesel are comparable with diesel. Major advantage of biodiesel is renewable, easily available, production of biodiesel won't affect environment like diesel, and biodiesel can be extracted from any type of vegetable oil, lower emission rates compared to diesel [1]. All over the world many researches are going on biodiesel aiming to replace the fossil fuel to reduce the dependency on the fossil fuels. Many non-edible oils are available for the production of biodiesel fuel. Some of the oils are Neem Oil, Mahua oil, cotton seed, Karanja oil, Jatropha oil, Castor oil, Rape seed oil, linseed oil, punnai oil, Acid oil, jojoba oil, rice bran oil, rubber seed oil, grape seed oil etc.

Even though biodiesel has comparable properties to be used in engines there is significant difference in properties like viscosity of the oil which lead to poor spray of fuel during injection and incomplete combustion resulting in unwanted emissions. To overcome the issues related to viscosity methods like transesterification, pyrolysis, micro emulsion, de-polymerization is used to produce biodiesel. Generally, transesterification process method is used to produce biodiesel. Based on the free acid content of the oil, esterification and transesterification or direct transesterification is carried out to extract biodiesel. Use of biodiesel extracted from vegetable oils such as karanja, jatropha, neeem, Karanja, rapeseed, soybean as fuel resulted in lower efficiency and higher smoke for diesel.

Karanja (Pongamia Pinnata) and jatropha are abundantly available in India and researchers study the performance characteristics of karanja oil in diesel engines. Moreover, Karanja oil is non edible and finds no food versus scarcity problem. Many literatures indicate the use of biodiesel from karanja oil and its effective use in direct injection diesel engines. Yehliu et al. Yehliu et al. reported that the BSFC of biodiesel is increased by 15% compared to diesel which is due to 15% lower calorific value and observed in some cases increase in particulate matter emission due to unburned hydrocarbon emissions [2]. Godhiganur et al. discussed the effect of mahua biodiesel and its blends in diesel engine. It is found that for maximum blend of mahua biodiesel resulted in reduced emissions of hydrocarbon and Carbon monoxide emissions but nitrous oxide emission and BSFC were higher [3].

Muralidharan et al. recorded the performance of waste cooking oil with diesel. They observed that brake thermal efficiency of B40 blend is marginally higher than diesel. Also, biodiesel and its blends resulted in higher combustion pressure and lower HRR in comparison with diesel [4]. The observation of other researchers indicates that use of biodiesel as fuel in diesel engines results in increased BSFC due to lower calorific value and NOx emission due to higher oxygen content and better combustion. [5-8]. In this present work karanja oil biodiesel is used for running the single cylinder diesel engine and to study the performance and emissions by running the fuel.

## 2. Materials and Methodology

### *2.1 Materials and experimental design*

Karanja oil was purchased from a dealer shop, 99.9% pure analytical grade methanol, Sulphuric acid and potassium hydroxide in lead shot form of above 85% pure were used in the production of biodiesel. The apparatus required for the experiment are half a litre spherical glass beaker, a speed controlled mechanical stirrer, a temperature-controlled heater and a digital thermometer.

### *2.2 Transesterification process*

Karanja biodiesel is extracted by two stage process. The first stage is esterification to produce ester and the next stage is alkaline transesterification process to produce biodiesel. The conversion process is done to reduce the viscosity of the oil. If the viscosity is higher than it may lead to poor atomization and improper combustion [9]. If the FFA content of the oil is minimal of 2.5%, then one step transesterification process with a base catalyst should be done and if it exceeds 2.5%, then esterification and transesterification processes should be preferred. In this study as the FFA content of Karanja oil was 5.45%, thus two stage transesterification method is employed.

The FFA content in oil is decided by titration test. For this test, oil sample of 2-5g is taken in a conical flask and 25ml of ethanol is added to the oil and 4-5 drops of phenolphthalein indicator was added to the flask. It was heated for about 4-5min and then allowed to cool down. Following it is titrated against 0.1N of KOH solution. The KOH solution is added until permanent pink colour appears. The FFA content is calculated based on the burette solution during titration. In Karanja oil, the FFA content is around 5.45%. Thus, two step transesterification process is carried out. In acid transesterification process methanol to oil ratio of 8:1 and 2% of concentrated sulphuric acid to the volume of oil is taken. The esterification process carried on for a period of 90 minutes at a temperature of 60°C. After the process the FFA is checked to move on with the alkaline esterification.

In alkaline transesterification process, oil is pre-heated and the solution is mixed by stirrer for constant mixing 500rpm is maintained. By dissolving KOH pellets in methanol (molar ratio 8:1) the methoxide solution is prepared. When the oil reaches certain temperature, in the reactor the prepared methoxide solution is poured slowly. The reaction was maintained at 60°C with a time period of 60 minutes. The by-product glycerine is removed and then to remove the soap content from the biodiesel, it is washed with distilled water.

## 3. Experimental setup and Measurements

The experiment was performed in Kirloskar AV1 make four stroke water cooled diesel engine. The rated power is 5.20 kW at a constant speed of 1500 rpm. Engine is connected to the swinging field electrical generator with the load bank. The engine experimental setup is shown in figure 1 and specification is given in the table 1. AVL-444 gas analyser is used to measure the amount HC, CO and NOx emissions. The pressure at in-cylinder is measured using Piezoelectric pressure sensor at every crank angle using a charge amplifier transducer. Variables such as engine speed, fuel flow, and emission characteristics were acquired from the data acquisition system connected with the engine. Specific fuel consumption, Brake thermal efficiency, and brake power were evaluated from study of performance of the engine.

**Table 1.** Engine Specifications

| Engine Parameters | Specifications |
|---|---|
| Make | Kirloskar AV 1, Water cooled, Four Stroke |
| Number of cylinders | Single |
| Bore | 87.5 mm |
| Compression ratio | 17.5:1 |
| Maximum power | 5.20 kW |
| Speed | 1500 rpm |
| Dynamometer | Electrical |
| Injection pressure | 200 bar |

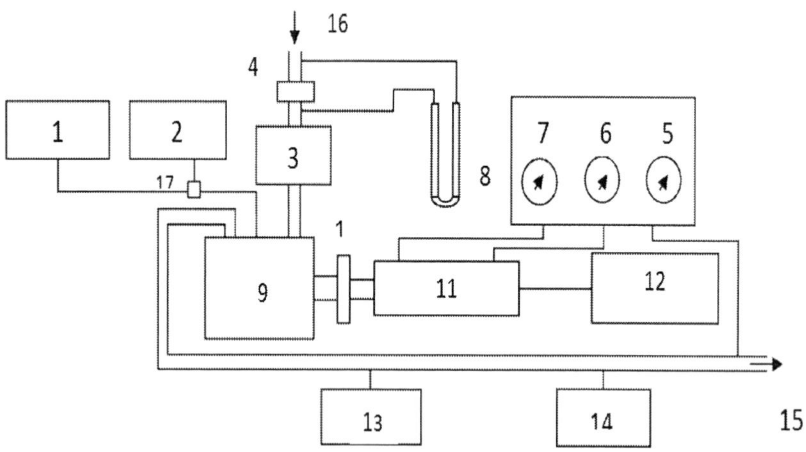


1.Diesel Tank 2.Alternate Fuel Tank 3.Air Tank 4.Orifice 5.Exhaust Gas Temperature 6.Voltmeter 7.Ammeter 8.U-Tube Manometer 9.Engine 10.Coupling 11.Electric Generator 12.Electrical Loading 13.AVL Smoke Meter 14.AVL Exhaust Gas Analyzer 15.Exhaust Gas 16.Air Intake 17.Two way valve.

Figure 1: Experimental Setup of test engine

## 4. Result and Discussions

### *4.1 Properties of Karanja oil*

Karanja oil property has been studied before carrying out the esterification and transesterification process for biodiesel production. Its physicochemical properties have been studied to find the suitability as feed stock for biodiesel production. Table 3 depicts the physicochemical properties of Karanja oil. The properties like density, free fatty acid content (FFA), kinematic viscosity, Acid value, Iodine value, molecular weight was studied.

**Table 2.** Physicochemical properties of karanja oil

| PARAMETERS | VALUES |
|---|---|
| Density at 15 °C (Kg/m$^3$) | 930 |
| Kinematic viscosity at 40 °C (mm$^2$/s) | 12.5 |
| Free fatty acid (% FFA as oleic acid) | 1.5% |
| Acid value (mg KOH/g) | 5.06 |
| Iodine value (g Iodine/100 g) | 86.5 |
| Molecular weight (g/mol) | 876.16 |

**Table 3.** Properties of karanja oil methyl ester

| PROPERTIES | DIESEL | BIODIESEL |
|---|---|---|
| Density(15 °C) (Kg/m$^3$) | 833 | 886 |
| Kinematic viscosity(40 °C) in cST | 2.67 | 4.3 |
| Flash point(°C) | 65 | 157 |
| Calorific Value (MJ/kg) | 43.06 | 41.07 |

### *4.2 Performance Characteristics*

#### *4.2.1 Specific fuel consumption*

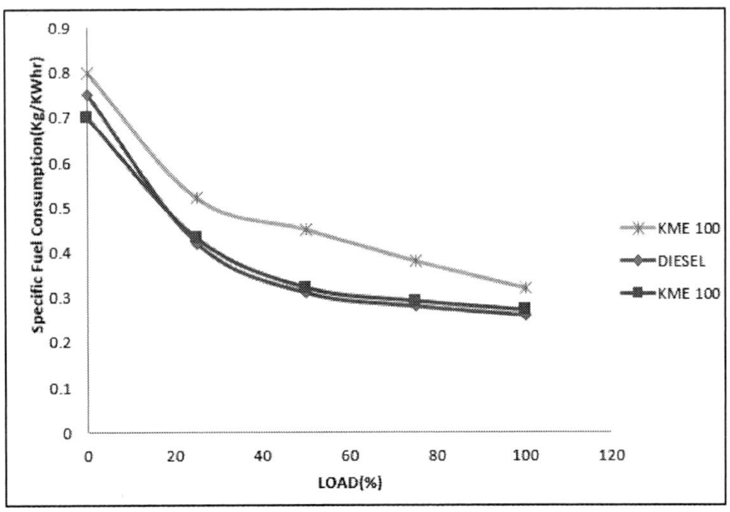


Figure 2. Specific fuel consumption against load

The specific fuel is defined as the ratio of fuel consumed to the brake power. consumption of diesel, biodiesel and its blend are shown in the figure 2. The Karanja oil biodiesel has higher specific fuel consumption compared to diesel for all load conditions. Karanja biodiesel has low calorific value compared to diesel, hence more amount of fuel was consumed to produce same power output [10]. It is observed that specific fuel consumption increases by 19% and 12.85% for biodiesel and B20 blend compared to that of diesel at full load.

#### *4.2.2 Brake Thermal Efficiency*

The brake thermal efficiency of diesel, biodiesel and its blend are shown in the figure 3. The Karanja oil biodiesel has lower BTE compared to that of diesel. The drop in brake thermal efficiency for biodiesel and diesel blends is due to reduced calorific value and higher viscosity [10]. The higher viscosity of the fuel affects the atomization and vaporization of biodiesel. It is observed that BTE decreases by 16% and 9% for biodiesel and B20 blend compared to that of diesel at full load.

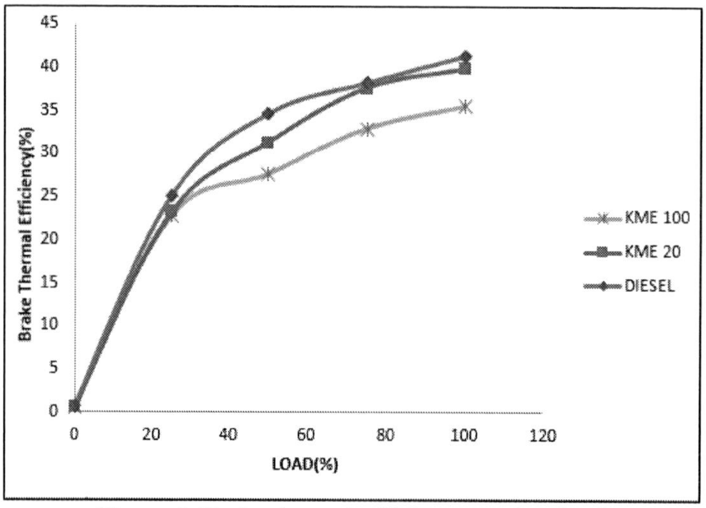


Figure 3. Brake thermal efficiency against load

### *4.3 Emission Characteristics*

#### *4.3.1 Carbon monoxide*

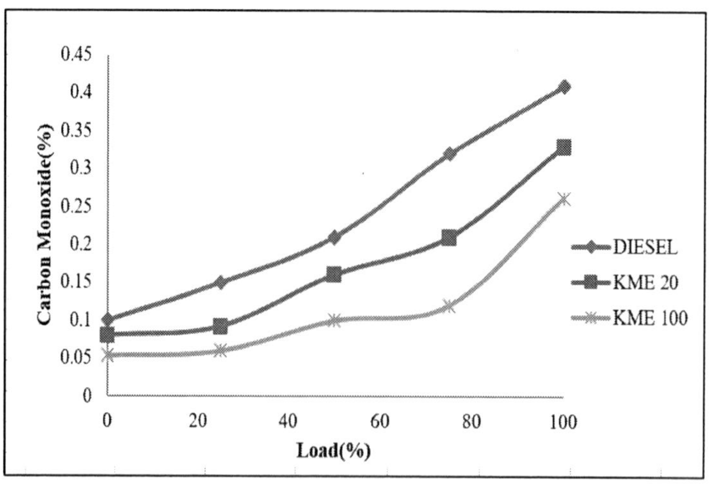


Figure 4. Carbon monoxide against load

The figure 4 shows the variation of carbon monoxide (CO) emission of diesel, biodiesel and its blend with respect to load. By using biodiesel, reduction in CO is observed which is due to the increased proportion of oxygen in biodiesel compared to diesel which promotes oxidation and results in complete combustion [11]. Thus, there is a decrease of 33.5% and 20.2% for biodiesel and B20 blend compared to that of diesel at full load

### *4.3.2 Hydrocarbon (HC)*

The figure 5 shows the variation of Hydrocarbon (HC) emission of diesel, biodiesel and its blend with respect to load. Since, biodiesel has higher oxygen content which results in complete combustion of fuel. Thus, HC is less compared to that of diesel. The increased gas temperature and higher cetane number of biodiesel and their blends were responsible for this decrease [11]. Thus, there is a decrease of 26.5% and 14.7% for biodiesel and B20 blend compared to that of diesel at full load.

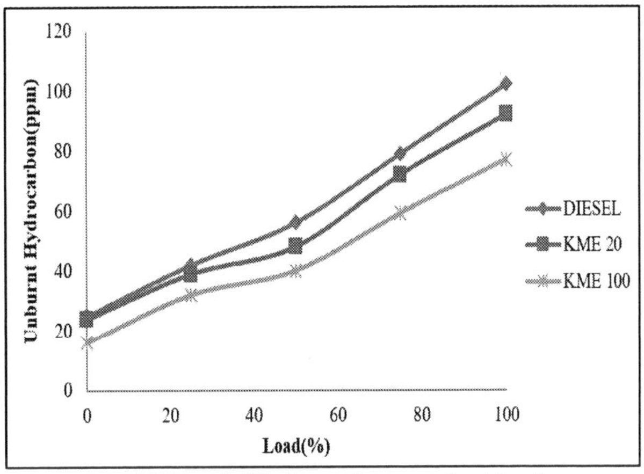


Figure 5. Hydrocarbon against load

### *4.3.3 Oxides of Nitrogen (NOx)*

The figure 6 shows the variation of Oxides of Nitrogen (NOx) emission diesel and biodiesel blends with respect to load. The NOx from biodiesel is higher compared to that diesel. This is mainly due to better combustion and improved cetane number of biodiesel blends. Hence the ignition delay is reduced which leads to better combustion resulting higher operating temperature. The inert nitrogen reacts with oxygen at higher temperature to form NOx [12]. Thus, there is an increase of 16.2% and 11.5% for biodiesel and B20 blend compared to that of diesel at full load.

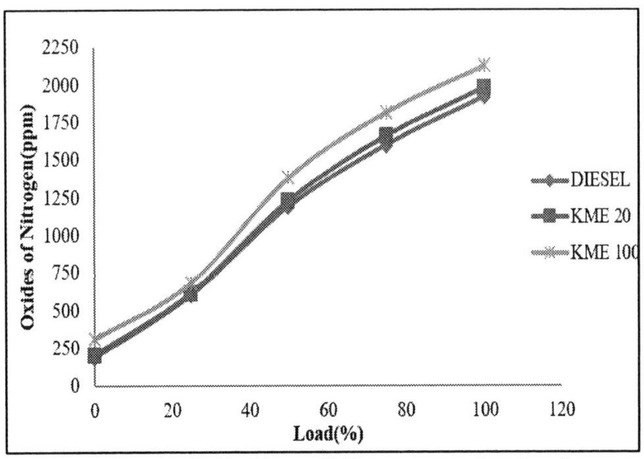


Figure 6. Oxides of Nitrogen against load

## 5. Conclusions

This paper investigates on the production of Karanja oil Methyl Ester (KOME) and its performance characteristics in diesel engine.

- The parameter condition used for production of Karanja Methyl Ester (KOM) are for esterification 8:1 Molar ratio of Methanol to oil, 90 min Reaction time, 60°C Reaction temperature and 2% ($H_2SO_4$) (w/w) Catalyst concentration.
- For transesterification 8:1 Molar ratio, 60 min Reaction time, 60°C Reaction temperature and 1% (KOH) (w/w) Catalyst concentration and obtained yield is 92.36%.
- There is a slight increase of specific fuel consumption for B20 blend and biodiesel than that of diesel
- Brake thermal efficiency decreases by 16% and 9% for biodiesel and B20 blend than that of diesel
- There is a significant reduction in CO and HC emission than that of diesel
- There is a slight increase in NOx emission due to high oxygen content in biodiesel.

From the conclusion it is clear that the karanja oil biodiesel and its blends can be used as a substantial fuel for direction injection diesel engines

## 6. References

[1] Dubey P, Gupta R, (2017) Effects of dual bio-fuel (Jatropha biodiesel and turpentine oil) on a single cylinder naturally aspirated diesel engine without EGR. Applied Thermal Engineering **115:**1137–47.

[2] Yehliu K, Boehman AL, Armas O. (2010) Emissions from different alternative diesel fuels operating with single and split fuel injection. Fuel;**89** (2):423–37.

[3] Godiganur S, Murthy CS, Reddy RP.( 2009) 6BTA 5.9 G2-1 Cummins engine performance and emission tests using methyl ester mahua (Madhuca indica) oil/diesel blends. Renewable energy;**34** (10):2172–7.

[4] Muralidharan K, Vasudevan D. (2011) Performance, emission and combustion characteristics of a variable compression ratio engine using methyl esters of waste cooking oil and diesel blends. Applied Energy: **88** (11):3959–68.

[5] Kalam MA, Masjuki H. (2004) Emissions and deposit characteristics of a small diesel engine when operated on preheated crude palm oil. Biomass Bioenergy: **27** (3):289–97.

[6] Ozsezen AN, Canakci M, Turkcan A, Sayin C. (2009) Performance and combustion characteristics of a DI diesel engine fueled with waste palm oil and canola oil methyl esters. Fuel; **88** (4):629–36.

[7] Sanjid A, Masjuki H, Kalam M, Rahman SA, Abedin M, Palash S. (2014) Production of palm and jatropha based biodiesel and investigation of palm-jatropha combined blend properties, performance, exhaust emission and noise in an unmodified diesel engine. Journal of Cleaner Production; **65**:295–303.

[8] Jiaqiang E, Liu T, Yang W, Deng Y, Gong J. (2016) A skeletal mechanism modeling on soot emission characteristics for biodiesel surrogates with varying fatty acid methyl esters proportion. Applied Energy; **181**:322–31.

[9] Gopinath. S , Devan P.K (2019) Optimization and Prediction of reaction parameters of plastic Pyrolysis oil production using Taguchi Method, Iranian Journal of Chemistry and Chemical Engineering; **39** 021-9986, 2020

[10] C. Bibin, P. Seeni Kannan, and P. K. Devan (2018) Performance, emission and combustion characteristics of a DI diesel engine using blends of punnai oil biodiesel and diesel as fuel. Thermal Sciences 233–23.

[11] PK Devan, NV Mahalakshmi (2009) A study of the performance, emission and combustion characteristics of a compression ignition engine using methyl ester of paradise oil–eucalyptus oil blends, Applied Energy **86** (5), 675-680

[12] PK Devan, NV Mahalakshmi (2009) Performance, emission and combustion characteristics of poon oil and its diesel blends in a DI diesel engine, Fuel **88** (5), 861-867

IOP Conf. Series: Materials Science and Engineering **923** (2020) 012069 doi:10.1088/1757-899X/923/1/012069

# INVESTIGATION ON PERFORMANCE AND EMISSION OF PONGAMIA BIODIESEL USING DIETHYL ETHER AND ZINC OXIDE AS ADDITIVE IN DIESEL ENGINE

**Rajesh K, Devan P K, Lokesh B, Rajesh Kumar K, Vignesh R, Yogesh Kumar R**

Department of Mechanical Engineering, RMK College of Engineering and Technology, Thiruvallur - 601206, Tamil Nadu, India.

E-mail: rajeshmechh@gmail.com

**Abstract.** Biodiesel is one of the better replacement for diesel fuel because of the depleting fuel resources. The cost of the biodiesel compared to diesel is major setback in commercialization. As mass production is encouraged the cost involved in production will reduce and may available at the cost less than diesel. Biodiesel is one of the promising source of fuel which stands as one of the better alternative diesel fuel. This study represents the production of biodiesel by using esterification and transesterification of pongamia oil which is obtained from pongamia pinnata. Diethyl ether and zinc oxide used as additive to enhance cetane number, properties of biodiesel. The DI diesel engine is tested for its Performance and emission characteristics using pure biodiesel, biodiesel with 5%DEE biodiesel with 50ppm of zinc oxide, diesel on four stroke direct injection diesel engine were studied. And found that biodiesel with 5%DEE gives better result than biodiesel. Biodiesel with added additives gives 5% higher brake thermal efficiency than diesel at full load condition, hydrocarbon emission less than 10%, carbon monoxide emission 40.35% less than B100, NOx is 0.7%, hydrocarbon 11.36% were lower than biodiesel. Thus, pongamia biodiesel with 5%diethyl ether can be used as one of the alternative source of energy for diesel.

## 1. Introduction

The biodiesel is gaining more importance in the recent years because of its suitability in comparison to diesel in compressed ignition engines. Lot of research is going on biodiesel due to the price of the conventional fuel, environmental concerns due to pollution from the burning of fossil fuels, scarcity due to increasing population and regular use of transportation [1]. The government also create awareness in using substitute for petroleum based fuels. As a result many companies are focusing more on alternate fuels. Focus on alternate fuels not only solve the problems related to scarcity, pollution and environmental effects, it also will help farmers in their growth. Due to price and environmental concern biodiesel is considered as better substitute to diesel fuel and it is concentrated more for its advantages.

There are alternative ways of production, with different sorts of raw materials: refined crude or frying oils. The main advantages of using biodiesel are it is safe to use, handling and transportation is easy, due to rich oxygen content better combustion, cetane number is high, it is biodegradable, higher flash and fire points, self lubricating, etc [2]. The main disadvantage of using vegetable oils directly in diesel engines is its viscosity and volatility. To overcome the drawback that is to reduce the viscosity and to improve volatility, transesterification process is carried out. It is the process of elimination of glycerides from oil and mixing the esters of oil with alcohol to form methyl or ethyl esters. This process improves the suitability of oil to use in diesel engine and hence the performance and combustion is improved. Also biodiesel on combustion emits lesser pollutants compared to diesel. Nano particles have gained special importance because of their chemical and physical properties [3]. To improve the performance of biodiesel in diesel engine, Nano additives were added and tested. In this work the pongamia biodiesel and blended with additives were tested in diesel engine and tested for its performance and emission characteristics to compare with diesel.

## 2. ENGINE PERFORMANCE AND EXHAUST EMISSION TESTS

The performance of the diesel engine is tested with various blends of biodiesel with additives in a four stroke water cooled, vertical, direct injection single cylinder diesel engine test rig which is given in the figure 1. The load to the engine is applied by means of coupling eddy current dynamometer. The rated power of the engine is 5.2 kW at 1500 rpm. The engine is mounted on the test bed with all necessary connections. The emission like CO, HC, NOx were measured using AVL exhaust gas analyser. The specification of the test engine is as given below:

**Table 1 Engine Specification**

| | |
|---|---|
| Engine Type | Kirloskar TV 1 |
| Stroke Length of Piston | 0.11m |
| Cylinder diameter | 0.0875m |
| No. of Cylinder | 1 |
| No. of Stroke | 4 |
| Type of cooling | Water Cooled |
| Rated Power | 5.2 kw at1500RPM |
| Compression ratio | 17.5:1 |
| Fuel used | Diesel |
| Eddy current dynamometer Dynamometer arm length | 0.185m |

*Figure1. Engine test rig*

## 3. Result and Discussions

### *3.1 Performance Characteristics*

#### *3.1.1 Brake thermal efficieny (BTE)*

The brake thermal efficiency is low for diesel at full load condition compared to biodiesel. The brake thermal efficiency at 25% load is almost equal to diesel,B100,B100+5%DEE and B100+ZNO 50PPM.The brake thermal efficiency at 50% load is higher for B100+5%DEE.After 75% of load, the brake thermal efficiency of diesel follows a flat line till 100% of load. The brake thermal efficiency curve of B100,B100+5%DEE and B100+ZNO 50ppm follows a gradually increasing path after 50% of load. The mixture of diethyl ether increases the premixed combustion phase which may be the reason for increased brake thermal efficiency of B100+5%DEE [4]. Diethyl ether is cetane improver which is primary reason for all the other consequent effects of B100+5%DEE. Thus we can get to a conclusion that the biodiesel obtained from pongamia oil has a higher brake thermal efficiency than the diesel.

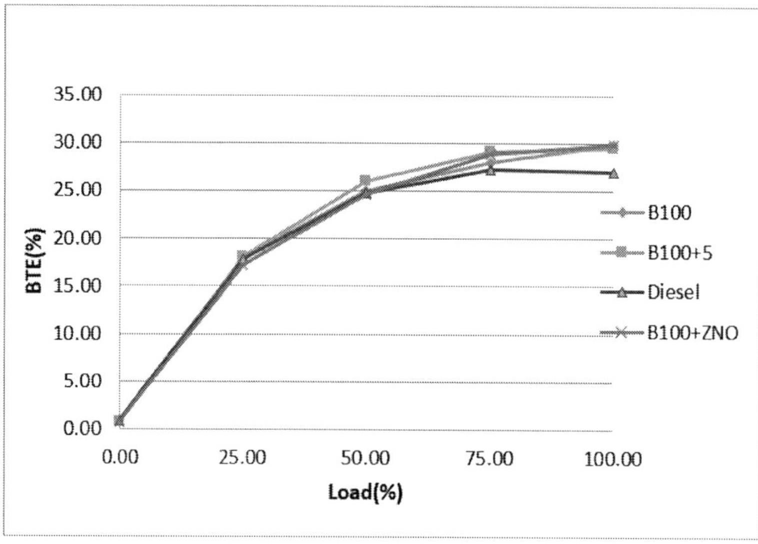


*Figure 2. Load Vs BTE*

#### *3.1.2 Specific Fuel Consumption (SFC)*

The BSFC of diesel is less than the pure biodiesel,B100+5%DEE and less than B100+ZNO+50ppm.This increase in BSFC of biodiesel is mainly due to increase in density which is caused by addition of diethyl ether and zinc oxide [5]. Due to lower calorific value of biodiesel there is increase in BSFC which makes it less economical than the diesel. The difference in BSFC of diesel and biodiesel is higher at 25% load. The variation of BSFC between biodiesel and diesel has decreased at 50%load and maintains the same difference till 100% load. Thus it clearly shows that diesel is economical than the biodiesel obtained from pongamia oil.

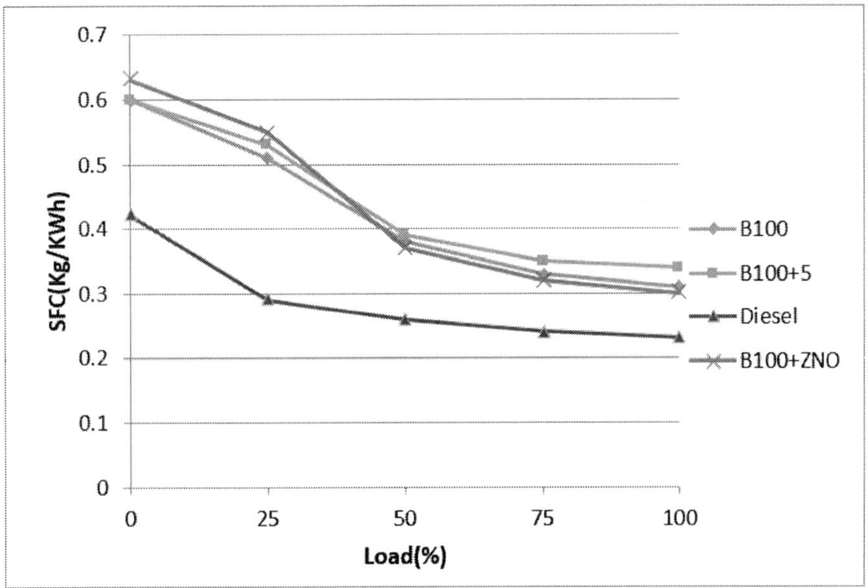


*Figure 3. Load Vs SFC*

### *3.1.3 Exhaust Gas Temperature (EGT)*

The exhaust gas temperature of diesel is lower than that of B100+5%DEE.At the initial stage the exhaust gas temperature is similar except B100 which is due to low Calorific value of the biodiesel. But the exhaust gas temperature increases after the addition of additives to the biodiesel. This increase in exhaust gas temperature may be due to the combustion during the latter diffusion phase [6] . At full load the exhaust gas temperature of B100+ZNO 50ppm is higher than B100,B100+5%DEE and diesel. This increase in EGT after addition of ZNO is because metal based additives are combustion improvers which increases the exhaust gas temperature.

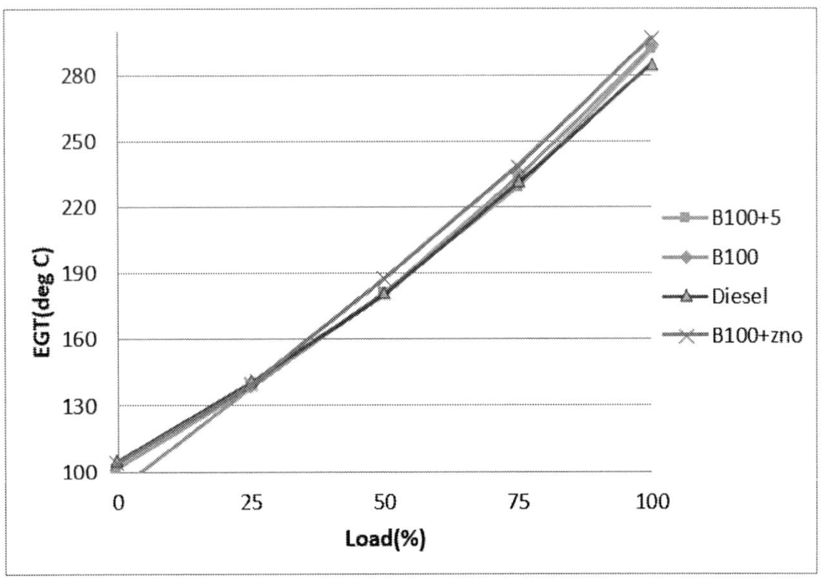


*Figure 4. Load Vs EGT*

### *3.2 Emission Analysis*

*3.2.1 Hydro Carbon*

Engine when tested with diesel recorded highest hydrocarbon compared to biodiesel. B100+5%DEE has less hydrocarbon compared to diesel base line. This is due to the free oxygen available in biodiesel because of the presence of esters. After addition of zinc oxide the hydrocarbon increases rapidly higher than the diesel. This is due to the increased cetane number and no change in compression ratio and less time available for fuel burning [7]. At 50% load the hydrocarbon released by diesel and B100+5%DEE is almost equal. After that the trend of B100+5%DEE is always less than diesel base line.

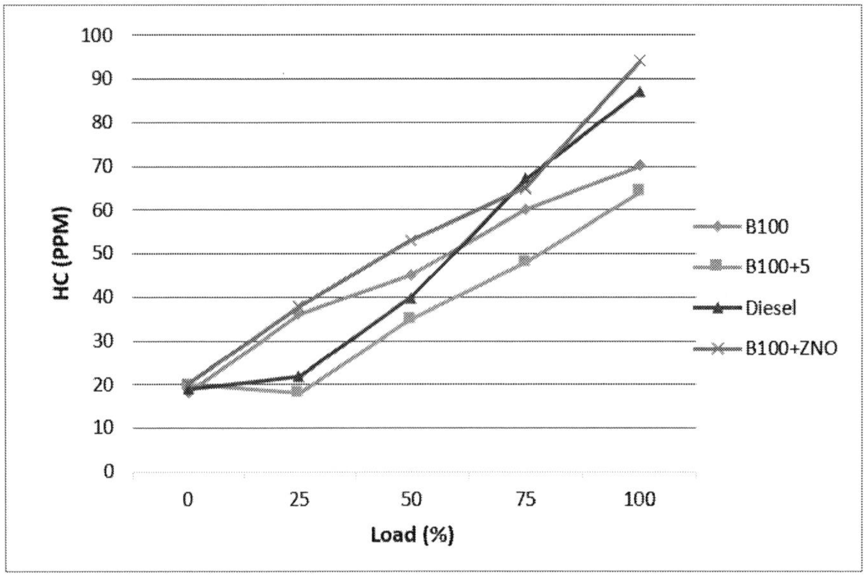


*Figure 5. Load Vs HC*

*3.2.2 Carbon monoxide*

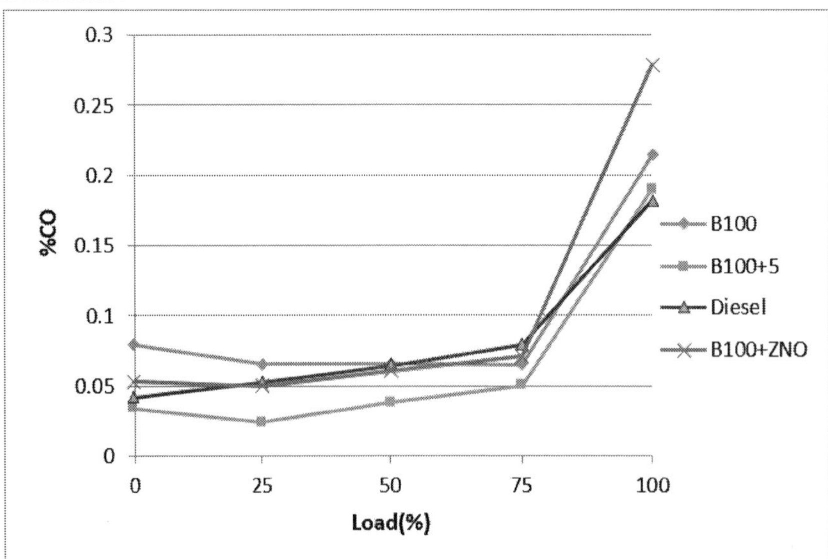


*Figure 6. Load Vs CO*

IOP Conf. Series: Materials Science and Engineering **923** (2020) 012069 doi:10.1088/1757-899X/923/1/012069

The carbon monoxide emission of biodiesel is lower than that of diesel base line. The addition of diethyl ether reduced the carbon dioxide emission till 90% of the load. After that carbon mooxide emission decreases slightly than diesel base line. This decrease in carbon monoxide emission is due to catalytic activity and complete combustion of the fuel [8]. Metal based additives owing to their catalytic activity enhance the oxidation process of hydrocarbons leading to higher carbon monoxide emission. This is the reason that B100+ZNO 50ppm has higher carbon monoxide emission than diesel and B100.

*3.2.3 Nitrogen oxides (NOx)*

NOx emission is higher in diesel base line. The addition of zinc oxide which is a primary NOx reducer reduces the NOx emission at 100%load. But the NOx emission of B100+ZNO 50ppm at 50% load higher than that of diesel. The addition of zinc oxide tends to reduce NOx only at 50ppm and 100ppm. If zinc oxide is added at higher volume of 50ppm it causes high reactivity and by triggering the micro explosion phenomenon which consequently led to decreased NOx formation [4]. The variation in NOx after adding diethyl ether is due to less complete combustion which intum reduces its emission.s

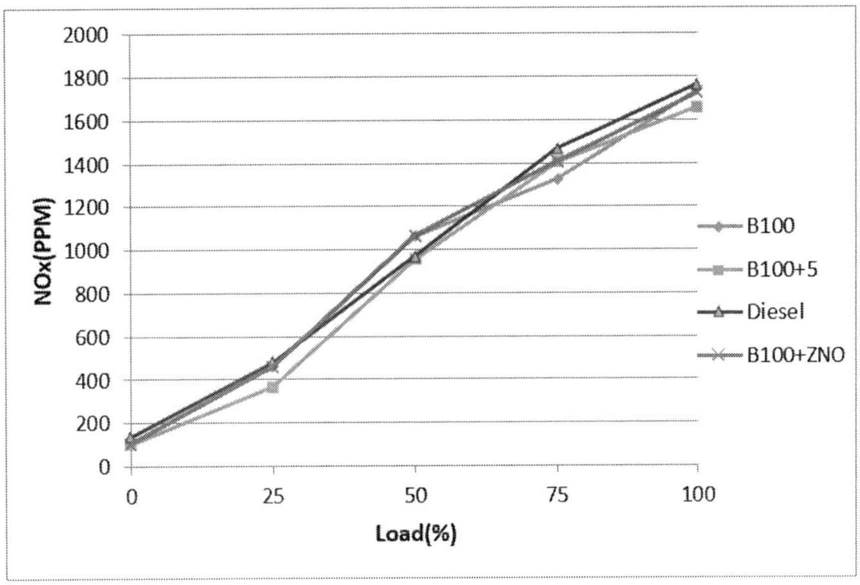


*Figure 7. Load Vs NOx*

*3.2.4 Smoke Opacity*

Diesel has very less opacity compared to the biodiesel obtained from pongamia oil. This increased smoke opacity is due to the less time available for the fuel to be burnt completely [9]. The smoke was highest at 50% load for B100+50%DEE.The variation in smoke opacity is also highest at 50%load. The increase in smoke opacity cannot be compensated further and complete burning of fuel is the only way to avoid it. Thus we can conclude that diesel has less opacity than pongamia biodiesel at all load conditions.

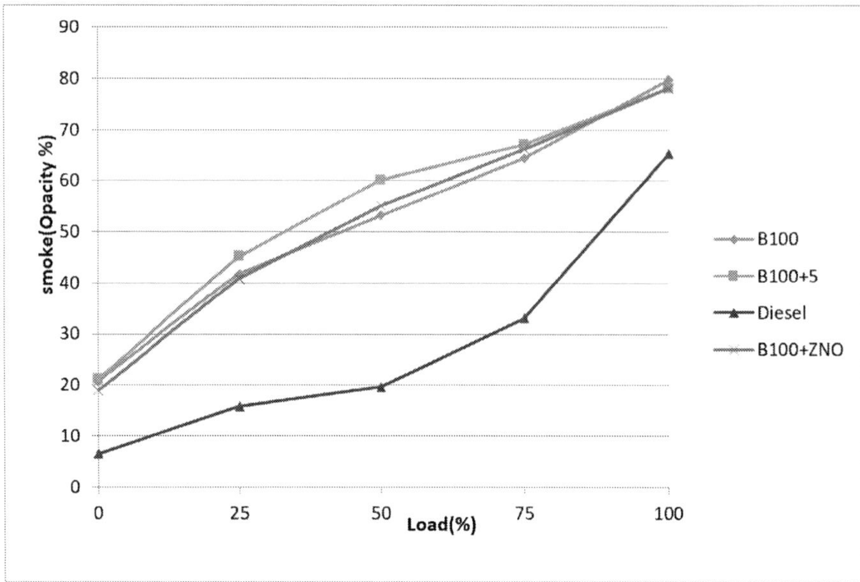


Figure 8. Load Vs Smoke Opacity

## CONCLUSION

From this work “Pongamia biodiesel production using methanol and evaluating it’s performance using diethyl ether and zinc oxide as additives”, the following conclusions were made.

It was noted that 5% of sulphuric acid gives high yield of 90.15% of biodiesel than the other concentration of sulphuric acid when maintained at 60deg celcius and 60 min of reaction time with methanol to oil molar ratio of 6:1.

The performance and emission characteristics of biodiesel was compared with diesel. Pongamia biodiesel has BTE of 7.56% higher than diesel, NOx emission is 5% less than diesel and 15% higher smoke opacity than diesel. Other characteristics like hydrocarbon emission is 25.42%, carbon monoxide emission is 27.85% lower than diesel.

Added zinc oxide and diethyl ether as additive to enhance the biodiesel property and emission of biodiesel.

The performance, emission and combustion characteristics of biodiesel with additive. Additive added biodiesel gives brake thermal efficiency 5% higher than diesel at full load condition, hydrocarbon emission less than 10%, carbon monoxide emission 40.35% less than B100, NOx is 0.7%, hydrocarbon 11.36% were lower than biodiesel.

From these results it is conclude that Pongamia biodiesel with diethyl ether is one of the best alternative fuel to diesel.

## References

[1] C. Bibin, P. Seeni Kannan, and P. K. Devan, “Performance, emission and combustion characteristics of a DI diesel engine using blends of punnai oil biodiesel and diesel as fuel,” Therm. Sci., vol. 00, no. 00, pp. 233–233, 2018.

[2] PK Devan, NV Mahalakshmi, A study of the performance, emission and combustion characteristics of a compression ignition engine using methyl ester of paradise oil–eucalyptus oil blends, Applied Energy 86 (5), 675-680
[3] P.K.Devan, S.Gopinath, K.Rajesh, S.Madhu, "Improving the characteristics of engine oil using nanofluid as coolant in combat vehicles", Materialstoday: proceedings, Volume 22, Part 3, 2020, Pages 1130-1134
[4] Qi DH, Chen H, Geng LM, Bian YZ. "Effect of diethyl ether and ethanol additives on the combustion and emission characteristics of biodiesel-diesel blended fuel engine" Renew Energy 2011;36:1252–8.
[5] B. Prabakaran, A. Udhoji, Experimental investigation into effects of addition of zinc oxide on performance, combustion and emission characteristics of diesel-biodiesel-ethanol blends in CI engine, Alexandria Eng. J. 55 (2016) 3355–3362.
[6] S.Senthur Prabu, M.A.Asokan, S.Prathiba, ShakkeelAhmed, GeorgePuthean, "Effect of additives on performance, combustion and emission behavior of preheated palm oil/diesel blends in DI diesel engine" Renewable Energy, Volume 122, July 2018, Pages 196-205
[7] Harish Venu, Venkataramanan Madhavan , "Effect of Al2O3 nanoparticles in biodiesel-diesel-ethanol blends at various injection strategies: Performance, combustion and emission characteristics", Fuel 186 (2016) 176–189 Contents
[8] A.Prabu, "Nanoparticles as additive in biodiesel on the working characteristics of a DI diesel engine", Ain Shams Engineering Journal, Volume 9, Issue 4, December 2018, Pages 2343-2349
[9] Harish Venu, Venkataramanan Madhavan, "Influence of diethyl ether (DEE) addition in ethanol-biodiesel-diesel (EBD) and methanol-biodiesel-diesel (MBD) blends in a diesel engine", Fuel, Volume 1891 February 2017Pages 377-390

IOP Conf. Series: Materials Science and Engineering **923** (2020) 012070 doi:10.1088/1757-899X/923/1/012070

# Numerical analysis of energy savings due to the use of PCM integrated in lightweight building walls

R.Sheeja[a], S. Kumar[b*], P. Chandrasekar[b], Abraham J S Jospher[b], Sai Krishnan[c]
[a] Department of Civil Engineering, Easwari Engineering College, Chennai, 600089
[b] Department of Mechanical Engineering, Rajalakshmi Engineering College, Chennai, 602 105.
[c] Department of Mechanical Engineering, Rajalakshmi Institute of Technology, Chennai.

[*]E-mail: kumarsubarayan@gmail.com

Abstract. A new type of PCM layer integrated into an exterior building wall is to reduce the amplitude of the indoor temperature and maintaining the better thermal comfort for human beings. A numerical simulation has been carried out with and without integration of PCM in building walls. In this paper, the effect of different configuration of external building walls were studied and analyzed by the integration of PCM building walls. Using finite difference method, one dimensional transient heat transfer model was developed and solved. The indoor temperature was also analyzed for the summer and winter climatic conditions of Chennai city, India (13.1 Latitude and 79.9 Longitude) in the PCM and Non-PCM buildings. The building wall consists of clay brick with filled with PCM. The numerical simulation result shows that, PCM filled with clay brick wall improves the thermal inertia significantly. Location of PCM layer in the building wall is one of the important factors to reduce the heat gain in the building wall before it enters the room.

*Keywords:* Phase Change materials, Energy efficient buildings, Building energy Conservation, Thermal Performance, DESIGNBUILDER

## 1. Introduction

The building sector is one of the largest energy consumers around the world with a total of 30–40% of the global energy consumption in which a substantial amount of energy is used to enhance indoor thermal comfort [1]. Also, the energy consumption in this sector has been increased rapidly in recent years due to high living standards, increased population growth, and people spending more time indoors. The energy used by the buildings was mainly generated by the heat gain through building envelopes. Among it, the building wall was consumed the energy around 25%, followed by the roofs for 22%, windows for 23% and others for 30%. The massive construction of the enveloped building wall is to reduce the energy consumption of the buildings. Increasing thermal mass of building walls is to increase the thermal performance, knowing as passive cooling energy saving methods [2].The building thermal system maintains at the building envelope, in which the thermal balance between energy consumption, energy production and storage of energy within the building. This thermal balance considerably affected by the factor's external thermal solicitations and the external building envelopes act as a barrier to external factors[3].The construction of building envelope plays a major role in determining the cooling energy demands of a building. Naturally available cool thermal energy is utilized to minimize the use of mechanically assisted air conditioning systems in buildings[4]. However, the construction of traditional building material gives thermal mass to a building bricks or blocks of construction make it heavy construction and high volume. The choice of suitable construction method provides an opportunity to enhance the thermal comfort, reducing the energy demands for space conditioning system and diminishes indoor temperature swings by absorbing heat and gradually releasing it when the decrease the ambient temperature, leading to a reduction and a shift the peak load[5].The energy storage system can be used with high storage density and high thermal capacity. The dynamic building envelope materials could be used to control the heat flow into and out of buildings. The innovative macro encapsulated Phase change materials (PCMs) have been considered as a way to increase the energy efficiency, ability of storing and releasing large amounts of heat within a small volume and thermal inertia of buildings [6]. PCM has been integrated in building envelopes to improve the thermal comfort and reducing the cooling/heating loads [7]. B. James and P. Delaney studies stated that, PCMs incorporated buildings in various climatic weather zones in the USA, can potentially save 10–30% of the annual heating loads [8]. An experimental work was carried out in USA showed the results, 10–26% of the energy savings were attained in summer months by using PCMs in building walls [9].The plaster incorporating micro encapsulated PCM was coated on interior building walls which is reduced by 2 K indoor temperature and maintained the thermal comfort for longer periods in German [10]. Experimental analysis of PCM integrated building was investigated and results found that, the PCM building walls could make the interior comfort temperature and decreasing the cooling load of air-conditioning system[11, 12].Also, during off- peak period in the winter season, heat pump could be used to melt the PCM in wallboards above the transition temperature. The stored heat energy could be released in a room at peak hours in which the heating demand might be reduced [13, 14] thermal performance. The conventional residential building results has been compared and evaluated with the PCM residential building.

Macro encapsulated PCM integrated in clay hollow brick in Mediterranean countries was investigated and discussed using temperature, relative humidity, solar radiation and climatic weather condition. The west and south facing walls equipped with thin layer of PCM was analyzed under different climatic weather conditions. The heat flux reductions were 29.7% and 51.3% for the west and south wall respectively. The maximum heat flux time delays were 2.3 hour for west wall and 1 hour for south wall. The maximum daily heat transfer reduction were 3.6% in the west wall and 27.1% in south wall [15]. The energy and economic analysis of the integration of PCM in typical multistory office building located in five different cities of China, was simulated for a whole year. The simulated results showed that the energy savings resulting from PCM integration were more prominent for office buildings located in cold regions as well as in cold winter and hot summer regions [16]. The economic analysis of PCM integrated office buildings were not sufficient to recover the investment which is located in different cities of China. During summer and winter season, the energy savings was found that the integration of PCM building in Zhengzhou, Changsha and Shenyang has high economic value and the investment appear to be attractive. However, the investment of PCM buildings cannot be recovered in Hong Kong and Kunming and do not offer economic benefit. The potential of PCM in minimize the energy consumption at different climatic condition of Australia and the effectiveness of PCM have been analyzed using EnergyPlus building simulation software. Five different PCM melting temperature ranges were used to identify the optimum melting range of PCM for each city. The potential of PCM building has reduce the energy consumption in Australian cities under warm temperate, cold temperate and mild temperate zones. The PCM building has very minor effect on hot and humid climate zone. The effectiveness of PCM depends on PCM layer thickness, surface area, thermostat range, location of application and local climate in building. The Melting point of PCM was selected outside the temperature range does not provide efficient energy reduction irrespective of the maximum reduction in temperature fluctuations [17].

Using Energy Plus building simulation software to simulate the PCM building wallboards on summer day in Auckland. The indoor temperature fluctuation was reduced daily up to 4 ◦C [18]. PCM effect was estimated using Enthalpy method through Energy Plus software. The verification and validation study [19] was conducted recently using Energy Plus software. The thermal performance of PCM buildings was accurately calculated using this same software and met the several building guidelines. Hence, the present study has been investigated and evaluated using "Energy plus".

From the above literature reviews indicated that, the integration of PCM in building materials is increase the energy storage in buildings in which reduces energy reduction and indoor temperature fluctuation. To the best of the author's knowledge, no published literature review on PCM integrated in building wall in warm and humid climatic conditions of Chennai city, India and no data is available about potential cooling load reduction by using PCMs in such climate. The aim of the present paper, a detailed study of thermal performance of PCM integrated in building wall was investigated through numerical simulation using Energy Plus software under warm and humid weather conditions of Chennai city (13.1 Latitude and 79.9 Longitude), India. The effects of phase change temperature, location, melting range of PCM, amount of PCM and shape of enthalpy curve on PCM building wall were reduced the heat gain reduction in PCM buildings were studied and reported.

## 2. Numerical methods and model description

### *2.1 Numerical method*

Simulation of PCM building was carried out using "Energyplus" building simulation software which is developed by U.S Department of Energy. It is a strong tool for simulation of thermal buildings and also has various option and modulus with an enormous data library. Additionally, almost all of its parts have been validated with experimental studies since the first release of the software [20]. The one-dimensional heat conduction finite difference (CondFD) solution is working as heat balance algorithm in EnergyPlus to simulate the thermal performance of PCM in building walls. EnergyPlus with the CondFD algorithm solved by a fully implicit finite difference scheme which is coupled with an enthalpy–temperature function to account for phase change energy was used in this study. The specific heat capacity ($C_p$) of PCMs is updated in each iteration according to the following Equation (1)

$$C_p = \frac{h_i^j - h_i^{j-1}}{T_i^j - T_i^{j-1}} \tag{1}$$

Where

T = node temperature (k),

J = actual time step for simulation,

j-i = previous time step,

I = modeled node,

h = user defined function of the specific enthalpy of PCMs as a function of temperature

The iteration scheme assures the correct $C_p$ is used in each time step based on the user defined enthalpy–temperature function [21]. The validation and verification studies for heat transfer calculation and CondFD solution algorithm in EnergyPlus have been performed many researchers [7, 22]. In order to ensure the accuracy of the CondFD model and the simulation, the dissertation space was 3 and the time step was set to one minute for simulation [20]. Moreover, the strong hysteresis PCMs cannot be accurately simulated, so that, negligible hysteresis PCM can be used to attain suitable results [20].

The PCM (HS 29) used for this research has been selected directly from the Design Builder simulation program which is having the melting temperature of 29°C. The phase change energy was considered into account by applying enthalpy-temperature function. The formulation of a node of PCM is given by Equation (1).

### *2.2 Validation*

Many verification and validation has been carried out on ConFD models and PCM of Energy Plus different test in which consist of comparative testing, analytical verification and empirical validation [23].

In addition, EnergyPlus PCM algorithm was validated against experimental data by other researchers [7, 24].The CondFD solution and PCM algorithms of EnergyPlus were verified and validated against comparative testing, empirical validation and an analytical verification [25]. Furthermore, the EnergyPlus PCM building model was validated against the experimental work [24] which was powerful agreement between numerical simulation results and experimental data.

*2.3. Building model*

A single-zone building with single floor prototype building was selected with floor area of 16 $m^2$ (4 m width x 4 m breadth x 3.5 m height) as shown in figure 1. Thee window was located in eastern and northern side of the building with size of (0.9 m x 1.2 m) size. The 3 mm thickened window glass having solar transmittance of 0.45 whereas the transmittance of visible light and the coefficient of the thermal conductivity of the glass were 0.7 and 0.9 W/m K, respectively. The property of the window glass has given in Table 5. The distance between floor and each window's bottom was 1 m. The ratio of window to wall area in the northern and southern side was 26% within the range of civil building thermal design code recommended. Based on the solar radiation magnitude, the PCM was arranged in all side of the walls. The sides of the walls covered by PCM and the Schematic diagram of PCM and Non-PCM wall as shown in figure 2 and 3. The details of selected PCM thermophysical properties and details building materials properties with building envelopes details are given in the Table 1 to 4. The PCM was added on the interior surface of exterior building walls. The building model is considered for office use. The effects of PCM cooling energy performance, thermal comfort was investigated with modelling and analysis of PCM and Non-PCM building with same orientation, same materials and same dimensions. The climatic weather data files for building simulations were mostly attained from the EnergyPlus Weather (EPW) database in which weather data provided in EnergyPlus format from 20 sources.

**Table 1.** Properties of phase change materials (PCMs)

| | |
|---|---|
| Melting point (° C) | 29 |
| Conductivity (W/mK) | 0.172 |
| Density (kg/m3) | 976 |
| Specific heat (J/kg K) | 1390 |
| Latent heat (J/kg) | 194 |

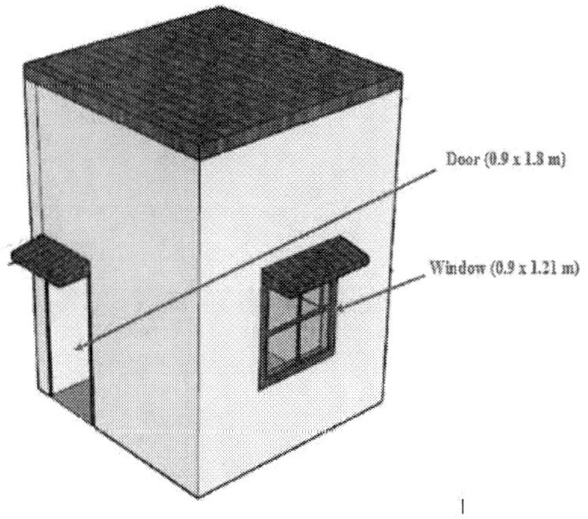


**Figure 1.** Building model geometry

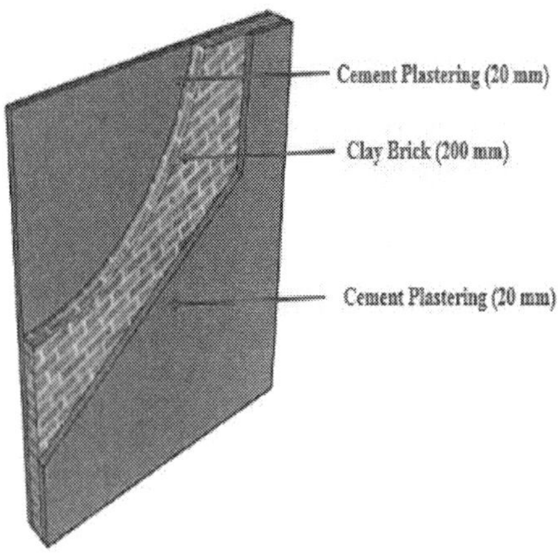


**Figure 2.** Schematic of wall

**Table 2.** Details of building envelope

| Envelope | Construction (Inside to Outside) |
|---|---|
| Floor | 120 mm reinforced concrete |
| wall | 20 mm mortar, 200 mm brick, 20 mm mortar |
| Wall with PCM | 15 mm mortar, 200 mm brick, 25 mm PCM,15 mm mortar |
| Roof | 15 mm mortar,100 mm reinforced concrete, 15 mm mortar |

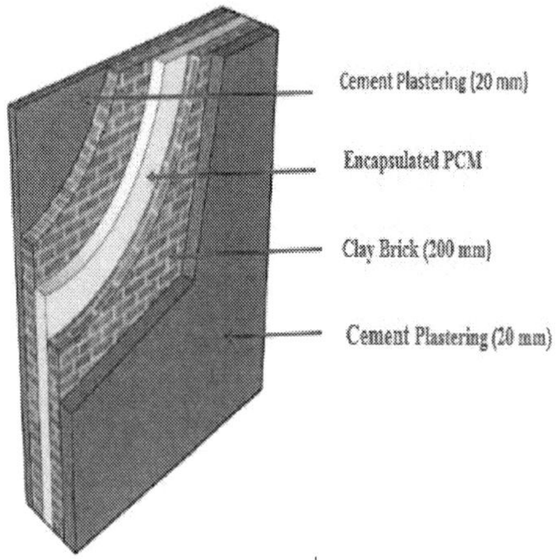


**Figure 3.** Schematic of PCM wall

**Table 3.** Physical properties of door

| Materials | Thickness of the door (m) | Thermal Conductivity 'K' (W/mK) | Density ($kg/m^3$) | Specific heat (J/kg K) |
|---|---|---|---|---|
| Plywood door | 0.0381 | 0.12 | 510 | 1380 |

**Table 4.** Physical properties of building envelope

| Building Materials | Density ($kg/m^3$) | Specific Heat (J/kg K) | Thermal conductivity (W/mK) | Thermal Resistance ($m^2K/W$) |
|---|---|---|---|---|
| Cement | 1600 | 1050 | 0.84 | 0.15 |
| Mortar | 1800 | 1.05 | 0.93 | 0.01 |
| Reinforced concrete | 2500 | 0.92 | 1.74 | 0.02 |
| Thermal insulating materials | 1000 | 0.92 | 0.085 | 0.07 |

**Table 5.** Window construction

| Optical data type | Spectral average |
|---|---|
| Thickness (m) | 0.0035 |
| Front Side Solar Reflectance at Normal Incidence | 0.078 |
| Visible Transmittance at Normal Incidence | 0.9 |
| Back Side Solar Reflectance at Normal Incidence | 078 |
| Solar Transmittance at Normal Incidence | 0.8 |
| Front Side Visible Reflectance at Normal Incidence | 0.081 |
| Front Side Infrared Hemispherical Emissivity | 0.85 |
| Infrared Transmittance at Normal Incidence | 0 |
| Back Side Infrared Hemispherical Emissivity Thermal | 0.85 |
| Thermal conductivity [W/m K] | 0.9 |

## 3. Results and discussion

The simulated PCM and Non-PCM room temperatures for the months of January, February, May and June as shown in figure 4 to 7. The simulated results show that, the PCM room temperature profile provides maintained within a thermal human comfort for winter season. During summer season, the PCM room temperature always less than that of Non-PCM room temperature as shown in figure 5 to 8. The PCM layer added in the building wall was reduced the interior PCM room temperate and increasing the energy saving possibilities. The great temperature difference between day and night time will help PCM play its role in solid–liquid phase change cycles transformation and improve the energy saving possibilities. From figure 5, shows that, the higher values of minimum temperature suggest that the PCM walls are discharging more heat than the controller model during cooling process. During day time, the PCM wall can absorbed heat and release it at night time when there is no heat source inside the room. The PCM effectiveness was evaluated by comparing the interior and exterior wall temperature with and without addition of PCM in buildings. During the sunshine period, the PCM can prevent the heat gain enter into interior area whereas at night time the PCM can decrease the interior heat flow losses. In the month of January and February the variation of PCM and Non-PCM room temperature as shown in figure 5 and 6.In the month of January and February the temperature of PCM room Varies from 25 to 29°C

The Non-PCM room temperature varies from 29 to 33°C. The temperature drop for the PCM and Non-PCM room for the January and February month was 6°C. This is true for the entire month of January and February. This month during the early morning hours, the ambient air temperature was very less than that of melting of PCM temperature range for the period of 6 hours in which the temperature of range of 23°C to 26°C. The diurnal temperature variation in Chennai is about 7 °C.

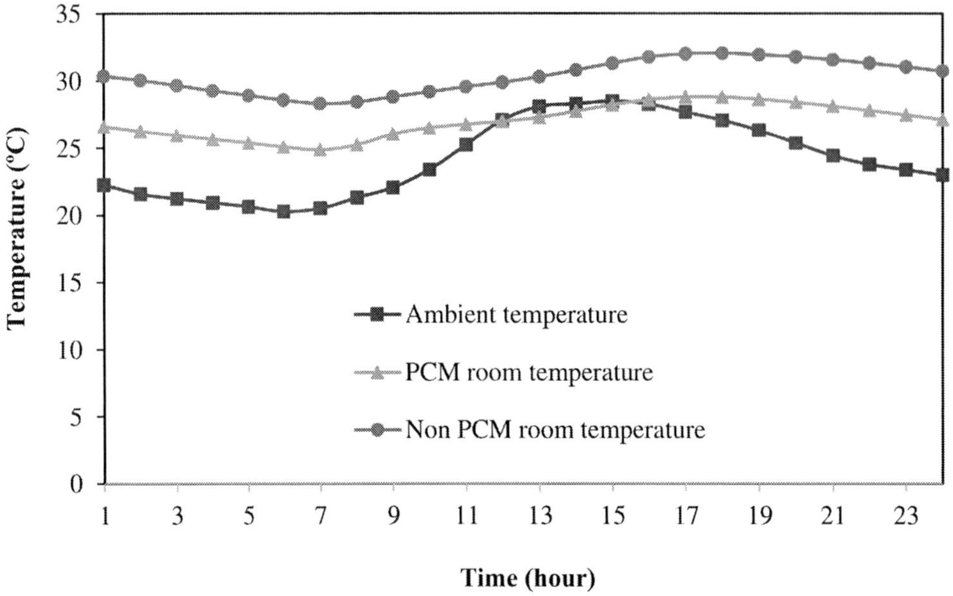


**Figure 4.** Variation PCM/Non-PCM room temperature for the month of January

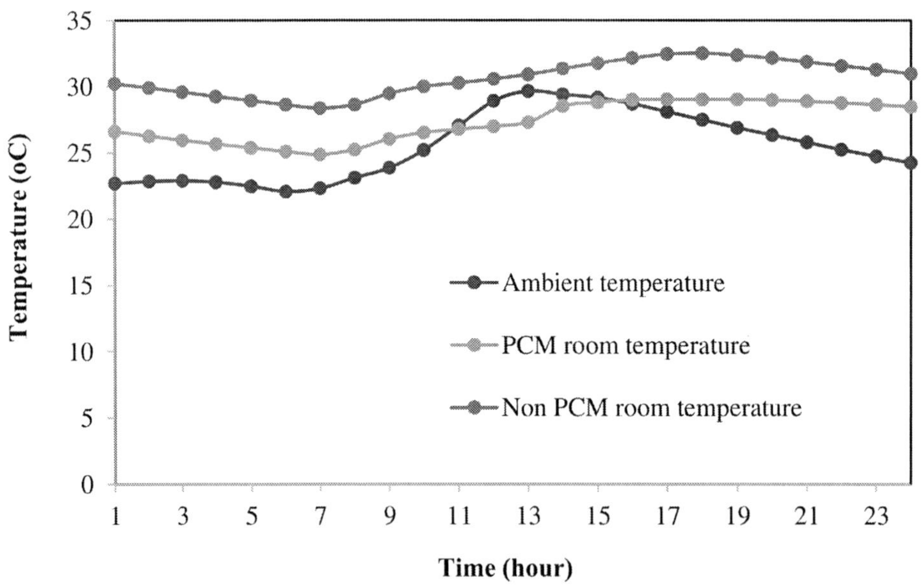


**Figure 5.** Variation PCM/Non-PCM room temperature for the month of February

During the sunshine period, the PCM can prevent the heat gain enter into interior area whereas at night time the PCM can decrease the interior heat flow losses. Figure 4 and 5 shows the variation of PCM and Non-PCM room temperature for the month of January and February. The PCM room temperature varies from 25 to 29°C for the period of 24 hours for the month of January and February months. The Non-PCM room temperature varies from 29 to 33°C. The temperature drop for the PCM and Non-PCM room for the January and February month was 6°C. This is true for the entire month of January and February. This month during the early morning hours, the ambient air temperature was very less than that of melting of PCM temperature range for the period of 6 hours in which the temperature of range of 23°C to 26°C. The diurnal temperature variation in Chennai is about 7 °C.

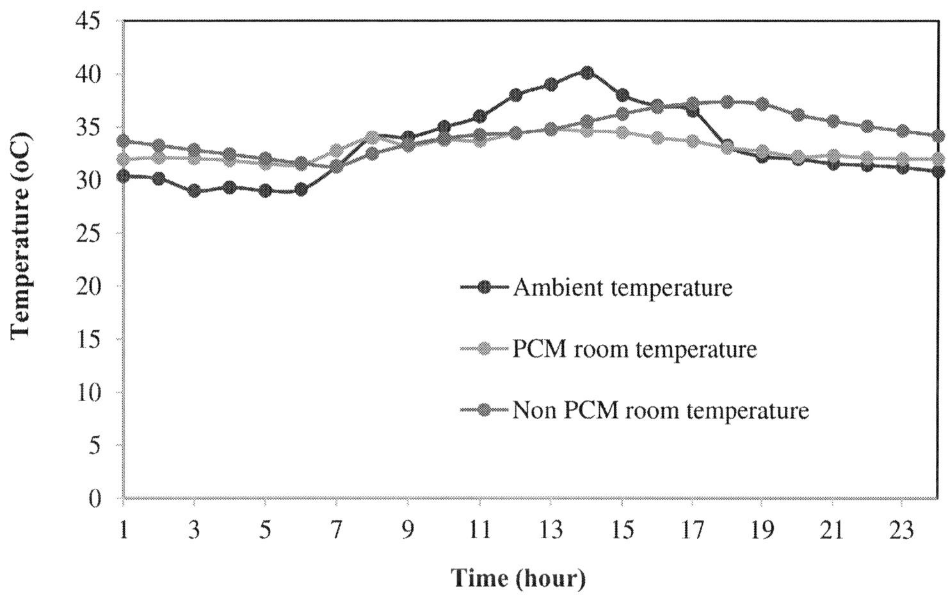


**Figure 6.** Variation PCM/Non-PCM room temperature for the month of April

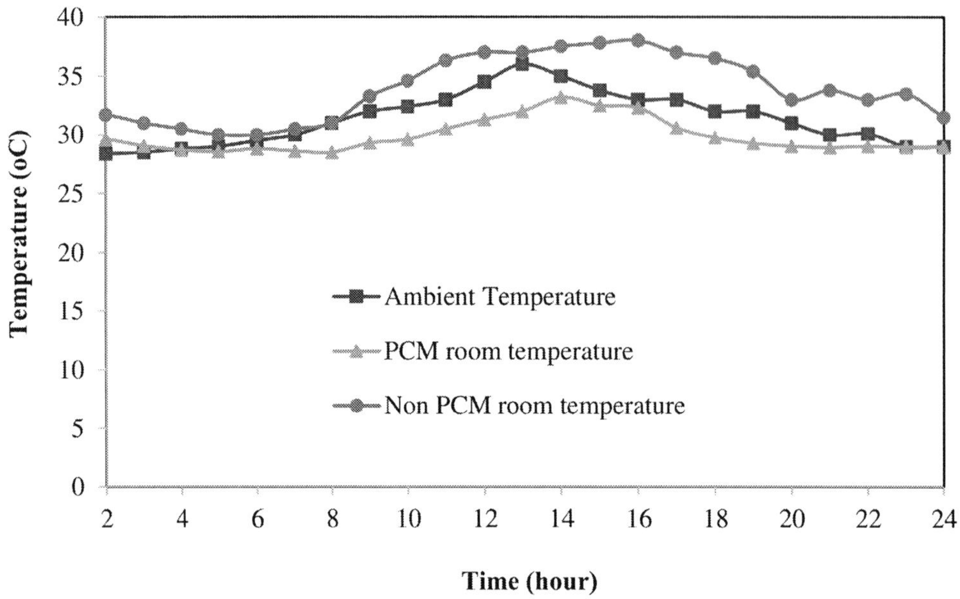


**Figure 7.** Variation PCM/Non-PCM room temperature for the month of July

The solar radiation heats up the exterior vertical wall surfaces of the building envelope in which larger temperature variation on the building envelope. This permit the PCMs can go through the full melting–freezing cycles leading to the temperature reduction of PCM building. The same temperature pattern is followed for the month of April and July as shown in figure 6 and 7. The ambient air temperature was increases during the night time in April and July months than in January and February. The freezing amount of PCM is less due to highest ambient temperature where it happens during night-time and during day time more melting happens. So that there will be a small rise in PCM temperature and the temperature drop 2-3°C compared to Non-PCM room temperature.

The April and July month with the temperature drop of 2-3°C, the same temperature pattern is followed by due to high ambient air temperature at day and night hours make it difficult in freezing of PCM and PCM change its phase in Small quantity. This temperature drop may be lower thermal conductivity of PCM material. The PCM integrated passive cooling building is good for January to March to due to lower ambient temperature. During summer season, this system underperforms due to higher ambient temperature.

The month wise minimum and maximum PCM and Non-PCM temperature variations as shown in figure 8. The maximum solar radiation reached to the Non-PCM room during noon time, so that it has reached maximum temperature. Again, the figure 8 shows that the Non-PCM room temperature is always greater than the PCM room temperature throughout the year. The PCM room, a constant temperature was maintained in the building walls at 26°C ± 3°C in January to March months, but Non-PCM wall was maintained at higher temperature. This is due to lower thermal conductivity of PCM and higher thermal mass in PCM room as the latent heat thermal energy storage system. The Non-PCM room stored as sensible heat energy storage and heat flow into the room in which increase the room temperature.

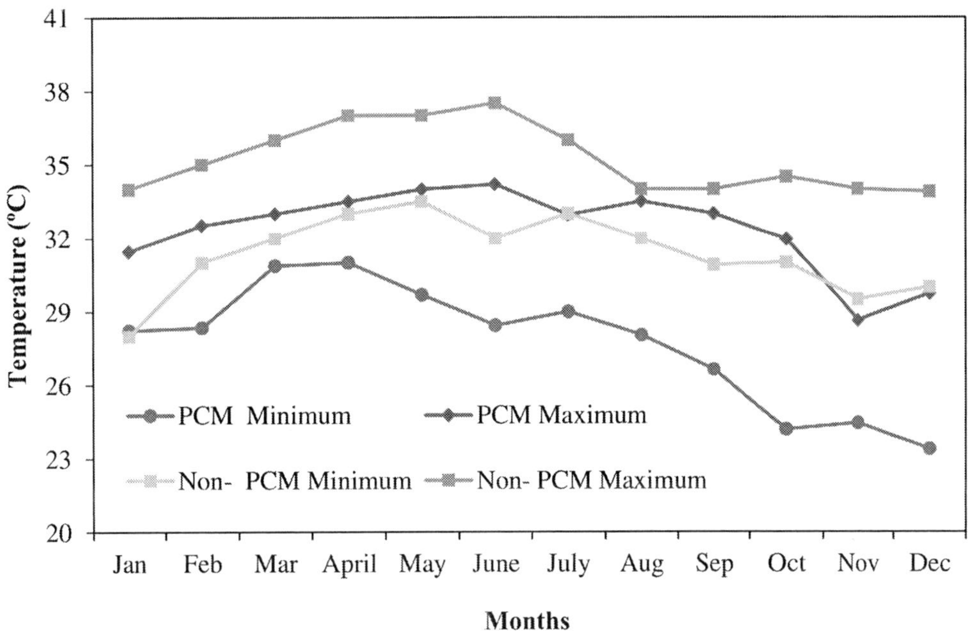


**Figure 8.** Monthly minimum and maximum experimental temperature variations of the PCM and non-PCM rooms

## 4. Conclusion

Several promising methods are taking place in the field of latent heat thermal energy storage system using PCMs integrated buildings. A detailed numerical simulation in PCM incorporation building architecture for space cooling was carried out using EnergyPlus software. It is the evident from the earlier studies that the thermal enhancement in a building due to the addition of PCMs depends on type of PCM, climatic condition, orientation of buildings, melting temperature, design. These parameters are optimized to determine the possibilities of successfully implementation PCM building materials. Therefore, the simulation analysis for the PCM building is definitely is a remarkable and quantifiable guidance for designing and of PCMs in building application. The selection of PCM for one climatic weather conditions will not be suitable for other climatic conditions.

The numerical simulation analysis was carried out with two identical buildings, both with same dimension, orientation and construction materials were used. One building integrated with PCM in building wall and another building was considered as non PCM building. The integrated PCM building wall was reduced the room temperature by 6°C in the month of January and March and a minimum temperature reduction is 3°C in May month. The selected PCM was performed well during the winter months and during summer months (April–July) a lesser temperature drop is observed. Hence, the PCM based passive cooling system is providing the reduction of indoor temperature and avoids high electrical consumption and reduces the greenhouse effect.

**Nomenclature:**

| | |
|---|---|
| C | : Specific heat capacity (J/kg °C) |
| T | : Nodal temperature, (°C) |
| $\rho$ | : Density of material |
| Q | : heat flux (W/m$^2$) |
| tm | : Phase change temperature (°C) |
| k | : Thermal conductivity (W/m °C) |
| t | : Thickness of the PCM layer (m) |
| L | : Liquid fraction temperature (°C) |
| LH | : Latent heat (J/kg) |

**References**

1. Ali, S 2014 Phase change materials integrated in building walls : A state of the art review *Renewable and Sustainable Energy Review.* **31** 870–906.

2. Liu, Y., Yang, L., Hou, L., Li, S., Yang, J., Wang, Q 2017 A porous building approach for modelling flow and heat transfer around and inside an isolated building on night ventilation and thermal mass *Energy* **141** 1914–1927.

3. Soares, N., Costa, J.J., Gaspar, A.R., Santos, P 2013 Review of passive PCM latent heat thermal energy storage systems towards buildings energy efficiency *Energy and Buildings* **59** 82–103.

4. Al-sanea, S.A., Zedan, M.F., Al-hussain S.N 2012 Effect of thermal mass on performance of

insulated building walls and the concept of energy savings potential *Applied. Energy* **89** 430–442.

5. Sadineni, S.B., Madala, S., Boehm R.F 2011 Passive building energy savings: A review of building envelope components *Renewable and Sustainable Energy Review* **15** 3617–3631.

6. Pasupathy, A., Athanasius, L., Velraj, R. 2008 Experimental investigation and numerical simulation analysis on the thermal performance of a building roof incorporating phase change material ( PCM ) for thermal management *Applied Thermal Engineering* **28** 556–565.

7. Kumar, S., Prakash, S.A., Pandiyarajan, V., Raj, V.A.A., Velraj, R. 2019 Effect of phase change material integration in clay hollow brick composite in building envelope for thermal management of energy efficient buildings *Journal of Building Physics.*

8. James, B., Delaney, P. 2012 Phase change materials: are they part of our energy efficient future *Proc. ACEEE summer study energy Effic. Build.* **24**, 160–172.

9. Athienitis, K. 1997 1-s2.0-S0360132397000097-*main.pdf* **32.**

10. Schossig, P., Henning, H., Gschwander, S. 2005 Micro-encapsulated phase-change materials integrated into construction materials *Solar Energy Materials and Solar Cell* **89** 297–306.

11. Menou, K., Castell, A., Farid, M.M., Boer, D., Cabeza, L.F. 2013 Life Cycle Assessment of experimental cubicles including PCM manufactured from natural resources ( esters ): A theoretical study *Energy and Buildings.* **51** 398–403.

12. Arce, P., Castellón, C., Castell, A., Cabeza, L.F. 2012 Use of microencapsulated PCM in buildings and the effect of adding awnings *Energy and Buildings* **44**, 88–93.

13. Pasupathy, A., Velraj, R. 2008 Effect of double layer phase change material in building roof for year round thermal management *Energy and Buildings* **40** 193–203.

14. Vicente, R., Silva, T. 2014 Brick masonry walls with PCM macrocapsules : An experimental approach *Applied Thermal Engineering* **67** 24–34.

15. Lee, K.O., Medina, M.A., Raith, E., Sun, X. 2015 Assessing the integration of a thin phase change material (PCM) layer in a residential building wall for heat transfer reduction and management *Applied Thermal Engineering* **137** 699–706.

16. Mi, X., Liu, R., Cui, H., Memon, S.A., Xing, F., Lo, Y.2016 Energy and economic analysis of building integrated with PCM in different cities of China *Applied Energy* **175** 324–336.

17. Alam, M., Jamil, H., Sanjayan, J., Wilson, J. 2014 Energy saving potential of phase change

materials in major Australian cities *Energy and Buildings* **78** 192–201.

18. Tardieu, A., Behzadi, S., Chen, J.J.J., Farid, M.M. 2011 Computer simulation and experimental measurments for an experimental PCM-impregnated office building. *Proc. Build. Simul. 2011 12th Conf. Int. Build. Perform. Simul. Assoc.* 56–63.

19. Tabares-Velasco, P.C., Christensen, C., Bianchi, M. 2012 Verification and validation of EnergyPlus phase change material model for opaque wall assemblies *Building and Environment* **54** 186–196.

20. Tabares-Velasco, P.C., Christensen, C., Bianchi, M. 2012 Verification and validation of EnergyPlus phase change material model for opaque wall assemblies *Building and Environment* **54** 186–196.

21. United States Department of Energy 2018 EnergyPlus Version 8.9.0 Documentation: *Engineering Reference*. 1716

22. Chan, A.L.S. 2011 Energy and environmental performance of building fac ¸ ades integrated with phase change material in subtropical Hong Kong *Energy and Buildings* **43** 2947–2955.

23. Zubillaga, O., Nogués, M., Medrano, M., Cabeza, L.F., Castellón, C., Leppers, R. 2006 Use of microencapsulated PCM in concrete walls for energy savings *Energy and Buildings* **39**, 113–119

24. Kuznik, F., Virgone, J.2009 Experimental assessment of a phase change material for wall building use *Applied Energy* **86**, 2038–2046.

25. Tabares-Velasco, P.C., Christensen, C., Bianchi, M.V.A. 2012 Validation methodology to allow simulated peak reduction and energy performance analysis of residential building envelope with phase change materials *ASHRAE Trans* **118** 90–97.

# Production and Characterization of Biodiesel from *Hevea Brasiliensis* and Diesel like Fuel from Waste Engine Oil

**Shreya Subedi[1], M Anand[2], Gokul Raghavendra S[3], Ranjitha J[3]***
[1]Department of Mechanical Engineering, School of Mechanical Engineering, Vellore Institute of Technology, Vellore-14
[2]Department of Science and Humanities, Kingston Engineering College, Vellore-59
[3]$CO_2$ Research and Green Technologies Centre, Vellore Institute of Technology, Vellore-14

*E-mail: ranjitha.j@vit.ac.in

**Abstract**. In today's application, it is essential to formulate the use of diesel and biodiesel in an environmentally generous manner. In this experimental study, an attempt was made to increase the performance and reduce the exhaust emission by blending biodiesel with diesel extracted from waste engine oil on their performance and emission characteristics. The bio-oil is extracted from the seeds of *Hevea Brasiliensis* using soxhlet extraction process. The crude bio-oil was converted into biodiesel using transesterification process. The percentage yield obtained was high, in the case of heterogeneous base catalyst CaO compared with homogenous base catalyst. Further, the trans-esterification reaction was optimized using the following condition reaction like temperature at 60°C, reaction time ~1.5 h, molar ratio 9:1 (oil: methanol) and 3% catalyst (w/w). The waste engine oil was distilled using fractional distillation process to produce fuels which is used in engines using inert silica granules. The bio-oil, bio-diesel and diesel samples were characterized using FT-IR and GC-MS spectroscopic techniques. Based on the spectral data, fatty acids, fatty acid methyl ester and hydrocarbon derivatives were identified. Similarly, physical properties were analysed for the produced fuel and biodiesel such as density, viscosity, flash and fire point and calorific value and the results are comparable with ASTM standards. Based on the fuel properties, biodiesel and diesel like fuel obtained from waste engine oil will be used in future to study their engine characteristics.

*Keywords*: *Hevea Brasiliensis*, waste engine oil, biodiesel, and GC-MS spectral analysis.

## 1. Introduction

As the world fossil fuel reserve is at a diminishing stage it is inevitable to find an alternative fuel, to obtain a proper solution for this impending problem researchers are forced to depend on self-sustaining energy resources. Among the globally available fuels, biodiesel has proved to be a remarkable fuel in the mark of history which was devised by Rudolf Diesel in 1990's to make his diesel engine work with the help of vegetable oil. Biodiesel has been acclaimed as an important substitute for conventional petroleum derived fuels, as concerned to its environmental impact and sustainability. Biodiesel is prepared by direct transesterification process of vegetable oil, animal fat or non-edible plant seed oils in which triglycerides reacts with the small molecules of alcohol such as methanol on behalf of catalyst. Relating to biodiesel production, the feedstock is of great importance that if they are grouped according to their degrees of purity, predominantly concerned to the level of free fatty acid. . Biodiesel is found to be a promising alternative fuel source as it has low content of Sox and Cox, organic as well as inorganic pollutants[1,2]. There are several methods adopted by researchers to convert bio-oil into biodiesel, which includes micro-emulsion, pyrolysis, transesterification etc. [3]. Transesterification is found to be the most effective method concerned to a main aspect that is biodiesel produced via this method is fully miscible with diesel in all kinds of proportion [4]. Recent studies has found that biodiesel produced from *Hevea brasiliensis* seed is a promising non edible plant seed which is widely available in the regions of south east Asia [5]. The average percentage weight of oil found in each kernel is about 40-50%[6] Homogeneous catalyst is usually adopted for biodiesel production in the case of vegetable oils, but the main disadvantages of using this type of catalyst includes high amount of soap formation ,large quantity of waste water production and also it results in corrosion of the reactor [7]. Numerous attempts has been made by researchers for the utilization of waste materials such as egg shells, animal bones, leaf extract, sea shells as the source of heterogeneous catalyst [8,9,10,11,12]. Lime stone based catalyst has shown efficient conversion of about 96.9% in the case of very high FFA *Hevea brasiliensis* seed oil into biodiesel [13]. Biodiesel produced from rubber seed oil in the presence of sodium metasilicate, calcium oxide based waste coral fragments has shown high amount of biodiesel yield in a single step transesterification reaction [14]. Several methods has been adopted in the previous studies for the extraction of rubber seed oil, includes solvent extraction and mechanical extraction methods(Santaso et.al) recent studies prevails that hydraulic press method is also efficient for oil extraction [15]. The hevea brasiliensis seed oil transesterification with the aid of homogeneous catalysts such as $H_2SO_4$, KOH has shown a yield of about 31-99% methyl esters depending on various conditions [16, 17]. In a study methyl propyl sulphonic acid – functionalized MCM-41 catalysts was used, which showed 96% of FAME yield [18]. Biodiesel produced from crude rubber seed oil via enzymatic transesterification using lipase immobilized on spherical silica aerogel under optimal condition yields about 93% of FAME [19]. The waste automotive engine oil produced throughout the world is concentrated to 24 million tons per year, the waste is hazardous hence it is a threat to the human society. A recent study has been conducted for the performance and emission characteristics in a diesel engine, biodiesel feedstock for this study was *moringa oleifra* and palm seed oil and this study reveals that the average emission of carbon monoxide for B5, B10 (13.7%, 17.3%) blend of biodiesel from palm oil was higher when compared to that of *moringa oleifra* (5.37%, 10.6%) [20]. The study conducted on cotton seed oil as the feedstock for biodiesel and the exhaust emission results obtained shows that there was a considerable reduction for particulate matter, carbon monoxide and as well as smoke emission [21]. The emission characteristics study conducted in a single cylinder diesel engine with Kapok oil biodiesel as the fuel can be compared to diesel for lower blends of B25 [22]. This paper focusses on Biodiesel production from Heavea brasiliensis seed and waste engine oil, and its performance emission characteristics on a four stroke diesel engine. The catalysts used for this research study is a CaO based heterogeneous catalysts named as Mytilus emulous.

## 2. Materials and Methods

2.1 The chemicals and reagents used for the extraction of biodiesel from the seeds of *Hevea Brasilinesis* are methanol, n-hexane, sulphuric acid.

2.2 Collection of seeds: *Hevea* brasiliensis seed were collected from Ernakulum district located in Kerala. The rubber shell was removed by manually and the seeds were cleaned. Approximately, 500 grams of rubber seeds were heated in an oven at 110°C for about 5 hours to remove moisture content for soxhlet extraction process. The rubber shell was removed by manually and the seeds were cleaned.

2.3 Extraction of bio-oil using Soxhlet Process: For this study, *Hevea* brasiliensis were taken and it was finely powdered. The equipment consist of a soxhlet extractor equipped with a round bottom flask with a capacity of 500 ml. About 250 g of finely powdered *hevea* seeds were taken ad it was weighed, and then it was transferred into a tea bag which functions as a filter media. After transferring the weighed powdered seed in the tea bag it was tightly tied to ensure that powder should not expel from it. Similarly, three bags were made and it was carefully placed in the soxhlet extraction tube. Then the cylindrical soxhlet extractor was taken carefully and was placed in the flask containing n- hexane as an organic solvent. Top of the soxhlet extraction tube is attached to a LB condenser, which operates on the circulation of water. The solvent n-hexane was heated to reflux the whole reaction mixture. When heated the solvent vapour passes up through the LB condenser and it falls into the chamber. The condenser setup ensures that it has zero loss of solvent vapour. The n-hexane solvent vapour gets cools and excluded back to the chamber. When the soxhlet extraction chamber gets fully filled, the chamber is emptied by symphonic action. After 8-10 hours of the reaction process and finally crude rubber seed oil was obtained.

2.4 Acid pre-treatment process: In this step one litre of crude rubber seed oil was taken, it requires 200ml of methanol to carry out acid treatment process. The crude rubber seed oil was carefully transferred into the conical flask and heated at a temperature of 50°c. After this step, methanol was added to the preheated rubber seed oil and it was carefully stirred for a few minutes. About 0.5% of sulphuric acid was added to the mixture, continuous heating and stirring was carried for 20-30 mints at atmospheric pressure. After completion of the reaction, and the product was poured into a separating funnel to separate the excess alcohol. The excess alcohol along with sulphuric acid and impurities get settled at the top layer and it was removed. The bottom layer was separated for alkaline transesterification.

2.5 Biodiesel transesterification: In this process, the lower layer separated during the acid pre-treatment process was taken in a conical flask and heated at a temperature of 55°c. Meanwhile, 3.5 % weight of prepared CaO based catalyst was dissolved in 300 ml methanol, mixed and it was poured into the flask. The mixture was heated and stirred about 30 mins, reaction was stopped and the mixture was allowed to separate into two layers. The bottom layer contained glycerol, then it was drawn off. The esters was formed in the upper layer, hot distilled water was sprayed over this layer and it was stirred gently.

2.6 Extraction of Waste Engine oil: The used waste engine oil was obtained and collected from an automobile Industry (Hyundai motors and services) located in Trivandrum district of Kerala. 5 litres of used automobile engine oil was collected .The oil was blackish in colour, the viscosity was very high and the obtained engine oil was used for refining process [23, 24].

## 3 Refining of waste engine oil

3.1. Removal of moisture content: Initially the crude waste automobile engine oil was taken in a cylindrical vessel and it was heated up to 100°C for a particular period of time in order to remove the moisture content.

3.2. Desulphurization: In order to remove the impurities, carbon content, and sulphur content in the moisture free waste engine oil, 8% of concentrated sulphuric acid was added and it was stirred well. This mixture was kept for 40 hours.

3.3. Fractional Distillation: A known amount of chemically treated crude oil was taken in a round bottom flask along with a few quantity of inert silica granules as shown in fig. The distillation column was set up, provided with a heat source at the end of the still pot. The chemically treated crude oil was heated to a temperature of 250-400°C. The condensate was cooled down using a graham condenser. When the desired distillation temperature was reached, the distillate was collected in a collection jar.

## 4. Results and Discussion

4.1 FT-IR spectra for bio-oil: The infrared spectra of the bio-oil sample shown in the figure 1. According to Figure 1, wave numbers of substantial functional groups in bio-oil were between 3288.99, 2976.92 and 2930.07 cm-1 as well as between 1690.32 cm-1, which indicated the existence of C =O bonds, C =C bonds, C–O bonds, and C–H bonds, etc. in bio-oil.

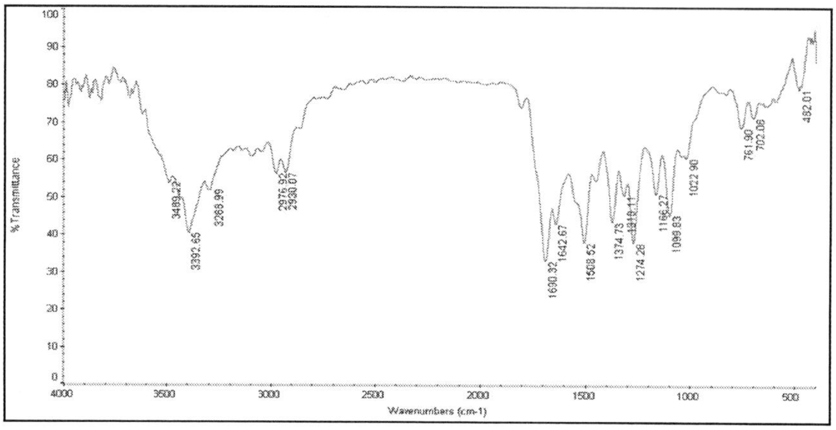


**Figure 1**: FT-IR for rubber seed oil

4.2 FT-IR spectra for waste engine oil: The infrared spectra of the bio-diesel sample shown in the figure 2. IR spectra clearly confirm the presence of functional groups in waste engine oils were –CH stretching of hydrocarbons in the region of 2975.64 and 2880.03 cm$^{-1}$. Similarly, on the side –CH bending of hydrocarbon derivatives were given peaks in the area of 1534.99 and 1410.93 cm$^{-1}$. Based on the FTIR results, clearly confirm the presence of hydrocarbon derivatives in the waste outboard engine oil.

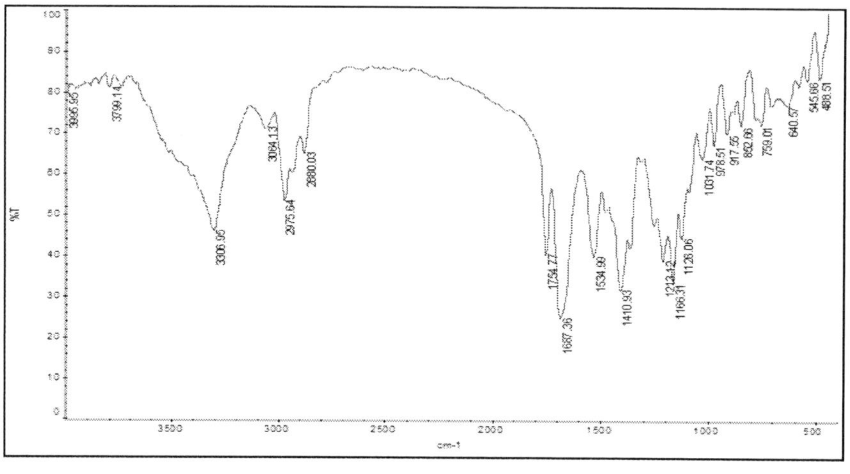


**Figure 2**: FT-IR for waste engine oil

4.3 Mass spectrometry library search: Identification of the components of the purified compound was matching their recorded spectra with the data bank mass spectra of NIST library V 11 provided by the instruments software. The rubber seed oil was analysed, and the composition of fatty acids identified. The individual peaks of the gas chromatogram were analysed, and the fatty acid components identified using MS database.

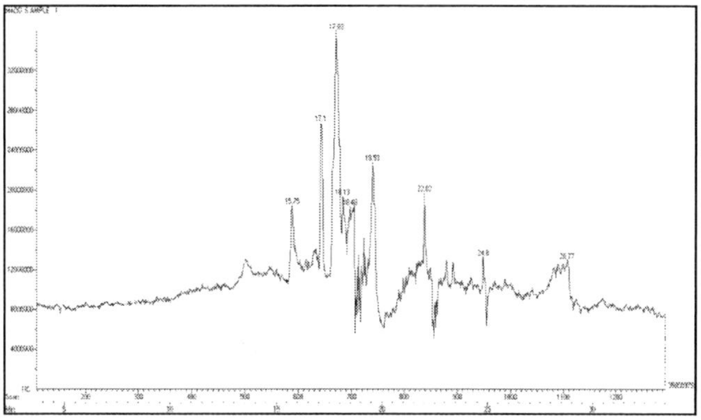

**Figure 3**: GC-MS for rubber seed oil

The relative percentage of fatty acid esters calculated from total ion chromatography by computerized integrator and results are presented in the Table 1. Totally five fatty acids were identified in the rubber seed oil. These fatty acids are only responsible for the production of bio-diesel using trans-esterification reaction. The fatty acids like Tetradecanoic acid, Pentadecanoic acid, n-hexadecanoic acid, 9-octadecanoic acid and hexadecanoic acid were identified using GC-MS NIST as shown in the figure 3.

**Table 1:** GC-MS Spectra for rubber seed oil

| Peak No. | RT (Min.) | Compound Name |
|---|---|---|
| 1 | 15.75 | Tetradecanoic acid |
| 2 | 17.10 | Pentadecanoic acid |
| 3 | 17.83 | n-hexadecanoic acid |
| 4 | 18.13 | hexadecanoic acid |
| 5 | 18.48 | Heptanoic acid |

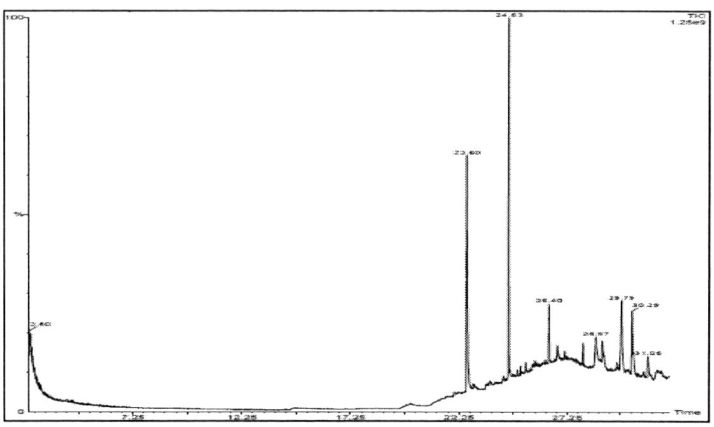

**Figure 4**: GC-MS for diesel like fuel

IOP Conf. Series: Materials Science and Engineering **923** (2020) 012071 doi:10.1088/1757-899X/923/1/012071

**Table 2:** GC-MS Spectra for diesel like fuel

| Peak No. | RT (Min.) | Compound Name |
|---|---|---|
| 1 | 22.60 | Tetradecanoic acid methyl ester |
| 2 | 24.53 | Pentadecanoic acid methyl ester |
| 3 | 26.40 | n-hexadecanoic acid methyl ester |
| 4 | 28.57 | hexadecanoic acid methyl ester |
| 5 | 30.29 | Heptanoic acid methyl acid |

The samples were analysed, and the composition of fatty acid methyl esters identified. The individual peaks of the gas chromatogram were analysed, and the fatty acid methyl ester components identified using MS database. The relative percentage of fatty acid esters calculated from total ion chromatography by computerized integrator and results are presented in the table 2. The biodiesel consists of Tetradecanoic acid methyl ester, Pentadecanoic acid methyl ester, n-hexadecanoic acid methyl ester, 9-octadecanoic acid methyl ester and hexadecanoic acid methyl ester were identified using GC-MS NIST as shown in the figure 4. The GC-MS scan also indicated that the concentration of fatty acid ethyl ester in the biodiesel is 98% due to the usage of heterogeneous catalyst in the transesterification. XRD patterns of synthesized CaO nanoparticles is shown in the Figure 5.

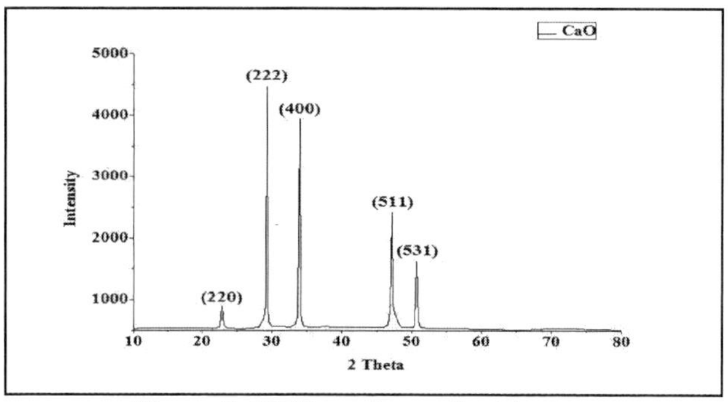


**Figure 5:** XRD spectrum of catalyst

XRD pattern shows that five different peaks were obtained at different 2 theta values. The strong peaks at 22.95, 29.31, 34.01, 47.10 and 50.78 are corresponded to the planes of CaO (2 2 0), (2 2 2), (4 0 0), (5 1 1) and (5 3 1), respectively (JCPDS card, 01-1160). These peaks corresponds to the standard peaks for CaO nanoparticles as mentioned in JCPDS card no 17-0912 and 01-1160.

**Figure 6**: SEM image of catalyst

The SEM result shows that synthesized CaO is crystalline in nature. The peak starts at 10 and ends at 50 nm values confirming the results of SEM analysis. The narrow peak indicates that the size range is not wider and all particles are similar in its dimension as shown in the figure 6.

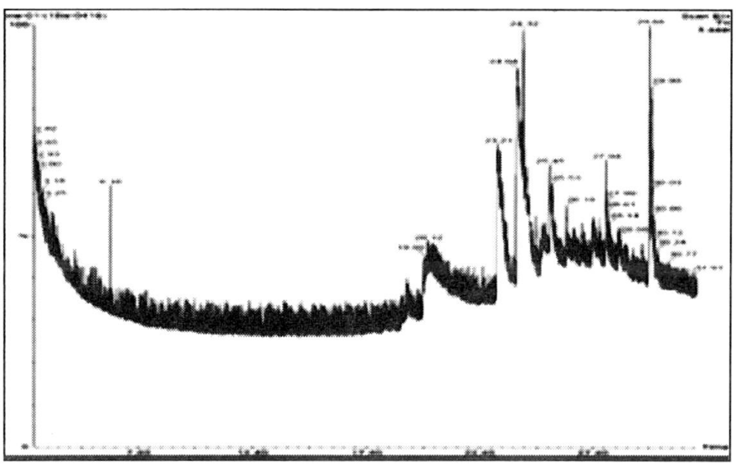

**Figure 7**: GC-MS Spectra of diesel like fuel

Waste is lubricating oil and distilled diesel samples were analysed using Perkin Elmer GC-MS in total scan mode to identify the hydrocarbon derivatives composition with the help of NIST library as shown in figure 7. The samples were analysed, and the composition of hydrocarbon derivatives identified. The individual peaks of the gas chromatogram were analysed, and the hydrocarbon derivative components identified using MS database. The relative percentage of hydrocarbons calculated from total ion chromatography by computerized integrator and results are presented in the Table 3.

**Table 3**: List of hydrocarbon compounds present in diesel like fuel

| S.No | Retention Time | Hydrocarbon compounds |
|---|---|---|
| 1 | 3.619 | Nonene |
| 2 | 4.519 | Decene |
| 3 | 6.010 | Cyclopropane |
| 4 | 6.730 | Tetradecene |
| 5 | 8.281 | Hexadecane |
| 6 | 8.441 | Undecane |
| 7 | 12.322 | Pentadecane |
| 8 | 13.503 | Hexadecane |
| 9 | 15.694 | Octacosane |
| 10 | 16.714 | Tridecane |

## 5. Conclusion

To provide sustainability and energy security, the renewable biodiesel was produced from the bio-oil of rubber seeds using heterogeneous catalyst. The synthesized calcium oxide (heterogeneous base catalyst) from Mytilus Emilious (CaO) showed effective conversion of crude bio-oil into biodiesel and enhanced the percentage yield of produced biodiesel 98%. The characterization studies revealed that the presence of FAME in the produced biodiesel from the rubber seed oil and diesel compounds lies in between the range of C9-C17 fatty acid composition. Physical properties for biodiesel from rubber seed and extracted diesel are within the ASTM standards limits. Diesel extracted from waste engine oil was found to have higher calorific value when compared to biodiesel. Based on the fuel properties, biodiesel and diesel like fuel obtained from waste engine oil will be used in future to study their engine characteristics.

## REFERENCES

[1] K. Neumann, K. Werth, A. Martín, A. Górak, Biodiesel production from waste cooking oils through esterification: Catalyst screening, chemical equilibrium and reaction kinetics, Chemical Engineering Research and Design, 107 (2016) 52-62.
[2] Meisam Tabatabaei, Keikhosro Karimi, Ilona Sárvári Horváth, Rajeev Kumar 5Recent trends in biodiesel production, Biofuel Research Journal 7 (2015) 258-267.
[3] M.A. Rajaeifar, A. Akram, B. Ghobadian, S. Rafiee, R. Heijungs, M. Tabatabaei, Environmental impact assessment of olive pomace oil biodiesel production and consumption: A comparative lifecycle assessment, Energy, 106 (2016) 87-102.
[4] M. Farooq, A. Ramli, A. Naeem, M. Saleem khan, Effect of different metal oxides on the catalytic activity of γ-Al2O3-MgO supported bifunctional heterogeneous catalyst in biodiesel production from WCO, RSC Advances, 6 (2016) 872-881.
[5] Yixibzhu, Jianchu Xu, Qiaohong Li, and Peter E Mortimer: Investigation of Rubber seed yield in Xishuanghanna and Estimation of Rubber seed oil Based Biodiesel potential in South Asia, Energy 69(2014) 837-842

IOP Conf. Series: Materials Science and Engineering **923** (2020) 012071 doi:10.1088/1757-899X/923/1/012071

[6] A S Ramdas, S Jayaraj and C. Muraleedharan: Biodiesel production from high FFA Rubber Seed oil Fuel, 84(2005)335-340
[7] S. Arumugam, Y.C. Kien, F. Paolo, K. Francis, Z. Sergey, M. Stanislav, Catalytic Applications in the Production of Biodiesel from Vegetable Oils, Sus., Chem, 2 (2009) 278 – 300.
[8] Y.H. Tan, M.O. Abdullah, C. Nolasco-Hipolito, Y.H. Taufiq-Yap, Waste ostrich- and chicken-eggshells as heterogeneous base catalyst for biodiesel production from used cooking oil: Catalyst characterization and biodiesel yield performance, Applied Energy, 160 (2015) 58-70.
[9] G. Joshi, D.S. Rawat, B.Y. Lamba, K.K. Bisht, P. Kumar, N. Kumar, S. Kumar, Transesterification of Jatropha and Karanja oils by using waste egg shell derived calcium based mixed metal oxides, Energy Conversion and Management, 96 (2015) 258-267.
[10] W. Roschat, T. Siritanon, T. Kaewpuang, B. Yoosuk, V. Promarak, Economical and green biodiesel production process using river snail shells-derived heterogeneous catalyst and co- solvent method, Bioresource Technology, 209 (2016) 343-350.
[11] A. Birla, B. Singh, S.N. Upadhyay, Y.C. Sharma, Kinetics studies of synthesis of biodiesel from waste frying oil using a heterogeneous catalyst derived from snail shell, Bioresource Technology, 106 (2012) 95-100.
[12] A. Obadiah, G.A. Swaroopa, S.V. Kumar, K.R. Jeganathan, A. Ramasubbu, Biodiesel production from Palm oil using calcined waste animal bone as catalyst, Bioresource Technology, 116 (2012) 512-516.
[13] Jolius Gimbun,, Shahid Ali Chitra Charan Suri Charan Kanwal, Liyana Amer Shaha, Nurul Hidayah Muhamad Ghazali, Chin Kui Cheng,, Said Nurdin :Biodiesel Production From Rubber Seed Oil Using Activated Cement Clinker As Catalyst, Procedia Engineering , 53 ( 2013 ) 13 –19.
[14] Wuttichai Roschat, Theeranum Siritanon, Boonyawan Yoosuk, Taweesak Sudyoadsul, and Vinich Promarak: Rubber seed oil as potential non-edible feedstock for biodiesel production using heterogeneous catalyst in Thailand, Renewable Energy 101(2017) 937-944.
[15] Santaso H, Iryanto, Inggrid M. Effects of temperature, pressure, preheating time and pressing time on rubber seed oil extraction using hydraulic press., Procedia Chemistry 2014; 9:248-56
[16] A.S. Silitonga, H.H. Masjuki, H. Chyuan, T. Yusaf, F. Kusumo, T.M.I. Mahlia, Synthesis and optimization of Hevea brasiliensis and Ricinus communis as feedstock for biodiesel production : A comparative study. Ind. Crop. Prod. 85 (2016) 274–286
[17] S.H. Dhawane, T. Kumar, G. Halder, Biodiesel synthesis from Hevea brasiliensis oil employing carbon supported heterogeneous catalyst: Optimization by Taguchi method. Renew.Energ.89 (2016) 506–514.
[18] S. Karnjanakom, S. Kongparakul, C. Chaiya, Biodiesel production from Hevea brasiliensis oil using SO3 H-MCM-41 catalyst. J. Environ. Chem. Eng. 4 (2016) 47–49.
[19] Arumugam. A, Thulasidharan. D, Gautham B. Jegadeesan: Process optimization of biodiesel Production from Hevea brasiliensis Oil using Lipase Immobilized on Spherical Silica Aerogel Renewable energy 2017.
[20] M. Mofijur, H.H. Masjuki, M.A Kalam , A.E. Atabani, I.M. Rizwanul Fattah, H.M. Mobarak :Comparative evaluation of performance and emission characteristics of Moringa oleifra and Palm oil based biodiesel in a diesel engine, Industrial crops and products 53(2014) 78-84.
[21] Md. Nurun Nabi, Md. Mustafizur Rahman, Md. Shamim Akhter: Biodiesel from cotton seed oil and its effect on performance and exhaust emissions, Applied Thermal Engineering 29(2009)2265-2270.
[22] S. Vedharaj, R. Vallinyagam, W.M. Yang, S.K Chou, KJE. Chau, P.S Lee: Experimental investigation of kapok (Ceiba pentandra) oil biodiesel as an alternative fuel for diesel engine, energy Conversion and Management 75(2013)773-779.
[23] Pirro, D.M. and Wessol, A.A., 2001, In-plant handling and purification for lubricant conservation, In: Lubrication Fundamentals, Second Edition, Marcel Dekker, Inc., New York, pp.19-1/19-24.
[24] Palani, A & Pandurangan A., 2005, Esterification of terephthalic acid with methanol over Mesoporous Al-MCM-41 molecular sieves, Journal of Molecular catalysis A: chemical 245 (2006) 101-105.

IOP Conf. Series: Materials Science and Engineering **923** (2020) 012072 doi:10.1088/1757-899X/923/1/012072

# Process Optimisation and Kinetics Study of Biodiesel Production from Dairy Waste Scum Using ZnO Heterogeneous Nanocatalyst

**Utsav Sharma[1], Muralidharan N.G[2], Vijayalakshmi S[2], Ranjitha J[2]***

[1]Department of Energy Engineering, School of Mechanical Engineering, Vellore Institute of Technology, India
[2]$CO_2$ Research and Green Technologies Centre, Vellore Institute of Technology, Vellore-14
***E-mail:** *ranjitha.j@vit.ac.in

**Abstract**. The present paper describes about the biodiesel production from diary waste using zinc oxide as a heterogeneous catalysts. The bio-oil from dairy waste scum was collected and used for biodiesel production. Zinc oxide nanostructures was synthesised by sol-gel method and characterised using particle size analyser, X-ray diffractometer and thermogravimetric analyser. The synthesised heterogeneous ZnO nanocatalyst was used for transesterification of dairy waste scum. The maximum biodiesel yield of 94.8% was obtained at the optimised reaction conditions, 0.8 wt. %, and methanol to oil molar ratio 9:1, reaction time 40 min, reaction temperature 65 °C, and mixing speed 300 rpm. The fatty acid methyl ester production reaction follows a first order reaction rate. This research work shows that synthesised ZnO nanocatalyst was an effective catalyst to produce biodiesel from dairy waste scum.

Key Words: Dairy waste, Scum oil, Transesterification, Kinetics, ZnO nanocatalyst.

## 1. Introduction

In the past decade, our environment has been polluted to a higher level, due to the increase in the consumption of non-renewable fuels. Consequently, Energy produced from non-renewable resources such as, coal and crude oil releases higher amount of toxic gases into the environment. These toxic gases can produce pollution creating activities resulted in global warming and ozone layer depletion [1]. The exhaust gases from vehicles and automobiles can also causes pollution in the environment. To minimize the pollution levels, alternative energy sources are preferred. These sources are locally available and causes lesser amount of gas emission. Such sources reduces pollution caused due to the chemical wastes and radioactive wastes [2]. Always, Renewable resources can provide clean and unlimited energy. Biodiesel is a long chain fatty acids comprising of mono methyl (or) ethyl esters. Biodiesel is a liquid fuel produced from oils, fats and lipids. These oils, fats and lipids are from renewable resources such as plants, animals and algae. Biodiesel can be used as a substitute for conventional petro diesel fuel. In a transesterification reaction, the triglycerides react with alcohol to form alkyl esters and glycerol as a byproduct. In the reaction mechanism, three moles of alcohol is required to react with one of triglyceride mole and form three moles of alkyl ester. It is a reversible reaction as shown in Eqn. 1. The reaction stoichiometry is 3:1 alcohol to oil. Usually alcohol is used in excess amount to shift the reaction equilibrium towards the product side [3]. Practically alcohol to oil molar ratio is more than 3:1 to increase the yield of the product.

$$TG + 3\,CH_3OH \xleftrightarrow{Catalyst} 3RCOOCH_3 + GL \qquad (1)$$

Biodiesel is non-toxic to the environment. It is a clean and completely burning fuel and also has high lubricity. The diesel engines can be directly fueled with biodiesel without any engine modifications. Biodiesel during combustion reaction produces lesser amount of air polluting substances compared to conventional diesel [4]. The word biodiesel itself implies that it is easily biodegradable and safer to the environment as it is from renewable resources. Biodiesel production activities encourages growth of local and rural economy. The production activity does not create any pollution. Several organizations promote the use of biodiesel. The physical and chemical properties of biodiesel matches with all international standards and it can be used directly (or) blended with diesel in any proportion from B0 to B100. The alternative fuels derived from oil from seeds, fat from animals and lipids from algae has a good scope in future due to the increasing pollution problems from petroleum resources. According to Knothe *et al.*, (2005), fueled an IC compression ignition engines with biodiesel proved that it is cleaner and safer than diesel [5]. The scarcity of fossil fuel motivates the researches to search a new biomass as a source of energy. The first generation feedstocks poses a threat to the food supply [6] and [7]. Anil Kumar Sarma *et al.*, (2005) reported that the preparation of biodiesel from Koroch (Pongamia glabra, another species under genus Pongamia) seed oil with 6:1 molar ratio and 1 g of $NaOCH_3$ [8]. Viscosity, density, cetane number, calorific value, IBP/FBP etc of resultant biodiesels was also evaluated. Vincent and Chakravarthi (2009) reported a novel method for improved manufacture of green biodiesel for which one or more solid metallic oxide base catalysts were selected from CaO, CaO-$Al_2O_3$, $CaTiO_3$, $BaTiO_3$, MgO-Al2$O_3$, Ti oxide, V oxide, Co oxide, Fe oxide, Hydrotalcite $Mg_6Al_2(CO_3)(OH)16.4H_2O$, Magnetite, Mg silicate and Ca silicate [9]. Kawashima *et al.*, (2009) reported that metal oxides of Ca, Mg and Zn are effective heterogeneous catalytic system for biodiesel preparation [10]. Research has been carried out in the field of CaO activity enhancement by pretreatment of CaO with methanol. The maximum yield of 90% was achieved when 0.1 g CaO was used as a catalyst. The most widely used heterogeneous catalysts for transesterification of triglycerides are metal oxides and mixed metal oxides. Kouzu *et al.*, (2008) investigated the catalyst activity of CaO, Ca $(OH)_2$ and $CaCO_3$ towards the transesterification of triglycerides of soybean oil and reported that CaO shows better activity [11]. Albuquerque *et al.*, (2008) synthesized MgO, CaO and tested for transesterification of ethyl butyrate [12]. Deepanraj *et al.*, (2017) reported biofuel from the blends of Jatropha (*Jatropha curcas*) and Karanja (*Pongamia pinnata*) oil were evaluated for their combustion properties. The biodiesel blends, the thermal efficiencies were 29.0% and 28.6%, respectively. The prolonged combustion was observed for Karanja oil blend in comparison to *Jatropha* oil blend [13]. Deepanraj *et al.*, (2012) reported performance and emission analyses for rice bran oil methyl ester (biodiesel) blends with diesel. The results reveal that blends of rice bran methyl ester with diesel up to 20% by volume (ZOBD) provide better engine performance and improved emission characteristics [14]. The present research work is mainly based on the biodiesel production from diary waste using a heterogeneous catalysts as ZnO.

## 2. Materials and Methods

2.1 Dairy Waste Scum Collection: The scum was collected from the area where the wash waters enter the effluent treatment section of the industry. Top layer is collected in glass beaker. Beaker was heated to a temperature of 60 °C to melt the fat into liquid phase. The heated oil was filtered to remove any suspended foreign particles from the sample. The Filtered oil was then centrifuged at 5000 rpm to remove water layer. The oil is then heated to remove any moisture content present in the oil.

2.2 Synthesis of ZnO nanocatalyst: The zinc Oxide nanoparticles was synthesized by sol-gel method. The zinc acetate and sodium hydroxide was used as precursors. 3 grams of $Zn(COOCH_3)_2$ was dissolved in 20 ml of deionized water. 15 ml of 0.1 M sodium hydroxide solution was prepared separately and poured dropwise into the zinc acetate solution under constant stirring for 30 minutes. After the addition of NaOH, 100 ml of ethanol was slowly added to the mixture and a white precipitate was formed. This precipitate was washed with distilled water and with ethanol. The ZnO nanoparticles were obtained after drying at 110 °C for 10 hours. The synthesis of ZnO nanoparticles is chemically given by Eqn. 2.

$$\boldsymbol{Zn(CH_3COO)_2.2H_2O + 2NaOH \longrightarrow ZnO + 2NaCH_3COO + H_2O} \qquad (2)$$

2.3 Characterization of ZnO nanocatalyst: A Mastersizer laser diffraction particle size analyzer (Malvern Instruments, Worchestershire, UK) was used. About 0.1 g of nanoparticle was analyzed at ambient temperature (20±2 °C). Crystallographic analysis of the nanoparticles was obtained by an X-ray powder diffractometer STADI P (STOE) with Ge (1 1 1) monochromatized Cu Kα radiation (λ = 1.54187A°). Diffraction data were collected at scanning speed of 2°/min over the 2θ range of 10-80°. XRD involves the diffraction of X-rays by the lattice of crystalline materials. In addition to the measurement of small crystallite sizes, X-ray diffraction can also be used to identify the crystalline phases, to measure the amount of those phases in the mixture, to determine crystal structures, and to determine the unit cell size of the crystalline material. Thermal stability of the nanoparticles was determined using TGA 7 (Perkin-Elmer) instrument operating in dynamic mode (heating rate = 10 °C/min under helium gas), was employed. The samples of less than 3 mg were placed.

2.4 Transesterification Reaction: A double necked one litre round bottomed reaction flask is placed over a magnetic stirrer with hot plate. One neck was used for reflux condenser and thermometer was inserted through the other neck to measure the reaction temperature. The heating mantle temperature can be adjusted as per the desired reaction temperature with a temperature controller of accuracy ± 0.5 °C. The magnetic pellet is taken inside the round bottom flask and the stirring speed of the reaction mixture can be adjusted with a control knob provided in the magnetic stirrer. The condenser is connected to a running water to reflux the methanol used. For each experimental study, the reaction flask was filled with 500 ml of dairy waste scum oil and then heated to the required reaction temperature. Transesterification reaction was conducted by the addition of a specified quantity of methanol and ZnO nanocatalyst. After addition of all reactants into the reactor, it was left undisturbed until the desired reaction time. The percentage of biodiesel yield was calculated by the volume of biodiesel produced divided by the volume of oil used, multiplied by 100.

## 3. Results and Discussion

3.1 Characterization Studies of ZnO Nanocatalyst

3.1.1 Nano Particle Size Analysis: The average particle size of the ZnO nanoparticle was found using the data from a particle size analyser. The size and intensity data was plotted as shown in the figure 1. The average particle size was 83.12 nm. This result shows that synthesised ZnO is crystalline in nature. The peak starts at 60 and ends at 100 nm which shows that the particle size is distributed between this specified ranges. The narrow peak indicates that the size range is not wider and all particles are similar in its dimensions and proves that the method used for synthesis process is good.

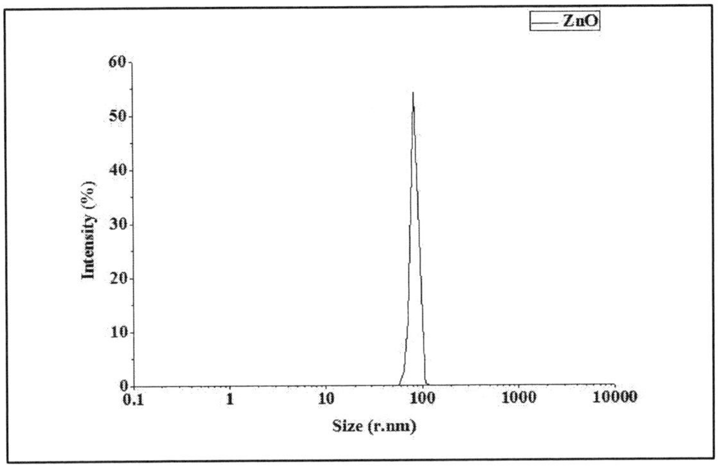


**Figure 1: Particle Size Analyser for ZnO nanocatalyst**

3.1.2 XRD Analysis: XRD patterns of the synthesized ZnO nanoparticles is shown in the figure 2. XRD pattern shoes that five different peaks were obtained at different 2 theta values. The highest peak is at angle 33.92° at (101) plane with 4781 intensity and at angle 36.40° with intensity 4085. These peaks corresponding to the standard peaks for ZnO nanoparticles as mentioned in JCPDS card no 36-1451. The sharper peaks shows that the particles are in nanosize as well as crystalline in nature. The crystalline structure of ZnO nanoparticles is again confirmed with these results.

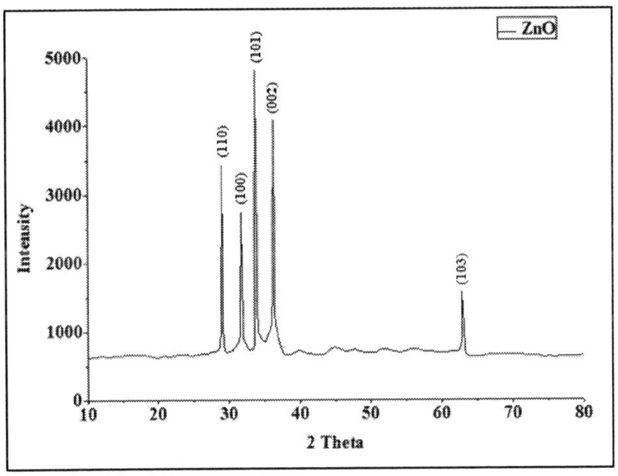


**Figure 2: XRD Patterns for ZnO**

3.1.3 Thermo-Gravimetric Analysis: The thermal stability of ZnO nanoparticles can be easily studied with the help of a thermogravimetric analysis. A plot between the temperature and the weight of the sample is shown in the figure 3. There is no weight loss till the temperature of 300 °C which indicates that the drying step was completely done and there is no moisture content in the nanoparticles. After 300 °C a sharp decrease of about 23% in the weight of the sample is due to the presence of zinc acetate precursor present in the sample. After this weight loss, the weight of the ZnO sample loaded is constant throughout the analysis which shows that it can withstand even at a temperature of 800 °C.

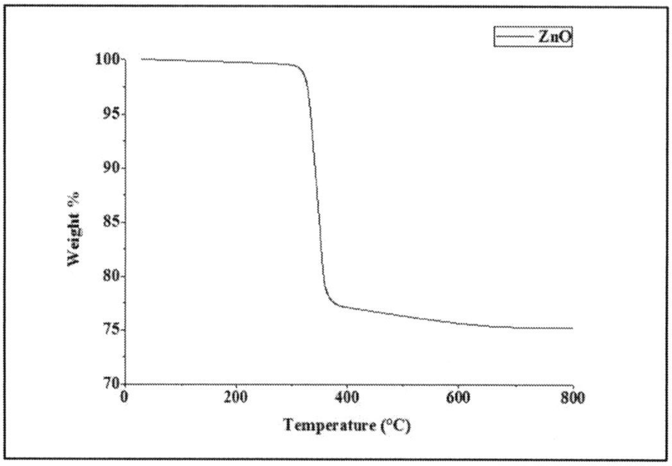


**Figure 3: TGA Plot for ZnO**

3.2 Biodiesel production

3.2.1 Catalyst optimisation: The catalyst quantity used in this reactions affects the reaction rate drastically though it doesn't alter the equilibrium conversion of the reversible reactions. If the catalyst is having more activity, less amount of the catalyst is required. In order to scrutinize the activity of ZnO catalyst, different weight ratios of catalyst to oil were used for the catalyst optimization studies of transesterification of a variety of mass ratios of catalyst to oil were used for transesterification process. Reactions were carried at constant parameters of reflux temperature, reaction time, molar ratio of methanol to oil and stirring speed. Figure 4 shows the effect of mass ratio of ZnO catalyst to oil on the transesterification reaction. 90.1% conversion was obtained when 0.8 wt. % of catalyst was used.

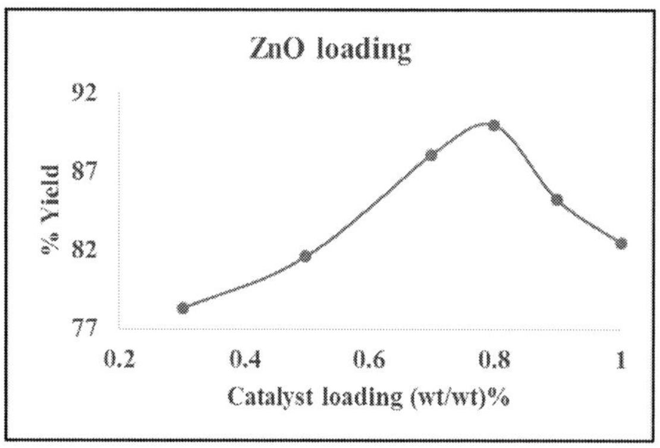


**Figure 4: Effect of catalyst loading**

3.2.2 Reaction temperature optimisation: The carbonyl activity of triglyceride have a tendency to improve its activity to initiate the reaction. The long chains of triglyceride restricts the carboxylic group activation. It is a complex reaction attained by the nucleophilic attack of the alcohol used. The carbonyl group activation can be done only at a temperature ranging the boiling point of methanol. Further, the reactions were conducted at different temperature using ZnO catalyst and with optimized values of catalyst loading. Figure 5 shows how the biodiesel yield varies at different reaction temperatures. From this figure, it is clear that biodiesel yield increases with the temperature. Higher yields of biodiesel 92.4 wt. % was achieved at 65 °C for ZnO catalysed reaction.

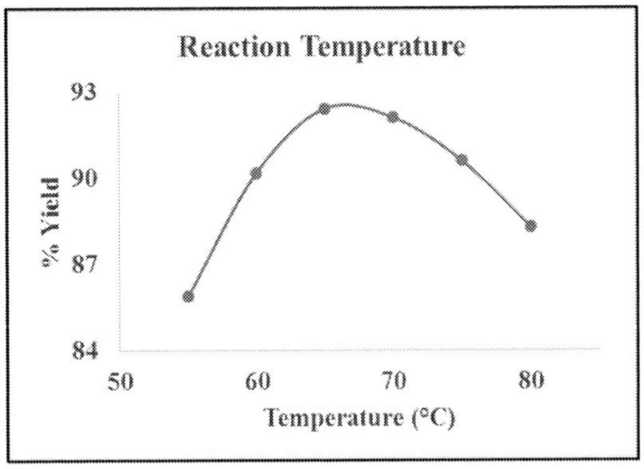


**Figure 5: Effect of reaction temperature**

3.2.3 Reaction time optimisation: The behaviour of reaction completion time was studied at different time intervals ranging from 20 to 70 min for ZnO catalysed reactions is shown in figure 6. The catalyst amount and reaction temperature values were previously optimized. The optimum conversion of 92.4 was observed at 40 min, when ZnO catalyst was used as a catalyst. When the reaction temperature increases, the rate of the reaction also increases and thus reducing the overall time consumed for the completion of reaction. Maximum biodiesel yield can be achieved if sufficient contact time is given for the reactants to collide [15].

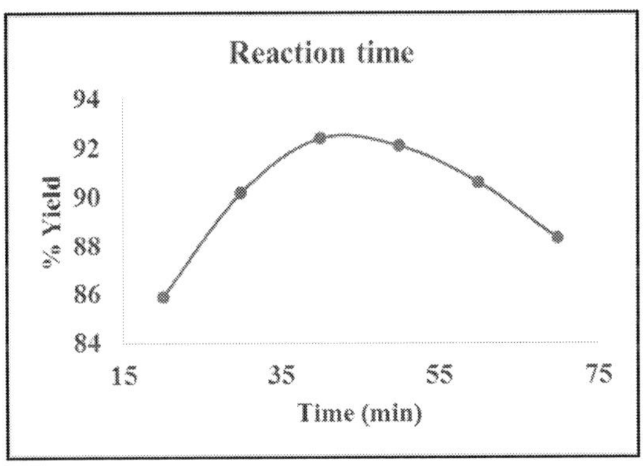


**Figure 6: Effect of reaction time**

3.2.4 Alcohol optimisation: The methanol to oil molar ratio is the most important parameter that influences the conversion of oil into ester. As per the stoichiometric calculations, 3 moles of methanol are required per mole of triglyceride. In practical, the molar relative amount of methanol is always maintained more than the stoichiometric ratio to drive the transesterification reaction to achieve complete conversion [16]. In this examination, the effect of molar ratio of methanol to dairy waste scum oil was investigated with different molar ratios of 3:1 to 15:1. Each reaction was run at individual optimized parameter values of catalyst concentration, reaction temperature and time for each catalyst. The changes in biodiesel yield according to the methanol oil molar ratio is shown in the figure 7. When ZnO catalyst was used, the conversion of triglyceride was observed as 94% at the 9:1 molar ratio of methanol to dairy waste scum oil with 40 min reaction time. The conversion of triglyceride was found to be increased with increase in molar ratio from 3:1 to 9:1. Further increase in molar ratio from 9:1 to 12:1 leads to decrease in the conversion of triglyceride. This is due to the presence of mass transfer limitation during the transesterification reaction.

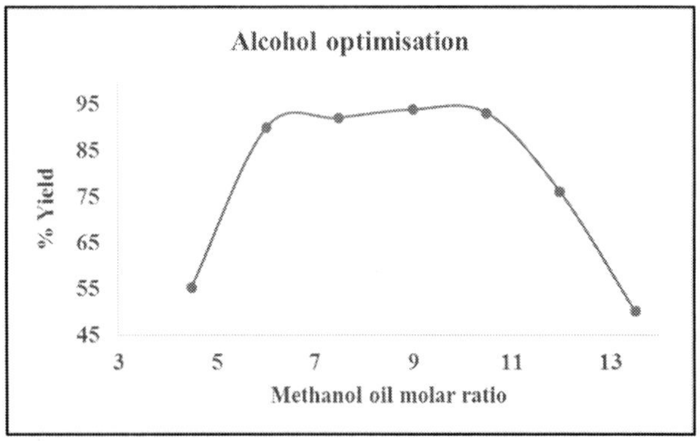


**Figure 7: Effect of methanol oil molar ratio**

3.2.5 Stirring speed optimisation: The experimental studies were done at different rpm, such as, 150, 200, 250, 300, 350 and 400 to optimize the stirring rate of the reactants. From figure 8 the mixing speed of 300 rpm produced 94.8% of FAME for ZnO catalyst. Therefore, 300 rpm mixing intensity can be maintained for dairy waste scum biodiesel production. Sivakumar et al. (2011) obtained similar results in stirring speed optimization studies [17]. They reported that 350 rpm mixing intensity was suitable for the production of dairy waste scum oil methyl ester.

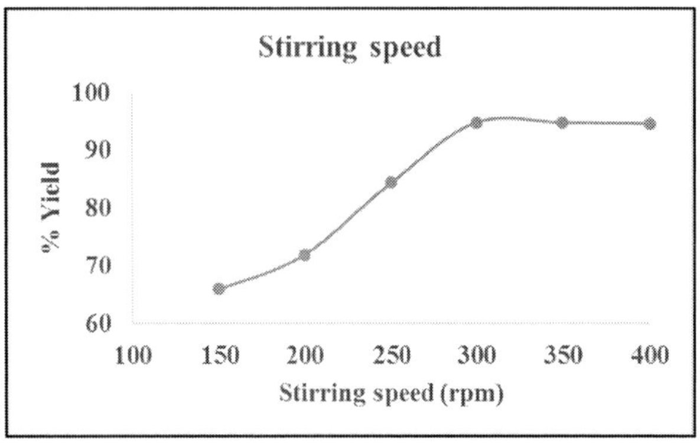


**Figure 8: Effect of stirring speed**

3.3 Kinetics of transesterification of dairy waste scum oil: The kinetic studies was carried out for the transesterification of dairy waste scum oil at the optimum reaction conditions. The reaction is based on first order and followed by the function of concentration of methyl ester. The rate constant of the reaction was determined based on the amount of the product take place at the reaction time interval [18]. In the present work, the increased amount of methyl ester reactant was selected for the kinetics study. Consequently the first order rate constant of the reaction was expressed in the Eqn. 3.

$$Rate = \frac{d(ME)}{dt} \quad (3)$$

Supplementary, a plot of dME/dt vs yield using log graph gives straight line involved in the transesterification reaction shows first order as shown in the figure 9. The $R^2$ value of the straight line found to be 94% good agreement and then additionally shows that the speed of reaction varies linearly with yield.

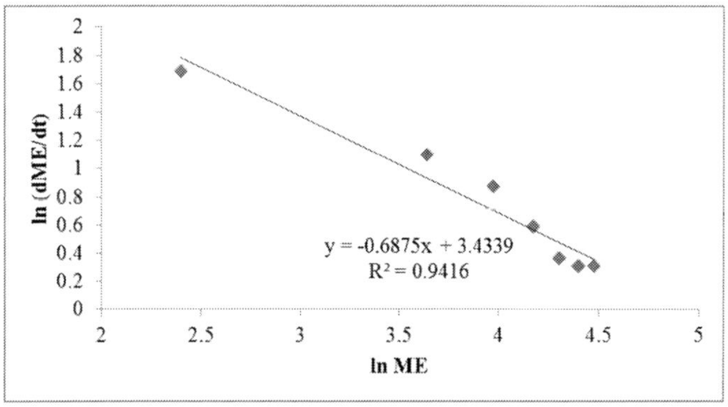


**Figure 9: A linear graph reaction rate vs yield on logarithmic graph (%) for transesterification of dairy waste scum oil.**

A plot relating ln k and inverse of temperature was plotted as shown in figure 10. This plot is used to determine the activation energy required using Arrhenius equation.

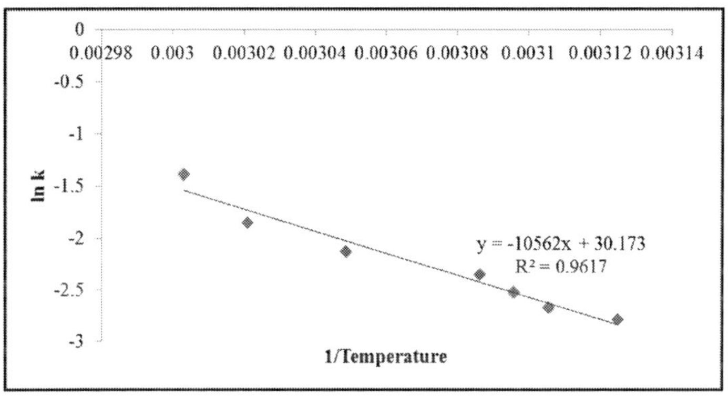


**Figure 10: A linear graph of Arrhenius equation (ln k v/s 1/T)**

Conclusion: The ZnO nanocatalyst was synthesised via sol-gel method and characterized using three different analysis. The synthesised heterogeneous catalyst was used to produce biodiesel from dairy waste scum oil. The transesterification reaction parameters were optimized and the optimum conditions are 0.8 wt. % ZnO nanocatalyst, 9:1 methanol to oil molar ratio, reaction completion time of 40 min, reaction temperature of 65 °C and stirring speed of 300 rpm. These conditions are similar to those reported previously [19] and [20]. The maximum yield of 94.8% was achieved when nano ZnO was used as a catalyst and the reaction follows first order kinetics.

## References

1. Adams, C., Peters, J. F., Rand M. C., Schroer, B. J. & Ziemke, M. C. (1983). Investigation of soybean oil as a diesel fuel extender: Endurance tests. Journal of the American Oil Chemists' Society 60, 1574-1579.

2. Coteron, A., Vicente, G., Martinez, M. & Aracil, J. (1997). Biodiesel production from vegetable oils. Influence of catalysts and operating conditions. Recent Research Developments in Oil Chemistry 1, 109-14.
3. Canakci, M., & Sanli, H. (2008). Biodiesel production from various feed stocks and their effects on the fuel properties. Journal of Industrial Microbiology and Biotechnology 35(5), 431-441.
4. Macedo, C. C. S., Abreu, F. R., Tavares, A. P., Alves, M. B., Zara, L. F., Rubim, J. C., & Suarez, A. Z. (2006). New heterogeneous metal- oxide based catalyst for vegetable oil trans-esterification. Journal of the Brazilian Chemical Society 17(7), 1291-1296.
5. Knothe, G., Karhl, J. & Van Gerpen, J. (2005). The biodiesel hand book. IL (USA): AOCS press Champaign.
6. Lam, M. K., Tan, K. T., Lee, K. T. & Mohamed, A.R. (2009). Malaysian palm oil: surviving the food versus fuel dispute for a sustainable future. Renewable & Sustainable Energy Reviews 13(6-7), 1456-1464.
7. Metzger, J. O. (2009). Fats and oils as renewable feedstock for chemistry. European Journal of Lipid Science and Technology 111(9), 865-876.
8. Anil Kumar Sarma, D. Konwer & P. K. Bordoloi (2005). A Comprehensive Analysis of Fuel Properties of Biodiesel from Koroch Seed Oil. Energy & Fuels 19, 656-657.
9. Vincent, M. J. & Chakravarthi, K. D. (2009). Transesterification catalysts for green biodiesel production. US patent 20060530.
10. Kawashima, A., Matsubara, K. & Honda, K. (2009). Acceleration of catalytic activity of calcium oxide for biodiesel production. Bioresource Technology 100, 696-700.
11. Kouzu, M., Kasuno, T., Tajika, M., Sugimato, Y., Yamanaka, S., & Hidaka, J. (2008). Calcium oxide as a solid base catalyst for transesterification of soybean oil and its application to biodiesel production. Fuel 87, 2798-2806.
12. Albuquerque, M. C. G., Azevedo, D. C. S., Cavalcante Jr, C. L., Gonzalez, J. S., Robles, J. M. M., Tost, R. M., Castellon, E. R., Lopez, A. J., & Torrres, P. M. (2008). Transesterification of ethyl butyrate with methanol using MgO/CaO catalyst. Journal of Molecular Catalysis A 300(1-2), 19-24.
13. Antonlin, G., Tinaut, F. V., Briceno, Y., Castano, V., Perez, C. & Ramirez, A. I. (2002). Optimization of biodiesel production by sunflower oil transesterification. Bioresource Technology 83(2), 111-114.
14. Kian Fei Yee, Keat Teong Lee, Riccardo Ceccato & Ahmad Zuhairi Abdullah (2011). Production of biodiesel from Jatropha curcas L. oil catalyzed by SO42-/ZrO2 catalyst: Effect of interaction between process variables. Bioresource Technology 102, 4285-4289.
15. Ramachandran, K., Sivakumar, P., Suganya, T. & Renganathan, S. (2011). Production of biodiesel from mixed waste vegetable oil using an aluminium hydrogen sulphate as a heterogeneous acid catalyst. Bioresource Technology 102, 7289-7293.
16. Sivakumar, P., Anbarasu, K. & Renganathan S. (2011). Bio-diesel production by alkali catalyzed transesterification of dairy waste scum. Fuel 90, 147-151.
17. Guo, F., Peng, Z.G., Dai, J.Y. & Xiu, Z.L. (2010). Calcined sodium silicate as solid base catalyst for biodiesel production. Fuel Processing Technology 91, 322-328.
18. B Deepanraj, M Srinivas, N Arun, G Sankaranarayanan, P Abdul Salam (2017). Comparison of jatropha and karanja biofuels on their combustion characteristics, International Journal of Green Energy,14(15) 1231-1237.
19. B Deepanraj, G Sankaranarayanan, P Lawrence 2012. Performance and emission characteristics of a Diesel engine fueled with rice bran oil methyl Ester blends, Daffodil International University Journal of Science & Technology, 7(2), 51-55.
20. Xiu, Z.L. & Guo, F. (2010). Application of novel solid acid and base catalysts in biodiesel synthesis. Journal of Biotechnology 150, S178.

# Performance and heat transfer evaluation of convex lens solar water heater with serrated, triangular cut and perforated twist inserts.

**S Thulasi[1], S Sivasubramaniam[2], V Thirumaran[3]**

[*1] *Assistant professor, Department of Mechanical Engineering, University College of Engineering, Thirukkuvalai, Nagapattinam- Dt, Tamilnadu, India. E-mail: thulasi_oxford@yahoo.com.* Tel.:+91 04366 245234.

[2] *Lecturer (Sr. Grade), Department of Foundry Technology, PSG Polytechnic College, Coimbatore- Dt, Tamilnadu, India.* Tel: +91 044-2572177

[3] *Final year Student, Department of Mechanical Engineering, University College of Engineering, Thirukkuvalai, Nagapattinam-Dt, Tamilnadu, India. E-mail:* Tel.:+91 04366 245234*;*

**Abstract:** This paper studies the performance and heat transfer analysis of Solar water heating systems incorporated by Convex lenses also with two types of twists CSPT(combined Serrated Perforated Twists) and CTPT(combined Triangular cut Perforated Twists) with twist ratios 3 and 6. Convex lenses are preferred to utilize the diffusible radiation and it focuses the solar rays in single point on the absorber plate. It helps the riser tube to get higher solar radiation at the instant. Nearly 45 lenses fixed over the glass and focusing radiations to 9 rows of riser tubes. Generally, twists give a swirling effect to the working fluid and it enables the water to catch up high heat from the riser tubes. In this case, CSPT and CTPT are introduced to examine the heat transfer characteristics. The experiments are carried out for plain and CTPT, CSPT collectors. Reynolds number and the Nusselt numbers are lying between 190 - 1266 and 4.104-16.7331. The friction factor is in the range of 0.035 to 0.1155. The implementation of convex lenses with the insertion of CSPT and CTPT helped to earn better heat transfer. CTPT with minimum twist ratio 3 has provided 1.9 times higher efficiency as compared to CSPT twist inserts. It enhances the overall efficiency by up to 9% than the conventional one.

**Keywords:** Convex lens, diffusible radiation, twist ratio, swirling effect, and Nusselt number

## 1.0 Introduction:

In the current scenario, solar energy has become widely used energy. Due to the depletion of fossil fuels now the world turns over toward the application of renewable energy. Renewable energy will play a huge role in the future. At present, the usage of renewable energy is growing rapidly. In this case, solar energy is one of the most important renewable energy worldwide. Solar energy contains a lot of benefits according to the usage. One of which is to convert the solar energy into thermal energy. It is the most effective and easy conversion process. Solar water heater works based on the above-said conversion process. This solar water heating system directly absorbs solar radiation and converts it into thermal energy for heating the water. Generally, solar water heater has been modified by many people according to design, placing of mirrors, construction of header, etc. those modifications have given better results. But in this paper, the convex lenses along with CTPT and CSPT swirl generators are used to boost the heat transfer in solar water heating system and the results are being discussed.

Kale et al. (1) investigated the performance of the convex lens CSP system for hot water and steam generation. They used six convex lenses with $0.0471m^2$ area and copper receiver of 5mm internal dia. and 10mm of outer dia. The results were found that it attains a maximum temperature of $81^0c$ at the mass flow rate of 1kg/hr and the efficiency obtained around 63-64%. He finalized that the presence of convex lenses will help to get the maximum result. Vinod Kumar Verma et al. (2) experimented to analyze the performance of convex lenses for heat transfer application. Six convex lenses and a copper receiver tube are used here. A two-axis manual tracking system is used to track the sun continuously. The results revealed that CSP by convex lenses is the best-suited one for solar water heating to get high heat transfer.

The above-said authors only published the articles according to the convex lens, but in this paper, the solar water heater with convex lenses is approached in different methods of design.

Twists are the best heat transfer improvement element in the heat augmentation technique. Murugan et al. (3) investigated the thermal performance of solar collectors using peripherally wing cut swirl generators (PSG). They used three types of swirl generators including PSG such as Triangular wings (Tri-PSG), Rectangular wings (Rec-PSG), Trapezoidal wings (Tra-PSG) in the solar water heater for enhancing heat transfer. Among the three types of wings the Trapezoidal wing cut swirl generator provided high thermal performance than the other two types. Hasanpour et al. (4), this paper explains the heat transfer and friction factor characteristics of double pipe heat exchanger with modified twisted tape such as perforated, V-cut, U-cut. The results were compared with the conventional heat exchanger. The results confirmed that the Nusselt number and friction factor for modified twists are higher than the typical twists except for the perforated twist. Kongaitpaiboon et al (5), introduced perforated conical ring (PCR) and analyzed the heat transfer characteristics of a heat exchanger. The results revealed that the heat transfer rate obtained in PCR increases 137% over the plain tube. Also, it was found that the PCR is more efficient than CR.

Li et al (6) depicted the heat transfer enhancement of triangular perforated fin by using CFD software and experimentally. The obtained results from the both are same while comparing. Finally, it is proved that the triangular perforated fin contributes to increase the heat transfer. Ammaralifarhan et al (7) experimented with three types of solar collectors. One has an absorber plate with perforation. The other two have a perforated absorber plate with 3 and 6mm holes. The results indicated that the solar collector of absorber plate with 3mm gained more energy than the absorber plate with 6mm perforation and the absorber plate of increasing number of holes with a small diameter is more efficient more than the absorber of decreasing number of holes with high diameter. Bodius Salam et al (8) researched heat transfer coefficient, friction factor, efficiency of water while twisted tape inserted into the tube of 26.6mm diameter. This twisted tape is having a rectangular cut of 8mm depth and 14mm width. The obtained heat transfer efficiencies vary from 1.9 to 2.3%. Eiamsaard et al. (9) investigated the performance of heat transfer, flow friction in tube assisted heat exchanger with oblique delta winglet twisted tape (O-DWT) and the results were compared with straight winglet twisted tape(S-DWT). Twist ratios of 3, 4 & 5 and wing cut ratios of 0.11, 0.21, and 0.32 were used. Results showed that the Nusselt number and friction factor increases with the decrease of twist ratio with the increase of depth of the wing cut ratio. O-DWT gives efficient results comparing with S-DWT.

Bhattacharya et al (10), analyzed the characteristics of friction factor and Nusselt number for laminar flow in a circular duct with the help of Integral transverse rib roughness along with Centre-cleared twisted tape. The results proved that the performance of circular duct with integral transverse rib roughness and Centre cleared twisted tape is more efficient than other enhancement techniques. Chang et al (11) studied the heat transfer in tube fitted with serrated twisted tape and smooth twisted tape. The Nusselt number and friction factor increases with the decrease of twisted ratio for both serrated and smooth twisted tape. The result of this experiment reveals that the heat transfer of water while using serrated twists is 1.2 or 1.3 times than the smooth twist. Sneha ponnada et al (12) conducted an experiment on the thermal performance of water in circular duct using with perforated twisted tape with alternate axis (PATT), perforated twisted tape (PTT), Regular twisted tape (RTT). The results found that the heat transfer rate in PATT, PTT, RTT is 48%, 43%, and 33%. Amar Raj Singh Suri et al (13) examined the heat transfer rate and friction in a heat exchanger using multiple square-cut twisted tapes. Results revealed that the heat transfer rate is considerably increased when using multiple square-cut twisted tape than without square-cut twisted tape. Eiamsa et al (14) carried out an experimental work about the serrated twisted tape on heat transfer enhancement against typical twisted tape (TT). Found out the results and the results showed that the use of serrated increases the heat transfer rate than the typical twist. Serrated twisted tape has a great influence on the enhancement of heat transfer rate.

Many research works had been conducted in heat exchangers by various authors in different geometry, but very few only have used different familiar shapes of twists in solar water heater. Hence, this research work has been carried out in a solar water heater with two different unworked geometries of twisted tapes such as CSPT and CTPT of two twist ratios of 3 and 6.

**2.0 Experimental set up:**

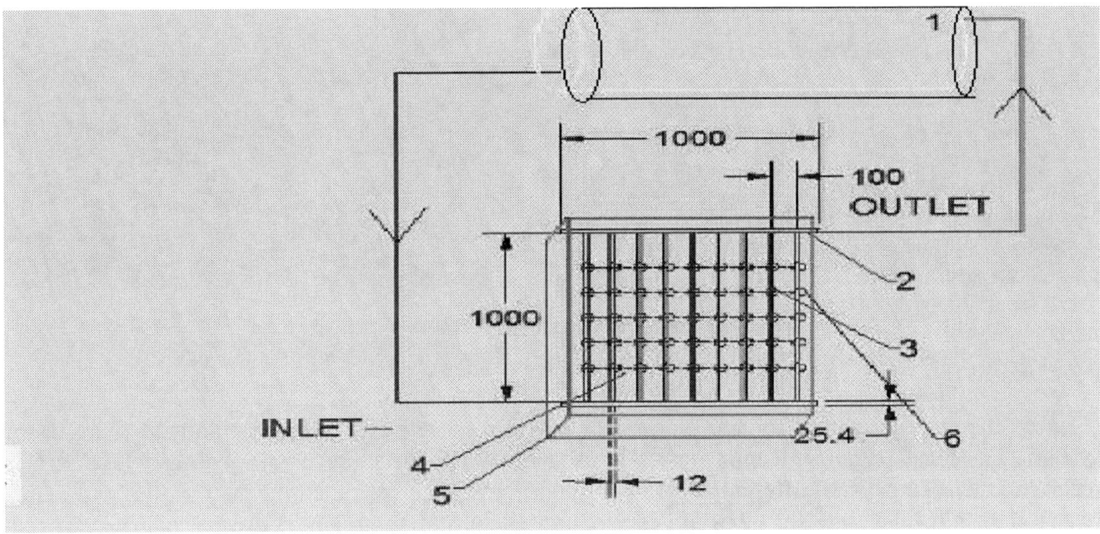


**Fig.1. Schematic diagram of the experimental setup.**

1. Water tank
2. Collector box
3. Riser tube
4. Absorber plate
5. Header
6. Convex lens

*2.1 Construction details:*

The entire collector set up is made up of wood with dimensions of 1000mm*1000mm. This wooden collector plays an important role as it prevents heat loss from the system. Here nine riser tubes of internal diameter 11mm and external diameter of 13mm are connected in parallel with Upper and Lower headers. Each riser tube is of 1m length and connected with headers at each end. Header is a main fluid distributional element having the uniform cylindrical shape of 25.4mm diameter and 1000mm length placed across the nine riser tubes. Both headers and riser tubes are made of copper. Twists of CTPT and CSPT are inserted within the tubes to provide swirling effect to the water. Besides with riser tubes are covered by an absorber plate made up of a copper sheet of 1000mm*1000mm*0.3mm and it is coated with black paint. The entire assembly inside the collector is closed by a transparent glass plate with convex lenses are in uniform intervals. The glass plate is of 5mm thick and nearly 45 convex lenses are placed on the glass plate of 300 mm distance from the top and bottom of the edge of the transparent glass. The distance between each convex lens is 100mm and the focal length of the lens is 13mm. Convex lenses mainly contribute to focus the solar radiation at a single point of the riser tubes. A storage tank of 100 liter capacity is used to store the water, which is made up of stainless steel. Two tubes of inlet and outlet are connected with the bottom and top header to feed and extract the water. Thermocouples are pinned at 35 points of the set up to take the readings. Here K-type thermocouple is suggested as it can measure up to $1000^0$C. Solar radiation will be gotten more if the setup is oriented with the earth`s latitude. For the Nagapattinam region, $18^0$ latitude is advisable to get maximum solar radiation.

Fig.2. Experimental setup.

*2.2 Concept of Convex lens:*

Usually, the convex is used for focusing. Here it secures itself as a very predominant thing in this system. The focal length of 13mm lenses is used here. The structure of the lens enables itself for receiving the solar rays of different angle and it concentrates all the radiations at a single point on the riser tubes. It enables the riser tubes to get maximum solar radiation. Normal plain solar water heater does not have this feature as its efficiency is also less. The presence of convex lenses help the system to attain maximum efficiency.

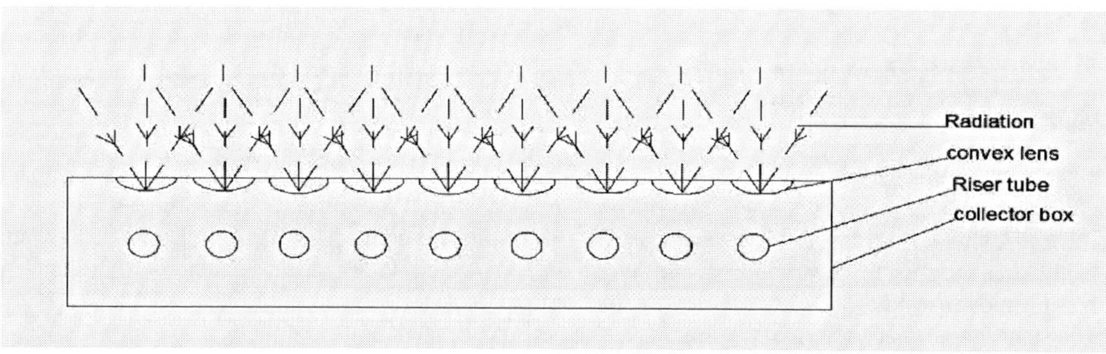


Fig.3. Top view of the solar water heater with convex lenses.

*2.3 Concept of CTPT and CSPT:*

Using twist inserts is an efficient augmentation method of heat transfer. The purpose of twists is for making swirl effect of water inside the tubes. It also reduces the flow speed of water as it stays a long time within the tubes. Hence the heat transfer rate between the riser tube and water will also be high. In this paper, we wish to give more heat transfer. So, we have introduced new modification twists such as CSPT (Combined Serrated Perforated Twists) and CTPT (Combined Triangular Perforated Twists). In CSPT, the combination of serrated and perforated has been made. In CTPT, the combination of a Triangular hole and perforated has been made. When water flows within the riser tubes, it will be allowed to flow along the helix of the twists and through the circular as well as triangular holes of the twist. It makes easier for the water particle to get collapse and mix with other particles. In this way, twists help the all water particle to carry an equal amount of heat along with itself. The inserting of CSPT and CTPT gives more swirl effect to the water than the existing twists.

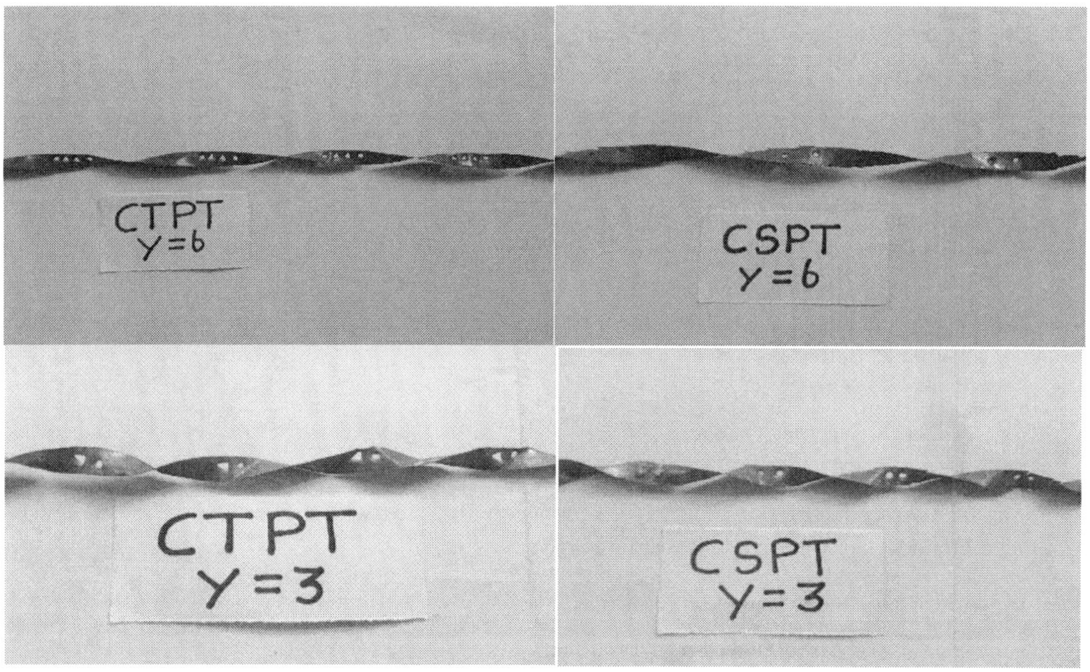


Fig.4. Pictorial view of CTPT (Combined serrated and perforated twist) and CTPT (Combined Triangular cut and perforated twist).

## 3.0 Data Processing:

The Normalized Nusselt number ($Nu_t/Nu$), Normalised friction factor ($f_t/f$) and thermal performance factor ($\eta_t/\eta$) are evaluated for the uniform header solar water heating system with helical twisted tape inserts of ratio's Y =3,4,5,6 and without twisted tape inserts.

*3.1 Heat Transfer:*

The Heat Transfer rate of the water in the single riser tube is estimated through the following basic equation (Liao Q, Xin MD, 2000) (15).

$$Q = mc_p(T_{out} - T_{in}) = U_0 A_0 (T_{w0} - T_m) \quad (1)$$

$$\frac{1}{(U_0 A_0)} = \frac{1}{(h_i A_i)} + \frac{\ln(D_0/D_i)}{(2\pi k_w L)} \quad (2)$$

$$h_i = mc_p (T_{out} - T_{in}) / A_0 (T_{w0} - T_m) \quad (3)$$

The internal convective heat transfer co-efficient, $h_i$ is determined by the equation (3). The experimental Nusselt Number is calculated using the following equation

$$Nu = h_i D_i / k. \quad (4)$$

All the fluid thermophysical properties are determined at the bulk mean temperature, $T_m$.

*3.2 Pressure drop:*

The differential pressure transducers are attached to the riser tubes through proper tapping for measuring the pressure drop of each riser tube over the length of the riser tube.

$$f = \frac{\Delta P}{\left(\frac{L}{D_i}\right)\left(\frac{\rho u_m^2}{2}\right)} \quad (5)$$

Where ΔP is the pressure drop over the length L.

*3.3 Thermal Performance:*

The thermal performance of the solar water heater is calculated by using the Hottel-Whillier-Bliss [16] equation as follows

$$\eta = F_R(\tau\alpha) - F_R U_l \frac{T_{in} - T_a}{H_t} \quad (6)$$

The following equations are helped to calculate the heat loss Co-efficient ($U_t$), Transmittance – absorbance Product $(\tau\alpha)$ and heat removal factor ($F_R$).

$$Q = H_t (\tau\alpha) - U_l (T_P - T_a) \quad (7)$$

$$U_l = Q - H_t(\tau\alpha) / (T_a - T_p) \quad (8)$$

$$F_R = Q / H_t(\tau\alpha) - U_l (T_{in} - T_a) \quad (9)$$

Every 15 minutes once the Instantaneous thermal performance is calculated.

## 4.0 Results and Discussion:

*4.1 Verification of Nusselt number and Friction factor for convex lens plain collector:*

The experimental Nusselt number and friction factor of the plain collector has validated with fundamental equations 9 & 10 and deviations fall within ±6.74 and ± 9.03% respectively. The increase of solar radiation increases the mass flow rate, Nusselt number and Friction factor in Phase1.In Phase 2 the decrease of solar radiation decreases the mass flow rate, Nusselt number and friction factor.

$$Nu = 1.86\,(GZ)^{1/3}\,(\mu/\mu w)^{0.14} \text{ for } GZ > 10. \quad (10)$$

$$f = 16/Re \quad (11)$$

*4.2 Effect of CTPT and CSPT twist inserts in heat transfer and Friction factor:*

The combined triangular cut and circular perforation provided more impact and centrifugal forces radially towards the next helix than combined serrated and circular perforation geometry. In this research, CTPT-3 gives a higher Nusselt number and Friction factor for the twist ratio 3.Figure 5(a) and (b) shows the Nusselt number variations between CSPT-3 &6, CTPT-3&6 and plain collector. Figure 6(a) and 6(b) shows the Friction factor variations for the above said, collectors. The results revealed that the Nusselt number and friction factor for CTPT -3 has 8.98% and 5.23% which is higher than CSPT-3 respectively.

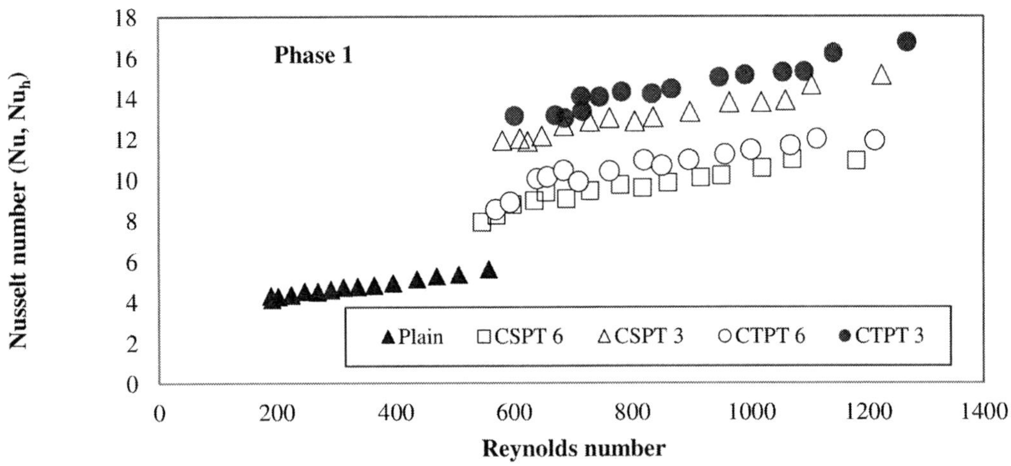


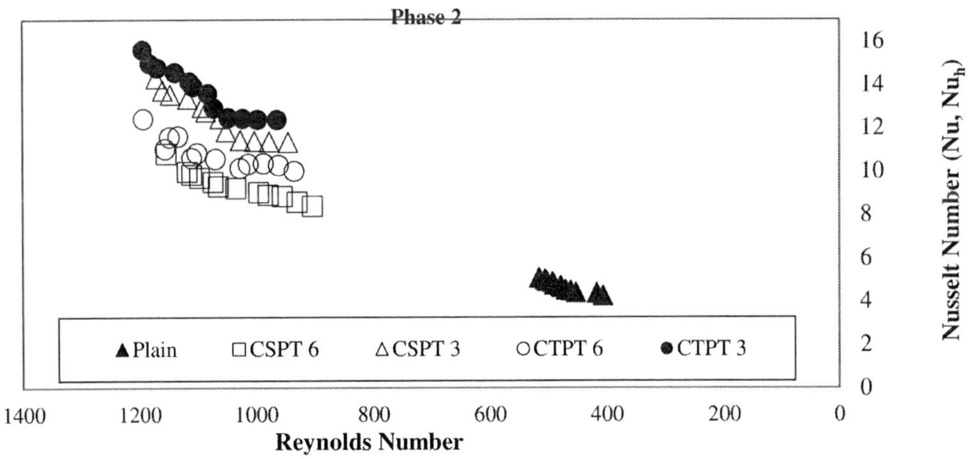


Fig.5 (a & b) Nusselt number verification for CTPT and CSPT (Phase1&phase2).

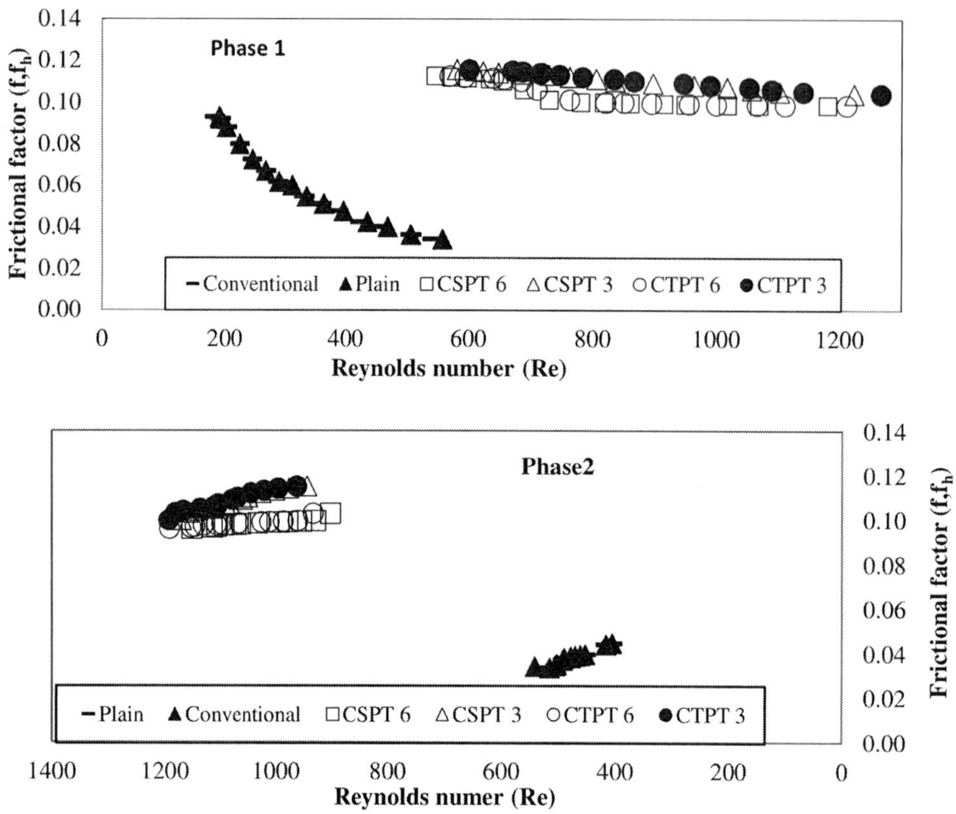


Fig.6 (a & b) Frictional factor verification for CSPT & CTPT (Phase1&phase2).

*4.3 Effect of CTPT and CSPT in Thermal performance:*

Fig.7 shows the thermal performance characteristics of CTPT, CSPT, and plain collectors. The efficiency of the collectors increases with increase of solar intensity in phase1 and it reduces with decrease of solar intensity in phase2. The results confirmed that the CTPT with minimum twist ratio 3 exerts optimum thermal performance than CSPT with twist ratio 3and 6 and is 1.9%

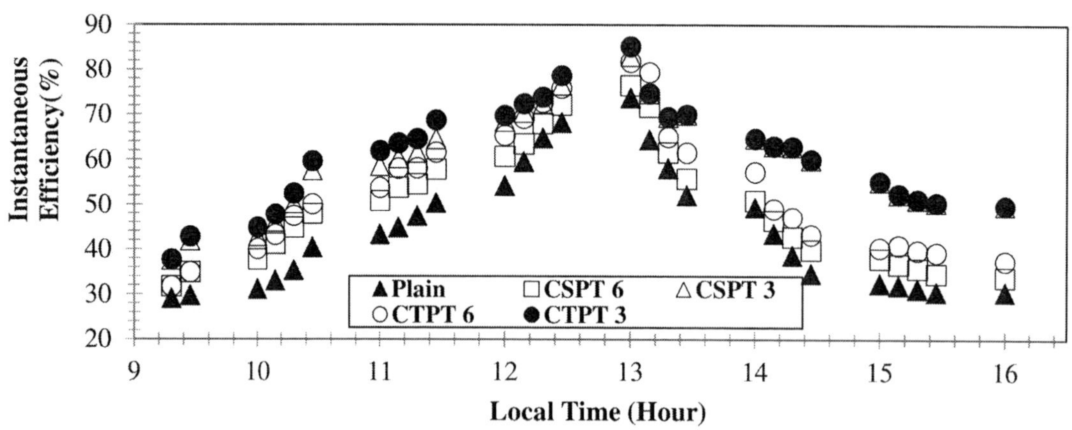


Fig.7. Relationship between Time and Efficiency.

**5.0 Conclusion:**

The results of the Convex lens solar water heating system operated with CSPT and CTPT of twist ratio3 & 6 are experimentally evaluated and compared with each other.

The average Nusselt no obtained from CSPT-6, CTPT-6, CSPT-3, and CTPT-3 and plain are 9.46, 10.74, 12.866, 14.13 and 4.70 respectively. The mass flow rate attained at CSPT-6, CTPT-6, CSPT-3, CTPT-3 are 0.0361, 0.0363, 0.037, 0.0379kg/s respectively.

In addition with the mean efficiency achieved by CSPT-6, CTPT-6, CSPT-3 & CTPT-3 are 49.77, 53.47, 59.8 & 61.03% respectively.

From the above results, it is confirmed that the CTPT with minimum twist ratio 3 provides higher efficiency in all parameters than the other twist.

The convex lens collector inserted with CTPT of twist ratio 3 has achieved 1.9 times higher efficiency than the CSPT collector.

**6.0 References:**

[1] Kale S.S and Shinde N.N, 2012. Design and Development of CSP Using Convex Lenses for Domestic Water Heating and Steam Generation; *International Journal of Mechanical Engineering and Research,* **1,** 74-78.

[2] Vinod Kumar Verma Er, Vipin Tripathi, Vivek Kumar Verma and Muhaid Huzaifa. Development of CSP using convex lenses for domestic water heating; *International Journal of Research in Engineering and Technology,* ISSN: 2319-1163 | pISSN: 2321-7308

[3] Murugan M, Vijayan R and Saravanan A, 2019. Improving the performance of a trapezoidal-trough Thermosiphon solar collector using peripherally wing-cut swirl generator; *International Journal of Green Energy*, **16**, 1196-1209.

[4] Hasanpour A, Farhadi M and Sedighi K, 2016. Experimental heat transfer and pressure drop Study on typical, perforated, V-cut and U-cut twisted tapes in a helically corrugated heat exchanger; *International Communications in Heat and Mass Transfer*, **71,** 126-136.

[5] Kongkaitpaiboon V, Nanan K and Eiamsa-ard S, 2010. Experimental investigation of heat transfer and turbulent flow friction in a tube fitted with perforated conical-rings; *International Communications in Heat and Mass Transfer*, **37** 560-567

[6] Juan li, Xiang ling and Hao peng, 2013. Field synergy analysis on convective heat transfer and Fluid flow of a novel triangular perforated fin; *International Journal of Heat and Mass Transfer*, **64**, 526-535.

[7] Ammar Ali Farhan, Hana Abdul Hussein Sahi, 2017. Energy Analysis of Solar Collector With perforated Absorber Plate; *Journal of Engineering,* **23**.

[8] Bodius Salam, Sumana Biswas, Shuvra Saha, Muhhamad Mostafa K and Bhuiya, 2013. Heat Transfer enhancement in a tube using rectangular-cut twisted tape insert; *5th BSME International Conf. on Thermal Engineering, Procedia Engineering.* **56**, 96 – 103.

[9] Eiamsa ard S, Wongcharee K, Eiamsa-ard P and Thianpong C, 2010. Heat transfer enhancement in a tube using delta-winglet twisted tape inserts; *Applied Thermal Engineering* **30**, 310–318.

[10] Suvanjan Bhattacharya, Subhankar Saha, Sujoy Saha, 2013. Laminar flow heat transfer enhancement in a circular tube having integral transverse rib roughness and fitted with Center-cleared twisted tape; *Experimental Thermal and Fluid Science,* **44,** 727–735.

[11] Shyy woei chang, Yih jena jan and Jin shuen liou, 2007. Turbulent heat transfer and pressure drop in a tube fitted with serrated twisted tape; *International Journal of Thermal Sciences*, **46**, 506-518.

[12] Sneha ponnada, Subrahmanyam T and Naidu S V, 2019. An experimental investigation on heat transfer and friction factor of Silicon Carbide/water Nanofluids in a circular tube; *Energy Procedia*, **158**, 5156-5161.

[13] Amar Raj Singh Suri, Anil Kumar and Rajesh maithani, 2017. Heat transfer enhancement of heat exchanger tube with multiple square perforated twisted tape inserts; *Chemical Engineering and Processing: Process intensification,* **116**, 76-96.

[14] Smith Eiamsa-ard and Pongjet pomvonge, 2010. Thermal characteristics in a round tube fitted with serrated twisted tape; *Applied Thermal Engineering*, **30**, 1673-1682

[15] Liao Q and Xin M D, 2000. Augmentation of convective heat transfer inside tubes with three Dimensional internal extended surfaces and twisted-tape inserts; *Chemical Engineering Journal*, **78**, 95-105.

[16] Hottle H C., Whiller A, 1958. Evaluation of flat-plate solar collector performance; in *Trans. Of conference on the use of Solar Energy,* **2**, 74-78.

IOP Conf. Series: Materials Science and Engineering **923** (2020) 012074 doi:10.1088/1757-899X/923/1/012074

# Solar Energy based low cost Converter for Rural Houses Electrification

**T D Sudhakar[1], K N Srinivas[2] and S Kaja Mohideen[3]**

[1]Professor / DEEE, St. Joseph's College of Engineering, Chennai, India

[2]Professor & Head / DEEE, SRM University (Ramapuram Campus), Chennai, India

[3]Senior Professor / DECE & Director (PG Admissions), B.S.Abdur Rahman Crescent Institute of Science & Technology, Chennai INDIA

t.d.sudhakar@gmail.com

**Abstract.** A Rectified Sinusoidal based Pulse generation method for a DC – DC converter is proposed in this discussion. To exhibit this, the methodology is applied to a Self-Lift topology based SEPIC converter. In this paper a two stage grid connected system is used. In stage I the said converter is used for boosting and DC rectification. This converter is considered, can provide the required voltage from a renewable source. The DC rectification is done by using the RSPWM technique. Stage II uses an inverter which inverts the DC rectified signal to produce AC signal. The usage of the bulky transformers are avoided by using the proposed modulation technique. The notable advantages of the RSPWM technique are : (i) the switches of the inverter are functioned at low frequency (50Hz), (ii) harmonic content of the output waveform is reduced because of RSPWM, (iii) decrease of the filter size and (iv) inverter switches life time are increased as complicated control techniques are avoided. These features are validated through simulations.

## 1. Introduction

Conventional power generation methodologies using the fossil fuel sources are depleted every day and in future might get exhausted. In this view, renewable energy sources which are non exhaustible are getting popular [1]; in particular the solar energy based power generation is the widely popular one. The solar system produces relatively low voltage as an output and it requires very high gain step up converters for integrating it with the grid. There are two stages of a solar grid connected system [2]. In stage I a DC – DC conversion is done and in stage II DC – AC conversion is done. The required high gain is brought in either at stage I, stage II or in both the stages. If the required gain is not attained in both the stages then transformers are used. Hence the solar based grid connected systems are generallycategorized as transformer and transformerless

systems. The transformerless systems are further classified into high gain converters which include (i) DC – DC converters [3] and (ii) DC – AC converters [4]. The DC – DC converters which have a high gain are further categorized [5], [6]as converters with additional components, and the components are
(A): coupled inductor
(B): switched capacitors
(C): inductor and switched capacitors
(D): coupled inductor and switched capacitors
(E): Interleaved boost converters
In this proposed work self-lift SEPIC converter is used which comes under the category of converters with Inductor and Switched Capacitor. The output of the DC – DC converter is modulated to follow a rectified sine waveform [12]. In this paper the converter output is modulated to follow as DC Rectified Output voltage. The output of the converter is modulated by applying Rectified Sinusoidal PWM (RSPWM) technique based pulse generation technique.

## 2. High Gain DC – DC Converter

The self-lift SEPIC converter [13] is shown in Figure 1. The circuit is designed as explained below. The designed values of various components are shown in Table 1. Using these values, the self-lift SEPIC converter fed by solar energy source is simulated in MATLAB for its operation as shown in Figure 2.

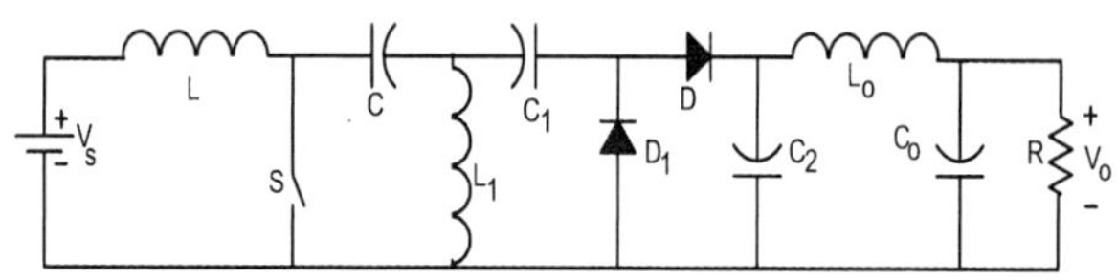


Figure 1 : Basic Diagram of self-lift SEPIC converter

The input to the converter ($V_S$) is 48V DC and the voltage ($V_0$) and the current ($I_0$) at the output side is 230V DC and 2.3A DC for a 100Ω resistive load.
By substituting the input and output voltages the gain (K) value is calculated using the formula $1-K=\frac{V_S}{V_0}$, then K= 0.7913.
The converter switching frequency is 50KHZ.
$I_S$=11.02083A, by $V_S I_S = V_0 I_0$
It is assumed that 5% is the ripple content in the output current.
As a result $\Delta I_S$=5% of $I_S$= 0.55A.
Then, $L=\frac{K \times V_S}{f_{sw} \times \Delta I_S}$ = 0.001379H.
Pretentious 5% of disturbance in the voltage at the output side, then 5% of $V_S$ is the $\Delta V_C$value that is 2.4V.
Then $C=\frac{I_0}{f_{sw} \times \Delta V_C}$= $1.9167\times 10^{-5}$F.
Similarly value of $C_1$ is obtained as $1.9167\times 10^{-5}$F using $C_1=\frac{I_0}{f_{sw} \times \Delta V_{C1}}$, where$\Delta V_{C1}$ is 5% of $V_S$.
The value of $C_2$ is calculated using the formula $C_2=\frac{I_0 K}{f_{sw} \times \Delta V_{C1}}$.
On substitution the value is $3.165 \times 10^{-6}$F.
The expression of $L_0$ is given by $L_0=\frac{V_0 K}{8\times f_{sw}^2 \times \Delta I_0 \times C_2}$.On substitutions, $L_0$=0.025H.
$C_0$is calculated using $C_0=\frac{I_0 K}{64\times f_{sw}^3 \times L_0 C_2 \Delta V_0}$,
which gives $C_0$= $0.25\times 10^{-9}$F.

Table 1 : Design values of self-lift SEPIC converter

| Components | Values |
|---|---|
| $V_S$ | 48V |
| $V_o$ | 230V |
| L | 0.001379H |
| $L_1$ | 0.001379H |
| $L_0$ | 0.025 H |
| C | $1.9167\times10^{-5}$F |
| $C_0$ | $0.25\times10^{-9}$F |
| $C_1$ | $1.9167\times10^{-5}$F |
| $C_2$ | $3.165\times10^{-6}$F |
| R | 100Ω |

**2.1 PWM based self-lift SEPIC converter with inverter:**

The converter steps up the input voltage of 48 V from the solar panel to an output voltage of 230 V. For a load of 100 Ω, the converter voltage at input and output side then output current obtained are revealed in Figure 3, Figure 4 and Figure 5 respectively.

The converter DC output voltage of 230 V is inverted, to give an AC voltage of 230 V (Figure 6) and 2.3 A (Figure 7). The Total Harmonic Distortion (THD) value in the waveform is 53.09% (figure 8).

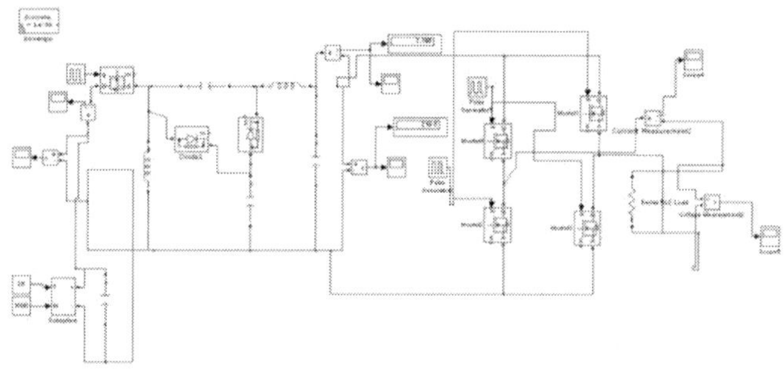

Figure 2 : Solar panel with basic self-lift SEPIC converter and inverter

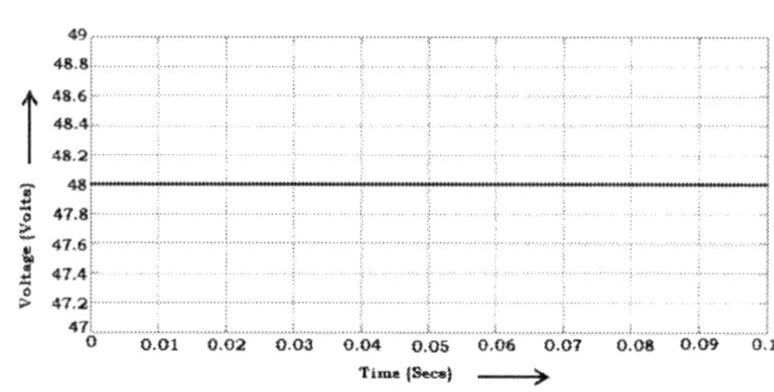


Figure 3 : Input side Voltage

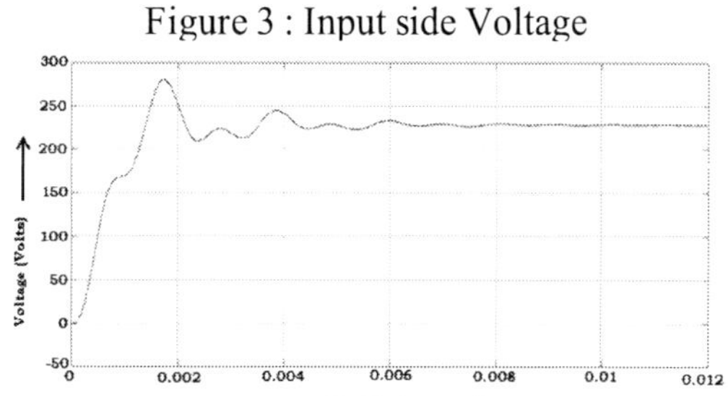

Figure 4 : Output side Voltage

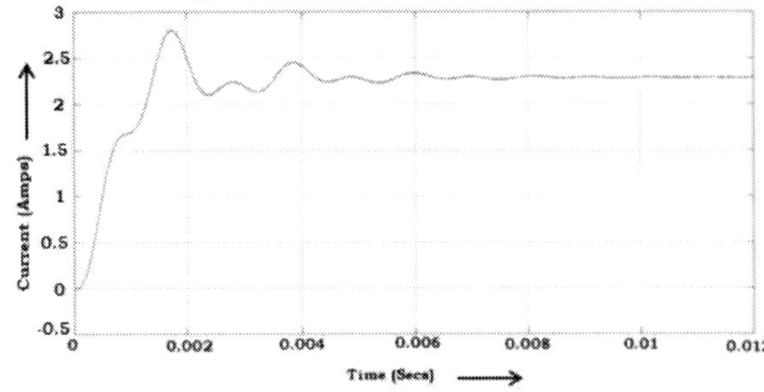


Figure 5 : Output side Current

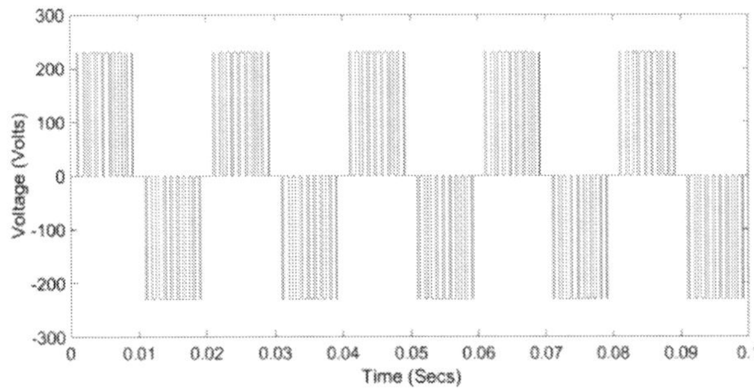


Figure 6 : Inverter side output voltage

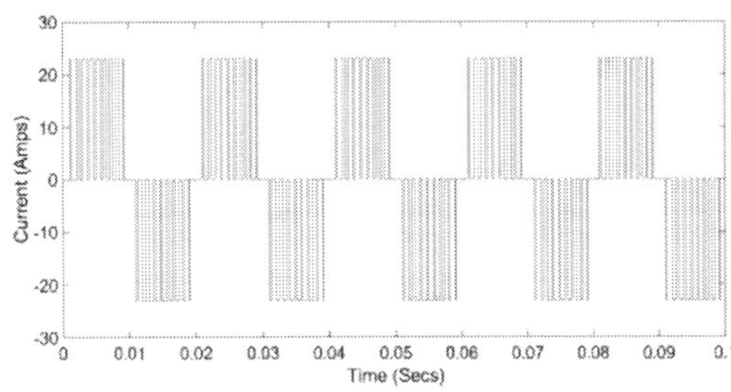


Figure 7 : Inverter side output current

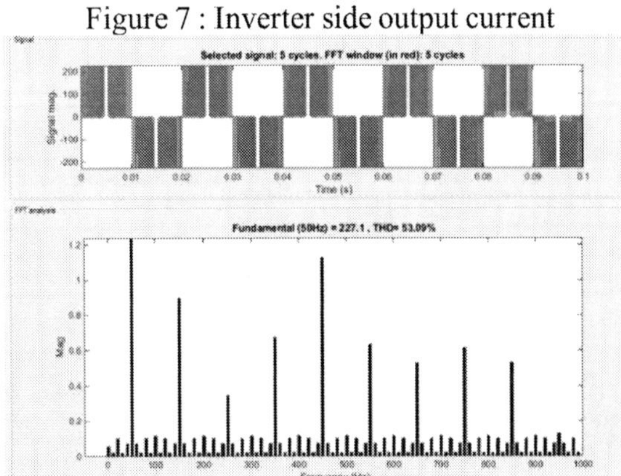

Figure 8 : FFT analysis of the Output voltage waveform of Solar panel with basic self-lift SEPIC converter and inverter

## 3. Proposed Methodology

The block diagram of the proposed work is shown in the Figure 9. This includes majorly the self-lift SEPIC converter for grid based applications. In this diagram a PV Module is used as an input

IOP Conf. Series: Materials Science and Engineering **923** (2020) 012074 doi:10.1088/1757-899X/923/1/012074

for the self-lift SEPIC converter. The converter is triggered by a Rectified Sinusoidal Pulse Width Modulation (RSPWM), which produces an output as 'Negative Cycle Converted' DC rectified output voltage. This output, is given as an input to the inverter. The inverter converts the 'Negative Cycle Converted' DC Rectified Output voltage to a sinusoidal AC voltage and feeds it to the grid.
The proposed RSPWM procedure is same as that of Sinusoidal Pulse Width Modulation (SPWM), but there is a difference in operation. The SPWM is used in inverters; instead, RSPWM is used in DC – DC converters. In the case of SPWM method a sinusoidal modulating signal, which is the desired fundamental component of the inverter, is related with a high frequency triangular carrier waveform to produce the pulsating signal; whereas in RSPWM the desired fundamental component is a rectified sine waveform, which is related with a high frequency triangular carrier waveform to produce the pulsating signal.
In SPWM technique, the DC voltage is converted to an AC voltage using conventional SPWM method and a step up transformer is used to step up the AC sine voltage to the magnitude of mains AC, for possible grid connected applications. In the Rectified SPWM method the amplitude of the DC voltage is controlled to look like a rectified sinusoidal voltage. The step up characteristic is established with duty cycle control and the modulation in the amplitude is achieved by dynamically changing the duty cycle at a high frequency to match the rectified sinusoidal voltage.

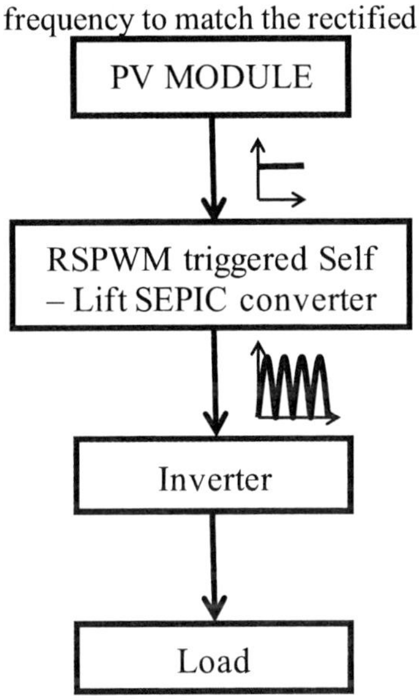


Figure 9 Block Diagram of the Proposed Work

### 3.1 Rectified SPWM based self-lift SEPIC converter:

The RSPWM technique based self-lift SEPIC converter is shown in the figure 10. This is done in order to get a rectified sine waveform as the output of the DC – DC converter. By applying the RSPWM technique modelled to the self-lift SEPIC converter the output voltage obtained as shown in the figure 11. The output voltage waveform is a rectified waveform. If this voltage is connected to an inverter operating at a low frequency (say 50Hz) a sinusoidal waveform can be obtained.

IOP Conf. Series: Materials Science and Engineering **923** (2020) 012074 doi:10.1088/1757-899X/923/1/012074

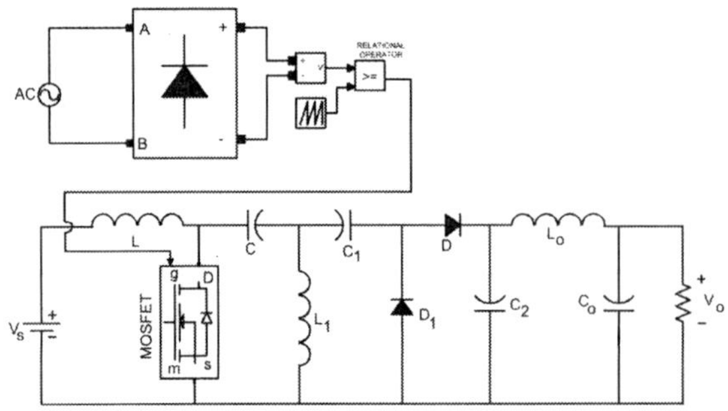


Figure 10 RSPWM Technique based self-lift SEPIC converter

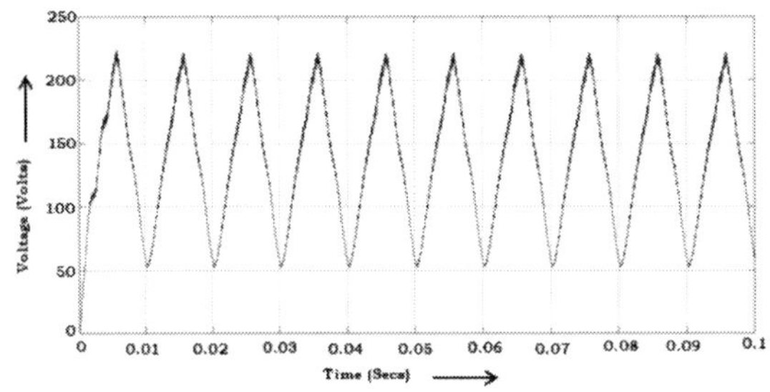


Figure 11 Output voltage of RSPWM Technique based self-lift SEPIC converter

**3.2 Rectified SPWM based self-lift SEPIC converter with inverter:**

Figure 12 shows applying the RSPWM technique to the self-lift SEPIC converter with inverter. The output voltage and output current are obtained are shown in the figure 13 and figure 14 respectively. The circuit shown in Figure 11 is used along with an H bridge inverter. The converter output voltage is connected to an inverter circuit to obtain a sinusoidal waveform. The $V_{max}$ of the $V_{ac}$ is 230 V, whereas the load current is 2.3 A for a load of 100 Ω.

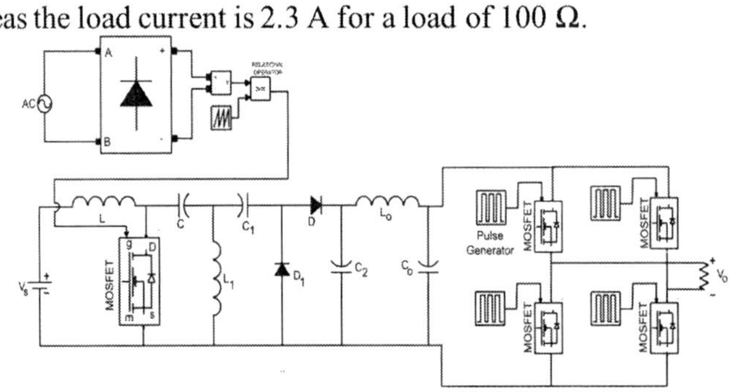


Figure 12 RSPWM Technique based self-lift SEPIC converter with inverter

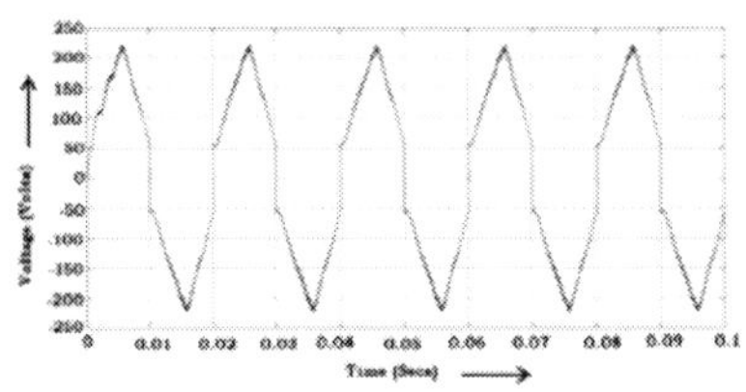


Figure 13 Output voltage of RSPWM Technique based self-lift SEPIC converter with inverter

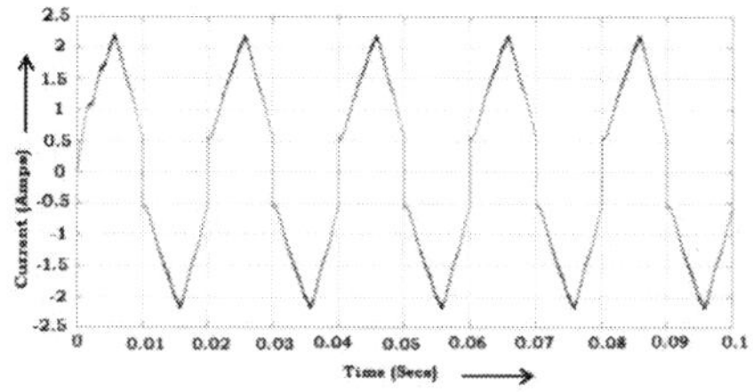


Figure 14 Output current of RSPWM Technique based self-lift SEPIC converter with inverter

### 3.3 Solar Power fed RSPWM based self-lift SEPIC converter with inverter:

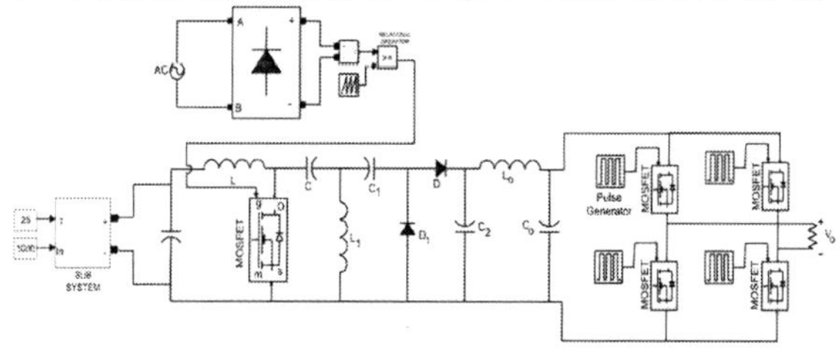


Figure 15 Solar panel with RSPWM based self-lift SEPIC converter and inverter

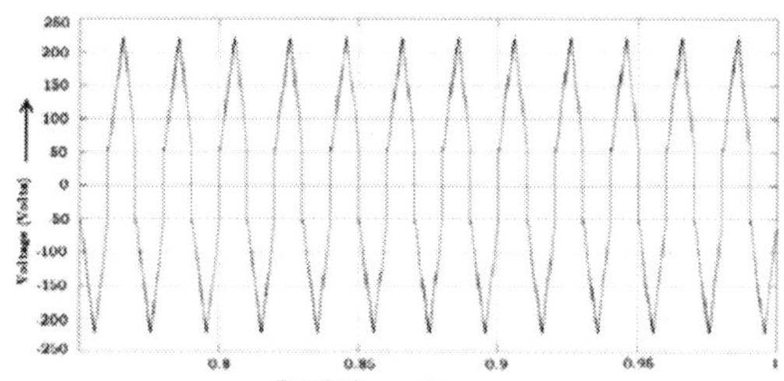


Figure 16 Output voltage waveform of Solar panel with SPWM based self-lift SEPIC converter and inverter

Figure 15 represents the PV system with RSPWM technique based self-lift SEPIC converter along with inverter implemented in the MATLAB / Simulink. Figure 16 and figure 17 show the output voltage and output current representation of this full fledged system. The value of $V_O$ is 220 V and $I_O$ is 2.2 A.

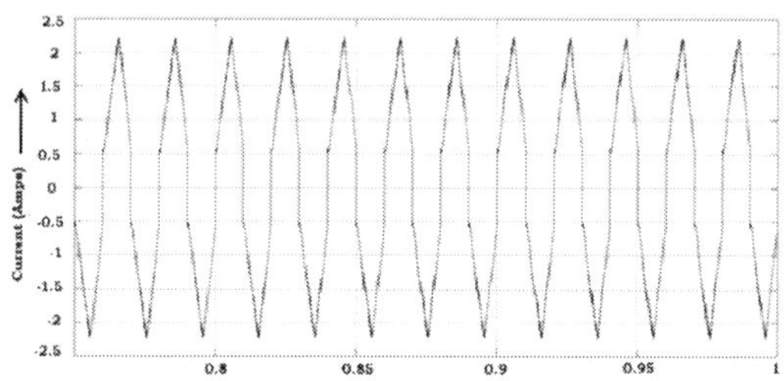


Figure 17 Output current waveform of Solar panel with SPWM based self-lift SEPIC converter and inverter

IOP Conf. Series: Materials Science and Engineering **923** (2020) 012074 doi:10.1088/1757-899X/923/1/012074

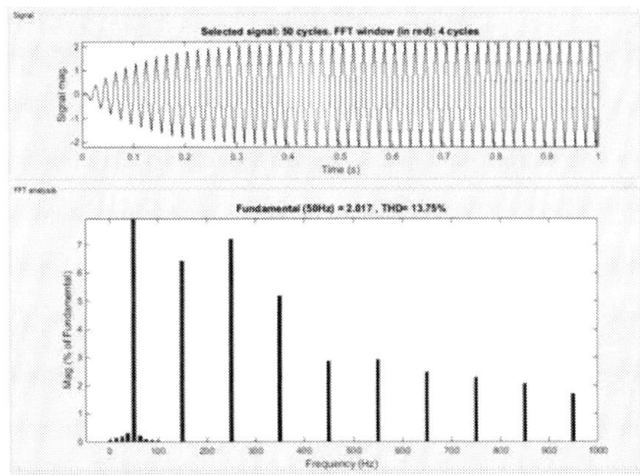

Figure 18 FFT analysis of the Output voltage waveform of Solar panel with SPWM based self-lift SEPIC converter and inverter

FFT is done on the output voltage waveform obtained using RSPWM technique based solar powered self-lift SEPIC converter – inverter system. The THD of the waveform is 13.75% as shown in figure 18. FFT analysis is done to the output voltage waveform using normal pulse given to the solar power self-lift converter – inverter system. The Total Harmonic Distortion (THD) of the waveform is 53.09% as shown in figure 8. Both these THD values are obtained without filter. The evaluated values are tabulated in Table 2.

Table 2: Comparison of THD values of the above circuits.

| SI. No | Circuit | Voltage | | Current | |
|---|---|---|---|---|---|
| | | Value | THD% | Value | THD% |
| 1. | PV Panel connected to the Self-Lift SEPIC Converter with Inverter | 230 | 53.09 | 2.3 | 53.09 |
| 2. | PV Panel connected to the RSPWM Technique based Self-Lift SEPIC Converter with Inverter | 230 | 13.75 | 2.3 | 13.75 |

The solar panel produces 48V DC voltage, which is boosted to 230V DC by the self–lift SEPIC. This boosted DC voltage is transformed to output sine wave voltage 230V AC with an inverter
The closed loop control of RSPWM triggered self–lift SEPIC converter with inverter as shown in Figure 19. In this technique, the output of the converter ($V_{actual}$) is taken and compared with the required rectified sinusoidal waveform ($V_{Ref}$) and fed to the comparator. The output of the comparator is compared with the saw tooth waveform to generate the gate pulse for the switch. Based on the error signal the switch's gate pulse is varied to get the required rectified sinusoidal voltage waveform.
The output voltage and its THD values are shown in figure 20 and figure 21 respectively are the simulation results of the closed loop self–lift SEPIC converter using the design values for a output voltage of 320V. For this voltage waveform the obtained THD value is 2.49% without any filter.

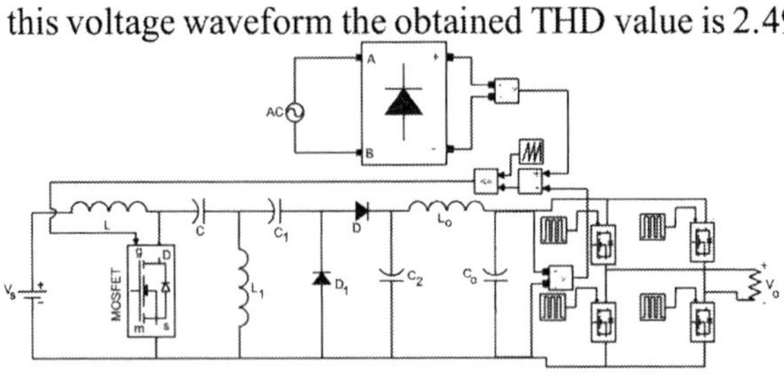

Figure 19 VMC based closed loop control of RSPWM triggered self–lift SEPIC converter

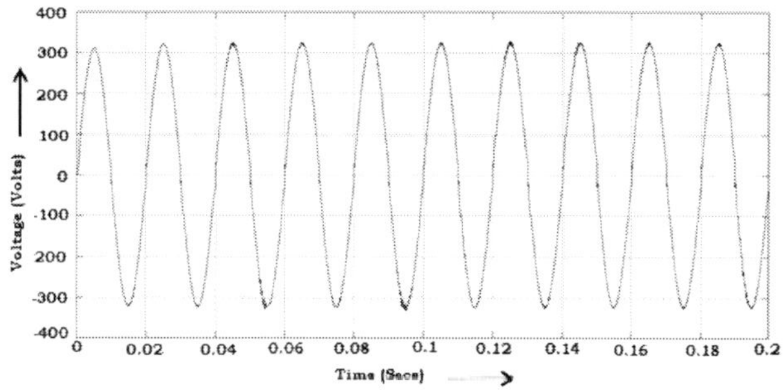


Figure 20 Output voltage waveform of VM control based closed loop control of RSPWM triggered self–lift SEPIC converter

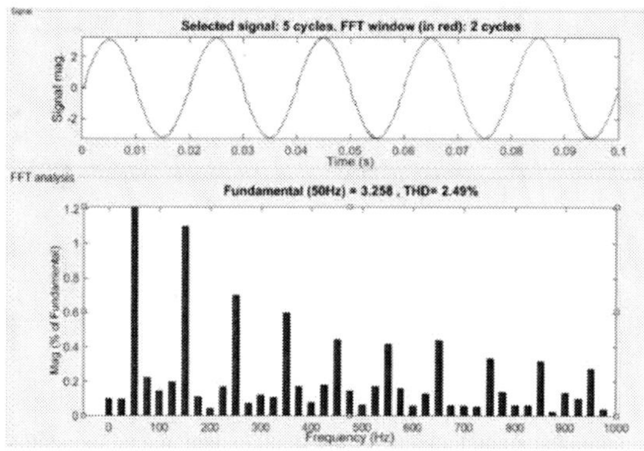


Figure 21 THD value for output voltage waveform of VM control based closed loop control of RSPWM triggered self–lift SEPIC converter fed inverter

## 4. CONCLUSION

This paper proposes a transformer less single stage DC-DC converter for a grid connected system. A RSPWM based Self-lift SEPIC converter is modelled to produce a 'negative-cycle-converted' DC rectified output voltage as output which is verified. This methodology reduces the switching stress of the four switches of the inverter. Hence the life time of the inverter is improved. The Self-lift SEPIC converter switch is operated at high frequency; the inverter switches are operated at low frequency. Therefore the control circuit required for the overall network is very small. The error rectification process is easier in these types of control circuitry.

The advantage of the proposed RSPWM based converter is reduction of the filter size connected to the inverter side. Hence cost can be reduced and stresses on the inverter switches are reduced. As the inverter is operated at a low frequency the noise from the inverter also gets reduced. Hence, this model proves to be safer, as only one switch is operated at higher frequency, when compared to the conventional circuit model where four switches of the inverter are operated at high frequency.

## 5. REFERENCE

[1]. Tsengenes G, Adamidis G. Investigation of the behavior of a three phase grid connected photovoltaic system to control active and reactive power. Electr Power Syst Res 2011; 81(1):177–84.

[2]. Om PrakashMahela, Abdul Gafoor Shaik Comprehensive overview of grid interfaced solar photovoltaic systems Renewable and Sustainable Energy Reviews 68 (2017) 316–332

[3]. Nousiainen L, Suntio T. Dual-mode current-fed semi-quadratic buck-boost converter for transformerless modular photovoltaic applications. In: Proceedingsof the 2011-14th European Conference on Power Electronics and Applications (EPE 2011); 2011, p. 1–10.
[4]. Chinnaiyan VK, Jerome J, Karpagam J. An experimental investigation on a multilevel inverter for solar energy applications.Int J Electr Power Energy Syst 2013; 47(0):157–67.
[5]. Wuhua Li, Xiaodong Lv, Yan Deng, Jun Liu, Xiangning He A Review of Non-Isolated High Step-Up DC/DC Converters in Renewable Energy Applications2009 Twenty-Fourth Annual IEEE Applied Power Electronics Conference and Exposition Pages: 364 - 369,
[6]. WuhuaLi; Xiangning HeReview of Nonisolated High-Step-Up DC/DC Converters in Photovoltaic Grid-Connected Applications IEEE Transactions on Industrial Electronics (Volume: 58, Issue: 4, April 2011 ) Page(s): 1239 - 1250
[7]. T. F. Wu, Y. S. Lai, J. C. Hung, and Y. M. Chen, "Boost converter with coupled inductors and buck–boost type of active clamp," *IEEE Trans. Ind. Electron.*, vol. 55, no. 1, pp. 154–162, Jan. 2008.
[8]. K. K. Law, K. W. Cheng, and Y. P. Yeung, "Design and analysis of switched-capacitor-based step-up resonant converters," *IEEE Trans. Circuits Syst. I, Fundam. Theory Appl.*, vol. 52, no. 5, pp. 1998–2016, May 2005.
[9]. E. H. Ismail, M. A. Al-Saffar, and A. J. Sabzali, "High conversion ratio dc–dc converters with reduced switch stress," *IEEE Trans. Circuits Syst. I, Reg. Papers*, vol. 55, no. 7, pp. 2139–2151, Aug. 2008.
[10]. R. J. Wai and R. Y. Duan, "High-efficiency bidirectional converter for power sources with great voltage diversity," *IEEE Trans. Power Electron.*, vol. 22, no. 5, pp. 1986–1996, Sep. 2007.
[11]. W. Li and X. He, "A family of interleaved dc/dc converters deduced from a basic cell with winding-cross-coupled inductors (WCCIs) for high step-up or step-down conversions," *IEEE Trans. Power Electron.*, vol. 22, no. 4, pp. 1499–1507, Jul. 2008.
[12]. Soren Baekhoj Kjaer, John K. Pederson and F. Blaabjerg, "A Review of Single–Phase Grid–Connected Inverters for Photovoltaic modules", IEEE Transactions on Industry Applications, Vol. 41, Issue 5, October 2005, pp 1292 – 1306.
[13]. T. Jones Basil, M. Chilambarasan, M. Ramesh Babu "Design and Simulation of Fused Self-Lift Converter Topology ", International Journal of Applied Engineering Research, ISSN 0973-4562, Vol. 10 No.20 (2015) pp. 18533 – 18537.

# Computational analysis of triglyceride molecule involving in transesterification reaction

**Gokul Raghavendra Srinivasan[1,3], Yashasvi Sharma[2], Vijayalakshmi S[1], Ranjitha J[1,*]**
[1]$CO_2$ Research and Green Technologies Centre, VIT University, Vellore, India
[2]Department of Mechanical Engineering, VIT University, Vellore, India
[3]Department of Mechanical Engineering, Bharath Institute of Higher Education and Research, Chennai, India

***E-mail:** ranjitha.j@vit.ac.in

**Abstract.** This present study deals with the quantum computational analysis of palmitic-oleic-stearic acid based triglyceride molecule available in waste animal fat, which involves in transesterification reaction during biodiesel production. From stereospecific analysis, stearic acid, oleic acid and palmitic acid, and were found at SN1, SN2 and SN3 positions respectively. The proposed triglyceride model was developed using Gaussian 09 and computational analyses were carried out using Density functional theory (DFT) method. Based on geometric optimization, the average bond length between C-C bond, C=C bond, C-H bond, C-O bond and C=O bond was measured as 1.53Å, 1.33Å, 1.09Å, 1.43Å and 1.23Å respectively. Average dipole moment of the triglyceride molecule was measured as 1.3010 Debye while electronic dislocation was noted in X+Z direction. In addition, the overall molecular energy was found to be -2522.7425 a.u. and optimization energy as -58.95 kJ/mol. Also, vibrational frequencies were analysed using the IR spectrum whereas electrostatic potential were mapped for the triglyceride molecule.

Key Words: Triglyceride, Quantum computational chemistry, Activation Energy, HOMO & LUMO, Electrostatic Potential, Vibration Frequencies and optimization energy.

## 1. Introduction

The increase in the demand for energy has been making the researchers to rely on self-sustaining biodiesel, which are renewable with lower emission characteristics and robust combustion features [1].To globalize this biofuel in world, the need for understanding its molecular properties is very primary need as these properties decides the overall nature of the fuel. But the market demands forces the experts to optimize the transesterification reaction to evaluate the production economy alone. Accordingly numerous researches have been carried out to optimize the transesterification reaction using various statistical & analytical techniques. The optimization of scum oil as feedstock for biodiesel was done using response surface methodology [2], whereas Taguchi method was implemented for evaluating the feasible production of biodiesel from *Manilkara zapota* (L.) seed oil [3]. Similarly, other high end tools like artificial neural network has been trained in MATLAB for auto optimization for any level of complexity involved in production. However, an effective optimization can be achieved only if the molecular properties of a molecule are analysed prior to the statistical analysis. Various molecular properties include molecular activation energy, electrostatic potential, molecular orbitals, vibrational frequencies etc. [4,5] which are decided based upon the bond length &

angle, strength of the nucleus, electrons in the outermost orbital, spin quantum number etc. Understanding these properties helps one in calculating parameters which influences the rate of transesterification reaction, to a greater extent. However, very less work has been carried out to study these properties for a triglyceride molecule involved in transesterification reaction. Among these few, Asukama et al., 2009 has conducted a theoretical study on triglyceride molecule involved in transesterification reaction which predicted the reaction order for an ideal reaction [6]. The computational study of the above work was influenced based upon the quantum study done on a KDP crystal for its colouring using dye, where electrostatic potential distribution was studied to understand its colouring mechanism [4]. Similarly, thermochemical analysis was done on oleic acid using quantum computation approach to analyses its behavior during a transesterification reaction [7].

In this study, the molecular properties of triglycerides derived from waste animal fat has been studied using Gaussian View software tool V5.0 [8, 9], where the molecular activation energy, electrostatic potential distribution, vibrational frequencies were analysed completely.

## 2. Materials and Methods

2.1 Identification of fatty acids: Waste animal fat was collected from the animal wastes and was extracted using Soxhlet's extraction technique. The extracted fat was degummed using orthophosphoric acid, water washed and dried. Following that, the fat was mixed with hexane and was analysed using gas chromatography [10]. GC-MS spectra identified the dominant fatty acids present in the waste fat thereby helping in designing the triglyceride molecule.

2.2 Design of triglyceride molecule: The triglyceride molecule was designed using Avogadro software and was imported to Gaussian tool using mol.2 file format. Arrangement of fatty acids in the triglyceride molecule was based on the Stereoscopic Number. The molecular model was initially optimized using auto optimization tool for proper alignment of atoms in their respective positions. The optimized model was then submitted for molecular analysis.

2.3 Optimization and frequency Analysis of triglyceride molecule: The actual optimization and determination were calculated using Density Functional Theory (DFT) method of quantum mechanics. The molecule was considered to be in ground state and the default spin was assumed for spin state [11]. The parameters involved in optimization and frequency determination of triglyceride molecule is tabulated below in table 1.

2.4 Electrostatic potential of triglyceride molecule: The electrostatic potential distribution was determined using the Density Functional Theory (DFT) method of quantum mechanics under the job type- energy. The molecule was considered to be in ground state and the default spin was assumed for spin state [11]. The basis set used for determining the energy of molecule was 6-31G*, known for computing larger organic molecules. The parameters involved in determination of energy and electrostatic potential distribution of triglyceride molecule is tabulated below in table 1.

**Table 1: Gaussian calculation setup for computational analyses**

| | Optimization and Frequency Analysis | Electrostatic Potential charge distribution | Activation Energy (Ground State) | Activation energy (Transition State) |
|---|---|---|---|---|
| Job Type | Opt + Freq | Energy | Energy | Energy |
| State | Ground State | Ground State | Ground State | Transition State |

| Method | DFT | DFT | DFT | DFT |
|---|---|---|---|---|
| Spin State | Default Spin | Default Spin | Default spin | Default spin |
| | B3LYP | B3LYP | B3LYP | B3LYP |
| Basis Set | 6-31G* | 6-31G* | 6-31G* | 6-31G* |
| Charge | | | 0 | 1 |

2.5 Activation energy of triglyceride molecule: The activation energy was determined using the Density Function Theory (DFT) Method in quantum mechanics under the job type- energy. The molecule was considered to be in ground state and the default spin (B3LYP) was assumed for spin state. The basis set used for determining the energy of molecule was 6-31G*, also known for computing larger organic molecules. The charge of the molecule in ground state was found to 0 whereas charge in excited state was found to be +1[11].

## 3. Results and Discussion

The waste animal fat was rendered from the subcutaneous wastes and was characterized for fatty acids using GC-MS spectra ,which identified Stearic Acid (C18),Oleic Acid (C18:1) and Palmitic Acid (C16) as the most dominant fatty acids present in it. Figure 1 shows the GC spectra of waste animal fat. The position and order of the fatty acids in the triglyceride molecule were decided based upon the Stereoscopic Number System (SN) and degree of saturation [12]. Accordingly, unsaturated oleic acid was well placed at SN2 position. Meanwhile, saturated Stearic Acid and Palmitic Acid among the identified fatty acids, was positioned in SN1 and SN3 position.

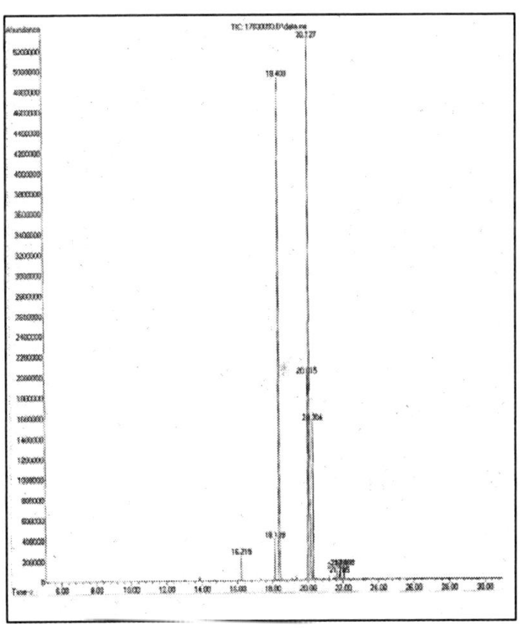

**Figure 1: GC Spectra of waste animal fat based triglyceride molecule**

Figure 2 shows the initial design model of the triglyceride molecule with no optimized geometry and ground state energy level. The model consisted of 168 atoms and 490 electrons in total and was arranged according to the tetravalency of the carbon with respect to hydrogen and oxygen and corresponding chain length of respective fatty acids.

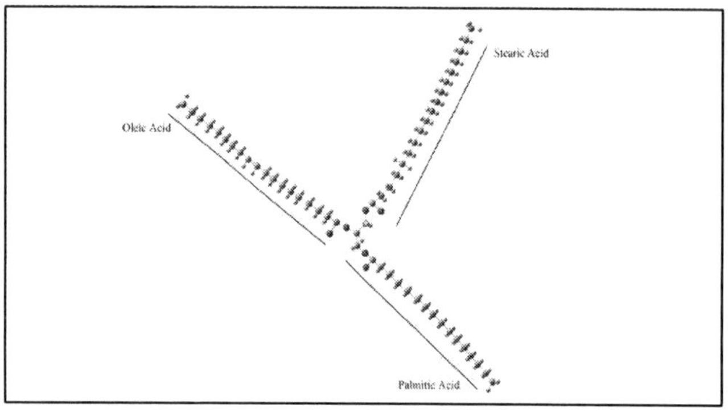


**Figure 2: Initial molecular model of the triglyceride molecule (pre-optimization)**

Figure 3 shows the optimized model of waste animal fat based triglyceride. The initial optimization was done using auto optimization tool which uses Merck Molecular Force Field (MMFF94(s)) method [13] where every individual atom in the molecule was considered as charged spheres with predetermined diameter and charges ,and were optimized based upon their interactions [11]. The overall energy required for optimizing the triglyceride molecule including the bond stretching energy, bond bending energy and torsional energy was estimated to be -58.95 kJ/mol. The final optimization of the triglyceride was optimized through Gaussian calculation setup using semi empirical method which works on successive iteration.

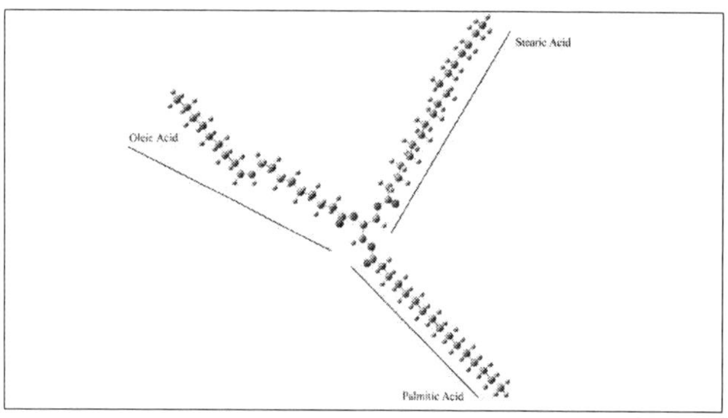


**Figure 3: Optimized model of the triglyceride molecule (post-optimization)**

Figure 4 and 5 depicts the variation in the geometry between the auto optimized model and mathematically optimized model. The latter model was computed based upon the Gaussian calculation setup→optimization job type, where it was solved until the interatomic distances between two carbon atoms were brought to the closer range of accuracy. This was achieved using semi empirical method of quantum mechanics, where all the physical dimensions related to the model was optimized based on successive iterations. Total number of iterations taken to solve this model was 67 and the energy was found to be -0.770 Hartee (2021.64 kJ/mol). The RMS gradient norm was bought nearly to 0 after 67 iterations and in addition, the deviation was with negligible difference post 50th iteration, thereby indicating that any value considered beyond 50th iteration was accurate with minimum variation. Figure 6 and 7 shows the energy curve and RMS Gradient Curve.

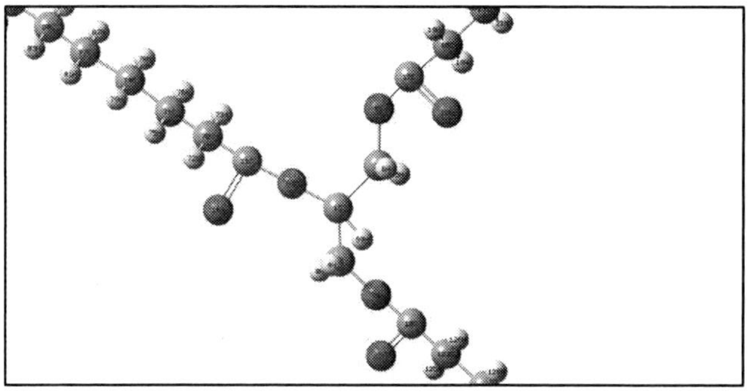

**Figure 4: Auto Optimized model of triglyceride molecule (Glyceride spine)**

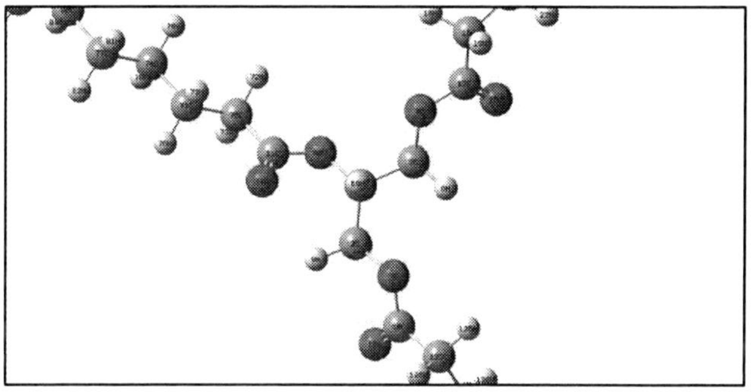

**Figure 5: Mathematically Optimized model of triglyceride molecule (Glyceride spine)**

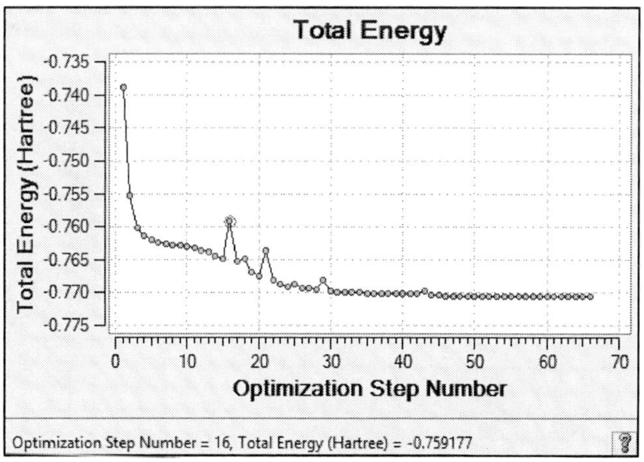


**Figure 6: Energy Curve of Optimized Triglyceride molecule (in steps)**

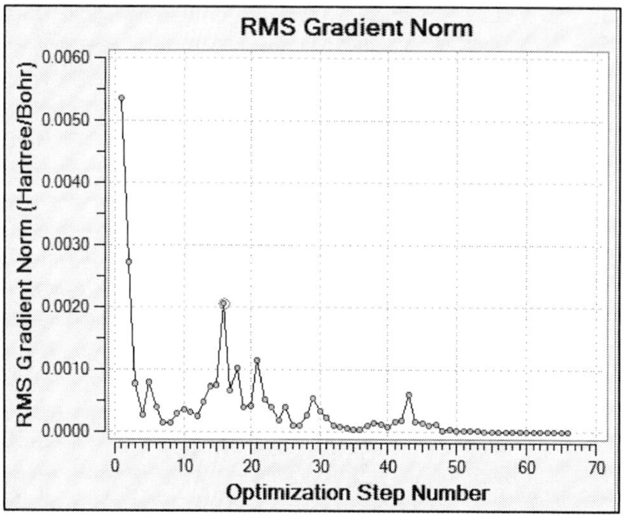


**Figure 7: RMS Gradient Norm Curve of Optimized Triglyceride molecule (in steps)**

Figure 8 and 9 depicts the modification in the spatial geometry of the double bond in fatty acid at SN2 position in the triglyceride molecule. The bend in the optimized model was because of the Cis-Configuration type isomerism, which is usually recorded for an ideal optimized oleic acid. The angle between C92 and C89 atoms was found to be 127.321° and the same was measured between C97 and C95 atoms because of its geometrical symmetricity. Similarly, the other carbon atoms were aligned based upon their spatial alignment with another carbon atom and their respective hydrogen atoms.

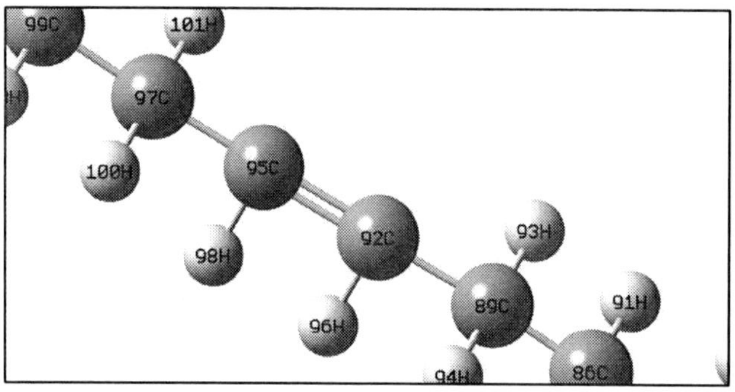


**Figure 8: Double Bond in Oleic Acid (Pre-optimization)**

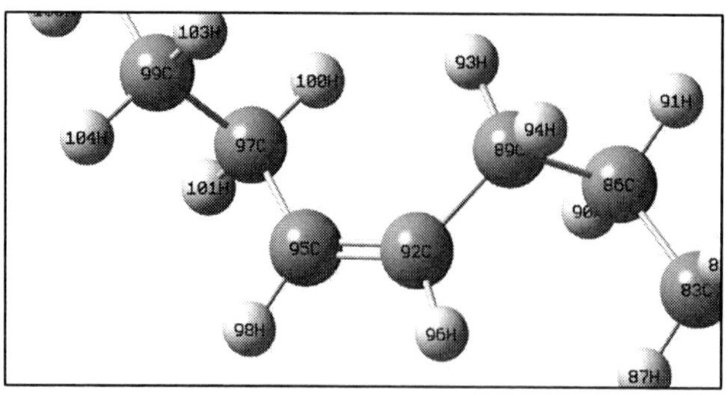

**Figure 9: Double Bond in Oleic Acid (Post-optimization)**

Figure 10-15 explains the optimization of majorly concentrated bonds which influences the transesterification reaction carried out later. The optimized bond length between C95-C92, C1-C2 and C1-C3 was found to be 1.3382Å, 1.533361Å and 1.53178Å respectively, whereas C2-O7, C1-O68 and C3-O4 were found to be 1.45096Å, 1.46109Å and 1.45141Å respectively. On average, it can be concluded the bond length between two saturated carbon atoms were found to be 1.53Å whereas for an unsaturated carbon bond it was found to be 1.33Å. The average bond length between the saturated bond between carbon and oxygen was found to be in order of 1.45Å. These bond lengths impart major role in transesterification reaction, where increased bond length imparts higher activity in reaction. Table 2 summarizes the bond length before and after Gaussian optimization.

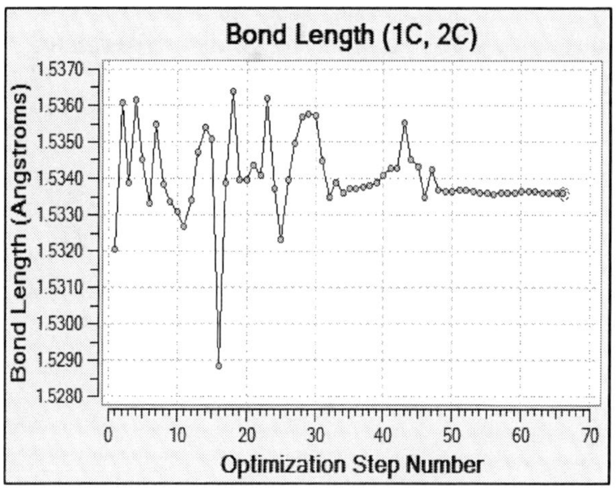


**Figure 10: Optimization of Bond length for C1-C2 bond using Gaussian iteration**

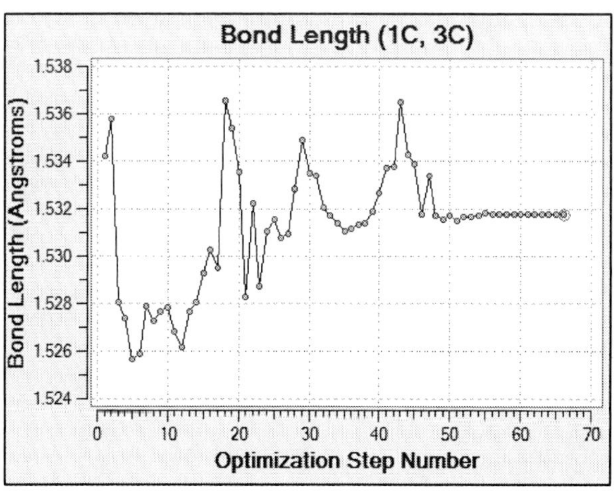


**Figure 11: Optimization of Bond length for C1-C3 bond using Gaussian iteration**

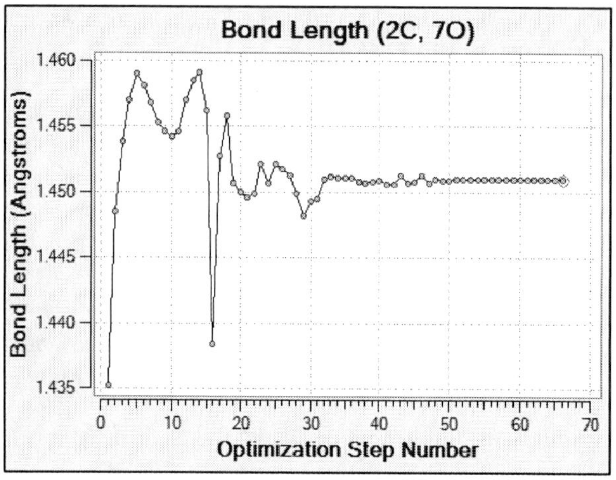


**Figure 12: Optimization of Bond length for C2-O7 bond using Gaussian iteration**

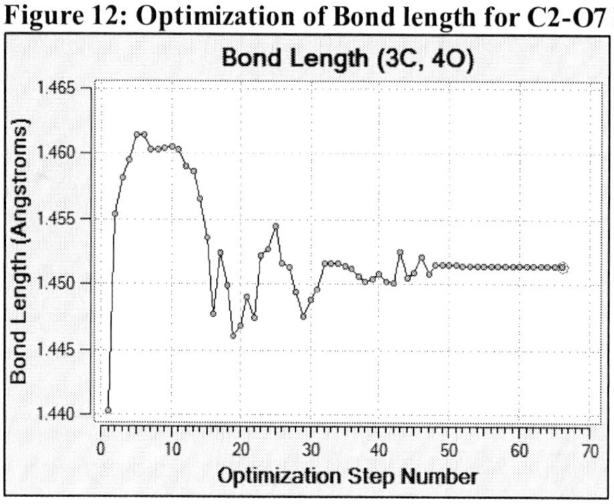


**Figure 13: Optimization of Bond length for C3-O4 bond using Gaussian iteration**

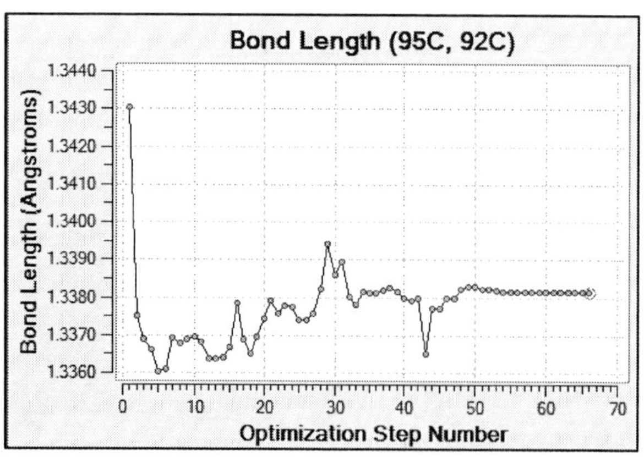


**Figure 14: Optimization of Bond length for C92-C95 bond using Gaussian iteration)**

IOP Conf. Series: Materials Science and Engineering **923** (2020) 012075 doi:10.1088/1757-899X/923/1/012075

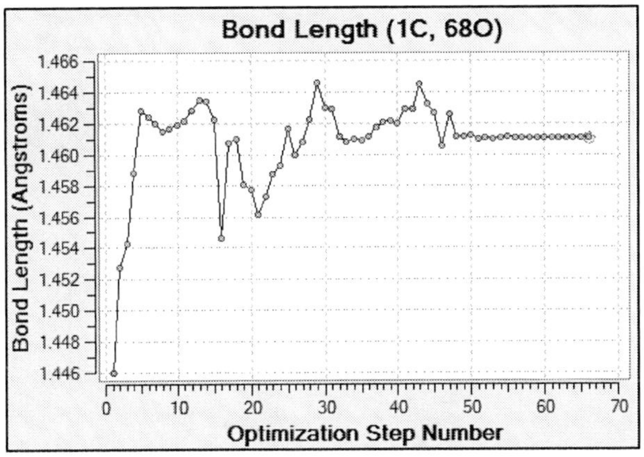


**Figure 15: Optimization of Bond length for C1-O68 bond using Gaussian iteration**

**Table 2: Bond length before and after Gaussian optimization**

| s.no | Interested bond | Bond length (Å) Pre-optimization | Bond length (Å) Post-optimization |
|---|---|---|---|
| 1 | C95-C92 | 1.3430 | 1.3382 |
| 2 | C1-C2 | 1.5320 | 1.5334 |
| 3 | C1-C3 | 1.5342 | 1.5318 |
| 4 | C2-O7 | 1.4352 | 1.4510 |
| 5 | C1-O68 | 1.4460 | 1.4611 |
| 6 | C3-O4 | 1.4403 | 1.4514 |

Figure 16 a&b depicts the electrostatic potential mapping of the triglyceride molecule. The electrostatic map shows the overall charge distribution throughout the triglyceride molecule and their related properties as well. The net charge of the molecule was found to be 0 since all atoms were paired and electrons shared covalently, in other words the model was a molecule . The spin state was considered as singlet and the whole iteration was computed based upon the DFT with 6-31G* as its basis set. Overall energy for the molecule was calculated to be -2522.7425 a.u.(-10999.1573*10-18 J) and the dipole moment was found to be 1.3875 debaye (0.287 eÅ).

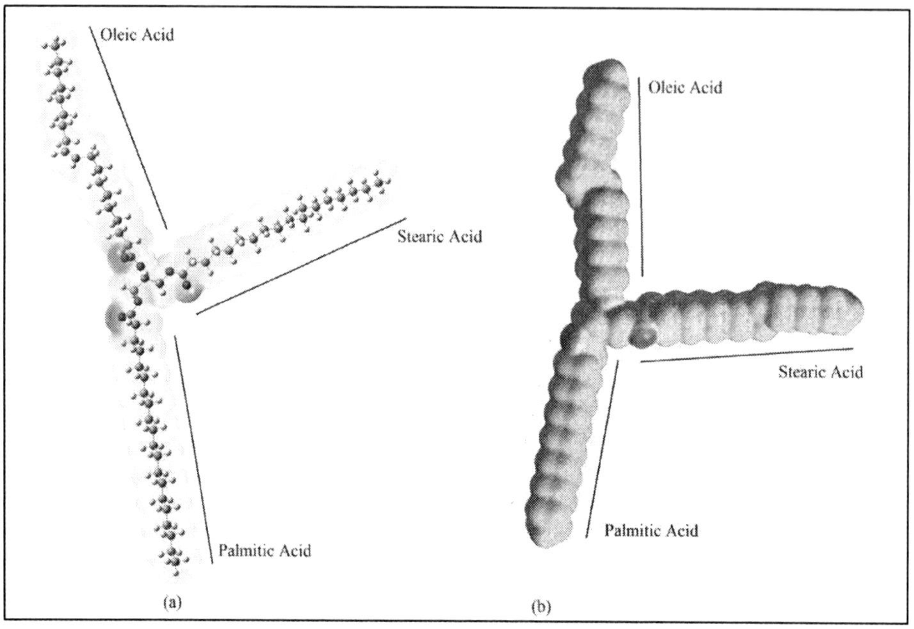


**Figure 16: Electrostatic potential mapping of the triglyceride molecule (a) transparent surface (b) wireframe surface**

Figure 17a & b shows the electrostatic potential map in form of transparent surface and meshed surface at glycerol spine. The variation in charge distribution at molecule from electrostatic potential map was identified using the colour code variation. The energy scale ranged from -0.273846 a.u. (718.99 KJ/mol) to 0.318700 a.u. (836.746914 KJ/mol), where the oxygen molecule, which readily involves in reaction is in the range of -0.27 a.u. (-718 KJ/mol) exhibiting negative charge whereas the inert carbon atoms is in the range of 0.31 a.u. (837 KJ/mol) carries the positive charges. The O68 atom has the highest negative charge, thereby making the positively charged alkyl ion to attack it first, followed by the remaining oxygen molecules based on magnitude of their charges. Table 3 shows the charge of individual atoms that involves in reaction.

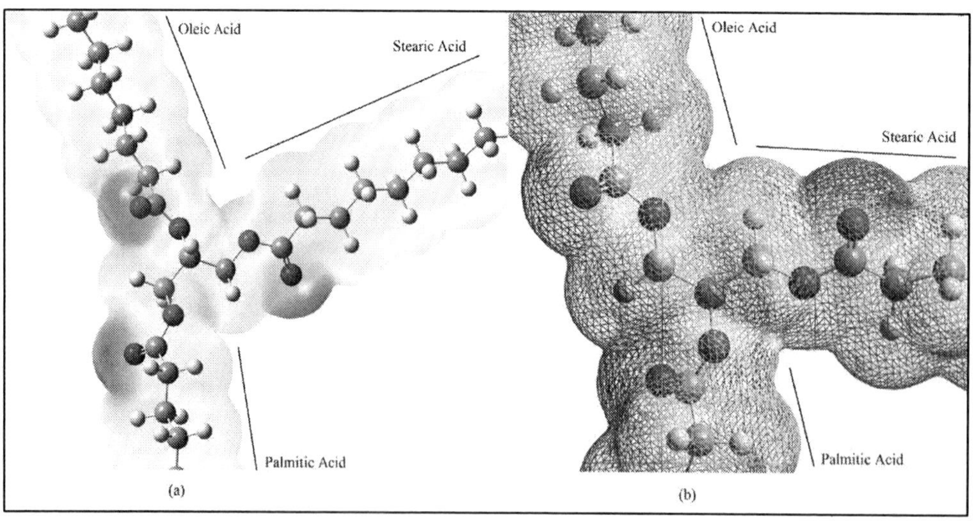

**Figure 17: Electrostatic Potential Mapping at Glycerol Spine (a) Transparent surface (b) Wireframe Surface**

**Table 3: Atomic charge of individual atoms involving in transesterification reaction**

| Atom Number | Atomic Charge (in a.u.) | Atomic Charge (in KJ/mol) |
|---|---|---|
| O15 | -0.273846 | -7.4517 |
| O4 | -0.262573 | -7.1449 |
| C12 | 0.316875 | 8.6226 |
| C10 | 0.318700 | 8.6722 |
| O7 | -0.263718 | -7.1761 |
| O13 | -0.271112 | -7.3773 |
| C11 | 0.318486 | 8.6664 |
| O14 | -0.273373 | -7.4388 |
| O68 | -0.263721 | -7.17621 |

The activation energy of the triglyceride molecule was studied by computing the energy levels at the ground state and excited state of the molecule. The energy level in the ground state was found to be -2539.06 a.u. (6666302.54 KJ/mol) whereas for the excited state, it was found to be -2538.860 a.u. (6665777.44 KJ/mol). Similarly, the dipole moment varied from 1.3010 to 18.0651 Debaye.

$\Delta E$= 2539.06 -2538.860 a.u. = 0.2 a u or 525.1 KJ/mol

The variation in energy predicts the amount of energy required to activate the triglyceride molecule for the reaction from it ground state. The data related to the activation energy is tabulated in table 4.

**Table 4: Energy and dipole moment of molecule in ground and transition state**

| Ground state | | Transition state (charge:1, spin: doublet) | |
|---|---|---|---|
| Energy | -2539.06 a.u. | Energy | -2538.860 a.u. |
| Dipole moment | 1.3010 debaye | Dipole moment | 18.0651 debaye |

Figure 18 represents the HOMO and LUMO plot of triglyceride molecule. It can be concluded that the HOMO mapping was higher at SN2 position, and then followed by SN3 position and SN1 position. The LUMO position was occupied at the unsaturated bond between carbon-carbon atoms belonging to fatty acids in SN2 position.

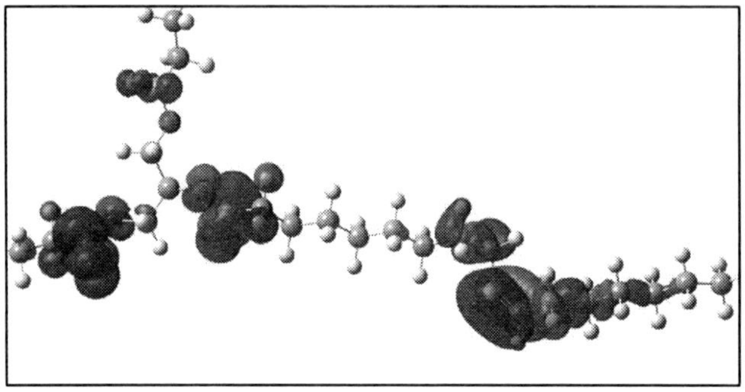

**Figure 18: HOMO & LUMO Mapping of Triglyceride Molecule**

Figure 19 shows the Infrared (IR) Spectrum of the triglyceride molecule. The major dominant peaks are 1819.08 $cm^{-1}$,1821.88 $cm^{-1}$ and 2694.50$cm^{-1}$ whose corresponding Dipole strengths were 588.4613, 301.5978, 541.2702 $10^{-40}esu^2cm^2$. The Bond with a wavelength of 1819.08 $cm^{-1}$ represents the carbon and oxygen (C=O) stretching, 1821.88 $cm^{-1}$ represents the C=O stretching and 2694.50$cm^{-1}$ represents carbon and hydrogen (C-H) stretching [14]. The table 5 consolidates the IR spectra for corresponding frequencies and Dipole strength.

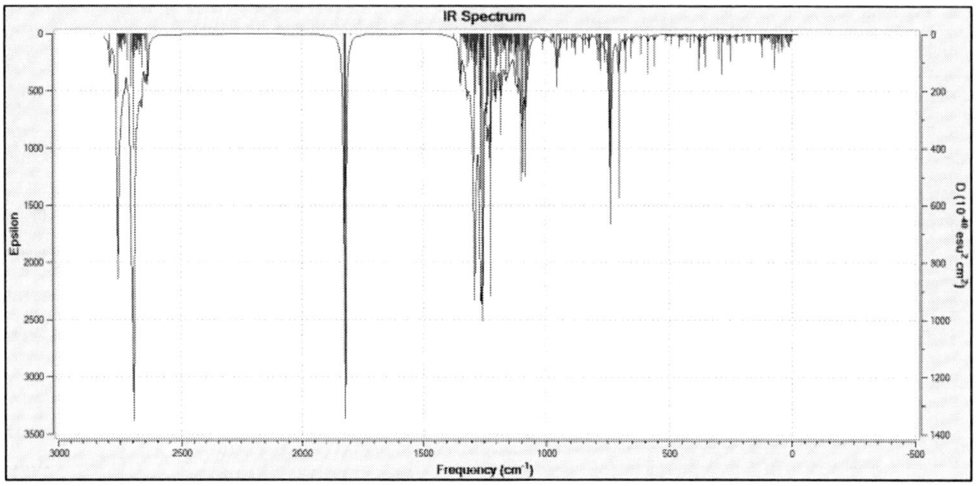


**Figure 19: Infrared (IR) Spectrum of the Triglyceride molecule**

**Table 5: Spectral data of Dominant peaks and their corresponding strengths**

| Frequency ($cm^{-1}$) | Dipole strength (10-40 $esu^2cm^2$) |
|---|---|
| 1819.08 | 588.4613 |
| 1821.88 | 301.5978 |
| 2694.98 | 541.2702 |
| 1254.85 | 276.9249 |
| 1256.19 | 232.5961 |
| 2758.27 | 126.5725 |
| 2759.06 | 144.9754 |
| 2759.37 | 152.9581 |

## 4. Conclusions

The quantum computational study of the triglyceride molecule extracted from waste animal fat was studied successfully. The most dominant fatty acids present in it were found as stearic acid, oleic acid and Palmitic acid. The energy of the triglyceride molecule was found as -2522.7425 a.u. (10999.1573*10-18 J) and the dipole moment was found to be 1. 3010 debaye (0.287 eÅ). The IR spectrum revealed the bond behaviour between the carbon and hydrogen atoms and carbon and oxygen atoms in addition with carbon and carbon atoms. The activation energy required for the triglyceride to respond a reaction was found to be 0.2 a.u. Understanding these molecular properties of the molecules helps one in understanding properties and its role in reaction.

## References

1. Srinivasan, G.R. and Jambulingam, R., 2018. Comprehensive Study on Biodiesel Produced from Waste Animal Fats-A Review. Journal of Environmental Science and Technology, 11(3), pp.157-166.

2. Yatish, K. V., Lalithamba, H. S., Suresh, R., Arun, S. B., & Kumar, P. V. (2016). Optimization of scum oil biodiesel production by using response surface methodology. Process Safety and Environmental Protection, 102, 667-672.

3. Kumar, R. S., Sureshkumar, K., & Velraj, R. (2015). Optimization of biodiesel production from Manilkara zapota (L.) seed oil using Taguchi method. Fuel, 140, 90-96.

4. Asakuma, Y., Nishimura, M., Li, Q., Ang, H. M., Tade, M., Maeda, K., & Fukui, K. (2007). Colouring mechanism of dyed KDP crystal by quantum chemistry. Journal of Molecular Structure: THEOCHEM, 810(1-3), 7-13.

5. Asakuma, Y., Li, Q., Ang, H. M., Maeda, K., & Fukui, K. (2006, April). The Effects of Impurities on Crystal Morphology Formation: an experimental and molecular modelling investigation. In WCPT5 (5thWorld Congress Powder Technology, 2006).

6. Asakuma, Y., Maeda, K., Kuramochi, H., & Fukui, K. (2009). Theoretical study of the transesterification of triglycerides to biodiesel fuel. Fuel, 88(5), 786-791.

7. Silva, M. V., Hori, C. E., & Reis, M. H. (2015). Thermochemical data of the oleic acid esterification reaction: A quantum mechanics approach. Fluid Phase Equilibria, 406, 168-174.

8. Stöckigt, D. (1996). Application of density functional theory/Hartree-Fock hybrid methods. Geometries and bond dissociation energies of Al+ complexes. Chemical physics letters, 250(3-4), 387-392.

9. Gokul, R.S. and Jambulingam, R., 2019. Computational analysis of methyl oleate in biodiesel produced from waste beef tallow. Research Journal of Chemistry and Environment, 23(2), 51-59.

10. B Deepanraj, G Sankaranarayanan, P Lawrence 2012. Performance and emission characteristics of a Diesel engine fueled with rice bran oil methyl Ester blends, Daffodil International University Journal of Science & Technology, 7(2), 51-55.

11. Atkins, P. W., & Friedman, R. S. (2011). Molecular quantum mechanics. Oxford university press.

12. Berry, S. E. (2009). Triacylglycerol structure and in transesterification of palmitic and stearic acid-rich fats: an overview and implications for cardiovascular disease. Nutrition research reviews, 22(1), 3-17.

13. Halgren, T. A. (1999). MMFF VI. MMFF94s option for energy minimization studies. Journal of Computational Chemistry, 20(7), 720-729.

14. Srinivasan, G.R., Shankar, V., Chandra Sekharan, S., Munir, M., Balakrishnan, D., Mohanam, A. and Jambulingam, R., 2020. Influence of fatty acid composition on process optimization and characteristics assessment of biodiesel produced from waste animal fat. Energy Sources, Part A: Recovery, Utilization, and Environmental Effects, pp.1-19.

IOP Conf. Series: Materials Science and Engineering **923** (2020) 012076 doi:10.1088/1757-899X/923/1/012076

# Physicochemical Analysis, Elicitation and Charaterisation of Bio-oil from *Aloe barbadensis*

**Jagadesan P[1], S Prithivi Rajan[1], Aditya Singh Chauhan[1], Chandana S Narayana[1], Vijayalakshmi S[2], Ranjitha J[2,*]**

[1]*Department of Mechanical Engineering, School of Mechanical Engineering, Vellore Institute of Technology, India*
[2]*$CO_2$ Research and Green Technologies Centre, Vellore Institute of Technology, Vellore-14
**Email ID: ranjitha.j@vit.ac.in**

**Abstract**: Biodiesel from plants can compete economically with petroleum diesel fuels as a future prospective fuel. The bio-oil was extracted from the plant of *Aloe barbadensis* by soxhlet extractor using hexane as an organic solvent, and then further, the oil was analyzed for moisture content, pH, specific gravity, density, viscosity, saponification value, refractive index, peroxide value, acid number, free fatty acid, and iodine value. The results showed that four major fatty acids, like nonadecanoic acid, n-hexadecanoic acid, octadecadienoic acid and octadecenoic acid were present in the bio-oil of *Aloe barbadensis* and characterized by GC-MS, which showed that the bio-oil could be used for producing high quality of biodiesel. In furture studies, converted crude biooil into biodiesel using hetergenous catalyst will be anlaysed for their engine characteristic studies.

**Keywords:** Physicochemical Analysis, Bio-oil, Biodiesel.

## 1. INTRODUCTION

The demand for energy around the world is influenced by many factors like urbanization, increased population, and better living standards. People have realized that the present energy reserves of fossil fuels have been declined with many environmental factors. An alternative of energy with a better economy which can provide a sustainable future in both domestic and industrial point is bio-diesel[1&2].Bio diesel remarked as "Cardon dioxide neutral" since most of the carbon produced in combustion is retrieved during cultivation of crops is produced from plant and animal fats raw or used has methyl ester as its main chemical content. Biodiesel has favorable characteristics like non-toxic, biodegradable, regenerative, and does not pollute the water, which makes it an economically friendly product. Lower sulfate and particulate matters and lesser hydrocarbon emissions make it a future fuel to beat current petroleum fuels economically. Feedstock rich in fatty acids like food waste, animal fats, and vegetable oil both edible and inedible seeds and byproduct obtained in the refining process is much cheaper, thereby reducing the production cost and are more preferred[3].The production of biodiesel from animal and plant fats and alcohol production by fermentation under microbes are the pronounced source of renewable energy[3]. The critical source of biodiesel is the oils extracted from plant seeds. While the edible seed oils met the demand as a source of food, the vegetable oil extracted from non-edible seeds acts as a secondary source of feedstock in biodiesel production[4]. Demand as food source and cost of producing the edible seed oil into fuel in higher. This is an added advantage of letting us prefer non-edible oilseed in the production of biodiesel. Aloe Vera, an herbaceous family plant, also called Aloe Barbadensis, Barbados Aloe, etc. has the scientific name Aloe Vulgarize. It is cultivated in many types of soil. It consumes less water and can be seen easily in places having a tropical climate. The leaves of the plant are thick, juicy, and green, which grows up to 30 to 60 cm long. A tasteless and colorless gel called aloe gel, which is highly viscous in nature, is found in the leaves of the plants. As the age of the leaves of the aloe vera increases, the viscosity increases too[6]. Enzymes, vitamins, minerals, sugar, saponins, lignin, amino acid, and salicylic acid are the constituents present about more than 98% along with water polysaccharides like cellulose, hemicellulose, acemannan, glucomannan, pectin, and mannose derivates. The infection of fungal disease on the plant may affect the biomedical medicals mentioned above [9]. They were widely used in medical, skin elated cosmetics, and agricultural purposes. It is also used in the production of bio-beverages and can be used in the treatment of polluted water[7&11]. Processing of the aloe vera for such purpose includes dehydration and heating, followed by grinding of the leaves. Pyrolysis treatment of aloe vera is found to be low cost, eco-friendly, manage biomass, and economically feasible by research studies[6]. Aloe vera farming has seena commercial reputation in medicinal products[8]. A sample dehydrated at 70 and 80 when heated shows changes in average molecular weight in bioactive polysaccharide in fresh aloe with an increase from 45kDa to 75kDa[9]. Dirt and bacterial at surface level are removed when the leaves are washed with an antimicrobial agent. This process is then culled, trimmed, and the rind is removed. This final leaf material looked similar to a proceeded fillet fish and called as "fillet". This filleting can be done either manually or mechanically, which involves the removal of the tip of the leaf, top and bottom rinds, the lower part of the leaf base, and the spines along the leaf margins. Water is used to rinse the inner part of the leaf fillets, followed by depulping and pasteurization. Further processing is treatment using enzymes, deaeration by filtration and using activated carbon to decolorize (see the aloe vera leaf processing section), and preservative addition [10 & 11]. Final purification includes filtration followed by the stabilization of liquid. The obtained final fluid is clean and is comparable to the organoleptic characteristic of inner leaf juice, but generally holding three times more bio-active constituents than juice produced by hand filleted as discussed in the next section of this document [12-16].

## 2. MATERIALS AND METHODS:

### 2.1 Collection of raw material and extraction of bio-oil from *Aloe barbadensis*

The fresh plant leaves were collected from VIT University, Tamil Nadu, and India. The leaves obtained from the *Aloe barbadensis* plant were washed properly and were cut into small pieces for drying purposes. For three weeks, the cut pieces of aloe vera leaves were air-dried in the shade so that loss of phytoconstituents from direct sunlight can be avoided. After proper drying, it was finely powdered using the pulverizer and sieved. It was then stored in an air-tight container after homogenization for further analysis as shown in the figure 1.The plant leaves were washed with water, and kernels were separated from leaves. The leaves were left for drying in the Hot-Air Oven for about 3 hours. After drying, grains were separated from leaves. The separated kernels were finely grounded.The extraction of bio-oil was done using Soxhlet extractor. The crude bio-oil is extracted with 200mL of hexane as a solvent for about 72 hours at 80 °C. Then the extract was concentrated using a rotary evaporator at 50°C. The extracted crude oil were stored for further use as shown in the figure 2.

Cleaning
↓
Edges of the leaves are removed
↓
Leaves are cleaved into different sections
↓
Dried in shades of sunlight
↓
Finely powdered
↓
Storing (air-tight container)
↓
Washing using kernels
↓
Dried using hot air oven
↓
kernels are separated
↓
Bio oil extraction

*Fig 1. Process flow diagram for the extraction of bio-oil from Aloe vera leaves*

IOP Conf. Series: Materials Science and Engineering **923** (2020) 012076 doi:10.1088/1757-899X/923/1/012076

*Figure 2: Extraction process of Aloe barbadensis*

## 2.2 Characterization of Oil

The percentage (%) yield of the leaf oil was calculated gravimetrically. Odor, color, and physical state of the oil were estimated by sensory evaluation. They were characterized for specific gravity using specific gravity bottle, and P.H. was determined using pH meter, moisture content by the oven-dry method, ash content by heating to dryness in a furnace, refractive index using Abbe refractometer, kinematic viscosity using a viscometer [17]. Other properties analyzed were the saponification values determined by titrimetry method. Acid value, iodine value, and peroxide value were determined by titrimetry, according to FAO.

### Gas Chromatography

An Agilent 6890 gas chromatograph was equipped with a straight deactivated 2 mm direct injector liner and had a 15m Alltech EC-5 column (250μ I.D., 0.25μ film thickness). For sample introduction, a split injection was used, and the split ratio was set to 10:1. The oven temperature program was programmed to begin at 35°C, hold for 2 minutes, then ramp at 20°C per minute to 300°C and hold for 5 minutes. The flow rate of helium carrier gas was set to 2 ml/minute(constant flow mode).

### Mass Spectrometry

A JEOL GCmate II benchtop double-focusing magnetic sector mass spectrometer operating in electron ionization (E.I.) mode with TSS-2000[1] software was used for all analyses. Low resolution mass spectra were acquired at a resolving power of 1000 (20% height definition) and scanning from m/z 25 to m/z 700 at 0.3 seconds per scan with a 0.2-second inter-scan delay. High-resolution mass spectra were acquired at a resolving power of 5000 (20% height definition) and scanning the magnet from m/z 65 to m/z 750 at 1 second per scan. Qualitative analyses were performed on an H.R.; high-resolution mass spectrometer Samples were introduced into the electron ionization source using a drawn micro capillary at a flow rate of 1 ml/min. The mass spectrometer was operated in positive ion mode with a spray voltage of 2.4 kV, a capillary temperature of 250, a capillary voltage of 29.0 V, a 1.5-u ion isolation window, and a 100-ms maximum inject time. The average Scans of the M.S. spectra approximately 100 scans were averaged for the $MS^2$ and $MS^3$ spectra.

**Mass spectrometry library search**

Identification of the components of the purified compound was matching their recorded spectra with the info bank mass spectra of NIST library V 11 provided by the instrument's software. Based on the GC-MS spectral data, six fatty acids were identified, and further, the bio-oil will be converted into biodiesel using the trans-esterification process. The Bio-oil from aloe vera was converted into biodiesel using the NaOH catalyst. The produced biodiesel yield was calculated gravimetrically.

## 3.RESULTS AND DISCUSSION:

The bio-oil content of *Aloe barbadensis* was calculated, and it can be used for biodiesel production would be highly economical. The oil content was found to be 21.57%. The oils had agreeably oily and very dark in color is an advantage to use as biodiesel feedstock. The physical state of the oil was liquid at room temperature. The specific gravity of the oil was 0.86, which is close to the standard range of 0.87–0.90 for biodiesel [18-21]. Density and gravity are other important parameters for diesel fuel injection systems. The values are supposed to be maintained within tolerable limits to permit optimal air to fuel ratios for complete combustion. The moisture content of the oil was in the limit to the ASTM Standards. The results of the physical properties are tabulated in Table 1.

**Table 1: Physical properties of *Aloe barbadensis* leaf oils**

| S.No | Chemical properties | *Aloe vera* |
|---|---|---|
| 1. | pH | 4.5 |
| 2. | Iodine value (g/100g) | 105.02 |
| 3. | Acid value (mgKOH/g) | 0.05 |
| 4. | Saponification value (mgKOH/g) | 220 |
| 5. | Peroxide value (meq/kg) | 7.89 |
| 6. | Refractive index | 1.447 |

The results of the chemical properties are tabulated in Table 2. The pH value of the oil was acidic. A measure of the unsaturation of fats and oils is the Iodine value. High iodine value indicatesthe presence of highly unsaturated fats and oils. Aloe vera leaf oil iodine values are within the limits of standard. The saponification value is used in checking adulteration. The saponification values are lesser than the ASTM standard. The refractive index of oils depends on properties like molecular weight, fatty acid chain length, degree of unsaturation, and degree of conjugation. Triglycerols have a higher refractive index than do their constituent free acids. The values of the refractive index for different oils generally vary between 1.447 and 1.482 [22-24]. All these physicochemical properties suggest that the leaf oil adoptable for biodiesel production.

**Table 2: Chemical properties of *Aloe barbadensis* leaf oil**

| S.No | Physical properties | *Aloe vera* |
|---|---|---|
| 1. | Colour | Light yellow |
| 2. | Odor | Agreeably Oily |
| 3. | Moisture (%) | 98.82 |
| 4. | Oil Content (%) | 42 % |
| 5. | Specific Gravity | 1.002 |
| 6. | Physical State at R.T. | oil |
| 7. | Density (g/cm$^3$) | 0.93 |
| 8. | Viscosity, at 40$^0$C, centistokes | 28.4 |

The fatty acid composition of aloe veraleaf analyzed by GC-MS was shown in table 3. The fatty acid profile shows that the *Aloe barbadensis* leaf oil can be used for quality biodiesel. It was clear that most of the fatty acids like nonadecanoic acid, n-hexadecanoic acid, octadecadienoic acid and octadecenoic acid present were essential fatty acids that can be converted into biodiesel [25 & 26]. Biodiesel was produced using the oil extracted from *Aloe barbadensis,* and the yield was 82%.

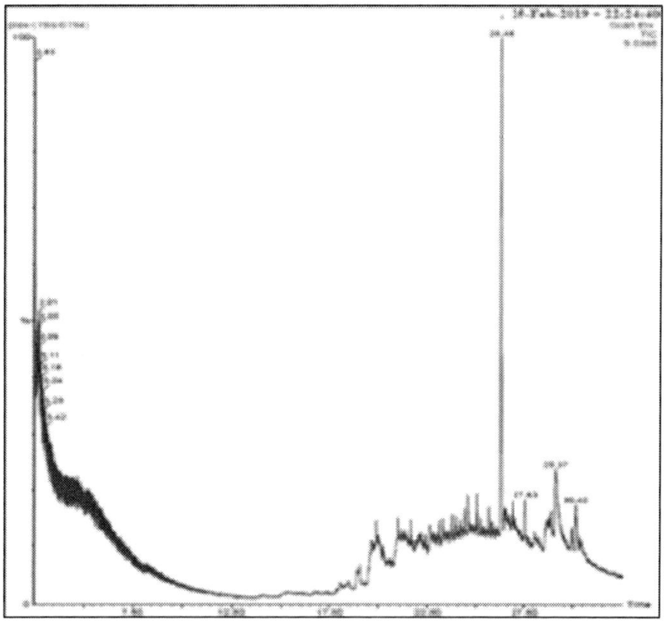

***Figure 3: GC-MS Spectrum for crude oil obtained from Aloe barbadensis***

| S.No | RETENTION TIME | FATTY ACIDS | STRUCTURE |
|---|---|---|---|
| 1 | 26.48 | Nonadecanoic acid | 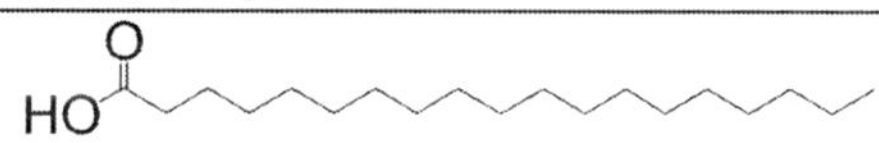 |
| 2 | 27.83 | n-Hexadecanoic acid | 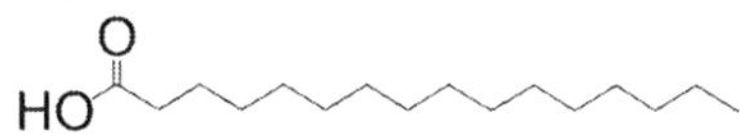 |
| 3 | 29.37 | Octadecadienoic acid | 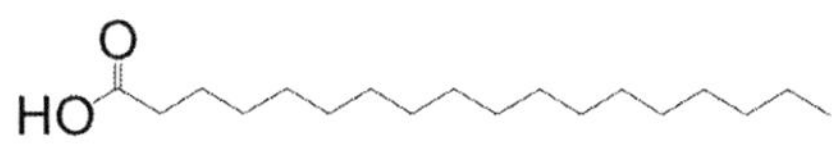 |
| 4 | 30.42 | Octadecenoic acid | 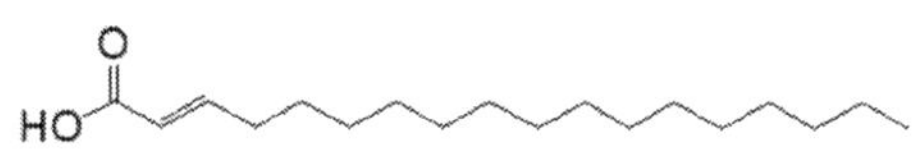 |

**Table 3: Fatty acids composition of *Aloe barbadensis* leaf oil**

## CONCLUSION:

The following conclusions were made by biodiesel production; (i) The physicochemical properties of bio-oil distinctly confirmed that the bio-oil from *Aloe vera* was used for biodiesel production; (iii) Overall four fatty acids were identified as nonadecanoic acid, n-hexadecanoic acid, octadecadienoic acid and octadecenoic acid using GC-MS spectral data, and these fatty acids were only responsible for biodiesel production; (iii) The extracted crude oil was converted into biodiesel and achieved 82% yield using hetergenous catalytst; (iv) Finally, the produced biodiesel will be used to check their performance and emission characteristics in diesel engine in furture work.

IOP Conf. Series: Materials Science and Engineering **923** (2020) 012076 doi:10.1088/1757-899X/923/1/012076

**REFERENCES:**

[1] Padmanabhan, S., Ganesan, S., Masen, A., Anvesh, V., Pradeep, B., Ganesan, S., Pradeep, B. (2018). Influence of Cerium Oxide additive on Aloe Vera Biodiesel on CI Engine Influence of cerium oxide additive and aloe vera biodiesel on a CI engine. *International Journal of Ambient Energy*, *0*(0), 1–5.

[2] S, A. N., Narasegowda, P. N., & Road, K. R. (2016). Physico-chemical and phytochemical evaluation of Aloe acutissima leaves, 4633–4640.

[3] Scala, K. D. I., Vega-galvez, A., Ah-hen, K., Nunez-mancilla, Y., Tabilo-munizaga, G., Pérez-won, M., & Giovagnoli, C. (2013). Chemical and physical properties of aloe vera (Aloebarbadensis Miller) gel stored after high hydrostatic pressure processing, *33*(1), 52–59.

[4] Silva, M., Machado, A. E., Castro, A. C. R. De, & Carvalho, B. (2015). Biodiesel Production Process Starting From Aloe Vera, 33–35. https://doi.org/10.14684/ICECE.9.2015.33-35

[5] Vindhya, P., Reddy, S. S. K., & Naik, V. G. (2014). Investigation on the Performance and Emissions of Aloe Vera Blends with EGR System, *3*(10), 16858–16865.

[6] Soares, E. J., Siqueira, R. N., Leal, L. M., Barbosa, K. C. O., Cipriano, D. F., & Freitas, J. C. C. (2019). Journal of Non-Newtonian Fluid Mechanics The role played by the aging of aloe vera on its drag reduction properties in turbulent flows. *Journal of Non-Newtonian Fluid Mechanics*, *265*(August 2018).

[7] Giannakoudakis, D. A., Hosseini-bandegharaei, A., Tsafrakidou, P., Triantafyllidis, K. S., Kornaros, M., &Anastopoulos, I. (2018). Aloe vera waste biomass-based adsorbents for the removal of aquatic pollutants : A review. *Journal of Environmental Management*, *227*(July), 354–364.

[8] Chagas, N., Alves, S., Guedes, G., Luiz, R., Albuquerque-júnior, C., Santana, D., Ribeiro, G. (2018). Journal of Photochemistry &Photobiology, B : Biology Analysis of Aloe Vera cytotoxicity and genotoxicity associated with endodontic medication and laser photobiomodulation☆. *Journal of Photochemistry & Photobiology, B: Biology*, *178*(November 2017), 348–354.

[9] Avasthi, S., Gautam, A. K., &Bhaduaria, R. (2018). Journal of Applied Research on Medicinal and Aromatic Plants Changes in biochemical components in Aloe vera (L.) Burm .f. leaves infected with Fusarium proliferatum (Matsushima) Nirenberg. *Journal of Applied Research on Medicinal and Aromatic Plants*, (October), 1–6.

[10] Farrugia, C. E., Burke, E. S., Haley, M. E., Bedi, K. T., & Gandhi, A. (2019). S.C. *Complementary Therapies in Clinical Practice*. https://doi.org/10.1016/j.ctcp.2019.01.013

[11] Hussain, S. A., Yadav, V., Reddi, S., Patil, G. R., & Singh, R. R. B. (2019). Journal of Applied Research on Medicinal and Aromatic Plants Thermal processing conditions affect in vitro immunostimulatory activity of Aloe vera juice. *Journal of Applied Research on Medicinal and Aromatic Plants*, *12*(November 2018), 73–77.

[12] El Fawal, G. F., Omer, A. M., & Tamer, T. M. (2019). Evaluation of antimicrobial and antioxidant activities for cellulose acetate films incorporated with Rosemary and Aloe Vera essential oils. *Journal of Food Science and Technology*, *56*(3), 1510–1518.

[13] Fardsadegh, B., &Jafarizadeh-Malmiri, H. (2019). Aloe vera leaf extract mediated green synthesis of selenium nanoparticles and assessment of their in vitro antimicrobial activity against spoilage fungi and pathogenic bacteria strains. *Green Processing and Synthesis*, *8*(1), 399–407.

[14] Tuong Phuoc Tho and Nguyen Truong Thinh, "A study of Aloe vera peeling and dicing system," *2017 International Conference on System Science and Engineering (ICSSE)*, Ho Chi Minh City, 2017, pp. 44-47.

[15] P. Yuvasree, K. Nithya and N. Neelakandeswari, "Biosynthesis of silver nanoparticles from Aloe vera plant extract and its antimicrobial activity against multidrug resistant pathogens," *International Conference on Advanced Nanomaterials & Emerging Engineering Technologies*, Chennai, 2013, pp. 84-86.

[16] D. Ismik, M. S. Findikoglu, D. S. Mansuroglu, E. Bulus and Y. M. Şahin, "Synthesis of Aloe Vera Loaded Chitosan Nanoparticle with Ionic Gelling Method," *2019 Scientific Meeting on Electrical-Electronics & Biomedical Engineering and Computer Science (EBBT)*, Istanbul, Turkey, 2019, pp. 1-4.
[17] Padmanabhan, S., Vinod Kumar, T., Chandrasekaran, M., & Ganesan, S. (2018). Investigation of Sapindus seed biodiesel with nano additive on single cylinder diesel engine. *International Journal of Ambient Energy*, 1–8.
[18] Shah, S., Sharma, S., & Gupta, M. N. (2004). Biodiesel preparation by lipase-catalyzed transesterification of Jatropha oil. *Energy and Fuels*, *18*(1), 154–159.
[19] Otterbach, A., & Wenclawiak, B. W. (1999). Ultrasonic/Soxhlet/supercritical fluid extraction kinetics of pyrethrins from flowers and allethrin from paper strips. *Fresenius' Journal of Analytical Chemistry*, *365*(5), 472–474.
[20] S. Boonyagul, P. Thunyakitpisal and W. Tachaboonyakiat, "Physical properties and biocompatibility of acemannan scaffold for biomedical applications," *The 7th 2014 Biomedical Engineering International Conference*, Fukuoka, 2014, pp. 1-5.
[21] S. Boonyagul, P. Thunyakitpisal and W. Tachaboonyakiat, "Physical properties and biocompatibility of acemannan scaffold for biomedical applications," *The 7th 2014 Biomedical Engineering International Conference*, Fukuoka, 2014, pp. 1-5.
[22] S. C. Pillai, P. K. K. Namboori, N. C. A. Kumar, K. V. Gopal, A. S. Kumar and P. Bharath, "Investigating multi target drug action of Aloe vera using computational analysis," *2011 2nd National Conference on Emerging Trends and Applications in Computer Science*, Shillong, 2011, pp. 1-4.
[23] Arun B. chhetri, Martin S tango, Suzane M budge, K. Chris Watts and M Rafiqul Islam, "Non-Edible Plant Oils as New Source of Biodiesel Production". *Int. J. Mol. Sci.* 2008, *9*(2), 169-180
[24] Faisal Abnisa, W.M.A. Wan Daud, J.N. Sahu "Optimization and characterization studies on bio-oil production". YU Jun1, YUE Peng-xiang1, YU Hong2, SUN Ji-ping1 & ZHENG Yuan-bin
[25] Karthikeyan, S., A. Elango and A. Prathima. 2016. "The Effect of Cerium Oxide Additive on the Performance and Emission Characteristics of a CI Engine Operated with Rice Bran Biodiesel and Its Blends." *International Journal of Green Energy* 13 (3): 267–273.
[26] Z. X. Lim *et al.*, "Switching Dynamics and Conductance Quantization of Aloe Polysaccharides-Based Device," in *IEEE Transactions on Electron Devices*, vol. 66, no. 7, pp. 3110-3117, July 2019.

IOP Conf. Series: Materials Science and Engineering **923** (2020) 012077 doi:10.1088/1757-899X/923/1/012077

# Hybrid energy harvesting system using IOT

Jayarama Pradeep*, S.Krishnakumar, M.Sowmiya

Department of Electrical and Electronics Engineering,

St. Joseph's College of Engineering, Chennai, Tamilnadu, India

*Email id : jayapradeepkp@gmail.com.

**ABSTRACT:**
Global warming, environmental pollution and power production costs are the major issues that the world is facing and many research work is done in the above areas to reduce the adverse effects arises due to fore mentioned issues. In particular, many research has been carried in power generation without $CO_2$ emission and polluting the environment. This can be achieved by using renewable sources for electrical energy production in optimized manner. This paper presents a hybrid energy harvesting by combining thermoelectric Generators (TEG) along with solar energy to develop electrical energy. The proposed framework has been intended to harvest energy to an extent that it can gather both solar energy and the heating effect in the panel at the same time. The solar radiations are made to fall on the PV panel, which is productive enough to collect the light radiations falling on it and convert into electrical energy. Meanwhile, these radiations cause warming up of the PV panel, prompting the decrease in its effectiveness of production. Hence in order to increase the efficiency of the system energy harvesting, this heat production is utilized using TEG. The thermoelectric generators are used which retain this warmth and create an electrical energy in the form of DC based on seebeck effect. This project utilize a PV panel and a thermoelectric generator to incorporate both light and warmth for harvesting electrical energy. This output energy is needed to regulate with the help of dc-dc converter. The contribution of energy from PV panel and thermoelectric generator are given to the buck boost converter. The output of the converter is stored in the battery which can be used for direct dc load or through inverter for ac load. This project also implements internet of things (IOT) technology to monitor and manage the electric energy harvested through this technology from anywhere.

## 1.INTRODUCTION

Recently the energy consumption is in the range of 10 terawatts per year , and in 2050 it may reach about 40 terawatts. This energy requirement using conventional source like fossil fuel, nuclear etc., will produce $CO_2$ emission and pollutes the environment [1]. In order to overcome this problem renewable energy sources are being used widely for the production of non $CO_2$ electrical energy. The simplest approach to control $CO_2$ by mid-century is one in which photovoltaics (PV) and other renewables are used for electricity generation, for transportation usage of hydrogen fuel, and fossil fuels can be used for residential and industrial heating [2]. Thus, solar energy conversaion will play a vital role in satisfying the world energy demand in future [3]. Due to wide research happened in solar cells, extraction of maximum energy from PV panel using multi junction cells are used nowadays. Even while using advanced technology in PV cells, a huge amount of the solar irradiance is not converted into electricity and is dissipated as heat. Hence to increase the system efficiency in a better manner harvesting of this heat produced in PV panel can be used which is considere as complementary "green" technology, namely, thermoelectricity [4]. According to this methodology, a thermoelectric (TE) generator (TEG) can be attached to the back side of the PV panel to harvest the e heat for generation of additional electricity, resulting from the Seebeck effect (the thermoelectric effect [5-11]). The Internet of Things is a constantly-evolving area of research, with many new applications and ideas being generated almost daily. From intelligent farming to smart devices in our homes, the IoT has affected almost every industry in our modern world [12]. The IoT has been evolving fastly over the last few years. Billions of devices and systems are already connected, with more being added every day. The IoT communicates with sensor data and not with words which makes it much reliable. This makes it possible to continuously monitor and control devices, systems, processes and infrastructure. There are already countless practical applications, from smart homes to smart lighting to smart roads. Interconnected production has become a major trend in the industrial sector, too [17] [18]. Hence, interconnection of solar and TEG will make the energy production effectively and efficiently by including IoT for proper regulating and management of electrical energy.

## 2. EXISTING SYSTEM

Nowadays solar energy is becoming popular and consumed commercially in huge proportion. As it is natural source of energy the production cost of power using solar is very much less compared to other conventional sources. The concept of TEG is being used in panel whose temperature difference can be converted as electrical energy which is used in miniscule range for various loads. Mainly, in the existing system, the solar panel and the thermal power plants are separately handled. This results in a large amount of losses due to energy conversion from solar energy to electrical energy. The cost for power generation by this method is huge. The load imbalance which occur in this system cannot be managed very easily. Fig 1 shows the existing system of energy conversion using solar energy where each energy conversion system works independently and not economical.

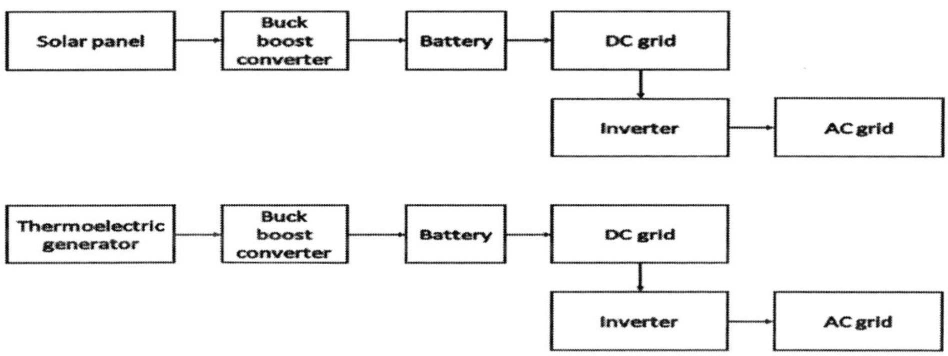


**Figure 1 Block diagram of the Existing system**

## 3. PROPOSED SYSTEM

In proposed system as per figure 2, the energy is harvested using solar and TEG which similar to the existing system, but the harvested energy is hybridized which can be properly regulated and managed through microcontroller and IoT mechanism. The produced energy by hybrid technique are used for DC load through battery and Buck boost converter. The same energy is converted to ac through inverter and it is used for AC load. These two operations are exactly same as existing system, but in proposed system the output of PV and TEG is hybridized. Therefore, the solar panel output and thermoelectric generator output is given to the buck-boost converter which provides a regulated output voltage to match the battery charging voltage. Power supply for the DC grid will be taken from the battery terminals directly. The power supply for AC grid will be taken from the battery terminals via inverter. Apart from hybrid technique, microcontroller and Arduino nano is used to measure the power generation by solar and thermoelectric generator for proper management through IoT. The same will be monitored on the LCD.

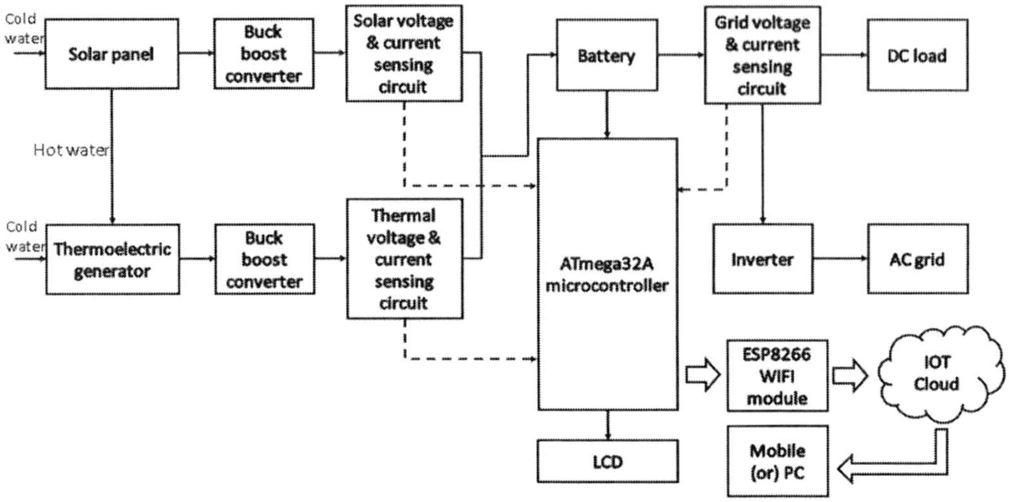


**Figure 2 Block diagram of Proposed Hybrid Energy Harvesting system**

In the proposed system, the load balancing problem due to individual source supply has been overcome due to hybrid technology i.e., the integration of heat and light concept is introduced. The major advantage of the proposed system is that the system consist of various protections like reverse current protection. The bypass switch helps to decide whether the battery or the load is to be connected.

## 4. HARDWARE SETUP

The overall hardware setup can be divided into two units namely, power generation unit and power regulating and management unit. Figure 3 shows the power generation unit which comprises of solar panel and thermo electric generator unit which poses of temperature sensors and conversion unit.

**Figure 3 Solar panel and TEG set up**

Figure 4 represents the hardware setup of power regulating and management unit. The hardware setup consists of Solar Panel, Thermoelectric Generator, Buck Boost Converter, Digital Temperature Sensor, 5V DC Regulator, ATMEGA32A Microcontroller, ESP8266 Wi-Fi Module,LM358,16X2 LCD as the major components. This unit gets the input from the power generation unit and converts into dc or ac voltage based upon the requirement of load with the help of buck boost converter and inverter.

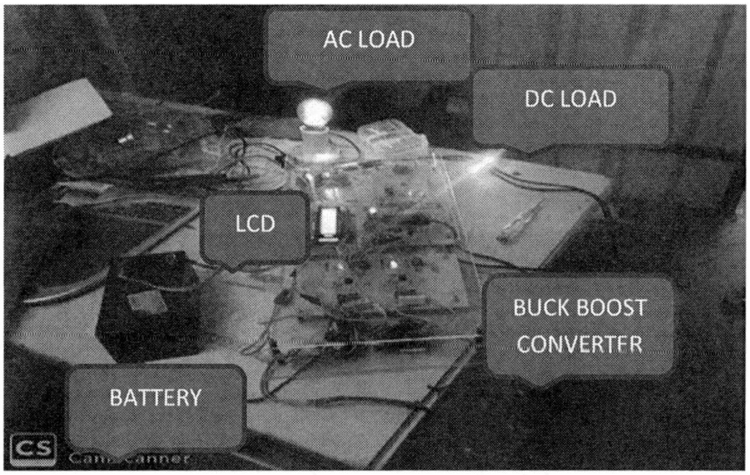


**Figure 4 Power regulating and Management hardware setup**

***a. BUCK BOOST CONVERTER***

Figure 5 and 6 shows the hardware part of buck boost converter and circuit diagram of it. The working operation of the DC to DC converter is the inductor in the input resistance has the unexpected variation in the input current. If the switch is ON then the inductor feed the energy from the input and it stores the energy of magnetic energy. If the switch is closed it discharges the energy. The output circuit of the capacitor is assumed as high sufficient than the time constant of an RC circuit is high on the output stage. The huge time constant is compared with the switching period and make sure that the steady state is a constant output voltage Vo(t) = Vo(constant) and present at the load terminal.

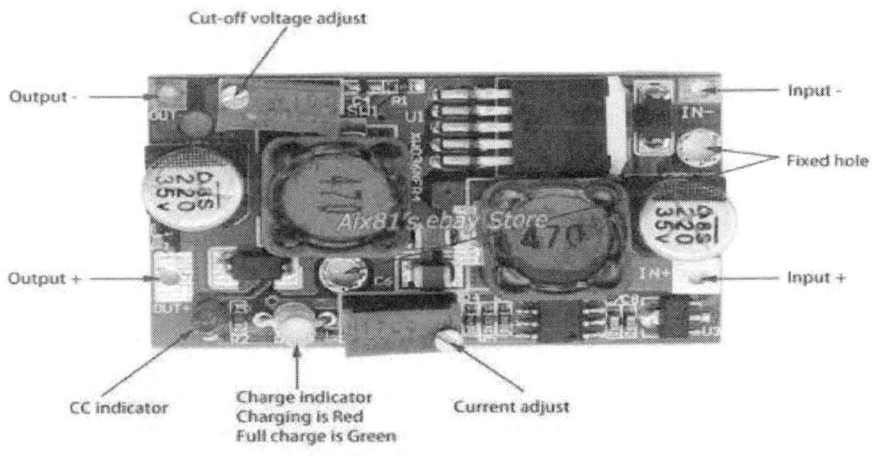


**Figure 5. BUCK BOOST CONVERTER**

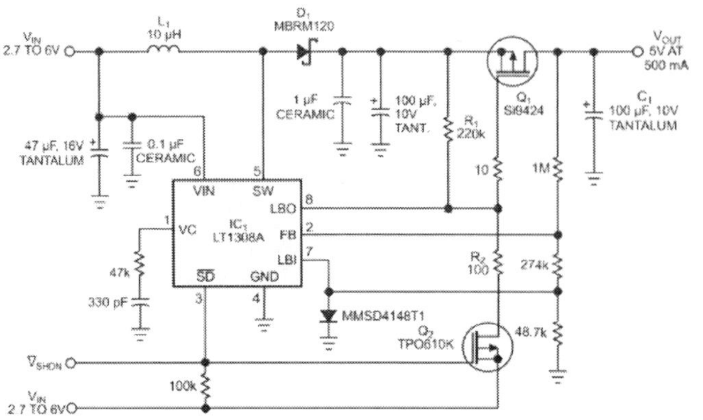


**Figure 6.CIRCUIT DIAGRAM OF BUCK BOOST CONVERTER**

***b. DIGITAL TEMPERATURE SENSOR:***

The DS18B20 digital thermometer provides 9-bit to 12-bit Celsius Temperature measurements and has an alarm function with non-volatile user programmable upper and lower trigger points. The DS18B20 communicates over a 1-Wire bus that by definition requires only one data line for Communication with a central microprocessor. It can derive power directly from the data line, eliminating the need for an external power supply.

**Figure 7 DIGITAL TEMPERATURE SENSOR**

They work great with any microcontroller using a single digital pin, andyou can even connect multiple ones to the same pin, each one has a unique 64-Bit ID burned in at the factory to differentiate them. When using withMicrocontroller put a 4.7k resistor to sensing pin, which is required as a pull-upFrom the DATA to VCC line.

## 5. HARDWARE RESULTS

Figure 8 represents the temperature of the solar panel which works according to seebeck principle. The PV panel when exposed certain radiance the temperature difference on the panel is measure which is as indicated.

**Figure 8 Panel Temperature**

IOP Conf. Series: Materials Science and Engineering **923** (2020) 012077 doi:10.1088/1757-899X/923/1/012077

Figure 9 indicates the voltage output from the solar panel which 15 volts and Figure 10 indicates the output voltage of 9V from TEG due to the temperature difference as mentioned in Figure 5

**Figure 9 Output voltage due to PV panel**

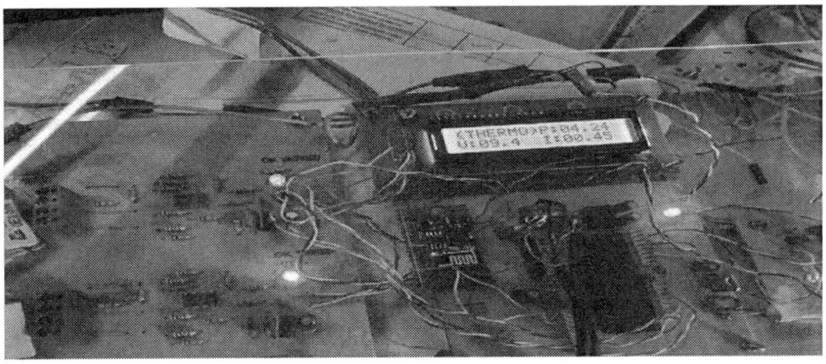

**Figure 10 Output Voltage due to TEG**

Figure 11 shows the working of complete circuit board which consists of two section namely, buck boost converter whose output is connected to LED and the output of buck boost converter which is in the form of dc is converted to ac with the help of inverter section and that output is given to bulb.

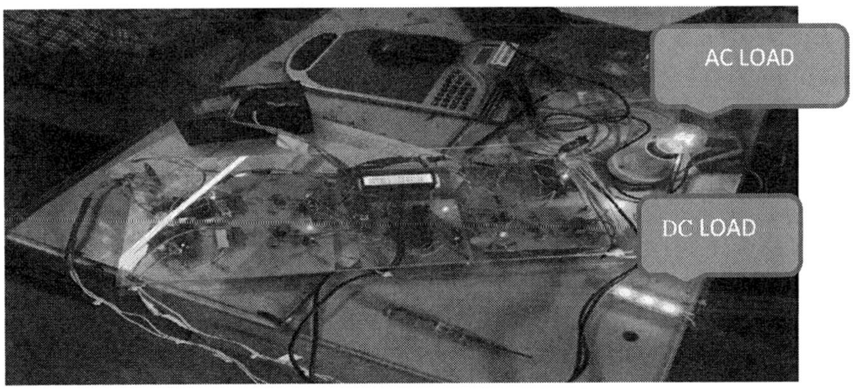


**Figure 11 Output of ac and dc load**

## 6.CONCLUSION AND FUTURE ENHANCEMENT

With the fast growing demand of smart devices like IoT systems, the energy management and sustainable power supply research is being conducted widely. This paper proposes integration of IoT with the hybrid energy technology namely solar energy and TEG. Eventhough, TEG are able to harvest enough energy, the voltage produced is not enough to operate various loads. Hence Integration of renewal energy source along with TEG is proposed, especially in commercial areas where need of electricity is more. It causes no effect on nature i.e. pollution free, at the same time not proneness any kind of accident due to lightning. It is also useful to minimize power supply load i.e. cut short power charge. By using this system, electricity charge is very less and requires less maintenance charge for the equipment.

The designing of this equipment is done in such a way that it is very compact and acts as user friendly. The cost of energy harvesting through this hybrid source can be reduced if it is manufactured in large scale. And becomes much affordable. Due to this technology, the power shedding or power failure problems can be easily rectified at all times. Hence, it is the most reliable renewable power or electricity resources with less expenditure.

In this paper, not only the energy harvesting by solar energy and thermoelectricity is discussed but also proper voltage regulation using a DC–DC converter to boost the output voltage and also proper energy management using IoT is also discussed. This project is at an intermediate stage in its research and soon will have the potential to explore in market with enabling technology.

## REFERENCES

[1] T. M. Razykov, C. S. Ferekides, D. Morel, E. Stefanakos, H. S. Ullal, and H. M. Upadhyaya, 2011 Solar photovoltaic electricity Current status and future prospects *Solar Energy* **85** doi.org/10.1016/j.solener.2010.12.002.

[2] Zweibel, K 2005 The Terawatt challenge for thin film PV. In: *Poortmans, J., Archipov, V. (Eds.)*, Thin Film Solar Cells: Fabrication, Characterization and Application John Wiley pp. 18–22.

[3] Kazmerski, L. 2005 Solar photovoltaics R&D: at the tipping point: a 2005 technology overview *Journal of Electron Spectroscopy and related Phenomena* **150 (2–3)** pp. 105–135

[4] Ofer Beeri, , Oded Rotem, Eden Hazan, Eugene A. Katz, Avi Braun, , and Yaniv Gelbstein 2015 Hybrid photovoltaic-thermoelectric system for concentrated solar energy conversion: Experimental realization and modeling *AIP. Journal of Applied Physics* **118** doi: 10.1063/1.4931428.

[5] D. Kraemer, L. Hu, A. Muto, X. Chen, G. Chen, and M. Chiesa 2008 Photovoltaic-thermoelectric hybrid systems: A general optimization methodology," *Applied Physics Lett*er **92** 243503 .

[5] W. G. J. H. M. Van Sark, 2011 Feasibility of photovoltaic–Thermoelectric hybrid modules,Appl. Energy **88**, pp.2785-2792.

[6]Yu. Vorobiev, J. Gonzalez-Hernandez, P. Vorobiev, and L. Bulat 2006 Thermal-photovoltaic solar hybrid system for efficient solar energy conversion *Solar Energy* **80** pp.170 -178.

[7]E. A. Chavez-Urbiola, Yu. V. Vorobiev, and L. P. Bulat, 2012 Solar hybrid systems with thermoelectric generators *Solar Energy* **86** pp. 369-379.

[8] T. Liao, L. Bihong, and Y. Zhimin 2014 Performance characteristics of a low concentrated photovoltaic–thermoelectric hybrid power generation device *International Journal of Thermal Science* **77**, pp. 158-169.

[9] N. Wang, L. Han, H. He, N. H. Park, and K. Koumoto 2011 A novel highperformance photovoltaic–thermoelectric hybrid device *Energy Environmnet and Science*. **4**, 3676.

[10]D. Yang and H. Yin 2011 Energy conversion efficiency of a novel hybrid solar system for photovoltaic, thermoelectric, and heat utilization *IEEE Transactions in Energy Converters.* **26**(2), 662

[11] Hajji, M.; Naimi, S.E.; Hajji, B.; El Hafyani, M.L. 2014 A numerical Modelling of hybrid photovoltaic/thermal (PV/T) collector, *In Microelectronics (ICM), 2014 26th International Conference* pp.152-155,

[12] Jiju K. et. al., 2014. "Development of Android based online monitoring and control system for Renewable Energy Sources." *Computer, Communications, and Control Technology (I4CT), International Conference on. IEEE*

[13] Yoshihiro G. et. al., 2007 Integrated management and remote monitoring system for telecommunications power plants with fully DC-powered center equipment INTELEC 07-29th *International Telecommunications Energy Conference. IEEE*
[14] Limpahan, K.S.; Vina, R.R.; Alagao, F.B. 2014 Performance Evaluation of direct-contact cooling tower used on thermoelectric Module and parabolic trough collector for power generation In Humanoid, Nanotechnology *Information Technology, Communication and Control, Environment and Management (HNICEM), 2014 International Conference* pp.1-6,
[15] Hongfei Wu; Kai Sun; Min Chen; Zhe Chen; Yan Xing 2013 Hybrid Centralized-distributed power conditioning system for thermoelectric Generator with high energy efficiency *in Energy Conversion Congress and Exposition (ECCE)* pp.4659-4664
[16] Kabalci, Ersan, Gorgun A. and Kabalci Y. 2013 Design and implementation of a renewable energy monitoring system. Power Engineering, Energy and Electrical Drives (POWERENG), Fourth International Conference on. IEEE, doi: 10.1109/PowerEng.2013.6635759
[17] Pallavi Ravindra Joshi1 and Prof. M S khan 2017 IOT Based Smart Power Management System Using WSN *International Journal of Advanced Research Trends in Engineering and Technology* **04** pp783-786
[18] Leo Louis 2016 Working Principle of Andruino and Using It as a tool for study and research", *International Journal of Control, Automation, Communication and Systems (IJCACS)* **1**, No.2. pp.21-29.

# AUTHOR INDEX